General Circulation Model Development

This is Volume 70 in the
INTERNATIONAL GEOPHYSICS SERIES
A series of monographs and textbooks
Edited by RENATA DMOWSKA, JAMES R. HOLTON,
    and H. THOMAS ROSSBY

A complete list of books in this series appears at the end of this volume.

General Circulation Model Development

Edited by

David A. Randall

Colorado State University
Fort Collins, Colorado

ACADEMIC PRESS

A Harcourt Science and Technology Company

San Diego   San Francisco   New York   Boston   London   Sydney   Tokyo

The cover contains three graphics based on work by Akio Arakawa. The
top left graphic is adapted from Figure 4 in Chapter 1 of this volume.
It illustrates the development of nonlinear computational instability.
The top right graphic is adapted from Figure 1 of the 1974 paper
by Arakawa and Schubert. (From Arakawa, A., and W. H. Schubert
(1974). Interaction of a cumulus cloud ensemble with the large-scale
environment, Part I. *Journal of the Atmospheric Sciences*, **31**,
674-701. With permission.) It illustrates an ensemble of cumulus
clouds embedded in a large-scale environment. The bottom left graphic
is taken from Figure 13 of Chapter 1 of this volume. (Redrawn
from Arakawa, 1975). It illustrates the relationships among several
cloud regimes and the large-scale circulation in the tropics and subtropics.

This book is printed on acid-free paper. ♾

Academic Press
A Harcourt Science and Technology Company
525 B Street, Suite 1900, San Diego, California 92101-4495, USA
http://www.academicpress.com

Academic Press
Harcourt Place, 32 Jamestown Road, London NW1 7BY, UK
http://www.academicpress.com

Library of Congress Catalog Card Number: 00-102252

International Standard Book Number: 0-12-578010-9

PRINTED IN THE UNITED STATES OF AMERICA
00 01 02 03 04 05 SB 9 8 7 6 5 4 3 2 1

*Professor Akio Arakawa lecturing at
the AAFest in January 1998.*

Contents

Chapter 1

A Personal Perspective on the Early Years of General Circulation Modeling at UCLA

Akio Arakawa

Chapter 2

A Brief History of Atmospheric General Circulation Modeling

Paul N. Edwards

Chapter 3

Clarifying the Dynamics of the General Circulation: Phillips's 1956 Experiment

John M. Lewis

I. Introduction   **91**

Chapter 4

Climate Modeling in the Global Warming Debate

J. Hansen, R. Ruedy, A. Lacis, M. Sato, L. Nazarenko, N. Tausnev, I. Tegen, and D. Koch

Chapter 5

A Retrospective Analysis of the Pioneering Data Assimilation Experiments with the Mintz–Arakawa General Circulation Model

Milton Halem, Jules Kouatchou, and Andrea Hudson

Chapter 6

A Retrospective View of Arakawa's Ideas on Cumulus Parameterization

Wayne H. Schubert

Chapter 7

On the Origin of Cumulus Parameterization for Numerical Prediction Models

Akira Kasahara

Chapter 8

Quasi-Equilibrium Thinking

Kerry Emanuel

Chapter 9

Application of Relaxed Arakawa–Schubert Cumulus Parameterization to the NCEP Climate Model: Some Sensitivity Experiments

Shrinivas Moorthi

Chapter 10

Solving Problems with GCMs: General Circulation Models and Their Role in the Climate Modeling Hierarchy

Michael Ghil and Andrew W. Robertson

Chapter 11

Prospects for Development of Medium-Range and Extended-Range Forecasts

Anthony Hollingsworth

Chapter 12

Climate Services at the Japan Meteorological Agency Using a General Circulation Model: Dynamical One-Month Prediction

Tatsushi Tokioka

Chapter 13

Numerical Methods: The Arakawa Approach, Horizontal Grid, Global, and Limited-Area Modeling

Fedor Mesinger

Chapter 14
Formulation of Oceanic General Circulation Models

James C. McWilliams

Chapter 15
Climate and Variability in the First Quasi-Equilibrium Tropical Circulation Model

Ning Zeng, J. David Neelin, Chia Chou, Johnny Wei-Bing Lin, and Hui Su

Chapter 16

Climate Simulation Studies at CCSR

Akimasa Sumi

Chapter 17

Global Atmospheric Modeling Using a Geodesic Grid with an Isentropic Vertical Coordinate

David A. Randall, Ross Heikes, and Todd Ringer

Chapter 18

A Coupled GCM Pilgrimage: From Climate Catastrophe to ENSO Simulations

Carlos R. Mechoso, Jin-Yi Yu, and Akio Arakawa

Chapter 19

Representing the Stratocumulus-Topped Boundary Layer in GCMs

Chin-Hoh Moeng and Bjorn Stevens

Chapter 20

Cloud System Modeling

Steven K. Krueger

Chapter 21
Using Single-Column Models to Improve Cloud-Radiation Parameterizations

Richard C. J. Somerville

Chapter 22
Entropy, the Lorenz Energy Cycle, and Climate

Donald R. Johnson

Chapter 23

Future Development of General Circulation Models

Akio Arakawa

Contributors

Numbers in parentheses indicate the pages on which the authors' contributions begin.

Akio Arakawa (1, 539, and 721), Department of Atmospheric Sciences, University of California, Los Angeles, California 90095

Chia Chou (457), Department of Atmospheric Sciences and Institute of Geophysics and Planetary Physics, University of California, Los Angeles, California 90095

Paul N. Edwards (67), School of Information, University of Michigan, Ann Arbor, Michigan 48109

Kerry Emanuel (225), Program in Atmospheres, Oceans, and Climate, Massachusetts Institute of Technology, Cambridge, Massachusetts 02139

Michael Ghil (285), Department of Atmospheric Sciences and Institute of Geophysics and Planetary Physics, University of California, Los Angeles, California 90095

Milton Halen (165), NASA Goddard Space Flight Center, Greenbelt, Maryland 20771

James Hansen (127), NASA Goddard Institute for Space Studies, New York, New York 10025

Ross P. Heikes (509), Department of Atmospheric Science, Colorado State University, Fort Collins, Colorado 80523

Andrea Hudson (165), NASA Goddard Space Flight Center, Greenbelt, Maryland 20771

Anthony Hollingsworth (327), European Centre for Medium-Range Weather Forecasts, Shinfield Park, Reading, Berks RG2 9AX, United Kingdom

Donald R. Johnson (659), Space Science and Engineering Center, University of Wisconsin–Madison, Madison, Wisconsin 53706

Akira Kasahara (199), National Center for Atmospheric Research, Boulder, Colorado 80307

Dorothy Koch (127), NASA Goddard Institute for Space Studies, New York, New York 10025

Jules Kouatchou (165), School of Engineering, Morgan State University, Baltimore, Maryland 21239

Steve Krueger (605), Department of Meteorology, University of Utah, Salt Lake City, Utah 84112

Andrew Lacis (127), NASA Goddard Institute for Space Studies, New York, New York 10025

John M. Lewis (91), National Severe Storms Laboratory, Norman, Oklahoma 73069; and Desert Research Institute, Reno, Nevada 89512

Johnny Wei-Bing Lin (457), Department of Atmospheric Sciences and Institute of Geophysics and Planetary Physics, University of California, Los Angeles, California 90095

James C. McWilliams (421), Department of Atmospheric Sciences/IGPP, University of California, Los Angeles, California 90095

Fedor Mesinger (373), NCEP Environmental Modeling Center, Camp Springs, Maryland 20746

Carlos R. Mechoso (539), Department of Atmospheric Sciences, University of California, Los Angeles, California 90095

Chin-Hoh Moeng (577), MMM Division, NCAR, Boulder, Colorado 80307

Shrinivas Moorthi (257), Environmental Modeling Center, National Centers for Environmental Predictions, National Weather Service, NOAA, Camp Springs, Maryland 20746

Larissa Nazarenko (127), NASA Goddard Institute for Space Studies, New York, New York 10025

J. David Neelin (457), Department of Atmospheric Sciences and Institute of Geophysics and Planetary Physics, University of California, Los Angeles, California 90095

David A. Randall (509), Department of Atmospheric Science, Colorado State University, Fort Collins, Colorado 80523

Todd D. Ringler (509), Department of Atmospheric Science, Colorado State University, Fort Collins, Colorado 80523

Andrew W. Robertson (285), Department of Atmospheric Sciences and Institute of Geophysics and Planetary Physics, University of California, Los Angeles, California 90095

Reto Ruedy (127), NASA Goddard Institute for Space Studies, New York, New York 10025

Minino Sato (127), NASA Goddard Institute for Space Studies, New York, New York 10025

Wayne H. Schubert (181), Department of Atmospheric Science, Colorado State University, Fort Collins, Colorado 80523

Richard C. J. Somerville (641), Scripps Institution of Oceanography, University of California, San Diego, La Jolla, California 92037

Bjorn Stevens (577), Department of Atmospheric Sciences, University of California, Los Angeles, California 90095

Hui Su (457), Department of Atmospheric Sciences and Institute of Geophysics and Planetary Physics, University of California, Los Angeles, California 90095

Akimasa Sumi (489), Center for Climate System Research, University of Tokyo, Meguro-ku, Tokyo, Japan

Nicholas Tausnev (127), NASA Goddard Institute for Space Studies, New York, New York 10025

Ina Tegen (127), NASA Goddard Institute for Space Studies, New York, New York 10025

Tatsushi Tokioka (355), Japan Meteorological Agency, 1-3-4 Otemachi, Chiyoda-ku, Tokyo, Japan

Jin-Yi Yu (539), Department of Atmospheric Sciences, University of California, Los Angeles, California 90095

Ning Zeng (457), Department of Atmospheric Sciences and Institute of Geophysics and Planetary Physics, University of California, Los Angeles, California 90095

Foreword

The volume you now hold in your hands is not the usual collection of miscellaneous papers from an office drawer, hastily collected by friends anxious to honor an admired and respected fellow scientist. Instead, this book consists of papers especially prepared for presentation at the January 1998 Arakawa Retirement Symposium. This favors us with substantial and appreciative assessments of Akio Arakawa's contributions and their applications. But we also have his personal account of how he approached the two subjects in which his contributions were of the first rank: his physically guided formulation of the finite-difference methods used in numerical simulations of fluid motions, and a ground-breaking treatment of the mutual interaction between convective clouds and the surrounding atmosphere. I think this volume is especially valuable for these insights.

Arakawa was one of the many young gifted Japanese scientists who emigrated to the United States in the late 1950s and early 1960s and who were greatly influenced by their esteemed teacher, Professor Shigekata Syono of Tokyo University. An exciting account of this group has been written by John Lewis (1993).

Akio was the youngest of three brothers, but the only one to choose science as a career. He credits his mother, a schoolteacher, as having the most intellectual influence on his life. During World War II, he was enrolled in a special seven-year high school in which the last three years were devoted to an elective subject, among which he chose physics. Here he was stimulated by a young physics teacher, who, according to Arakawa, "gave rather advanced lectures." He, like most of the other Japanese entrants into the field of U.S. meteorology, then entered Tokyo University, the most prestigious among the Japanese universities. In the university entrance examination he chose science over engineering, for which the meteorological world must be grateful. But he was not exposed to geophysical subjects until after receiving his bachelor degree in physics, when the lack of employment opportunities in that area led him to enter the Japanese Meteorological Agency. His initial introduction to practical meteorology over the stormy western Pacific and his later involvement in theoretical aspects of the atmosphere are described in the first paper in this volume.

The influx of young Japanese meteorologists into the United States in the early 1960s was a boon to American meteorology. This was true not only for Arakawa's field of numerical weather prediction and general circulation modeling, but also for turbulence theory, hurricane modeling, and the study of severe storms, where the names of Ogura, Ooyama, Kurihara, Murakami, and Fujita stand out. Then, in the area chosen by Akio Arakawa, the names of Gambo, Manabe, Miyakoda, Kasahara, Yanai, and Sasaki must be recognized for their strong contributions to these subjects. (Among these and other Japanese emigrants, Fujita is perhaps the only one whose career was not shaped from its early days by Professor Syono.)

As someone who also worked in his field, I was of course greatly interested in and reasonably aware of Arakawa's ideas. But the two chapters by Arakawa in this volume have great human and scientific interest that was not evident to me at the time. I draw your attention especially to Akio's recounting of his initial efforts in Los Angeles where Professors Jacob Bjerknes and Yale Mintz had spent several years gathering statistics on the zonal budgets of momentum and temperature from radiosonde data. Then Yale met Akio in 1959 at the International Symposium on Numerical Weather Prediction in Tokyo. He quickly hired Akio to work with him at the University of California–Los Angeles (UCLA) in developing a numerical model of the atmospheric general circulation. Here we have a young man, with a fresh doctoral degree, newly arrived in a strange land, and faced with the strong desire of an established and energetic professor to jump quickly into the new field of numerical simulation of the atmosphere. The professor counts on Akio to do all of the computer design and programming. But the young man realizes that progress in this field requires that before the first "Do loop" is written for a general circulation code, an important technical problem must be solved: the formulation of a physically meaningful finite-difference scheme that will simulate the nonlinear processes in the equations of motion. What conviction this must have required to convince the professor! And, what patient understanding and faith were required in turn of the professor to accept this argument—*and then get his funding agencies to agree!* This collaboration of Akio Arakawa and Yale Mintz was certainly fortuitous for meteorology.

Arakawa's work on numerical methods has been glowingly described by Douglas Lilly (1997), who also did pioneer work in this area. The wisdom of Arakawa's strategy became apparent as early as 1965, when the Committee on Atmospheric Sciences of the National Research Council prepared its planning study, *The Feasibility of a Global Observation and Analysis Experiment.* This document (National Research Council, 1966) was a prelude to the Global Weather Experiment (FGGE) 12 years later. An important aspect of the committee's task was to estimate the predictability of the atmosphere with respect to large-scale motions; that is, for how long could forecasts be made if it were possible to observe the atmosphere in reasonable detail?

Under the guidance of Jule Charney, three general circulation models designed by Cecil Leith at Livermore Laboratories, Joseph Smagorinsky at the Geophysical Fluid Dynamics Laboratory of NOAA, and Akio Arakawa at UCLA were used for parallel integrations of several weeks to determine the growth of small initial errors. Only Arakawa's model had the aperiodic behavior typical of the real atmosphere in extratropical latitudes, and his results were therefore used as a guide to predictability of the real atmosphere. This aperiodic behavior was possible because Arakawa's numerical system did not require the significant smoothing required by the other models, and it realistically represented the nonlinear transport of kinetic energy and vorticity in wave number space.

But an equally striking result that I remember from Arakawa's efforts was his presentation in the early 1970s of his parameterization of cloud convection and its interaction with the environment. His latitudinal cross section showed in dramatic fashion how small cumulus clouds in the trade winds changed into towering cumulonimbus clouds as air approached the intertropical convergence zone in an environment that had been shaped by the clouds themselves. Somehow this pictorial representation was more convincing than a thousand maps of predicted rainfall! And knowing something of the hard and almost painful steps necessary to obtain this result made it all the more valuable as a milestone in understanding the atmosphere.

It is now time to study and enjoy the wisdom contained in this volume. You will return often. Thank you, Akio Arakawa!

Norman Phillips

REFERENCES

1. Lewis, J. (1993). Meteorologists from the University of Tokyo: Their exodus to the United States following World War II. *Bull. Am. Meteorological Soc.* **74**, 1351–1360.
2. Lilly, D. (1997). Testimonial to Akio Arakawa. *J. Comp. Physics* **135**, 101–102.
3. National Research Council (1966). The feasibility of a global observation and analysis experiment, Publication 1290. National Academy of Sciences, Washington, DC.

Preface

On January 20–22, 1998, a symposium was held at the Northwest Campus Auditorium on the University of California–Los Angeles (UCLA) campus. The official title of the symposium was "General Circulation Modeling, Past Present and Future: A Symposium in Honor of Akio Arakawa," but we informally called it the "AA Fest." The AA Fest was organized as a celebration of the career of UCLA's Professor Akio Arakawa, who has been among the leaders in the field of atmospheric general circulation model (GCM) development from its beginning.

Akio Arakawa obtained his B.Sc. in physics from Tokyo University in 1950 and his D.Sc. in meteorology from the same institution in 1961. In the early 1950s, he served for one year on a weather ship in the North Pacific, an experience that made a strong impression on him. Subsequently, still during the 1950s, he conducted forecasting research at the Meteorological Research Institute, which is operated by the Japan Meteorological Agency. Norman Phillips's first numerical simulation of general circulation inspired Arakawa to develop his own model of global atmospheric circulation, and during 1961–1963 he was an assistant research meteorologist at UCLA, working with Professor Yale Mintz on the development of what was to become the Mintz–Arakawa GCM. After returning to Japan for two years, he joined the faculty at UCLA in 1965, and has remained there ever since, conducting his wide-ranging research on GCM development and related scientific issues.

Akio Arakawa's two best known papers are his 1966 exposition of the energy- and enstrophy-conserving finite-difference Jacobian operator and his 1974 paper with Wayne Schubert in which they presented their theory of the interactions of a cumulus cloud ensemble with large-scale circulations. These are both remarkably insightful papers, but the most amazing thing about them is that the same person led both studies. The world of numerical methods for the solution of the partial differential equations of geophysical fluid dynamics is mathematically "clean"; the much more down-to-earth world of cumulus parameterization is mired in the fantastically intricate and "dirty" phenomena of moist convection. The problem of general circulation model development spans both of these seemingly disparate fields, and Akio Arakawa has made major contributions to both.

In doing so, he has demonstrated that the two kinds of problems are not as different as they appear. He creates differencing schemes that mimic the key physical properties of the exact equations, thus transforming mathematical problems (discretize the momentum equation) into physical problems (find a discrete analog of the momentum conservation principle from which it is possible to derive discrete analogs of the kinetic energy and potential enstrophy conservation principles). He distills parameterizations that capture the essence of an infinitely detailed physical system into the simplest possible mathematical statement. He seeks perfection, and he is getting closer every year.

Several decades ago, Syukuro Manabe gave a lecture on global atmospheric modeling before a large audience, which included me as a graduate student. He began by briefly summarizing the status of the relatively few GCM development projects under way at that time. He outlined the recent successes and ongoing struggles of several modeling groups, and then he came to the UCLA effort: "Oh," he said, "Arakawa will have a *perfect* model—but it's not quite finished yet."

As demonstrated by his contributions to this volume, Akio Arakawa is still working on the perfect model, and all of us who are privileged to know him are very glad that he is not quite finished yet.

This book and the symposium on which it is based were made possible by the efforts of many people. The symposium was generously supported by the National Science Foundation, the National Aeronautics and Space Administration, the Office of Naval Research, the Department of Energy, the Center for Climate System Research at the University of Tokyo, the World Climate Research Program, and the College of Natural Sciences at UCLA. Dr. Kayo Ide of UCLA made heroic contributions to the organization and planning of the symposium, and her efforts were absolutely crucial to its success. Professor Roger Wakimoto of UCLA provided both institutional support and sage advice in the planning of the symposium. I am grateful to Dr. Frank Cynar of Academic Press, whose enthusiasm for this project and useful suggestions along the way have made my work easier and more enjoyable. Finally, Cindy Carrick of Colorado State University ably assisted me in my efforts to put this book together.

David Randall
Fort Collins, Colorado

A Personal Perspective on the Early Years of General Circulation Modeling at UCLA

Akio Arakawa
Department of Atmospheric Sciences
University of California, Los Angeles, California

I. INTRODUCTION

This chapter is based on the lecture I presented at the beginning of the AA Fest: Symposium on General Circulation Model Development: Past, Present, and Future, held at UCLA, January 20–22, 1998. As the title indicates, this chapter is primarily my memoir on the evolution of ideas in

General Circulation Model Development

the history of general circulation modeling at UCLA, with references to the history of numerical modeling of the atmosphere in general.

Section II presents a review of the pre-GCM periods in the general history, covering its prelude and the "epoch-making" first phase. Section III then presents my personal history during the pre-UCLA period, and Section IV describes my first work at UCLA on the so-called "Arakawa Jacobian." Section V presents a brief review of the development of the "Mintz–Arakawa model," which is the first generation of the UCLA general circulation model. Section VI then describes the "magnificent" second phase in the general history of numerical modeling of the atmosphere and an outline of the different generations of the UCLA general circulation model developed during that phase. Sections VII through X discuss the evolution of those generations in more detail for selected modeling aspects: vertical differencing, horizontal differencing, formulation of planetary boundary layer processes and formulation of moist processes, respectively. Closing remarks are given in Section XI.

II. EARLY HISTORY OF NUMERICAL MODELING OF THE ATMOSPHERE

I divide the general history of numerical modeling of the atmosphere into four phases: the prelude, the "epoch-making" first phase, the "magnificent" second phase, and the "great-challenge" third phase (Fig. 1). The beginnings of the first, second, and third phases roughly correspond to the development of the early numerical weather prediction (NWP) models, that of the early general circulation models (GCMs), and that of the recent coupled atmosphere–ocean GCMs, respectively. In the rest of this section, I present a brief historical review of the prelude and the first phase.

A. THE PRELUDE (–1950)

V. Bjerknes (1904) is considered the first advocate of NWP. He pointed out:

> If it is true, as every scientist believes, that subsequent atmospheric states develop from the preceding ones according to physical law, then it is apparent that the necessary and sufficient conditions for the rational solution of forecasting problems are the following:
>
> 1. A sufficiently accurate knowledge of the state of the atmosphere at the initial time,

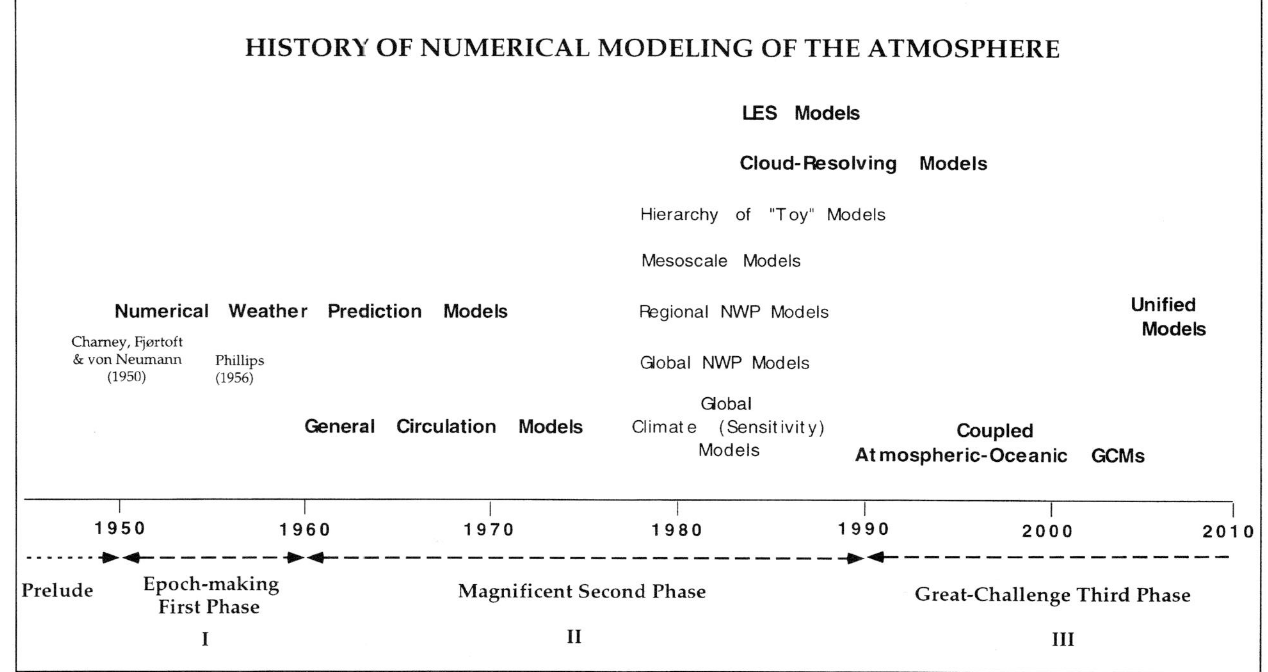

Figure 1 Chart showing the history (and near future) of numerical modeling of the atmosphere.

2. A sufficiently accurate knowledge of the laws according to which one state of the atmosphere develops from another.

These statements now sound obvious. In pointing out the second condition, however, Bjerknes distinguished the laws for changes "from degree to degree in meridian and from hour to hour in time" from those "from millimeter to millimeter and second to second," indicating that we did not know the laws for the former well enough to forecast weather. I would say that we still do not know such laws, and the history of numerical modeling of the atmosphere is that of the struggle to establish such laws. Recognizing the complexity of the problem, Bjerknes said:

> The problem is of huge dimensions. Its solution can only be the result of a long development.... I am convinced that it is not too soon to consider this problem as the objective of our researches.

Bjerknes (1914) further expressed his enthusiasm for weather prediction as a scientific problem:

> I shall be more than happy if I can carry on the work so far that I am able to predict the weather from day to day after many years of calculation. If only the calculations shall agree with the facts, the scientific victory will be won. Meteorology would then have become an exact science....

Apparently stimulated by Bjerknes, Richardson (1922) attempted actual forecasts. (For an excellent review of Richardson's work, see Platzman, 1967.) Interestingly, the basic structure of his model is not very different from that of the models we now use, including the use of finite-difference methods (in contrast to a graphical method Bjerknes had in mind). Richardson said:

> The scheme is complicated because the atmosphere is complicated. But it has been reduced to a set of computing forms. These are already to assist anyone who wishes to make partial experimental forecasts.... In such a way it is thought that our knowledge of meteorology might be tested and widened and concurrently the set of forms might be revised and simplified.

Richardson again recognized the forecasting problem as a problem in science, in which "our knowledge of meteorology might be tested and widened."

In spite of the imaginative and laborious work, Richardson's 6-hr forecast of the surface pressure at two points over Europe turned out to be a complete failure, mainly because he extrapolated the calculated initial tendencies over a period of 6 hr. With the equations he used, which we now call the primitive equations, instantaneous tendencies are strongly influenced by the existence of high-frequency oscillations such as those due to the Lamb wave and internal inertia-gravity waves and, therefore, the calculated tendencies should not have been extrapolated over such a long time interval.

Richardson was a perfectionist, reluctant to introduce even a minor approximation. The modern history of numerical modeling of the atmosphere, however, followed a completely different path. A particularly important event on this path was Rossby's (Rossby *et al.*, 1939) recognition of the relevance of absolute vorticity advection to large-scale wave motions (the Rossby wave).

In 1946, von Neumann called a conference of meteorologists to tell them about the general-purpose electronic computer he was building and to seek their advice and assistance in designing meteorological problems for its use. Jule Charney, who attended the conference, said in an unpublished manuscript coauthored with Walter Munk:

> To von Neumann, meteorology was par excellence the applied branch of mathematics and physics that stood the most to gain from high-speed computation.

According to Charney, the established figures attending the conference, however, were interested but less than enthusiastic. Rossby perhaps best voiced their feeling by stating, "The mathematical problem is not yet defined: there are more unknowns than equations." This can be interpreted as a statement of the existence of what we now call the "parameterization" problem. The problem arises from the fact that model equations explicitly deal with only the large-scale portion of the broad spectrum shown in Fig. 2, while the effect of smaller scales on the large scale cannot be totally ignored. Rossby further said, again according to Charney,

> Computation could not be successful before observation, experiment and analysis had led to a better understanding of fundamental atmospheric processes, in particular of atmospheric turbulence.

Although Rossby was absolutely right in foreseeing the problems involved in numerical modeling of the atmosphere, it is ironic that the model used for the first successful NWP was basically Rossby's own, which entirely bypasses such problems.

The prelude of the history reached its climax in the late 1940s through the rapid development of theories on large-scale motions in the atmosphere. These theories include the baroclinic and barotropic instability theories by Charney (1947), Eady (1949), and Kuo (1951), scale analysis for quasi-geostrophic approximation by Charney (1948), and introduction of the concept of equivalent-barotropy and its application by Charney (1949) and Charney and Eliassen (1949).

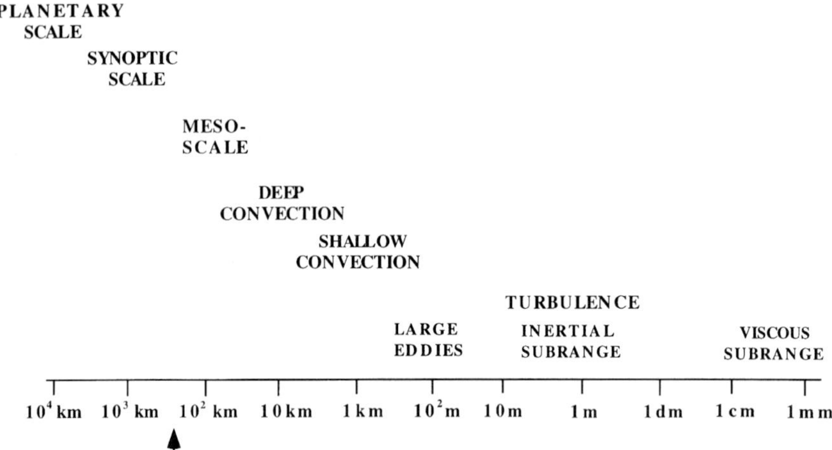

Figure 2   Chart showing the spectrum of atmospheric phenomena. The arrow shows a scale representing typical distances between weather stations and a typical resolution of the early and present GCMs.

B. The "Epoch-Making" First Phase (1950–1960)

The "epoch-making" first phase of numerical modeling of the atmosphere (see Fig. 1) began with the successful 24-hr numerical weather prediction of 500-mb geopotential height by Charney, Fjørtoft, and von Neumann (1950) using the adiabatic quasi-geostrophic equivalent-barotropic model, in which the absolute geostrophic vorticity is horizontally advected by the geostrophic velocity.

Naturally, there were some criticisms against this work. Perhaps the most common and obvious criticism was "500 mb geopotential is not weather." According to Charney, Norbert Wiener at MIT, who proposed to employ linear "black box" prediction methods based on long time series of past data, stated:

> Von Neumann and Charney were misleading the public by pretending that
> the atmosphere was predictable as deterministic system.

In spite of these criticisms, I consider the 10-year period that immediately followed this work "epoch making," not only in the history of numerical modeling of the atmosphere but also in the history of meteorology in general. Through this work, the relevance of such a simple dynamical model for daily changes of weather was demonstrated for the first time in history, and thus dynamic meteorologists began to be directly involved in the practical problem of forecasting. In this way, dynamic meteorology and synoptic meteorology began to merge during this phase.

Realistically including three-dimensional baroclinic effects, but with only a few degrees of freedom in the vertical, then became the main target of NWP model development (Phillips, 1951; Eady, 1952; Eliassen, 1952; Charney and Phillips, 1953; Bolin, 1953). In particular, Charney and Phillips (1953) presented a foundation for vertical discretization of quasi-geostrophic multilevel models. Using an adiabatic three-level quasi-geostrophic model, they reported a successful prediction of the rapid development of a storm observed over the United States in November 1950.

Perhaps at least partly stimulated by this success, NWP became operational first in the United States in 1955 and later in many other countries; but naturally forecasts were not always successful. This disappointment, however, led to encouragement rather than discouragement that the forecasting problem be viewed as a broader scientific problem. That direction can be seen from the issues discussed at the International Symposium on Numerical Weather Prediction held in Tokyo, November 1960, the proceedings for which were published by the Japan Meteorological Society in 1962. Those issues included, among many others,

- Nongeostrophic effects
- Diabatic effects, condensation effects in particular
- Control of discretization errors, especially in long-term integrations

At the end of the symposium, Charney emphasized, "The scientific problems of numerical weather prediction are inseparable from the scientific problems of meteorology in general." Again, this is an obvious statement. However, it was not always remembered when the pragmatic aspects of numerical modeling had to be emphasized.

Another "epoch-making" development during the early part of the first phase was the recognition of the close relation between the dynamics of "cyclones" and that of "general circulation" through

- Observational studies on the meridional transports of angular momentum and heat (e.g., Starr and White, 1954; Bjerknes and Mintz, 1955) and establishment of the concept of available potential energy for understanding the energy cycle of the atmosphere (Lorenz, 1955)
- Identification of the symmetric and wave regimes in laboratory experiments (e.g., Fultz, 1956; Hide, 1956)
- The numerical general circulation experiment by Philips (1956)

Phillips's numerical experiment highlighted these developments. He used the quasi-geostrophic two-level model applied to a middle-latitude β plane with friction and heating terms included. The heating term was a pre-scribed function of latitude. First, to obtain a sufficiently strong meridional temperature gradient starting from a state of rest, a zonally symmetric preliminary experiment lasting 130 days was performed. Small random perturbations were then added to start the zonally asymmetric main experiment, during which the large-scale components of the perturbations grew through baroclinic instability, modifying the general circulation from the zonally symmetric regime to the wave regime. In the latter regime, waves transported heat poleward across the middle latitudes and trans-ported westerly momentum into the middle latitudes, where the westerly jet at the upper level and the meridional temperature gradient below became stronger. At the same time, the zonally averaged meridional circulation in the middle latitudes changed from the Hadley type to the Ferrel type, producing the midlatitude surface westerlies. In this way, the experiment simulated the very basic features of the observed general circulation of the atmosphere, whose causes had been more or less a matter of speculation. Unfortunately, the experiment could not be contin-ued sufficiently long to reach a statistical quasi-equilibrium due to compu-tational instability, later interpreted by Phillips (1959) as nonlinear compu-tational instability (see Section IV). For more details about Phillips's experiment, see Lewis (1998) and the chapter by Lewis, Chapter 3, in this book.

Phillips's work excited many people, including my later collaborator, Yale Mintz of UCLA. He said (Mintz, 1958):

> Although there are details that are wrong, the overall remarkable success achieved by Phillips in using the hydrodynamical equations to predict the mean zonal wind and mean meridional circulations of the atmosphere must be considered one of the landmarks of meteorology.

Charney's statement given at the end of the Tokyo symposium quoted earlier and this statement by Mintz symbolized the dawn of the next phase of numerical modeling of the atmosphere, which I call the "magnificent" second phase (see Section VI).

III. AA'S PERSONAL PRE-UCLA HISTORY

At this point I would like to change the subject to my personal history (Fig. 3). My pre-UCLA period (1950–1960) roughly corresponds to the epoch-making first phase and the UCLA period (1961–) to the magnifi-cent second phase and the beginning of the great-challenge third phase. I have been fortunate to witness and experience all of these phases and to

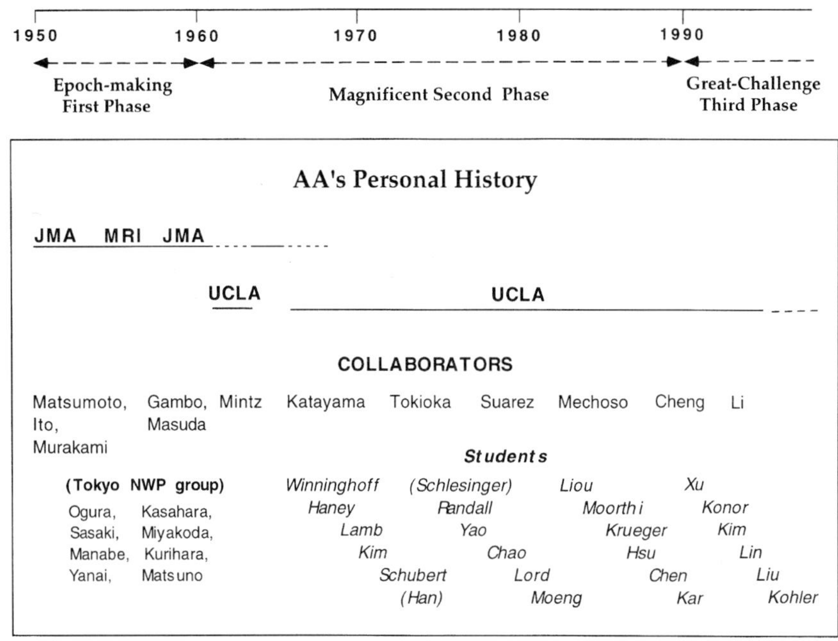

Figure 3 Chart showing the personal history of Akio Arakawa.

have a number of outstanding collaborators, students, and friends (including the members of the Tokyo Numerical Weather Prediction Group during the late 1950s) throughout my entire career, as listed in Fig. 3.

Although I published 11 papers during the pre-UCLA period, it was basically a learning period for me. In 1950, I received my B.S. degree in physics from the University of Tokyo and entered the Japan Meteorological Agency (JMA) with almost no knowledge of meteorology. I did so because, at that time in Japan, the jobs available for physics graduates were extremely limited. JMA administrators must have recognized that my real interest might not be in meteorology, so they put me on a weather ship for a 1-year period to experience severe weather. Another victim of this policy was Vic Ooyama. At least in my case, the policy worked. The weather ship made routine surface, aerological, and marine observations, and analyzed daily surface maps for its own safety. I enjoyed participating in all of these activities and developed a curiosity for weather, especially the laws behind its change.

The locations of the two ship stations maintained by JMA in the early 1950s were X-RAY (152°E, 39°N) and TANGO (135°E, 29°N). Weather is especially tough at X-RAY in the winter season when cold air flows out from the continent to the warmer ocean. During such an event, air becomes moist-convectively unstable as its lower part is warmed and moistened by the sea surface. Because the air above is still relatively dry, the resulting cumulus convection is usually not associated with extensive anvils. Thus we could often see clear sky between cumulus towers. I was fascinated as I watched such clouds from the weather ship. Of course I did not know at that time that I was going to do research on cumulus convection.

In 1951, I was transferred to the forecast research division of the Meteorological Research Institute (MRI), which belongs to JMA. This determined the direction of my entire career. The division was headed by Hidetoshi Arakawa, one of the most famous dynamic meteorologists then in Japan. Although I am not related to him, I owe him very much for giving me the opportunity to work at MRI.

The first research I was involved in was cooperative research with Seiichi Matsumoto and Hiroshi Itoo on observed troughs and frontal systems in the westerlies over Japan. Through this research, which naturally involved interpretation of observations, I gradually developed a desire to fill the great gap between linear theories and nature. Luckily, this period coincided with the beginning of the epoch-making first phase of numerical modeling of the atmosphere. I was particularly inspired by the concept of quasi-geostrophy and then fascinated by the fact that even highly simplified dynamical models such as the quasi-geostrophic barotropic model have some relevance to extremely complicated day-to-day weather changes. Because there was no electronic computer available in Japan at that time, my colleagues and I practiced and used Fjørtoft's graphical method (Fjørtoft, 1952) to apply the barotropic and simple baroclinic models to experimental forecasts.

In the mid-1950s, I also developed a curiosity about why the general circulation of the atmosphere is as observed. This seemed to me one of the most fundamental questions in meteorology. I was excited about the new view on the general circulation developing at that time. I was stimulated by papers on the observed angular momentum and heat balance of the atmosphere, and I made my own small effort to better understand the dynamical constraints on eddy transports in the steady wave regime (Arakawa, 1957a, 1957b).

My excitement about the new developments reached its climax when Phillips's paper appeared in 1956. To share this excitement with other Japanese meteorologists, I published a monograph on the general circulation of the atmosphere through the Meteorological Society of Japan (Arakawa, 1958), which contained an extensive review cf the subject and my own interpretation of why general circulation of the atmosphere is as observed. The monograph also included results from experimental medium-range forecasts of index cycles using a second-order closure model for a barotropic atmosphere. This work was performed in collaboration with K. Gambo (Gambo and Arakawa, 1958; also in Arakawa, 1961) using an electronic computer with only 128 words of memory.

In 1959, JMA started operational NWP and I became a member of the team responsible for developing operational models and objective analysis methods. The first model used for operational forecasts was the barotropic model. I watched and examined vorticity fields predicted by the model everyday. "Noodling" (Platzman, 1961; Miyakoda, 1962) of the vorticity field, as schematically shown in Fig. 4, was usually apparent in predicted maps, and the model sometimes blew up even within 48 hr. The problem occurred where the deformation of the flow was large, not in a region of

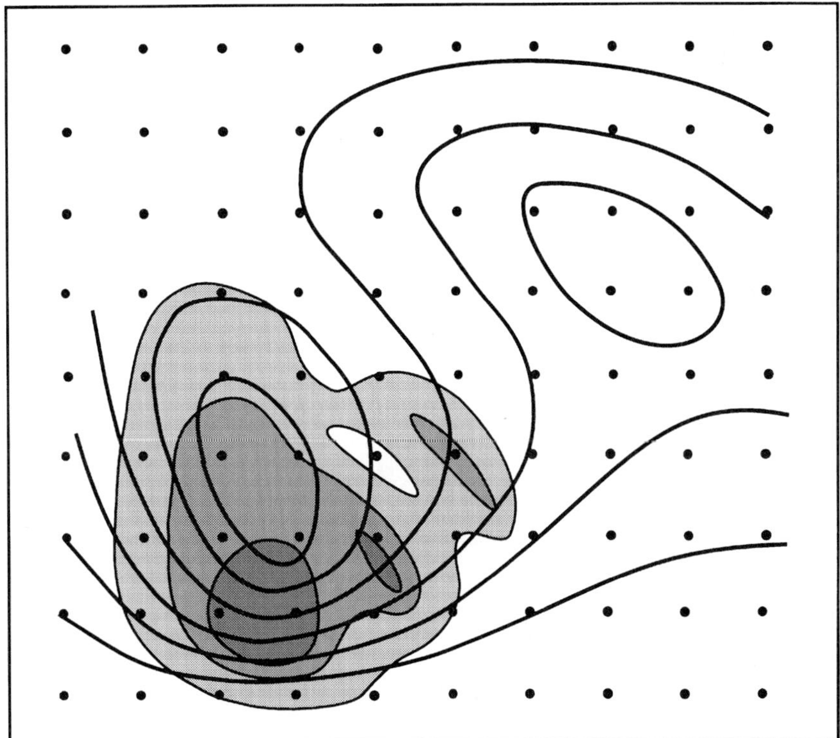

Figure 4 Schematic map showing an early stage of nonlinear computational instability with "noodling" of predicted vorticity. Contours without shading are for geopotential height and those with shading are for vorticity.

strong wind as anticipated from the Courant–Friedrich–Levy stability criterion. Also, growing disturbances did not have an eigenmode-like characteristic spatial structure. Moreover, shortening the time step only postponed the catastrophe. All of these symptoms suggested that the problem was not due to the usual linear computational instability, but was instead due to the "nonlinear computational instability," which made Phillips's model blow up about 30 days after the introduction of random disturbances.

It was fortunate for me that JMA produced maps of the predicted vorticity field operationally. If I had only looked at the predicted geopotential field, I might not have noticed the way in which "noodling" of the vorticity grows. By watching those maps, I began to feel that something was fundamentally different between the dynamics of the continuous system and that of the discrete system. My experience with Fjørtoft's graphical method contributed to the feeling, which later led me to the work of the Arakawa Jacobian (see Section IV).

The Tokyo symposium in 1960 was one of the most important international events in the early history of NWP. This is partly because the timing of the symposium coincided with the dawn of the magnificent second phase. Practically all the important figures in the field, including Bolin, Charney, Döös, Eliassen, Fjørtoft, Gates, Kuo, Lorenz, Mintz, Namias, Pfeffer, Phillips, Platzman, Shuman, Smagorinsky, and Wurtele, attended the symposium. For me, of course, this was the first opportunity to appear on the international scene. As mentioned in Section I, one of the subjects discussed at the symposium was how to include nongeostrophic effects in NWP models. I presented a paper on this subject, which was later published in the proceedings of the symposium (Arakawa, 1962). This paper discussed energetically consistent pairs of the vorticity and divergence equations, which include the balanced model (Charney, 1962; Lorenz, 1960) with Charney's (1955) balance equation; the self-stabilizing effect of a baroclinically unstable wave through the increase of static stability as it develops; and the effect of nongeostrophic effects on frontogenesis.

In 1961, shortly after I received my D.Sc. degree in meteorology, I took a leave of absence from JMA and came to UCLA as a visitor to Mintz's project for a 2-year period. The possibility of this visit was broached before the Tokyo symposium during a conversation between Mintz and Shigekata Shyono, professor at the University of Tokyo and the leader of the Tokyo Numerical Weather Prediction Group. At that time, Mintz had an ambitious plan to develop a general circulation model of the atmosphere based on the primitive equations and was looking for someone to help him. Even from the beginning, Mintz was interested in simulating the geographical distribution of heating over the entire globe season by season with a realistic land–sea distribution and topography. I was excited about the possibility of participating in this task, which was an ideal opportunity for me to combine my interests in NWP and the general circulation of the atmosphere.

When I came to UCLA, however, I was determined to look into the problem of nonlinear computational instability before developing a GCM. Thus, my first role after joining Mintz's project was to persuade him to slow the development, giving first priority to designing model dynamics suitable for long-term integrations. At first, Mintz was irritated by the slower progress. However, he quickly became the strongest supporter of my effort. The main product during this early part of my visit to UCLA was the so-called "Arakawa Jacobian," which is described in the next section.

IV. THE "ARAKAWA JACOBIAN"

As Phillips (1959) showed (and as I experienced with the JMA operational NWP model), nonlinear computational instability may occur in solutions of the nondivergent barotropic vorticity equation, which is perhaps the simplest nonlinear dynamical equation applicable to the real atmosphere. Although motions contributing to general circulation of the atmosphere are usually divergent, they are to a good approximation quasi-nondivergent as far as the horizontal advection terms are concerned. I therefore believed that a finite-difference scheme adequate for use in a GCM should be adequate for nondivergent motions. This motivated me to consider first the nondivergent barotropic vorticity equation, which can be written as

$$\partial \nabla^2 \psi / \partial t = J(\nabla^2 \psi, \psi), \tag{1}$$

where ψ is the streamfunction, $\nabla^2 \psi (\equiv \zeta)$ is the vorticity, ∇^2 and J are the Laplacian and Jacobian operators given by

$$\nabla^2 = \partial^2 / \partial x^2 + \partial^2 / \partial y^2 \tag{2}$$

and

$$J(p, q) = (\partial p / \partial x)(\partial q / \partial y) - (\partial p / \partial y)(\partial q / \partial x), \tag{3}$$

respectively, and x and y are the horizontal Cartesian coordinates.

Phillips (1959) presented a two-wave-component example of nonlinear computational instability, which may appear in finite-difference solutions of Eq. (1) as a consequence of aliasing error. Let us assume, initially, that

$$\psi = \psi_1 + \psi_2, \tag{4}$$

where ψ_1 and ψ_2 are the streamfunctions for the two wave components with

$$\nabla^2\psi_1 = -k_1^2\psi_1 \quad \text{and} \quad \nabla^2\psi_2 = -k_2^2\psi_2. \tag{5}$$

Using Eqs. (4) and (5) with a finite-difference Jacobian $\mathbf{J}$ that satisfies $\mathbf{J}(p, q) = -\mathbf{J}(q, p)$, including $\mathbf{J}(p, p) = 0$, we obtain

$$\partial\nabla^2\psi/\partial t = (k_2^2 - k_1^2)\mathbf{J}(\psi_1, \psi_2). \tag{6}$$

As an example, consider these cases:

$$(\psi_1)_{i,j} = C[\sin(\pi i/2) + \cos(\pi i/2)]\sin(2\pi j/3) \tag{7}$$

and

$$(\psi_2)_{i,j} = U\cos(\pi i)\sin(2\pi j/3). \tag{8}$$

Using the simplest centered finite-difference approximations to the Laplacian and Jacobian operators in Eq. (6), we find, after aliasing $\sin(3\pi i/2) \to -\sin(\pi i/2)$, $\cos(3\pi i/2) \to +\cos(\pi i/2)$, and $\sin(4\pi j/3) \to -\sin(2\pi j/3)$,

$$\partial\psi_{i,j}/\partial t = (\sqrt{3}\,U/10d^2)C[\sin(\pi i/2) + \cos(\pi i/2)]\sin(2\pi j/3). \tag{9}$$

Comparing Eq. (9) with Eqs. (7) and (8), we see that no new wave components are generated in this system by aliasing so that Eq. (4) with Eqs. (7) and (8) is valid for all t. If $U > 0$, however, the amplitude of ψ_1 given by C exponentially increases with time, whereas the amplitude of ψ_2 given by U remains constant. [If $U < 0$, consider $(\sin\pi i/2 - \cos\pi i/2)$ instead of $(\sin\pi i/2 + \cos\pi i/2)$ in Eq. (7).] This result, which is a simplified version of Phillips's (1959), demonstrates that aliasing error can in fact cause computational instability. While it is convincing, this conclusion might have generated a pessimistic outlook on the control of nonlinear computational instability since in practice we cannot avoid the existence of aliasing error in any finite-difference scheme for the Jacobian. Yet I believed that there was no inherent reason that the aliasing error must always *grow* in time.

In the preceding example, ψ_1 grows when $U > 0$ because Eqs. (7) and (9) are then in phase and, therefore, they are positively correlated. I thought this might not be the case in another discretization of the Jacobian. Here we note that Eq. (9) is for an aliased wave with no counterpart in the true solution and, therefore, its phase has nothing to do with the accuracy of the solution. I then thought that, by redesigning the finite-difference Jacobian $\mathbf{J}$ without sacrificing its accuracy, it might be possible to make the phase of the aliased wave tendency $90°$ out of phase with the existing wave so that the wave does not grow in time. For the two-wave-component system, the nongrowth condition is given by the no correlation requirements,

$$\overline{\psi_1 \mathbf{J}(\psi_1, \psi_2)} = 0 \quad \text{and} \quad \overline{\psi_2 \mathbf{J}(\psi_1, \psi_2)} = 0. \tag{10}$$

The problem then becomes the construction of a finite-difference Jacobian for arguments p and q that satisfies discrete analogs of

$$\overline{p\mathbf{J}(p, q)} = 0 \quad \text{and} \quad \overline{q\mathbf{J}(p, q)} = 0. \tag{11}$$

If a finite-difference Jacobian satisfying Eq. (11) can, in fact, be constructed, the impact of its use in solving Eq. (1) is tremendous because it implies that conservations of discrete analogs of enstrophy, $\overline{\zeta^2}/2 = \overline{(\nabla^2\psi)^2}/2$, and kinetic energy, $\overline{\mathbf{v}^2}/2 = \overline{(\nabla\psi)^2}/2$, where $\mathbf{v}$ is the velocity, are formally guaranteed *regardless of the initial condition*. If either of these quadratic quantities is conserved, the solution at each grid point must be bounded and, therefore, there is no room for nonlinear computational instability. Moreover, if both are conserved, there would be no systematic energy cascade to smaller scales, as in the continuous system discussed by Fjørtoft (1953). Then a relatively small amount of energy would accumulate in small scales, for which numerical errors are large, and thus the overall numerical error of the solution would remain relatively small. In this way, the solution would also approximately maintain other statistical properties of the exact solution, such as conservation of the higher moments of the statistical distribution of vorticity.

Once this objective was defined, the actual design of the Arakawa Jacobian satisfying Eq. (11) became rather straightforward if we look at finite-difference Jacobians in terms of interactions between grid points (Arakawa, 1966; see also Arakawa, 1970; Arakawa and Lamb, 1977; or Arakawa, 1988). The trick for conserving ζ^2, for example, is to formulate $\mathbf{J}(\zeta, \psi)$, written in the flux convergence form $-\nabla \cdot (\mathbf{v}\zeta)$ using $\nabla \cdot \mathbf{v} = 0$, in such a way that $\zeta\mathbf{J}(\zeta, \psi)$ can also be written in the flux convergence form $-\nabla \cdot (\mathbf{v}\zeta^2/2)$. This can be achieved through expressing the vorticity flux $\mathbf{v}\zeta$ from one grid point to another as the corresponding mass flux times the arithmetic mean of ζ at those two grid points. Here it is important for the mass flux to satisfy a discrete analog of $\nabla \cdot \mathbf{v} = 0$. Conservation of

$\overline{\mathbf{v}^2} = \overline{(\nabla\psi)^2}$ can then be achieved by requiring the condition $\mathbf{J}(\zeta, \psi) = -\mathbf{J}(\psi, \zeta)$.

The possibility of conserving a squared quantity through the choice of an appropriate expression for the flux was shown by Lorenz (1960) for the vertical discretization of the potential temperature equation (see Section VII). Thus, the work of the Arakawa Jacobian can be considered an application of such a choice to the advection of vorticity by two-dimensional nondivergent velocity, while the antisymmetric property of the Jacobian operator is maintained.

It is interesting to see that, as later recognized by Sadourny *et al.* (1968) and Williamson (1968), the simplest and most straightforward centered finite differencing with a hexagonal grid automatically gives a scheme that has the same properties as the Arakawa Jacobian. It is also interesting to see that the Arakawa Jacobian can be derived using a finite-element method (Jespersen, 1974).

I essentially finished this work in early 1962 and presented it at various conferences in 1962 and 1963. It is rather embarrassing that I did not publish the work until 1966. My strongest motivation then was for development of a comprehensive GCM, which requires a generalization of the approach to the primitive equations. This was not easy, especially when curvilinear orthogonal coordinates are used. It is also embarrassing that the paper (Arakawa, 1966) is called "Part I"—and Part II has never been published. The material I originally had in mind for Part II, which numerically demonstrates the performance of the Arakawa Jacobian, was later published with a different title (Arakawa, 1970; see also Arakawa, 1972; Arakawa and Lamb, 1977; Arakawa, 1988). One of the important results included in the material is that long-term solutions with finite-difference Jacobians conserving energy but not enstrophy (e.g., $\mathbf{J}_3$ given below) can be meaningless, even though they are stable, due to spurious energy cascade to smaller scales. For additional historical introduction of the Arakawa (1966) paper, see Lilly (1997).

Although most people in the atmospheric modeling community almost immediately recognized the importance of this work, I had difficulty convincing some other people because the approach was not standard in numerical analysis. Another type of objection to this approach was "why require conservation while nature does not conserve?" This kind of argument mixes up the mathematical problem of formulating the advection term with the physical problem of formulating the effect of small scales on large scales. The objective of this approach is not conservation itself in the solution of the entire equation (or the entire system of equations); it is the problem of choosing a discrete expression for the advection term that leads to conservation in the solution if there are no other terms. Through such a choice, we gain more freedom in formulating the dissipation terms so that the formulation can be done based on physical considerations. In addition, while conservation is a global constraint, what we gain from this approach can be seen in local solutions as well. As Takacs (1988) nicely put it:

> ...it is not the absolute conservation that is important, but rather the manner in which conservation is obtained. As pointed out by Arakawa (1966), Sadourny (1975) and others, global conservation of nonlinear quantities is a result of correct representation of the local dynamic interaction of a triad of waves. The reverse, however, is not true, i.e., global conservation in itself does not imply correct local nonlinear dynamics.

Conservation, however, should not be overemphasized for general problems since its importance is problem dependent. Obviously, it is not important for low Reynolds number problems, in which the viscosity term dominates over the advection term, *as long as the scheme is stable.*

Through the work described in this section, I recognized that some of the standard concepts in numerical analysis, such as those of truncation error and order of accuracy, do not necessarily provide a sufficient guide for constructing a satisfactory finite-difference scheme. This is especially true for motions such as those in the atmosphere, in which viscosity plays only secondary roles for the majority of its domain. A higher order accuracy is of course desirable as long as solutions remain smooth. For a nonlinear system such as Eq. (1), however, the smoothness of solutions is scheme dependent. This is why a nonconservative higher order scheme can be worse than a conservative lower order scheme.

Finding the truncation error (and order of accuracy) consists of substituting a continuous function (or functions) into the finite-difference expression and expanding the result into a Taylor series with respect to the grid size. Arakawa (1966) presented the truncation error for each of the finite-difference Jacobians, $\mathbf{J}^{++}(\zeta, \psi)$, $\mathbf{J}^{\times+}(\zeta, \psi)$, and $\mathbf{J}^{+\times}(\zeta, \psi)$, which are the simplest centered finite-difference analogs of $J(\zeta, \psi)$ based on the following differential forms, respectively:

$$J(\zeta, \psi) = (\partial\zeta/\partial x)(\partial\psi/\partial y) - (\partial\zeta/\partial y)(\partial\psi/\partial x) \qquad (12)$$

$$= -(\partial/\partial x)(\psi\partial\zeta/\partial y) + (\partial/\partial y)(\psi\partial\zeta/\partial x) \qquad (12a)$$

$$= (\partial/\partial x)(\zeta\partial\psi/\partial y) - (\partial/\partial y)(\zeta\partial\psi/\partial x). \qquad (12b)$$

[Following Lilly (1965), the notations $\mathbf{J}_1(\zeta, \psi)$, $\mathbf{J}_2(\zeta, \psi)$, and $\mathbf{J}_3(\zeta, \psi)$ are used in my later publications in place of $\mathbf{J}^{++}(\zeta, \psi)$, $\mathbf{J}^{\times+}(\zeta, \psi)$, and $\mathbf{J}^{+\times}(\zeta,\psi)$, respectively.] Arakawa (1966) showed that $\mathbf{J}^{++}(\zeta, \psi)$, which was the most commonly used including Phillips (1956, 1959), conserves neither enstrophy nor energy, $\mathbf{J}^{\times+}(\zeta, \psi)$ conserves enstrophy but not energy, $\mathbf{J}^{+\times}(\zeta, \psi)$ conserves energy but not enstrophy, and the Arakawa Jacobian defined by

$$\mathbf{J}_{\mathrm{A}}(\zeta, \psi) \equiv [\mathbf{J}^{++}(\zeta, \psi) + \mathbf{J}^{\times+}(\zeta, \psi) + \mathbf{J}^{+\times}(\zeta, \psi)]/3 \qquad (13)$$

conserves both. [In my later publications, the notation $\mathbf{J}_7(\zeta, \psi)$ is also used for $\mathbf{J}_{\mathrm{A}}(\zeta, \psi)$.] Note that, in spite of these differences in conservation properties, all the finite-difference Jacobians defined above share second-order accuracy. The forms of truncation errors are different between the schemes, but they do not immediately reveal the differences in conservation properties of the finite-difference Jacobians.

When nonlinearity is not dominant, on the other hand, solutions with these finite-difference Jacobians are not very different since they share the same order of accuracy. For example, with a uniform flow in the x direction, U, all of them reduce to the centered second-order finite-difference scheme $-U(\zeta_{i+1} - \zeta_{i-1})/2d$ for $-U\partial\zeta/\partial x$. Here i is the integer index identifying grid points and d is the grid size. Thus, the use of $\mathbf{J}_{\mathrm{A}}$ does not eliminate or reduce any deficiencies the centered second-order scheme may have, such as the computational dispersion of short waves (see Mesinger and Arakawa, 1976, and Chapter 23, Section IV.B, of this book). A higher order scheme can reduce such deficiencies as long as the solution remains smooth. For example, the fourth-order accurate version of $\mathbf{J}_{\mathrm{A}}$, which was also presented in Arakawa (1966), is generally superior to the original second-order version.

The following paragraphs, which are an edited excerpt from Arakawa and Lamb (1977), summarize the view I had on the merit of the approach discussed in this section:

> As the grid size approaches zero, the finite-difference solution obtained with any "convergent" scheme will eventually approach the true solution. If the grid size is sufficiently small, the order of accuracy determines how rapidly its solution approaches the true solution. Although many schemes can share the same order of accuracy, the solutions of those schemes approach the true solution along different paths in a function space, generally with different statistics ... One of the basic principles used in the design of the finite difference scheme for the GCM is to seek a finite difference scheme whose solutions approach the true solution along a path on which the statistics are analogous to those of the true solution.

V. DEVELOPMENT OF THE MINTZ–ARAKAWA MODEL

After finishing the derivation of the Arakawa Jacobian in late 1961, I began to work on designing dynamics for the primitive equation model.

Influenced by the successful work of Phillips (1956), and for economic reasons, a two-level model excluding the stratosphere was an almost obvious choice. Yet developing a primitive equation model for a global domain with surface topography was an extremely challenging task in many ways.

For the vertical coordinate, we chose the σ coordinate proposed by Phillips (1957a), and modified it to give $\sigma = 0$ at the model top placed at 200 mb. The prognostic variables were the surface pressure and the temperatures and velocities at the upper and lower levels (Fig. 5). Moisture was not predicted so that the model was "dry." Although the choice of the σ coordinate was a reasonable one, some difficulties were anticipated over steep topography, where the horizontal pressure gradient force is a difference of two large terms of about equal magnitude. I tried to minimize the error in the difference, at least its systematic part, through a careful design of the vertical discretization of the hydrostatic equation. For more details on this subject, see Section VII.

The first task in the horizontal discretization was to derive a finite-difference scheme for the momentum equation that is equivalent to the use of the Arakawa Jacobian in the vorticity equation when the motion is nondivergent. This work required some laborious manipulations, but overall it was rather straightforward as far as nondivergent flow is concerned. Real difficulties appeared in generalizing the scheme to divergent flow over a sphere. I will come back to this problem in Section VIII.

It was also difficult to choose a grid or a grid system covering the entire globe. Influenced by Phillips's (1957b) idea, in which a grid on the polar stereographic projection for high latitudes and a grid on the Mercator projection for low latitudes are coupled in middle latitudes through interpolation between the grid points, the first thing I tried was coupling two polar stereographic grids near the equator. I almost immediately abandoned this idea due to tremendous computational difficulties, and decided

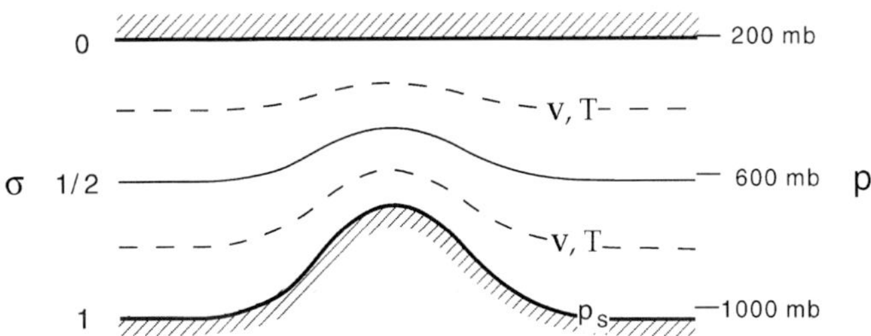

Figure 5 Vertical structure of the two-level model constructed by Mintz and Arakawa in the early 1960s.

to use a grid based on the spherical coordinates with uniform grid intervals in both longitude and latitude. To avoid the use of the extremely short time step required for computational stability due to converging meridians near the poles, each polar cap poleward of 77° was represented by a single point located at the pole.

In the meantime, Mintz proceeded to design and elaborate model physics. His formulation of model physics, which more or less followed his earlier ideas (Mintz, 1958), are described by Mintz (1965) and summarized by Johnson and Arakawa (1996). See also Sections IX and X of this chapter for the formulation of surface heat flux and convective heating. The model included seasonal changes of solar radiation, without diurnal change in the standard version. The long-wave cooling for each layer was given as a function of the temperature at the lower level of the model. These functions were empirically determined using the mean cooling rates calculated by Takahashi *et al.* (1960) for a model atmosphere with average cloud distribution. In this way, development of the first generation of the UCLA GCM, which later became known as the "Mintz-Arakawa model," was completed by the middle of 1963 when I left UCLA for Japan.

The following paragraphs are an excerpt from Johnson and Arakawa (1996):

> Arakawa learned from Mintz throughout the entire period of their association at UCLA, especially during the period of developing physics for the initial GCM just described. Still the most valuable experience for Arakawa was to observe and study Mintz's approach to research. Mintz was a perfectionist, untiringly seeking to understand nature in physical terms. When the GCM was being developed, he attempted to interpret almost every detail of the simulations. When the integrity of the results failed, he suspected erroneous coding. In this way, he found a number of code errors, although he himself never prepared a single FORTRAN statement.

While I was in Japan, Mintz performed simulation studies with the two-level GCM. Mintz (1965) presented the results of a January simulation with 7° latitude by 9° longitude horizontal resolution. Although the simulations of lows over the tropical continents and highs over the Southern Hemisphere subtropical oceans and the Northern Hemisphere subtropical Pacific are poor, presumably due to the lack of explicit calculation of latent heat release, the simulations of highs and lows in higher latitudes, especially the Siberian high and the circumpolar trough around Antarctica, are very good even by present standards (see Fig. 5 of Johnson and Arakawa 1996).

After spending 2 years in Japan, I returned to UCLA in 1965 as a member of the faculty and resumed working on general circulation modeling.

VI. SECOND PHASE OF NUMERICAL MODELING OF THE ATMOSPHERE AND THE EVOLUTION OF DIFFERENT GENERATIONS OF THE UCLA GCM

A. THE "MAGNIFICENT" SECOND PHASE (1960–1990)

Before describing the second and later generations of the UCLA GCM, I would like to go back to the general history of numerical modeling of the atmosphere. As mentioned earlier, the beginning of the magnificent second phase of numerical modeling of the atmosphere (see Fig. 1) roughly corresponds to the development of early GCMs. Besides the Mintz–Arakawa model just described, these GCMs include those developed at Geophysical Fluid Dynamics Laboratory (GFDL; Smagorinsky, 1963; Smagorinsky *et al.*, 1965; Manabe *et al.*, 1965), Lawrence Livermore Radiation Laboratory (LRL; Leith, 1964), and the National Center for Atmospheric Research (NCAR; Kasahara and Washington, 1967).

Development of the early GCMs stimulated the meteorological community to look into the feasibility of a global observation and analysis experiment, as reported by the National Academy of Sciences (1965). This report included the results of Charney's famous predictability experiments using three GCMs, one of which was the Mintz–Arakawa model. An international research program of unprecedented scale in the atmospheric sciences, the Global Atmospheric Research Program (GARP), then followed. The program tremendously stimulated and widely supported the worldwide efforts in general circulation modeling almost throughout the magnificent second phase. Major changes from the first phase to the second phase are listed here:

- The primitive equation approach became standard.
- Computational difficulties associated with the use of the primitive equations and those with long-term integrations were essentially overcome.
- Heating was made the result of motion as well as the cause of motion.
- Water–vapor mixing ratio became a standard prognostic variable.
- The importance of the cumulus parameterization problem was recognized.
- Comprehensive GCMs were developed and extensively used.
- Numerical models with diversified objectives were developed (see Fig.1) and became indispensable tools for predicting and investigating the broad spectrum of atmospheric phenomena.

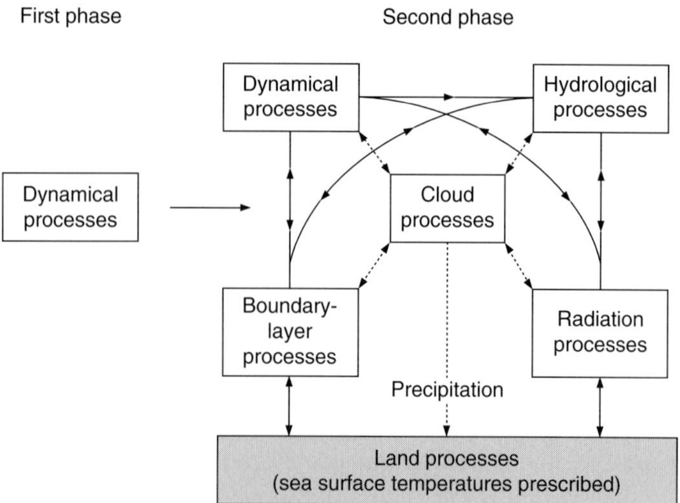

Figure 6   Expansion of processes included in typical GCMs from the epoch-making first phase to the magnificent second phase.

The third item from the top distinguishes the GCMs developed during the second phase from the model used by Phillips (1956), in which heating was a prescribed function of latitude. The right panel of Fig. 6 illustrates processes typically included in those GCMs. Since most models developed during the first phase included only dynamical processes, as shown in the left panel of Fig. 6, the expansion of the scope from the first to second phases was truly magnificent.

Reflecting on the last item in the preceding list, one of the trends of the second phase was the diversification of numerical models of the atmosphere, as shown in Fig. 1. To study specific problems, a hierarchy of idealized "toy" models has also been found useful (see Chapter 10).

B. Evolution of Different Generations of the UCLA GCM

UCLA's major contributions to the magnificent second phase were through almost continuous development of new generations of the GCM, which were made available to other institutions for their own further development and applications. Figure 7 shows the evolution of different generations of the UCLA GCM, with the number of levels, the pressure at model top, the names of the institutions that further developed and applied each generation, and the names of the major contributors to the evolution. The Generation I GCM is the Mintz–Arakawa model described in Section V. In the rest of this section, I outline the model characteristics of Generations II, III, and IV, primarily using Fig. 8, leaving more details of selected modeling aspects to Sections VII through X.

The early version of the Generation II GCM had the same two-level vertical structure as that of the Generation I GCM shown in Fig. 5. The

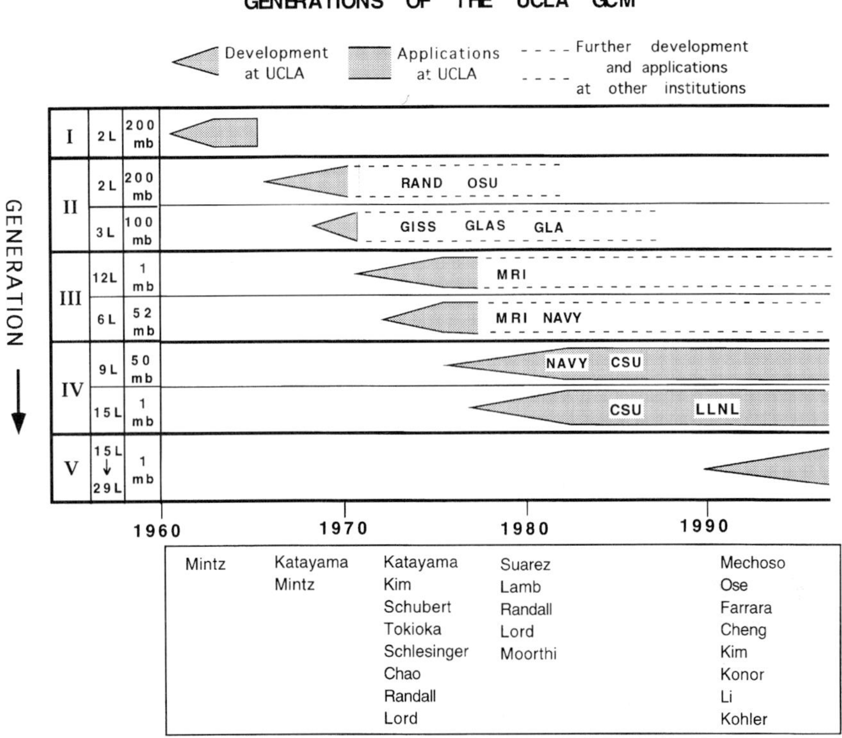

Figure 7 Chart showing the evolution of different generations of the UCLA GCM, with names of the institutions that further developed and applied the GCM and names of the major contributors to the development. The number of levels and the pressure at model top are shown in the second and third columns from left. RAND, Rand Corporation; OSU, Oregon State University; GISS, Goddard Institute for Space Studies, NASA; GLAS, Goddard Laboratory for Atmospheric Science, NASA; GLA, Goddard Laboratory for Atmospheres, NASA; MRI, Meteorological Research Institute, Japan; NAVY, U.S. Navy Fleet Numerical Oceanographic Center & Environmental Prediction Research Facility; CSU, Colorado State University; CWB, Central Weather Bureau, Republic of China; LLNL, Lawrence Livermore National Laboratory.

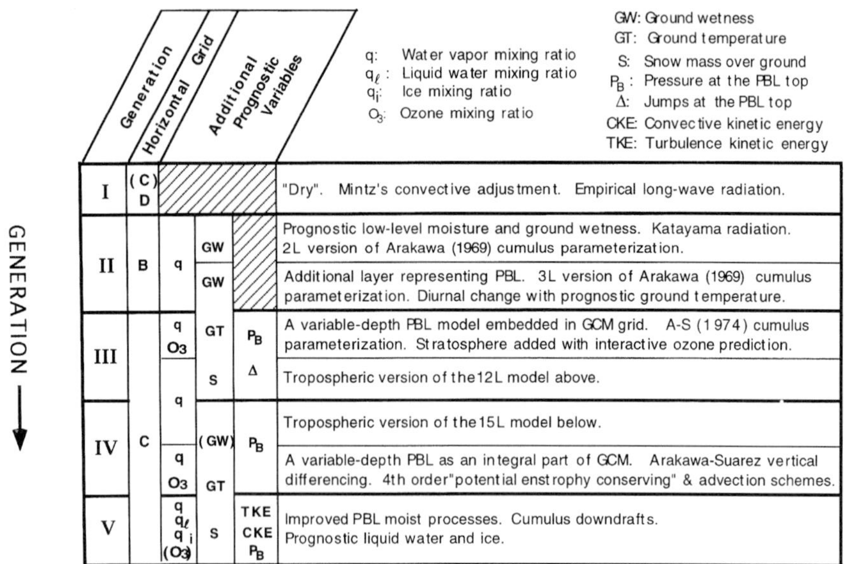

Generation	Horizontal Grid	Additional Prognostic Variables			Description
I	(C) D				"Dry". Mintz's convective adjustment. Empirical long-wave radiation.
II	B	q	GW		Prognostic low-level moisture and ground wetness. Katayama radiation. 2L version of Arakawa (1969) cumulus parameterization.
II	B	q	GW		Additional layer representing PBL. 3L version of Arakawa (1969) cumulus parameterization. Diurnal change with prognostic ground temperature.
III		q O3	GT	P_B	A variable-depth PBL model embedded in GCM grid. A-S (1974) cumulus parameterization. Stratosphere added with interactive ozone prediction.
III			S	Δ	Tropospheric version of the 12L model above.
IV	C	q	(GW)	P_B	Tropospheric version of the 15L model below.
IV	C	q O3	GT		A variable-depth PBL as an integral part of GCM. Arakawa-Suarez vertical differencing. 4th order "potential enstrophy conserving" & advection schemes.
V		q q_ℓ q_i (O3)	S	TKE CKE P_B	Improved PBL moist processes. Cumulus downdrafts. Prognostic liquid water and ice.

Figure 8 Chart outlining the model characteristics of different generations of the UCLA GCM. Prognostic variables in parentheses are temporally frozen.

horizontal domain now covers the entire globe with uniform grid intervals in both longitude and latitude. This became possible after introducing the technique of zonal smoothing of selected terms in the prognostic equations near the poles, which relaxes the Courant–Friedrich–Levy stability criterion by making the effective grid interval in longitude longer (see Section VIII). The horizontal grid structure was also changed from the D Grid to the B Grid (see also Section VIII), with an increased horizontal resolution of 4° latitude by 5° longitude, which became standard for the UCLA GCM. Modifications of the model physics include the addition of water–vapor mixing ratio (for the lower layer) and ground wetness to the set of prognostic variables, the explicit calculation of radiative transfer using the scheme developed by Akira Katayama (Katayama, 1969, 1972; see also Schlesinger, 1976), and the implementation of a two-level version of Arakawa's early cumulus parameterization (Arakawa, 1969). This two-level version of the Generation II GCM, which was still called the "Mintz–Arakawa model," was documented by a group at IBM (Langlois and Kwok, 1969) and in more detail by Gates's group at Rand Corporation (Gates *et al.*, 1971).

Further development of the Generation II GCM included the addition of another layer of 100 mb depth next to the lower boundary, the addition of the ground temperature and snow mass over land to the set of prognostic variables, and the inclusion of the diurnal change in solar insolation. This three-level version of the Generation II GCM was briefly described by Arakawa *et al.* (1969) and later described in detail by Arakawa (1972) and Katayama (1972) who also explained the rationale for its development.

The Generation III GCM is the first multilevel model developed at UCLA. The dynamical aspects of this generation, including the change of horizontal grid structure from the B Grid to the C Grid, are described in detail by Arakawa and Lamb (1977). Major changes in model physics include the implementation of a bulk model for the planetary boundary layer (PBL) based on Deardorff (1972). This PBL model is embedded in the vertically discrete GCM, with explicit prediction of the pressure and the jumps of the prognostic variables at PBL top (Randall, 1976). When the PBL top is higher than the condensation level, the PBL has a sublayer of stratocumulus clouds. Another important change was the inclusion of the Arakawa–Schubert cumulus parameterization (Arakawa, 1972; Schubert, 1973; Arakawa and Schubert, 1974). The 12-level version of the Generation III GCM also includes prediction of ozone mixing ratio with interactive photochemistry (Schlesinger, 1976; Schlesinger and Mintz, 1979).

In the Generation IV GCM, the variable-depth PBL is made an integral part of the vertically discrete model, becoming the lowest layer of the GCM (Suarez *et al.*, 1983). For the troposphere above the PBL, the vertical discretization follows Arakawa and Suarez (1983). The horizontal differencing of the momentum equation is based on the scheme presented by Takano and Wurtele (1981; see Appendix A), which is the fourth-order version of the "energy and potential enstrophy conserving scheme" for the shallow water equations, as designed by Arakawa and Lamb (1981). The horizontal advection scheme for the scalar variables also has fourth-order accuracy (see Appendix B). The Arakawa–Schubert cumulus parameterization was further refined in this generation following the work of Chao (1978), Lord (1978), Lord and Arakawa (1980), Lord (1982), and Lord *et al.* (1982).

VII. VERTICAL DIFFERENCING IN THE UCLA GCM

A. BACKGROUND: LORENZ'S MODEL

Constructing a GCM based on the primitive equations was a challenging task in many ways during the early 1960s. Throughout the first phase of numerical modeling of the atmosphere, 1950–1960, it was standard to use quasi-geostrophic models both in operational forecasts and research, and going beyond quasi-geostrophic models was only done experimentally as far as baroclinic models were concerned. An early example of such experiments is the work by Charney, Gilchrist, and Shuman (1956), who reported that inclusion of some nongeostrophic effects in the prognostic equations did not produce better forecasts, presumably due to the loss of some kind of consistency. Efforts to improve this situation included an extension of the quasi-geostrophic scale analysis to the analysis of the balanced system of equations (e.g., Charney, 1962) and the derivation of energetically consistent sets of vorticity and divergence equations (e.g., Lorenz, 1960; Arakawa, 1962).

In an earlier paper, Lorenz (1955) discussed the total potential energy, $P + I$, the available potential energy, $A \equiv (P + I) - (P + I)_{\min}$, and the gross static stability, $S \equiv (P + I)_{\max} - (P + I)$. Here P is the potential energy, I is the internal energy, and $(P + I)_{\min}$ and $(P + I)_{\max}$ are the minimum and maximum values of $(P + I)$, respectively, that can be obtained by adiabatic mass redistribution from the state in question. Isentropic surfaces are purely horizontal for state $A = 0$ and purely vertical for state $S = 0$. Under adiabatic frictionless processes, these quantities satisfy

$$d(K + A)/dt = 0, \tag{14}$$

$$d(K + P + I)/dt = 0, \tag{15}$$

$$d(K - S)/dt = 0. \tag{16}$$

Here K is the kinetic energy. In the balanced and primitive equation models, these conservation laws hold exactly (though the definition of K differs between the models). In quasi-geostrophic models, on the other hand, Eq. (14) holds approximately but Eqs. (15) and (16) do not. This may be considered as one of the important deficiencies of quasi-geostrophic models. Arakawa (1962), for example, pointed out that the self-stabilization of a developing baroclinic disturbance due to the associated increase of static stability does not operate in quasi-geostrophic models.

Lorenz (1960) further discussed the problem of maintaining important integral constraints in a vertically discrete balanced model with the p coordinate. He stated:

> Our problem is to do this (vertical differencing) in such a way that reversible adiabatic processes still have numerically equal effects upon kinetic energy, total potential energy, and gross static stability. To this end, we define θ and ψ (streamfunction for horizontal velocity) within each layer. At this point we depart from many of the currently used models in which the wind field is

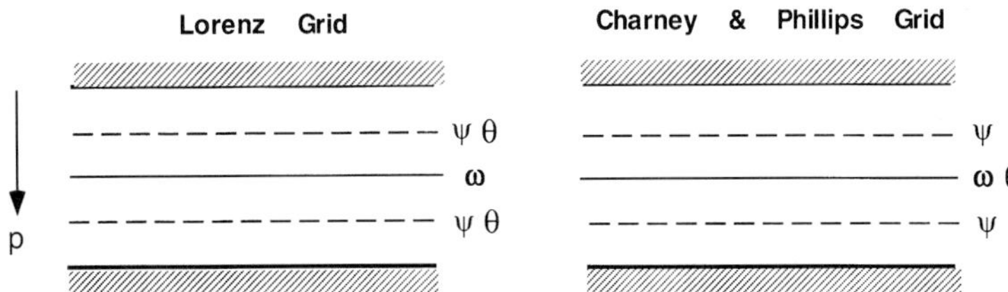

Figure 9 The Lorenz grid and the Charney–Phillips grid applied to the two-level model with the pressure coordinate. Here ψ is the streamfunction for horizontal velocity, θ is the potential temperature, and ω is vertical p velocity, Dp/Dt.

defined at n levels and the temperature field $n-1$ levels (see Charney and Phillips, 1953).

We refer to this type of vertical grid, introduced by Lorenz (1960), as the *Lorenz grid*. Figure 9 compares the Lorenz grid and the Charney–Phillips grid for the case of a two-level model with the p coordinate.

The integral constraints maintained by Lorenz's model included conservation of total energy, $K + P + I$, under adiabatic and frictionless processes, and conservation of the average values of θ and θ^2 under adiabatic processes, where θ is the potential temperature. For a two-level model, conservation of the average values of two functions of θ, such as θ itself and θ^2, is sufficient to constrain the adiabatic mass redistributions necessary to define the available potential energy A and the gross static stability S.

It is interesting that conservation of the mean of θ^2 is achieved in the Lorenz model by formulating the vertical flux of θ at an interface of two layers as the product of the corresponding vertical mass flux and the arithmetic mean of the potential temperatures for the two layers above and below. Recall that a similar formulation is used in the Arakawa Jacobian to conserve $\overline{\zeta^2}$.

B. Evolution of Vertical Differencing in the UCLA GCM

When constructing the Generation I GCM, I was greatly influenced by Lorenz's approach as just described, and determined to extend his approach to the vertical discretization of the primitive equations with the σ coordinate. The Lorenz grid was also attractive from a practical point of view since the grid allowed two temperatures to be predicted even for the two-level model and thus static stability could be predicted, rather than prescribed as in the case of a two-level model with the Charney–Phillips grid.

In retrospect, however, the departure from the Charney–Phillips grid was not a good decision (see Arakawa and Moorthi, 1988; Arakawa and Konor, 1996; see also Chapter 23 in this volume). The Charney–Phillips grid is almost unquestionably the best choice for quasi-geostrophic models (Arakawa and Moorthi, 1988). Although great advantages are reaped from using the Lorenz grid for nongeostrophic models, discretization of such models should be a generalization of, rather than a departure from, the best discretization for quasi-geostrophic models as long as quasi-geostrophy is a good first approximation.

In any case, the Lorenz grid with a modified σ coordinate was an almost obvious choice for the Generation I GCM and became a tradition throughout the history of the UCLA GCM (as is the case for most of the existing large-scale models.) Using that grid, vertical discretizations in Generations I, II, and III satisfy the following integral constraints:

1. That the pressure gradient force generate no circulation of vertically integrated momentum along a contour of the surface topography
2. That the finite-difference analogs of the energy conversion term have the same form in the kinetic energy and thermodynamic energy equations and thus the mass integral of the total energy be conserved under adiabatic and frictionless processes
3. That the mass integral of the potential temperature θ be conserved under adiabatic processes
4. That the mass integral of some $f(\theta)$ other than θ itself, such as θ^2 or $\ln \theta$, be conserved under adiabatic processes.

Constraints 2, 3, and 4 with $f(\theta) = \theta^2$ follow Lorenz (1960). Constraint 1 was added to constrain the horizontal pressure gradient force in a σ-coordinate model, whose error can be serious near steep topography. When the p coordinate is used, the horizontal pressure gradient force is a gradient vector. Then a line integral of its tangential component taken along an arbitrary closed curve on a coordinate surface vanishes. Thus, error in computing geopotential ϕ does not matter for the generation of vorticity as long as the coordinate surface does not intersect the Earth's surface. A similar situation exists for the θ coordinate. When $\sigma = p/p_s$ is used as the vertical coordinate, on the other hand, the pressure gradient force is given by $-[\nabla_\sigma \phi + \sigma \alpha \nabla p_s]$, where α is the specific volume satisfying the hydrostatic equation $\partial \phi / \partial \sigma = -\alpha p_s$. Then the above constraint on the line integral generally does not hold for individual coordinate surfaces. Yet we can show that the horizontal pressure gradient force generates no circulation of *vertically integrated* horizontal momentum along a contour of the surface topography or along a surface isobar, or along an arbitrary curve if p_s is a single-valued function of ϕ_S. A global consequence of this constraint is that the horizontal pressure gradient force generates no angular momentum of the atmosphere except through the mountain torque.

Constraint 2 also can reduce the systematic error of the horizontal pressure gradient force, because it requires the work done by the force to be consistent with the conversion of total potential energy to kinetic energy anticipated from the discrete thermodynamic energy equation. Constraints 3 and 4, on the other hand, are of the form of the discrete thermodynamic equation.

The Generation I and II GCMs satisfy all of these constraints with $f(\theta) = \theta^2$ for constraint 4 (Arakawa, 1972). The Generation III GCM also satisfies all of these with $f(\theta) = \theta^2$ for the troposphere and $f(\theta) = \ln \theta$ for the stratosphere (see Arakawa and Lamb, 1977, for motivation). In addition, the vertical differencing scheme for the stratosphere was designed by Tatsushi Tokioka to eliminate any false computational internal reflections of the wave energy propagating in a resting isothermal atmosphere (Arakawa and Lamb, 1977; Tokioka, 1978).

When all four of the above constraints are imposed, however, the discrete hydrostatic equation becomes nonlocal, and the way in which the nonlocality appears can seriously affect the local accuracy. Arakawa and Suarez (1983), therefore, abandoned constraint 4 and instead required that the discrete hydrostatic equation be local. They derived a family of vertical difference schemes that satisfies constraints 1 and 2 and the locality requirement, and showed that the scheme proposed by Simmons and Burridge (1981) at ECMWF is a member of the family. They further showed that another member of the family satisfies constraint 3 as well. The Generation IV GCM uses this scheme for the troposphere, whereas the vertical difference scheme for the stratosphere remains the same as that for the Generation III GCM.

The vertical difference scheme for advection of water vapor and ozone mixing ratios in Generation III and later generations conserves the mass integral of both q and $\ln q$, where q is the mixing ratio, except when water vapor is saturated (see Section X.D).

C. Further Remarks on Vertical Differencing

The conservation of enstrophy achieved by the Arakawa Jacobian for two-dimensional nondivergent flow is an effective computational constraint on local solutions, as well as global solutions, because the enstrophy is the square of second-order derivatives of the streamfunction, which is sensitive to the amount of energy existing in small scales. The conservation properties discussed in this section, on the other hand, can be physically important, but they usually do not provide effective computational constraints on local solutions (possibly except constraints 1 and 2 near steep topography when the σ coordinate is used). For example, if the finite-difference scheme for the momentum equation is unstable, K can increase almost without limit, while satisfying the total energy conservation Eq. (15), since $P + I$ is an almost infinite energy source for K. Also, satisfying those conservation properties does not eliminate or reduce any inherent deficiencies the Lorenz grid may have, such as the existence of a vertical computational mode (Arakawa and Konor, 1996; see also Section III.B of Chapter 23, this volume).

I should also mention that discretization of the vertical advection of water vapor (and other atmospheric constituents) is not quite a settled problem in my mind and I consider it as one of the most important issues in future development of numerical models of the atmosphere. See Section IV of Chapter 23 for further discussion of these problems.

VIII. HORIZONTAL DIFFERENCING IN THE UCLA GCM

A. HORIZONTAL DIFFERENCING IN THE GENERATION I GCM

As mentioned earlier, the first task in horizontal discretization for constructing the Generation I GCM was to derive a finite-difference scheme for momentum advection that is equivalent to the use of the Arakawa Jacobian for vorticity advection when the motion is nondivergent. This approach was followed almost throughout the entire history of the UCLA GCM, with the exception of Generation II.

The first step in this task was to decide the way in which

$$\zeta = (1/a \cos \varphi)[\partial v/\partial \lambda - \partial(u \cos \varphi)/\partial \varphi] \tag{17}$$

is finite differenced, where ζ is the vorticity, u and v are the zonal and meridional components of velocity, respectively, λ is the longitude, and φ is the latitude. Finite differencing of Eq. (17) in turn depends on the way in which u and v are distributed over the grid points.

Given the grid for the vorticity equation shown in Fig. 10a, the simplest and most straightforward way of distributing u and v is the one shown in Fig. 10b. With this grid, the finite-difference expression of ζ consists of the

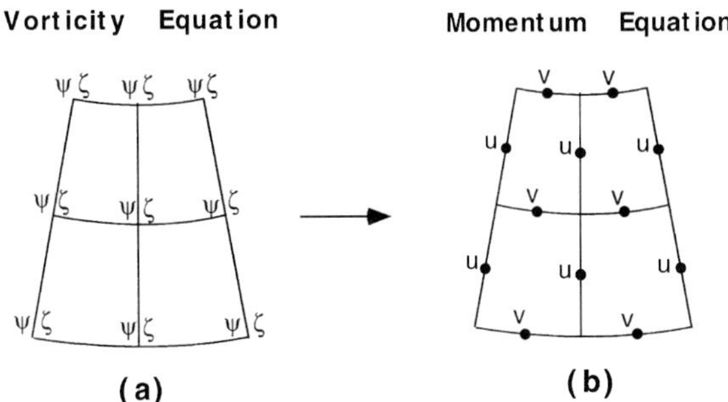

Figure 10 (a) Horizontal grid for the vorticity advection on which the Arakawa Jacobian is based. (b) Corresponding grid for velocity components.

differences of u and v over single grid intervals without any space averaging. If the expression involves space averaging, the smallest scale checkerboard patterns in the u and v fields do not contribute to the magnitude of ζ and, therefore, to that of enstrophy, $\overline{\zeta^2}/2$. Then, even when the enstrophy is conserved, this does not computationally constrain the evolution of a checkerboard pattern.

Having decided on the grid for velocity components, the second step was to derive a finite-difference scheme for momentum advection that is equivalent to the Arakawa Jacobian for vorticity advection when the motion is nondivergent. As I mentioned earlier, this work required some laborious manipulations, but overall it was rather straightforward as far as a scheme for nondivergent motions is concerned. Obviously, the result is not unique since the irrotational part of the momentum advection is not constrained, and a discrete analog of the nondivergence condition can be used to rewrite any part of the scheme. This nonuniqueness does not matter as long as the motion is nondivergent, but it does matter in generalizing the scheme to divergent flow.

The positions of scalar points must also be decided in the grid for the primitive equations. In the Generation (I) GCM, which is a very preliminary version of the Generation I GCM, I used the upper left grid in Fig. 11, which was later given the name "C Grid." The results from this preliminary version, however, were very strange since an initially strong westerly jet rapidly (within several days) broke down. In retrospect, this was due to a special type of computational instability, which I will discuss later in Section VIII.F. At that time I thought that the problem was in C Grid and I almost immediately switched to the left lower grid in Fig. 11, which was later given the name "D Grid." My reasoning was that this grid

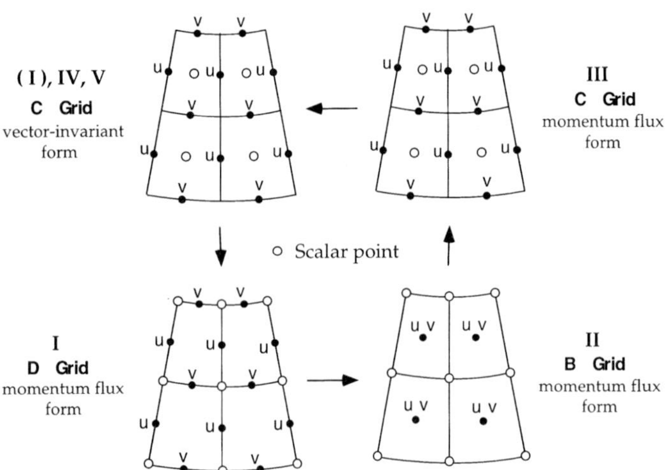

Figure 11   Horizontal grids used in different generations of the UCLA GCM. (I) denotes a preliminary version of Generation I.

should be better for quasi-geostrophic motions since the expression for a geostrophic balance is the most straightforward with this grid. In fact, the scalar points in D Grid coincide with the streamfunction points in Fig. 10a. This is certainly reasonable for quasi-geostrophic motions. The Generation I GCM was then constructed using this grid.

The Generation I GCM was successful in many ways, and any deficiencies of D Grid were hidden until either an artificial lateral boundary was introduced for an experimental purpose or moisture was introduced as a prognostic variable to construct the Generation II GCM. Especially with moisture, the results with D Grid were almost disastrous at times. It is natural that predicted fields tend to be noisier when the condensation process is included. An inadequate formulation of condensation might even produce runaway convection. It was strange to me, however, that the noisy pattern in the predicted pressure and temperature fields rapidly propagated in space with an obviously nonphysical large group velocity.

B. Geostrophic Adjustment in Discrete Systems

Motions associated with the noise generated with D Grid were obviously not in a geostrophic balance even approximately. I then recognized that there was no point in using a grid structure suitable for *describing* quasi-geostrophic motions if the model is not capable of *producing* those motions through geostrophic adjustment. As Arakawa and Lamb (1977) later stated,

> ... there are two main computational problems in the simulation of large-scale motions with the primitive equations. One is the proper simulation of the geostrophic adjustment. The other is the proper simulation of the slowly changing quasi-geostrophic (and, therefore, quasi-nondivergent) motion after it has been established by geostrophic adjustment.

Collaborating with Frank Winninghoff, who was working on a data assimilation problem, I looked into geostrophic adjustment mechanisms in discrete analogs of the shallow water equations. We considered five ways of distributing the dependent variables, A Grid through E Grid, with the simplest centered finite-difference scheme for each. A Grid is the standard nonstaggered grid; B Grid, C Grid, and D Grid are staggered grids as shown in Fig. 11; and E Grid is similar to B Grid but rotated by 45° while the directions of u and v remain unchanged. As presented in Winninghoff (1968), Arakawa (1972), and in later publications, including Arakawa and Lamb (1977), geostrophic adjustment is greatly affected by the choice of grid due to the different dispersion properties for inertia-gravity waves. For the discrete pressure gradient force to be centered with grids other than the C Grid, the pressure difference over one grid interval must be averaged in space. Due to this averaging, the force vanishes for a pattern characterized by either a one-dimensional grid-to-grid oscillation or a two-dimensional checkerboard pattern in the pressure field. Thus, geostrophic adjustment does not operate for these patterns. From this point of view, the C Grid is the best and the D Grid is the worst.

With the C Grid, however, as well as with the D Grid, the Coriolis force involves space averaging because u and v are not defined at the same point. Then the Coriolis force vanishes for a pattern characterized by either a one-dimensional grid-to-grid oscillation or a two-dimensional checkerboard pattern, in the velocity field this time. Geostrophic adjustment again does not operate for these patterns.

More specifically, we can show (see Arakawa and Lamb, 1977, for example) that geostrophic adjustment can be best simulated with the C Grid when d/λ is sufficiently smaller than 2. Here d is the grid size, $\lambda = (gh)^{1/2}$ is the radius of deformation, and $(gh)^{1/2}$ is the speed of gravity wave. When d/λ is near or larger than 2, on the other hand, geostrophic adjustment is poorly simulated with the C Grid.

Unlike the case of the shallow water equations, however, h is not a prescribed constant in an atmospheric model and, therefore, λ depends on the equivalent depth of the wave. For the inertia-Lamb wave, the equivalent depth is such that λ is approximately 3000 km and, therefore, $d/\lambda < 2$ is well satisfied by typical horizontal grid sizes of atmospheric models. Similarly, $d/\lambda < 2$ can easily be satisfied by λ of the gravest vertical mode of internal inertia-gravity waves. These situations indicate that the C Grid is the best for the geostrophic adjustment between the surface pressure and the vertically integrated wind velocity, and for the thermal wind adjustment between the vertical mean temperature and the vertical mean wind shear. The C Grid, however, has difficulties in the thermal wind adjustment for higher vertical modes.

In any case, the D Grid, which was used in the Generation I GCM, was the worst choice for a primitive equation model, in which the geostrophic adjustment mechanism must be explicitly simulated. The C Grid is not perfect, however, as pointed out earlier. I thus chose the B Grid for the Generation II UCLA GCM as a compromise.

C. Horizontal Differencing in the Generation II GCM

The decision to use the B Grid for the Generation II UCLA GCM immediately raised two problems. One problem is that, with the B Grid, the finite-difference scheme for the momentum equation cannot be made formally equivalent to the use of the Arakawa Jacobian for vorticity advection, even when the motion is nondivergent, since the B Grid is different from the grid shown in Fig. 10b. Therefore I required in the Generation II GCM that $\overline{(\partial u/\partial y)}^2$ and $\overline{(\partial v/\partial x)}^2$ be separately conserved *during the advection* by nondivergent velocity. This can be done by designing the finite-difference Jacobian representing the momentum advection, $\mathbf{J}(u, \psi)$ for example, to satisfy $\overline{(\partial u/\partial y)\partial \mathbf{J}(u, \psi)/\partial y} = 0$. [Note that $\partial \mathbf{J}(u, \psi)/\partial y = \mathbf{J}(\partial u/\partial y, \psi)$ when $\mathbf{J}(u, \partial\psi/\partial y) = 0$ is satisfied in the finite-difference Jacobian.] Since momentum is not simply advected in the momentum equation, mainly due to the existence of the pressure gradient force, these constraints are not analogous to any physical constraints on nondivergent flow such as enstrophy conservation; they are, however, equally effective computational constraints on the advection terms.

The other problem associated with the use of the B Grid comes from the discrete forms of the pressure gradient force in the momentum equation and the horizontal convergence term in the continuity equation. With this grid, centered finite-difference expressions for these terms inevitably involve space averaging: meridional averaging of zonal difference and zonal averaging of meridional difference (see the lower right grid in Fig. 11). Due to this averaging, a checkerboard pattern in the mass field does not contribute to the pressure gradient force and thus such a pattern is decoupled from the dynamics of the model. Similarly, a checkerboard pattern in the velocity field does not contribute to the horizontal convergence term and thus such a pattern is decoupled from the mass budget of the model. To avoid this situation, the Generation II GCM uses a one-sided difference at one time level. However, to obtain an overall accuracy comparable to the centered difference, the one-sided difference on the opposite side is used at the next time level. [This is the time-alternating space-uncentered (TASU) scheme.]

Problems similar to those discussed here for the B Grid exist for the E Grid. See Mesinger (1973) and Janjic (1974, 1984) for different approaches to dealing with these problems (also see Chapter 13 in this volume).

D. ZONAL SMOOTHING OF SELECTED TERMS NEAR THE POLES

Another major computational feature introduced into the Generation II GCM is the zonal smoothing of selected terms near the poles, which include the pressure gradient force in the zonal component of the momentum equation and the convergence of zonal mass flux in the continuity equation. This smoothing became necessary as the horizontal domain of the GCM expanded to cover the entire globe with equal grid intervals in latitude as well as in longitude.

It is well known that, when an explicit time-differencing scheme is used for an oscillation equation, $|\nu|\Delta t < 1$ or a similar condition must be satisfied for computational stability (see Mesinger and Arakawa, 1976, for example). Here ν is the frequency (of the continuous solution) and Δt is the time interval. For a hyperbolic partial differential equation, the stability condition becomes the Courant–Friedrich–Levy (CFL) condition. In a one-dimensional case, the condition becomes

$$C_{\max}\Delta t/\Delta x < \text{constant}, \tag{18}$$

where $C_{\max}$ is the maximum phase speed due to either advective or wave-propagation processes in the model and Δx is the grid size in space. The grid size appears in the denominator because, due to the truncation error in space, the larger the grid size the slower the effective maximum frequency.

The zonal smoothing of selected terms in the Generation II and later generations of the UCLA GCM was designed to reduce the effective maximum frequencies in high latitudes, which are otherwise too high compared with those in lower latitudes due to the small Δx near the poles.

Zonal smoothing of the selected terms can do the reduction by increasing the effective Δx. In this way, the use of an extremely short time interval necessary to satisfy the CFL condition can be avoided. The smoothing operation does not smooth the prognostic variables themselves. It is simply a generator of multiple-point differences in the space finite-difference scheme. For more details, see Arakawa and Lamb (1977) and Takacs and Balgovind (1983).

E. Horizontal Differencing in the Generation III GCM

The Generation III UCLA GCM returned to the C Grid (see Fig. 11) to further pursue the possibility of making the finite-difference scheme for momentum advection equivalent to the Arakawa Jacobian for vorticity advection when the motion is nondivergent. As pointed out earlier, such a scheme cannot be uniquely determined. Although this lack of uniqueness does not matter as long as the motion is nondivergent, it does matter in generalizing the scheme to divergent flow.

We can think of two principles that can guide us in generalizing the momentum advection scheme to a divergent flow: (angular) momentum conservation and potential enstrophy conservation. In either case, kinetic energy conservation under advective processes may be simultaneously considered. The Generation III GCM followed the former by using the flux convergence form of the momentum equation.

For the C Grid (see Figs. 11 and 10), Arakawa and Lamb (1977) showed that, for a nondivergent flow, the use of $\mathbf{J}_A(u, -\overline{\psi}^y)$ for $-\mathbf{v} \cdot \nabla u$ at u points and $\mathbf{J}_A(v, \overline{\psi}^x)$ for $-\mathbf{v} \cdot \nabla v$ at v points is equivalent to the use of $\mathbf{J}_A(\zeta, \psi)$ for $-\mathbf{v} \cdot \nabla \zeta$ at ζ points. Here $\mathbf{v}$ is the horizontal velocity, $\mathbf{J}_A$ is the Arakawa Jacobian as previously defined, and $\overline{(\ )}^x$ and $\overline{(\ )}^y$ are the averages over two neighboring grid points in x and y, respectively. For nondivergent flow with $\nabla \cdot \mathbf{v} = 0$, these schemes for $-\mathbf{v} \cdot \nabla u$ and $-\mathbf{v} \cdot \nabla v$ immediately give schemes for the momentum flux convergence, $-\nabla \cdot (\mathbf{v} u)$ and $-\nabla \cdot (\mathbf{v} v)$. Arakawa and Lamb then generalized these schemes to the case of divergent flow while conserving energy. In this generalization, however, exact conservation of the enstrophy for nondivergent flow on a sphere, based on the vorticity given by Eq. (17), was sacrificed.

F. Horizontal Differencing in the Generation IV GCM

The horizontal differencing of the momentum equation in the Generation IV UCLA GCM is based on the scheme for the shallow water equations derived by Takano and Wurtele (1981), which is the fourth-order version of the energy and potential enstrophy conserving scheme of Arakawa and Lamb (1981). Potential enstrophy conservation rather than (angular) momentum conservation guided construction of these schemes. A description of the Takano and Wurtele scheme is given in Appendix A, since it has never been formally published.

Considering the shallow water equations, Arakawa and Lamb (1981) demonstrated the importance of conserving the potential enstrophy, $\overline{hq^2}/2$ $\equiv (1/2)(f + \zeta)^2/h$, in simulating flow over steep orography. Here $q \equiv (f + \zeta)/h$ is the potential vorticity for shallow water, ζ is the vorticity, and h is the depth of the fluid layer. Based on (the zonal and meridional components of) the vector invariant form of the momentum equation given by

$$\partial \mathbf{v}/\partial t = -q\mathbf{k} \times h\mathbf{v} - \nabla(K + \phi), \tag{19}$$

where $K \equiv \mathbf{v}^2/2$, $\phi \equiv g(h + h_s)$ and h_s is the height of the lower boundary, Arakawa and Lamb (1981) derived a family of second-order schemes that conserve potential enstrophy and energy when the mass flux $h\mathbf{v}$ is nondivergent. The scheme derived by Sadourny and subsequently used by the European Centre for Medium Range Forecasts (Burridge and Haseler, 1977) is a member of this family (see also Arakawa and Hsu, 1990). Arakawa and Lamb (1981) further showed that another member of this family conserves those quantities even when the mass flux is divergent. Since potential enstrophy reduces to the usual enstrophy when h is constant, the use of any of these schemes for momentum advection is equivalent to the use of the Arakawa Jacobian for vorticity advection when the motion is nondivergent.

Another important advantage of using schemes based on Eq. (19) is that the derivatives of velocity or its components appear only in q and $\nabla\mathbf{v}^2/2$, both of which are well defined even at the poles. The flux convergence form used in the Generation III GCM, on the other hand, includes the longitudinal convergence of the momentum flux. This flux is multivalued at the poles and, therefore, its convergence is generally infinite. The momentum equation also includes the metric term involving $u \tan\varphi/a$, where a is the Earth's radius, which also becomes infinite at the poles. Thus, the total inertia effect near the poles generally involves a difference of two large terms.

Solutions with this family of schemes generally behave well for the shallow water equations (e.g., Arakawa and Lamb, 1981). For a three-dimensional flow, however, the governing equations are analogous to the shallow water equations only when material surfaces are used as coordinate surfaces, as in the θ coordinate under an adiabatic process. The use of such schemes with the θ coordinate then guarantees conservation of the potential enstrophy based on the (quasi-static version of) Ertel's potential vorticity $(2\mathbf{\Omega} + \nabla \times \mathbf{v}) \cdot \nabla\theta/\rho$, where $\mathbf{\Omega}$ is the Earth's angular velocity vector. When the p or σ coordinate is used, however, this analogy breaks down for the baroclinic (or internal) modes.

At least partly for this reason, a formal application of such schemes to a model with the p or σ coordinate, with the replacement of h by the mass of the model layer, can cause an "internal symmetric computational instability," as pointed out by Hollingsworth and Kållberg (1979) and Hollingsworth *et al.* (1983). This instability, which is also called "symmetric instability of computational kind" (SICK), is characterized by spurious energy conversion from the zonal kinetic energy to the energy of meridionally propagating internal inertia-gravity waves. This instability is unique in the sense that it is a linear computational instability although it originates from space differencing, not from time differencing. The existence of this computational instability also gives us the lesson that the results of testing a space-differencing scheme with the shallow water equations does not necessarily apply to a three-dimensional model.

This instability, however, can be eliminated by a proper formulation (a SICK-proof formulation) of the term $\mathbf{v}^2/2$ in Eq. (19), which generally violates the strict conservation of energy unless the mass flux is nondivergent. For more details on this subject, see Hollingsworth *et al.* (1983). Arakawa and Lamb (1981) and Appendix A of this chapter include the SICK-proof formulations for the Arakawa–Lamb and Takano–Wurtele schemes, respectively.

The horizontal differencing of the advection equation for scalar variables in the Generation IV GCM is also based on a fourth-order scheme on the C Grid. Because the scheme has not been published anywhere, its description is included in this chapter as Appendix B.

IX. FORMULATION OF PBL PROCESSES IN THE UCLA GCM

A. FORMULATION OF PBL PROCESSES IN THE GENERATION I GCM

The formulation of convective processes in the Generation I UCLA GCM (Mintz, 1965) was based on moist-convective adjustment assuming that a sufficient amount of water vapor was available for condensation whenever conditional instability existed. In this sense, the model is "fully moist" rather than "dry," although no grid-scale condensation was included.

The idea of moist-convective adjustment was partially used even in determining the surface heat flux by considering "boundary layer convection." In this formulation, the temperature lapse rate between the model's lower level and the ground was adjusted toward its critical value, which is the moist adiabatic lapse rate when conditionally unstable. An empirically determined coefficient is used for the adjustment. Over oceans, the adjustment modified the lower level temperature T_3 while the ground temperature Tg was fixed. Over land, on the other hand, both T_3 and Tg were modified in such a way that the sum of the upward convective heat flux and the surface long-wave radiation flux (prescribed) is equal to the surface insolation, and thus no heat is stored in the ground.

This formulation is of course very crude. For example, there is no room for the Monin–Obukhov similarity theory (Monin and Obukhov, 1954) for the surface layer or even use of the bulk aerodynamical formula for surface fluxes. The formula applied to the total heat flux (sensible heat flux plus latent heat flux) can be written as

$$(F_h)_s = \rho C_H |\mathbf{v}_a|(h_g^* - h_a), \tag{20}$$

where F denotes the turbulent flux, h is the moist static energy defined by $h \equiv s + Lq \equiv c_p T + gz + Lq$, s is the dry static energy, q is the mixing ratio of water vapor, L is the latent heat per unit mass of water, the subscript s denotes the surface, the subscript a denotes the air at a height within the surface layer such as the anemometer level, and h_g^* denotes the saturation moist static energy of the ground. Other symbols are standard. Besides the direct dependence of the flux on wind speed $|\mathbf{v}_a|$, the coefficient C_H is a function of the ground wetness, surface Richardson number, and surface roughness. None of these "details" matters in the above formulation.

Nevertheless, the formulation used in the Generation I GCM is in the right direction at least conceptually. For models without an extremely high vertical resolution, h_a in Eq. (20) is unknown so that the formula simply relates the two unknowns, $(F_h)_s$ and h_a, to each other. (To simplify the argument, here I pretend $|\mathbf{v}_a|$ and C_H are known.) This situation, which can easily be forgotten when we are concerned with the "details," is in sharp contrast to the usual application of the formula to observations.

Because the surface layer is approximately a constant flux layer, we may replace the left-hand side of Eq. (20) by $(F_h)_{s+}$, which is F_h at the top of the surface layer. If $(F_h)_{s+}$ can be determined from the processes occurring above the surface layer, the bulk aerodynamical formula simply diagnoses h_a, if it is needed, from the known flux. The formulation in the Generation I GCM bypasses this diagnosis.

More generally, it is important to remember that what really determines the time-averaged surface heat flux is the rate of removal of heat from the surface layer. The same can be said for sensible heat and latent heat separately and for momentum. I will emphasize this again in Chapter 23 of this book.

For the surface stress the Generation I GCM uses the bulk aerodynamic formula for momentum with a constant drag coefficient based on $\mathbf{v}_a$ determined by a linear extrapolation of $\mathbf{v}$ from above.

B. FORMULATION OF PBL PROCESSES IN THE GENERATION II GCM

The most important change in the model physics from Generation I to Generation II was the inclusion of (low-level) moisture and ground wetness as prognostic variables. Formulation of the PBL processes in Generation II followed the approach of Generation I, but as further elaborated by Katayama. It now calculated the surface air temperature, which is necessary to determine the drag coefficient since it depends on the temperature difference between the ground and surface air. The calculation used h_a, as determined by Eq. (20), with the surface relative humidity diagnosed from the relative humidity at the model's lower level and the ground wetness.

Further development of the Generation II GCM included the addition of the ground temperature and snow mass over land as prognostic variables. It also introduced an additional model layer of approximately 100-mb depth next to the lower boundary to explicitly represent the outer PBL (the PBL above the surface layer). The addition of this layer was especially important for the implementation of a cumulus parameterization scheme being developed around that time (see Section X). This version of the Generation II GCM was briefly described by Arakawa *et al.* (1969) and later described in detail by Arakawa (1972) including the rationale for its development.

C. BACKGROUND FOR THE PBL FORMULATIONS FOR LATER GENERATIONS

While working on the Generation II UCLA GCM, Mintz and I began to feel that a more sophisticated PBL formulation was badly needed. Mintz therefore invited James Deardorff to UCLA for a 6-month period to look into what we could do for PBLs in GCMs. The result is described by Deardorff (1972).

The first step of Deardorff (1972) was the application of the standard Monin–Obukhov similarity theory, as formulated by Businger *et al.* (1971), to obtain relations between the surface fluxes and the profiles of temperature, moisture, and wind in the surface layer. Here two length scales (other than the height itself) appear: the Monin–Obukhov length L and the surface roughness length z_0. Deardorff then made an additional similarity assumption for the outer PBL. Here he assumed that the relevant length scales are the Monin–Obukhov length and the PBL depth, the latter of which constrains the length scale of turbulence in the outer PBL, replacing the surface roughness length for the surface layer. Matching these two formulations at a height typical of the surface layer top gives the desired relations between the surface fluxes and bulk properties of the outer PBL. In these relations, which are formally similar to the bulk aerodynamical formula, the mean values of the outer PBL appear instead of the anemometer-level values, and the stability dependence of the surface transfer coefficients is expressed with a bulk Richardson number that depends on the PBL depth.

The PBL depth, however, is highly variable in space and time, and depends on the history of the PBL. Deardorff (1972) proposed using a rate equation representing the mass budget for the PBL to prognostically determine the PBL depth. In this equation, turbulent mass entrainment through the PBL top plays a leading role. The equation also includes the mass sink due to the upward cumulus mass flux originating from the PBL. Regarding this point, it was fortunate that the Arakawa and Schubert (1974) cumulus parameterization, which can provide the cumulus mass flux at the PBL top for given large-scale conditions, was being developed at about the same time.

Another attractive aspect of including a variable-depth PBL in a GCM is that stratocumulus clouds can be explicitly treated as a saturated sublayer of the PBL when the PBL top is higher than the condensation level. When such a sublayer exists, the PBL top becomes a cloud top, across which the radiation and turbulent fluxes and the thermodynamic properties of air rapidly change. Furthermore, the radiative cooling concentrated near the cloud top drives in-cloud turbulence, making the character of the PBL dramatically different from that of a cloud-free PBL.

In 1974, an international study conference to discuss the second objective of GARP—the physical basis for climate and climate change—was held in Stockholm. The following is an excerpt from the position paper on cloud processes that I presented at the Conference (Arakawa, 1975):

> The air in a cloud is almost always highly turbulent and the cloud is the product of complicated interactions of moist-convective turbulence with larger-scale circulations, radiation, and the microphysical cloud processes.
>
> Because of the variety of cloud regimes and the complexity of the controlling mechanisms, the modeling of time-dependent clouds is perhaps the weakest aspect of the existing general circulation models and may be the most difficult task in constructing any reliable climate model. A considerable effort should be made to improve the cloud parameterizations in general circulation models... .

The possibility of treating the cloud-free, cumulus-topped and strato-cumulus-topped PBLs using the unified framework of a variable-depth PBL model, as illustrated in Fig. 12a, was extremely attractive to me in view of the importance of simulating the observed distributions of cloud regimes, such as those shown schematically in Fig. 13. (Recent observations show that this figure should be slightly modified. See Section V.C of Chapter 23.)

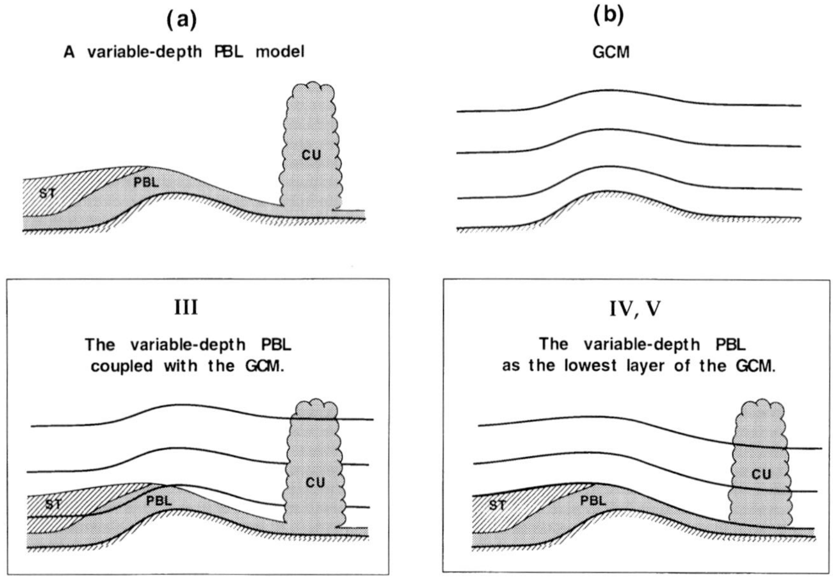

Figure 12   Implementation of the variable-depth PBL into a vertically discrete GCM.

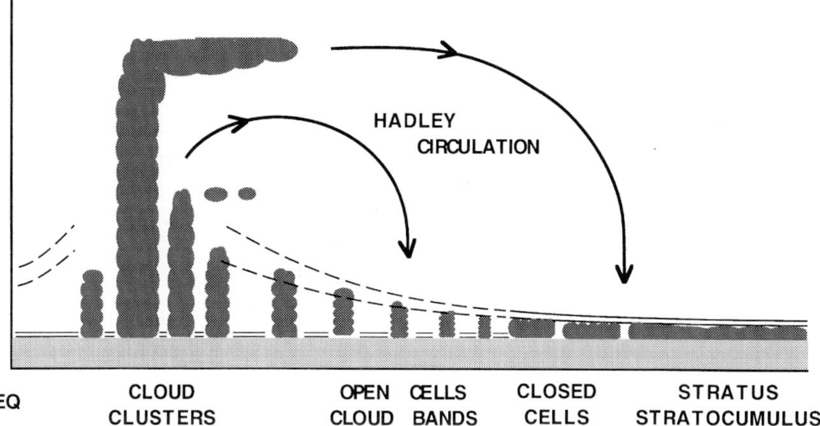

Figure 13 Schematic NE–SW cross section over the northeastern Pacific, summarizing typical observed cloud regimes. From right to left, the sea surface temperature increases and subsidence decreases. The stippled area is the PBL, the top of which is shown by the continuous and discontinuous double-stroked lines. The dashed lines above the cumulus clouds show an inversion layer, which is principally the trade wind inversion. (Redrawn from Arakawa, 1975.)

D. Formulation of PBL Processes in the Generation III and IV GCMs

A variable-depth PBL model, which follows Deardorff (1972) (but not entirely), was implemented into the Generation III UCLA GCM by Randall (1976), with an emphasis on the formulation of the stratocumulus cloud sublayer and its instability through evaporative cooling of air entrained from above.

The variable-depth PBL model had a vertical structure like Fig. 12a, while the GCM had its own vertical structure based on the σ coordinate as illustrated in Fig. 12b. When these two models were coupled, sharing the lower part of the model atmosphere as in the lower left panel of Fig. 12, maintaining consistencies between the two models became computationally very demanding. To decrease the chance of introducing inconsistencies, the Generation III GCM predicted "jumps" at the PBL top (denoted by Δ in Fig. 8), which are the differences between the PBL and free-atmosphere values of the prognostic variables, rather than directly predicting the PBL values.

Although this approach produced some encouraging results, it is very complicated in practice, largely because the GCM layer containing the PBL top can vary from one grid point to the next, and from one time step to the next. This experience led to the decision in 1977 to introduce the variable-depth PBL model as the lowest layer of the GCM (Suarez *et al.*, 1983), which is an integrated part of the GCM's vertical structure. This was done with a generalized σ coordinate, in which the PBL top is a coordinate surface, as illustrated in the lower right panel of Fig. 12.

The major advantage of using such a coordinate is that the PBL properties are expected to be "similar" along a coordinate surface, making the formulation of processes concentrated near the PBL top much more tractable. The recent improvement in simulating stratocumulus incidence (Li *et al.*, 1999) with the Generation V UCLA GCM is made possible by these advantages of the PBL formulation. For further discussion of this type of coordinate, see Section V.B of Chapter 23.

X. FORMULATION OF MOIST PROCESSES IN THE UCLA GCM

A. FORMULATION OF "MOIST PROCESSES" IN THE GENERATION I GCM

As mentioned earlier, the formulation of convective processes in the Generation I UCLA GCM (Mintz, 1965) was based on moist-convective adjustment assuming that a sufficient amount of water vapor is available for condensation whenever conditional instability exists. Besides "boundary layer convection" described in Section IX.A, the two-level GCM includes "internal convection," through which temperatures at the upper and lower levels of the model, T_1 and T_3, are adjusted toward a moist adiabat without changing the temperature linearly extrapolated to the surface (i.e., $\Delta T_1 = 3\Delta T_3$, where Δ denotes the change due to the adjustment). The coefficient of the adjustment was determined empirically from observed mean rainfall and temperature lapse rates.

B. STRUGGLE TO FIND THE PHYSICAL BASIS FOR CUMULUS PARAMETERIZATION

The Generation II GCM introduced water vapor mixing ratio at the lower level as a new prognostic variable. This was my first exposure to a moist model. I was amazed to see how different the performance of a moist model could be from that of a dry model mainly because the heat of condensation, which is the dominant part of heating in a moist model, is motion dependent. Also, as pointed out in Section VIII.A, simulation of geostrophic adjustment is more important in a moist model due to the frequent occurrence of heating concentrated in a single grid point, which locally breaks down the thermal wind balance.

The first attempt to include cumulus effects in the Generation II GCM was the continued use of the adjustment scheme used in Generation I, but applying it only when the (low-level) relative humidity exceeds a critical value, i.e., when

$$RH \geq (RH)_{crit}. \tag{21}$$

When $(RH)_{crit}$ is a prescribed constant, however, the amount of adjustment can be discontinuous (or "spiky") in time and space. This happens because, even when the lapse rate Γ is considerably steeper than the moist-adiabatic lapse rate Γ_m, no adjustment takes place if RH is even slightly smaller than $(RH)_{crit}$. As RH reaches $(RH)_{crit}$ or slightly exceeds it, however, a sudden adjustment of the lapse rate from Γ to Γ_m takes place. A quick remedy of this situation is to let $(RH)_{crit}$ depend on $\Gamma - \Gamma_m$. This was done in an early version of the Generation II GCM and worked reasonably well. However, the physical basis for determining $(RH)_{crit}$ was not sufficiently clear to me at that time.

The period of developing the Generation II GCM was still in the early part of the magnificent second phase of the general history of numerical modeling. This period was also in the middle of an epoch-making phase of tropical meteorology. In this phase, the observational and modeling studies by Riehl and Malkus (1958, 1961), Yanai (1961), Ooyama (1964), Charney and Eliassen (1964), Kuo (1965), and Ooyama (1969) recognized the importance of collective effects of cumulus convection in the tropical atmosphere and formulated those effects in tropical cyclone models. These studies and the formulation of moist-convective adjustment by Manabe *et al.* (1965) in the GFDL GCM, as well as my own experience with the early version of the Generation II GCM, stimulated me to seriously consider the problem of cumulus parameterization, the physical basis for parameterizability in particular. I thought that the question of parameterizability was more than a matter of curiosity since the logical structure of a cumulus parameterization should reflect one's understanding of parameterizability.

There were two specific questions to be answered:

1. How can cumulus clouds modify their environment while condensation takes place only inside clouds?
2. What quantity can be assumed to be in a quasi-equilibrium without loss of predictability?

Question 1 must be answered to formulate the effect of a cloud ensemble on the large-scale environment. Do clouds modify the environment only through mixing of cloud air with the environment as they decay? Alternatively, can mature clouds continuously modify the environment even when they are steady? Regarding question 2, there was no doubt in my mind that parameterizability means the existence of some kind of quasi-equilibrium between cumulus-convective and large-scale processes. The core of the parameterization problem is to explicitly formulate such a quasi-equilibrium to close the problem without loss of predictability of day-to-day changes. Any assumed balance in the large-scale budgets, for example, cannot be used for a quasi-equilibrium since the large-scale budget equations are the model's prognostic equations. Obviously we cannot use the same equations twice, one for finding the unknown cumulus terms based on a balance and the other for model prediction based on an imbalance. Instead, for example, a quasi-equilibrium may be assumed for a measure of the overall intensity of cumulus convection. Then free fluctuations of cumulus activity not modulated by the large-scale processes would be eliminated. This situation is analogous to the filtering problem in large-scale dynamics, as Wayne Schubert discusses in Chapter 6 of this book. In Arakawa (1969) and Arakawa and Schubert (1974), a bulk measure of the cloud buoyancy, which represents the temperature difference between clouds and the environment, is chosen to be in a quasi-equilibrium.

C. Formulation of Moist-Convective Processes in the Generation II GCMs: Cumulus Parameterization by Arakawa (1969)

After considerable struggling, I arrived at preliminary answers to the two questions raised above, as outlined in Arakawa (1969). The logic of this parameterization consists of three steps.

Step 1: Construction of a Cloud Ensemble Model

The parameterization was designed for the three-layer (or three-level) version of the Generation II GCM, in which the lowest layer represents the PBL. Corresponding to this vertical structure, three types of clouds are considered: deep and shallow clouds originating from the PBL and "middle-level" clouds originating from the middle level. Despite its simple vertical structure, the parameterization explicitly formulates the effects of cloud-induced subsidence and cloud air detrainment on the large-scale environment.

One of the three cloud types considered in Arakawa (1969) is shown in Fig. 14. In this figure, solid arrows show large-scale mass fluxes and open arrows show superposed cumulus-induced mass fluxes. In choosing this cloud model, I was influenced by Ooyama's (1964, 1969) two-level tropical cyclone model. As in his model, the factor η is determined by the nonbuoyancy condition at cloud top for each cloud type. The GCM, however, had two vertical degrees of freedom for the temperature above the PBL, while Ooyama's model had only one interface of homogeneous layers, which is equivalent to having only one temperature. This led me to close the cumulus-induced mass circulation as shown in Fig. 14 to separate the detrainment and cumulus-induced subsidence effects on the cloud environment. Considering the budgets for each layer of the environment, I expressed $(\partial s/\partial t)_c$, $(\partial q/\partial t)_c$, and, therefore, $(\partial h/\partial t)_c$ in terms of the cumulus mass flux at cloud base (C in Fig. 14), where the subscript c denotes the cloud effect and, as previously defined, s is the dry static energy, q is the water-vapor mixing ratio, and $h \equiv s + Lq \equiv c_p T + gz + Lq$ is the moist static energy.

The use of this highly idealized cloud model greatly simplifies the parameterization problem. Since the number of unknowns is decreased to only one, cloud base mass flux C, the cloud model contributes to the closure of the problem (called "Type II" closure by Arakawa and Chen, 1987; Arakawa, 1993).

Step 2: Determination of the Condition for the Existence of Clouds

What remains to be formulated is the determination of cloud base mass flux C. The initial step toward this objective is to determine the condition

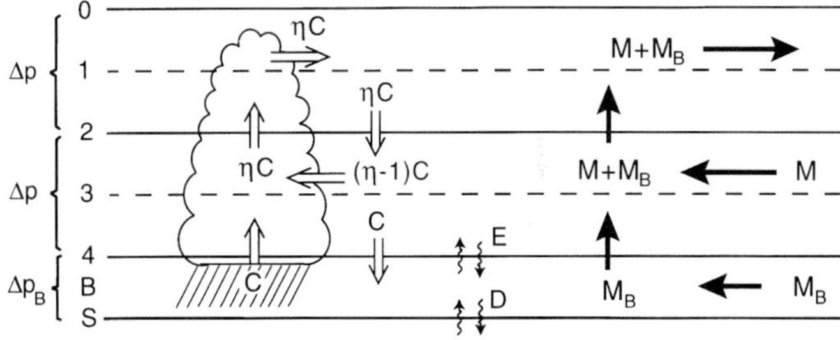

Figure 14 One of the three cloud types considered in Arakawa's (1969) parameterization for a three-level model. Solid and open arrows show large-scale and superposed cumulus-induced mass fluxes, respectively.

for the existence of clouds for each cloud type. Let h_k^* be the saturated value of moist static energy of the environment at level k and h_B be the moist static energy of the PBL. We then find that $h_B \geq (h_1^*, h_3^*)$ is necessary for the cloud type shown in Fig. 14 not to be negatively buoyant at levels 3 and 1. If $h_1^* > h_3^*$, as is usually the case, the necessary condition becomes

$$h_B - h_1^* \geq 0. \tag{22}$$

The left-hand side is a measure of moist-convective instability defined for this type of cloud. Condition (22) can be further rewritten as

$$(\text{RH})_B \geq 1 - c_p[(T_B - T_1) - (T_B - T_1)_m]/Lq_B^*, \tag{23}$$

where $(\text{RH})_B \equiv q_B/q_B^*$ is the relative humidity of the PBL and the subscript m represents the value corresponding to the moist-adiabatic vertical structure defined by $h_B^* - h_1^* = 0$. The right-hand side of Eq. (23) is the expression for $(\text{RH})_{\text{crit}}$ (for the PBL), which I had been looking for.

Step 3: Identification of Adjustment and Introduction of a Principal Closure

With $(\partial s/\partial t)_c$, $(\partial q/\partial t)_c$ and, therefore, $(\partial h/\partial t)_c$ formulated in terms of C, we see that a positive C tends to decrease h_B mainly through drying and increase h_1^* through warming. Then we can write

$$\frac{\partial}{\partial t}(h_B - h_1^*) = -SC + F, \tag{24}$$

where the first and second terms on the right-hand side represent the contributions of cumulus and large-scale processes to $\partial(h_B - h_1^*)/\partial t$, respectively. The coefficient S is a combined measure of $-\partial h/\partial z$ and $\partial h^*/\partial z$, and it is usually positive. Then the cumulus term tends to decrease $h_B - h_1^*$ as long as $C > 0$ and represents the self-stabilizing effect of clouds, which may be called "adjustment." When positive, on the other hand, the term F tends to increase $h_B - h_1^*$. This destabilizing effect by large-scale processes may be called "large-scale forcing." Recall that Eq. (24) is a consequence of using the cloud model, in which cloud effects on the environment are explicitly formulated (Type II closure).

Recognizing that the parameterizable part of cumulus activity is the part forced by large-scale processes, we now exclude free fluctuations of cumulus activity from the objective of parameterization. Then the time scale of the *net* change of $h_B - h_1^*$ is the same as the time scale of the large-scale forcing, F. If we further *hypothesize* that the cumulus adjustment occurs sufficiently rapidly compared to the time scale of F, we can neglect the left-hand side of Eq. (24) compared with the terms on the right-hand side, leading to the quasi-equilibrium of $h_B - h_1^*$. Then we have

$$C \approx F/S. \tag{25}$$

This hypothesis is the "principal closure" (called Type I closure by Arakawa and Chen, 1987; Arakawa, 1993) of this parameterization. The parameterization is now fully closed.

Summary

Here I summarize the logical structure of Arakawa (1969) using terms that are more general for later convenience.

Step 1: Relate $(\partial T/\partial t)_c$ and $(\partial q/\partial t)_c$ to a single variable, cloud base mass flux, m_B (denoted by C in Fig. 14)—Type II closure.

Step 2: Express the condition for the existence of clouds as $\mathbf{A}(T, q) > 0$, where $\mathbf{A}$ is a discrete version of a differential-integral operator involving vertical differences. In the above example, $\mathbf{A}(T, q) \equiv h_B - h_1^*$.

Step 3: Introduce a hypothesis on the quasi-equilibrium of $\mathbf{A}(T, q)$—Type I closure.

Figure 15 schematically illustrates the equilibrium, large-scale forcing and adjustment in an idealized $\Gamma - (\mathrm{RH})_B$ space, where Γ is the mean lapse rate and $(\mathrm{RH})_B$ is the relative humidity of the PBL.

It is very important to note that the hypothesis introduced above is on the self-stabilizing effect of cumulus activity and, therefore, the quasi-equilibrium assumption is applied to $\mathbf{A}(T, q)$, which is a measure of moist-convective instability. Other variables, including temperatures and humidities at individual levels, are not necessarily in a quasi-equilibrium even in an approximate sense.

D. Vertical Advection of Moisture in the Generation III and IV GCMs

The Generation III GCM is the first multilevel model developed at UCLA. In the earlier Generation II GCM, especially in its two-level

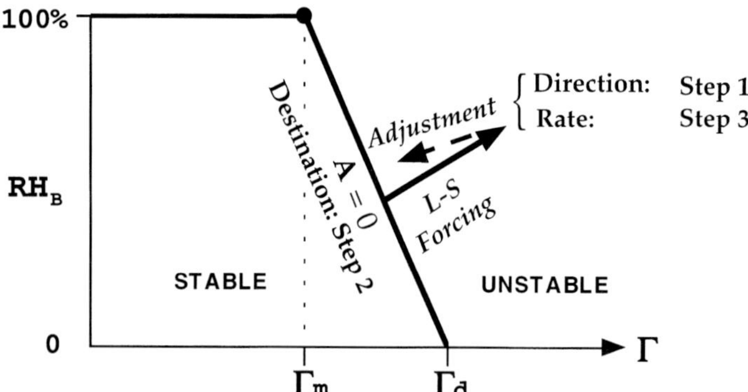

Figure 15   A schematic illustration of the equilibrium, large-scale forcing and adjustment in an idealized $\Gamma - (RH)_B$ space, where Γ is the mean lapse rate, Γ_d is the dry-adiabatic lapse rate, Γ_m is the moist-adiabatic lapse rate, $(RH)_B$ is the relative humidity of the PBL, $A = 0$ represents the marginal moist-convective instability defined in step 2 in the text, which is the destination of the adjustment. Step 1 and step 2 define the direction and rate of the adjustment, respectively, responding to the large-scale forcing. The dot represents the equilibrium state used in the moist-convective adjustment scheme of Manabe *et al.* (1965).

version, we did not have to worry about the vertical redistribution of moisture due to either advective or moist-convective processes. One of the first things I noticed during the development of the Generation III GCM was that simulating the vertical redistribution of moisture is a difficult problem both computationally and physically.

As far as the computational aspect is concerned, a part of the difficulty comes from the large fractional change of the mixing ratio in the vertical direction, covering a wide range of values. The same problem can exist for other atmospheric constituents if their three-dimensional distributions are explicitly predicted. As mentioned in Section VII.B, our solution to overcome this difficulty was to use a vertical difference scheme that conserves the mass integral of both q and $\ln q$, where q is the mixing ratio, except when water vapor is saturated.

When water vapor is saturated, the problem can become even trickier since heat of condensation is involved. If the vertical flux of q at an interface of model layers is calculated using an inadequately interpolated value of q, a spurious growth of condensation may occur even when there is no conditional instability between the model layers. I called this instability *conditional instability of computational kind* (CICK). In Generation III and in later generations of the UCLA GCM, we use a vertical interpolation of q that is free from this instability, when at least one of the two layers involved in the flux calculation is saturated (CICK-proof interpolation; Arakawa and Lamb, 1977). The basic idea is to interpolate h^* and RH to the interface separately, so that no spurious conditional instability and no spurious supersaturation are generated, and then diagnose h there. From this $h(\equiv s + Lq)$ and the value of s at the interface specified by the vertical differencing of the dynamics, q at the interface is diagnosed.

I should again mention that discretization of the vertical advection of water vapor (and other atmospheric constituents) is not quite a settled problem in my mind. (See Chapter 23 for further discussion.)

E. FORMULATION OF MOIST-CONVECTIVE PROCESSES IN THE GENERATION III AND IV GCMS: CUMULUS PARAMETERIZATION BY ARAKAWA AND SCHUBERT (1974)

The cumulus parameterization described in Section X.C was designed exclusively for a three-level model and could not be directly generalized to multilevel models. Development of a cumulus parameterization for such models required the following questions to be answered, in addition to questions 1 and 2 already raised in Section X.B:

3. Since the vertical distribution of cumulus effects depends on cloud type, cumulus parameterization should determine the spectral distribution of clouds. What is an appropriate framework for doing this?
4. How does the subcloud layer control cumulus activity? What is the nature of the feedback in this link?

The original objective of the paper by Arakawa and Schubert (1974) (hereafter AS) was not necessarily to present a readily usable cumulus parameterization; instead, the objective was to construct a theoretical framework that could be used for understanding the physical and logical basis for cumulus parameterization. Parameterizability was still the major concern of the paper. In developing the framework, we were especially careful not to miss explicitly stating assumptions and idealizations we had to introduce into the framework.

More specifically, AS attempted to answer questions 3 and 4, and then modify, extend and elaborate the preliminary answers Arakawa (1969) had for questions 1 and 2. In particular, AS introduced a spectral cumulus ensemble model for step 1 and the cloud work function defined for each cloud type for $\mathbf{A}(T, q)$ in steps 2 and 3. (For these steps, see the summary given near the end of Section X.C.) The cloud work function is an integral measure of cloud buoyancy and replaces $h_B - h_i^*$ in Arakawa (1969). For deep clouds, the cloud work function is similar to the convective available potential energy (CAPE), but generally not the same. It is more related to an instability criterion, or the process of releasing energy that depends on cloud type, rather than the amount of energy available for *all* clouds.

The spectral cumulus ensemble model relates the *vertical distributions* of cumulus heating, Q_1, and cumulus drying, Q_2, to the *spectral distribution* of cloud base mass flux into different types of clouds. Suppose that the model has N levels above the cloud base. Then there are $2N$ unknowns for the cloud layer: Q_1 and Q_2 at N levels. On the other hand, the model allows N cloud types since there are N levels to identify cloud top. Thus, the spectral cloud ensemble model decreases the number of degrees of freedom for the unknowns from $2N$ to N, providing one-half of the necessary closures. This is the closure of Type II, which is due to the coupling of the vertical profiles of Q_1 and Q_2 through the cumulus mass flux. This constrains the direction of the adjustment shown in Fig. 15. The rest of the necessary closures are provided by the quasi-equilibrium of the cloud work function applied to N cloud types. This is the closure of Type I and constrains the destination and rate of the adjustment shown in Fig. 15.

The combination of the variable-depth PBL model discussed in Section IX.C and the AS cumulus parameterization discussed here was an almost perfect marriage. To calculate the vertical profiles of the thermodynamic properties of cloud air, the cumulus parameterization must know the thermodynamic properties of the PBL air and the PBL depth. To calculate the time change of the PBL depth, on the other hand, the PBL model must know the mass flux into clouds through the PBL top. The original paper by AS distinguished cloud base level and PBL top and discussed their mutual regulations, interpreting why those two levels are usually very close in nature. In the actual implementation of this parameterization, the difference between these levels is ignored.

For more details and discussion of the AS cumulus parameterization, see the chapter by Wayne Schubert in this book, Arakawa and Chen (1987), Arakawa (1993), Arakawa and Cheng (1993), Randall *et al.* (1997), and the original papers referenced at the end of Section VI.

There have been a number of criticisms against the AS paper. These criticisms can be roughly classified into two groups. The statement "too complicated" represents one group. The paper is in fact complicated; for example, it includes 200 equations! In addition, the implementation of the parameterization into a GCM involves many technical details (e.g., Lord *et al.*, 1982; Cheng and Arakawa, 1997b) and, perhaps most importantly, it is computationally expensive. The statement "too simple" represents the other group. The cloud ensemble model used in the paper is in fact physically simple; for example, there is no downdraft, no ice phase, etc.

Moreover, the paper's focus is only on parameterization of cumulus convection in a quasi-equilibrium with large-scale processes.

Perhaps the real criticism against the AS paper is "It does a simple thing in a complicated way." While I basically admit that this statement is true, my response to it would depend on what "complication" means.

If "complication" means the large number of equations in the paper, I would say that there is a great gap in the link between the first principles and the cumulus parameterization problem. Filling this gap is an important scientific issue that should not be ignored. Apparently, it is a complicated task although the answer we expect from the parameterization is a relatively simple one.

If "complication" means a large amount of computation, I would say that an entire GCM "does a simple thing in a complicated way." To an outsider, what a GCM does to find the global warming a century from now, say, finding one number, must seem to be "doing a simple thing in a complicated way." I would further say that nature also "does a simple thing in a complicated way" in the sense that nature is complicated, but our research is focused on relatively simple aspects of nature's behavior. GCMs try to mimic nature's own complicated way of doing simple things.

XI. CLOSING REMARKS

So far, I haven't said much about the currently developing Generation V UCLA GCM. This is because the focus of this article is on the *early* years of general circulation modeling at UCLA. Here I simply list some of the recent or ongoing revisions for the Generation V: change of the radiation scheme following Harshvardan *et al.* (1987, 1989), inclusion of an orographic gravity wave drag parameterization following Kim and Arakawa (1995), inclusion of convective downdraft effects in the cumulus parameterization (Cheng and Arakawa, 1997a), revision of PBL moist processes following Li *et al.* (1999), implementation of a prognostic closure in the AS cumulus parameterization following Randall and Pan (1993), and inclusion of explicit prediction of ice and liquid clouds following Köhler *et al.* (1997).

Finally, I would like to emphasize that development of a GCM involves much more than the subjects I covered in this article, and the present UCLA GCM is an accumulated product of the ingenuity and hard work of many people. Development of a GCM suggests a number of unique research topics, which were addressed by Ph.D. dissertations of many of my former students. Frequently I have found that a brilliant idea may not work as it is in a GCM due to the existence of negative feedbacks. It may influence the results, however, in an unexpected way. Thus the GCM has been an excellent teacher for me, hard to cheat and at times too demanding, but ultimately appreciated.

The magnificent second phase of numerical modeling of the atmosphere is now over and we have entered the great-challenge third phase (see Fig. 1), in which many challenging problems are waiting for us. I will discuss some of those issues in the closing chapter of this book.

ACKNOWLEDGMENTS

I would like to thank all participants of the AA Fest and those who initiated the idea of the symposium and pursued it to become such a big event. I especially thank Kayo Ide, David Randall, and Roger Wakimoto for the tremendous amount of their time and energy spent organizing the symposium. I also thank the Center for Climate System Research of the University of Tokyo, Department of Energy, National Aeronautics and Space Administration, National Science Foundation, Office of Naval Research, UCLA College of Letters and Sciences, and World Climate Research Program for their financial support of the Symposium and UCLA's Department of Atmospheric Sciences and Institute of Geophysics and Planetary Physics for their administrative support.

Taking this opportunity, I would like to extend my thanks to all of my former and present collaborators, students and friends, especially those listed in Fig. 3, for their stimulation, encouragement, and valuable assistance. I also acknowledge the funding agencies who have generously supported our research on general circulation modeling for many years, UCLA for providing the excellent research and teaching environment, and JMA for my early years of operational and research experience, and finally my wife Kazuko for her understanding, patience, and collaboration for almost 45 years.

Preparation of this article is supported by the NSF grant ATM-96139, NASA grant NAG 5-4420, and DOE grant DE-FG03-91ER61214. I greatly appreciate the help provided by Prof. David Randall and Drs. John Farrara and Celal Konor in revising the manuscript.

APPENDIX A

A FOURTH-ORDER ENERGY AND POTENTIAL ENSTROPHY CONSERVING SCHEME FOR THE SHALLOW-WATER EQUATIONS BY TAKANO AND WURTELE (1982)

Let i and j be the longitude and latitude indices, respectively, shown in Fig. A.1. We define

$$(\delta_i A)_i \equiv (A_{i+1/2} - A_{i-1/2}),$$

$$(\delta_i A)_{i+1/2} \equiv (A_{i+1} - A_i), \tag{A.1}$$

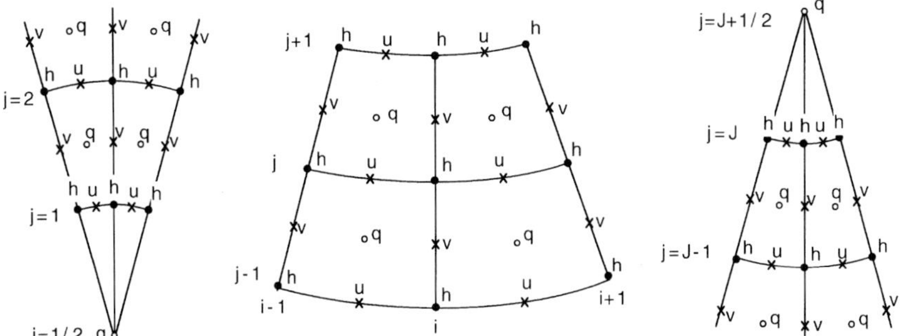

Figure A1

$$(\overline{A}^{i})_i \equiv \frac{1}{2}(A_{i+1/2} + A_{i-1/2}),$$

(A.2)

$$(\overline{A}^{i})_{i+1/2} \equiv \frac{1}{2}(A_{i+1} + A_i), \text{ etc.},$$

$$(\delta x)_j = a \cos\varphi_j \Delta\lambda, \ j = 1,\dots,J,$$

(A.3)

$$(\delta y)_{j-1/2} = a\Delta\phi, \ j = 1,\dots,J+1,$$

$$(\delta x)_{1/2} = 0, \quad (\delta x)_{j+1/2} = \left(\overline{\delta x}^{j}\right)_{j+1/2}, \ j = 1,\dots,J-1,$$

$$(\delta x)_{J+1/2} = 0,$$

(A.4)

$$(\delta y)_1 = \frac{3}{2}(\delta y)_{1/2}, \quad (\delta y)_j = \left(\overline{\delta y^{j}}\right)_{j}, \ j = 2,\dots,J-1,$$

$$(\delta y)_J = \frac{3}{2}(\delta y)_{J+1/2},$$

(A.5)

$$(\Delta^2)_1 = (\delta y)_{1/2}(\delta x)_1, \quad (\Delta^2)_j = (\delta y)_j\left(\overline{\delta x}^{j}\right)_j, \ j = 2,\dots,J-1,$$

$$(\Delta^2)_J = (\delta y)_{J+1/2}(\delta x)_J.$$

(A.6)

The discrete continuity equation used is

$$\frac{\partial h_{i,j}}{\partial t} = -\frac{1}{(\Delta^2)_j}(\delta_i u^* + \delta_j v^*)_{i,j},$$

(A.7)

where the mass fluxes are defined by

$$u^*_{i+1/2,j} = (\bar{h}^i)_{i+1/2,j} u_{i+1/2,j} (\delta y)_j, \qquad j = 1,\ldots,J \qquad (A.8)$$

and

$$v^*_{i,j+1/2} = (\bar{h}^j)_{i,j+1/2} v_{i,j+1/2} (\delta x)_{j+1/2}, \qquad j = 0,\ldots,J. \qquad (A.9)$$

The zonal and meridional components of the first term on the right-hand side of Eq. (19) are discretized as

$$(-qhv)_{i+1/2,j}$$

$$= -\frac{1}{(\delta x)_j} [\, \alpha_{i+1/2,j} v^*_{i+1,j+1/2} + \beta_{i+1/2,j} v^*_{i,j+1/2} + \gamma_{i+1/2,j} v^*_{i,j-1/2}$$

$$+ \delta_{i+1/2,j} v^*_{i+1,j-1/2} - \lambda_{i+1/2,j+1/2} u^*_{i+1/2,j+1}$$

$$+ \lambda_{i+1/2,j-1/2} u^*_{i+1/2,j-1} - \varepsilon_{i+1,j} u^*_{i+3/2,j} + \varepsilon_{i,j} u^*_{i-1/2,j}\,], \qquad (A.10)$$

$$(qhu)_{i,j-1/2} = \frac{1}{(\delta y)_{j-1/2}} [\, \gamma_{i+1/2,j} u^*_{i+1/2,j} + \delta_{i-1/2,j} u^*_{i-1/2,j}$$

$$+ \alpha_{i-1/2,j-1} u^*_{i-1/2,j-1} + \beta_{i+1/2,j-1} u^*_{i+1/2,j-1}$$

$$+ \mu_{i+1/2,j-1/2} v^*_{i+1,j-1/2} - \mu_{i-1/2,j-1/2} v^*_{i-1,j-1/2}$$

$$+ \phi_{i,j} v^*_{i,j+1/2} - \phi_{i,j-1} v^*_{i,j-3/2}\,], \qquad (A.11)$$

$$\alpha_{i+1/2,j} = \frac{1}{24}(2q_{i+3/2,j+1/2} + 3q_{i+1/2,j+1/2} + 2q_{i+1/2,j-1/2}$$

$$+ q_{i+3/2,j-1/2} - q_{i+1/2,j+3/2} - q_{i-1/2,j+1/2}), \qquad (A.12)$$

$$\beta_{i+1/2,j} = \frac{1}{24}(3q_{i+1/2,j+1/2} + 2q_{i-1/2,j+1/2} + q_{i-1/2,j-1/2}$$

$$+ 2q_{i+1/2,j-1/2} - q_{i+3/2,j+1/2} - q_{i+1/2,j+3/2}), \qquad (A.13)$$

$$\gamma_{i+1/2,j} = \frac{1}{24}(2q_{i+1/2,j+1/2} + q_{i-1/2,j+1/2} + 2q_{i-1/2,j-1/2}$$

$$+ 3_{i+1/2,j-1/2} - q_{i+1/2,j-3/2} - q_{i+3/2,j-1/2}), \qquad (A.14)$$

$$\delta_{i+1/2,j} = \frac{1}{24}(q_{i+3/2,j+1/2} + 2q_{i+1/2,j+1/2} + 3q_{i+1/2,j-1/2}$$

$$+ 2q_{i+3/2,j-1/2} - q_{i-1/2,j-1/2} - q_{i+1/2,j-3/2}), \qquad (A.15)$$

$$\varepsilon_{i,j} = \frac{1}{24}(q_{i+1/2,j+1/2} + q_{i-1/2,j+1/2} - q_{i-1/2,j-1/2} - q_{i+1/2,j-1/2})$$

(A.16)

$$\phi_{i,j} = \frac{1}{24}(-q_{i+1/2,j+1/2} + q_{i-1/2,j+1/2} + q_{i-1/2,j-1/2} - q_{i+1/2,j-1/2}),$$

(A.17)

$$\lambda_{i+1/2,j+1/2} = \frac{1}{24}(q_{i+3/2,j+1/2} - q_{i-1/2,j+1/2}), \qquad \text{(A.18)}$$

$$\mu_{i+1/2,j+1/2} = \frac{1}{24}(q_{i+1/2,j-1/2} - q_{i+1/2,j+3/2}). \qquad \text{(A.19)}$$

These are applied at all i, j except

$$\alpha_{i+1/2,J} = 0, \beta_{i+1/2,J} = 0, \gamma_{i+1/2,1} = 0,$$
$$\delta_{i+1/2,1} = 0, \quad \lambda_{i+1/2,1/2} = 0, \quad \lambda_{i+1/2,J+1/2} = 0. \qquad \text{(A.20)}$$

Here $q_{i+1/2,j+1/2}$, is given by

$$q_{i+1/2,j+1/2} = \left[\frac{\overline{\Delta^2}^j f + (\delta y)(\delta_i v) - \delta_j((\delta x)u)}{\overline{\overline{h}^i \Delta^2}^j} \right]_{i+1/2,j+1/2} \qquad \text{(A.21)}$$

and, at the poles,

$$q_{i+1/2,1/2} = \frac{\frac{1}{2}(\Delta^2)_1 f_{1/2} - \frac{1}{I}\Sigma_i(\delta x)_1 u_{i+1/2,1}}{(\Delta)_1^2 \Sigma_i h_{i,1}/2I}, \qquad \text{(A.22)}$$

$$q_{i+1/2,J+1/2} = \frac{\frac{1}{2}(\Delta^2)_J f_{J+1/2} + \frac{1}{I}\Sigma_i(\delta x)_J u_{i+1/2,J}}{(\Delta^2)_J \Sigma_i h_{i,J}/2I}, \qquad \text{(A.23)}$$

where I denotes the number of grid points in longitude.

The terms involving K in Eq. (19) are finite differenced as

$$\left(-\frac{1}{a\cos\varphi}\frac{\partial K}{\partial\lambda} \right)_{i+1/2,j} = -\frac{1}{(\delta x)_j}(\delta_i K)_{i+1/2,j}, \qquad \text{(A.24)}$$

$$\left(-\frac{1}{a}\frac{\partial K}{\partial\varphi} \right)_{i,j-1/2} = -\frac{1}{(\delta y)_{j-1/2}}(\delta_j K)_{i,j-1/2}, \qquad \text{(A.25)}$$

where

$$K_{i,j} = \frac{1}{(\Delta^2)_j}\left[(\delta x)(\delta y)\overline{K^u}^i + \overline{(\delta x)(\delta y)K^v}^j\right]_{i,j}. \qquad (A.26)$$

In the original Takano–Wurtele scheme,

$$K^u_{i+1/2,j} = u^2_{i+1/2,j}/2, \; K^v_{i,j+1/2} = v^2_{i,j+1/2}/2, \qquad (A.27)$$

while the SICK-proof expressions used in the current UCLA GCM are given by

$$K^u_{i+1/2,j} = \frac{1}{6}\left(\left(\frac{u_{j+1} + u_{j-1}}{2}\right)^2 + 2u^2_j\right)_{i+1/2} \qquad (A.28)$$

and

$$K^v_{i,j+1/2} = \frac{1}{6}\left(\left(\frac{v_{i+1} + v_{i-1}}{2}\right)^2 + 2v^2_i\right)_{j+1/2}. \qquad (A.29)$$

APPENDIX B

A Fourth-Order Horizontal Difference Scheme for the Thermodynamic Equation

Consider a square grid in orthogonal coordinates, ξ and η. In Fig. B.1, the circles show the scalar points such as the θ points. We define the difference and average operators as in Eqs. (A.1) and (A.2). In addition, as in Eq. (A.8) and (A.9), we define the mass fluxes

$$u^* \equiv \overline{\pi}^i\frac{u}{n}\Delta\xi, \qquad v^* \equiv \overline{\pi}^j\frac{v}{m}\Delta\eta, \qquad (B.1)$$

where π is the pseudo-density of the vertical coordinate, m and n are the map factors for the ξ and η directions, respectively. For this scheme, the term $(\Delta\xi\Delta\eta/mn)\nabla \cdot (v^*\theta)$ is finite differenced as

$$\frac{1}{12}\left[\left(6u^* + 3\overline{\overline{u^*}}^\xi - \overline{\overline{u^*}}^\eta\right)_{10}(\theta_{20} + \theta_{00})\right.$$

$$\left. - \left(6u^* + 3\overline{\overline{u^*}}^\xi - \overline{\overline{u^*}}^\eta\right)_{-10}(\theta_{00} + \theta_{-20})\right.$$

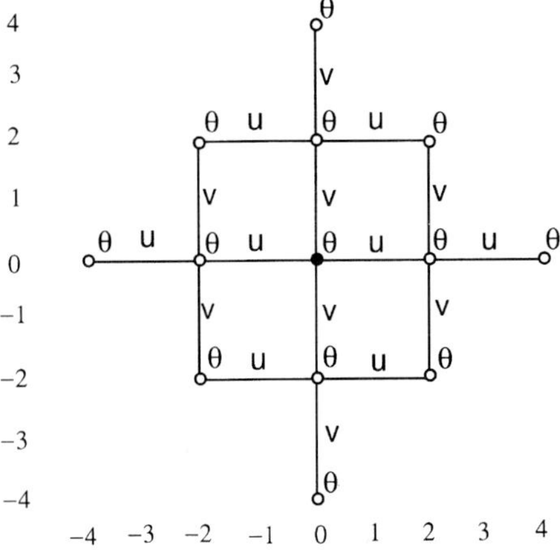

Figure B1

$$+\left(6v^* + 3\overline{\overline{v^*}}^{\eta}_{\eta} - \overline{\overline{v^*}}^{\xi}_{\xi}\right)_{01}(\theta_{02} + \theta_{00})$$

$$-\left(6v^* + 3\overline{\overline{v^*}}^{\eta}_{\eta} - \overline{\overline{v^*}}^{\xi}_{\xi}\right)_{0-1}(\theta_{00} + \theta_{0-2})$$

$$-\left(\overline{u^*}^{\xi}\right)_{20}(\theta_{40} + \theta_{00}) + \left(\overline{u^*}^{\xi}\right)_{-20}(\theta_{00} + \theta_{-40})$$

$$-\left(\overline{v^*}^{\eta}\right)_{02}(\theta_{04} + \theta_{00}) + \left(\overline{v^*}^{\eta}\right)_{0-2}(\theta_{00} + \theta_{0-4})$$

$$+\frac{1}{8}\left\{(\delta_\xi u^* - \delta_\eta v^*)_{02} - (\delta_\xi u^* - \delta_\eta v^*)_{20}\right\}(\theta_{22} + \theta_{00})$$

$$-\frac{1}{8}\left\{(\delta_\xi u^* - \delta_\eta v^*)_{-20} - (\delta_\xi u^* - \delta_\eta v^*)_{0-2}\right\}(\theta_{00} + \theta_{-2-2})$$

$$+\frac{1}{8}\left\{(\delta_\xi u^* - \delta_\eta v^*)_{02} - (\delta_\xi u^* - \delta_\eta v^*)_{-20}\right\}(\theta_{-22} + \theta_{00})$$

$$-\frac{1}{8}\left\{(\delta_\xi u^* - \delta_\eta v^*)_{20} - (\delta_\xi u^* - \delta_\eta v^*)_{0-2}\right\}(\theta_{2-2} + \theta_{00})\Bigg] = 0.$$

$$(B.2)$$

REFERENCES

Arakawa, A. (1957a). On the maintenance of zonal mean flow. *Pap. Met. Geophys.*, **8**, 39–54.

Arakawa, A. (1957b). On the mean meridional circulation in the atmosphere *J. Meteor. Soc. Japan, 75th Anniversary Volume*, 230–236.

Arakawa, A. (1958). Modern theory of general circulation of the atmosphere. *Kisho Kenkyu Note*, **9**, No. 4, Meteor. Soc. Japan (in Japanese).

Arakawa, A. (1961). The variation of general circulation in the barotropic atmosphere *J. Meteor. Soc. Japan*, **39**, 49–58.

Arakawa, A. (1962). Non-geostrophic effects in the baroclinic prognostic equations. *In* "Proceedings of the International Symposium on Numerical Weather Prediction," Tokyo, 1960, pp. 161–175, Meteor. Soc. Japan.

Arakawa, A. (1966). Computational design for long-term numerical integration of the equations of fluid motion: Two-dimensional incompressible flow. Part I. *J. Comp. Phys.*, **1**, 119–143. Reprinted in *J. Comp. Phys.*, **135**, 103–114.

Arakawa, A. (1969). Parameterization of cumulus clouds. *In* "Proceedings of the WMO/IUGG Symposium on Numerical Weather Prediction, Tokyo, 1968, pp. IV-8-1–IV-8-6. Japan Meteorological Agency.

Arakawa, A. (1970). Numerical simulation of large-scale atmospheric motions. *In* "Numerical Solution of Field Problems in Continuum Physics, Proceedings of a Symposium in Applied Mathematics," Durham, NC, 1968, SIAM-AMS Proceedings (G. Birkhoff and S. Varga, eds.) Vol. 2, pp. 24–40. American Mathematical Society.

Arakawa, A. (1972). Design of the UCLA general circulation model, Technical Report 7, Numerical simulation of weather and climate. Department of Meteorology, UCLA.

Arakawa, A. (1975). Modelling clouds and cloud processes for use in climate model. *In* "The Physical Basis of Climate and Climate Modelling," GARP Publication Series No. 16, pp. 183–197, WMO.

Arakawa, A. (1988). Finite-difference methods in climate modeling. *In* "Physically-Based Modeling and Simulation of Climate and Climate Change," (M. Schlesinger, ed.), Part I, pp. 79–168. Kluwer Academic Publishers, New York.

Arakawa, A. (1993). Closure assumptions in the cumulus parameterization problem. *In* "The Representation of Cumulus Convection in Numerical Models of the Atmosphere" (K. A. Emanuel and D. J. Raymond, eds.), pp. 1–16. Am. Meteor. Soc.

Arakawa, A., and J.-M. Chen (1987). Closure assumption in the cumulus parameterization problem. *In* "Short- and Medium-Range Numerical Weather Prediction" (T. Matsuno, ed.), Special Volume, pp. 107–131. *J. Meteor. Soc. Japan*.

Arakawa, A., and M.-D. Cheng (1993). The Arakawa-Schubert cumulus parameterization. *In* "The Representation of Cumulus Convection in Numerical Models of the Atmosphere" (K. A. Emanuel and D. J. Raymond, eds.), pp. 123–136. Am. Meteor. Soc.

Arakawa, A. and Y.-J. G. Hsu (1990). Energy conserving and potential-enstrophy dissipating schemes for the shallow water equations. *Mon. Wea. Rev.*, **118**, 1960–1969.

Arakawa, A., and C. S. Konor (1996). Vertical differencing of the primitive equations based on the Charney-Phillips grid in hybrid σ-p vertical coordinates. *Mon. Wea. Rev.*, **124**, 511–528.

Arakawa, A., and V. R. Lamb (1977). Computational design of the basic dynamical processes of the UCLA general circulation model. *In* "General Circulation Models of the Atmosphere," (J. Chang, ed.), Methods in Computational Physics, Vol. 17, pp. 173–265. Academic Press, San Diego.

Arakawa, A., and V. R. Lamb (1981). A potential enstrophy and energy conserving scheme for the shallow water equations. *Mon. Wea. Rev.*, **109**, 18–36.

Arakawa, A., and Y. Mintz, with the participation of A. Katayama, J.-W. Kim, W. Schubert, T. Tokioka, M. Schlesinger, W. Chao, D. Randall, and S. Lord (1974). The UCLA general circulation model. Notes distributed at the workshop, March 25–April 4, 1974, Department of Meteorology, UCLA.

Arakawa, A., and S. Moorthi (1988). Baroclinic instability in vertically discrete systems. *J. Atmos. Sci.*, **45**, 1688–1707.

Arakawa, A., and W. H. Schubert (1974). Integration of a cumulus cloud ensemble with the large-scale environment. Part I. *J. Atmos. Sci.*, **31**, 674–701.

Arakawa, A., and M. J. Suarez (1983). Vertical differencing of the primitive equations in sigma-coordinates. *Mon. Wea. Rev.*, **111**, 34–45.

Arakawa, A., A. Katayama, and Y. Mintz (1969). Numerical simulation of the general circulation of the atmosphere. (Appendix I, A. Arakawa: Parameterization of cumulus convection. Appendix II, A. Katayama: Calculation of radiative transfer). *In* "Proceedings of the WMO/IUGG Symposium on Numerical Weather Prediction, Tokyo, 1968, pp. IV-7–IV-8-12. Japan Meteorological Agency.

Bjerknes, V. (1904). Das Problem der Wettervorsage, betrachtet vom Standpunkte der Mechanik und der Physik. *Meteor. Z.*, **21**, 1–7. (English translation by Yale Mintz, Los Angeles, 1954.)

Bjerknes, V. (1914). Die Meteorologie als exakte Wissenshaft [Meteorology as an exact science]. *Mon. Wea. Rev.*, **42**, 11–14.

Bjerknes, J., and Y. Mintz (1955). Investigation of the general circulation of the atmosphere. Final report, General Circulation Project AF 19(122)-48, sponsored by Geophysical Research Directorate, Department of Meteorology, UCLA.

Bolin, B. (1953). Multiple-parameter models of the atmosphere for numerical forecasting purposes. *Tellus*, **5**, 207–218.

Burridge, D. M., and J. C. Haseler (1977). A model for medium range forecasting, Tech. Report 4. ECMWF, Reading, UK.

Businger, J. A., J. C. Wyngaard, Y. Izumi, and E. F. Bradley (1971). Flux-profile relationships in the atmospheric surface layer. *J. Atmos. Sci.*, **28**, 181–189.

Chao, W. C.-W. (1978). A study of conditional instability of the second kind and a numerical simulation of the intertropical convergence zone and easterly waves, Ph.D. Thesis. Department of Atmospheric Sciences, UCLA.

Charney, J. G. (1947). The dynamics of long waves in a baroclinic westerly current. *J. Meteor.*, **4**, 135–162.

Charney, J. G. (1948). On the scale of the atmospheric motions. *Geodes. Publ.* **17**, No. 2.

Charney, J. G. (1949). On a physical basis for numerical prediction of large-scale motions in the atmosphere. *J. Meteor.*, **6**, 371–385.

Charney, J. G. (1955). The use of the primitive equations in numerical weather prediction. *Tellus*, **7**, 22–26.

Charney, J. G. (1962). Integration of the primitive and balance equations. *In* "Proceedings of the International Symposium on Numerical Weather Prediction," Tokyo, 1960, pp. 131–152. Meteor. Soc. Japan.

Charney, J. G., and A. Eliassen (1949). A numerical method for predicting the perturbations of the middle latitude westerlies. *Tellus*, **1**, 38–54.

Charney, J. G., and A. Eliassen (1964). On the growth of the hurricane depression. *J. Atmos. Sci.*, **21**, 68–75.

Charney, J. G., and N. A. Phillips (1953). Numerical integration of the quasi-geostrophic equations for barotropic and simple baroclinic flows. *J. Meteor.*, **10**, 71–99.

Charney, J. G., B. Gilchrist, and F. G. Shuman (1956). The prediction of general quasi-geostrophic motions. *J. Meteor.*, **63**, 489–499.

Charney, J. G., R. Fjørtoft, and J. von Neumann (1950). Numerical integration of the barotropic vorticity equation. *Tellus*, **2**, 237–254, 1950.

Cheng, M.-D., and A. Arakawa (1997a). Inclusion of rainwater budget and convective downdrafts in the Arakawa–Schubert cumulus parameterization. *J. Atmos. Sci.*, **54**, 1359–1378.

Cheng, M.-D., and A. Arakawa (1997b). "Computational procedures for the Arakawa–Schubert cumulus parameterization," Tech. Report 101. General Circulation Modeling Group, Department of Atmospheric Sciences, UCLA.

Deardorff, J. W. (1972). Parameterization of the planetary boundary layer for use in general circulation models. *Mon. Wea. Rev.*, **100**, 93–106.

Eady, E. T. (1949). Long waves and cyclone waves. *Tellus*, **1**, 35–52.

Eady, E. T. (1952). Note on weather computing and the so-called $2\frac{1}{2}$-dimensional model. *Tellus*, **4**, 157–167.

Eliassen, A. (1952). Simplified models of the atmosphere, designed for the purpose of numerical weather prediction. *Tellus*, **4**, 145–156.

Fjørtoft, R. (1952). On a numerical method of integrating the barotropic vorticity equation. *Tellus*, **4**, 179–194.

Fjørtoft, R. (1953). On the changes in the spectral distribution of kinetic energy for two-dimensional non-divergent flow. *Tellus*, **5**, 225–230.

Fultz, D. (1956). A survey of certain thermally and mechanically driven systems of meteorological interest. *In* "Fluid models in Geophysics, Proc. 1st Symposium on the Use of Models in Geophys. Fluid Dynamics," Baltimore, MD, 1953, pp. 27–63.

Gambo, K., and A. Arakawa (1958). Prognostic equations for predicting the mean zonal current, Tech. Report 1. Numerical Weather Prediction Group, Tokyo.

Gates, W. L., E. S. Batten, and A. B. Nelson (1971). A documentation of the Mintz–Arakawa two-level atmospheric general circulation model, R-877-ARPA. Rand Corp.

Harshvardan, R. D., D. A. Randall, and T. G. Corsetti (1987). A fast radiation parameterization for atmospheric circulation models. *J. Geophys. Res.* **92**, 1009–1016.

Harshvardan, R. D., D. A. Randall, T. G. Corsetti, and D. A. Dazlich (1989). Earth radiation budget and cloudiness simulations with a general circulation model. *J. Atmos. Sci.* **46**, 1922–1942.

Hide, R. (1956). Fluid motion in the earth's core and some experiments on thermal convection in a rotating liquid. *In* "Fluid Models in Geophysics, Proc. 1st Symposium on the Use of Models in Geophys. Fluid Dynamics," Baltimore, MD, 1953, pp. 101–116.

Hollingsworth, A., and P. Kållberg (1979). Spurious energy conversions in an energy-enstrophy conserving scheme, Internal Report 22. ECMWF, Reading, UK.

Hollingsworth, A., P. Kållberg, V. Renner, and D. M. Burridge (1983). An internal symmetric computational instability. *Quart. J. Roy. Meteor. Soc.*, **109**, 417–428.

Hsu, Y.-J. G., and A. Arakawa (1990). Numerical modeling of the atmosphere with an isentropic vertical coordinate. *Mon. Wea. Rev.*, **118**, 1933–1959.

Janjic, Z. I. (1974). A stable centered difference scheme free of two-grid-interval noise. *Mon. Wea. Rev.*, **102**, 319–323.

Janjic, Z. I. (1984). Nonlinear advection schemes and energy cascade on semi-staggered grids. *Mon. Wea. Rev.*, **112**, 1234–1245.

Jespersen, D. C. (1974). Arakawa's method is a finite element method. *J. Comp. Phys.*, **16**, 383–390.

Johnson, D., and A. Arakawa (1996). On the scientific contributions and insight of Professor Yale Mintz. *J. Climate*, **9**, 3211–3224.

Kasahara, A., and W. M. Washington (1967). NCAR global general circulation model of the atmosphere. *Mon. Wea. Rev.,* **95,** 389–402.

Katayama, A. (1969). Calculation of radiative transfer. *In* "Proceedings of the WMO/IUGG Symposium on Numerical Weather Prediction," Tokyo, 1968, pp. IV-8-7–IV-8-10. Japan Meteorological Agency.

Katayama, A. (1972). A simplified scheme for computing radiative transfer in the troposphere, Technical Report 6, Numerical simulation of weather and climate. Department of Meteorology, UCLA.

Kim, Y. J., and A. Arakawa (1995). Improvement of orographic gravity wave parameterization using a mesoscale gravity wave model. *J. Atmos. Sci.,* **52,** 1875–1902.

Köhler, M., C. R. Mechoso, and A. Arakawa (1997). Ice cloud formulation in climate modeling. *In* "7th Conference on Climate Variations," Long Beach, CA, February 2–7, 1997, pp. 237–242. American Meteorological Society.

Kuo, H. L. (1951). Dynamic aspects of the general circulation and the stability of zonal flow. *Tellus,* **3,** 268–284.

Kuo, H. L. (1965). On formation and intensification of tropical cyclones through latent heat release by cumulus convection. *J. Atmos. Sci.,* **22,** 40–63.

Langlois, W. E., and H. C. W. Kwok (1969). Description of the Mintz–Arakawa numerical general circulation model. Technical Report 3, Numerical simulation of weather and climate. Department of Meteorology, UCLA.

Leith, C. E. (1964). Numerical simulation of the Earth's atmosphere. Report under contract W-7405-eng-48, Lawrence Radiation Laboratory, Livermore, CA.

Lewis, J. M. (1998). Clarifying the dynamics of the general circulation: Phillips's 1956 experiment. *Bull. Am. Meteor. Soc.,* **79,** 39–60.

Li, J.-L. F., C. R. Mechoso, and A. Arakawa (1999). Improved PBL moist processes with the UCLA GCM. *In* "10th Symposium on Global Change Studies," Dallas, Texas, January 10–15, 1999, pp. 423–426. American Meteorological Society.

Lilly, D. K. (1965). On the computational stability of numerical solutions of time-dependent non-linear geophysical fluid dynamical problems. *Mon. Wea. Rev.,* **93,** 11–26.

Lilly, D. K. (1997). Introduction to "Computational design for long-term numerical integration of the equations of fluid motion: Two-dimensional incompressible flow. Part I." *J. Comp. Phys.,* **135,** 101–102.

Lord, S. J. (1978). Development and observational verification of a cumulus cloud parameterization, Ph.D. Thesis. Department of Atmospheric Sciences, UCLA.

Lord, S. J. (1982). Interaction of a cumulus cloud ensemble with the large-scale environment. Part III. *J. Atmos. Sci.,* **39,** 88–103.

Lord, S. J., and A. Arakawa (1980). Interaction of a cumulus cloud ensemble with the large-scale environment. Part II. *J. Atmos. Sci.,* **37,** 2677–2692.

Lord, J. S., W. Chao, and A. Arakawa (1982). Interaction of a cumulus cloud ensemble with the large-scale environment. Part IV. *J. Atmos. Sci.,* **39,** 104–113.

Lorenz, E. N. (1955). Available potential energy and the maintenance of the general circulation. *Tellus,* **7,** 157–167.

Lorenz, E. N. (1960). Energy and numerical weather prediction. *Tellus,* **12,** 364–373.

Manabe, S., J. Smagorinsky, and R. F. Strickler (1965). Simulated climatology of a general circulation model with a hydrological cycle. *Mon. Wea. Rev.,* **93,** 769–798.

Mesinger, F. (1973). A method for construction of second-order accuracy difference schemes permitting no false two-grid-interval wave in the height field. *Tellus,* **25,** 444–458.

Mesinger, F., and A. Arakawa (1976). Numerical methods used in atmospheric models. GARP Publication Series 17, **1,** WMO.

Mintz, Y. (1958). Design of some numerical general circulation experiments. *Bull. Res. Counc. Isr. Geosci.*, **7G**, 67–114.

Mintz, Y. (1965). Very long-term global integration of the primitive equations of atmospheric motion: An experiment in climate simulation. WMO Tech. Notes 66, 141–167; and *Meteor. Monogr.* **8**, No. 30, 1968, 20–36.

Miyakoda, K. (1962). A trial of 500 hour barotropic forecast. *In* "Proceedings of the International Symposium on Numerical Weather Prediction," Tokyo, 1960, pp. 221–240. Meteor. Soc. Japan.

Monin, A. S., and A.M. Obukhov (1954). Basic laws of turbulent mixing in the ground layer of the atmosphere. *Akad. Nauk SSR Geofiz. Inst. Tr.*, **151**, 163–187.

National Academy of Sciences (1965). The *feasibility of a global observation and analysis experiment*. Report of the Panel on International Meteorological Cooperation to the Committee on Atmospheric Sciences. National Research Council, October 1965. (See *Bull. Amer. Meteor. Soc.*, **47**, 1966, 200–220.)

Ooyama, K. (1964). A dynamical model for the study of tropical cyclone development. *Geofisica Internacional*, **4**, 187–198.

Ooyama, K. (1969). Numerical simulation of the life-cycle of tropical cyclones. *J. Atmos. Sci.*, **26**, 3–40.

Phillips, N. (1951). A simple three-dimensional model for the study of large-scale extratropical flow patterns. *J. Meteor.* **8**, 381–394.

Phillips, N. A. (1956). The general circulation of the atmosphere: A numerical experiment. *Quart. J. Roy. Meteor. Soc.*, **82**, 123–164.

Phillips, N. A. (1957a). A coordinate system having some special advantages for numerical forecasting. *J. Meteor.*, **14**, 184–185.

Phillips, N. A. (1957b). A map projection system suitable for large-scale numerical projection. *J. Meteor. Soc. Japan*, **56**, 175–186.

Phillips, N. A. (1959). An Example of non-linear computational instability. *In* "The Atmosphere and the Sea in Motion," pp. 501–504. Rockefeller Institute Press, New York.

Platzman, G. W. (1961). An approximation to the product of discrete functions. *J. Meteor.*, **18**, 31–37.

Platzman, G. W. (1967). A retrospective view of Richardson's book on weather prediction. *Bull. Am. Meteor. Soc.*, **48**, 514–550.

Randall, D. A. (1976). The interaction of the planetary boundary layer with large-scale circulations, Ph.D. Thesis. Department of Atmospheric Sciences, UCLA.

Randall, D. A., and D.-M. Pan (1993). Implementation of the Arakawa–Schubert cumulus parameterization with a prognostic closure. *In* "The Representation of Cumulus Convection in Numerical Models of the Atmosphere" (K. A. Emanuel and D. J. Raymond, eds.), pp. 137–144. Am. Meteor. Soc.

Randall, D. A., P. Ding, and D. M. Pan (1997). The Arakawa–Schubert parameterization. *In* "The Physics and Parameterization of Moist Convection" (R. T. Smith, ed.), pp. 281–296, Kluwer Academic Publishers, New York.

Richardson, L. F. (1922). "Weather Prediction by Numerical Processes." Cambridge University Press, Cambridge, MA.

Riehl, H., and J. S. Malkus (1958). On the heat balance in the equatorial trough zone. *Geophysica*, **6**, 503–538.

Riehl, H., and J. S. Malkus (1961). Some aspects of hurricane Daisy, 1958. *Tellus*, **13**, 181–213.

Rossby, C.-G., and Collaborators (1939). Relation between the intensity of the zonal circulation of the atmosphere and the displacement of the semipermanent centers of action. *J. Mar. Res.*, **2**, 38–55.

Sadourny, R. (1975). The dynamics of finite-difference models of the shallow water equations. *J. Atmos. Sci.,* **32,** 680–689.

Sadourny, R., A. Arakawa, and Y. Mintz (1968). Integration of the nondivergent barotropic vorticity equation with an icosahedral-hexagonal grid for the sphere. *Mon. Wea. Rev.,* **96,** 351–356.

Schlesinger, M. E. (1976). A numerical simulation of the general circulation of atmospheric ozone, Ph.D. Thesis. Department of Atmospheric Sciences, UCLA.

Schlesinger, M. E., and Y. Mintz (1979). Numerical simulation of ozone production, transport and distribution with a global atmospheric general circulation model. *J. Atmos. Sci.,* **36,** 1325–1361.

Schubert, W. H. (1973). The interaction of a cumulus cloud ensemble with the large-scale environment, Ph.D. Thesis. Department of Meteorology, UCLA.

Simmons, A. J., and D. M. Burridge (1981). An energy and angular momentum conserving vertical finite-difference scheme and hybrid vertical coordinates. *Mon. Wea. Rev.,* **109,** 758–766.

Smagorinsky, J. (1963). General circulation experiments with the primitive equations. *Mon. Wea. Rev.,* **91,** 99–164.

Smagorinsky, J., S. Manabe, and J. L. Holloway, Jr. (1965). Numerical results from a nine-level general circulation model of the atmosphere. *Mon. Wea. Rev.,* **93,** 727–768.

Starr, V. P., and R. M. White (1954). Balance requirements of the general circulation. *Geophys. Res. Papers,* **35.** Geophysical Research Directorate, Cambridge, MA.

Suarez, M. J., and A. Arakawa (1979). Description and preliminary results of the 9-level UCLA general circulation model. *In* "Proceedings of the Fourth Conference on Numerical Weather Prediction," pp. 290–297. Am. Meteor. Soc.

Suarez, M. J., A. Arakawa, and D. A. Randall (1983). The parameterization of the planetary boundary layer in the UCLA general circulation model: Formulation and results. *Mon. Wea. Rev.,* **111,** 2224–2243.

Takacs, L. L. (1988). On the effects of using *a posteriori* methods for the conservation of integral invariants. *Mon. Wea. Rev.,* **116,** 525–545.

Takacs, L. L., and R. C. Balgovind (1983). High latitude filtering in global grid point models. *Mon. Wea. Rev.,* **111,** 2005–2015.

Takahashi, K., A. Katayama, and T. Asakura (1960). A numerical experiment of the atmospheric radiation. *J. Meteor. Soc. Japan,* **38,** 175–181.

Takano, K., and M. G. Wurtele (1981). A fourth order energy and potential enstrophy conserving difference scheme. Final Report, Sep. 1978–Sept. 1981, AFGL-TR-82-0205. Air Force Geophysics Laboratory, Boston, MA.

Tokioka, T. (1978). Some considerations on vertical differencing. *J. Meteor. Soc. Japan,* **56,** 98–111.

Williamson, D. L. (1968). Integration of the barotropic vorticity equation on a spherical geodesic grid. *Tellus,* **20,** 642–653.

Winninghoff, F. J. (1968). On the adjustment toward a geostrophic balance in a simple primitive equation model with application to the problems of initialization and objective analysis, Ph.D. Thesis. Department of Meteorology, UCLA.

Yanai, M. (1961). A detailed analysis of typhoon formation. *J. Meteor. Soc. Japan,* **39,** 187–214.

Chapter 2

A Brief History of Atmospheric General Circulation Modeling

Paul N. Edwards

Program in Science, Technology & Society, Stanford University, Stanford, California

I. INTRODUCTION

This article presents preliminary results of an attempt to trace the history of atmospheric general circulation modeling, focusing on the period through 1985. *Important caveats*: This is *not* intended as a definitive account. Rather, it is an exploratory study that will be revised and corrected over the next 2 years, as I prepare a book-length history of climate modeling (Edwards, in press). More information about this project is provided at the end of the essay. This chapter certainly contains mistakes and incomplete coverage, for which I apologize in advance. I encourage anyone who finds significant omissions or errors to let me know

General Circulation Model Development

about them, so that the final version of this history can be accurate and complete.

Finally, I should stress that what follows is written from the perspective of a historian of science, rather than that of a scientist.

II. BEFORE 1955: NUMERICAL WEATHER PREDICTION AND THE PREHISTORY OF GCMs

In the early 20th century, the Norwegian Vilhelm Bjerknes argued that atmospheric physics had advanced sufficiently to allow weather to be forecast using calculations. He developed a set of seven equations whose solution would, in principle, predict large-scale atmospheric motions.

Bjerknes proposed a "graphical calculus," based on weather maps, for solving the equations. Although his methods continued to be used and developed until the 1950s, both the lack of faster calculating methods and the dearth of accurate observational data limited their success as forecasting techniques (Nebeker, 1995).

A. RICHARDSON'S "FORECAST FACTORY"

In 1922, Lewis Fry Richardson developed the first numerical weather prediction (NWP) system. His calculating techniques—division of space into grid cells, finite difference solutions of differential equations—were the same ones employed by the first generations of general circulation model (GCM) builders. Richardson's method, based on simplified versions of Bjerknes's "primitive equations" of motion and state (and adding an eighth variable, for atmospheric dust) reduced the calculations required to a level where manual solution could be contemplated. Still, this task remained so large that Richardson did not imagine it as a weather forecast technique. His own attempt to calculate weather for a single 8-hr period took 6 weeks and ended in failure.

His model's enormous calculation requirements led Richardson to propose a fanciful solution he called the "forecast factory." The "factory"—really more like a vast orchestral performance—would have filled a vast stadium with 64,000 people. Each one, armed with a mechanical calculator, would perform part of the calculation. A leader in the center, using colored signal lights and telegraph communication, would coordinate the forecast.

Yet even with this fanciful apparatus, Richardson thought he would probably be able to calculate weather only about as fast as it actually happens. Only in the 1940s, when digital computers made possible automatic calculation on an unprecedented scale, did Richardson's technique become practical (Richardson, 1922).

B. Computers, Weather, and War in the 1940s

The Princeton mathematician John von Neumann was among the earliest computer pioneers. Engaged in computer simulations of nuclear weapons explosions, he immediately saw parallels to weather prediction. (Both are nonlinear problems of fluid dynamics.) In 1946, soon after the ENIAC became operational, von Neumann began to advocate the application of computers to weather prediction (Aspray, 1990). As a committed opponent of Communism and a key member of the WWII-era national security establishment, von Neumann hoped that weather modeling might lead to weather control, which might be used as a weapon of war. Soviet harvests, for example, might be ruined by a U.S.-induced drought (Kwa, 1994, in press).

Under grants from the U.S. Weather Bureau, the Navy, and the Air Force, he assembled a group of theoretical meteorologists at Princeton's Institute for Advanced Study (IAS). If regional weather prediction proved feasible, von Neumann planned to move on to the extremely ambitious problem of simulating the entire atmosphere. This, in turn, would allow the modeling of climate. Jule Charney, an energetic and visionary meteorologist who had worked with Carl-Gustaf Rossby at the University of Chicago and with Arnt Eliassen at the University of Oslo, was invited to head the new Meteorology Group.

The Meteorology Project ran its first computerized weather forecast on the ENIAC in 1950. The group's model, like Richardson's, divided the atmosphere into a set of grid cells and employed finite-difference methods to solve differential equations numerically. The 1950 forecasts, covering North America, used a two-dimensional grid with 270 points about 700 km apart. The time step was 3 hr. Results, while far from perfect, were good enough to justify further work (Charney *et al.*, 1950; Platzman, 1979).

C. The Swedish Institute of Meteorology

The Royal Swedish Air Force Weather Service in Stockholm was first in the world to begin routine real-time numerical weather forecasting (i.e., with broadcast of forecasts in advance of weather). The Institute of Meteorology at the University of Stockholm, associated with the eminent meteorologist Carl-Gustaf Rossby, developed the model. Forecasts for the North Atlantic region were made three times a week on the Swedish BESK computer using a barotropic model, starting in December 1954 (Bergthorsson *et al.*, 1955; Institute of Meteorology, 1954).

D. The Joint Numerical Weather Prediction Unit

About 1952, Von Neumann, Charney, and others convinced the U.S. Weather Bureau and several research and forecasting agencies of the Air Force and Navy to establish a Joint Numerical Weather Prediction (JNWP) Unit. The JNWP Unit opened in Suitland, Maryland, in 1954, under the directorship of George Cressman. It began routine real-time weather forecasting in May 1955 (Nebeker, 1995). Yet it was more than a decade before numerical methods began to outstrip in accuracy the "subjective method" employed by human forecasters.

Initially, the computer models used for NWP employed simplifying assumptions. Only in the 1960s did models based on the Bjerknes/ Richardson primitive equations replace barotropic and baroclinic models.

III. 1955–1965: ESTABLISHMENT OF GENERAL CIRCULATION MODELING

In the mid-1950s, the weather models used by forecasters were still regional or continental (versus hemispherical or global) in scale. Calculations for numerical weather prediction were limited to what could be accomplished in a couple of hours on then-primitive digital computers. In addition, the time constraints of analog-to-digital data conversion and long-distance communication imposed limitations on the scale of operational weather forecasting.

Yet for theoretical meteorologists—unconcerned with real-time forecasting—general circulation modeling became a kind of holy grail.

By mid-1955 Normal Phillips had completed a two-layer, hemispheric, quasi-geostrophic computer model of the general circulation (Phillips, 1956). Despite its primitive nature, Phillips's model is now often regarded as the first working GCM.

As computer power grew, the need for simplifying assumptions (such as barotropy and quasi-geostrophy) diminished. Many individuals throughout the world, including Phillips, began experiments with primitive equation models in the late 1950s (Hinkelmann, 1959). Between the late 1950s and the early 1960s, four separate groups began—more or less independently—to build many-leveled, three-dimensional GCMs based on the primitive equations of Bjerknes and Richardson. Details of these efforts are given in the four following sections.

IV. THE GEOPHYSICAL FLUID DYNAMICS LABORATORY

The first laboratory to develop a continuing program in general circulation modeling opened in 1955. In that year, at von Neumann's instigation, the U.S. Weather Bureau created a General Circulation Research Section under the direction of Joseph Smagorinsky. Smagorinsky felt that his charge was to continue with the final step of the von Neumann/Charney computer modeling program: a three-dimensional, global, primitive equation GCM of the atmospheric (Smagorinsky, 1983). The General Circulation Research Section was initially located in Suitland, Maryland, near the Weather Bureau's JNWP unit. The lab's name was changed in 1959 to the General Circulation Research Laboratory (GCRL), and it moved to Washington, D.C.

In 1955–1956, Smagorinsky collaborated with von Neumann, Charney, and Phillips to develop a two-level, zonal hemispheric model using a subset of the primitive equations (Smagorinsky, 1958). Beginning in 1959, he proceeded to develop a nine-level primitive equation GCM, still hemispheric (Smagorinsky, 1963). Smagorinsky was among the first to recognize the need to couple ocean models to atmospheric GCMs; he brought the ocean modeler Kirk Bryan to the GCRL in 1961 to begin this research (Smagorinsky, 1983).

The General Circulation Research Laboratory was renamed the Geophysical Fluid Dynamics Laboratory (GFDL) in 1963. In 1968, GFDL moved to Princeton University, where it remains.

A. Manabe and the GFDL General Circulation Modeling Program

In 1959, Smagorinsky invited Syukuro Manabe of the Tokyo NWP Group to join the General Circulation Research Laboratory. (Smagorinsky had been impressed by Manabe's publications in the *Journal of the*

Meteorological Society of Japan.) He assigned Manabe to the GCM coding and development.

By 1963, Smagorinsky, Manabe, and their collaborators had completed a nine-level, hemispheric primitive equation GCM (Manabe, 1967; Manabe *et al.*, 1965; Smagorinsky *et al.*, 1965). Manabe was given a large programming staff. He was thus able to focus on the mathematical structure of the models, without becoming overly involved in coding.

In the mid-1960s, as Smagorinsky became increasingly involved in planning for the Global Atmospheric Research Program (GARP), Manabe became the *de facto* leader of GFDL's GCM effort, although Smagorinsky remained peripherally involved. Until his retirement in 1998, Manabe led one of the most vigorous and longest lasting GCM development programs in the world.

Manabe's work style has been highly collaborative. With his colleagues Strickler, Wetherald, Holloway, Stouffer, and Bryan, as well as others, Manabe was among the first to perform carbon dioxide doubling experiments with GCMs (Manabe, 1970, 1971), to couple atmospheric GCMs with ocean models (Manabe and Bryan, 1969), and to perform very long runs of GCMs under carbon dioxide doubling (Manabe and Stouffer, 1994). Another characteristic of Manabe's work style is a focus on basic issues rather than on fine-tuning of model parameterizations. He retired in 1998, but remains active.

B. The GFDL Atmospheric GCMs

Note that the names given in the following section are informal terms used by GFDL members, who do not always agree on their interpretation.

1. MARKFORT

The MARKFORT series began with Smagorinsky's nine-level, 3-D hemispheric model. It was used well into the 1960s. Initially, the model was run on the IBM STRETCH. A number of GFDL's most influential publications resulted from the MARKFORT model.

2. Zodiac

The Zodiac finite-difference model series was the second major GFDL GCM. The chief innovation was the use of a new spherical coordinate system developed by Yoshio Kurihara (Kurihara, 1965). This model remained in use throughout the 1970s.

3. Sector

The Sector series was not an independent GCM, but a subset of the GFDL global models. To conserve computer time (especially for coupled ocean–atmospheric modeling), integrations were performed on a 60-deg longitudinal "slice" of the globe, with a symmetry assumption for conversion to global results. In the early sector models, highly idealized land–ocean distributions were employed (Manabe *et al.* 1975).

4. SKYHI

Work on SKYHI, a high-vertical-resolution GCM covering the troposphere, stratosphere, and mesosphere, began in 1975 (Mahlman *et al.*, 1978).

5. GFDL Spectral Model

In the mid-1970s, GFDL imported a copy of the spectral GCM code developed by W. Bourke at the Australian Numerical Meteorological Research Centre (Bourke, 1974; Gordon, 1976; Gordon and Stern, 1974). Interestingly, Bourke and Barrie Hunt had originally worked out the spectral modeling techniques while visiting GFDL in the early 1970s.

6. Supersource

Beginning in the late 1970s, Leith Holloway began to recode the GFDL spectral model to add modularity and user-specifiable options. The result was Supersource, the modular, spectral atmospheric GCM that remains in use at GFDL today. "Holloway fit the physics from Manabe's grid model (Zodiac and relatives) into the spectral model. Holloway then unified all the versions of this new spectral model into one Supersource" (Ron Stouffer, personal communication, 1997).

Users can specify code components and options. Among these options is a mixed-layer ocean model, but Supersource itself does not contain an ocean GCM. Supersource code has frequently been used as the atmospheric component in coupled OAGCM studies (Manabe and Stouffer, 1988, 1994). It will be replaced by a new model in 2000.

V. THE UCLA DEPARTMENT OF METEOROLOGY

Jacob Bjerknes, who founded the UCLA Department of Meteorology in 1940, had a strong interest in the problem of the atmospheric general circulation. This tradition continued with Yale Mintz, a graduate student of Bjerknes's who received his Ph.D. in 1949. He continued to work at UCLA, becoming associate project director with Bjerknes. In the late 1950s, Mintz began to design numerical general circulation experiments (Mintz, 1958).

A. MINTZ AND ARAKAWA

Like Smagorinsky, Mintz recruited a Japanese meteorologist, Akio Arakawa, to help him build GCMs. Arakawa, known for his mathematical wizardry, was particularly interested in building robust schemes for the parameterization of cumulus convection. Mintz and Arakawa constructed a series of increasingly sophisticated GCMs beginning in 1961. "Ironically, Arakawa's first role after joining the project was to persuade him to slow the development, giving first priority to designing model dynamics suitable for long-term integrations" (Johnson and Arakawa, 1996). The first-generation UCLA GCM was completed in 1963. Arakawa then went back to Japan, but Mintz persuaded him to return to UCLA permanently in 1965.

In the latter half of the 1960s, IBM's Large Scale Scientific Computation Department in San Jose, California, provided important computational assistance and wrote a manual describing the model (Langlois and Kwok, 1969).

B. WIDESPREAD INFLUENCE

Of all the GCM groups in the world, the UCLA laboratory probably had the greatest influence on others, especially in the 1960s and 1970s. This was due not only to continuing innovation (particularly in cumulus parameterization), but also to the openness of the UCLA group to collaboration and sharing. Whereas GFDL, and to a lesser extent the National Center for Atmospheric Research (NCAR), were pure-research institutions, UCLA operated in the mode of an academic graduate program. The Department of Meteorology's graduates carried the UCLA model with them to other institutions, while visitors from around the world spent time at the group's laboratories (Arakawa, 1997, personal communication to Paul N. Edwards).

C. THE UCLA MODELS

The key characteristics of the UCLA model series and its spinoffs are neatly pictured in a chart made by Arakawa (see Fig. 7 in Chapter 1). Until the 1980s, UCLA typically focused on model development, leaving "production" of the models (i.e., use in experimental studies) to other institutions. Generation numbers given here are my own.

1. UCLA I (Prototype)

The first Mintz–Arakawa model was a two-level global, primitive equation GCM at a 7° latitude × 9° longitude horizontal resolution. It included realistic land–sea distributions and surface topography. Mintz never learned to program computers; Arakawa carried out all the model coding. This prototype model was abandoned about 1965.

2. UCLA II

When Arakawa returned to UCLA from Japan in 1965, he and Mintz began work on the first-generation "production" UCLA GCM. It increased model resolution to 4° latitude × 5° longitude, although it still had only two vertical levels, and introduced a new horizontal grid structure—the Arakawa–Lamb B Grid (Arakawa and Lamb, 1977). This was an extremely influential GCM. About 1970, Lawrence Gates, a UCLA graduate, carried the model with him to the RAND Corporation, where he used it in a series of studies sponsored by the Advanced Research Projects Agency of the U.S. Department of Defense. The RAND version of the model was eventually carried to Oregon State University (Gates, 1975).

3. UCLA II (3-level)

The second-generation UCLA model essentially extended the vertical resolution of the second-generation model to three levels. This model was carried to three NASA laboratories. In 1972, a nine-level version was begun at the Goddard Institute for Space Studies (GISS) in New York, whose current model is a direct descendant. Later in the 1970s it traveled to the Goddard Laboratory for Atmospheric Sciences and the Goddard Laboratory for Atmospheres (A. Del Genio, 1998, personal communication).

4. UCLA III

This 6- and 12-level model used the Arakawa–Lamb C Grid, a finite-difference horizontal grid. All subsequent UCLA models have also employed this scheme. In the mid-1970s, versions of this model, with slightly different sets of prognostic variables, were built. One version was exported to the U.S. Naval Environment Prediction Research Facility and the Fleet Numerical Oceanographic Center, both in Monterey, California. This model evolved into the operational NOGAPS forecasting system (Hogan and Rosmond, 1991). It was also given to the Meteorological Research Institute in Tsukuba, Japan, where it continues to be used in a wide variety of forecasting and climate studies.

5. UCLA IV

Work on the fourth-generation UCLA model began in the late 1970s. The chief innovation of this model generation was a new vertical coordinate system, which used the top of the planetary boundary layer as a coordinate surface. A version of this model remains in use at UCLA into the present, although a fifth-generation model was built in 1990.

UCLA IV was also adopted by the Navy research centers mentioned earlier. In addition, it was taken to the Goddard Laboratory for Atmospheres in the early 1980s. Code for this model was extensively rewritten (Randall, 2000, personal communication). In 1988, the model was brought to Colorado State University by David Randall, another former student of Arakawa.

Versions of this model made their way to Lawrence Livermore National Laboratory and also to the Central Weather Bureau of the Republic of China.

VI. THE LIVERMORE ATMOSPHERIC MODEL

In 1960, Cecil E. "Chuck" Leith began work on a GCM at Lawrence Livermore National Laboratories (LLNL). Trained as a physicist, Leith became interested in atmospheric dynamics and received the blessing of LLNL director Edward Teller for a project on the general circulation. Teller's approval stemmed from his long-term interest in weather modification.

After receiving encouragement from Jule Charney, Leith spent a summer in Stockholm at the Swedish Institute of Meteorology. There he coded a five-level GCM for LLNL's newest computer, the Livermore Automatic Research Calculator (LARC), due to be delivered in the fall of 1960. Leith wrote the code based solely on the manual for the new machine.

Although aware of the Smagorinsky–Manabe and Mintz–Arakawa efforts, Leith worked primarily on his own. He had a working five-level model by 1961. However, he did not publish his work until 1965 (Leith,

1965). Nevertheless, by about 1963 Leith had made a film showing his model's results in animated form and had given numerous talks about the model.

Leith ceased work on his model—known as LAM (Leith atmospheric model or Livermore atmospheric model)—in the mid-1960s, as he became increasingly interested in statistical modeling of turbulence. In 1968, he went to NCAR, where he was instrumental in a number of climate modeling projects.

The initial LAM model was based on the Bjerknes–Richardson primitive equations. It had five vertical levels and used a $5° \times 5°$ horizontal grid. It covered only the Northern Hemisphere, with a "slippery wall" at 60°N. To damp the effects of small-scale atmospheric waves, Leith introduced an artificially high viscosity, which caused serious problems and helped to stimulate Leith's career-long interest in turbulence.

VII. THE NATIONAL CENTER FOR ATMOSPHERIC RESEARCH

The National Center for Atmospheric Research, established in 1960, began a GCM effort in 1964 under Akira Kasahara and Warren Washington. Two different model series were eventually constructed, designated here as NCAR 1–3 and CCM 0–1.

A. THE KASAHARA–WASHINGTON MODELS (NCAR 1–3)

The first-generation NCAR GCM was developed starting in 1964, with first publication in 1967. It was a simple two-layer global model with a 5° horizontal resolution.

The second-generation model, completed around 1970, added a great deal of flexibility. The basic model had a 5° horizontal, six-layer resolution, but it could also be run at resolutions as fine as 0.625° horizontal over a limited domain, with up to 24 vertical layers.

NCAR 3, finished about 1973, also allowed multiple resolutions, including a user-specifiable vertical increment. The most significant changes, however, involved improved finite-difference schemes.

The Kasahara–Washington group focused a great deal of attention on numerical schemes for finite-difference approximations. In addition, a great deal of work was done on problems of computational error arising from round-off (Kasahara and Washington, 1967).

B. The Community Climate Model

In the latter part of the 1970s, NCAR gradually abandoned the Kasahara–Washington model. In its place, NCAR developed a community climate model (CCM), intended to serve not only modelers working at NCAR, but the large constituency of affiliated universities associated with NCAR's parent organization, the University Corporation for Atmospheric Research. The CCM was initially based on the Australian Numerical Meteorological Research Centre model and an early version of the European Centre for Medium Range Weather Forecasts (ECMWF) model. It also incorporated elements of the GFDL models.

The NCAR CCM series of models was especially important because of the relatively large community of researchers who were able to use it. Versions of the model were adopted by a number of other groups in the late 1980s. This was made possible by NCAR's strong focus on documentation and modularity. User manuals and code documentation were made available for all elements of the models starting with CCM-0B.

1. CCM-0A

The initial version of the community climate model was based on the spectral model of the Australian Numerical Meteorological Research Centre (Bourke *et al.*, 1977). One member of the ANMRC team (K. Puri) brought the model to NCAR during an extended visit. Later, it was extensively revised.

2. CCM-0B: A Combined Forecast and Climate Simulation Model

A second version of the community climate model was developed in 1981. This model's guiding purpose was "NCAR's decision to utilize the same basic code for global forecast studies (both medium- and long-range) and for climate simulation. Economy and increased efficiency could then be achieved by documenting and maintaining only one set of codes. Changes from one application to the other could be relatively straightforward in a model with modular design. The use of one basic model for both forecasting and climate studies has potential scientific value since a major part of long-range (one- to two-week) forecast errors is due to the drift toward a model climate which differs from that of the atmosphere. Thus, improvements in the climate aspects of the model should lead to improvements in forecasts" (Williamson *et al.*, 1987).

CCM-0B was designed to include the best elements of other existing models. Initial code for CCM-0B came from an early version of the ECMWF model. Physical parameterizations, including the radiation and cloud routines of Ramanathan, and numerical approximations were added from CCM-0A (Ramanathan *et al.*, 1983). Energy balance and flux prescriptions from the early GFDL models were incorporated, while vertical and temporal finite differences matched from the Australian spectral model that was the basis for CCM-0A (Williamson *et al.*, 1987).

3. CCM-1

CCM-1 evolved from CCM-0B in the mid-1980s. The primary differences were changed parameterizations, new horizontal and vertical diffusion schemes, and changes to moisture adjustment and condensation schemes.

VIII. 1965–1975: SPREAD OF GCMs

By 1965, then, three groups in the United States had established ongoing efforts in general circulation modeling:

- Geophysical Fluid Dynamics Laboratory
- UCLA Department of Meteorology
- National Center for Atmospheric Research

In addition, a small group at the UK Meteorological Office had begun work on a GCM, under Andrew Gilchrist, but published very little until the 1970s. At this point, GCMs and modeling techniques began to spread by a variety of means. Commonly, new modeling groups began with some version of another group's model. Some new groups were started by post-docs or graduate students from one of the three original GCM groups. Others built new models from scratch.

The GCM family tree, shown in the Appendix at the end of this chapter, offers a visual map of these relationships.

A. Modeling Groups Proliferate

Among the important GCM groups established in 1965–1975 were these:

- RAND Corporation (Santa Monica, California)

- Goddard Institute for Space Studies (New York, New York)
- Australian Numerical Meteorological Research Centre (Melbourne, Australia; later this became the Bureau of Meteorology Research Centre)

Each group initially borrowed an existing model, but subsequently made significant modifications of its own.

B. Modeling Innovations

Two important innovations of the 1965–1975 decade were coupled atmosphere–ocean models and spectral transform techniques.

1. Coupled Atmosphere–Ocean Models

GFDL was among the first groups to attempt coupling of an atmospheric GCM to an ocean model. Initially, highly simplified ocean models (one-layer "swamp" oceans) were used. These were succeeded by two-level "mixed-layer" ocean models. In 1969, Manabe and Bryan published the first results from a coupled ocean–atmosphere general circulation model (OAGCM). However, this model used a highly idealized continent–ocean configuration. Results from the first coupled OAGCM with more realistic configurations were published in 1975 (Manabe *et al.*, 1975).

2. Spectral Transform Techniques

Spectral methods are an alternative to finite-difference schemes, the method used by all of the first-generation primitive equation GCMs. They express the horizontal variation of dynamic model fields in terms of orthogonal spherical harmonics. The technique simplifies the solution of many of the nonlinear partial differential equations used in general circulation modeling. Its utility had been explored as early as 1954 (Platzman, 1960; Silberman, 1954).

Heavy calculational demands made spectral methods unsuitable for use in early GCMs. Faster computers, and improvements in algorithms for spectral methods that reduced their calculational intensity, led to their adoption in GCMs around 1970 (Bourke, 1974; Eliasen *et al.*, 1970; Orszag, 1970; Robert, 1969).

C. Research on Carbon Dioxide and Climate

The important role of carbon dioxide, water vapor, and other "greenhouse" gases in the atmosphere's heat retention capacity had been recognized in the 19th century by the Swedish scientist Svante Arrhenius, who had also speculated—with remarkable prescience—on the possibility of anthropogenic climate change from the combustion of fossil fuels (Arrhenius, 1896).

Little further work on the greenhouse effect was done until the late 1940s, when radioactivity in the atmosphere stimulated interest in "tracer" studies of various atmospheric constituent gases (Callendar, 1949; Suess, 1953). This gradually led to a revival of interest in the possibility of anthropogenic influences on climate (Plass, 1956). During the International Geophysical Year (1957–1958), Revelle and Suess (1957) proposed monitoring the carbon dioxide content of the atmosphere. This led to the establishment of Keeling's station at Mauna Loa in the same year, which soon established the regular annual increases in the carbon dioxide concentration (Keeling, 1960).

During 1965–1975, studies of the effect of changing carbon dioxide concentrations on the Earth's radiative equilibrium began in earnest, as data from Mauna Loa continued to show steady CO_2 increases. The first studies used simpler one- and two-dimensional models, rather than GCMs (Manabe and Wetherald, 1967). Responses to CO_2 doubling became the standard form of this experiment. The first use of a GCM to study the effects of carbon dioxide doubling came in 1975 (Manabe and Wetherald, 1975).

D. Early Climate Politics and GCMs

During this period, anthropogenic effects on climate were usually considered under the rubric of weather modification, which had been among the stimuli for early efforts in weather modeling. Literature on the subject frequently uses the phrase "inadvertent climate modification" when discussing anthropogenic climate change, to make the parallel (National Research Council, 1966; Study of Man's Impact on Climate, 1971).

1. SCEP and SMIC

With the rise of the environmental movement in the early 1970s came early interest in world-scale environmental problems. Two important studies, both prepared as input to the 1972 United Nations Conference on the Human Environment, noted the possibility of "inadvertent climate modification." The Study of Critical Environmental Problems (SCEP) focused on pollution-induced "changes in climate, ocean ecology, or in large terrestrial ecosystems." It cited GCMs as "indispensable" in the study of possible anthropogenic climate change.

The Study of Man's Impact on Climate (SMIC) also endorsed GCMs. (Its section on this subject was drafted by Manabe.) Both SCEP and SMIC recommended a major initiative in global data collection, new international measurement standards for environmental data, and the integration of existing programs to form a global monitoring network. These reports are widely cited as the origin of public policy interest in anthropogenic climate change (Study of Critical Environmental Problems, 1970; Study of Man's Impact on Climate, 1971).

2. Other Issues

In the early 1970s, several other large-scale atmospheric issues rose to public awareness. Notable among these were stratospheric ozone depletion, acid rain, and upper atmosphere pollution problems raised by the controversial supersonic transport.

IX. 1975–1985: GCMs MATURE

In this decade, more modeling groups were established. Research programs consisted primarily of improving existing modeling techniques through higher resolution, better parameterizations, and coupling ocean and atmospheric GCMs. Increasingly, modelers began to perform GCM-based experiments. Longer models runs, made possible by faster computers, were an important part of experimental strategies. Increasing political attention to the climate change issue, especially in the United States, raised the visibility of GCMs both inside and outside climate science.

A. COMPUTER POWER

The rapid growth of computer power during this period is illustrated by the following in Table I. Most groups building GCMs either owned or had access to large, fast supercomputers. Greater computer power allowed longer runs, smaller grids, and larger numbers of runs.

B. Spread of Modeling Capacity

New GCM modeling groups established during this period include these:

- Max Planck Institut (Hamburg, Germany)
- NASA Goddard Laboratory for Atmospheric Sciences
- NASA Goddard Laboratory for Atmospheres
- Colorado State University
- Oregon State University
- National Meteorological Center
- Lawrence Livermore National Laboratory
- European Centre for Medium-Range Weather Forecasts (Reading, UK)

By the end of this period, European modeling groups—especially the ECMWF—had begun to mount a significant challenge to U.S. dominance in general circulation modeling.

C. Modeling Innovations and Experiments

The decade from 1975 to 1985 was marked by steady improvement in existing techniques, rather than major innovation. Increasingly sophisticated and computationally efficient schemes were developed for these areas of interest:

- Spectral transforms
- Hydrological cycles

Table I

Computers in Use at GFDL, 1956–1982

Computer	Time period	Relative power
IBM 701	1956–1957	1
IBM 704	1958–1960	3
IBM 7090	1961–1962	20
IBM 7030	1963–1965	40
CDC 6600	1965–1967	200
UNIVAC 1108	1967–1973	80
IBM 360/91	1969–1973	400
IBM 360/195	1974–1975	800
Texas Instruments X4ASC	1974–1982	3000

From Geophysical Fluid Dynamics Laboratory (1981).

- Coupled OAGCMs
- Radiative transfer, including atmospheric chemistry
- Moist convection
- Continental surfaces
- Boundary layer turbulence

Carbon dioxide doubling experiments became commonplace.

D. Climate Politics

During 1975–1989, the possibility of global warming became a policy issue within scientific agencies both in the United States and internationally. Studies were conducted by the National Academy of Sciences, the Council on Environmental Quality, the U.S. Department of Energy, the World Meteorological Organization, and others. Congressional hearings called for action, and funding for climate research grew steadily. In 1985, at Villach, Austria, an influential climate science conference recommended policy studies of climate change mitigation techniques, including international treaties.

In the early 1980s, the effects of smoke and dust from a superpower nuclear exchange were tested with climate models, leading to the issue of "nuclear winter" (Covey *et al.*, 1984; Sagan, 1983; Thompson and Schneider, 1986). Action on the ozone depletion issue—sparked by observations of an Antarctic ozone "hole"—produced the Montreal Protocol on the Ozone Layer in 1985. Transboundary pollution problems, notably acid rain, were also high on the political agenda. All of these raised public awareness of global atmospheric problems, but the issue of climate change did not achieve the status of mass politics until about 1988 (Schneider, 1989).

X. CONCLUSION

By the 1980s, computer models of atmosphere and ocean general circulation had become the primary tool in studies of climate. This marked a major historical transformation from a previous era, in which virtually the only tool for climate studies was the statistical record.

Perhaps the most important aspect of this shift was the ability to perform model-based "experiments" to project possible causes of climatic change. This led to the remarkable visibility of GCMs in political debates over anthropogenic climate change, which continues into the present with the work of the Intergovernmental Panel on Climate Change and the Conferences of Parties to the Framework Convention on Climate Change, signed at Rio de Janeiro in 1992.

Another major product of the shift to numerical models was the development of vast global data networks, from many different instrument modalities. These were built to supply the information necessary to predict weather, but the data record is now very nearly sufficient in length and global coverage to allow accurate studies of climate as well. Without the availability of computer models, these data networks would probably not have been constructed, since they could not have been processed or understood in any other way. The pioneering GCM builders have now retired, turning over their monumental project to a large and growing generation of successors. This volume of essays dedicated to Akio Arakawa is a fitting tribute to one of the major scientific achievements of the 20th century.

APPENDIX

The GCM Family Tree

A "family tree" that describes important relations among the major modeling groups is shown in Fig. 1. While the GCM Family Tree captures only the most direct relationships among GCM groups, it can serve a useful heuristic purpose in tracing the main lines of institutional affiliation.

Participating in GCM IIistory

The GCM Family Tree is part of an evolving WWW-based project in "participatory history." We hope to collect archival materials—including documents, informal memoirs, and any other information related to the history of GCMs—and make them available on-line to historians, scientists, and anyone interested in this fascinating story.

The group building the site—funded by the Alfred P. Sloan Foundation and sponsored by the American Institute of Physics and the American Geophysical Union—is posting materials that (like this article) are *still in draft form*. The Web address is www.aip.org/history/gcm. Anyone interested in participating in the project can be added to a notification list by contacting the author at pne@umich.edu.

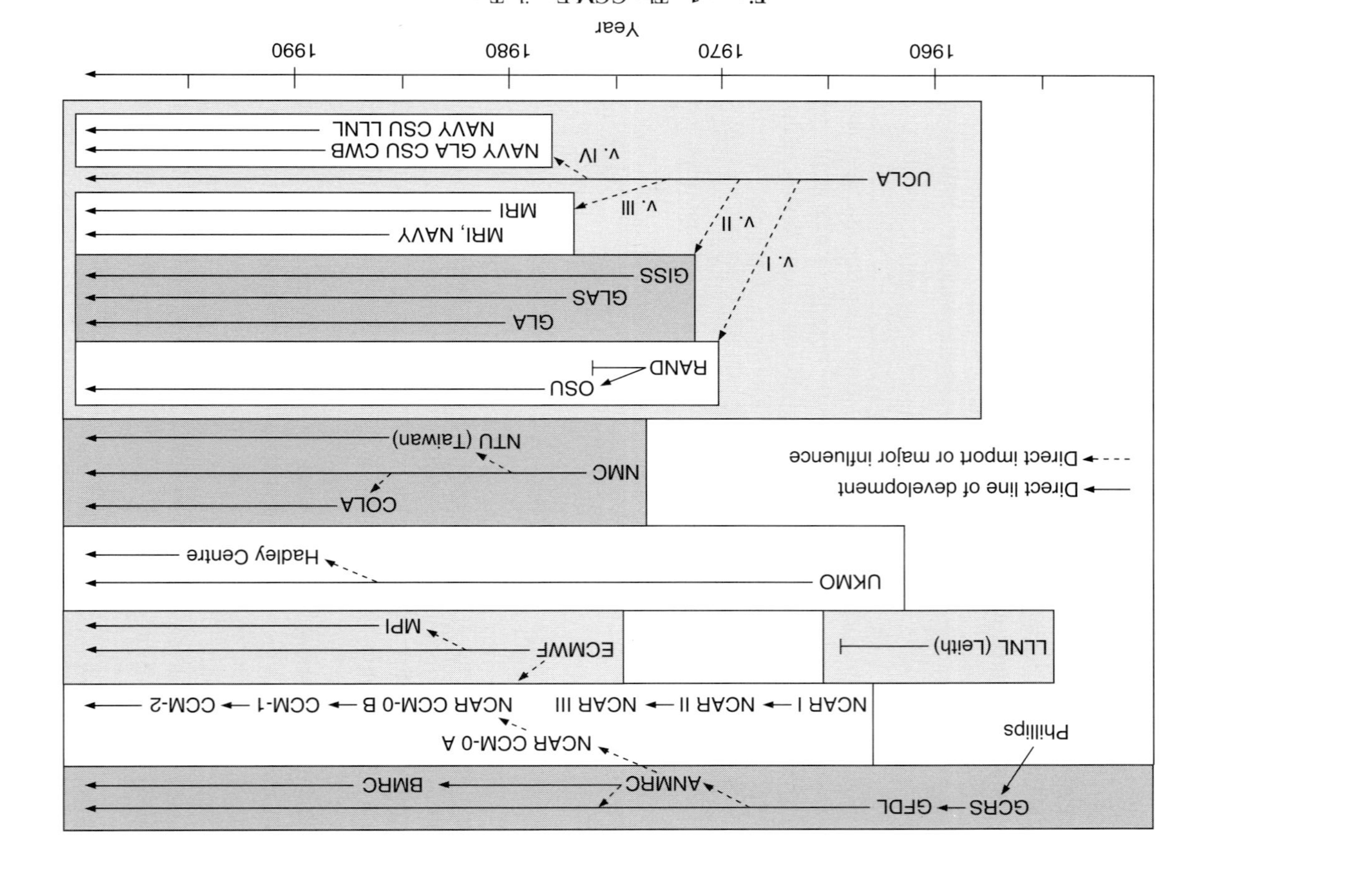

Figure 1 The GCM Family Tree.

Why Contribute to the Archive?

The purpose of the project is to see if the interactive capability of the World Wide Web can be used not only to present information, but also to collect it. We are especially interested in information that might not otherwise be preserved or that researchers would not easily be able to find.

We would like to gather information that would not be part of any official record while it is still relatively fresh in participants' memories. We seek physical material related to the development of GCMs, such as model documentation, memoirs, and correspondence. We are also interested in learning about existing collections of material related to this history.

All contributions will become part of a public archive on the history of atmospheric GCMs. For the life of the Web site, e-mail contributions will be posted there. Eventually, they will be preserved in an electronic archive, along with the physical material donated to us.

REFERENCES

Arakawa, A. Interviewed by Paul N. Edwards, July 17–18, 1997, University of California, Los Angeles.

Arakawa, A., and V. R. Lamb (1977). Computational design of the basic dynamical processes of the UCLA General Circulation Model. *In* "General Circulation Models of the Atmosphere" (J. Chang, ed.), pp. 173–265. Academic Press, San Francisco.

Arrhenius S. (1896). On the influence of carbonic acid in the air upon the temperature of the ground. *Philos. Mag. J. Sci.* **41**, 237–276.

Aspray, W. (1990). "John von Neumann and the Origins of Modern Computing." MIT Press, Cambridge, MA.

Bergthorsson, P., B. R. Doos, S. Frylkund, O. Haug, and R. Lindquist (1955). Routine forecasting with the barotropic model. *Tellus* **7**, 272–274.

Bourke, W. (1974). A multi-level spectral model. I. Formulation and Hemispheric integrations. *Monthly Weather Rev.* **102**, 687–701.

Bourke, W., B. McAvaney, K. Puri, and R. Thurling (1977). Global modeling of atmospheric flow by spectral methods. *In* "General Circulation Models of the Atmosphere" (J. Chang, ed.), pp. 267–324. Academic Press, San Francisco.

Callendar, G. S. (1949). Can carbon dioxide influence climate? *Weather* **4**, 310–314.

Charney, J. G., R. Fjörtoft, and J. von Neumann (1950). Numerical integration of the barotropic vorticity equation. *Tellus* **2**, 237–254.

Covey, C., S. H. Schneider, and S. L. Thompson (1984). Global atmospheric effects of massive smoke injections from a nuclear war: Results from general circulation model simulations. *Nature* **308**, 21–25.

Edwards, P. N. (in press). "The World in a Machine: Computer Models, Data Networks, and Global Atmospheric Politics." MIT Press, Cambridge, MA.

Eliasen, E., B. Machenhauer, and E. Rasmussen (1970). "On a numerical method for integration of the hydrodynamical equations with a spectral representation of the horizontal fields," Report 2. Institut for Teoretisk Meteorologi, Köbenhavns Universitet, Denmark.

Gates, W. L. (1975). "A Review of Rand Climate Dynamics Research." Report WN-9149-ARPA. Rand Corporation, Santa Monica, CA.

Geophysical Fluid Dynamics Laboratory (1981). "Geophysical Fluid Dynamics Laboratory: Activities—FY80, Plans—FY81." U.S. Department of Commerce, Princeton, NJ.

Gordon, C. T. (1976). Verification of the GFDL spectral model. *In* "Weather Forecasting and Weather Forecasts: Models, Systems, and Users. Notes from a Colloquium, Summer 1976" (D. L. Williamson, L. Bengtsson, and A. H. Murphy, eds.), Vol. 2. Advanced Study Program, National Center for Atmospheric Research, Boulder, CO.

Gordon, T., and B. Stern (1974). Spectral modeling at GFDL. Report of the International Symposium on Spectral Methods in Numerical Weather Prediction, GARP Programme on Numerical Experimentation.

Hinkelmann, K. (1959). Ein numerisches Experiment mit den primitiven Gleichungen. *In* "The Atmosphere and the Sea in Motion: Scientific Contributions to the Rossby Memorial Volume" (B. Bolin and E. Eriksson, eds.), pp. 486–500. Rockefeller Institute Press, New York.

Hogan, T. F., and T. E. Rosmond (1991). The Description of the Navy Operational Global Atmospheric Prediction System's Spectral Forecast Model. *Monthly Weather Rev.* **119,** 1786–1815.

Institute of Meteorology, University of Stockholm (1954). Results of forecasting with the barotropic model on an electronic computer (BESK). *Tellus* **6,** 139–149.

Johnson, D. R., and A. Arakawa (1996). On the scientific contributions and insight of Professor Yale Mintz. *J. Climate* **9,** 3211–3224.

Kasahara, A., and W. M. Washington (1967). NCAR global general circulation model of the atmosphere. *Monthly Weather Rev.* **95,** 389–402.

Keeling, C. D. (1960). The concentration and isotopic abundances of carbon dioxide in the atmosphere. *Tellus* **12,** 200–203.

Kurihara, Y. (1965). Numerical integration of the primitive equations on a spherical grid. *Monthly Weather Rev.* **93,** 399–415.

Kwa, C. (1994). Modelling technologies of control. *Sci. as Culture* **4,** 363–391.

Kwa, C. (in press). The rise and fall of weather modification. *In* "Changing the Atmosphere: Science and the Politics of Global Warming" (P. N. Edwards and C. A. Miller, eds.). MIT Press, Cambridge, MA.

Langlois, W. E., and H. C. W. Kwok (1969). Description of the Mintz–Arakawa numerical general circulation model, Technical Report 3. Dept. of Meteorology, University of California, Los Angeles.

Leith, C. E. (1965). Numerical simulation of the earth's atmosphere. *In* "Methods in Computational Physics" (B. Alder, S. Fernbach, and M. Rotenberg, eds.), pp. 1–28. Academic Press, New York.

Mahlman, J. D., R. W. Sinclair, and M. D. Schwarzkopf (1978). Simulated response of the atmospheric circulation to a large ozone reduction. *In* "Proceedings of the WMO Symposium on the Geophysical Aspects and Consequences of Changes in the Composition of the Stratosphere," Toronto, Canada, June 26–30, 1978, pp. 219–220.

Manabe, S. (1967). General circulation of the atmosphere. *Trans. Am. Geophys. Union* **48,** 427–431.

Manabe, S. (1970). The dependence of atmospheric temperature on the concentration of carbon dioxide. *In* "Global Effects of Environmental Pollution" (S. F. Singer, ed.), pp. 25–29. D. Reidel, Dallas, TX.

Manabe, S. (1971). Estimates of future change of climate due to the increase of carbon dioxide. *In* "Man's Impact on the Climate" (W. H. Matthews, W. W. Kellogg, and G. D. Robinson, eds.), pp. 250–264. MIT Press, Cambridge, MA.

Manabe, S., and K. Bryan (1969). Climate calculations with a combined ocean–atmosphere model. *J. Atmos. Sci.* **26,** 786–789.

Manabe, S., and R. J. Stouffer (1988). Two stable equilibria of a coupled ocean–atmosphere model. *J. Climate* **1,** 841–865.

Manabe, S., and R. J. Stouffer (1994). Multiple-century response of a coupled ocean–atmosphere model to an increase of atmospheric carbon dioxide. *J. Climate* **7,** 5–23.

Manabe, S., and R. Wetherald (1967). Thermal equilibrium of the atmosphere with a given distribution of relative humidity. *J. Atmos. Sci.* **24,** 241–259.

Manabe, S., and R. T. Wetherald (1975). The effects of doubling the CO_2 concentration on the climate of a general circulation model. *J. Atmos. Sci.* **XXXII,** 3–15.

Manabe, S., J. Smagorinsky, and R. F. Strickler (1965). Simulated climatology of general circulation with a hydrologic cycle. *Monthly Weather Rev.* **93,** 769–798.

Manabe, S., K. Bryan, and M. J. Spelman (1975). A global ocean–atmosphere climate model: Part I. The atmosphere circulation. *J. Phys. Oceanog.* **5,** 3–29.

Mintz, Y. (1958). Design of some numerical general circulation experiments. *Bull. Res. Council of Israel* **76,** 67–114.

National Research Council (1966). Weather and climate modification, Publication 1350. National Academy of Sciences, Washington, DC.

Nebeker, F. (1995). "Calculating the Weather: Meteorology in the 20th Century." Academic Press, New York.

Orszag, S. A. (1970). Transform method for calculation of vector-coupled sums: Application to the spectral form of the vorticity equation. *J. Atmos. Sci.* **27,** 890–895.

Phillips, N. A. (1956). The general circulation of the atmosphere: A numerical experiment. *Quart. J. Roy. Meteorolog. Soc.* **82,** 123–164.

Plass, G. N. (1956). The carbon dioxide theory of climatic change. *Tellus* **8,** 140–154.

Platzman, G. W. (1960). The spectral form of the vorticity equation. *J. Meteorol.* **17,** 653–644.

Platzman, G. W. (1979). The ENIAC computations of 1950—gateway to numerical weather prediction. *Bull. Am. Meteorolog. Soc.* **60,** 302–312.

Ramanathan, V., E. J. Pitcher, R. C. Malone, and M. L. Blackmon (1983). The response of a spectral general circulation model to refinements in radiative procesess. *J. Atmos. Sci.* **40,** 605–630.

Randall, D. (n.d.). Colorado State University general circulation model: Introduction. http://kiwi.atmos.colostate.edu/BUGS/BUGSintro.html.

Revelle, R., and H. E. Suess (1957). Carbon dioxide exchange between the atmosphere and ocean and the question of an increase of atmospheric CO_2 during the past decades. *Tellus* **9,** 18–27.

Richardson, L. F. (1922). "Weather Prediction by Numerical Process." Cambridge University Press, Cambridge, UK.

Robert, A. J. (1969). The integration of a spectral model of the atmosphere by the implicit method. *In* "Proceedings of the WMO IUGG Symposium on Numerical Weather Prediction in Tokyo," Japan, November 26–December 4, 1968 (World Meteorological Organization and International Union of Geodesy and Geophysics, eds.), pp. VII-9–VII-24. Meteorological Society of Japan, Tokyo.

Sagan, C. (1983). Nuclear war and climatic catastrophe: Some policy implications. *Foreign Affairs* **62,** 257–292.

Schneider, S. H. (1989). "Global Warming: Are We Entering the Greenhouse Century?" Vintage Books, New York.

Silberman, I. S. (1954). Planetary waves in the atmosphere. *J. Meteorol.* **11,** 27–34.

Smagorinsky, J. (1958). On the numerical integration of the primitive equations of motion for baroclinic flow in a closed region. *Monthly Weather Rev.* **86,** 457–466.

Smagorinsky, J. (1963). General circulation experiments with the primitive equations. *Monthly Weather Rev.* **91,** 99–164.

Smagorinsky, J. (1983). The beginnings of numerical weather prediction and general circulation modeling: Early recollections. *Adv. Geophys.* **25,** 3–37.

Smagorinsky, J., S. Manabe, and J. L. Holloway (1965). Numerical results from a nine-level general circulation model of the atmosphere. *Monthly Weather Rev.* **93,** 727–768.

Study of Critical Environmental Problems (1970). "Man's Impact on the Global Environment." MIT Press, Cambridge, MA.

Study of Man's Impact on Climate (1971). "Inadvertent Climate Modification." MIT Press, Cambridge, MA.

Suess, H. E. (1953). Natural radiocarbon and the rate of exchange of carbon dioxide between the atmosphere and the sea. *In* "Nuclear Processes in Geologic Settings" (National Research Council Committee on Nuclear Science, ed.), pp. 52–56. National Academy of Sciences, Washington, D.C.

Thompson, S. L., and S. H. Schneider (1986). Nuclear winter reappraised. *Foreign Affairs* **64,** 981–1005.

Williamson, D. L., J. T. Kiehl, V. Ramanathan, R. E. Dickinson, and J. J. Hack (1987). Description of NCAR community climate model (CCM1), NCAR/TN-285 + STR. National Center for Atmospheric Research, Boulder, CO.

Chapter 3

Clarifying the Dynamics of the General Circulation: Phillips's 1956 Experiment

John M. Lewis

National Severe Storms Laboratory
Norman, Oklahoma
and Desert Research Institute
Reno, Nevada

I. INTRODUCTION

One housand years ago, the Viking colonizer Erik the Red knew of the stiff westerly winds that resided over the North Atlantic. These persistent winds hindered his passage from Iceland to Greenland in 990 A.D. Fourteen out of the 25 ships under his command failed to make the pilgrimage because of the gales and associated rough seas (Collinder, 1954). Christopher Columbus was more fortunate, finding the northeast trades on his first voyage to the West. By the time Queen Elizabeth founded the East Indian Trade Company in 1600, ocean traders knew full well that once their ships reached the mouth of the Mediterranean, sails could be continuously set and yards braced for a following wind (Fig. 1; see color insert).

When these surface observations over the Atlantic were coupled with Newton's system of dynamics (available by the 1680s), the stage was set for a rational study of the atmosphere's general circulation. Astronomer Edmund Halley (1656–1742), knowledgeable of Newtonian mechanics before the publication of *Principia* in 1687, attempted a systematic study of the low-latitude wind systems, namely, the trades and the monsoon (Halley, 1686). In Louis More's biography of Issac Newton (1642–1727), written correspondence between Halley and Newton is presented (More, 1934). Based on the information in these letters, it is clear that Halley was familiar with the material in Newton's monumental treatise, *The Mathematical Principles of Natural Philosophy* (Newton, 1687) or simply *Principia* (Principles). In fact, Halley was a driving force behind publication of *Principia*.

Nearly 50 years passed before the first conceptual model of the atmosphere's circulation emerged, and the honor of discovery fell to a relatively unknown English scientist—George Hadley (1685–1768). In his essay of ~ 1300 words, free of equations, Hadley (1735) used arguments based on the conservation of angular momentum to explain the trades:

> From which it follows, that the air, as it moves from the tropics towards the equator, having a less velocity than the parts of the earth it arrives at, will have a relative motion contrary to that of the diurnal motion of the earth in those parts, which being combined with the motion towards the equator, a NE. wind will be produced on this side of the equator and a SE. on the other. (p. 59)

Lorenz (1967) has carefully traced the development of ideas associated with the atmosphere's general circulation from the time of Halley and Hadley to the mid-20th century. His historical research shows that advances appeared to fall into time blocks of approximately a half century. Typically, an idea gained credibility and was published in the leading texts of the day, only to be challenged by the avant garde. New theoretical ideas emerged, often concurrent with observational facts, only to suffer the same fate as the precedent theory.

By the 1930s–1940s, conceptual models began relying on an ever increasing set of upper air observations—pilot balloon observations from early century later complemented by observations from radiosondes and instrumented aircraft. The picture was nevertheless incomplete, suffering from a lack of simultaneous measurements over latitudinal swaths commensurate with the pole-to-equator distance. The hint and hope for a global observational view, however, came with the heroic study by Scandinavian meteorologists Jacob Bjerknes and Erik Palmén (1937). Bjerknes coordinated the simultaneous release of radiosondes ("swarm ascents") from 11 European countries to study the evolution of a midlatitude depression (extratropical cyclone). Data from 120 radiosondes were used to analyze the storm. As recalled by Palmén, "It was most exciting to see that we were able to construct maps for different isobaric levels over practically the whole of Europe for a period of about two days" (Palmén, 1980, p. 28). The cross sections in this paper spanned ~ 3500 km and featured a pronounced sloping frontal zone as well as a bifurcation in the tropopause height that was linked to the front. The wind structure normal to the sections could be inferred from the isotherm pattern in conjunction with the thermal wind relation.

Coupled with these improvements in the atmospheric observation system, the vicissitudes of World War II spurred the development of high-speed computation. In 1946–1947, this computational power was brought to bear on two challenging problems in physics—both formulated by scientists at Los Alamos Scientific Laboratory. The first was the numerical solution to a hydrodynamics-radiative transfer problem associated with the explosive release of energy from thermonuclear reaction, and the second was the simulation of neutron diffusion in fissionable materials (Ulam, 1964). Both experiments used the ENIAC (Electronic Numerical Integrator and Computer), a computer ostensibly designed for the computation of artillery firing tables, but rewired for the physics experiments. John von Neumann was a central figure in these experiments, and in spring of 1946 he contemplated a numerical weather prediction (NWP) experiment. This project, labeled the Meteorology Project at Princeton's Institute for Advanced Study (IAS), officially started on July 1, 1946. Three years later, after a fitful start linked to staffing problems, a team led by Jule Charney made the celebrated short-range forecasts on the ENIAC (Charney *et al.*, 1950). Nebeker (1995) has carefully examined events associated with the Meteorology Project, and eyewitness accounts are also available (Platzman, 1979; Thompson, 1983; Smagorinsky, 1983).

Steady improvements to short-range NWP accrued during the early 1950s, in large part due to more realistic models that accounted for energy conversion in extratropical cyclones. Encouraged by the success of these forecasts, IAS team member Norman Phillips began to contemplate longer range prediction using the IAS computer. His work took the form of a numerical simulation of the atmosphere's general circulation for a period of ~ 1 month. The work was completed in 1955 and Phillips communicated the results to von Neumann, who immediately recognized their significance. Von Neumann hastily arranged a conference in October 1955, Application of Numerical Integration Techniques to the Problem of the General Circulation, held at Princeton University. In his opening statement at the conference, von Neumann (1955) said:

> I should like to make a few general remarks concerning the problem of forecasting climate fluctuations and the various aspects of the general circulation that cause such fluctuations. Specifically, I wish to point out that the

hydrodynamical and computational efforts which have been made in connection
with the problem of short-range forecasting serve as a natural introduction to
an effort in this direction [Following a discussion of prediction partitioned by
time scale, von Neumann continues].... With this philosophy in mind, we held
our first meeting nine years ago at the Institute for Advanced Study to discuss
the problem of short-range weather prediction. Since that time, a great deal of
progress has been made in the subject, and we feel that we are now prepared to
enter into the problem of forecasting the longer-period fluctuations of the
general circulation. (pp. 9–10)

Following this conference, which highlighted his numerical experiment,
Phillips entered the research into competition for the first Napier Shaw
Memorial Prize, a prize honoring England's venerated leader of meteorology, Sir Napier Shaw (1854–1945), on the occasion of the centenary of his
birth (the competition was announced in April 1954). The subject for the
first competition was "the energetics of the atmosphere." On June 20,
1956, "... the adjudicators recommended that the prize be given to Norman A. Phillips of the Institute of Advanced Study, Princeton, U. S. A. for
his essay 'The general circulation of the atmosphere: a numerical experiment,' which had been published in the *Quarterly Journal* [*of the Royal
Meteorological Society*] (82, p. 1230) [April 1956]..." (Prize, 1956).[1]

This numerical experiment is retrospectively examined; furthermore, an
effort is made to trace the steps that led Phillips to undertake the
research. We begin by reviewing the state of knowledge concerning atmospheric general circulation in the 1940s and early 1950s, with some
attention to the underlying controversies.

II. GENERAL CIRCULATION: IDEAS AND CONTROVERSIES, 1940s TO EARLY 1950s

To appreciate the momentous changes that took place in general
circulation theory between ~ 1940 and 1955, one has only to read Brunt's
classic text (Brunt, 1944, Chap. 19), and follow this with a reading of
Eady's (1957) contribution 13 years later, "The General Circulation of the
Atmosphere and Oceans." From Brunt, the reader is left feeling that a
consistent theory of the atmosphere's general circulation is out of reach:
"It has been pointed out by many writers that it is impossible to derive a
theory of the general circulation based on the known value of the solar
constant, the constitution of the atmosphere, and the distribution of land
and sea.... It is only possible to begin by assuming the known tempera-

---

[1] The adjudicators also commended the excellence of the entry "On the dynamics of the
general circulation" by Robert Fleagle (1957).

ture distribution, then deriving the corresponding pressure distribution, and finally the corresponding wind circulation" (Brunt, 1944, p. 405).

Eady's discussion, on the other hand, promotes a sense of confidence that the general circulation problem, albeit complicated, was yielding to new theoretical developments in concert with upper air observations. His final paragraph begins "If from this incomplete survey, the reader has gained the impression that general circulation problems are complicated, this is as it should be. The point is that mere complication does not prevent their being solved. Much of the complication shows itself when we attempt to give precise answers instead of vague ones.... To answer problems in any branch of geophysics we need vast quantities of observations but we also need precise, consistent, mathematical theory to make proper use of them" (Eady, 1957, p. 151).

Certainly the 10-year period prior to Phillips's numerical experiment was one of ferment as far as general circulation was concerned. A brief review of the major issues and themes during this period follow.

A. ROSSBY: LATERAL DIFFUSION

Rossby's interest in the general circulation problem can be traced to his review paper on atmospheric turbulence (Rossby, 1927). In this paper, the work of Austrian meteorologists Wilhelm Schmidt and Albert Defant was highlighted. Defant (1921) had suggested that traveling midlatitude cyclones and anticyclones could be viewed as turbulent elements in a quasi-horizontal process of heat exchange between air masses, and he quantified the process by calculating an *austausch* or exchange coefficient following Schmidt (1917). Rossby was attracted by this concept (especially in the context of momentum transfer), and he applied it to the gulf stream and tropospheric westerlies (Rossby, 1936, 1937, 1938a,b, respectively).

Rossby summarized his ideas in a wide-ranging review article in *Climate and Man* (*Yearbook of Agriculture*), a compendium of meteorology that was shaped by a diverse committee headed by Chief of the Weather Bureau Francis Reichelderfer (Rossby, 1941). Rossby relied on the three-cell model of circulation that emanated from the work of 19th-century scientists William Ferrel and James Coffin (Ferrel, 1859; Coffin, 1875). This conceptual model, as it appeared in Rossby's article, is shown in Fig. 2. Here we see two direct cells: the equatorial cell (called the "Hadley cell") and the polar cell. The indirect cell in the midlatitudes is called the "Ferrel cell."

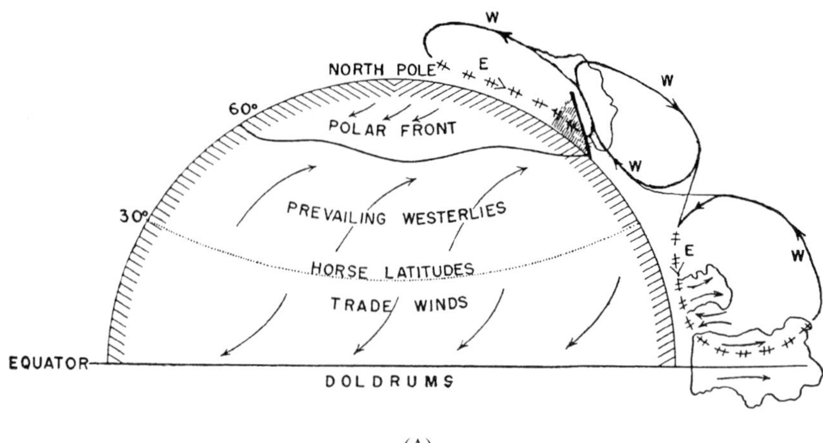

(A)

(B)

Figure 2 (A) Three-cell conceptual model of global circulation (extracted from Fig. 4 of Rossby, 1941). Deep cumulus cloud is indicated in the equatorial zone, clear sky is associated with descending air in the subtropics ($\sim 30°$N), and precipitation occurs in association with ascent of air over the polar front zone. Westerly/easterly winds are indicated along the meridional circulation circuits by the solid lines/"hatched" symbols. (B) Rossby is shown sitting at his desk in the U. S. Weather Bureau building in Washington, DC (ca. 1940). (Rossby photo courtesy of K. Howard and the Library of Congress.)

Regarding the westerlies, Rossby (1941) argued as follows:

> In the two direct circulation cells to the north and to the south, strong westerly winds are continuously being created at high levels. Along their boundaries with the middle cell, these strong westerly winds generate eddies with approximately vertical axes. Through the action of these eddies the momentum of the westerlies in the upper branches of the two direct cells is diffused toward middle latitudes, and the upper air in these regions is dragged along eastward. The westerlies observed in middle latitudes are thus frictionally driven by the surrounding direct cells...the air which sinks in the horse latitudes spreads both polewards and equatorwards. The poleward branch must obviously appear as a west wind.... (p. 611)

Rossby modified his ideas by the late 1940s—vorticity becoming the transferable property rather than momentum (Rossby, 1947).

B. Jeffreys –Starr –Bjerknes –Priestley –Fultz: Asymmetric Eddies

Tucked away near the end of a paper that explored atmospheric circulation by analogy with tidal theory, Harold Jeffreys argued that asymmetric eddies (cyclones/anticyclones) "...not unlike that described by Bjerknes...." were an essential component of the atmosphere's general circulation (Jeffreys, 1926). Quantitative arguments based on the conservation of angular momentum led him to state that a steady meridional (axially symmetric) circulation could not be maintained. Balance could only be achieved when the frictional torque was balanced by angular momentum transport due to asymmetric eddies. The governing equation for this transport is the integral (around a latitude circle) of the product of horizontal wind components. Quoting Jeffreys (1926, p. 99), "Considering any interchange of air across a parallel of latitude, then uv [the product of horizontal winds] must be negative both for the air moving north and for that moving south. This corresponds to the observed preponderance of south-westerly and north-easterly winds over those in the other two quadrants." (Jeffreys chose a coordinate system where u was directed southward and v eastward. Thus, the sign of uv in Jeffreys's coordinate system is opposite to that found in the more conventional system where u points eastward and v northward.)

Jeffreys came to this conclusion after grappling with the frictional formulation in his theory. The paper conceals this battle, but his reminiscence exposes it:

> ...the point was that you could solve the [atmospheric] problem when you had adopted the hydrodynamical equations to a compressible fluid...you could solve that for a disturbance of temperature of the right sort, and you could

> solve it in just the same way as you did for the tides—and it just wouldn't work!
> At least it worked all right when you didn't put in any friction. When you put
> friction in, it turned out that the friction in the result would stop the circulation
> in about a fortnight, and I had to start again, and I found that the only way to
> do it was to have a strong correlation between the easterly and northerly
> components of wind. (Jeffreys, 1986, p. 14)

Jeffreys's theory laid dormant for ~ 20 years. It was rejuvenated in the
late 1940s by Victor Starr (1948), Bjerknes (1948), and Charles Priestley
(1949). In the second paragraph of Starr's paper, he says "In reality, this
essay may be construed as a further extension of the approach to the
problem initiated by Jeffreys." Starr, who had exhibited his prowess with
mathematical physics applied to the geophysical system (see, e.g., Starr,
1939, 1945), displayed another aspect of his skill as a researcher in this
essay—namely, a clarity of expression and an expansive research vision. In
essence, the essay became the blueprint for Starr's research plan at MIT
during the next decade.[2] The upper air observations collected in the
postwar period made it clear that there was a decidedly NE–SW tilt to the
horizontal streamlines, "...so common on meteorological maps, [it] is a
necessary automatic adjustment to provide for the poleward transfer of
atmospheric angular momentum" (Starr, 1948, p. 41). Dave Fultz's hydrodynamical laboratory experiments confirmed the tilted streamline patterns
and became an independent source of support for Jeffreys's theory. (Photographs from Fultz's experiment are shown in Starr, 1956.)

The initial investigations by Starr and Bjerknes led to independent,
long-term efforts (at MIT and UCLA, respectively) to collect and archive
upper air data on a global scale. These assidious efforts led to sets of
general circulation "statistics"—measures of the temporally and/or spatially averaged terms in the heat and angular momentum budget equations
(see the contributions by Starr and White, 1951, and Mintz, 1951, 1975).
Priestley's work is notable, however, because his calculations relied on
observed winds rather than geostrophic approximations to the wind. Priestley continued his work on these problems until the early 1950s "...before
yielding to the greater resources of the two American pairs, Bjerknes–
Mintz and Starr–[Robert] White..." (Priestley, 1988, p. 104).

Photographs of the scientists who were instrumental in studying the
asymmetric aspects of the general circulation are shown in Fig. 3.

---

[2] Starr was the second recipient of the Ph.D. in meteorology from the University of
Chicago (Summer 1946) [The first recipient was Morris Neiberger (Autumn 1945).] Starr
accepted a faculty position at MIT in 1947.

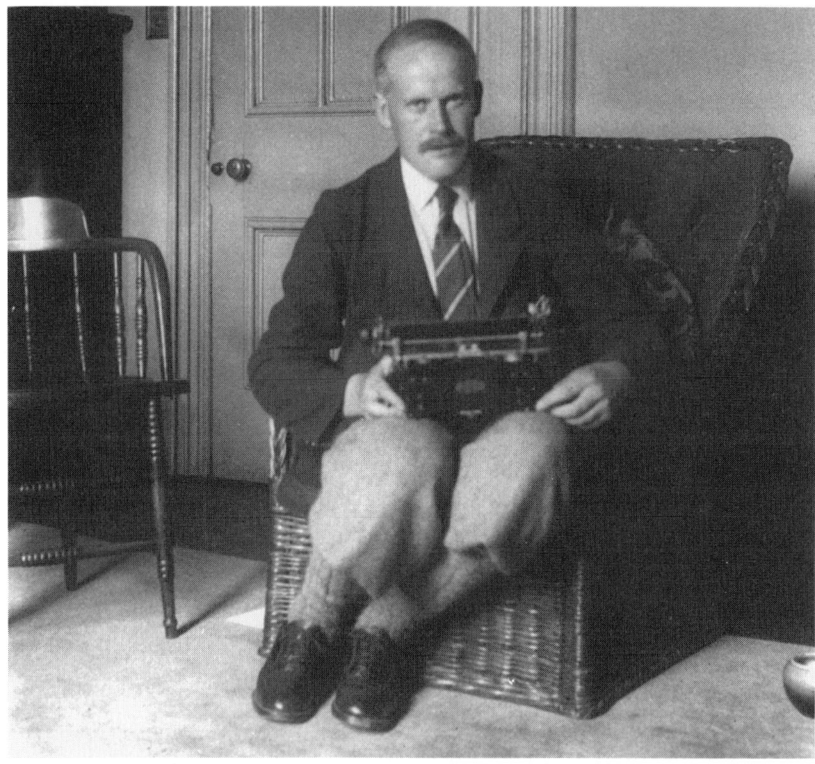

(A)

Figure 3 (A) Harold Jeffreys sits in his office at Cambridge (ca. 1928). (B) C. H. B. Priestley (ca. 1980). (C) Jacob Bjerknes (in the foreground) and Dave Fultz at the University of Chicago's Hydrodynamics Laboratory (1953). (D) Victor Starr (ca. 1965). (Courtesy of Lady Jeffreys, Dave Fultz, Constance Priestley, and the MIT archives.)

C. PALMÉN AND RIEHL: JET STREAMS

The existence of the strong and narrow band of upper level westerlies, labeled the jet stream, was established by forecasters in Germany (late 1930s) and the United States (early 1940s) (see Seilkopf, 1939, and Flohn, 1992; Riehl *et al.*, 1954; and Plumley, 1994, respectively). Following World War II, Rossby obtained funding from the Office of Naval Research (ONR) for a comprehensive study of atmospheric general circulation (including the dynamics of the jet stream). He invited Erik Palmén to assume a leadership role in this research. Palmén had spent his early career at Finland's Institute for Marine Research, and was named director of the institute in October 1939, just 2 months before Russia invaded

(B)

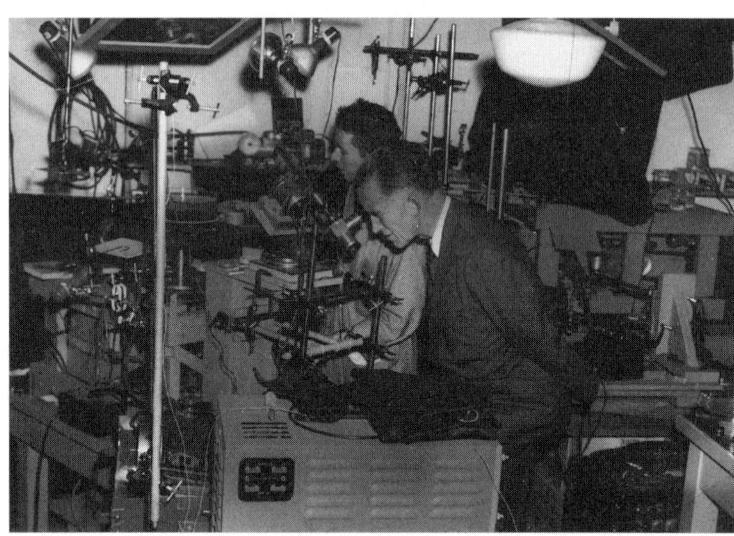

(C)

Figure 3 (*Continued*)

(D)

Figure 3 (*Continued*)

Finland. Throughout the remainder of WWII, Palmén's scientific work was severely curtailed. "He [Palmén] was born again in the setting of the general circulation project at the U of C [University of Chicago]" (C. Newton, personal communication, 1990). He remained at Chicago for 2 years (1946–1948), returning to Finland in late 1948 as chair professor of meteorology at the University of Helsinki. His frequent long-term visits to Chicago during the next decade, however, made him a fixture at the U of C's Institute of Meteorology.

In June 1947, the expansive report on the ONR project appeared under the authorship of staff members of the Department of Meteorology (Staff Members, 1947). Salient features of the jet stream were enumerated in the Summary section of the paper. Notable were the following: (1) The jet is located in or just south of a zone in which a large fraction of the middle and upper troposphere temperature contrast between polar and equatorial regions is concentrated; and (2) below the jet stream, it is possible to identify a well-defined frontal zone, intersecting the ground south of the jet stream.

Palmén became convinced that the concept of a single circumpolar jet was questionable, and he proposed the existence of a second jet, which he called the subtropical jet. "He [Palmén] thought that the great mass of air convected to the upper troposphere in the tropics could not all then descend in the subtropics. As evidence kept mounting, one began to speak of the "subtropical jet stream" found mainly above 500 mb and not undergoing the many violent north–south oscillations of the northern, soon called "polar jet stream" (Riehl, 1988).

Following Palmén's return to Finland in 1948, Herbert Riehl became the scientific leader of the jet stream project. Through the continued sponsorship of ONR, research flights across the circumpolar jet stream were initiated in 1953 (Riehl, personal communication, 1994).

D. Controversies

Amid such rapid advancement in meteorology, along with the slate of competing ideas, there is little wonder that this period had its share of controversies. A considerable amount of heated debate occurred at the daily map briefings at University of Chicago in the late 1940s. George Cressman offered daily discussions and forecasts with all the available maps (from mid-Pacific Ocean to the Ural Mountains in Russia—240° of longitude in the Northern Hemisphere). There was no end to the arguments about general and cyclone circulations that followed Cressman's briefings. The "reverse cell" of midlatitudes created fuel for the verbal exchanges. The abrupt transition from equatorward westerlies at high level in this middle cell to the neighboring easterlies in the equatorward or Hadley cell was conceptually difficult to understand (see Palmén and Newton, 1969, Chap. 1, for a summary of research that established the existence of the upper level easterlies). In Riehl's words, " . . . [why should] the equatorward westerlies, virtually friction-free in high atmosphere, . . . quickly diminish and go over into easterlies, just where the maximum west wind is observed" (Riehl, 1988, p. 4).

One of the most celebrated scientific exchanges occurred in the Correspondence section of the *Journal of Meteorology*. Starr and Rossby (1949) wrote a short article reconciling their differences on the role of angular momentum conservation in the atmosphere's general circulation. Their "differences" were minor, essentially related to the interpretation of terms in the equation of angular momentum conservation. One of the statements in the article, however, created an uproar. This cardinal statement reads: *"Most of the classic theories for the general circulation were based upon the assumption that it is this effect of meridional circulations which maintains the angular momentum of the zonal motions in the atmosphere.* It is this assumption that *both of us* call into question for reasons enumerated by Rossby [1941]."* They go on to say that, in their opinion, it is the advective transport of relative angular momentum—the *uv* term in Jeffreys's formulation—that is of prime importance in the mechanics of the general circulation.

Four months after the appearance of the Rossby–Starr article, Palmén wrote a letter to the editor that adamantly questioned the conclusion stated above (Palmén, 1949). He argued that the mean meridional circulation term could not be discounted; furthermore, Palmén made order of magnitude estimates of the meridional transport and found them comparable to the eddy transport term. The verbiage was strong and it elicited an ordered yet acerbic response from Starr (1949). Quoting Starr, p. 430 "Apparently Palmén suspects me of highest heresy lest I suggest that the energy production process may also be accomplished without the aid of meridional circulations. This I have indeed proposed... the hypothesis that meridional cells are of small importance seems to be bearing fruit. Indeed if such are the fruits of heresy, then I say let us have more heresy."

Although more stimulating than controversial, the general circulation statistics generated by the research teams at UCLA and MIT were demanding explanation. For example, the work of Bjerknes (and Mintz) at UCLA showed that the poleward eddy heat flux had its maximum at 50° latitude and was strongest near the ground. On the other hand, the poleward eddy angular momentum flux had its maximum near 30° and was strongest near the tropopause (Bjerknes, 1955).

Thus, by the mid-1950s, major questions related to the atmosphere's general circulation begged for answers. Among the issues were the respective roles of the mean meridional circulation and transient eddies in the momentum and energy budgets, mechanism for the maintenance of the westerlies (jet streams), and the dynamical basis for alternating wind regimes at the surface.

III. THE EXPERIMENT

Norman Phillips had been exposed to much of the controversy on general circulation theory while a graduate student at the University of Chicago in the late 1940s and early 1950s. During this same period,

Phillips's interest in dynamic meteorology was awakened through a careful reading of Charney's paper on the scale of atmospheric motions (Charney, 1948). He became convinced that simple baroclinic models (in particular, models that stratified the troposphere into two or three layers) could faithfully depict the principal features of cyclogenesis. His early work with these models, both theoretically and numerically, proved to be fundamentally important for the subsequent work on numerical simulation of the atmospheric general circulation.

Although Phillips's doctoral and postdoctoral research concentrated on the short-range prediction problem (Phillips, 1951; Charney and Phillips, 1953), he had an abiding interest in the general circulation problem that came in part from his exposure to the debates at Chicago, but also from his own practical experience as a research assistant on the ONR general circulation research project. These two research themes or components, the theoretical investigation of baroclinic motions and the phenomenological view of global circulation, came together for Phillips in early 1954. He was employed by the Institute for Advanced Study at this time, but was on leave to the International Institute of Meteorology in Sweden. As he recalls:

> From graduate school days at Chicago we had a pretty good idea of what the leading theoreticians and synopticians thought about how the general circulation worked. So it was not too difficult for me to first do this study in the '54 paper [Phillips, 1954] to see what baroclinic unstable waves might do—force an indirect circulation and then ... the lateral north and south boundary conditions would require direct circulation further out towards the pole and equator. And that this indirect circulation, in middle latitudes was the process, turbulent process that Rossby always referred to vaguely as giving rise to the surface westerlies. The explanation of surface westerlies had been the main challenge in the general circulation for centuries. They all knew that a direct circulation with the equator flow would not produce westerlies. So they had to put in little extra wheels, to end up creating polar flow in mid-latitudes. This seemed to all fit together so it encouraged me to go back to Princeton [in April 1954] and convince Jule [Charney] with that paper that yeah, that should be a logical thing to spend my time on. He was my boss. (Phillips, 1989, p. 25)

Figure 4 shows Phillips and others at IAS in the early 1950s.

Another key factor or ingredient in Phillips's strategy for designing the general circulation experiment was the success of the laboratory simulations of hemispheric flow by Fultz and English geophysicist Raymond Hide. Phillips (1955) writes:

> In spite of the unavoidable dissimilarities between the laboratory experiments and the atmosphere, certain experimental flow patterns are remarkably like those to be seen on weather maps. Thus, one is almost forced to the conclusion that at least the gross features of the general circulation of the atmosphere can be predicted without having to specify the heating and cooling in great detail. (p. 18)

Figure 4 Some of the members of the Meteorology Project at the Institute for Advanced Study in 1952. From left to right: Jule Charney, Norman Phillips, Glenn Lewis, N. Gilbarg, and George Platzman. The IAS computer, MANIAC I, is in the background. This picture was taken by Joseph Smagorinsky, another member of the Meteorology Project. (Courtesy of J. Smagorinsky.)

(See Hide, 1969, for a comprehensive review of research on laboratory simulations of the atmosphere's general circulation.)

A. MODEL AND COMPUTATIONAL CONSTRAINTS

Phillips adopted a set of dynamical constraints not unlike those used in short-range forecasting of the large-scale tropospheric flow—a two-level quasi-geostrophic model with horizontal winds specified at the 750- and

250-mb levels and mean temperature defined at 500 mb. Net radiation and latent heat processes were empirically parameterized by a heating function—a linear asymmetric function of the north–south distance (denoted by coordinate y, $-W \le y \le +W$), vanishing at $y = 0$.

The salient features of the model follow: quasi-geostrophic and hydrostatic constraints on the beta plane,[3] where lateral diffusion of vorticity is included at both levels and frictional dissipation is parameterized at the lower level. Following Phillips, subscripts are used as follows: 1, 250 mb; 2, 500 mb; 3, 750 mb; and 4, 1000 mb. The vorticity (ζ) at 1000 mb is found by linear (in pressure) extrapolation of vorticity from the 750- and 250-mb levels, i.e., $\zeta_4 = (3\zeta_3 - \zeta_1)/2$. Streamfunction, geopotential, and wind components are found by using an equivalent extrapolation formula.

To model a "hemispheric" region, the north–south dimension (y direction) of the domain was set to 10^4 km (approximate equator-to-pole distance on the Earth's surface). The east–west dimension (x direction) was chosen to accommodate one large baroclinic disturbance [$\sim (5\text{--}6) \times 10^3$ km]. Phillips cleverly allowed for the life cycle of the eddies by postulating periodic boundary conditions in the x direction—thus the disturbances typically moved out of the domain on the eastern boundary and entered along the western boundary.

The discretized arrays of variables shared computer memory with the stored program, and this was the limiting factor on the dimensionality of the problem. The IAS computer had 1024 words of internal memory and 2048 words of slower magnetic drum memory. This dictated arrays of (17×16) in the y and x directions, respectively. The associated grid intervals were $\Delta x = 375$ km and $\Delta y = 625$ km. Since the mean temperature (level 2, 500 mb) is proportional to the difference in geopotential between levels 1 and 3, the dependent variables for the problem are the geopotential arrays (streamfunctions); thus, the instantaneous state of the modeled atmosphere is determined by roughly 500 numbers. The horizontal domain (with the grid spacing shown on the inset) is displayed in Fig. 5.

As might be expected in those early days of computer modeling, execution time for the model run was long and the associated coding was tedious and laborious. Using nominal time steps of 1 hr, the 31-day forecast required 11–12 hr on the IAS machine. As recalled by Phillips (personal communication, 1997):

> Code was written in what would now be called "machine language" except
> that it was one step lower—the 40 bits of an instruction word (two instructions)

---

[3] The beta plane was introduced by Rossby *et al.* (1939) to simplify the equations of motion on the sphere. In this formulation, the kinematic effects of the earth's curvature are ignored but the dynamical effects are retained through the inclusion of the variation of the Coriolis parameter. Phillips assumed the beta plane's point of tangency was 45°N.

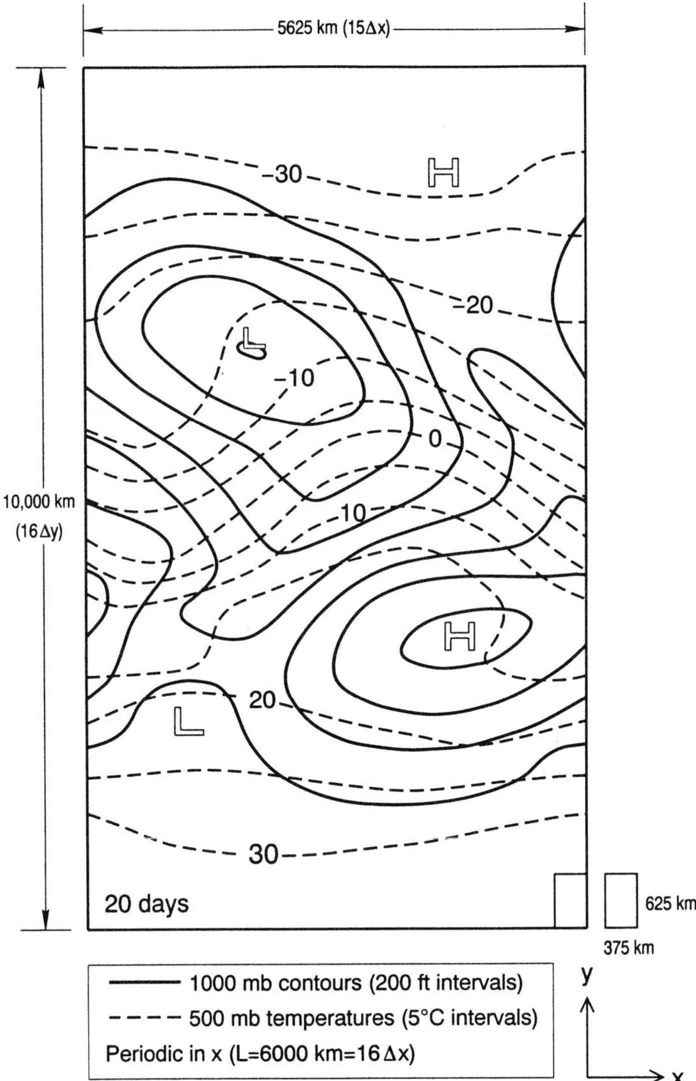

Figure 5   On day 20 of the simulation, the synoptic scale disturbance exhibits the character-istics of a developing cyclone with attendant frontogenesis. The mesh size is shown beside the model's horizontal domain. (From Phillips, 1956. With permission.)

were written by us in a 16-character (hexadecimal) alphabet 0, 1—9, A, B, C, D, E, F instead of writing a series of 0's and 1's; e.g., "C" represented the four bits "1100."

There was no automatic indexing—what we now call a "DO-LOOP" was programmed explicitly with actual counting. Subroutines were used, but calls to them had to be programmed using explicitly stored return addresses. In the first year or so of the IAS machine, code and data were fed in by paper tape. Von Neumann eventually got IBM to allow one of their card readers to be modified so that punched cards could be used for input and output.

B. THE BASIC STATE

In accord with studies of baroclinic instability via analytical dynamics, Phillips established a basic state solution on which perturbations could be superimposed. To derive this basic state, he started with an isothermal atmosphere at rest and used the model constraints to incrementally march forward in units of 1 day. The net heating gradually built up a latitudinal temperature gradient and associated zonal wind structure. The empirical heating/cooling rate of $0.23°C/day$ (at $y = \pm W$) led to a latitudinal temperature gradient of $60.2°C/10^4$ km after 130 days. At this stage of the integration, the meridional circulation consisted of a single weak direct cell (as Hadley had envisioned) superimposed on a zonal circulation that was independent of x. The latitudinal temperature gradient gave rise to a vertical wind shear of ~ 2 $m\,s^{-1}\,km^{-1}$, sufficient for the growth of small-amplitude perturbations in the zonal flow.

Charney (1959) and, more recently, Wiin-Nielsen (1997) have investigated steady-state solutions to Phillips's model. It is clear from their investigations that Phillips's basic state was not the steady-state solution. Quoting Wiin-Nielsen: "From the values of the zonal velocities in [Phillips's basic state] it is obvious that the model at this stage did not make a good approximation to the steady state derived here. His [Phillips's] purpose was only to obtain a zonal state where the vertical wind shear (or equivalently, the horizontal temperature gradient) was sufficiently large to be a state which was unstable for small perturbations. It is, however, of interest to see what the spin-up time is for the model to approximate the derived steady zonal state .... It is seen that the asympotic level is almost reached after $t = 4.32 \cdot 10^8$ seconds which is equivalent to 5000 days (13.7 years)" (Wiin-Nielsen, 1997, p. 6).

C. THE DISTURBED STATE

A random number generating process was used to introduce perturbations into the geopotential field, where the perturbations were identical at levels 1 and 3. Incremental steps of 1 hr were used to march forward in time and the following events took place:

1. A disturbance developed with wavelength of ~ 6000 km (similar to the disturbance shown in Fig. 5), and the flow pattern tilted westward with height; the wave moved eastward at ~ 21 m s^{-1} (1800 km day^{-1}).)

2. Transport of zonal momentum into the center of the region by horizontal eddies created a jet of ~ 80 m s^{-1} at 250 mb, and at the same time a pattern of easterly–westerly–easterly zonal winds was established at the 1000-mb level.

D. ZONAL-MEAN WINDS

The time evolution of the zonal-mean fields is displayed in Fig. 6. (zonal-mean implies an average over the x coordinate and is denoted by an overbar). Time (in days) is shown along the abscissa, where $t = 0$ (days $= 0$) is the time when disturbances were introduced into the simulation and the total period of simulation is 31 days.

The zonal-mean wind components at 250 mb are shown in the top panels of Fig. 6: $\overline{V}_1$ (meridional component in cm s^{-1}) and $\overline{u}_1$ (zonal component in m s^{-1}). The extrapolated zonal wind at 1000 mb, u_4, is shown in the lower left panel. The $\overline{V}_1$ field shows two sign reversals as one moves along the y axis (j index) on days ~ 10–25. The magnitude of this meridional component is greatest in the middle zone, reaching values of ~ 60–80 cm s^{-1}.

The $\overline{u}_1$ pattern shows westerly winds at all latitudes for approximately the first 25 days of simulation. The strongest winds are in the middle zone where speeds are ~ 40–60 m s^{-1} (days 10–25). At the 1000-mb level, the zonal winds ($\overline{u}_4$) exhibit an alternating pattern of easterly, westerly, and easterly winds.

The summary of the mean-zonal flow has been depicted in the lower right panel of Fig. 6, where the extremities of the y coordinate have been labeled "equator" and "pole" (to indicate that the north and south limits of the beta plane have been chosen commensurate with the pole-to-equator distance). Here the W and w indicate the strongest and weaker westerly flow at the upper level, respectively; these westerlies overlie the alternating pattern of easterlies and westerlies at the 1000-mb level. The arrows at the upper level, directed toward the strongest midlatitude westerlies (the jet stream), are indicative of the flux of eddy momentum into the jet (as discussed in the next subsection).

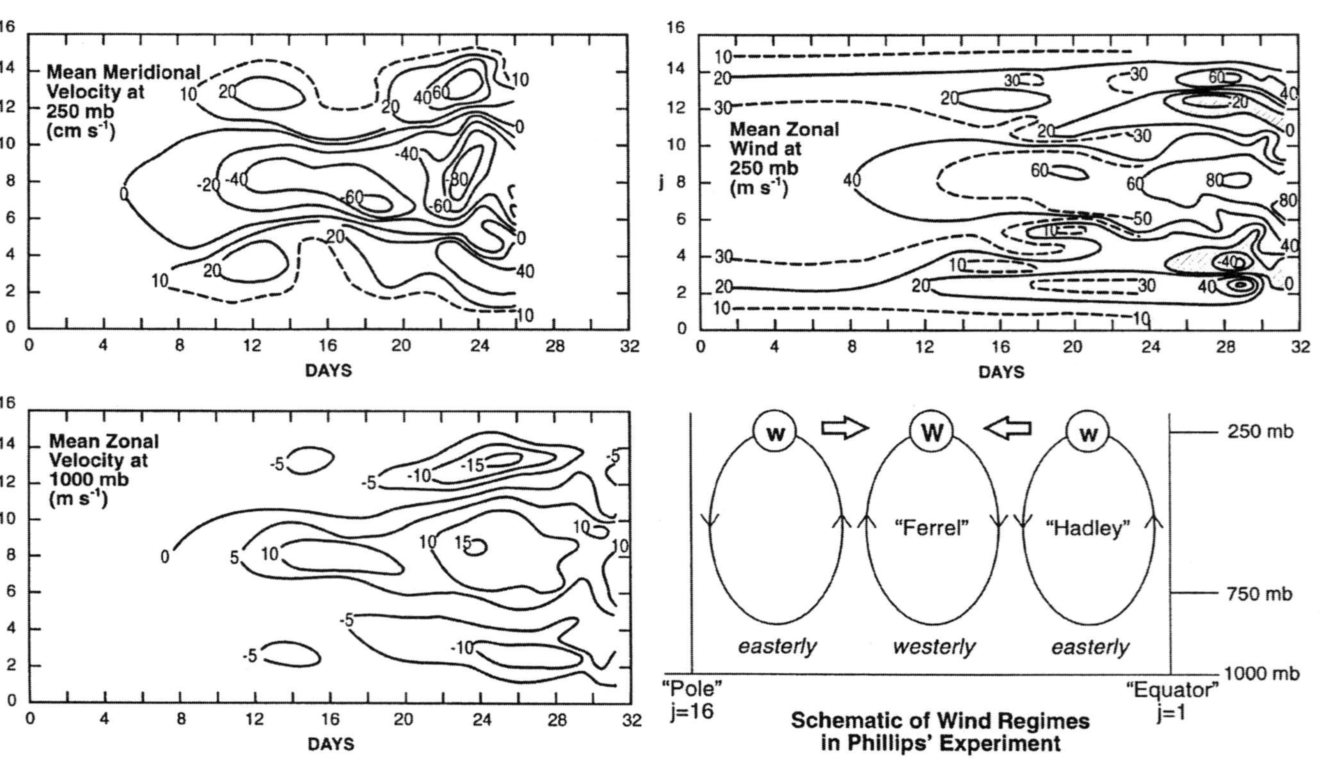

Figure 6 Latitudinal distribution of the mean meridional and zonal winds over the 31-day period of simulation. (From Phillips, 1956. With permission.)

Because the zonal-mean meridional flow at 750 mb is equal and opposite to that at 250 mb, a three-cell pattern can be inferred. Because of the similarity between this three-cell structure and that postulated from earlier studies, the labels "Ferrel" and "Hadley" have been added. Phillips, however, did not use these terms in the discussion of his results, only "…we see the appearance of a definite three-cell circulation, with an indirect cell in middle latitudes and two somewhat weaker cells to the north and south. This is a characteristic feature of the unstable baroclinic waves in the two-level model, as has been shown previously by the writer (Phillips, 1954). After 26 days, the field of $\overline{V}$ became very irregular owing to large truncation errors, and is therefore not shown" (Phillips, 1956, pp. 144–145).

E. MOMENTUM BUDGET

To clarify the processes that give rise to the jet, Phillips tabulated the momentum budget based on statistics over the 11-day period, days 10–20 inclusive. Information found in Phillips (1956, Tables 4 and 5) is graphically represented in Fig. 7. At the upper level, the tendency $(\overline{\partial u_1 / \partial t})$ in midlatitudes is mainly determined by the meridional circulation $(\alpha \, \overline{V_1})$ and the eddy transport $[-\frac{\partial}{\partial y}(\overline{u_1' v_1'})]$, the latter being the larger. The contribution from the meridional circulation is in general opposite to the observed changes in $\overline{u_1}$, so as to reduce the effect of the eddy term at 250 mb. As stated by Phillips, "The resulting picture is thus very much like that postulated by Rossby as existing during the building up of a zonal wind maximum (Staff Members, 1947)" (Phillips, 1956, p. 152).

The profiles at level 3 indicate that the midlatitude westerlies form in response to the meridional circulation, the $(f_0 \cdot \overline{V_3})$ term. Thus, the meridional circulation tends to balance both the large values of $[-\frac{\partial}{\partial y}(\overline{u' v'})$ in the upper atmosphere and the effect of surface friction on the lower atmosphere. As retrospectively examined by Phillips,

> Thus Palmén and Starr had missing features in their respective views, Starr could not explain the low level westerlies without the indirect meridional circulation, and Palmén could not explain the upper level westerlies without the eddies. (Phillips, personal communication, 1997)

F. THERMODYNAMIC BUDGET

As a complement to the zonally averaged momentum budget, Phillips tabulated the terms in the thermodynamic energy equation. These results

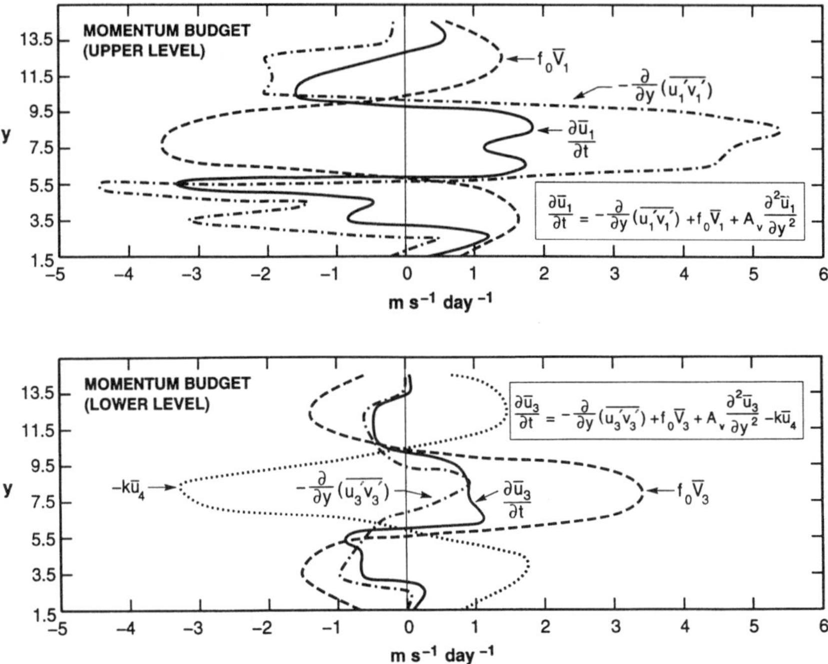

Figure 7 Latitudinal distribution of the various terms in the momentum budget equations at the upper and lower levels. The equations were averaged over the 11-day period, days 10–20 inclusive. Parameterized coefficients of lateral diffusion and friction are denoted by A_v and k, respectively. The diffusion terms at both levels were negligibly small and have not been plotted. (From Phillips, 1956. With permission.)

are displayed in Fig. 8. Here, the net radiation term heats the atmosphere in low latitudes and cools it at high latitudes. The convergence of eddy heat transport, $[-\frac{\partial}{\partial y}(\overline{v_1' T_2'})]$, opposes the net radiation, tending to destroy the latitudinal temperature gradient, especially in midlatitudes. The meridional circulation term, $(\alpha\ \omega_2)$, on the other hand, tends to increase the latitudinal temperature gradient $(-\partial T_2/\partial y)$ due to the reverse circulation of the Ferrel cell.

G. ENERGETICS

Because the heating function is a linear and asymmetric function about $y = 0$ (45°N latitude), the total amount of energy added or subtracted from the system is zero. However, there is a positive correlation between the heating and mean meridional temperature (i.e., the heating is positive/negative in the region of higher/lower mean temperature). This generates *available potential energy*. In Phillips's model, this energy is expressed as the spatial integral of the squared deviation of the 500-mb temperature (a deviation from the standard atmosphere). It is derivable from the governing equations of the two-level quasi-geostrophic model, first appearing in Phillips (1954). Lorenz's (1955) systematic treatment of available potential energy is acknowledged by Phillips: "…in a beautiful attempt to reconcile the synoptic meteorologist's intuitive association of *available* potential energy with temperature gradients, [Lorenz] has recently shown how a similar expression can be approximated from the usual definition of the potential plus internal energy…." (Phillips 1956, p. 135). It is clear from information in Phillips's oral history interview that he was unaware of Lorenz's contribution until the general circulation experiment was completed (Phillips, 1989, p. 27).

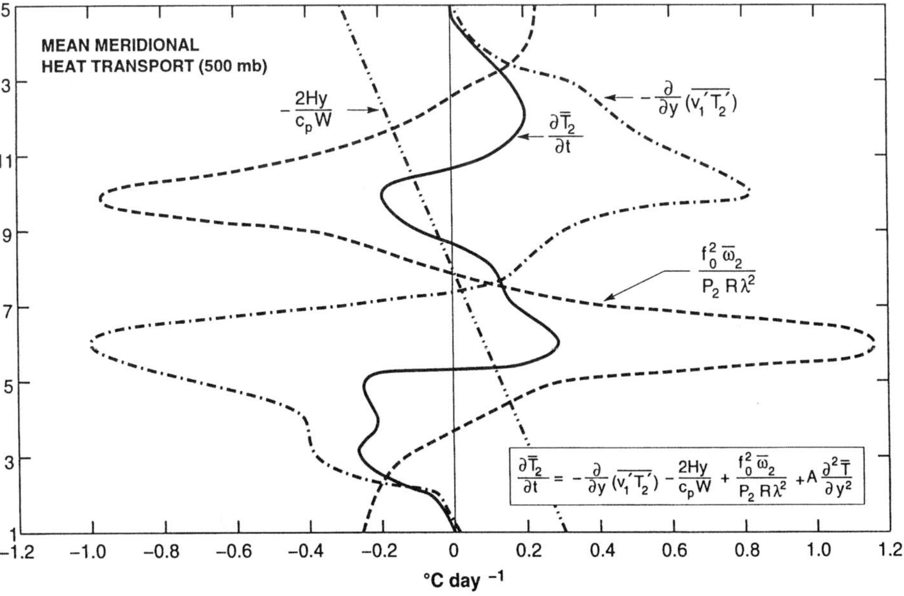

Figure 8 Latitudinal distribution of the various terms in the thermodynamic equation, averaged over the 11-day period, days 10–20 inclusive. The lateral diffusion coefficient is denoted by A; P_2 is 500 mb, R is the gas constant, and λ^2 is a positive parameter related to the static stability (assumed constant). The diffusion term is of negligible magnitude and has not been plotted. (From Phillips, 1956. With permission.)

The energy, both kinetic (K) and the available potential (P), are partitioned into zonal-mean components ($\overline{K}$ and $\overline{P}$) and perturbations about this mean, referred to as eddy components (K' and P'). At each step of the model integration, the various energy components are calculated along with the energy transformations. Phillips then found the *temporal* average of these quantities over a 22-day period of simulation (days 5–26).

The results are presented in Fig. 9 (patterned after the diagram found in Oort, 1964). The generation of mean meridional available potential energy is represented by the symbol $\overline{G}$, and it is shown in the upper left corner of the schematic diagram. This generation term is theoretically balanced by the dissipation of energy D, which takes the form of lateral diffusion and surface friction in Phillips's model. As indicated by Phillips's results and subsequent studies, the energy cycle generally proceeds from $\overline{P}$ to P' and K' and finally to $\overline{K}$—a counterclockwise movement around the diagram (Wiin-Nielsen and Chen, 1993, Chap. 7). The transformation rates

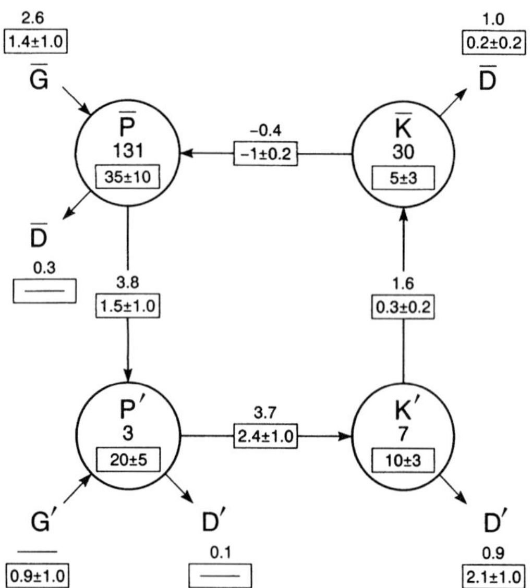

Figure 9 Energy diagram showing the reservoirs of kinetic (K) and available potential energy (P), where zonal-mean and eddy components are denoted by $(\overline{\ldots})$ and $(\ldots)'$, respectively. The transformation rates between the various components are indicated along the lines connecting the reservoirs. If positive, the energy is transferred in the direction indicated. Energy generation/dissipation is denoted by G/D, respectively. Oort's observationally based statistics are shown in the rectangular boxes, and Phillips's simulated statistics are written above these boxes. The energy units are (1) reservoirs, $\mathrm{J\,m^{-2}} \times 10^{5}$; and (2) transformation rates, $\mathrm{W\,m^{-2}}$.

are indicated along lines connecting the various energy reservoirs, where a positive value indicates transfer in the direction of the arrow.

Phillips, of course, had little basis for validation of his calculated energy exchanges (the top set of values at the various junctions in the energy diagram). He nevertheless appeared to be pleased that the generation and dissipation almost balanced and that the generation term was "...about half of the estimated rate of energy dissipation in the real atmosphere (Brunt 1944)...the model is undoubtedly too simple to expect any closer agreement" (Phillips, 1956, p. 154). He is circumspect when he writes "It is of course not possible to state definitively that this diagram is a complete representation of the principal energy changes occurring in the atmosphere, since our equations are so simplified, but the verisimilitude of the forecast flow patterns suggests quite strongly that it contains a fair element of truth. Further computations with more exact equations will presumably refine the picture considerably, as will an extension of observational studies using real data" (Phillips, 1956, p. 154).

When the first comprehensive set of general circulation statistics became available in the next decade (Oort, 1964), Phillips's cautious optimism was rewarded. Oort had judiciously combined results from various observational studies (with limited information from Phillips, 1956, and Smagorinsky, 1963) to make mean annual estimates of the terms in the energy budget of the Northern Hemisphere. Oort's mean annual statistics are displayed in the rectangular boxes of Fig. 9. Phillips did not account for the generation of eddy available potential energy (a very difficult component of the system to determine because it depends on the heating in the atmospheric waves). On the other hand, Oort's study made no attempt to calculate the dissipation associated with the available potential energy (a modeled term that tends to smooth the temperature gradients). The sense of Phillips's energy transformations, as well as their magnitudes, is quite consistent with Oort's. The absolute values of the energy components in the reservoirs, however, are significantly different. The variability of these statistics on seasonal, let alone monthly, time scales could account for part of the difference, but the simplified model dynamics also shared responsibility for this discrepancy. It would be nearly 10 years before more complete models of the general circulation would begin to faithfully represent this aspect of the energetics (Smagorinsky *et al.*, 1965).

IV. REACTION TO THE EXPERIMENT

Fortunately, some of the discussion that followed Phillips's oral presentation of his work has been preserved. Excerpts from these discussions are presented, and they are followed by vignettes that feature retrospective viewpoints from several prominent scientists who worked on the general circulation problem in the 1950s.

A. SIR NAPIER SHAW LECTURE

As the recipient of the first Napier Shaw Prize in 1956, Phillips was invited to deliver a seminar on his paper to the Royal Meteorological Society. The state of affairs in the English meteorological establishment at this time was not far removed from that in the United States. Both countries were in the throes of initiating operational numerical weather prediction, and both had strong traditions in synoptic meteorology. Reginald Sutcliffe, director of research at the British Meteorological Office (BMO), had a strong team of researchers at the BMO, including John Sawyer, and England possessed a bonafide world-class theoretician in Eric Eady of Imperial College. These scientists, along with other members of England's meteorological elite, were in attendence at Phillips's presentation. Questions and replies that followed the talk were recorded in the *QJRMS* (1956). Broad issues and questions that arose are paraphrased as follows, where the author(s) of the questions are identified in parentheses:

1. Unrealistic initial condition (i.e., starting the simulation from a state of rest) (P. Sheppard and R. Sutcliffe)
2. Excessive strength of the indirect cell (P. Sheppard)
3. Absence of condensation processes that could possibly explain the "slow rate of baroclinic development" (B. Mason and Sutcliffe)
4. Questionable physical significance of the transformation of energy between K' and $\bar{K}$ (G. Robinson)
5. Question regarding the secondary jets to the north and south of the main jet. Can these jets be considered similar to the subtropical jet and can we deduce that these are established by different mechanisms than the main jet stream? (J. Sawyer)

Phillips's responses to these questions (and others) are detailed in the *QJRMS* (1956). He seemed to be particularly stimulated by the question posed by Sheppard on the indirect circulation and Sawyer's question related to the subtropical jet. He sided with Sheppard and agreed that the indirect circulation of the middle cell was probably overestimated (citing evidence from observational studies at UCLA); furthermore, he was reluctant to claim that the secondary jets in the simulation were manifestations of the subtropical jet (as postulated and studied by Palmén).

The most encouraging remark came from Eric Eady:

> I think Dr. Phillips has presented a really brilliant paper which deserves detailed study from many different aspects. I am in complete agreement with the point of view he has taken and can find no fault with his arguments, either in the paper or in the presentation. With regard to the statement by Prof. Sheppard and Dr. Sutcliffe, I think Dr. Phillips' experiment was well designed. Numerical integrations of the kind Dr. Phillips has carried out give us a unique opportunity to study large-scale meteorology as an experimental science. By using a simple model and initial conditions which never occur in the real atmosphere he has been able to isolate, and study separately, certain fundamental properties of atmospheric motion—the kind of procedure adopted by all good experimenters…. An experiment which merely attempted to ape the real atmosphere would have been very poorly designed and very much less informative.

B. PRINCETON CONFERENCE

The issue that received the most attention at the Symposium on the Dynamics of Climate at Princeton University in October 1955 was truncation error in the numerical experiment (See "Discussions" in *Dynamics of Climate*; Pfeffer, 1960). During the last 10 days of the 31-day period of simulation, there was a steady deterioration of the flow pattern. There appeared to be little doubt in Phillips's mind and in the opinion of the others at the symposium that the culprit was truncation error, i.e., numerical error that accrued from finite-difference approximations to the governing equations. Quoting Phillips (1956, p. 157): "It was thought initially that the introduction of a lateral eddy-viscosity into the equations would eliminate some of the bad effects of truncation errors, by smoothing out the small-scale motions. To some extent this was true…but evidently a still more fundamental modification of the equations is required." Phillips (1959) would later identify nonlinear computational instability as a contributor to this noise problem.

C. VIGNETTES

Norman Phillips visited Stockholm in early 1956 and presented his research results at the International Meteorological Institute. Rossby, director of the institute, was especially interested in Phillips's experiment because it addressed issues related to cyclogenesis (and associated frontogenesis). The Bergen school model of cyclone development had placed emphasis on instabilities that formed on existing fronts (see, e.g., Solberg,

1928; Kotschin, 1932; Eliassen, 1962), whereas the work of Charney (1947) and Eady (1949) discussed cyclogenesis in terms of the upper level tropospheric wave. (Figure 5 shows an upper level Charney–Eady wave and the associated surface pressure pattern.) Following the seminar, Rossby held forth and had an elongated discussion with Phillips on the numerical simulation of the cyclogenesis process (Wiin-Nielsen, personal communication, 1993). Wiin-Nielsen reconstructs this discussion where Rossby's and Phillips's statements are denoted by R and P, respectively:

> *R*: Norman, do you really think there are fronts there?
>
> *P*: Yea, look at the temperature fields packed up very nicely.
>
> *R*: But Norman, what's the process that creates these fronts? Where do they come from?
>
> *P*: Well, they come out of a very simple dynamics.
>
> *R*: And what is that?
>
> *P*: I have very simple linear heating between the equator and pole, simple dissipation, but of course there is no water vapor or no precipitation, no clouds, totally dry model.
>
> *R*: Yes, Norman, and it should be that! Because here we are getting this front—and it has nothing to do with clouds/rising motion, it is a sheer dynamic effect that comes as a result of the development.

Charney discussed this same issue in a paper commemorating the work of Jacob Bjerknes. Quoting Charney (1975):

> His [Phillips's] experiment also helped to resolve an apparent inconsistency that I had found in my own and Bjerknes' work on the cyclone. I had been struck by the fact that while there was a one-to-one correspondence between long upper air waves and the primary surface cyclones—which always seemed to form fronts—there was no such correspondence between the upper wave and the secondary and tertiary frontal waves in a cyclone family... In Phillips' experiment... the dominantly unstable baroclinic wave mode appeared and, in its nascent stage, very much resembled the theoretical prediction from small perturbation analysis; but when the wave developed to finite amplitude, it exhibited the typical concentration of isotherms of a frontal wave. Thus the deformation field in the developing baroclinic wave produce frontogenesis in the form of the frontal wave, so that the primary cyclone wave does not form on a preexisting front, rather it forms at the same time as the front and appears as the surface manifestation of the upper wave... once the front has formed, it may permit frontal instabilities of the type analyzed by Solberg [1928] and Kotschin [1932] and later more completely by Eliassen [1962] and Orlanski [1968]. It would seem that the latter type is the "cyclone wave" of Bjerknes and Solberg [1922], whereas the former is the "cyclone wave" of Bjerknes, [and] Holmboe [1944], Eady [1949], and Charney [1947]. (p. 12)

Phillips's experiment had a profound effect outside the United States, especially in the strongholds of dynamic meteorology such as the International Meteorological Institute at Stockholm and in Tokyo, Japan, at the University of Tokyo's Geophysical Institute and at the Japan Meteorological Agency (JMA). Akio Arakawa, a scientist at JMA in the mid-1950s, recalls his reaction to Phillips's work:

> I myself was also extremely inspired by Phillips' work. My interest around the mid-50s was in general circulation of the atmosphere, mainly those aspects as revealed by observational studies on the statistics of eddy transports by Starr and White at MIT and Bjerknes and Mintz at UCLA, and laboratory experiments by Fultz [at University of Chicago] and Hide at MIT. At the same time, I was also interested in numerical weather prediction, through which dynamical meteorologists began to be directly involved in actual forecasts. Phillips' work highlighted the fact, which people began to recognize around that time, that the dynamics of cyclones and that of general circulation are closely related. I was so excited about these findings that I published a monograph through Japan Meteorological Society (Arakawa, 1958)... to let Japanese meteorologists recognize the important ongoing progress in our understanding of general circulation of the atmosphere. (Arakawa, personal communication, 1997)[4]

V. EPILOGUE

George Hadley presented the first consistent theory of the general circulation of the atmosphere. A little over 200 years later, with the advent of high-speed computation, Norman Phillips would blend theory and observations in the design of a numerical experiment—an experiment that he hoped would clarify the interaction between synoptic scale eddies and the general circulation.

The experiment design was bold. The simplicity of the model dynamics exhibited an almost irreverent disregard for the complexities of the real atmosphere—the governing equations were quasi-geostrophic, there were no mountains, no land/sea contrast, and water vapor was only indirectly accounted for in the empirically derived heating function. The strength of the contribution rested on (1) the reasonable simulation of the energy transformation rates and (2) the explanation of interacting physical processes (the mean meridional circulation and the eddy transport) that gave rise to the midlatitude westerlies and the alternating surface wind regimes. The experiment also demonstated the linkage between surface frontogenesis and evolution of the planetary wave—in essence, it showed that fronts grow on the broad north–south temperature gradient field according to the Charney–Eady theory. This result inspired a cadre of young theoreti-

---

[4] In collaboration with Donald Johnson, Arakawa has coauthored a tribute to Yale Mintz (Johnson and Arakawa, 1996). The influence of Phillips's work on Mintz has been carefully documented in their paper.

cians to analytically and numerically examine the dynamics of frontogenesis in the next decade (see the review by Orlanski *et al.*, 1985, Sect. 2).

From a politico–scientific viewpoint, Phillips's work led to the establishment of an institutional approach to deterministic extended-range forecasting. Von Neumann was the champion of this effort. As recalled by Joseph Smagorinsky (1983):

> Phillips had completed, in the mid-1950s, his monumental general circulation experiment. As he pointed out in his paper, it was a natural extension of the work of Charney on numerical prediction, but Phillips' modesty could not obscure his own important contributions to NWP. The enabling innovation by Phillips was to construct an energetically complete and self-sufficient two-level quasi-geostrophic model which could sustain a stable integration for the order of a month of simulated time.... A new era had been opened...von Neumann quickly recognized the great significance of Phillips' paper and immediately moved along two simultaneous lines.... One was to call a conference on "The Application of Numerical Integration Techniques to the Problem of General Circulation" in Princeton during October 26–28, 1955...[and] the other initiative by von Neumann was stimulated by his realization that the exploitation of Phillips' breakthrough would require a new, large, separate, and dedicated undertaking...[he] drafted a proposal to the Weather Bureau, Air Force, and Navy justifying a joint project on the dynamics of the general circulation.... The proposal, dated August 1, 1955, was more or less accepted the following month as a joint Weather Bureau–Air Force–Navy venture. I was asked to lead the new General Circulation Research Section, and reported for duty on October 23, 1955. (pp. 25–29)

This research unit, initially a companion project alongside the short-range numerical forecasting unit in Washington, DC, soon attained a separate identity with the help of Weather Bureau Chief Robert White and became known as the Geophysical Fluid Dynamics Laboratory in 1965. And within the 10-year period from 1955 to 1965, major institutional efforts in global numerical simulation were started at the National Center for Atmospheric Research, Lawrence Livermore Laboratory, and UCLA (in the United States), and at the Meteorological Office–United Kingdom (abroad).

The experiment helped resolve the Starr–Palmén controversy, but it accomplished this goal in concert with a series of diagnostic studies of the general circulation that accrued from the late 1950s through the 1960s (see Palmén and Newton, 1969, Chaps. 1 and 2). Palmén, Riehl, and disciples eventually came to accept Starr's thesis regarding the primacy of the eddies in transporting momentum poleward, while Starr, Rossby, and company accepted the fact that mean meridional circulations are essential ingredients in the global balance requirements.

In his oral history interview, Phillips makes it clear that he greatly benefited from Rossby, Palmén, Platzman, and Charney—these scientists stimulated and challenged him at pivotal junctures on his path to the experiment. As he said, "I began to learn more about how fortunate I was to have Platzman as a thesis advisor.... George, as you know, has a characteristic of being accurate as well as being right. And I think I've, I hope I've learned some of that from him..." (Phillips, 1989, p. 13). The experiment certainly contained that element of verity that we all search for in our research, and by example, Phillips inspired his contemporaries and a host of us in the succeeding generations.

ACKNOWLEDGMENTS

I am grateful for a series of letters from Norman Phillips over the past several years. In this correspondence, he clearly presented his scientific experiences at Chicago, Princeton, and Stockholm. I gained perspective on the experiment by meshing these personal reminiscences with his scientific contributions and the wealth of information contained in Phillips's oral history interview. Tony Hollingsworth and colleagues (Akira Kasahara, Joseph Tribbia, and Warren Washington) are congratulated for their superlative work in the collection of this oral history. Hollingsworth's knowledge of Phillips's oeuvre was encyclopedic.

Bulletin appointed reviewers offered valuable suggestions for revision that were followed, and the result was a significantly improved manuscript. Throughout, Aksel Wiin-Nielsen shared his knowledge of general circulation theory with me. Additionally, his unpublished notes, *Lectures in Dynamic Meteorology* (University of Michigan, ca. 1965) served as a pedagogical guide as I worked my way through the literature on atmospheric general circulation.

Eyewitness accounts concerning early developments in numerical simulation of atmospheric motion have been provided by the following scientists, where "O" and "L" denote oral history or letter-of-reminiscence, respectively, and where the date of the communication is noted within parentheses:

Akio Arakawa, L (4-14-97)	Joseph Smagorinsky, L (5-28-97)
Fred Bushby, L (10-29-97)	Larry Gates, L (4-15-97)
Phil Thompson, O (5-18-90)	Brian Hoskins, O (10-25-95)
Warren Washington, L (4-17-97)	Akira Kasahara, L (8-20-93)
Aksel Wiin-Nielsen, O (4-22-93)	Syukuro Manabe, L (4-14-97)
Terry Williams, L (9-3-93)	

Photo acquisition credit goes to the following people: Dave Fultz, Kenneth Howard, Bertha Jeffreys, Constance Priestley, Joseph Smagorinsky, Athelstan Spilhaus, Monika Stutzbach-Michelsen; and the following institutions: Cambridge University, Library of Congress, Massachusetts Institute of Technology, and Seewarte (Bundesamt für Seeschiffahrt und Hydrographie, Hamburg, Germany). Finally, I want to thank Joan O'Bannon, graphics specialist at the National Severe Storms Laboratory, for her faithful reproduction of data from Phillips's 1956 paper.

REFERENCES

Arakawa, A. (1958). Modern theory of general circulation. *Kisho Kenkyu* **9**, 4, (in Japanese).
Bjerknes, J. (1948). Practical application of H. Jeffreys' theory of the general circulation. *In Résumé des Mémoires Réunion d' Oslo*, pp. 13–14.

Bjerknes, J. (1955). The transfer of angular momentum in the atmosphere. *In* "Sci. Proc. Int. Assoc. Meteor.," pp. 407–408.

Bjerknes, J., and J. Holmboe (1944). On the theory of cyclones. *J. Meteorol.* **1**, 1–22.

Bjerknes, J., and E. Palmén (1937). Investigations of selected European cyclones by means of serial ascents. *Geofys. Publikasjoner* **12**, 1–62.

Bjerknes, J., and H. Solberg (1922). Life cycle of cyclones and the polar front theory of atmospheric circulation. *Geofys. Publikasjoner* **3** (1), 1–18.

Brunt, D. (1944). "Physical and Dynamical Meteorology," 2nd ed. Cambridge University Press, Cambridge, MA.

Charney, J. (1947). The dynamics of long waves in a baroclinic westerly current. *J. Meteor.* **5**, 135–162.

Charney, J. (1948). On the scale of atmospheric motions. *Geofys. Publikasjoner*, **17**, 2.

Charney, J. (1959). On the general circulation of the atmosphere. *In* "The Atmosphere and Sea in Motion" (B. Bolin, ed.), pp. 178–193. Rockefeller Institute Press and Oxford University Press, New York.

Charney, J. (1975). Jacob Bjerknes—An appreciation. *In* "Selected Papers of Jacob Aall Bonnevie Bjerknes" (M. Wurtele, ed.), pp. 11–13. Western Periodicals, North Hollywood, CA.

Charney, J., and N. Phillips (1953). Numerical integration of the quasi-geostrophic equations for barotropic and simple baroclinic flow. *J. Meteorol.* **10**, 71–99.

Charney, J., R. Fjørtoft, and J. von Neumann (1950). Numerical integration of the barotropic vorticity equation. *Tellus* **2**, 237–254.

Coffin, J. (1875). The winds of the globe: Or the laws of the atmospheric circulation over the surface of the earth. *In* "Smithsonian Contribution to Knowledge 268," Vol. 20. Smithsonian Institution, Washington, DC.

Collinder, P. (1954). Chap. 4 *in* "A History of Marine Navigation." Trans. from Swedish by M. Michael. Batsford, Ltd., London.

Defant, A. (1921). Die Zirkulation in der Atmosphäre in den Gemässigten Breiten der Erde [The circulation of the atmosphere in the temperate latitudes of the earth]. *Geografiska Ann.* **3**, 209–266.

Eady, E. (1949). Long waves and cyclone waves. *Tellus* **1**, 33–52.

Eady, E. (1957). The general circulation of the atmosphere and oceans. *In* "The Earth and Its Atmosphere" (D. Bates, ed.). pp. 130–151. Basic Books, New York.

Eliassen, A. (1962). On the vertical circulation in frontal zones. *Geofys. Publikasjoner*, **24**, 147–160.

Ferrel, W. (1859). The motions of fluids and solids relative to the earth's surface. *Math. Mon.* **1**, 140–147, 210–216, 300–307, 366–372, 397–406.

Fleagle, R. (1957). On the dynamics of the general circulation. *Quart. J. Roy. Meteorolog. Soc.* **83**, 1–20.

Flohn, H. (1992). "Meteorologie im Übergang Erfahrungen und Erinnerungen (1931–1991) [Meteorology in Transition (1931–1991), Experience and Recollection], pp. 6–8 Ferd Dümmlers, Bonn.

Hadley, G. (1735). Concerning the cause of the general trade-winds. *Phil. Trans. London* **39**, 58–62.

Halley, E. (1686). An historical account of the trade-winds and monsoons observable in the seas between and near the tropicks with an attempt to assign the physical cause of said winds. *Phil Trans.* **26**, 153–168.

Hide, R. (1969). Some laboratory experiments on free thermal convection in a rotating fluid subject to a horizontal temperature gradient and their relation to the theory of the global atmospheric circulation. *In* "The Global Circulation of the Atmosphere" (G. Colby, ed.), pp. 196–221. Royal Meteorological Society, Berkshire, UK.

Jeffreys, H. (1926). On the dynamics of geostrophic winds. *Quart. J. Roy. Meteorolog. Soc.* **52**, 85–104.

Jeffreys, H. (1986). Oral history. Transcription of an interview by M. McIntyre. (Available from the Royal Meteorological Society History Group, 104 Oxford Rd., Reading, Berkshire, RG1 7LL, England.)

Johnson, D., and A. Arakawa (1996). On the scientific contributions and insight of Professor Yale Mintz. *J. Climate* **9**, 3211–3224.

Kotschin, N. (1932). Über die Stabilität von Margulesschen Diskontinuitäts-flächen [On the stability of Margules' discontinuity surface]. *Beiträge Phys. Atmos.* **18**, 129–164.

Lorenz, E. (1955). Available potential energy and the maintenance of the general circulation. *Tellus* **7**, 157–167.

Lorenz, E. (1967). "The Nature of the Theory of the General Circulation of the Atmosphere." WMO No. 218.TP.115. World Meteorological Organization, Geneva. (Available from World Meteorological Organization, 33 Ave. de Bude, Geneva, Switzerland, 1202.)

Mintz, Y. (1951). The geostrophic poleward flux of angular momentum in the month of January 1949. *Tellus* **3**, 195–200.

Mintz, Y. (1975). Jacob Bjerknes and our understanding of the atmospheric general circulation. *In* "Selected Papers of Jacob Aall Bonnevie Bjerknes" (M. Wurtele, ed.), 4–15. Western Periodicals, North Hollywood, CA.

More, L. (1934). "Isaac Newton (a biography)." Charles Scribner's Sons, New York.

Nebeker, F. (1995). "Calculating the Weather (Meteorology in the 20th Century)." Academic Press, San Diego.

Newton, I. (1687). "Philosophiae naturalis principia mathematica" [A. Koyre and I. Cohen (with A. Whitman), eds.]. Harvard University Press, Boston, 1972.

Oort, A. (1964). On estimates of the atmospheric energy cycle. *Mon. Wea. Rev.* **22**, 483–493.

Orlanski, I. (1968). Instability of frontal zones. *J. Atmos. Sci.* **25**, 178–200.

Orlanski, I., B. Ross, L. Polinsky, and R. Shaginaw (1985). Advances in the theory of atmospheric fronts. *Adv. in Geophys.* **28B**, 223–252.

Palmén, E. (1949). Meridional circulations and the transfer of angular momentum in the atmosphere. *J. Meteor. (Correspondence)* **6**, 429–430.

Palmén, E. (1980). Oral history. Transcript of an interview by H. Taba. *In* "The 'Bulletin' Interviews." World Meteorological Organization Report 708, pp. 25–33. (Available from World Meteorological Organization, 33 Ave. de Bude, Geneva, Switzerland, 1202.)

Palmén, E., and C. Newton (1969). "Atmospheric Circulation Systems (Their Structure and Physical Interpretation)." Academic Press, San Diego.

Pfeffer, R., ed. (1960). "Dynamics of Climate—Proceedings of a Conference on the Application of Numerical Integration Techniques to the Problem of the General Circulation," October 26–28, 1955. Pergamon Press, New York.

Phillips, N. (1951). A simple three-dimensional model for the study of large-scale extratropical flow patterns," Ph. D. dissertation. Department of Meteorology, University of Chicago.

Phillips, N. (1954). Energy transformations and meridional circulations associated with simple baroclinic waves in a two-level, quasi-geostrophic model. *Tellus* **6**, 273–286.

Phillips, N. (1955). The general circulation of the atmosphere: A numerical experiment. Presented at the Conference on Applications of Numerical Integration Techniques to the Problem of the General Circulation. *In* "Dynamics of Climate" (R. Pfeffer, ed.), 18–25. Pergamon Press, New York, 1960.

Phillips, N. (1956). The general circulation of the atmosphere: A numerical experiment. *Quart. J. Roy. Meteor. Soc.* **82**, 123–164, 535–539.

Phillips, N. (1959). An example of non-linear computational instability. *In* "Atmosphere and Sea in Motion (Rossby Memorial Volume)" (B. Bolin, ed.), pp. 501–504. Rockefeller Press, New York.

Phillips, N. (1989). Oral history. Transcribed interview by T. Hollingsworth, W. Washington, J. Tribbia, and A. Kasahara. [Available from NCAR Archives, P. O. Box 3000, Boulder, CO, 80303.)

Platzman, G. (1979). The ENIAC computations of 1950—gateway to numerical weather prediction. *Bull. Am. Meteor. Soc.* **48**, 514–550.

Plumley, W. (1994). Winds over Japan. *Bull. Am. Meteor. Soc.* **75**, 63–68.

Priestley, C. (1949). Heat transport and zonal stress between latitudes. *Quart. J. Roy. Meteor. Soc.* **75**, 28–40.

Priestley, C. (1988). Oral history. Transcript of an interview by H. Taba. *In* "The 'Bulletin' Interviews." World Meteorological Organization Report 708, p. 21. (Available from World Meteorological Organization, 33 Ave. de Bude, Geneva, Switzerland, 1202.)

Prize (1956). The Napier Shaw Memorial Prize. *Quart. J. Roy. Meteor. Soc.* **82**, 375.

Riehl, H. (1988). General circulation studies in Chicago from the 1940's into the 1950's. In "Palmén Mem. Symp. on Extratropical Cyclones" Helsinki, Finland, 29 Aug.–2 Sep., 1988, pp. 4–5. Amer. Meteor. Soc.

Riehl, H., M. Alaka, C. Jordan, and R. Renard (1954). "The Jet Stream," *Meteor. Monogr.*, No. 7. Amer. Meteor. Soc.

Rossby, C.-G. (1927). The theory of atmospheric turbulence—A historical resumé and an outlook. *Mon. Wea. Rev.* **55**, 1–5.

Rossby, C.-G. (1936). Dynamics of steady ocean currents in light of experimental fluid mechanics. *Papers Phys. Oceanogr. Meteor.* **5** (1), 43.

Rossby, C.-G. (1937). On the mutual adjustment of pressure and velocity distributions in certain simple current systems. *J. Mar. Res.* **1**, 15–28.

Rossby, C.-G. (1938a). On the role of isentropic mixing in the general circulation of the atmosphere. *In* "Proc. Fifth Congress on Applied Mechanics," Cambridge, MA, pp. 373–379. Harvard University and Massachusetts Institute of Technology, Cambridge, MA.

Rossby, C.-G. (1938b). Aerological evidence of large scale mixing in the atmosphere. *Trans. Am. Geophys. Union*, I, 130–136.

Rossby, C.-G. (1941). The scientific basis of modern meteorology. *In* "Yearbook of Agriculture, Climate and Man." Department of Agriculture, Govt. Printing Office, Washington, DC.

Rossby, C.-G. (1947). On the distribution of angular velocity in gaseous envelopes under the influence of large-scale horizontal mixing processes. *Bull. Am. Meteor. Soc.* **28**, 53–68.

Rossby, C.-G., and Collaborators (1939). Relation between variations in the intensity of the zonal circulation of the atmosphere and the displacements of the semi-permanent centers of action. *J. Mar. Res.* **2**, 38–55.

Schmidt, W. (1917). Der Massenaustausch bei der ungeordneten Strömung in freier Luft und seine Folgen [Mass exchange by disorderly (turbulent) motion in the free air and its consequences]. *Wiener Sitzber.* **II**, 126–142.

Seilkopf, H. (1939). "Maritime Meteorologie: Handbuch der Fliegenwetterkunde, II" (Maritime Meteorology: Handbook for Aviation Weather), Berlin, (R. Habermehl, ed.), Vol. 2, pp. 142–150.

Smagorinsky, J. (1963). General circulation experiments with the primitive equations I. The basic experiment. *Mon. Wea. Rev.* **91**, 99–164.

Smagorinsky, J. (1983). The beginnings of numerical weather prediction and general circulation modeling: Early recollections. *Adv. Geophysics* **25**, 3–37.

Smagorinsky, J., S. Manabe, and J. Holloway (1965). Numerical results from a nine-level general circulation model of the atmosphere. *Mon. Wea. Rev.* **93**, 727–768.

Solberg, H. (1928). Integrationen der atmosphärischen Störungsgleichungen [Integration of the atmospheric perturbation equations]. *Geofys. Publikasjoner*, **5**, 9, 1–120.

Staff Members (1947). On the general circulation of the atmosphere in middle latitudes (A preliminary summary report on certain investigations conducted at the Univ. of Chicago during the academic year 1946–47). *Bull. Am. Meteor. Soc.* **28,** 255–280.

Starr, V. (1939). The readjustment of certain unstable atmospheric systems under conservation of vorticity. *Mon. Wea. Rev.* **67,** 125–134.

Starr, V. (1945). A quasi-Lagrangian system of hydrodynamical equations. *J. of Meteor.* **2,** 227–237.

Starr, V. (1948). An essay on the general circulation of the earth's atmosphere. *J. Meteor.* **5,** 39–43.

Starr, V. (1949). Reply to Palmén (1949). *J. Meteor. Correspondence* **6,** 430.

Starr, V. (1956). The circulation of the atmosphere. *Sci. Am.* **195,** 40–45.

Starr, V., and C.-G. Rossby (1949). Interpretations of the angular-momentum principle as applied to the general circulation of the atmosphere. *J. Meteor.* **6,** 288.

Starr, V., and R. White (1951). A hemispheric study of the atmospheric angular-momentum balance. *Quart. J. Roy. Meteor. Soc.* **77,** 215–225.

Thompson, P. (1983). A history of numerical weather prediction in the United States. *Bull. Am. Meteor. Soc.* **84,** 755–769.

Ulam, S. (1964). Computers in mathematics. *Sci. Am.* **203,** 203–217.

von Neumann, J. (1955). Some remarks on the problem of forecasting climate fluctuations. In "Dynamics of Climate" (R. Pfeffer, ed.), pp. 9–11. Pergamon Press, New York, 1960.

Wiin-Nielsen, A. (1997). On the zonally-symmetric circulation in two-level quasi-geostrophic models, unpublished manuscript.

Wiin-Nielsen, A., and T.-C. Chen (1993). "Fundamentals of Atmospheric Energetics." Oxford University Press, New York.

Chapter 4

Climate Modeling in the Global Warming Debate

J. Hansen, R. Ruedy, A. Lacis, M. Sato, L. Nazarenko, N. Tausnev,
I. Tegen, and D. Koch
NASA Goddard Institute for Space Studies, New York, New York

I. INTRODUCTION

Akio Arakawa played a key role in the development of the Goddard Institute for Space Studies (GISS) global climate models (GCMs). Along with Jule Charney, Arakawa also motivated us to use those models to analyze climate sensitivity and processes involved in global warming. The current suite of GISS models, ranging from the global ocean to the Earth's mesosphere and Mars, continues to have dynamical cores that are fundamentally based on Arakawa's numerical methods.

We summarize the origins of climate modeling at GISS in the 1970s and later extension into a family of global models. Our first model application was to the fundamental question of how sensitive the Earth's climate is to external forcings, such as changes of atmospheric composition and solar irradiance. We also discuss climate predictions based on models driven by realistic transient climate forcings. The topical question of "missing atmospheric absorption" is considered in the penultimate section. Finally, we present a summary perspective of global warming issues. For the sake of informality, this chapter is written mainly in the first person by the first author, Jim Hansen.

II. GISS GLOBAL CLIMATE MODELS

A. WEATHER MODEL PRELUDE

When I came to GISS as a postdoctoral candidate in the late 1960s my primary interest was in planetary atmospheres, especially the clouds of Venus, and I focused on radiative transfer theory as a tool to study the Venus clouds. But at about that time the director of GISS, Robert Jastrow, concluded that the days of generous NASA support for planetary studies were numbered, and he thus began to direct institutional resources toward Earth applications.

The principal upshot was a concerted effort for GISS to get involved in testing the value of space observations for improving weather forecasts. Jule Charney of MIT, serving as a scientific consultant to GISS, provided the intellectual underpinnings, arguing that daily global measurements of atmospheric temperature profiles, if inserted continuously in a global weather prediction model, could sufficiently constrain the temperature, pressure, and wind fields in the model and hence lead to more accurate weather forecasts.

The first requirement for testing this hypothesis was a good weather prediction model, i.e., a computer program solving the fundamental equations for atmospheric structure and motion: the conservation equations for energy, mass, momentum and water substance, and the ideal gas law. That is where Akio Arakawa came in. Charney recommended that GISS import the UCLA two-layer atmospheric model of Yale Mintz and Arakawa and increase the model's vertical resolution, thus making full use of the temperature profiles measured by satellites and presumably increasing the model's forecast capability. Because Arakawa was the architect of the model, it was only through his enthusiastic cooperation that the model could be adapted for the GISS project. Milt Halem was the project director, Richard Somerville led the meteorological analysis of model capabilities, and Peter Stone was the principal consultant on atmospheric dynamics.

I had only a minor responsibility in the GISS modeling project, specifically to calculate the solar radiative heating, a term in the energy equation that is of little importance for weather forecasts. But this project, together with a Venus spacecraft project, provided resources that permitted hiring someone to work with me, and I used that opportunity to bring Andy Lacis, who was just completing his Ph.D. thesis in astrophysics at the University of Iowa, to GISS. Although our main interest was in planetary studies, our involvement with the weather model made it practical for us to initiate a climate modeling effort several years later.

Andy soon became the GISS expert in modeling of atmospheric radiation. We developed a method for calculating solar heating of the atmosphere (Lacis and Hansen, 1974) that used a crude eight-point k distribution to represent water vapor absorption over the entire spectrum. We also parameterized ozone absorption and cloud and molecular scattering, using analytic formulas fit to off-line radiative transfer calculations. This parameterization was cited by Paltridge and Platt (1976) as "a classic example of the derivation of a parameterization scheme whose validity has been tested by comparison with the results of complex but precise numerical solutions" (p. 91) and it was adopted in a number of GCMs and regional models. Although this parameterization of solar heating was sufficiently accurate for weather models, and was used in the GISS weather model (Somerville *et al.*, 1974), it did not include aerosols and was not designed for or ever used in any of our climate models. Decades later it became inadvertently involved in the current issue about "missing atmospheric absorption," but we argue in Section V that this missing absorption is primarily a misunderstanding.

Perhaps our main (inadvertent) contribution during the weather modeling era was to improve the lighting in the GISS building. Andy and I always worked until about 9 P.M., by which time everyone else had gone home. Just before leaving we would have a contest of hardball Frisbee standing at opposite ends of the hallway. The object was to throw the Frisbee so hard that the opponent would fail to catch it. We soon became sufficiently skilled that the only good way to induce a miss was via the sudden change of direction that accompanied a skip off a light fixture. Unfortunately, these plastic fixtures were not always as strong as the Frisbee and cracks occasionally appeared in a light cover. Fortunately, the fixtures were identical throughout the building and it was easy to interchange them. Within several years there was more light getting through the fixtures throughout the building, which was good because they were grimy and fuliginous. And, fortunately, by the 1990s when the building was renovated and the lights replaced, we had retired from hardball Frisbee.

B. Initial GISS Climate Model

Our interest in global climate was an outgrowth of radiation calculations. Following the approach of Suki Manabe (Manabe and Moller, 1961;

Manabe and Strickler, 1964), we used a one-dimensional (1-D) radiative-convective model to estimate the effect of various human-made greenhouse gases (GHGs) on global mean temperature (Wang *et al.*, 1976). This 1-D modeling allowed us to be involved in climate studies while we were seeking support for 3-D climate modeling. In addition to greenhouse calculations, we used the 1-D model to test the climate effect of volcanic aerosols, simulating a cooling after the largest volcanic eruption of the previous 50 years, Mt. Agung in 1963, in reasonable agreement with observations (Hansen *et al.*, 1978).

The problem with 1-D models was that climate feedbacks were specified, rather than computed from first principles, so climate sensitivity was essentially prescribed. Realistic study of climate problems required a 3-D global climate model (GCM), so that physical processes involved in climate feedbacks could be modeled more explicitly. The need was for a model that could be run on climatic time scales, and it seemed to me that it could define the main features of the atmospheric general circulation without having a resolution as fine as that in a weather model. Peter Stone, referring to a paper by Merilees (1975), argued that the important large-scale eddies could be represented with resolution as coarse as about 1000 km.

That is where Arakawa's model came in, in a crucial way. Other studies suggested that fine resolution (a few hundred kilometers or less) was required in global models, but those studies used unrealistic horizontal viscosity that tended to damp out not only the numerical instabilities at which it was aimed, but also real atmospheric motions when the resolution was coarse (Merilees, 1975). However Arakawa had designed the finite-differencing schemes in his model to conserve fundamental integral properties, thus permitting stable integration of the equations with little artificial diffusion or smoothing. And because the computing time varies roughly in proportion to the cube of the horizontal resolution, the long simulations needed for climate studies are much more feasible with coarse resolution.

I presented a proposal to NASA in 1975 to develop a climate model from the GISS weather model. Although this first proposal was not supported, Kiyoshi Kawabata, a Venusian scholar in our planetary group, volunteered to test Arakawa's model at coarse resolution, as a part-time activity. We were delighted to find that the simulated general circulation looked reasonably realistic at $8° \times 10°$ resolution, and it was qualitatively similar at $4° \times 5°$, $8° \times 10°$, and even $12° \times 15°$ resolutions. This meant that Arakawa's model could provide the dynamical core that we needed for an efficient climate model, although we would need to provide "physics" required for climatic time scales.

Our practical need was for someone with complete command of the model, including the finite-differencing methods and model programming. As fate would have it, in 1977 Milt Halem moved his weather modeling group to the parent Goddard Center in Greenbelt, Maryland. That provided the opportunity for us to acquire from Halem's group a brilliant young mathematician, Gary Russell, who had been the principal programmer for the GISS weather model. Gary not only had the confidence and ability to completely overhaul parts of the model when necessary, but also an insight about the physics that is crucial for model development.

The other key player soon added to our group was David Rind, coming from Bill Donn's group at Columbia's Lamont Observatory. His background in atmospheric dynamics, including the upper atmosphere, was an essential complement to the others, particularly since many climate change mechanisms involve the stratosphere. David developed a broad interest in climate modeling, including paleoclimate studies, thus also providing a working connection with paleoclimate researchers and to their invaluable perspective on climate change. For more than a decade David has been the most effective person at GISS in spurring model development and applications, and he has been our most active researcher in the crucial area of evaluating model performance relative to observations.

This internal GISS climate group (Fig. 1) has been guided by regular consultations with Peter Stone from the time of our first musings about developing a model. Although Peter is best known as an atmospheric dynamicist, he advises on the entirety of the model and is a collaborator on many of the model applications. The other main contributors to our early modeling, all coauthors on the paper describing our first model (Hansen *et al.*, 1983), were Reto Ruedy, Larry Travis, and Sergej Lebedeff.

Tony Del Genio arrived at GISS at about the time we finished that paper, and since then he has been responsible for clouds and moist convection, leading to some of the most significant model improvements. Other important model improvements came from Greg Hartke for the planetary boundary layer, Michael Prather for quadratic upstream differencing for atmospheric tracers, Cynthia Rosenzweig and Frank Abramopoulos for ground hydrology, and Elaine Matthews for global vegetation properties.

The gestation period for our first 3-D climate model paper, published in 1983, was more than 5 years. In addition to model development being laborious (we included 61 sensitivity experiments in our first paper) and our innate tendency to be deliberate, other factors contributed to this long gestation. First, we were pursuing multiple objectives. Although my aim was to study global change, e.g., the greenhouse effect, the GISS director asked us to focus on the "farmer's forecast." Thus, in addition to model

Figure 1 *Left to right*: A. Lacis, J. Hansen, D. Rind, and G. Russell in the early 1980s.

development, we carried out experiments to test the influence of sea surface temperature and initial land surface and atmospheric conditions on 30-day forecasts. Second, we worked on simpler models that provided guidance for more detailed study, as exemplified by our 1981 paper "Climate impact of increasing atmospheric CO_2" based on a 1-D model (Hansen *et al.*, 1981). Third, it took us a long time to convince referees that a coarse resolution model was a legitimate climate model.

This last factor warrants a comment here, and it is touched on implicitly under our "philosophy" below and in the concluding section. It is inappropriate to equate model validity with resolution, in our opinion. Resolution should relate to science objectives and the phenomena to be represented. Our aim is to employ a resolution sufficient to define the general circulation, including transports by large-scale atmospheric eddies, to allow simulation of seasonal climate on global and regional scales. Although a weather prediction model must attempt to resolve and follow midlatitude synoptic storms precisely, that is not necessarily required of a climate model. Model intercomparisons indicate that our coarse model does a good job of simulating seasonal variation of precipitation over the United States (Boyle, 1998), for example. Improvements obtained with finer resolution must be weighed carefully against improvements obtained with better physics and against the advantages of an efficient model.

C. MODEL VARIATIONS AND PHILOSOPHY

The model that we documented in 1983, dubbed model II, was basically a tropospheric model. It was used for a number of climate studies in the 1980s, usually with a simple "Q-flux" treatment of the ocean, as described in Section III. The descendants of the original GISS climate model now form a family of models that can be used for more comprehensive investigations of climate change.

The most direct descendant of the original GISS model based on Arakawa's B Grid is the series of models SI95, SI97, SI99, which have been used and tested by students and faculty in the GISS Summer Institute on Climate and Planets (Hansen *et al.*, 1997c). These models, so far, have been run at $4° \times 5°$ resolution. Changes of model physics subsequent to model II include the moist convection parameterization (Del Genio and Yao, 1993), prognostic clouds (Del Genio *et al.*, 1996), the planetary boundary layer representation (Hartke and Rind, 1997), ground hydrology and evapotranspiration (Rosenzweig and Abramopoulos, 1997), numerical differencing schemes, including use of a quadratic upstream scheme (Prather, 1986) for heat and moisture, and various minor factors (Hansen *et al.*, 1997c). The SI95 model had the same 9 layers as model II, while the SI97 and SI99 models have 12 layers with 3 or 4 of these in the stratosphere. Current development gives priority to improved vertical resolution and better representation of physical processes.

The first major extension of the GISS model was to the stratosphere and mesosphere, with the development of the GISS global climate/middle atmosphere model (Rind *et al.*, 1988). That model is used with different choices for vertical resolution and model top, as high as about 80 km, and with increasingly sophisticated treatments of gravity wave drag. Recent applications of that model to solar cycle and ozone climate forcings (Shindell *et al.*, 1999a,b), including successful simulation of observed solar cycle changes, provide an incentive for improving the vertical resolution in other model versions. Inclusion of this model in the GISS stable allows testing of the model resolution and vertical extent required to simulate different climate phenomena.

Another variation of the GISS model is Gary Russell's coupled atmosphere–ocean model (Russell *et al.*, 1995). Both atmosphere and ocean use Arakawa's C Grid with the linear upstream method of Russell and Lerner (1981) for heat and water vapor. In addition Gary modified and simplified physics parameterizations, including replacement of the surface/boundary layer formulation with an extrapolation from the lowest model layer and replacement of the Del Genio *et al.* prognostic clouds with a simpler scheme having cloud optical thickness proportional to the square root of water vapor amount. The resulting model is faster and has an improved climatology for several climate diagnostics including sea level pressure distribution. A criticism that has been made is that the model yields an increasing cloud optical thickness with increasing temperature, contrary to observations at most places in the world (Tselioudis and Rossow, 1994; Del Genio and Wolf, 2000). But the model's efficiency has allowed it to be used for many climate studies and comparison of its results with other models has been valuable for model development and analysis of climate experiments. Also Russell's ocean model has been coupled with the B Grid atmosphere model, providing a useful comparison with the community ocean models used in most climate studies.

Still another variation is the Wonderland model (Hansen *et al.*, 1997b). This uses the physics of the 1983 model with $8° \times 10°$ resolution and an idealized cyclic geography, which makes the model fast enough for numerous century and millennium time scale simulations. The Wonderland model has been used for systematic analysis of the climate response to a wide range of radiative forcings (Hansen *et al.*, 1997c), and it has potential for paleoclimate studies. The Wonderland model has been temporarily abandoned because of its outdated physics, but, once we have model physics that we are satisfied with, we intend to revive it with the updated physical parameterizations.

Finally, I offer a few comments on our modeling philosophy. Our emphasis is on improved representation of the "physical" (including biological) processes. In our opinion, inadequate treatment of the physics is the primary restraint on understanding of long-term climate change. But better physics includes a need for higher vertical resolution in the atmosphere, where our present focus is on the planetary boundary layer and the upper atmosphere. Also Gary Russell emphasizes the need to handle nonlinear advection (the momentum equation) more accurately, which may require fundamental changes in the differencing schemes. Horizontal resolution in the atmosphere warrants continued examination, i.e., experimentation with finer grids. But, as we discussed in our 1983 paper, increased horizontal resolution is very expensive in resource requirements and relatively ineffective; when it is overemphasized, it limits the ability to attack fundamental issues. In comparison, there is a better justified need for improved resolution in ocean models. Along with the need for better physics in the atmosphere, this provides a primary drive for improved computer power.

A corollary of emphasis on model physics is the need to involve the research community in our model development and applications. GISS researchers can cover only a few topics in depth. But, if we can demonstrate that our model simulates characteristics of decadal climate change realistically and that it can help investigate the causes of long-term climate change, that should promote collaborations and interactions with leading researchers, and that in turn may provide a positive feedback advancing modeling capabilities.

Modeling philosophy must also relate to computing technology. It is commonly assumed that the fastest supercomputer is most productive for climate modeling. But the speed of a single run is only one consideration. Other factors include cost, the fraction of time available on the computer, the need for special programming, and especially how the computing approach meshes with the research objectives. We were among the first to emphasize the potential of workstations; for example, the ensembles of runs with the SI95 model (Hansen *et al.*, 1997c) were carried out on individual workstations. Now we have a 64-processor cluster that is well suited for ensembles of runs, but also, using a fraction of the processors in parallel, it permits use of a 32-layer $2° \times 2.5°$ model.

Ongoing technological advances in computing, data storage, and communications capabilities open new possibilities to advance modeling capabilities and understanding of long-term climate change. These advances will make it possible not only to include more realistic physics and higher model resolutions, but to systematically carry out ensembles of simulations and make the results readily available to the research community. This is an approach that we will pursue vigorously.

III. CLIMATE SENSITIVITY

A. CHARNEY REPORT

In 1979 the president's science advisor requested the National Academy of Science to study the carbon dioxide and climate issue. This resulted in the famous Charney (1979) report from a group of climate researchers, including Akio Arakawa, who met at Woods Hole in the summer of 1979.

Jule Charney, the panel chairman, decided to focus on a well-defined question: If the amount of atmospheric CO_2 were doubled, how much would the global average temperature increase by the time the system came to a new equilibrium? This question allowed use of the doubled CO_2 GCM studies of Suki Manabe that were already published (Manabe and

Wetherald, 1975) and in preparation (Manabe and Stouffer, 1980). The Charney panel also employed other tools, especially 1-D climate models, to analyze the topic.

Charney and Arakawa were interested personally in 3-D global models, which provided us opportunities for interactions with them. After Charney learned that we had initiated a doubled CO_2 experiment, we had several discussions with him and he asked Arakawa to visit GISS and work with us for a week. It was a good opportunity for us to talk with Akio not only about the doubled CO_2 results, but also about climate model development in general.

Our model result differed from the most recent model of Manabe, ours yielding a global warming of almost 4°C, while Manabe and Stouffer obtained 2°C. The conclusion that we reached with Arakawa, under the assumption that both models calculated the radiation accurately, was that differences between the models probably were caused by different strengths of climate feedback processes, especially sea ice and clouds. Specifically, there was relatively little Southern Hemisphere sea ice in the control run of Manabe and Stouffer, which would limit that positive feedback. Also their model used fixed clouds, while our model calculated reduced cloud cover with global warming, thus yielding more positive feedback.

Based on these model studies and their other deliberations, the Charney report estimated that equilibrium global climate sensitivity to doubled CO_2 was 3 ± 1.5°C. The range 1.5 to 4.5°C was broad and the stated uncertainty range was not meant to exclude the possibility of a sensitivity outside that range. Perhaps the best summary of the Charney report was their statement: "To summarize, we have tried but have been unable to find any overlooked or underestimated physical effects that could reduce the currently estimated global warming due to doubling of atmospheric CO_2 to negligible proportions" (p. 3).

The interactions with Charney and Arakawa stimulated us to analyze the contributions from each of the radiative feedbacks in our climate sensitivity experiments by inserting the changes (of sea ice, clouds, and water vapor) found in the GCM into a 1-D radiative model. This feedback analysis, developed by Andy Lacis, was used to help interpret our first published doubled CO_2 experiment (Hansen *et al.*, 1984). The separation of the climate response into that which would occur without feedbacks, ΔT_O, plus feedback contributions is the fundamental distinction between radiative forcing and climate response. ΔT_O measures the forcing in °C; the proportionality factor needed to convert this to a forcing in W/m^2 is 3.33. Thus the forcing for doubled CO_2 is $\Delta T_O \sim 1.25$°C or $\Delta F \sim 4.2\ W/m^2$.

B. Ice Age

Climate models by themselves can never yield an accurate and convincing knowledge of climate sensitivity. It is possible to change model parameters, e.g., in the cloud representation, that greatly alter the model sensitivity. And one can always think of climate feedbacks that may exist in the real world, but are entirely unrepresented in the model.

A more accurate measure of climate sensitivity can be obtained from analysis of empirical data with the help of climate models. Probably the best measure of climate sensitivity that we have now is that inferred from the last ice age, about 20,000 years ago. We now have a rather good knowledge of both the climate change between the last ice age and the current interglacial period as well as the change in the climate forcing that maintained the changed climate.

The important point is that, averaged over, say, 1000 years, the Earth had to be in near radiation balance with space during the middle of the last glacial period as well as during the current interglacial period. An imbalance of even 1 W/m^2 would have caused a rate of ocean temperature change or a change in the mass of glacial ice much greater than actually occurred.

The composition of the Ice Age atmosphere has been measured well from samples of air trapped in the polar ice sheets at the time of their formation (e.g., Lorius *et al.*, 1990). Planetary surface conditions, including the distribution of ice sheets, shorelines, vegetation, and surface albedo, have also been reconstructed (CLIMAP, 1981). The resulting radiative forcings that maintained the Ice Age cold were increased reflection of sunlight by the Earth's surface due mainly to larger ice sheets and altered vegetation distributions, decreased amounts of GHGs, and increased atmospheric aerosol loading (Hansen *et al.*, 1984, 1993; Hoffert and Covey, 1992). These surface and atmospheric changes caused a total forcing of $-6.6 \pm 1.5 \ W/m^2$ (Fig. 2).

This forcing maintained a global mean temperature change of about 5°C. CLIMAP (1981) reconstructions of ocean temperature, which had the last Ice Age being warmer than at present in much of the tropics, implied a global cooling of about 3.7°C during the last Ice Age. But recent data indicate that the tropics did cool by at least a few degrees (e.g., Guilderson *et al.*, 1994; Schrag *et al.*, 1996), so that a better estimate of the global mean Ice Age cooling is 5 ± 1°C.

Thus the climate sensitivity implied by the last Ice Age is about $5°C/(6.6 \ W/m^2) = 0.75°C$ per W/m^2, equivalent to 3 ± 1°C for doubled CO_2, in remarkable agreement with the analysis of Charney and Arakawa.

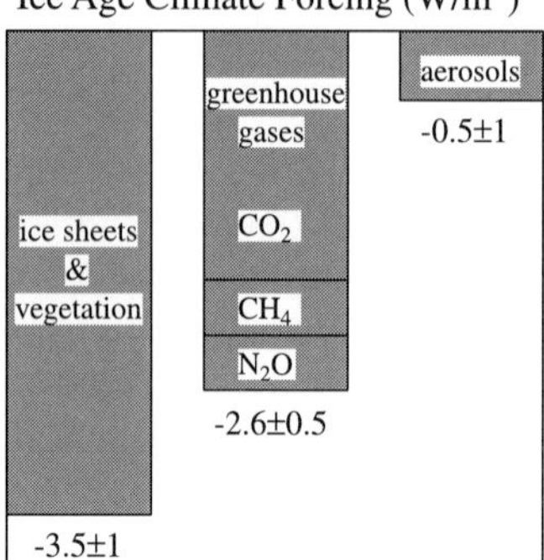

Figure 2   Climate forcings during the Ice Age 20,000 years ago relative to the current interglacial period. This forcing of -6.6 ± 1.5 W/m^2 and the 5°C cooling of the Ice Age imply a climate sensitivity of 0.75°C per 1 W/m^2.

The great thing about this empirical derivation is that it includes all climate feedbacks; any feedback that exists in the real world, whether we have thought about it yet or not, is incorporated, and that includes any changes of ocean heat transports.

A concern that can be raised about this empirical sensitivity is that climate sensitivity depends on the mean climate state. Variations of past climate and climate models both suggest that climate sensitivity is greater for a colder climate than for a warmer climate, and thus climate sensitivity inferred from comparison with the last Ice Age may not be accurate for the present climate. But, for several reasons, this concern is less substantial than it may appear. First, much of the higher sensitivity toward a colder climate is a consequence of increasing land ice cover with colder climate, and this factor is taken out in our present evaluation that treats land ice changes as a forcing, i.e., the inferred sensitivity refers only to the "fast" feedbacks, such as water vapor, clouds, and sea ice (Hansen *et al.*, 1984). Second, although the sea ice feedback is expected to increase toward colder climates, the nonlinearity should be moderate for small changes of the mean climate. Third, the sensitivity 0.75°C per W/m^2 if calculated to two decimals yields 3.2°C for our current estimate of doubled CO_2 forcing (Hansen *et al.*, 1998b) with this result representing the mean sensitivity between the last Ice Age and today. We conclude that 3 ± 1°C for doubled CO_2 is the appropriate estimate of climate sensitivity for today's global temperature.

IV. TRANSIENT CLIMATE: CLIMATE PREDICTIONS

A. CLIMATE RESPONSE TIME: SIMPLE OCEAN MODELS

The Charney report discussed only briefly the issue of how long it takes the climate system to more or less fully respond to a climate forcing. Charney realized that it was necessary to account for the ocean heat capacity beneath the mixed layer, and I recall him suggesting that the response time to increased CO_2 could be a few decades, on the basis of overturning times for near surface ocean layers in the tropics and subtropics. What was not realized at that time was that the climate response time is a function not only of the ocean's overturning rate, but of climate sensitivity itself. In fact, it is a very strong function of climate sensitivity. This issue does not alter Charney's analysis, because he focused on the equilibrium response to doubled CO_2. But climate sensitivity and response time become intimately connected if one attempts to infer climate sensitivity from observed transient climate change, and the climate response time raises a severe problem for policy makers.

I became especially interested in climate response time with the publication of the Carbon Dioxide Assessment Committee report (CDAC, 1983). This report seemed to be aimed at damping concern about anthropogenic climate change; at any rate, that was a likely effect of their conclusion that climate sensitivity was probably near the lower end of the range that Charney had estimated ($1.5°C$ for doubled CO_2). But their conclusion was based on the magnitude of observed global warming in the past century and the assumption that most of the warming due to human-made GHGs should already be present. Specifically, their analysis assumed that the climate response time could be approximated as being 15 years and that the response time was independent of climate sensitivity.

The fact that climate response time is a strong function of climate sensitivity is apparent from the following considerations. First, note that climate feedbacks, such as melting sea ice or increasing atmospheric water vapor, come into play only in conjunction with temperature change, not in conjunction with the climate forcing. Thus, even if the ocean's heat capacity could be represented as that of a simple slab mixed layer ocean, the response time would increase in proportion to the feedbacks (and thus in proportion to climate sensitivity). And, second, while the feedbacks are coming into play, the heat perturbation in the ocean mixed layer can mix into the deeper ocean, further delaying the surface response to the forcing.

Investigation of this issue requires a realistic estimate of the rate of heat exchange between the ocean surface (well-mixed) layer and the deeper ocean. Our approach to this problem in the early 1980s was to attach a simple representation of the ocean to our atmospheric GCM. We used this ocean representation for our transient climate predictions, described in the next section, as well as for investigation of climate response time. The objectives of the ocean representation were (1) to obtain a realistic climate response time at the Earth's surface and (2) to achieve a realistic distribution of surface climate in the model's control run despite the absence of a dynamic simulation of the ocean.

One part of the ocean representation was vertical exchange of heat anomalies beneath the ocean mixed layer. For our 1-D radiation model we had used a vertical diffusion coefficient based on observed global penetration of transient tracers. For the 3-D model Inez Fung determined local diffusion coefficients by using transient ocean tracer observations to establish a relationship between the vertical mixing rate and the local stability at the base of the winter mixed layer. This relationship and the Levitus ocean climatology were then used to obtain effective mixing coefficients beneath the mixed layer for the entire ocean, as described in our Ewing symposium paper (Hansen *et al.*, 1984).

The second part of the ocean representation was a specification of horizontal heat transports in the ocean suggested by Peter Stone and developed by Gary Russell, as described briefly in our Ewing paper and in more detail by Russell *et al.* (1985). Specifically, we employed the ocean heat transports implied by the energy balance at the ocean surface in our GCM when the model was driven by observed sea surface temperatures. This approach of specifying the horizontal ocean heat transport has come to be known as the Q-flux ocean model and is used with the mixed layer model alone as well as with the mixed layer attached to a diffusive ocean.

The upshot of our climate simulations was that climate response time is a strong function of climate sensitivity. The response time is only about 15 years if climate sensitivity is near the lower limit estimated by Charney (1.5°C for doubled CO_2), but more than 100 years if climate sensitivity is 4.5°C for doubled CO_2. The climate sensitivity inferred from paleoclimate data, about 3°C for doubled CO_2, suggests that the climate response time is at least 50 years.

Such a long response time raises a severe conundrum for policy makers. If, as seems likely, GHGs are the dominant climate forcing on decadal time scales, there may be substantial warming "in the pipeline" that will occur in future decades even if GHGs stop increasing. Such yet to be realized warming calls into question a policy of "wait and see" for dealing with the uncertainties in climate prediction. The difficulty of halting climate change once it is well under way argues for commonsense measures that slow down the climate experiment while a better understanding is developed.

B. GLOBAL CLIMATE PREDICTIONS

We used the model described above, with Q-flux horizontal ocean transports and spatially variable diffusive mixing of temperature anomalies beneath the mixed layer, for the first transient climate predictions with a 3-D global climate model (Hansen *et al.*, 1988). Climate change in this model was driven by observed and projected GHG changes and secondarily by aerosols from volcanic eruptions.

Figure 3 compares observed global surface temperature with the simulations, which were carried out in 1987. The large interannual variability makes it difficult to draw inferences based on only 11 years of data subsequent to the calculations. But so far the world has been warming at a rate that falls within the range of scenarios considered.

Scenarios A, B, and C differed in their growth rates of GHGs and in the presence or absence of large volcanic eruptions. Scenario A assumed that GHGs would increase exponentially at rates characteristic of the preceding 25 years and that there would be no large volcanic eruptions. Scenario

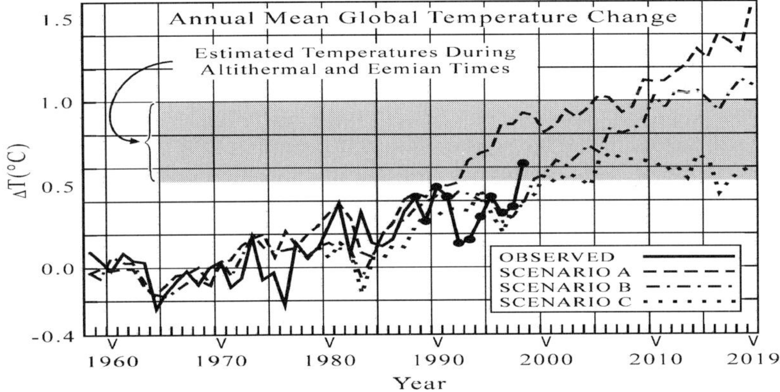

Figure 3 Global surface air temperature computed with GISS model in 1987 (Hansen *et al.*, 1988) and observed global temperature based on meteorological station measurements (Hansen *et al.*, 1999), including update subsequent to model predictions ($\cdots\cdots$).

A was designed to reach the equivalent of doubled CO_2 by about 2030, consistent with the estimate of Ramanathan *et al.* (1985). Scenario B had an approximately linear growth of GHGs, reaching the equivalent of doubled CO_2 at about 2060. Scenario B included occasional cooling from volcanic eruptions in 1995 and 2015. Scenario C had a still slower growth rate of GHGs with a stabilization of GHG abundances after 2000 and the same volcanos as in scenario B.

What is the climate forcing in the real world? Both GHGs and volcanic aerosols have been well measured in recent decades. The observed GHG changes and volcanic aerosols both correspond closely to scenarios B and C (Hansen *et al.*, 1998a,b), which are practically the same until year 2000. The main difference is that the large volcano in the 1990s occurred in 1991 in the real world, while in the model it occurred in 1995. Scenario C, with terminating GHG growth in 2000, is not expected to be realistic in the future. Thus scenario B is the most realistic.

The global temperature in scenario B increases by 1°C in 50 years (Fig. 3), with a rather steady warming rate of about 0.2°C/decade. This is in good agreement with observations of the past few decades, as described in detail by Hansen *et al.* (1999). But the absence of information on all climate forcings makes it difficult to draw substantive conclusions even from the 40-year record.

One important conclusion that can be drawn is that the rate of growth of GHGs in the real world is significantly less than in scenario A, the "business as usual" scenario with continued exponential growth of GHGs that is similar to the principal IPCC (1996) scenarios. The climate forcing due to observed growth rates of GHGs during the past several years is only about half of that in the scenarios commonly used by IPCC, such as IS92a or 1% CO_2 increase per year (Hansen *et al.*, 1998b). The slowdown in growth rates provides hope that the more drastic climate changes can be avoided.

Clarification of GHG scenarios is important for the global warming debate (Section VI) and for interpretation of present and future observed climate change. Although IPCC defines a broad range of scenarios, the full range is not emphasized. It is a common practice of modelers to employ a single scenario with a strong GHG growth rate. A strong forcing has the merit of yielding a large "signal-to-noise" ratio in the climate response. But use of a single scenario can be taken as a prediction in itself, even if that is not intended. Multiple scenarios are especially useful for problems that may involve nonlinear processes in a significant way. Thus we argue (Hansen *et al.*, 1998b) for use of a range of scenarios bracketing plausible rates of change, which was the intention of our scenarios A, B, and C.

C. FORCINGS AND CHAOS

We present an example of calculations with the current GISS GCM to bring the modeling discussion up to date. Specifically, we use the model version based on Arakawa's B Grid atmosphere that is employed by the Forcings and Chaos research team in the GISS Institute on Climate and Planets. Examples of recent results from the other principal variations of the GISS GCM are given by Shindell *et al.* (1999b) for simulated climate effects of solar cycle and ozone variability using the GISS climate/middle atmosphere model and by Russell *et al.* (2000) for simulated climate trends due to increasing CO_2 using the C Grid coupled atmosphere–ocean version of the GISS model.

The objective of the Forcings and Chaos group is to shed light on the roles of climate forcings and unforced climate variability ("chaos") in climate variability and change during recent decades. The approach is to make ensembles of simulations, adding various radiative forcings to the model one by one, and running the model with several different treatments of the ocean (Hansen *et al.*, 1997c). Initial simulations were made for the period 1979–1996 with the SI95 model, which was frozen during the Summer Institute of 1995. Trial simulations for the period 1951–1997 were made with the SI97 and SI99 models, and a larger array of simulations for 1951–1999 is planned for the SI00 model.

1. SI95 Simulations

The SI95 model, documented by Hansen *et al.* (1997c), had nine layers in the atmosphere with one or two layers in the stratosphere. This model was run with four representations of the ocean: (A) observed SST, (B) Q-flux ocean, (C) GISS ocean model (Russell *et al.*, 1995), and (D) an early GFDL ocean model (Bryan and Cox, 1972; Cox, 1984). The SI95 model was flawed by excessive absorption of solar radiation by sea ice, as illustrated by Fig. 1 of Hansen *et al.* (1997c). It was realized later that the excessive absorption was the result of a programming error that caused sea ice puddling to be active independent of surface temperature.

The SI95 simulations illustrated that most of the interannual variability of *regional* climate on an 18-year time scale at middle and high latitudes is chaotic, i.e., unforced. But a natural radiative forcing (volcanic aerosols) and an anthropogenic forcing (ozone depletion) were found to leave clear signatures in the simulated *global* climate that were identified in observations. The SI95 simulations were also used to infer a planetary radiation imbalance of about 0.5 W/m^2, leading to prediction of a new record global temperature that has subsequently occurred.

2. SI97 Simulations

Significant modifications in the SI97 model include the use of 12 atmospheric layers, changes to the planetary boundary layer (Hartke and Rind, 1997) and the clouds and moist convection (Del Genio *et al.*, 1996), correction of the programming error in the sea ice puddling, and addition of a parameterization for ice cover of lakes. The three additional layers increase the resolution in the tropopause and lower stratosphere region with the model top remaining at 10 mb. These modifications will be described in a future paper documenting the SI99 model and, in some cases, in future papers defining specific aspects of the model physics.

Improvements in the SI97 climatology over the SI95 model include (1) more realistic stratospheric temperatures, especially the longitudinal variations, although the stratosphere remains too warm at the winter pole and too cool at the summer pole; (2) more realistic poleward heat transports; (3) more accurate computations of stratospheric radiative forcings, especially due to stratospheric aerosol and ozone changes, resulting in accurate representation of stratospheric temperature change after large volcanos; (4) more accurate albedos for sea ice, improving the sea ice cover in coupled atmosphere ocean runs; and (5) more accurate winter temperatures in Canada.

Known outstanding problems with the SI97 model include (1) deficiencies in boundary layer stratus cloud cover off the west coast of the continents, resulting in a solar radiation flux at the ocean surface that is excessive by as much as 50 W/m^2 in the summer; (2) buildup of snow cover along the northeast coast of Siberia that fails to melt in the summer, a problem that was exacerbated by improved physical representations of the PBL and clouds; and (3) a still very crude representation of the stratosphere, including the rigid top at 10 mb and a sponge-layer drag in the top layer, resulting in errors in the stratospheric temperature distribution and circulation.

We carried out several simulations for the period 1951–1997 with the SI97 model that helped assess the model capabilities and deficiencies. Figure 4 (see color insert) shows the degree to which the SI97 model simulates observed surface temperature change during that 47-year period. Observed change of the surface temperature index, which consists of surface air temperature over land and SST over the ocean, is shown in Fig. 4b. The left column, Figs 4a, 4c, and 4e, shows climate model simulations of surface air temperature change driven only by observed changes of SST and sea ice, with the three cases providing an indication of the impact of uncertainties in these boundary conditions. Figures 4d and 4f add the two most accurately known radiative forcings, greenhouse gases (Hansen *et al.*, 1998b) and stratospheric aerosols (Sato *et al.*, 1993).

Two features in the observed climate change are of special interest: (1) high latitude warming over Siberia and the Alaska region, which is strongest in the winter, and (2) cooling over the contiguous United States, which is strongest in the summer. We discuss each of these briefly.

a. High-Latitude Warming

The model simulates the Alaska warming, but it does not simulate the Siberia warming well. The results may improve with the SI99 model, which eliminates the problem of growing glaciers in northeast Siberia, but that seems unlikely to be important in the winter. Additional climate forcings, including ozone, solar irradiance, and aerosol direct and indirect effects may be important. But it is likely that simulation of the winter warming in Siberia will require a better representation of the stratosphere. Shindell *et al.* (1999a) find that greenhouse gas forcing yields greater Siberian warming in the GISS climate/middle atmosphere model, associated with an intensification of the stratospheric polar vortex. This topic requires further study as the climate/middle atmosphere model has a sensitivity of 5.5°C for doubled CO_2, which may be larger than reality, and the climate forcing used by Shindell *et al.* (1999a) is similar to IPCC IS92a, which exceeds the observed greenhouse gas forcing. The Siberian warming is a part of the Arctic oscillation (Thompson and Wallace, 1998) that seems to be a natural mode not only of the real world but of climate models. Thus the stronger response in the experiment by Shindell *et al.* (1999a) might be in part a consequence of the bell being rung harder in that model. But the important point is the evidence that adequate representation of stratospheric dynamics is needed for simulating tropospheric climate.

This is an important practical matter for climate model development because the higher model top (80 km) and sophisticated gravity wave drag treatment in the climate/middle atmosphere model increase the computation time by a factor of 7. The plans for the SI model series, which is aimed at studies of surface climate, were to make moderate improvements in the representation of the stratosphere, perhaps increasing the model top to 50 km and including a simple representation of gravity wave effects. But if the suggestion of Shindell *et al.* (1999a), that even the mesosphere must be included to simulate the effects of solar variability on surface climate, is borne out, we will need to reconsider this strategy for model development.

b. United States Cooling

It is interesting that the GISS model driven by observed SST anomalies consistently simulates a cooling trend in the United States during the past 50 years. This cooling trend is not an accident, because it is captured by all of the five ensembles of SI97 model runs. All five ensembles yield greater cooling in the summer than in the winter, in agreement with observations. This suggests that the observed regional climate trend is a tropospheric phenomenon driven immediately by SST anomalies, and that the model can represent, at least in part, the immediate mechanisms for change. Although it will be a challenge to determine whether the SST anomalies are themselves forced or chaotic, it may be easier to make progress in partial understanding of this climate change by making simulations in which the SST anomalies are restricted to specific parts of the ocean. However, because of inherent limitations in the ability of specified SST experiments to deliver correct atmosphere to ocean flux changes, it will be necessary to also carry out experiments with other ocean representations that more realistically portray ocean-atmosphere interactions.

We point out elsewhere (Hansen *et al.*, 1999) the practical importance of understanding this climate change in the United States. During the past century, temperatures have increased slightly in the United States, but not as much as in most of the world, and the warmest temperatures in the United States occurred in the 1930s (Fig. 8 of Hansen *et al.*, 1999). Although long-term climate change in recent years seems to be reaching a level that is noticeable to the layperson in some parts of the world (Hansen *et al.*, 1998a), this is less so in the contiguous United States. However, if the SST patterns that are giving rise to the recent cooling tendency in the United States are a temporary phenomenon, there could be a relatively rapid change to noticeably warmer temperatures in the near future.

3. SI99 Simulations

The SI99 model was recently frozen to allow an array of simulations for 1951–1999 to be carried out. Principal changes in the SI99 model are (1) modification of the snow albedo parameterization to eliminate the growth of glaciers in northeast Siberia, (2) replacement of the tropospheric aerosol distribution of SI95 and SI97 with a new distribution based mainly on assumed aerosol sources and tracer transport modeling by Ina Tegen and Dorothy Koch, and (3) optional replacement of the fourth-order differencing scheme for the momentum equation with second-order differencing. The new aerosol distribution reduces solar heating of the surface by several watts per square meter, as shown in Section V. The second-order differencing eliminates excessive noise and model instability caused by the fourth-order scheme while reducing the computing time by about 25%. However, midlatitude storms move more slowly and do not cross the continents as realistically, so the fourth-order differencing is retained in the model coding and employed in many experiments.

The SI99 model will be documented in conjunction with a paper describing the array of simulations for 1951–1999. These experiments will differ from the array described by Hansen *et al.* (1997c) in several ways: (1) The period of simulation will be about five decades rather than two decades; (2) forcings each will be run individually rather than cumulatively, but some experiments will also include all or most of the forcings; (3) tropospheric aerosols will be included as a forcing; (4) dynamic ocean models are expected to include the GISS model, an up-to-date version of the GFDL MOM model, and the global isopycnal (MICOM) ocean model of Shan Sun and Rainer Bleck; and (5) access to model results will be provided via the GISS World Wide Web home page (www.giss.nasa.gov).

V. MISSING ATMOSPHERIC ABSORPTION

A prominent issue concerning climate models in the 1990s has been "missing atmospheric absorption." Surface, satellite, and *in situ* observations have been used to surmise that most climate models underestimate solar radiation absorbed in the atmosphere by 20–40 W/m^2 and overestimate solar radiation absorbed at the planetary surface by a similar amount. Such errors could affect the simulated atmospheric circulation and the drive for oceanic temperatures and motions.

Comprehensive review of this topic is beyond the scope of our paper. We refer instead to a few recent papers, which lead to many others. John Garratt and colleagues (1998) and Bob Cess and colleagues (1999) have been especially productive in providing observational data and interpretations in a series of papers going back at least to 1993. These scientists and others (cf. semipopular review by Kerr, 1995) deserve credit for stimulating discussions about atmospheric physics and verification of models, in the best spirit of scientific investigation.

The focus has been on identifying missing or underrepresented absorbers in the models. Arking (1996) argues that water vapor absorption is underestimated. Garrett *et al.* (1998) suggest that inaccurate water vapor calculations and aerosols contribute to the problem. Cess *et al.* (1999), however, present data that they interpret as indicating that the missing absorber is present only in cloudy skies, not clear skies. There has been much speculation about possible exotic mechanisms for absorption, such as water vapor dimers, that are not included in present models.

Not long ago Bob Cess presented a seminar at GISS summarizing evidence that he interpreted as requiring the presence of a missing absorber. He commented that Paul Crutzen not only agreed with this conclusion but stated that it was time to stop arguing about it. Although Bob took some solace in the support of a Nobel prize winner, somehow the thought that jumped to my mind on hearing this was one of Oscar Wilde's epigrams: "When people agree with me, I always feel that I must be wrong."

Observationally it is difficult, if not impossible, to obtain a clean separation of clear and cloudy skies, especially with satellite observations. For this reason, and because it is the total absorption that drives the atmosphere and ocean, it seems best to examine first the all-sky case. Martin Wild has presented extensive comparisons of modeled and "observed" solar radiation absorption (see Wild *et al.*, 1998, and references therein) that we will use for quantitative discussion.

We focus on three numbers: (1) the amount of solar radiation hitting the Earth's surface, $S\downarrow$, (2) the amount of solar radiation absorbed by the Earth's surface, $\alpha \times S\downarrow$, where α is the surface co-albedo, i.e., 1 minus the albedo), and (3) the amount of solar radiation absorbed by the atmosphere (A_{atm}). The debate in the literature has focused on atmospheric absorption, but we argue that A_{atm} is a tertiary quantity and is not observed. Thus it is better to consider the three quantities in the order listed here.

The solar radiation hitting the Earth's surface, $S\downarrow$, is a primary quantity, i.e., it can be measured and, indeed, has been measured at hundreds of stations around the world. The solar radiation absorbed by the Earth's surface, $\alpha \times S\downarrow$, is a secondary quantity. It cannot practically be measured with the needed accuracy, because it varies on small spatial scales. One must assume a global distribution of surface albedos, so $\alpha \times S\downarrow$ includes the uncertainties in both $S\downarrow$ and α. Similarly, the absorption in the atmosphere, A_{atm}, is a tertiary quantity and cannot be measured directly on a global scale, and its calculation requires additional input. That input can be an assumed (or measured) planetary albedo, which is often taken as 30%, or detailed information on clouds and other atmospheric properties required for radiative transfer calculations across the solar spectrum.

The GEBA (Global Energy Balance Archive) data for $S\downarrow$ are shown in Fig. 5a, top left (see color insert), based on measurements at about 700 stations (Ohmura *et al.*, 1998). Where there is more than one measurement within a $4° \times 5°$ gridbox, we average the results. The mean over all gridboxes having data, weighted by gridbox area, is 184 W/m^2, in agreement with Fig. 18 of Wild *et al.* (1998). The true global mean is uncertain due to the limited sampling, but this difficulty can be minimized by averaging the model results over the GEBA gridboxes (G Grid). In Table I we include the modeled $S\downarrow$ integrated over the G Grid and the true global average; these two ways of averaging over the world yield results that tend to differ by several W/m^2, not always in the same sense.

Table I compares the estimates of Wild *et al.* (1998) for global radiation quantities with values obtained in recent GISS global climate models. Model results are 5-year means, years 2–6 of 6-year runs. The SI95 model is described by Hansen *et al.* (1997c). One difference between SI99 and earlier models is more absorbing aerosols in the SI99 model, as quantified below. Another change that may affect these results is improvement in the cloud physics beginning with the SI97 model (Del Genio *et al.*, 1996). The radiation scheme is the same in all models: It uses the k distribution method for gaseous absorption and the adding method for multiple scattering with spectrally dependent aerosol and cloud scattering parameters to ensure self-consistency between solar and thermal regions. Clear comparisons can be made among the runs with the SI99 model, which differ only in atmospheric composition. Differences among the runs are meaningful only if they exceed a few W/m^2, because the cloud cover fluctuates from run to run, especially for the G Grid. The clearest demonstration of the aerosol effect is the run with all aerosols removed. This shows that the assumed 1990 aerosol distribution reduces $S\downarrow$ by 11 W/m^2 for the true global average and by 18 W/m^2 averaged over the GEBA gridboxes.

$S\downarrow$, as simulated in the GISS climate model, agrees well with the GEBA data, as summarized in Table I and Fig. 5. SI95 has 5–10 W/m^2

Table I

Global Radiation Quantities[a]

	$S\downarrow$ (W/m^2)		$\alpha \times S\downarrow$ (W/m^2)	A_{atm} (W/m^2)	Albedo (%)
	G Grid	Global			
Wild estimates	184	—	154	85	30
SI95 model, 1980 atmos.	194	190	167	66	30.8
SI99 model, 1950 atmos.	188	188	163	66	32.9
SI99 model, 1990 atmos.	179	182	159	70.4	33.0
SI99 model, no aerosols	197	193	168	63.5	32.3

[a] Estimated by Wild *et al.* (1998) and as calculated in recent versions of the GISS global climate model. Results are global, but for $S\downarrow$ results are also given for the GEBA network of stations.

more solar radiation hitting the surface than observed. But SI99, with its more absorbing aerosols, agrees even more closely with observations. Sulfate, black carbon, and organic aerosols are time dependent in the SI99 model, so results are given for both 1950 and 1990. The observations were taken over a few decades, so an average of 1950 and 1990 seems appropriate for comparison. With this choice the SI99 model agrees with GEBA data within 1 W/m^2 on average (Fig. 5, lower left), but if aerosols are removed there would be a significant discrepancy of 13 W/m^2 with GEBA (Fig. 5, lower right).

$\alpha \times S \downarrow$, the solar radiation absorbed by the Earth's surface, is at least 5 W/m^2 more in our current model than estimated by Wild *et al.* (1998), implying that our surface is slightly darker. Surface albedo in recent GISS models is specified in detail, with ocean albedo including effects of whitecaps as a function of wind speed (Gordon and Wang, 1994) and subsurface particulate scattering (Gordon *et al.*, 1988), while the land albedo varies seasonally with vegetation and snow cover and depends on soil properties (Matthews, 1983; Hansen *et al.*, 1983, 1997c). We believe that our largest error is an *underestimate* of surface absorption in the Himalayas in the summer. But the discrepancy with the estimate of Wild *et al.* (1998) for surface absorption is small in any case.

A_{atm}, the solar radiation absorbed in the atmosphere, is almost 15 W/m^2 less in our model than in the estimate of Wild *et al.* (1998). Much of this difference is associated with the planetary albedo in our model being higher (32–33%) than the observed albedo of 30%, which is based mainly on Earth Radiation Budget Experiment (ERBE) data (Barkstrom *et al.*, 1989).

In summary, there is no discrepancy between the model and observations of solar radiation reaching the Earth's surface. Our calculated atmospheric absorption of 70–71 W/m^2 is 14–15 W/m^2 less than that estimated by Wild. We argue below that absorbers omitted or underestimated in our model can increase atmospheric absorption to only about 75 W/m^2. Before considering the likely sources of the remaining 10 W/m^2 discrepancy with Wild's estimate for A_{atm}, we discuss how the near agreement of the GCM with GEBA observations can be reconciled with the conclusion that most models underestimate absorption by 20–40 W/m^2.

We believe, in agreement with Garrett (see above), that absorption by aerosols and water vapor has been underestimated in some models. That is why we said that the Lacis and Hansen (1974) parameterization for solar absorption may have inadvertently contributed to the "missing" atmospheric absorption issue. That parameterization, adopted by a number of GCM groups, does not include aerosols, and for that reason we never used it in our climate models. We use the more general correlated k distribution method (Lacis and Oinas, 1991) with explicit integration over the spectrum to achieve accurate scattering and absorption by clouds and aerosols. The water vapor parameterization of Lacis and Hansen, though quite accurate given its simplicity, underestimates absorption of solar radiation by 5–10% for typical water vapor amounts, as judged by the more general k distribution method or line-by-line calculations (Ramaswamy and Freidenreich, 1992). Especially when combined with the low water vapor amounts in many atmospheric models, this also contributes to underestimates of absorption of solar radiation.

The effect of aerosols is illustrated in Fig. 5 (and Table I), where we compare results from our SI99 model with and without aerosols. The aerosols in our SI99 model are a combination of sulfates, organics, black carbon, soil dust, and sea salt as summarized and compared with other aerosol climatologies in Table II. The sulfates, organics, and black carbon each contain a time-dependent anthropogenic component as well as a natural component. Time dependence is not included in either the soil dust or biomass burning (which contributes both organics and black carbon) because of insufficient available information. The aerosol distributions, based in part on aerosol transport models (Tegen *et al.*, 1997; Koch *et al.*, 1999), will be described in more detail elsewhere. The principal change of aerosols that has occurred in successive GISS climate models has been the addition of more absorbing aerosols, as illustrated in Fig. 6, which shows that the global mean aerosol single-scatter albedo decreased from 0.954 in the SI95 model to 0.927 in the SI99 model.

Absorption by SI99 aerosols is due principally to black carbon and soil dust, and only slightly to organics. The black carbon distribution, based on a transport model (Tegen *et al.*, 1997), is especially uncertain; if it is reduced by a factor of 2 the net single-scatter albedo increases from 0.927 to 0.943. The small absorption by organics, presumably occurring mainly at ultraviolet wavelengths, is based on measurements of Tica Novakov (private communication, 1999). Sea salt amount is very uncertain; we multiply the optical depth of Tegen *et al.* (1997) by 4 to account for submicron particles (Quinn and Coffman, 1999). But sea salt is nonabsorbing, so it has little effect on atmospheric absorption.

How realistic is the aerosol absorption in the SI99 model? Although we have concern that the black carbon amount could be exaggerated, other factors work the other way. Actual aerosols often are mixtures of compositions, which tends to decrease the net single-scatter albedo. Also satellite data (Nakajima, *et al.*, 1999) reveal greater aerosol amount in the tropical Western Pacific and Indian Ocean regions than in our model, perhaps in part a consequence of the fact that we did not have data to include

Table II
Aerosol Optical Depth and Single-Scatter Albedo

			Optical depth		Single-scatter albedo	
	Andreae	Seinfeld	SI95 model	SI99 model (1950/1990)	SI95 model	SI99 model
Sulfates						
Trop. natural	0.021	0.014	0.045	0.0067	1.00	1.00
Trop. anthro.	0.032	0.019	0.030	0.0090/0.0222	0.99	1.00
Black carbon						
Industrial	0.006[a]	0.003[a]	0.011	0.0021/0.0067	0.48	0.31
Biomass burning	—		—	0.0014	—	0.48
Organic carbon						
Natural	0.019	0.014	—	0.0032	—	0.98
Industrial	0.003	0.002	—	0.0086/0.0267	—	0.96
Biomass burning	0.027[a]	0.017[a]	—	0.0124	—	0.93
Soil dust	0.023	0.023	0.042	0.0324	0.96	0.89
Sea salt	0.003	0.003	0.012	0.0267	1.00	1.00
Other						
Volcanic	0.004	0.001	0.012	0.005 + variable (total = 0.0065/0.011)	1.00	1.00
NO_x	0.003	0.002	—	—	—	—
Industrial dust	—	0.004	—	—	—	—
Total	0.144	0.102	0.152	0.109/0.149	0.954	0.935 (1950) 0.927 (1990)

From Andreae, 1995, and Seinfeld, 1996.
[a] Black carbon included with organic aerosol optical depth.

time-dependent biomass burning and did not include a Western Pacific biomass source. Because of the complexity of aerosols, the best verification of aerosol absorption is probably field data for the net aerosol single-scatter albedo. Data from field campaigns off the eastern United States and near Europe and India suggest that absorption as great as that in Fig. 6 is not unrealistic, but more extensive and precise data are needed.

What about other possible absorption, besides aerosols? Several minor effects are not included in our present radiation calculations, for example, oxygen dimer (Newnham and Ballard, 1998) and nitrogen continuum (Boissoles *et al.*, 1994) absorption, but these are likely to produce at most a few W/m^2. A popular idea, championed by Bob Cess, is that clouds somehow absorb more sunlight than calculated. However, as a GCM experiment, we doubled the calculated absorption by liquid and ice cloud particles and found the effect to be negligible because of absorption by water vapor in the same spectral regions. Finite (horizontal) cloud extent

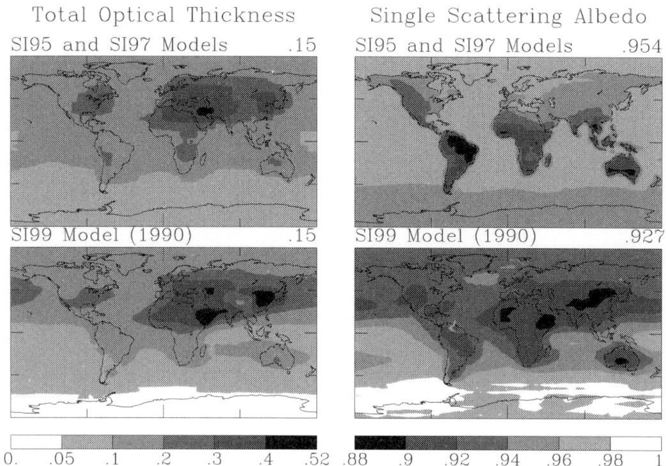

Figure 6 Optical depth and single-scatter albedo of aerosols in GISS GCM.

needs to be accounted for, but it does not introduce substantial absorption. Water vapor absorption is underestimated in our and many other models because the troposphere tends to be about 1–2°C cooler than observed, and thus also drier than observed, but at most this could produce a few W/m^2 of additional absorption. For these reasons we believe that atmospheric absorption is at most about 75 W/m^2.

Finally, assuming atmospheric absorption is not more than 75 W/m^2, how is the remaining 10 W/m^2 difference with Wild's estimate of 85 W/m^2 accounted for? In our present model 5 W/m^2 of this difference is in our larger surface absorption and the other 5 W/m^2 is in our planetary albedo being larger than 30% (our calculated albedo is about 31.5% if atmospheric absorption is 75 W/m^2). The ERBE planetary albedo of 30% is uncertain by at least 1% because it depends on uncertain models for the angular distribution of reflected sunlight and on detectors that do not have a uniform response over the solar spectrum. We suspect that an Earth albedo of 31–32% is possible. But the division of this 10 W/m^2 between surface absorption and planetary albedo can be shifted, and such detailed discussion pushes the data beyond current accuracy levels.

The bottom line is that we find no evidence for a 20–40 W/m^2 radiation mystery and no need for an exotic absorber. The solar radiation reaching the planetary surface is in good agreement between our climate model and observations. This does not mean that a better understanding of absorption of solar radiation, especially by atmospheric aerosols, is unimportant. On the contrary, we must have improved knowledge of aerosols and their changes to predict long-term climate change (Hansen *et al.*, 1998b), and the results and discussion in this section only reinforce the need for better aerosol observations.

VI. GLOBAL WARMING DEBATE

It has been 20 years since the global warming discussions of Charney and Arakawa in 1979. Is our understanding of this topic improving? The picture drawn by the media is one of opposing camps in perpetual fundamental disagreement. Opposing interpretations of the science seem likely to persist, because of the perceived economic stakes associated with energy policies.

The public debate is not as scientific as we would prefer. It can be difficult to find references for public statements or positions of participants. Publication of your own research does not ensure that it will be represented accurately. An egregious example, from my perspective, was congressional testimony of Patrick Michaels in 1998 in which he extracted from our Fig. 3 (see earlier section) the simulated global temperature for scenario A, compared this with observed global temperature, and concluded that my congressional testimony in 1988 had exaggerated global warming. If he had used the entire figure, and noted that real-world climate forcings have been following scenario B, he would have been forced to a very different conclusion.

Recently I had the opportunity to debate "global warming" with Richard Lindzen (Schlumberger, 1998), who has provided much of the intellectual underpinnings for global warming "skeptics." It seemed to me that it may aid future progress to delineate our fundamental differences of opinion, thus providing a way to pin each other down and a basis to keep tabs on progress in understanding. So I went through Dick's publications and made a list of our key differences, for use in my closing statement at the debate. As it turned out, closing statements were eliminated from the debate format at the last minute. But I used this list (Table III) in a debate with Patrick Michaels (AARST, 1998), and, with the same objective of pinning down key issues, I briefly discuss each of the six items here.

A. Reality of Warming

Lindzen (1989) and others have questioned the reality of global warming. Many "greenhouse skeptics" continue to argue that it is only an urban

Table III

Fundamental Differences with R. Lindzen, as Prepared for Schlumberger (1998) Discussion and Used in AARST (1998) Debate

---

1. Observed global warming: real or measurement problem?
   Hansen: Warming 0.5–0.75°C in past century; $\geq$ 0.3°C in past 25 years.
   Lindzen: Since about 1850 "more likely ... 0.1°C.
2. Climate sensitivity (equilibrium response to doubled CO_2).
   Hansen: 3 $\pm$ 1°C   *Lindzen*: $\leq$ 1°C
3. Water vapor feedback
   Hansen: Positive (upper tropospheric H_2O increases with warming)
   Lindzen: Negative (upper tropospheric H_2O decreases with warming)
4. CO_2 contributions to the $\sim$33°C natural greenhouse effect
   Lacis and Hansen: Removing CO_2 and trace gases with water vapor fixed would cool the Earth 5–10°C; with water vapor allowed to respond, it would remove most of the greenhouse effect.
   Lindzen: If all other GHGs (such as CO_2 and CH_4) disappeared, over 98% of the natural greenhouse effect would remain.
5. When will global warming and climate change be obvious?
   Hansen: With the climatological probability of a hot summer represented by two faces (say, painted red) of a six-faced die, judging from our model by the 1990s, three or four of the six die faces will be red. It seems to us that this is a sufficient "loading" of the dice that it will be noticeable to the man in the street
   Lindzen: I personally feel that the likelihood over the next century of greenhouse warming reaching magnitudes comparable to natural variability remains small.
6. Planetary disequilibrium
   Hansen: Earth is out of radiative equilibrium by at least 0.5 W/m^2.

---

effect. We summarize elsewhere (Hansen *et al.*, 1999) evidence that global surface temperature has risen sharply in recent decades and that there has been 0.5–0.75°C global warming since 1880. The warming is largest in remote ocean and high-latitude regions, where local human effects are minimal, and the geographical patterns of warming clearly represent climatic phenomena, not patterns of human development. The instrumental temperature measurements are supported by borehole temperature profiles from hundreds of locations around the world (Harris and Chapman, 1997; Pollack *et al.*, 1998) and by analysis of the near-global meltback of mountain glaciers during the past century (Oerlemans, 1994).

The issue of the reality of global warming survives only because tropospheric temperatures showed essentially no warming over the first 19 years of satellite measurements, 1979–1997. For such a brief period it is not expected that surface and tropospheric temperature changes must coincide, especially in view of measured and suspected changes of atmospheric ozone, aerosols, and clouds. Indeed, tropical surface temperatures hardly increased during 1979–1997, so we would not anticipate much increase of global tropospheric temperature (Hansen *et al.*, 1999). Because of the small temperature change during 1979–1997, small measurement errors can add to real differences in surface and tropospheric trends and cause a qualitative impact on their comparison. But tropospheric warming becomes obvious when one includes (radiosonde) data from several years preceding 1979 and as data following 1997 are added to the record. Temperature measurements deserve continued attention, but the reality of long-term warming is already apparent to most analysts and it is our expectation that this topic will recede as an issue as additional data are collected.

B. CLIMATE SENSITIVITY

Lindzen argues that climate sensitivity is less than or approximately 1°C for doubled CO_2 and may be as small as 0.3–0.5°C (Lindzen, 1997). We have presented an analysis of paleoclimate data (Hansen *et al.*, 1984, 1993, this paper) that we maintain not only confirms the climate sensitivity estimated by Charney and Arakawa, but sharpens it to 3 ± 1°C. It is our expectation that confidence in this high climate sensitivity will increase as paleoclimate data continue to improve and as their significance for analyzing climate sensitivity is more widely accepted. Climate models can contribute further to this discussion by showing that the details of paleoclimate changes can be simulated realistically.

The approach of attempting to infer climate sensitivity from the current rate of global warming, as discussed in CDAC (1983) and IPCC (1996), will remain fruitless as long as major climate forcings remain unmeasured (Hansen *et al.*, 1998b). A more meaningful constraint on climate sensitivity could be obtained from observations of ocean heat content, as discussed in Subsection F below, but full interpretation of changes in ocean heat content also requires that climate forcings be measured.

C. WATER VAPOR FEEDBACK

This feedback is related to climate sensitivity, but it is so fundamental that it deserves specific attention. Lindzen has argued that with global warming tropospheric water vapor will decrease at altitudes above 2–3 km (Lindzen, 1990). This contrasts sharply with our expectation based on global climate modeling that water vapor will increase through most of the troposphere with global warming (Hansen *et al.*, 1984).

Water vapor feedback has resisted definitive empirical assessment, because water vapor is not accurately measured and tropospheric temperature change in the past 20 years has been small. Ozone depletion, which cools the upper troposphere, complicates empirical assessment, because it tends to counteract upper tropospheric warming due to increasing carbon dioxide (Hansen *et al.*, 1997c). But ozone depletion is expected to flatten out, while the well-mixed greenhouse gases continue to increase. Thus it should be possible to verify this feedback empirically, if upper tropospheric water vapor is accurately monitored.

D. CO_2 Contribution to Natural Greenhouse

Lindzen (1992) has argued that "Even if all other greenhouse gases (such as carbon dioxide and methane) were to disappear, we would still be left with over 98% of the current greenhouse effect" (p. 88) and makes a similar statement elsewhere (Lindzen, 1993). We believe that this contention, also made in essence by other greenhouse skeptics, illustrates a lack of understanding of the basic greenhouse mechanism that in turn contributes to their expectation that climate should be stable. Although water vapor is the strongest greenhouse gas, the other greenhouse gases contribute a large portion of the present 33°C greenhouse effect on Earth.

Radiation calculations are straightforward, but they need to be made in the context of a climate model to be relevant. And because climate models are complex, results can be debated and obfuscated, which discourages any effort to invest time in addressing this somewhat academic issue per se. But the history of the Earth includes dramatic changes of both climate and atmospheric composition. Ongoing improvements in the knowledge of these changes will provide an opportunity to study the Earth's climate over a large range, and this will incidentally illuminate the contribution of CO_2 to the Earth's natural greenhouse effect.

E. When Will Climate Change Be Obvious?

Lindzen (1989) has said that he believes it unlikely that warming will reach magnitudes comparable to natural variability in the next century. On the contrary, we argue that global mean warming is already comparable to natural variability of global temperature and the warming should soon reach a level comparable to the natural variability of local seasonal mean temperature (Hansen *et al.*, 1988, 1998a). This topic is important because agreement on substantial efforts to curb global warming may require that climate change first be apparent to people.

We have examined practical measures of climate such as seasonal heating degree days, defining an index of change in units of the local standard deviation (Hansen *et al.*, 1998a). We find that in large parts of the world this index is at or near a level such that climate change should be noticeable to the perceptive layperson. If global warming continues as in our scenario B simulations, climate change should be more generally obvious in the next decade.

F. PLANETARY DISEQUILIBRIUM

The most fundamental measure of the state of the global greenhouse effect is the planetary "disequilibrium" (imbalance between incoming and outgoing radiation). Averaged over a few years, this imbalance is a simple measure of all climate forcings, measured and unmeasured. Specifically it is the integral over time of past forcings weighted by their exponential decay, with the decay constant being the ocean response time. But this imbalance is not a simple measure of the forcings, because the ocean response time, as discussed in Section IV.A, is not just a function of ocean mixing rates, but rather is a strong function of climate sensitivity. A planetary radiation imbalance must exist today, if climate sensitivity is as high (and thus the ocean response time as long) as we estimate and if increasing greenhouse gases are the dominant climate forcing.

Lindzen has not addressed specifically planetary radiation imbalance, as far as I know, but his positions regarding climate sensitivity and ocean response time would yield a negligible imbalance. We have inferred a planetary disequilibrium of at least approximately 0.5 W/m^2 based on climate simulations for 1979–1996 (Hansen *et al.*, 1997c). An imbalance of this magnitude has practical implications, implying that at least 0.4°C future global warming is still "in the pipeline."

It will be difficult to measure the radiation imbalance directly; we noted in Section V that the Earth's albedo presently is uncertain by at least 1% (3.4 W/m^2). But the imbalance can be deduced indirectly, because the only place the excess energy can go is into the ocean and into melting of ice. A global mean rate of even 0.1 W/m^2 used for melting ice would raise sea level by about 1 cm/year, well above observed rates. Thus most of the energy imbalance must raise the ocean temperature, which can be measured accurately.

White *et al.* (1998) find a substantial positive rate of heat storage between the sea surface and the top of the main pycnocline at latitudes

60°N–20°S for years 1955–1996. Our coupled atmosphere–ocean simulations (Plate 4 of Hansen *et al.*, 1997c) suggest that heat storage at higher latitudes may be large and that storage beneath the top of the main pycnocline is significant. Although temperature changes beneath the ocean mixed layer are small, the mass of water is so great that heat storage at depth can be important. Temperature measurements are needed globally for the full ocean depth.

The aim should be to measure the heat content with an accuracy sufficient to determine the rate of energy storage over a period as short as a year. Climate fluctuations such as El Niños cause a variability in the heat storage rate, but would not prevent use of it to infer information on climate forcings and the long-term energy imbalance. The rate of heat storage for the entire ocean would provide a crucial measure of the state of the planet, a measure that, in our opinion, is more fundamental than the mean global temperature.

VII. A CAUTIONARY CONCLUSION

Nostalgia can cloud perceptions, yet it is clear that the scientific approach of Arakawa and Charney, toward building of models and their application to climate problems, is a paragon for researchers. The essence of that approach is a focus on the relevant climate physics and design of models to represent that physics. A close corollary is use of the models to define needed observations, with continual iterations between data and models.

Technological advances in computing capabilities are opening the potential to advance our modeling capabilities and understanding of climate change. But achievement of that potential requires continued emphasis on the climate physics, on brainpower over megaflops. This may seem obvious, and any commentary perceived as criticism will be met with the response that the focus is on climate physics.

Yet it is difficult to witness current discussions of national climate research plans without concern. The most common measure of modeling prowess seems to be model resolution, or what is worse, the number of simulations that are added to the set of IPCC simulations for the 21st century. It is useful to have a number of such simulations, and we have argued for using and emphasizing a broad range of scenarios, yet with current uncertainties in the models and in the climate forcings driving the models, the law of diminishing returns with additional projections is reached quickly.

We are all pursuing the goal of understanding the climate system so that people and policy makers have information to help make the best decisions. The issue is how to get there. Moving in the direction of a centralized top-down approach is deleterious, in my opinion, because it opens too much of a danger of specification of what to compute and how to do it. That may be good for converging on a single answer, which might even be a goal of some people, but it is hardly in the interests of the best science and thus the long-term interests of the public.

These concerns should not mask an underlying optimism about the prospects for improved understanding of long-term climate change. The spectacular technical improvements in computing, data handling, and communication capability are ideal for increasing scientific cooperation and communication. At the same time there are improving capabilities for global observations that promise to make the modeling and scientific collaborations more productive.

Two topics of this chapter illustrate the potential for improved understanding of climate change: the cooling in the United States in the past 50 years and heat storage in the ocean. We found that models, notably of Arakawa's pedigree and with a relatively coarse resolution of 400–500 km, can simulate U.S. cooling. This provides the potential to investigate the mechanisms behind this regional climate trend, and in turn the possibility of anticipating future change. It should be straightforward to isolate the ocean regions driving the continental temperature change, but it may be more challenging to understand the causes of the ocean changes. A complete analysis will depend on having appropriate observations of climate forcings.

The rate of heat storage in the ocean is important for studies of regional climate change, and it is crucial for analysis of global climate change. An accurate current heat storage rate would provide an invaluable constraint on the net global climate forcing and climate sensitivity. Continued monitoring of heat storage, along with satellite monitoring of the major climate forcings, and preferably ice sheet and ocean topography, would serve as an integral measure of the state of the climate system and provide important data for analyzing mechanisms of long-term global climate change. Technology exists for the temperature measurements, but it must be deployed globally and measure the entire depth of the ocean.

ACKNOWLEDGMENTS

We thank Tica Novakov for providing absorption data for organic aerosols, Martin Wild for providing the GEBA data, David Randall for encouraging us to write this chapter, and Anthony Del Genio for critical review of the manuscript.

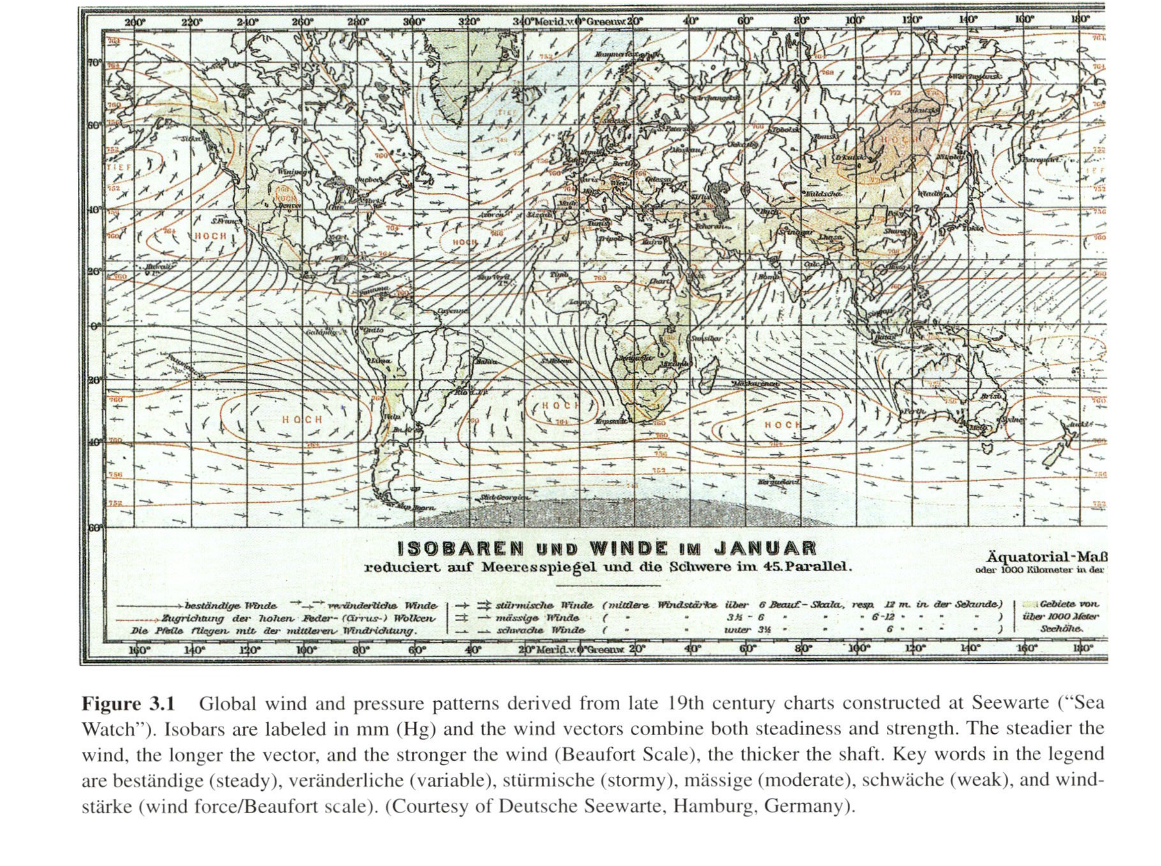

Figure 3.1 Global wind and pressure patterns derived from late 19th century charts constructed at Seewarte ("Sea Watch"). Isobars are labeled in mm (Hg) and the wind vectors combine both steadiness and strength. The steadier the wind, the longer the vector, and the stronger the wind (Beaufort Scale), the thicker the shaft. Key words in the legend are beständige (steady), veränderliche (variable), stürmische (stormy), mässige (moderate), schwäche (weak), and windstärke (wind force/Beaufort scale). (Courtesy of Deutsche Seewarte, Hamburg, Germany).

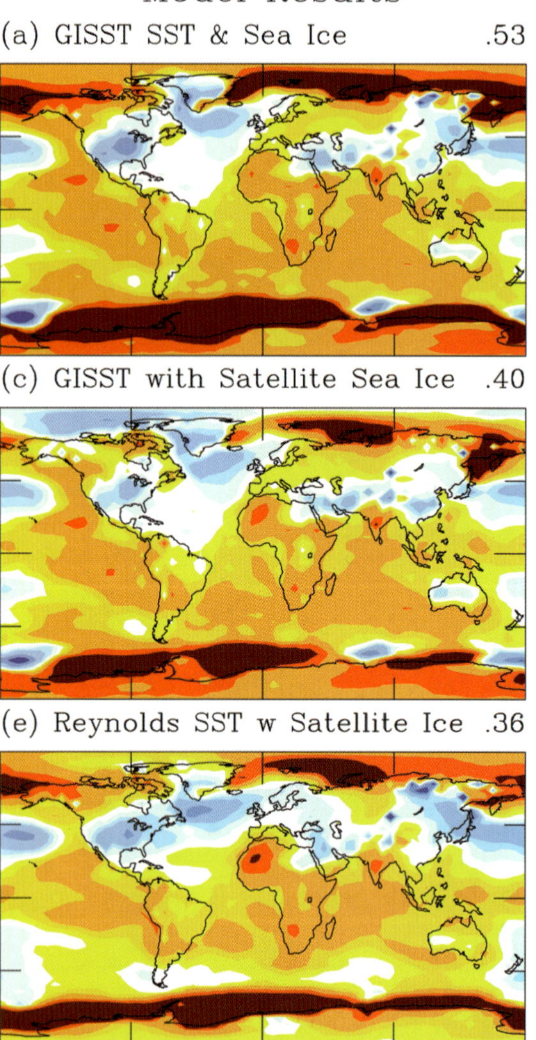

Figure 4.4 Part b: observed change of surface temperature index for 1951–1997 based on local linear trends using surface air temperature change over land and SST change over the ocean (Reynolds and Smith, 1994), with the latter measured for 1982–1998 and calculated based on ship measurements and an EOF analysis for 1950–1981 (Smith *et al.*, 1996). Other maps are simulated surface air temperature with the SST and sea ice specified in three ways as discussed in the text. In (d) and (f) the radiative forcings due to well-mixed greenhouse gases and stratospheric aerosols are included in addition to the surface forcings.

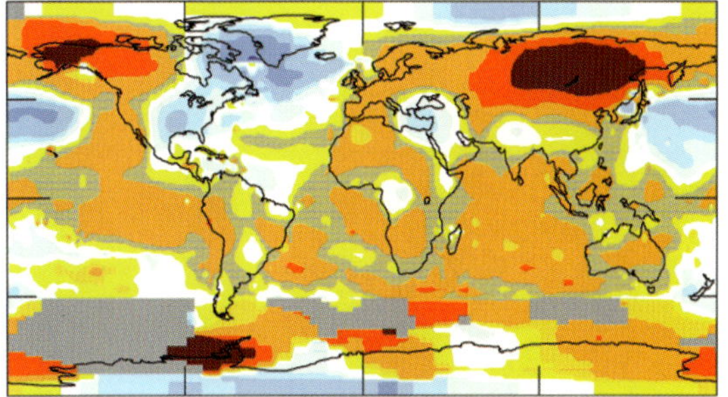

Model Results

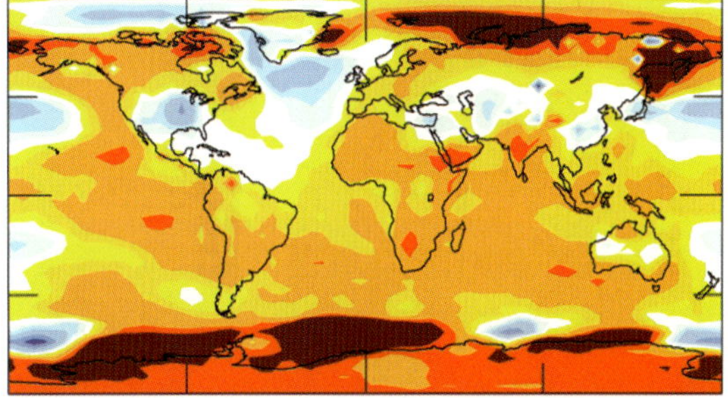

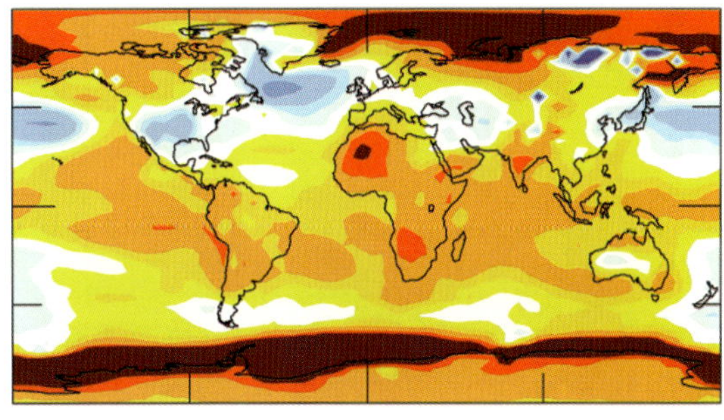

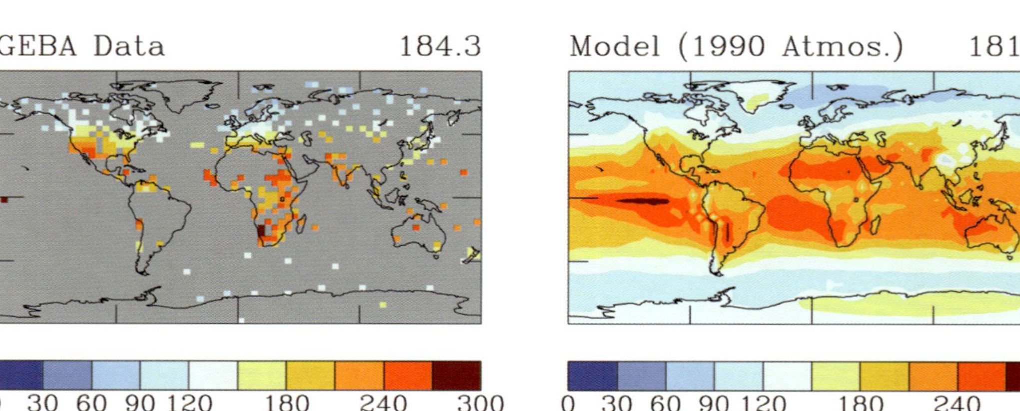

Figure 4.5 Comparison of solar radiation incident on planetary surface (W/m^2) in the SI99 model and in GEBA observations (Ohmura *et al.*, 1998). The four figures below are the difference between the model and observations, with the model containing aerosols estimated for 1950, 1990, the average of 1950 and 1990, and no aerosols, respectively. As discussed in the text, the case with 1950–1990 average aerosols is perhaps most relevant for comparison with the observations.

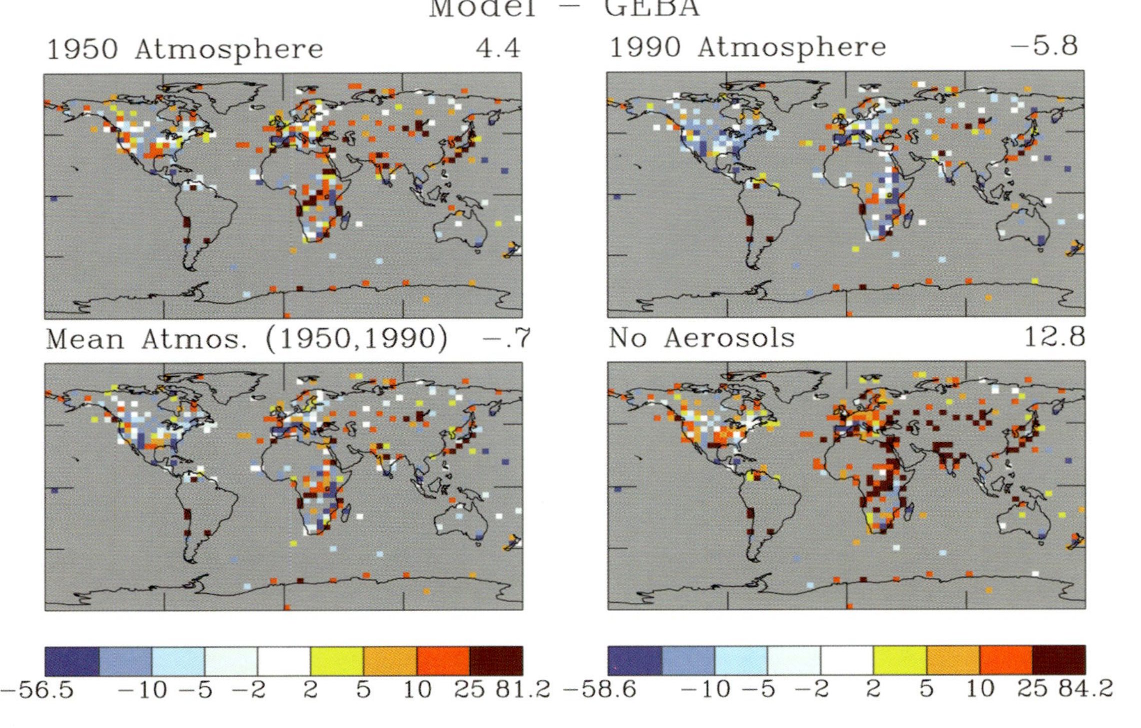

Model – GEBA
1950 Atmosphere   4.4
1990 Atmosphere   −5.8
Mean Atmos. (1950,1990)   −.7
No Aerosols   12.8
−56.5 −10 −5 −2 2 5 10 25 81.2
−58.6 −10 −5 −2 2 5 10 25 84.2

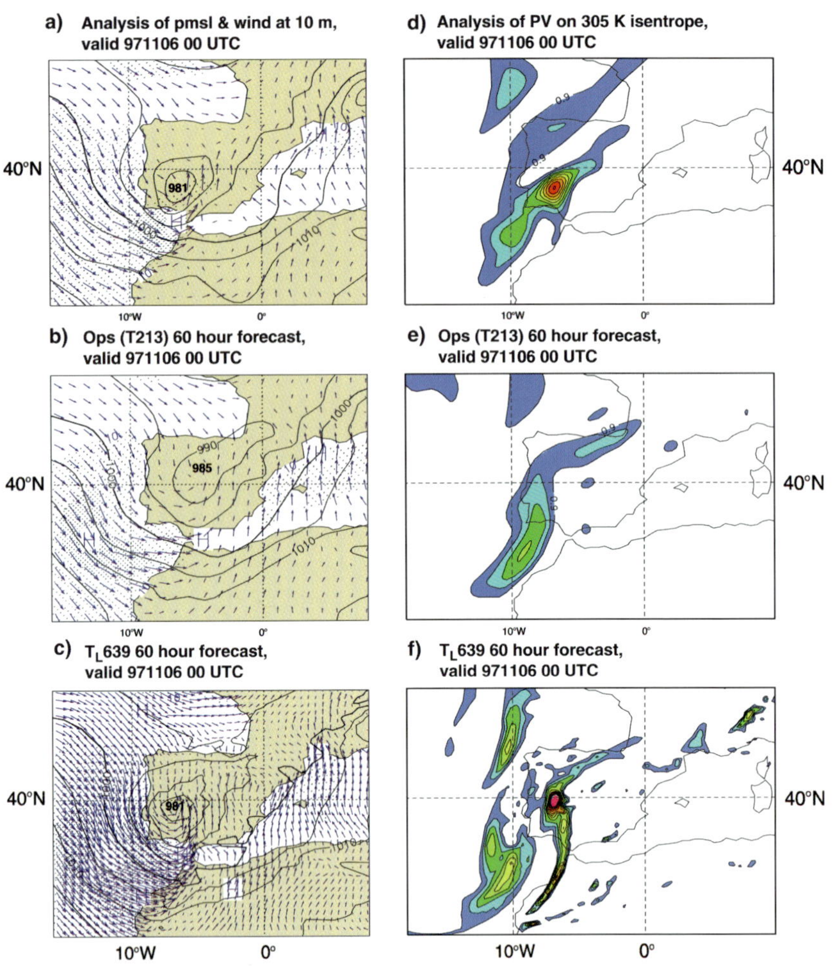

Figure 11.5 The figures on the left side show the analyzed 10-m wind and mean-sea-level pressure (Figure a) over the Iberian Peninsula at 0000UTC on November 6, 1997, together with two 60-h forecasts valid at the same time. Figure b shows the T213 60-h operational forecast while Figure c shows an experimental T$_L$639 60-h forecast. The figures on the right side show corresponding maps of the analyzed potential vorticity on the 305 K isentrope (Figure d) over the Iberian Peninsula at 0000UTC on November 6, 1997, together with two 60-h forecasts valid at the same time. Figure e shows the T213 60-h operational forecast while Figure f shows an experimental T$_L$639 60-h forecast.

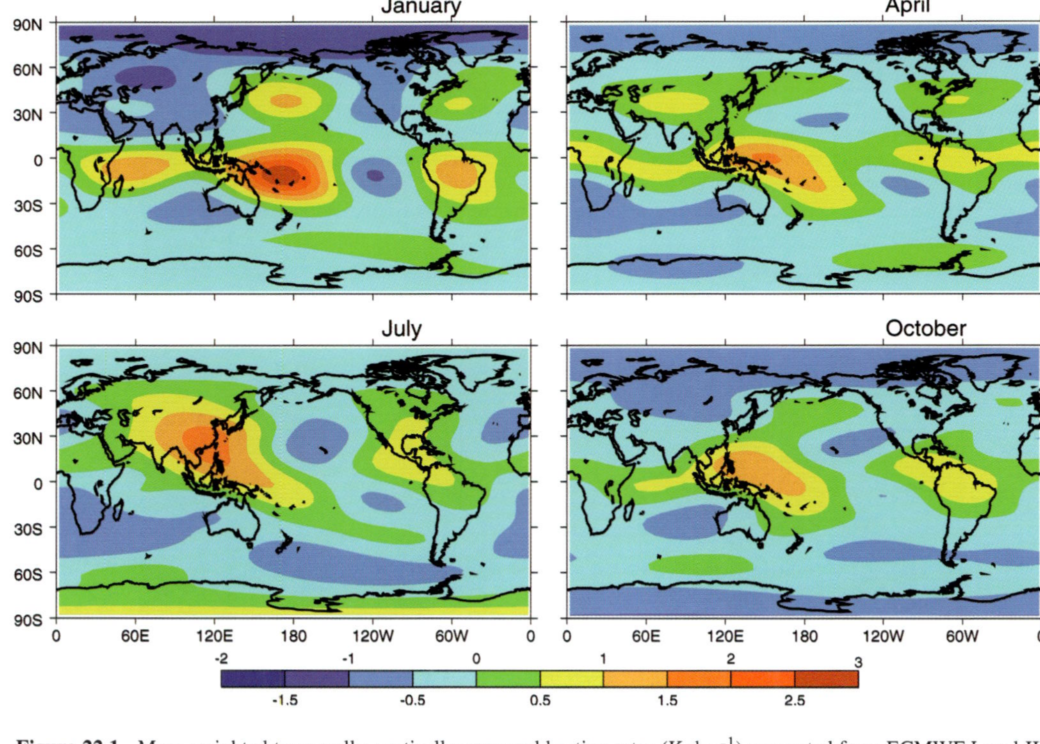

Figure 22.1 Mass-weighted temporally, vertically averaged heating rates (K day^{-1}) computed from ECMWF Level III analyses for January, April, July, and October 1979 (after Johnson, 1989). Heating distributions are filtered to emphasize heating from wavelengths greater than 10,000 km.

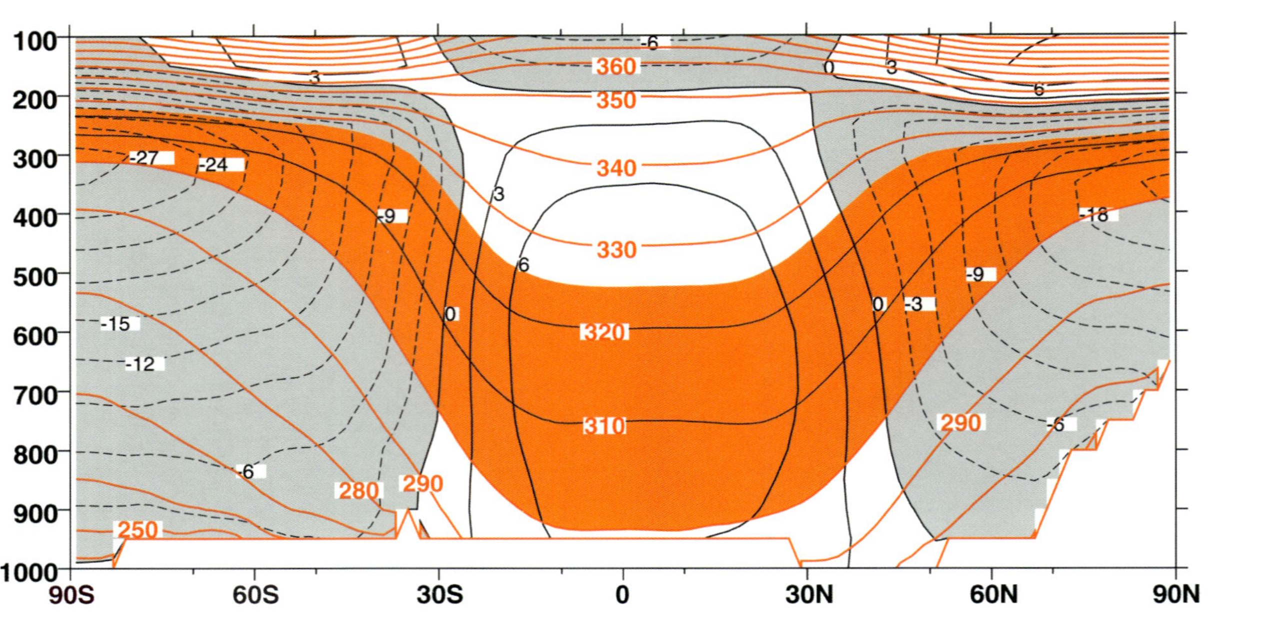

Figure 22.2 A vertical meridional distribution of mass weighted temporally, zonally averaged isentropic efficiency (units 10^{-2}) defined by $\hat{\varepsilon}$ equal to $< (1-T_\alpha/T) >$ [see Eq. (44)] and potential temperature $\theta[\phi, \overline{p}^{\lambda,t}(\phi,\theta)]$ (K) as determined from the isentropic temporally, zonally averaged pressure distribution, $\overline{p}^{\lambda,t}(\phi,\theta)$. Unshaded and shaded regions denote positive and negative efficiencies, respectively. The isentropic layers shaded red identify the atmospheric region within which the covariance of entropy sources and sinks with positive and negative efficiencies, respectively, are most effective in the generation for maintaining the atmosphere's circulation. The nonlinearity of this process is evident from the consideration that a cooling rate of 1 K day^{-1} in the high polar troposphere with $\varepsilon < 0.25$ is four times as effective in generation as a heating rate of 1 K day^{-1} in the low to mid tropical troposphere with $\varepsilon \approx 0.06$. Alternatively, the nondimensional efficiencies as indicated may be considered to be percentages of the heating/cooling that generates a reversible component. For example, in high polar troposphere in the northern hemisphere, the generation contribution εQ_m means that 27% of the cooling rate with $Q_m < 0$ and $\varepsilon < 0$ is a positive generation contribution to the reversible component equal to 27% of $|Q_m|$.

REFERENCES

AARST (American Association for the Rhetoric of Science and Technology), "Science Policy Forum," New York, Nov. 20, 1998, (G. R. Mitchell and T. M. O'Donnell, eds.), Univ. Pittsburgh.

Andreae, M. O. (1995). Climatic effects of changing atmospheric aerosol levels. *In* "World Survey of Climatology, Vol. 16: Future Climates of the World" (A. Henderson-Sellers, ed.), pp. 341–392. Elsevier, Amsterdam.

Arking, A. (1996). Absorption of solar energy in the atmosphere: Discrepancy between model and observations, *Science* **273**, 779–782.

Barkstrom, B., E. Harrison, G. Smith, R. Green, J. Kibler, and R. Cess (1989). Earth radiation budget experiment (ERBE) archival and April 1985 results. *Bull. Am. Meteor. Soc.* **70**, 1254–1262.

Boissoles, J., R. H. Tipping, and C. Boulet (1994). Theoretical study of the collision-induced fundamental absorption spectra of N_2–N_2 pairs for temperatures between 77 and 297K. *J. Quant. Spectrosc. Radiat. Transfer* **51**, 615–627.

Boyle, J. S. (1998). Evaluation of the annual cycle of precipitation over the United States in GCMs: AMIP simulations, *J. Climate* **11**, 1041–1055.

Bryan, K., and M.D. Cox (1972). An approximate equation of state for numerical model of ocean circulation. *J. Phys. Oceanogr.* **15**, 1255–1273.

CDAC (1983). "Changing Climate, Report of the Carbon Dioxide Assessment Committee." National Academy Press, Washington, DC.

Cess, R. D., M. Zhang, F. P. J. Valero, S. K. Pope, A. Bucholtz, B. Bush, C. S. Zender, and J. Vitko (1999). Absorption of solar radiation by the cloudy atmosphere: Further interpretations of collocated aircraft observations. *J. Geophys. Res.* **104**, 2059–2066.

Charney, J. (1979). "Carbon Dioxide and Climate: A Scientific Assessment." National Academy Press, Washington, DC.

CLIMAP Project Members (1981). Seasonal reconstruction of the Earth's surface at the last glacial maximum, *Geolog. Soc. Am.* Mmap and chart series, **MC-36**.

Cox, M. D. (1984). A primitive equation three-dimensional model of the ocean, GFDL Ocean Group Tech. Rep. 1. Geophys. Fuid Dyn. Lab., Princeton, NJ.

Del Genio, A. D., and M. S. Yao (1993). Efficient cumulus parameterization of long-term climate studies: The GISS scheme. *Am. Meteor. Soc. Monogr.* **46**, 181–184.

Del Genio, A. D., and A. Wolf (2000). Climatic implications of the observed temperature dependence of the liquid water path of low clouds in the Southern Great Plains. *J. Climate*, in press.

Del Genio, A. D., M. S. Yao, W. Kovari, and K. K. W. Lo (1996). A prognostic cloud water parameterization for global climate models, *J. Climate* **9**, 270–304.

Garratt, J. R., A. J. Prata, L. D. Rotstayn, B. J. McAvaney, and S. Cusack (1998). The surface radiation budget over oceans and continents, *J. Climate* **11**, 1951–1968.

Gordon, H. R., and M. Wang (1994). Influence of oceanic whitecaps on atmospheric correction of SeaWIFS, *Appl. Opt.* **33**, 7754–7763.

Gordon, H. R., O. B. Brown, R. H. Evans, J. W. Brown, R. C. Smith, K. S. Baker, and D. K. Clark (1988). A semi-analytic radiance model of ocean color. *J. Geophys. Res.* **93**, 10,909–10,924.

Guilderson, T. P., R. G. Fairbanks, and J. L. Rubenstone (1994). Tropical temperature variations since 20,000 years ago: Modulating interhemispheric climate change. *Science* **263**, 663–665.

Hansen, J. E., W. C. Wang, and A. A. Lacis (1978). Mount Agung eruption provides test of a global climate perturbation. *Science* **199**, 1065–1068.

Hansen, J., D. Johnson, A. Lacis, S. Lebedeff, P. Lee, D. Rind, and G. Russell (1981). Climatic impact of increasing atmospheric carbon dioxide. *Science* **213**, 957–966.

Hansen, J., G. Russell, D. Rind, P. Stone, A. Lacis, S. Lebedeff, R. Ruedy, and L. Travis (1983). Efficient three-dimensional global models for climate studies: Models I and II, *Mon. Wea. Rev.* **111**, 609–662.

Hansen, J., A. Lacis, D. Rind, G. Russell, P. Stone, I. Fung, R. Ruedy, and J. Lerner (1984). Climate sensitivity: Analysis of feedback mechanisms. *Geophys. Mono.* **29**, 130–163.

Hansen, J., I. Fung, A. Lacis, D. Rind, S. Lebedeff, R. Ruedy, G. Russell, and P. Stone (1988). Global climate changes as forecast by the Goddard Institute for Space Studies three-dimensional model. *J. Geophys. Res.* **93**, 9341–9364.

Hansen, J., A. Lacis, R. Ruedy, M. Sato, and H. Wilson (1993). How sensitive is the world's climate? *Natl. Geogr. Res. Explor.* **9**, 142–158.

Hansen, J. R. Ruedy, A. Lacis, G. Russell, M. Sato, J. Lerner, D. Rind, and P. Stone (1997a). Wonderland climate model. *J. Geophys. Res.* **102**, 6823–6830.

Hansen, J., M. Sato, and R. Ruedy (1997b). Radiative forcing and climate response. *J. Geophys. Res.* **102**, 6831–6864.

Hansen, J., and 42 Others (1997c). Forcings and chaos in interannual to decadal climate change. *J. Geophys. Res.* **102**, 25,679–25,720.

Hansen, J., M. Sato, J. Glascoe, and R. Ruedy (1998a). A common-sense climate index: Is climate changing noticeably? *Proc. Natl. Acad. Sci.* **95**, 4113–4120.

Hansen, J., M. Sato, A. Lacis, R. Ruedy, I. Tegen, and E. Matthews (1998b). Climate forcings in the industrial era. *Proc. Natl. Acad. Sci.* **95**, 12753–12758.

Hansen, J., R. Ruedy, J. Glascoe, and M. Sato (1999). GISS analysis of surface temperature change. *J. Geophys. Res.* **104**, 30997–31022.

Harris, R. N., and D. S. Chapman (1997). Borehole temperatures and a baseline for 20th-century global warming estimates. *Science* **275**, 1618–1621.

Hartke, G. J., and D. Rind (1997). Improved surface and boundary layer models for the GISS general circulation model. *J. Geophys. Res.* **102**, 16,407–16,442.

Hoffert, M. I., and C. Covey (1992). Deriving global climate sensitivity from paleoclimate reconstructions. *Nature* **360**, 573–576.

Intergovernmental Panel on Climate Change (1996). "Climate Change 1995" (J. T. Houghton, L. G. Meira, Filho, B. A. Callandar, N. Harris, A. Kattenberg, and K. Maskell, eds.). Cambridge Univ. Press, Cambridge, UK.

Kerr, R. A. (1995). Darker clouds promise brighter future for climate models. *Science* **267**, 454.

Koch, D., D. Jacob, I. Tegen, D. Rind, and M. Chin (1999). Tropospheric sulfur simulation and sulfate direct radiative forcing in the GISS GCM. *J. Geophys. Res.* **104**, 23799–23822.

Lacis, A. A., and J. E. Hansen (1974). A parameterization for the absorption of solar radiation in the Earth's. *J. Atmos. Sci.* **31**, 118–133.

Lacis, A. A., and V. Oinas (1991). A description of the correlated k distribution method for modeling nongray gaseous absorption, thermal emission, and multiple scattering in vertically inhomogeneous atmospheres. *J. Geophys. Res.* **96**, 9027–9063.

Lindzen, R. (1989). EAPS' Lindzen is critical of global warming prediction. *MIT Tech Talk*, **34**, No. 7, 1–6.

Lindzen, R. S. (1990). Some coolness concerning global warming. *Bull. Am. Meteorol. Soc.* **71**, 288–299.

Lindzen, R. S. (1992). Global warming: The origin and nature of the alleged scientific consensus. *Cato Rev. Bus. Govt.* **2**, 87–98.

Lindzen, R. (1993). Absence of scientific basis. *Nat. Geog. Res. Explor.* **9**, 191–200.

Lindzen, R. S. (1997). Can increasing carbon dioxide cause climate change? *Proc. Natl. Acad. Sci.* **94,** 8335–8342.

Lorius, C., J. Jouzel, D. Raynaud, J. Hansen, and H. Le Treut (1990). The ice-core record: Climate sensitivity and future greenhouse warming. *Nature* **347,** 139–145.

Manabe, S., and F. Moller (1961). On the radiative equilibrium and heat balance of the atmosphere. *Mon. Wea. Rev.* **89,** 503–532.

Manabe, S., and R. J. Stouffer (1980). Sensitivity of a global climate model to an increase of CO_2 concentration in the atmosphere. *J. Geophys. Res.* **85,** 5529–5554.

Manabe, S., and R. F. Strickler (1964). Thermal equilibrium of the atmosphere with a convective adjustment. *J. Atmos. Sci.* **21,** 361–385.

Manabe, S., and R. T. Wetherald (1975). The effects of doubling the CO_2 concentration on the climate of a general circulation model. *J. Atmos. Sci.* **32,** 3–15.

Matthews, E. (1983). Global vegetation and land use: New high resolution data bases for climate studies. *J. Clim. Appl. Meteor.* **22,** 474–487.

Merilees, P. E. (1975). The effect of grid resolution on the instability of a simple baroclinic model. *Mon. Wea. Rev.* **103,** 101–104.

Nakajima, T., A. Higurashi, N. Takeuchi, and J. R. Herman (1999). Satellite and ground-based study of optical properties of 1997 Indonesian forest fire aerosols. *Geophys. Res. Lett.* **26,** 2421–2424.

Newnham, D. A., and J. Ballard (1998). Visible absorption cross sections and integrated absorption intensities of molecular oxygen O_2 and O_4). *J. Geophys. Res.* **103,** 28,801–28,816.

Oerlemans, J. (1994). Quantifying global warming from the retreat of glaciers. *Science* **264,** 243–245.

Ohmura, A., and 13 coauthors (1998). Baseline Surface Radiation Network (BSRN/WCRP), a new precision radiometry for climate research. *Bull. Am. Meteor. Soc.* **79,** 2115–2136.

Paltridge, G. W., and C. M. R. Platt (1976). "Radiative Processes in Meteorology and Climatology." Elsevier, New York.

Pollack, H. N., H. Shaopeng, and P. Y. Shen (1998). Climate change record in subsurface temperatures: A global perspective. *Science* **282,** 279–281.

Prather, M. J. (1986). Numerical advection by conservation of second-order moments. *J. Geophys. Res.* **91,** 6671–6680.

Quinn, P. K., and D. J. Coffman (1999). Comment on "Contribution of different aerosol species to the global aerosol extinction optical thickness: Estimates from model results" by Tegen *et al. J. Geophys. Res.* **104,** 4241–4248.

Ramanathan, V., R. J. Ciccrone, H. J. Singh, and J. T. Kiehl (1985). Trace gas trends and their potential role in climate change. *J. Geophys. Res.* **90,** 5547–5566.

Ramaswamy, V., and S. M. Freidenreich (1992). A study of broadband parameterization of the solar radiative interactions with water vapor and water drops. *J. Geophys. Res.* **97,** 11,487–11,512.

Reynolds, R. W., and T. M. Smith (1994). Improved global sea surface temperature analyses. *J. Clim.* **7,** 929–948.

Rind, D., R. Suozzo, N. K. Balachandran, A. Lacis, and G. L. Russell (1988). The GISS global climate/middle atmosphere model, I. Model structure and climatology. *J. Atmos. Sci.* **45,** 329–370.

Rosenzweig, C., and F. Abramopoulos (1997). Land surface model development for the GISS GCM. *J. Climate* **10,** 2040–2054.

Russell, G. L., and J. A. Lerner (1981). A new finite-differencing scheme for the tracer transport equation. *J. Appl. Meteorol.* **20,** 1483–1498.

Russell, G. L., J. R. Miller, and L. C. Tsang (1985). Seasonal oceanic heat transports computed from an atmospheric model. *Dynam. Atmos. Oceans* **9,** 253–271.

Russell, G. L., J. R. Miller, and D. Rind (1995). A coupled atmosphere-ocean model for transient climate change studies. *Atmos. Oceans* **33,** 683–730.

Russell, G. L., J. R. Miller, D. Rind, R. A. Ruedy, G. A. Schmidt, and S. Sheth (2000). Comparison of model and observed regional temperature changes during the past 40 years. *J. Geophys. Res.,* in press.

Sato, M., J. E. Hansen, M. P. McCormick, and J. B. Pollack (1993). Stratospheric aerosol optical depth, 1850–1990. *J. Geophys. Res.* **98,** 22,987–22,994.

Schlumberger Research, Climate Change and the Oil Industry, A Debate, Ridgefield, CT, Oct. 15, 1998, available at www.slb.com/research/sdr50.

Schrag, D. P., G. H. Hampt, and D. W. Murray (1996). Pore fluid constraints on the temperature and oxygen isotopic composition of the glacial ocean. *Science* **272,** 1930–1932.

Seinfeld, J. H. (1996). "Aerosol Radiative Forcing of Climate Change." National Research Council, National Academy Press, Washington, DC.

Shindell, D. T., R. L. Miller, G. A. Schmidt, and L. Pandolfo (1999a). Greenhouse gas forcing of Northern Hemisphere winter climate trends. *Nature* **399,** 452–455.

Shindell, D. T., D. Rind, N. Balachandran, J. Lean, and P. Lonergan (1999b). Solar cycle variability, ozone and climate. *Science* **284,** 305–308.

Smith, T. M., R. W. Reynolds, R. E. Livesay, and D. C. Stokes (1996). Reconstruction of historical sea surface temperature using empirical orthogonal functions. *J. Clim.* **9,** 1403–1420.

Somerville, R. C. J., P. H. Stone, M. Halem, J. Hansen, J. S. Hogan, L. M. Druyan, G. Russell, A. A. Quirk, and J. Tenenbaum (1974). The GISS model of the global atmosphere. *J. Atmos. Sci.* **31,** 84–117.

Tegen, I., P. Hollrig, M. Chin, I. Fung, D. Jacob, and J. Penner (1997). Contribution of different aerosol species to the global aerosol extinction optical thickness: Estimates from model results. *J. Geophys. Res.* **102,** 23,895–23,915.

Thompson, D. W. J., and J. M. Wallace (1998). The Arctic oscillation signature in the wintertime geopotential height and temperature fields. *Geophys. Res. Lett.* **25,** 1297–1300.

Tselioudis, G., and W. B. Rossow (1994). Global, multiyear variations of optical thickness with temperature in low and cirrus clouds. *Geophys. Res. Lett.* **21,** 2211–2214.

Wang, W. C., Y. L. Yung, A. A. Lacis, T. Mo, and J. E. Hansen (1976). Greenhouse effects due to man-made perturbations of trace gases. *Science* **194,** 685–690.

White, W. B., D. R. Cayan, and J. Lean (1998). Global upper ocean heat storage response to radiative forcing from changing solar irradiance and increasing greenhouse gas/aerosol concentrations. *J. Geophys. Res.* **103,** 21,355–21,366.

Wild, M., A. Ohmura, H. Gilgen, E. Roeckner, M. Giorgetta, and J. J. Morcrette (1998). The disposition of radiative energy in the global climate system: GCM-calculated versus observational estimates. *Clim. Dynam.* **14,** 853–869.

Chapter 5

A Retrospective Analysis of the Pioneering Data Assimilation Experiments with the Mintz –Arakawa General Circulation Model

Milton Halem
NASA Goddard
Space Flight Center
Greenbelt, Maryland

Jules Kouatchou
School of Engineering
Morgan State University
Baltimore, Maryland

Andrea Hudson
NASA Goddard
Space Flight Center
Greenbelt, Maryland

I. INTRODUCTION

We have performed a retrospective analysis of a simulation study, published about 30 years ago, which had a profound impact on satellite meteorology. The paper had the strange title "Use of incomplete historical data to infer the present state of the atmosphere." It was authored by J. Charney, M. Halem, and R. Jastrow, and appeared in the *Journal of the Atmospheric Sciences*, in September 1969 (Charney *et al.* 1969). We decided that the numerical experiments which formed the basis of that paper should be repeated using a contemporary model, particularly in view of their relevance to upcoming satellite missions.

Secondly, by the end of 2000, NASA plans to launch the EOS PM platform, which will carry a new generation of temperature sounders, the Atmospheric Infra-Red Sounder (AIRS) and the Advanced Microwave Sounding Unit (AMSU). These sounders will have substantially increased spectral and spatial resolutions and are expected to produce an increase in accuracy over that of today, perhaps attaining 1 K accuracies throughout the column in clear and cloudy regions. AIRS will also provide greatly improved vertical humidity profiles, which really are not feasible with today's instruments. These expectations are reminiscent of the situation in July 1969, just after the launch of NIMBUS 3, which carried the first of a new class of remote sensors, namely, the Space Infra-Red Sounder (SIRS-A), which could acquire global vertical temperature profiles, with a potential accuracy of 1 K in clear tropical regions. Shortly thereafter, Dr. Morris Tepper, NASA program manager, visited Goddard Institute for Space Studies (GISS) to meet with Charney, Jastrow, and Halem to ask what impact such data could have in numerical weather prediction. It was then that Charney proposed that we conduct an experiment to assimilate complete temperature fields synoptically into a GCM, in order to infer the geostrophic winds. He called Mintz and Arakawa to ask them to lend GISS their model to perform such experiments, and they agreed to do so.

Those experiments produced some very interesting results that initially raised some skepticism in the community. Most modelers had expected that the insertion of "foreign" temperature fields without balancing would generate spurious disturbances in the model. Another conclusion which generated considerable discussion was that a knowledge of the temperature fields alone could lead to adjustments of the wind and pressure fields even in the tropics, where the geostrophic approximation is not accurate. The retrospective analysis reported here investigates the model dependencies of those results. At that time, the Mintz–Arakawa model had a very coarse spatial resolution by present standards, $7° \times 9°$ by two levels, and very crude physical parameterizations compared with today's models. Clearly, the simulation experiment of Charney *et al.* (1969) ignored the operational world weather observing system with hundreds of upper air radiosondes and thousands of ground surfaces observing systems and focused mainly on a conjecture that Charney (1969) had earlier presented at the 1968 International Numerical Weather Prediction Conference in Tokyo, Japan. The Charney conjecture was based on a simplified linear hydrodynamical model. In Chapter 6 of this volume, Schubert shows that the relevant system of first-order equations in several variables can be reduced to a single equation of higher order in a single unknown with a forcing term expressed in terms of higher order temporal and spatial derivatives. Initial conditions of state variables are replaced with higher order temporal derivatives of the single unknown variable. Such a linear higher order differential equation can be solved by the method of Green's functions, but Charney conjectured that the GCM would produce such a solution "automatically" if provided with the temperature history over a sufficiently long integration period. Although this conjecture was not at all obvious at the time, it is generally accepted today.

Ghil *et al.* (1977, 1979) analytically proved the Charney conjecture for certain simple atmospheric models. These results were extended by Ghil (1980). In practice, numerous problems with real data and with complexities of current atmospheric models render Ghil *et al.*'s theory not strictly applicable. However, the power of the process whereby continuous assimilation of temperature profiles can be used to infer complete global states or even just extratropical atmospheric states is still of considerable interest today. Thus, we set out to repeat the experiments of Charney *et al.* (1969) using a contemporary GCM.

II. DESCRIPTION OF EXPERIMENTS

In this retrospective study, we conduct a simulation experiment that is as nearly as possible identical to the original experiment of Charney *et al.* (1969), except that we employ the Goddard Earth Observing System (GEOS) GCM (Takacs *et al.*, 1994) in place of the Mintz–Arakawa GCM (Langlois and Kwok, 1969).

The satellite system configuration that the original Charney *et al.* (1969) experiments were designed to simulate consisted of one polar orbiting NIMBUS 3 satellite carrying infrared and microwave scanning sounders capable of providing temperature profiles throughout the atmosphere under clear and cloudy conditions, including the radiative surface temperatures. Based on today's NOAA operational satellite configuration, we assume for these experiments that two satellites can provide synoptic global coverage every 6 hr.

The original experiments consisted of generating a "history" record to represent the synoptic state of the atmosphere by conducting a long integration with a GCM. The Charney *et al.* (1969) experiment employed the Mintz–Arakawa GCM two-level model at 400 and 800 mb and $7° \times 9°$ grid spacing in latitude and longitude, respectively. A second integration was performed with the Mintz–Arakawa model starting with initial conditions from the "history" file at day 85 with a random perturbation error of 1 K added to the temperature field at all grid points. This integration was carried out to day 95 to produce an initial state that was considerably different from the history tape. Experiments all starting from this initial state of day 95 were then conducted assimilating the "history" temperature field with different random perturbation errors. The experiments tested a parametric range of assumed temperature accuracy and frequency of insertions.

Our current experiments used the GEOS GCM with 20 levels and $4° \times 5°$ grid spacing in latitude and longitude and much more detailed physical parameterizations (clouds, radiations, turbulence, surface processes, etc.); see Takacs *et al.* (1994). The "history" record was started from an atmospheric state provided by L. Takacs and integrated for 90 days. At day 30, a second integration was started for 60 days with a 1 K random perturbation introduced into the temperature field. The atmospheric state at day 60 was then used as the initial condition for two parametric temperature assimilation experiments. Synoptic temperature fields from the "history" record with random root mean square (rms) errors of 0, 1, and 2.5 K were assimilated into the GEOS GCM at different time intervals (every hour, 3, 6, and 12 hr) for 30 days. A fourth experiment assimilation was carried out with both the temperature field and the surface pressure field.

The following section compares the results of Charney *et al.* (1969) with those obtained by a contemporary model.

III. RESULTS OF GEOS SIMULATION EXPERIMENTS

As mentioned in the previous section, a "history" file was generated by carrying out the numerical integration of the GEOS GCM for 90 days. This file is treated throughout the remainder of the study as an exact measurement notwithstanding all of the limitations of the model. At day 30, a random perturbation or "error" of 1 K is introduced in the temperature fields at all grid points and all levels, and the flow is then recalculated from this initial state for 60 days. The resulting atmospheric state of the "perturbation" run will be compared with the "history" run to confirm that their respective fields are randomly correlated. Results are presented in terms of rms differences of the sea level pressure and 400-mb zonal winds. The results, summarized in Figs. 1 and 2, demonstrates that the sea level pressure and 400-mb wind differences between the perturbed integration and the unperturbed history files grow rapidly with time and then reach asymptotic error differences of 10 mb and 12 m s^{-1}, respectively. After 30 days, an examination of contour plotted differences shows that the sea

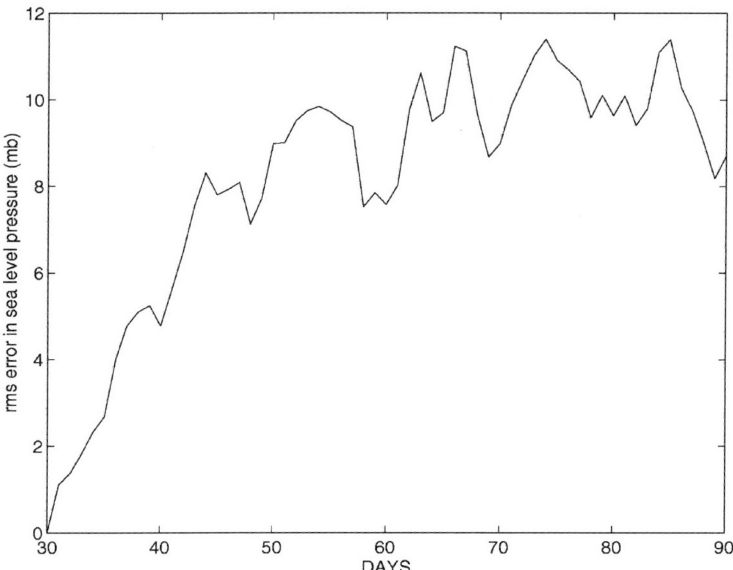

Figure 1 The rms differences in sea level pressure between the history and perturbed runs, from day 30 to day 90.

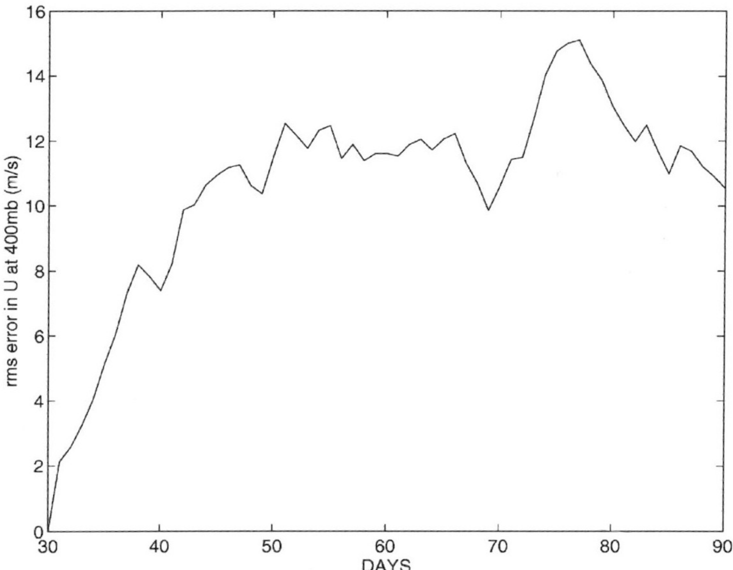

Figure 2 The rms error in the 400-mb zonal wind between history and perturbed files, from day 30 to day 90.

level pressure and the winds are meteorologically distinct and uncorre-
lated, with no remaining sign of their common parentage.

The next set of runs is designed to investigate the sensitivity of our
results to the frequency of data insertion. Charney *et al.* (1969) found that
a 12-hr insertion frequency was optimal, but we wanted to find out what
would be optimal for the GEOS GCM. The sensitivity experiments were
performed by starting from the perturbed file at day 60 and integrating the
GCM with exact temperatures inserted from the history file at specified
time intervals.

Figures 3 and 4 show the results of inferring the sea level pressure and
400-mb zonal wind fields by inserting data from the history temperature
file at intervals of 1, 3, 6, and 12 hr, respectively. It is seen that continuous
temperature insertions immediately arrest the growth in the sea level
pressure differences (Fig. 1), and reduce the differences to approximately 3
mb for insertions every 3 and 6 hr after 30 days. Insertions of temperature
fields every hour and every 12 hr produce asymptotic differences of 3.8 and
4.6 mb, respectively. The 400-mb zonal wind behaves similarly, reducing
the differences to about 3.5 m s^{-1} for insertions at intervals of 3 and 6 hr,
and to approximately 5.5 m s^{-1} for 1- and 12-hr insertion intervals. This is

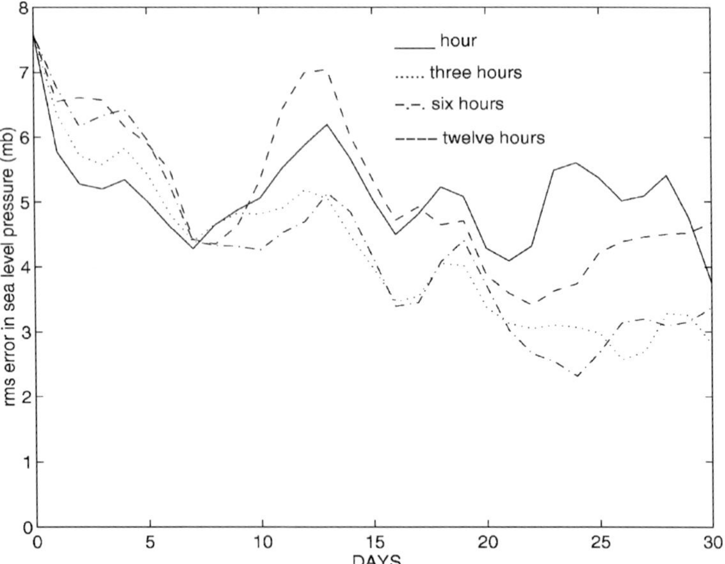

Figure 3 The rms error in sea level pressure in cases for which exact temperatures are
inserted every 1, 3, 6, and 12 hr at all grid points.

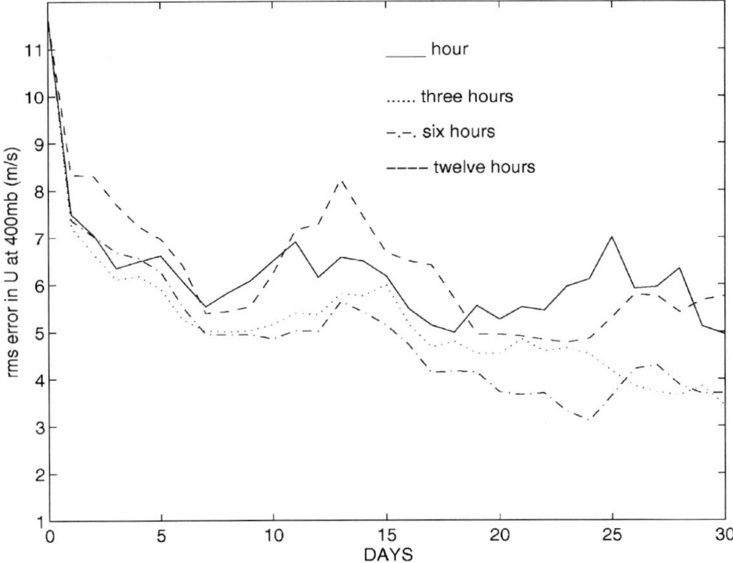

Figure 4   The rms error in 400-mb zonal wind (m s^{-1}), in cases for which exact temperatures are inserted every 1, 3, 6, and 12 hr at all grid points.

in contrast to the results without temperature corrections, given in Figs. 1 and 2, which show divergences from the history sea level pressure and 400-mb zonal wind, with amplitudes of 8 mb and 12 m s^{-1}, respectively, after 30 days.

The greatest reduction of rms error, i.e., the smallest rms error, was achieved when the "correct" temperatures were inserted every 3 or 6 hr. A more frequent insertion (every hour for instance) gives rise to oscillations in the wind field. The 6-hr interval was chosen for the experiment described below. This choice is consistent with an observing system consisting of two satellite overpasses a day. Operational weather forecasting systems today employ two satellites in this manner.

A second set of runs was performed in which temperatures were inserted at each grid point with random errors of 1 or 2.5 K, representing a range of observational errors, for comparison with exact temperature insertion. Figure 5 indicates that the insertion of temperatures with 1 K errors at 6-hr intervals reduces the global sea level pressure difference to approximately 3 mb, the same level as the insertion of exact temperatures. With temperature errors of 2.5 K, corresponding to the current estimated accuracies of today's operational sounders, the asymptotic differences are on the order of 4 to 5 mb. Figure 6 shows similar behavior with the global

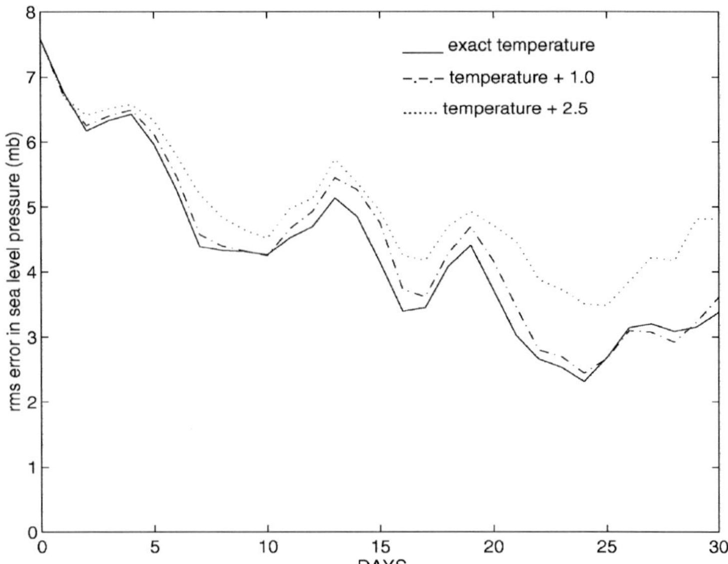

Figure 5   The rms error in sea level pressure, in cases for which temperatures with random error perturbations of 0, 1, and 2.5 K are inserted every 6 hr at all grid points.

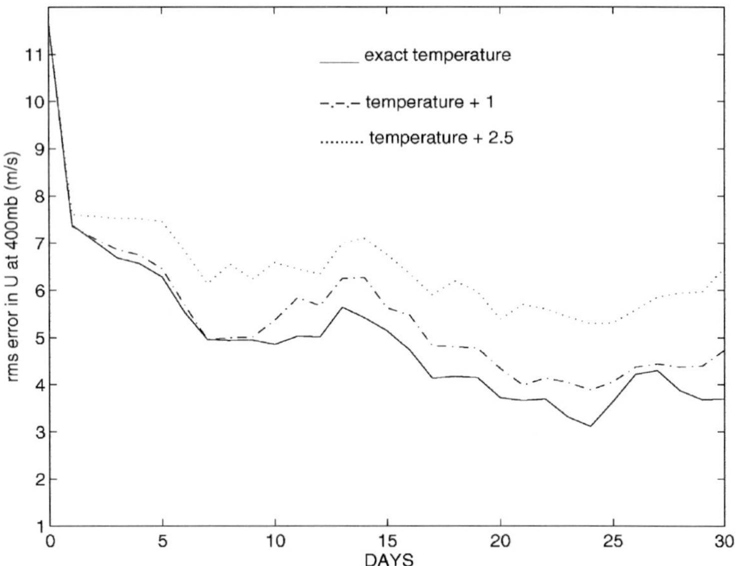

Figure 6   The rms error in 400-mb zonal wind ($m\ s^{-1}$), in cases for which temperatures with random error perturbations of 0, 1, and 2.5 K are inserted every 6 hr at all grid points.

wind adjustments, which reduce the wind errors to 4 and 6.5 m s^{-1}, respectively. Although this is a significant reduction of errors from the initial state, it falls somewhat short of the desired 3 m s^{-1} global wind errors.

We next wish to compare the results of the experiments described above with those derived earlier obtained by Charney *et al.* (1969). Figures 7 and 8, taken from Charney *et al.* (1969), show that the 400-mb extratropical and tropical zonal winds are reduced to below 1 m s^{-1} with 1 K temperature errors. These very favorable results, referred to earlier in the introduction, generated both skepticism and excitement over the prospective use of temperature sounders to infer the global wind fields.

Figure 9 shows that, for the GEOS GCM with 1 K sounding errors, the 400-mb wind differences at 48°N are reduced to about 4 m s^{-1}, while with 2.5 K temperature errors they are reduced to 6 m s^{-1}. These results are similar to those of Charney *et al.* (1969), but differ in the magnitude of the asymptotic errors. At the equator, shown in Fig. 10, the 1 K sounder errors lead to oscillatory wind adjustments ranging from 4 to 6 m s^{-1}, down from an uncorrected error of 7 m s^{-1}. Temperatures with 2.5 K errors also produce oscillations with magnitudes between 6 and 8 m s^{-1}, with a mean of 7 m s^{-1}, effectively showing no reduction relative to the uncorrected

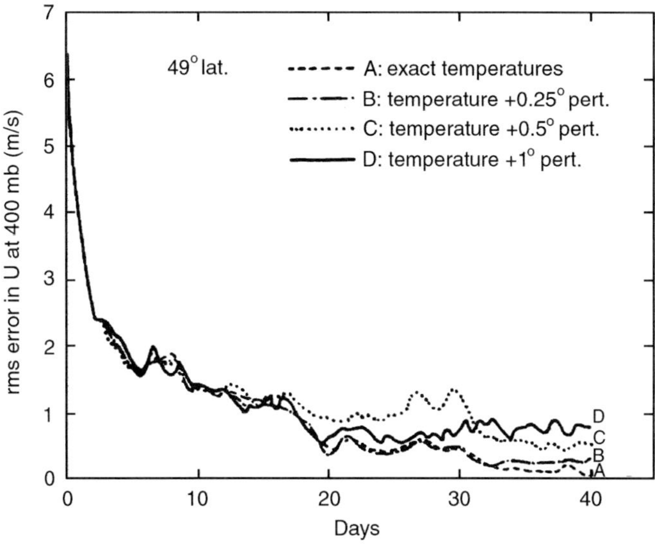

Figure 7 Charney *et al.* (1969) results with the two-level Mintz–Arakawa GCM: the rms error in 400-mb zonal wind (m s^{-1}) at 49°N, in cases for which temperatures with random error perturbations of 0, 0.25, 0.5, and 1 K are inserted every 12 hr at all grid points. (From Charney *et al.* (1969).)

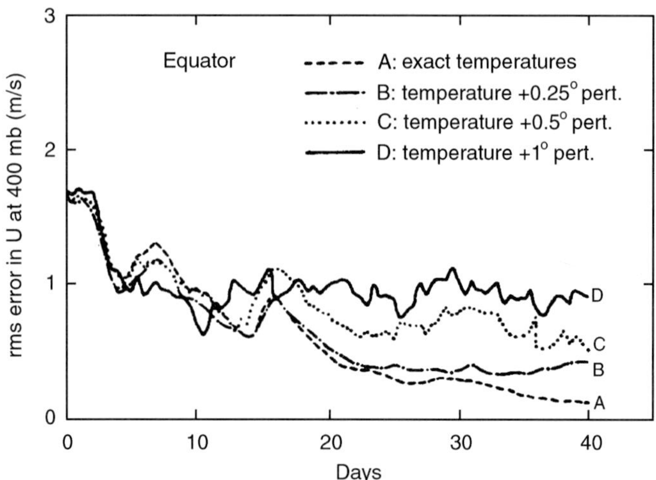

Figure 8   Charney *et al.* (1969) results with the two-level Mintz–Arakawa GCM: the rms error in the 400-mb zonal wind (m s^{-1}) at the equator, in cases for which temperatures with random error perturbations of 0, 0.25, 0.5, and 1 K are inserted every 12 hr at all grid points. (From Charney *et al.* (1969).)

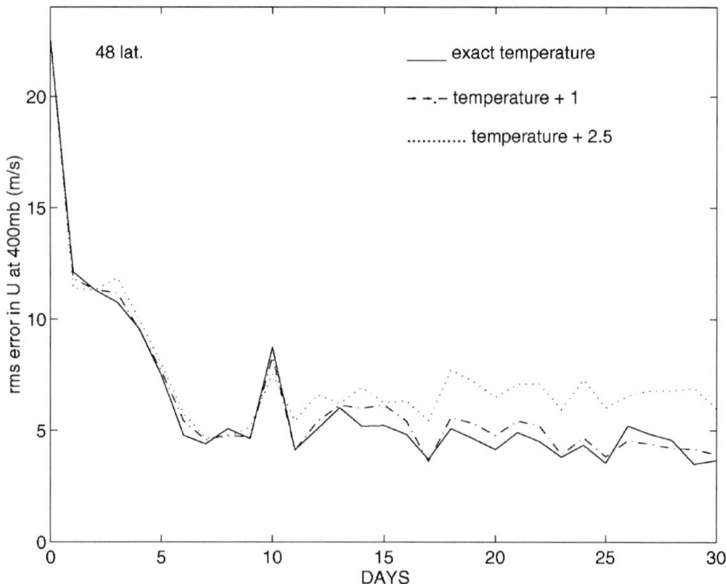

Figure 9   The rms error in 400-mb zonal wind (m s^{-1}) at 48°N, in cases for which temperatures with random error perturbations of 0, 1, and 2.5 K are inserted every 6 hr at all grid points.

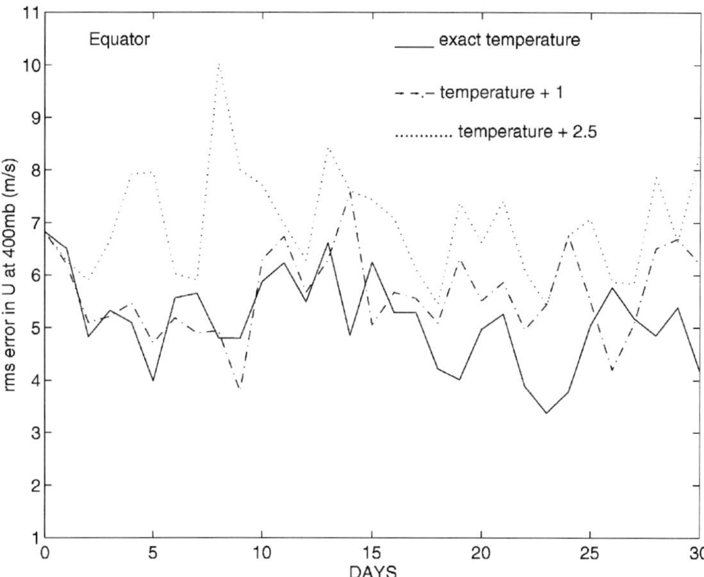

Figure 10 The rms error in 400-mb zonal wind (m s^{-1}) at the equator, in cases for which temperatures with random error perturbations of 0, 1, and 2.5 K are inserted every 6 hr at all grid points.

wind errors. This disagrees with the results of Charney *et al.*, which indicated that highly accurate tropical winds can be inferred from sounding data.

The last experiment was designed to explore whether combining surface pressure data together with temperature data helps in dynamical balancing, especially in the tropics. Figures 11, 12, and 13 compare the global zonal wind errors and meridional wind errors at 48°N and at the equator, for exact temperature insertions, with and without sea level pressure insertions. We see from Fig. 11 that the error reductions in the global winds are significantly greater when surface pressure fields are combined with temperature fields. A more noticeable reduction is achieved at 48°N (Fig. 12), in very close agreement with the results of Charney *et al.* (1969). However, even with exact observations of sea level pressure, there is very little improvement in the inferred equatorial zonal winds (Fig. 13).

IV. CONCLUSIONS

We have performed observing-system simulation studies whose basic objective is the determination of the relationship between the temperature

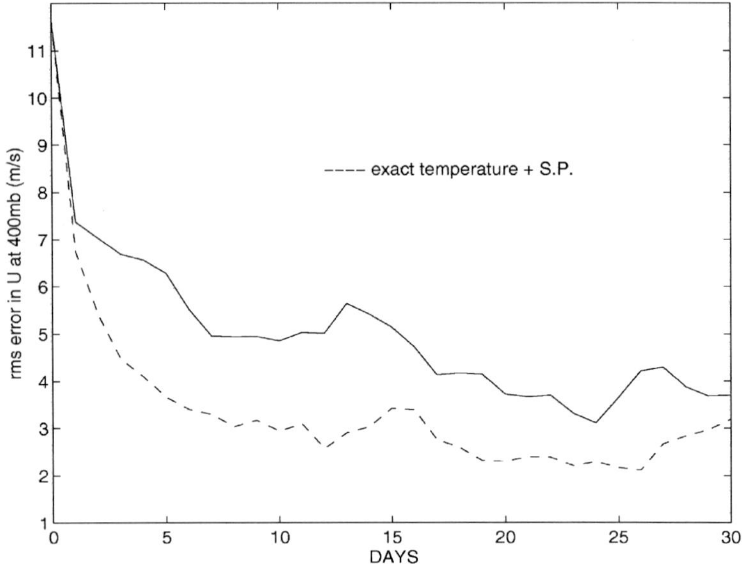

Figure 11   The rms error in 400-mb zonal wind (m s^{-1}), in cases for which exact temperatures are inserted with and without surface pressure every 6 hr at all grid points.

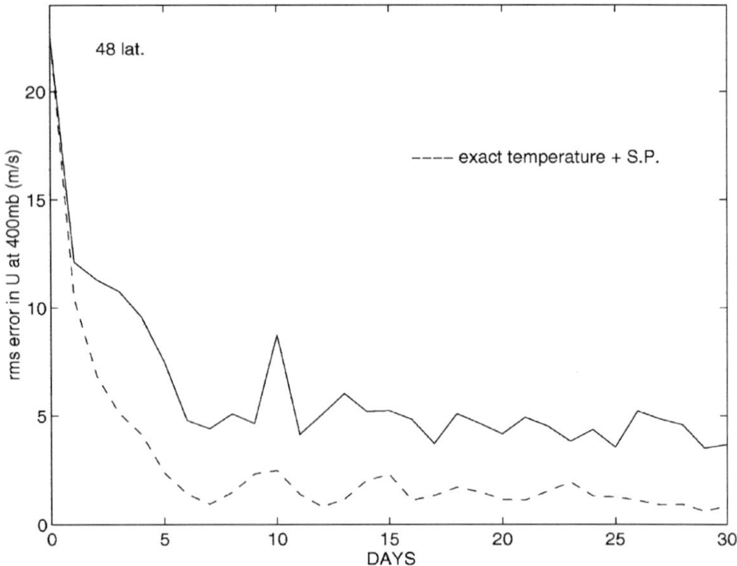

Figure 12   The rms error in 400-mb zonal wind (m s^{-1}) at 48°N, in cases for which exact temperatures are inserted with and without surface pressure every 6 hr at all grid points.

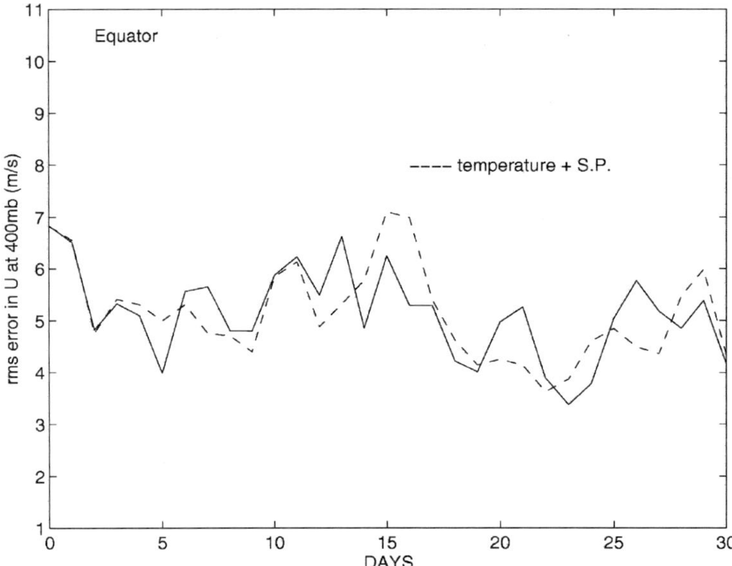

Figure 13   The rms error in 400-mb zonal wind (m s^{-1}) at the equator, in cases for which exact temperatures are inserted with and without surface pressure every 6 hr at all grid points.

errors and the inferred global winds and pressures, for realistic configurations of a proposed earth observing system with advanced vertical temperature sounders. Numerical results obtained with the GEOS GCM indicate that if a continuing day-by-day sequence or history of temperature profiles is inserted into the numerical integrations at appropriate time intervals, wind components and sea level pressures can be determined to a useful degree of accuracy. More precisely, we can draw the following conclusions:

- Based on limited idealized simulations with 1998 GEOS GCM, the gross accuracies of the inferred wind and sea level pressure fields are consistent with the findings of Charney *et al.* (1969), but with somewhat larger asymptotic errors.
- GCMs of higher spatial and vertical resolution assimilate temperature data to substantially improve the inferred winds and sea level pressure where no data are available.
- A system of two polar orbiting satellites with temperatures sounders of 1 K accuracy in clear and cloudy regions, combined with surface pressure observations, should be capable of inferring the global wind fields to the required accuracies of 3 m s^{-1}.

- The conclusion of Charney *et al.* (1969) that it is possible to infer tropical winds from temperature profiles may have been a model-dependent result.
- Assimilating surface pressure greatly improves the rate of adjustment and the asymptotic accuracies of the extratropical winds, but does not significantly improve the inferred tropical winds.

As mentioned earlier, the new integrations reported here were performed with a resolution of $4° \times 5°$ by 20 levels. We plan to carry out further simulations employing finer resolution versions of the same model, as well as additional experiments with other models, to assess the effects of model dependence.

ACKNOWLEDGMENTS

The study of Charney *et al.* (1969) was made possible by Profs. Arakawa and Mintz, who shared the Mintz–Arakawa GCM with our organization at NASA as early as 1964, and again in 1969, agreeing that we could conduct and publish independent research results based on the use of the model.

We also want to take this opportunity to acknowledge that we at NASA are deeply indebted to Professor Arakawa for encouraging so many of his students and colleagues at UCLA to visit the NASA Goddard Space Flight Center. Some have stayed on to become permanent members of our staff. Many have continued to work closely with Arakawa, in introducing his concepts into the NASA model-development effort. Arakawa has often shared with us at NASA his latest ideas and models, well before he publishes them. For example, in 1972 he provided to us an early version of his three-level model, which subsequently evolved into the GISS nine-level model.

We are grateful to R. Rood for making the GEOS GCM available for use in this study. We also wish to thank L. Takacs and S. Nebuda for implementing the GEOS GCM code on the NASA Center for Computational Science (NCCS) computing facilities. The computations presented were all performed on the SGI/CRAY J90 system at the NCCS at Goddard Space Flight Center. We are indebted to the NCCS for making their computing environment and resources available to the authors. We also thank J. Raymond, who provided support in the preparation of this document.

REFERENCES

Charney, J. G. (1969). "Proceedings 1968 WMO/IUGG Symp. on Numerical Weather Prediction," Tokyo, March 1969. Meteorological Society of Japan.

Charney, J., M. Halem, and R. Jastrow (1969). Use of incomplete historical data to infer the present state of the atmosphere. *J. Atmos. Sci.* **26,** 5, 1160–1163.

Ghil, M. (1980). The compatible balancing approach to initialization, and four-dimensional data assimilation. *Tellus* **32,** 198–206.

Ghil, M., B. Shkoller, and V. Yangarber (1977). A balanced diagnostic system compatible with a barotropic prognostic model. *Mon. Wea. Rev.* **105,** 1223–1238.

Ghil, M., M. Halem, and R. Atlas (1979). Time-continuous assimilation of remote-sounding data and its effect on weather forecasting. *Mon. Wea. Rev.* **107,** 140–171.

Langlois, W. E., and H. C. Kwok (1969). Numerical simulation of weather and climate, Technical Report 3. Dept. of Meterorology UCLA.

Takacs, L., A. Molod, and T. Wang (1994). Documentation of the Goddard Earth Observing System GEOS general circulation model, Version 1, Technical Memorandum 104606. NASA.

Chapter 6

A Retrospective View of Arakawa's Ideas on Cumulus Parameterization

Wayne H. Schubert

Department of Atmospheric Science
Colorado State University
Fort Collins, Colorado

I. INTRODUCTION

When Akio Arakawa graduated from Tokyo University with a B.Sc. degree in physics in 1950, the economy of postwar Japan was in a recovery phase and there were few job opportunities in physics. However, there were job opportunities in the Japan Meteorological Agency (JMA) and, fortunately for our field, Akio took one of them. His early duties in JMA included a stint on a weather ship during a typhoon passage and research into the synoptic meteorology of the East Asia region. These must have been times of intense study and self-education, with a developing excitement for numerical weather prediction and general circulation modeling.[1] In recognition of his many original research contributions in the 1950s, Arakawa was granted his D.Sc. degree from Tokyo University in 1961. One paper from this early period, entitled "Non-geostrophic effects in the baroclinic prognostic equations," showed that the limitations of quasi-geostrophic theory in describing frontogenesis and baroclinic wave occlusion could be overcome by the use of more accurate balanced models. That paper, like many of his, was way ahead of its time.

In 1961 Arakawa came to UCLA as a research meteorologist to work with Yale Mintz on the development of the Mintz–Arakawa general circulation model, later called the UCLA GCM. In 1965 Arakawa joined the academic faculty, much to the benefit of more than three decades of UCLA graduate students. The performance of long-term integrations with a global primitive equation model was not a well-established procedure in the early and mid-1960s. In perfecting the "dry dynamics core" of the UCLA GCM, Arakawa made important contributions to both vertical and horizontal discretization schemes. In particular, his 1966 paper on "Computational design for long-term numerical integration of the equations of fluid motion" is a classic.

To obtain a good simulation of the climate, a GCM must include an accurate treatment of the hydrological cycle, which includes cumulus convection. In the late 1960s, the parameterization of cumulus convection began to occupy more of Arakawa's thoughts. In this chapter, we examine some of Arakawa's early ideas on this problem, starting with those developed in 1968 for application in the three-level version of the UCLA GCM. For a broader perspective on the early history of cumulus parameterization (mainly the 1960s to the early 1970s) see the accompanying chapter by Akira Kasahara, Chapter 7.

One of the most difficult concepts in cumulus parameterization theory is the assumption involving quasi-equilibrium of the cloud work function. Because the quasi-equilibrium assumption leads to a diagnostic equation for the cloud base mass flux, it can be interpreted as a filtering approximation. This allows us to draw an analogy with quasi-geostrophic theory, which is a filtering approximation that leads to a diagnostic equation for the divergent part of the flow. Of course, the analogy is only partial, but it does help us understand the quasi-equilibrium assumption as a filtering approximation. It filters the transient adjustment of a cloud ensemble in the same sense that quasi-geostrophic theory filters transient inertia-gravity waves.

---

[1] For a personal history of this pre-UCLA period see Section III of Chapter 1 in this book.

Section II explores the half of the analogy associated with quasi-geostrophic theory, while Sections III and IV explore the half associated with quasi-equilibrium of the cloud work function. The heart of the analogy lies in Eqs. (15), (26), and (41), the former being a constraint on the tendencies of wind and pressure in quasi-geostrophic theory and the latter two constraints on the tendencies of temperature and moisture in cumulus parameterization theory.

II. PRIMITIVE EQUATION MODELS, QUASI-GEOSTROPHIC MODELS, AND THE CONCEPT OF FILTERING THE TRANSIENT ASPECTS OF GEOSTROPHIC ADJUSTMENT

Let us consider inviscid, forced, y-independent, small-amplitude motions (about a basic state of rest) in a shallow water fluid on an f plane. The nondimensional, linearized, shallow water primitive equations governing such motions are

$$\frac{\partial u}{\partial t} - v + \frac{\partial h}{\partial x} = 0, \tag{1}$$

$$\frac{\partial v}{\partial t} + u = 0, \tag{2}$$

$$\frac{\partial h}{\partial t} + \frac{\partial u}{\partial x} = Q\alpha^2 t e^{-\alpha t}. \tag{3}$$

We have nondimensionalized these equations by choosing $1/f$, c/f, H, and c as units of time, horizontal distance, vertical distance, and speed, where f is the constant Coriolis parameter, H the constant mean depth of the fluid, and $c = (gH)^{\frac{1}{2}}$ the pure gravity wave phase speed. The mass source/sink term on the right-hand side of Eq. (3) has been assumed to be separable in x and t, with the spatial dependence given by $Q(x)$ and the time dependence given by $\alpha^2 t e^{-\alpha t}$ where α is a specified constant. Small α (i.e., $\alpha \ll 1$) corresponds to slow forcing and large α (i.e., $\alpha \gg 1$) to rapid forcing, but the total forcing is independent of α, since $\int_0^\infty \alpha^2 t e^{-\alpha t}\,dt = 1$. As the initial condition for Eqs. (1)–(3), we assume $u(x,0) = 0$, $v(x,0) = 0$, and $h(x,0) = 0$, so that any flow field or pressure field is generated by the mass source/sink on the right-hand side of Eq. (3). Because of the assumed y-independent nature of the flow, all divergence is associated with u and all vorticity with v. Thus, we can refer to u as the divergent part of the flow and v as the rotational part of the flow.

By forming the combination $\partial(1)/\partial t + (2) - \partial(3)/\partial x$ we obtain

$$\frac{\partial^2 u}{\partial t^2} + u - \frac{\partial^2 u}{\partial x^2} = -\frac{\partial Q}{\partial x}\alpha^2 t e^{-\alpha t}, \tag{4}$$

which is the governing equation for the divergent flow $u(x,t)$. We can construct the solution for Eq. (4) by using Fourier transform methods. First, we introduce the Fourier transform pair

$$\hat{u}(k,t) = (2\pi)^{-\frac{1}{2}} \int_{-\infty}^{\infty} u(x,t) e^{-ikx}\, dx, \tag{5a}$$

$$u(x,t) = (2\pi)^{-\frac{1}{2}} \int_{-\infty}^{\infty} \hat{u}(k,t) e^{ikx}\, dk, \tag{5b}$$

where k is the horizontal wavenumber. A similar transform pair exists for $Q(x)$ and $\hat{Q}(k)$. We refer to $u(x,t)$ as the physical space representation of the divergent flow and $\hat{u}(k,t)$ as the spectral space representation of the divergent flow. Transforming Eq. (4) to spectral space via Eq. (5), we obtain the ordinary differential equation

$$\frac{d^2\hat{u}}{dt^2} + (1 + k^2)\hat{u} = -ik\hat{Q}\alpha^2 t e^{-\alpha t}. \tag{6}$$

As can easily be checked by direct substitution, the solution of Eq. (6) is

$$\hat{u}(k,t) = -\frac{ik\hat{Q}(k)}{\alpha^2 + \nu^2}\alpha^2 t e^{-\alpha t} - \frac{2ik\,\alpha^3\hat{Q}(k)}{(\alpha^2 + \nu^2)^2}e^{-\alpha t}$$

$$+ \frac{k\hat{Q}(k)}{2\nu}\left[\frac{\alpha^2(\alpha + i\nu)^2}{(\alpha^2 + \nu^2)^2}e^{-i\nu t} - \frac{\alpha^2(\alpha - i\nu)^2}{(\alpha^2 + \nu^2)^2}e^{i\nu t}\right], \tag{7}$$

where $\nu = (1 + k^2)^{\frac{1}{2}}$. Note that both $\hat{u}$ and $d\hat{u}/dt$ vanish at $t = 0$, as required by our initial condition. The first two terms on the right-hand side of Eq. (7) constitute a particular solution for Eq. (6), while the last two terms are the homogeneous solutions. The homogeneous solutions $e^{-i\nu t}$ and $e^{i\nu t}$ represent freely propagating inertia-gravity waves. If one wishes to plot the solution in physical space, Eq. (7) can be substituted into Eq. (5b) and the integral over k evaluated numerically. The solution $u(x,t)$ depends very much on whether the forcing is slow or rapid. In the case of a very slow mass source/sink, $\alpha \ll 1$ and the coefficients of $e^{-i\nu t}$ and $e^{i\nu t}$ are much less than unity, so that practically no freely propagating inertia-gravity waves are excited by the mass source/sink. In fact, for very slow forcing, only the first term on the right-hand side of Eq. (7) survives, and we obtain the approximate result

$$\hat{u}(k,t) \approx -\frac{ik\hat{Q}(k)}{1 + k^2}\,\alpha^2 t e^{-\alpha t} \qquad \text{if } \alpha \ll 1 \quad \text{(slow forcing)}. \tag{8}$$

We now show how result (8) is obtained directly from quasi-geostrophic theory. Thus, let us approximate the primitive equations, Eqs. (1)–(3), by the quasi-geostrophic equations:

$$-v + \frac{\partial h}{\partial x} = 0, \tag{9}$$

$$\frac{\partial v}{\partial t} + u = 0, \tag{10}$$

$$\frac{\partial h}{\partial t} + \frac{\partial u}{\partial x} = Q\alpha^2 t e^{-\alpha t}. \tag{11}$$

Under what conditions might we expect the solutions of Eqs. (9)–(11) to be nearly identical to the solutions of Eq. (1)–(3)? Obviously, we must limit ourselves to cases in which $|\partial u/\partial t| \ll |v|$. This tends to be true if the divergent flow is weak compared to the rotational flow (i.e., $|u| \ll |v|$) and the divergent flow is slowly changing (i.e., $|\partial/\partial t| \ll 1$, or in dimensional terms that the dimensional $|\partial/\partial t|$ is much less than f).

Following the same procedure used in deriving Eq. (4), we can combine Eqs. (9)–(11) to form a single equation for $u(x,t)$. Thus, by forming the combination $\partial(9)/\partial t + (10) - \partial(11)/\partial x$ we obtain

$$u - \frac{\partial^2 u}{\partial x^2} = -\frac{\partial Q}{\partial x}\,\alpha^2 t e^{-\alpha t}, \tag{12}$$

which is the quasi-geostrophic version of Eq. (4). The crucial difference between Eqs. (4) and (12) is that Eq. (12) is a diagnostic equation, so that the divergent flow at time t, as determined from Eq. (12), depends only on the mass source/sink at time t. The dependence of $u(x,t)$ on the past history of the mass source/sink has been lost. According to Eq. (12), a change in the mass source/sink at a certain spatial point has an immediate effect at all spatial points, as if information can propagate at infinite speed. Transforming Eq. (12) to spectral space via Eq. (5), we obtain the algebraic equation

$$(1 + k^2)\hat{u} = -ik\hat{Q}\alpha^2 t e^{-\alpha t}, \tag{13}$$

which is the quasi-geostrophic version of Eq. (6). Equation (13) can also be written as

$$\hat{u}(k,t) = -\frac{ik\hat{Q}(k)}{1 + k^2}\alpha^2 te^{-\alpha t}, \tag{14}$$

which allows easy comparison with Eqs. (7) and (8). Since Eq. (14) is identical to result (8), we conclude that the quasi-geostrophic equations predict the same flow evolution as the primitive equations in the special case of slow forcing.

To illustrate these concepts, we have prepared Fig. 1, which shows the h field for the shallow water primitive equation model (Figs. 1b and 1c) and the shallow water quasi-geostrophic model (Fig. 1a). The equations that were solved to produce Fig. 1 are identical to Eqs. (1)–(3) and Eqs. (9)–(11) except that polar coordinates and the axisymmetric assumption were used. The models are forced by a mass sink of the form $Q(r)\alpha^2 te^{-\alpha t}$, where $Q(r)$ is constant in the region $0 \leq r/a < 1$ and vanishes elsewhere. Figures 1b and 1c show $h(r,t)$ as determined by the shallow water primitive equation model for $\alpha = 1/2$ (slow mass removal) and $\alpha = 2$ (fast mass removal), respectively. Figure 1a shows $h(r,t)$ as determined by the shallow water quasi-geostrophic model. Figures 1a–1c have been constructed with a time axis of αt, and when $\alpha t = 5$, 96% of the eventual total mass removal has already occurred. Although the final states are the same in the three cases shown in Fig. 1, the transient states are different. In the case of the primitive equation model with rapid forcing ($\alpha = 2$) a large inertia-gravity wavefront is excited and propagates outward with time. This is in contrast to the case of the primitive equation model with slow forcing ($\alpha = 1/2$), where only a small amount of inertia-gravity wave activity is excited (Fig. 1b). The quasi-geostrophic model (Fig. 1a) filters all transient inertia-gravity wave activity, but is not significantly different than the slowly forced primitive equation model result. In fact, the quasi-geostrophic model result (Fig. 1a) can be considered a slightly smoothed version of the slow forcing case (Fig. 1b).

About the time he was developing quasi-geostrophic theory, Jule Charney visualized the atmosphere "as a musical instrument on which one can play many tunes." He thought of the high notes as the sound waves and gravity waves, and the low notes as the Rossby waves, with Mother Nature being "a musician more of the Beethoven than the Chopin type" in that she prefers "the low notes and only occasionally plays arpeggios in the treble and then only with a light hand." If Eq. (1)–(3) can be thought of as a whole piano, Eq. (9)–(11) might be thought of as a piano that has been

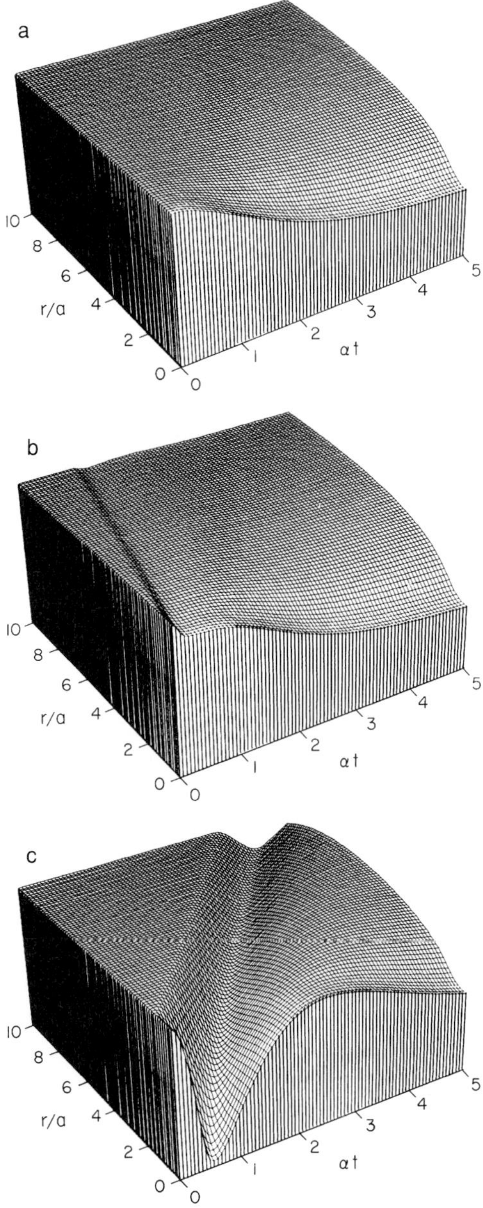

Figure 1 The height of the free surface $h(r, t)$ in the $(r/a, \alpha t)$ plane for the case of (a) the quasi-geostrophic model, (b) the primitive equation model with slow forcing ($\alpha = 1/2$), and (c) the primitive equation model with rapid forcing ($\alpha = 2$). The models are forced by a mass sink of the form $Q(r)\alpha^2 t e^{-\alpha t}$, where $Q(r)$ is constant in the region $0 \leq r/a < 1$ and vanishes elsewhere. Note that the quasi-geostrophic model result can be considered a slightly smoothed version of the primitive equation model result for the slow forcing case.

sawed in half, with only the low notes remaining usable. Even though its dynamic range is limited, it can still play some beautiful music.

To summarize this section, we emphasize that the reason we obtain the diagnostic equation, Eq. (12), for u in quasi-geostrophic theory is that the tendencies of v and h are constrained by $\partial(9)/\partial t$, i.e.,

$$\frac{\partial v}{\partial t} = \frac{\partial}{\partial x}\left(\frac{\partial h}{\partial t}\right). \tag{15}$$

Although constraint (15) is not satisfied for the high-frequency inertia-gravity waves, it is satisfied for the slowly evolving quasi-geostrophic flow. In the next section we draw the following analogy: Just as constraint (15) on the tendencies of v and h leads to a diagnostic equation for u in quasi-geostrophic theory, so the quasi-equilibrium constraint on the tendencies of temperature and moisture leads to a diagnostic equation for cloud base mass flux in cumulus parameterization theory.

III. ARAKAWA'S 1968 CUMULUS PARAMETERIZATION: LAYING THE CONCEPTUAL FOUNDATION FOR FUTURE WORK

In his 1968 paper (Proceedings of the WMO/IUGG Symposium on Numerical Weather Prediction, Tokyo) Arakawa considered a cumulus cloud ensemble that is in a statistically steady state.[2] He assumed that the thermodynamical features of the individual clouds within the ensemble are alike. In other words, he did not consider subensembles with different entrainment rates and different depths. At a particular longitude and latitude in the UCLA three-level GCM, one of three types of convection could occur. The three types of convection are shown in Fig. 2. Let us restrict our attention here to penetrating convection, also called type II convection. This is the type of convection most relevant to tropical cyclones and the intertropical convergence zone. Defining the dry static energy by $s = c_p T + gz$, the moist static energy by $h = s + Lq$, and the saturation moist static energy by $h^* = s + Lq^*$ (where q^* denotes the saturation mixing ratio), type II convection occurs when $h_B > (h_3^*, h_1^*) > h_3$, i.e., when conditional instability exists between the boundary layer and level 3, but not between level 3 and level 1. The mass fluxes associated with this convection are shown in Fig. 3, with C denoting the ensemble

---

[2] For a full discussion of the formulation of "moist processes" in the UCLA GCM, see Section X of Chapter 1 in this book.

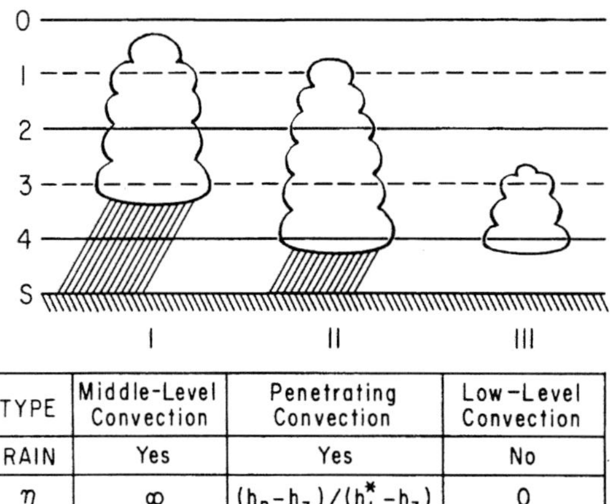

TYPE	Middle-Level Convection	Penetrating Convection	Low-Level Convection
RAIN	Yes	Yes	No
η	∞	$(h_B - h_3)/(h_1^* - h_3)$	0
CONDI-TION	$h_3 > h_1^*$	$h_B > (h_1^*, h_3^*) > h_3$	$h_1^* > h_B > h_3^*$

Figure 2 The three types of convection allowed in the first version (1968) of the three-level UCLA GCM.

mass flux at level 4 (cloud base), ηC the ensemble mass flux at level 2, and $(\eta - 1)C$ the entrainment. In pressure coordinates the budget equations for the dry static energy of each layer, when type II convection is occurring, are

$$\frac{\partial s_1}{\partial t} + \mathbf{v}_1 \cdot \nabla s_1 + \omega_2 \left(\frac{s_2 - s_1}{\Delta p} \right)$$

$$= \frac{g\eta C}{\Delta p} \left(\frac{1}{1 + \gamma_1} \right)(h_c - h_1^*) + g\eta C \left(\frac{s_1 - s_2}{\Delta p} \right), \qquad (16)$$

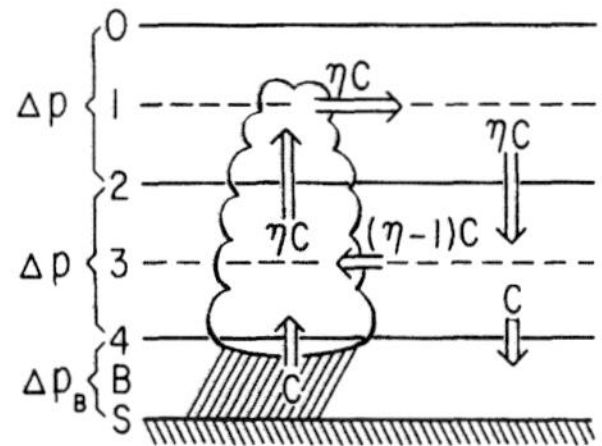

Figure 3 A schematic cloud for type II convection in the three-level UCLA GCM.

$$\frac{\partial s_3}{\partial t} + \mathbf{v}_3 \cdot \nabla s_3 + \omega_2 \left(\frac{s_3 - s_2}{\Delta p} \right) + \omega_4 \left(\frac{s_4 - s_3}{\Delta p} \right)$$

$$= g\eta C \left(\frac{s_2 - s_3}{\Delta p} \right) + gC \left(\frac{s_3 - s_4}{\Delta p} \right), \tag{17}$$

$$\frac{\partial s_{\mathrm{B}}}{\partial t} + \mathbf{v}_{\mathrm{B}} \cdot \nabla s_{\mathrm{B}} + \omega_4 \left(\frac{s_{\mathrm{B}} - s_4}{\Delta p_{\mathrm{B}}} \right) = gC \left(\frac{s_4 - s_{\mathrm{B}}}{\Delta p_{\mathrm{B}}} \right)$$

$$+ \frac{g}{\Delta p_{\mathrm{B}}} \rho_{\mathrm{S}} C_{\mathrm{E}} |\mathbf{v}_{\mathrm{S}}| (s_{\mathrm{S}} - s_{\mathrm{B}}), \tag{18}$$

where radiative processes have been neglected and where $\mathbf{v}_1$, $\mathbf{v}_3$, $\mathbf{v}_{\mathrm{B}}$ are the large-scale horizontal velocities at the three model levels, ω_2 and ω_4 are the large-scale vertical p velocities at the layer interfaces, ρ_{S} the surface air density, C_{E} the bulk aerodynamic coefficient for surface heat and moisture exchange, $|\mathbf{v}_{\mathrm{S}}|$ the surface wind speed, h_{c} is the moist static energy of the air inside the clouds in the upper layer, s_{S} the dry static energy of the surface, and γ_1 is defined as $\gamma = (L/c_p)(\partial q^*/\partial T)_p$, evaluated at level 1. Similarly, the budget equations for the water vapor mixing ratio of each layer, when type II convection is occurring, are

$$\frac{\partial q_1}{\partial t} + \mathbf{v}_1 \cdot \nabla q_1 + \omega_2 \left(\frac{q_2 - q_1}{\Delta p} \right)$$

$$= \frac{g\eta C}{\Delta p} \left[q_1^* + \frac{\gamma_1}{1 + \gamma_1} \frac{1}{L} (h_{\mathrm{c}} - h_1^*) - q_1 \right] + g\eta C \left(\frac{q_1 - q_2}{\Delta p} \right), \tag{19}$$

$$\frac{\partial q_3}{\partial t} + \mathbf{v}_3 \cdot \nabla q_3 + \omega_2 \left(\frac{q_3 - q_2}{\Delta p} \right) + \omega_4 \left(\frac{q_4 - q_3}{\Delta p} \right)$$

$$= g\eta C \left(\frac{q_2 - q_3}{\Delta p} \right) + gC \left(\frac{q_3 - q_4}{\Delta p} \right), \tag{20}$$

$$\frac{\partial q_{\mathrm{B}}}{\partial t} + \mathbf{v}_{\mathrm{B}} \cdot \nabla q_{\mathrm{B}} + \omega_4 \left(\frac{q_{\mathrm{B}} - q_4}{\Delta p_{\mathrm{B}}} \right)$$

$$= gC \left(\frac{q_4 - q_{\mathrm{B}}}{\Delta p_{\mathrm{B}}} \right) + \frac{g}{\Delta p_{\mathrm{B}}} \rho_{\mathrm{S}} C_{\mathrm{E}} |\mathbf{v}_{\mathrm{S}}| (q_{\mathrm{S}}^* - q_{\mathrm{B}}), \tag{21}$$

where q_{S}^* is the saturation mixing ratio at the pressure and temperature of the underlying surface. In Eqs. (16)–(21) the large-scale horizontal and vertical advection terms are on the left-hand side, while the surface flux and cumulus terms are on the right-hand side. The first term on the right-hand side of Eq. (16) and the first term on the right-hand side of

Eq. (19) are due to the detrainment of cloud air,[3] while the remaining terms proportional to ηC and C are due to cumulus-induced subsidence at levels 2 and 4.

Since the horizontal and vertical advection terms on the left-hand sides of Eqs. (16)–(21) are explicitly computed on the resolvable scales by the GCM, we can now define the cumulus parameterization problem as the determination of h_c, η, and C, the three unknowns on the right-hand sides of Eqs. (16)–(21). To determine h_c, Arakawa assumed that the detraining air at level 1 has vanishing buoyancy, i.e.,

$$h_c = h_1^*. \tag{22}$$

Note that this eliminates the detrainment term in Eq. (16), but not the detrainment term in Eq. (19). Since the cloud air in the upper layer is a mixture of air that has come from the boundary layer and air that has been entrained from layer 3, the ensemble budget equation for moist static energy is $\eta h_c = h_B + (\eta - 1)h_3$. When combined with Eq. (22), this can be solved for η to obtain

$$\eta = \frac{h_B - h_3}{h_1^* - h_3}. \tag{23}$$

With h_c determined by Eq. (22) and η determined by Eq. (23), the only remaining problem is the determination of the ensemble cloud base mass flux C. Arakawa has always felt that the real conceptual difficulty in parameterizing cumulus convection starts at this point. In his 1968 paper he argued as follows: Since the penetrating convection shown in Fig. 2 occurs when $h_B - h_1^* > 0$, first derive the equation for the tendency of $h_B - h_1^*$ from Eqs. (16)–(21). This results in

$$\frac{\partial(h_B - h_1^*)}{\partial t} + \mathbf{v}_B \cdot \nabla h_B - (1 + \gamma_1)\mathbf{v}_1 \cdot \nabla s_1$$

$$+ \omega_4\left(\frac{h_B - h_4}{\Delta p_B}\right) - (1 + \gamma_1)\omega_2\left(\frac{s_2 - s_1}{\Delta p}\right)$$

$$= -gC\left[\left(\frac{h_B - h_4}{\Delta p_B}\right) + (1 + \gamma_1)\left(\frac{s_1 - s_2}{\Delta p}\right)\eta\right]$$

$$+ \frac{g}{\Delta p_B}\rho_S C_E |\mathbf{v}_S|(h_S^* - h_B). \tag{24}$$

---

[3] The cooling and moistening effects of the detrainment of liquid water are not included in Arakawa's 1968 paper. They were included in later work.

Since the bracketed part of the first term on the right-hand side of Eq. (24) is positive and $C \geq 0$, cumulus convection acts to decrease $h_B - h_1^*$. If the surface flux term and the large-scale horizontal and vertical advective terms are constant in time and act to increase $h_B - h_1^*$, we expect a steady-state balanced mass flux C to be established, with $\partial(h_B - h_1^*)/\partial t = 0$. Even if the surface flux term and the large-scale horizontal and vertical advective terms are not constant in time, but are varying on a time scale that is longer than the adjustment time of the ensemble mass flux, a near balance will be maintained and $\partial(h_B - h_1^*)/\partial t$ will be negligible. Then, when the resulting diagnostic version of Eq. (24) is solved for C, we obtain

$$
C = \frac{-\mathbf{v}_B \cdot \nabla h_B + (1 + \gamma_1)\mathbf{v}_1 \cdot \nabla s_1 - \omega_4\left(\dfrac{h_B - h_4}{\Delta p_B}\right) + (1 + \gamma_1)\omega_2\left(\dfrac{s_2 - s_1}{\Delta p}\right) + \dfrac{g}{\Delta p_B}\rho_S C_E |\mathbf{v}_S|(h_S - h_B)}{g\left[\left(\dfrac{h_B - h_4}{\Delta p_B}\right) + (1 + \gamma_1)\left(\dfrac{s_1 - s_2}{\Delta p}\right)\eta\right]}
\tag{25}
$$

which shows how the ensemble mass flux C is controlled by large-scale horizontal and vertical advective processes, surface fluxes, and radiation [which would also appear in the numerator of Eq. (25) if we had included it in Eqs. (16)–(18)]. To summarize, with h_c determined by Eq. (22), η by Eq. (23), and C by Eq. (25), all the terms on the right-hand sides of Eqs. (16)–(21) are known, so that the cumulus parameterization theory for type II convection is closed.

As we have just seen, if the C given by Eq. (25) is used in the right-hand sides of Eqs. (16)–(21), then the tendencies of temperature and moisture in the column are constrained by $\partial(h_B - h_1^*)/\partial t = 0$, or equivalently

$$
\frac{\partial s_B}{\partial t} + L\frac{\partial q_B}{\partial t} - (1 + \gamma_1)\frac{\partial s_1}{\partial t} = 0,
\tag{26}
$$

a statement that the predictability of CAPE has been lost. Equation (26) is analogous to Eq. (15) in the sense that the rapid cloud ensemble adjustment process constrains the tendencies describing the time evolution of the temperature and moisture fields on the slower time scales, just as the rapid geostrophic adjustment process constrains the tendencies describing the time evolution of the balanced wind and mass fields on slower time scales. Of course, there are important conceptual differences between constraint (15) and constraint (26), one of which is that Eq. (15) is a full-time constraint (in the sense that it operates at all times and at all spatial points), whereas Eq. (26) is a part-time constraint (in the sense that it operates only when and where there is type II convection).

IV. GENERALIZATION TO THE SPECTRAL FORM OF CUMULUS PARAMETERIZATION THEORY

Just after Arakawa wrote his 1968 paper, an effort was begun to generalize the UCLA GCM to many more layers. About this time there occurred another event with very important long-term consequences— Michio Yanai left Tokyo University and accepted a faculty position at UCLA. Michio arrived with a knowledge that systematic differences in the vertical profiles of apparent heat source Q_1 and apparent moisture sink Q_2 held important information about cumulus activity. Using certain parts of parameterization theory, Michio was able to diagnostically compute cumulus mass fluxes from his Q_1 and Q_2 budgets. Although Akio's interest was in cumulus parameterization for the GCM, and Michio's interest was in the observational and diagnostic analysis of heat and moisture budgets, the daily interaction seemed to spur on both groups and led to rapid progress in both areas. Two classic papers from Michio's group investigate ITCZ convection (Yanai *et al.*, 1973) and trade cumulus convection (Nitta and Esbensen, 1974).

Returning to the parameterization problem, it was obvious that the old three-level parameterization needed a generalization that would allow for many cloud types. To allow clouds of many different depths at the same horizontal grid point simultaneously, the cloud ensemble was broken into subensembles of different entrainment rates. Using the z coordinate, the large-scale budget equations for $\bar{s}$ and $\bar{q}$ above the boundary layer were then written as

$$\rho\left(\frac{\partial \bar{s}}{\partial t} + \bar{\mathbf{v}} \cdot \nabla \bar{s} + \bar{w}\frac{\partial \bar{s}}{\partial z}\right) = D(\hat{s} - L\hat{\ell} - \bar{s}) + M_{\mathrm{c}}\frac{\partial \bar{s}}{\partial z} + \overline{Q}_{\mathrm{R}}, \quad (27)$$

$$\rho\left(\frac{\partial \bar{q}}{\partial t} + \bar{\mathbf{v}} \cdot \nabla \bar{q} + \bar{w}\frac{\partial \bar{q}}{\partial z}\right) = D(\hat{q} + \hat{\ell} - \bar{q}) + M_{\mathrm{c}}\frac{\partial \bar{q}}{\partial z}. \quad (28)$$

To make temperature and moisture predictions with Eqs. (27) and (28), we need to determine the total cumulus ensemble induced subsidence $M_{\mathrm{c}}(z)$, the detrainment $D(z)$, the dry static energy of the detraining air $\hat{s}(z)$, the water vapor mixing ratio of the detraining air $\hat{q}(z)$, and the liquid water mixing ratio of the detraining air $\hat{\ell}(z)$. If $m_{\mathrm{B}}(\lambda)\eta(z, \lambda)\,d\lambda$ is the vertical mass flux at level z due to all clouds with entrainment rates between λ and $\lambda + d\lambda$, then the total ensemble vertical mass flux at level z is

$$M_c(z) = \int_0^{\lambda_{max}} m_B(\lambda)\eta(z, \lambda)\, d\lambda, \tag{29}$$

and the detrainment is

$$D(z) = -m_B(\lambda_D(z))\eta(z, \lambda_D(z))\frac{d\lambda_D(z)}{dz}, \tag{30}$$

where $\lambda_D(z)$ is the entrainment rate of the subensemble, which just reaches level z. Later we shall need the function $z_D(\lambda)$, which is the inverse function of $\lambda_D(z)$, i.e., $z_D(\lambda)$ is the detrainment height of the subensemble with entrainment rate λ. The dry static energy, water vapor mixing ratio, and liquid water mixing ratio of the detraining air at level z are given by

$$\hat{s}(z) = s_c(z, \lambda_D(z)), \tag{31}$$

$$\hat{q}(z) = q_c(z, \lambda_D(z)), \tag{32}$$

$$\hat{\ell}(z) = \ell(z, \lambda_D(z)), \tag{33}$$

where $s_c(z, \lambda)$ and $q_c(z, \lambda)$ are given in terms of $h_c(z, \lambda)$ by

$$s_c(z, \lambda) = \bar{s}(z) + \frac{1}{1 + \gamma}\left[h_c(z, \lambda) - \bar{h}^*(z)\right], \tag{34}$$

$$q_c(z, \lambda) = \bar{q}^*(z) + \frac{1}{L}\frac{\gamma}{1 + \gamma}\left[h_c(z, \lambda) - \bar{h}^*(z)\right]. \tag{35}$$

The subensemble normalized mass flux $\eta(z, \lambda)$, the subensemble moist static energy $h_c(z, \lambda)$, and the subensemble total water content $q_c(z, \lambda) + \ell(z, \lambda)$ are determined by the subensemble budget equations

$$\frac{\partial \eta(z, \lambda)}{\partial z} = \lambda\eta(z, \lambda), \tag{36}$$

$$\frac{\partial[\eta(z, \lambda)h_c(z, \lambda)]}{\partial z} = \lambda\eta(z, \lambda)\bar{h}(z), \tag{37}$$

$$\frac{\partial\{\eta(z, \lambda)[q_c(z, \lambda) + \ell(z, \lambda)]\}}{\partial z} = \lambda\eta(z, \lambda)\bar{q}(z) - c_0\eta(z, \lambda)\ell(z, \lambda), \tag{38}$$

where c_0 is a constant of proportionality for the precipitation process [i.e., a constant of proportionality for the conversion of airborne liquid water droplets $\ell(z, \lambda)$ to precipitation]. The subensemble budget equations, Eqs. (36)–(38), require boundary conditions at the top of the boundary layer. These boundary conditions are $\eta(z_B, \lambda) = 1$, $h_c(z_B, \lambda) = h_M$, and $q_c(z_B, \lambda) + \ell(z_B, \lambda) = q_M$, where h_M and q_M are the moist static energy and water vapor mixing ratio of the subcloud mixed layer. The subensemble detraining at level z is assumed to have vanishing buoyancy there, i.e.,

$$h_c(z, \lambda_D(z)) = \bar{h}^*(z). \tag{39}$$

Except for the determination of the mass flux distribution function $m_B(\lambda)$, the parameterization is now closed. To see this, first solve the six equations (34)–(39) for the six functions $\eta(z, \lambda)$, $h_c(z, \lambda)$, $s_c(z, \lambda)$, $q_c(z, \lambda)$, $\ell(z, \lambda)$, and $\lambda_D(z)$. Use this result to determine $\hat{s}(z)$, $\hat{q}(z)$, and $\hat{\ell}(z)$ from Eqs. (31)–(33). Then, if $m_B(\lambda)$ is known, $M_c(z)$ and $D(z)$ can be determined from Eqs. (29) and (30). In other words, the parameterization problem has been reduced to the determination of the mass flux distribution function $m_B(\lambda)$.

The determination of the mass flux distribution function $m_B(\lambda)$ is based on the quasi-equilibrium of the cloud work function $A(\lambda)$, which is defined by

$$A(\lambda) = \int_{z_B}^{z_D(\lambda)} \frac{g}{(1 + \gamma)c_p\bar{T}} \eta(z, \lambda)\left[h_c(z, \lambda) - \bar{h}^*(z)\right] dz. \tag{40}$$

In analogy with the type II constraint in the three-level model [see Eq. (26)], the constraints on the large-scale tendencies of temperature and moisture, derived from the time derivative of Eq. (40), can be written as[4]

$$\int_{z_B}^{z_D(\lambda)} \frac{g}{(1 + \gamma)c_p\bar{T}} \eta(z, \lambda)\frac{\partial}{\partial t}\left[h_c(z, \lambda) - \bar{h}^*(z)\right] dz = 0. \tag{41}$$

Of course, to actually make Eq. (41) contain only large-scale tendencies we would need to express $h_c(z, \lambda)$ in terms of the large-scale fields through the solution of Eq. (37). If Eqs. (27) and (28) are now used in Eq. (41) to expresss the large-scale tendencies in terms of the large-scale advective terms, radiation terms, and cumulus terms, we could put all the cumulus terms on the left-hand side and all the large-scale advective and radiation terms on the right-hand side to obtain a diagnostic equation for $m_B(\lambda)$

---

[4] For simplicity, the time derivative of the lower limit of integration in Eq. (40) has been neglected.

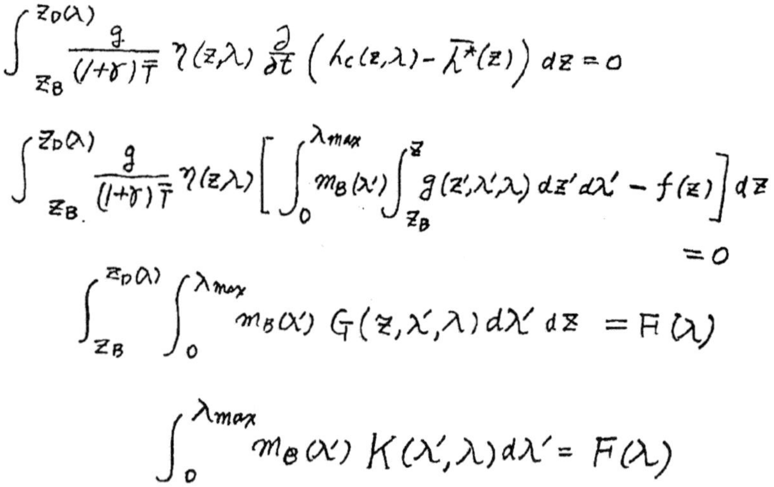

$$\int_{z_B}^{z_D(\lambda)} \frac{g}{(1+\gamma)\bar{T}}\, \eta(z,\lambda)\, \frac{\partial}{\partial t}\left(h_c(z,\lambda) - \bar{h}^*(z)\right) dz = 0$$

$$\int_{z_B}^{z_D(\lambda)} \frac{g}{(1+\gamma)\bar{T}}\, \eta(z,\lambda)\left[\int_0^{\lambda_{max}} m_B(\lambda')\int_{z_B}^{z} g(z',\lambda',\lambda)\, dz'\, d\lambda' - f(z)\right] dz$$
$$= 0$$

$$\int_{z_B}^{z_D(\lambda)} \int_0^{\lambda_{max}} m_B(\lambda')\, G(z,\lambda',\lambda)\, d\lambda'\, dz = F(\lambda)$$

$$\int_0^{\lambda_{max}} m_B(\lambda')\, K(\lambda',\lambda)\, d\lambda' = F(\lambda)$$

Figure 4 Reproduction of a handwritten transparency prepared by A. Arakawa in 1971. With only a few lines of mathematics, Arakawa deduced the form of the diagnostic equation for the mass flux distribution function $m_B(\lambda)$. The final line is the diagnostic equation for $m_B(\lambda)$, with kernel $K(\lambda, \lambda')$ and forcing $F(\lambda)$. At the time he produced this transparency, the exact forms of $K(\lambda, \lambda')$ and $F(\lambda)$ were not known.

analogous to Eq. (25), but obviously more general than Eq. (25). With remarkable insight, Arakawa deduced the form of this diagnostic equation with only a few lines of mathematics. His argument, from a handwritten transparency he prepared, is reproduced as Fig. 4. The final line,

$$\int_0^{\lambda_{max}} K(\lambda, \lambda') m_B(\lambda')\, d\lambda' = F(\lambda), \tag{42}$$

is the diagnostic equation for $m_B(\lambda)$, with kernel $K(\lambda, \lambda')$ and forcing $F(\lambda)$. At the time, the exact forms of $K(\lambda, \lambda')$ and $F(\lambda)$ were not known, and in fact it took some time to work them out, especially after the virtual temperature effects of water vapor and liquid water were added to the cloud work function of Eq. (40). It is important to note that physical considerations require that $m_B(\lambda)$ be nonnegative, and that Eq. (42) may not have a general solution under this constraint. Thus, some kind of "optimal" solution must be found. For a review of this problem and many subsequent developments, see Arakawa and Xu (1990), Moorthi and Suarez (1992), Arakawa and Cheng (1993), Randall and Pan (1993), Cheng and Arakawa (1994), Randall *et al.* (1997a, b), and Pan and Randall (1998) and references therein. For a perspective on the implications of "quasi-equilibrium thinking" about the physics of large-scale circulations in convecting atmospheres see Chapter 8 by Kerry Emanuel.

V. CONCLUSIONS

In the 1950s quasi-geostrophic models were used for both numerical weather prediction and general circulation modeling. Nowadays, NWP centers and GCM groups use global models based on the quasi-static primitive equations with the traditional approximation (i.e., the models use the hydrostatic equation and take advantage of the shallowness of the atmosphere to approximate the metric factors appearing in the gradient, divergence, and curl operators in spherical coordinates, a procedure that slightly distorts the Coriolis acceleration and the absolute angular momentum principle). The quasi-geostrophic model is not used much anymore. Even though the flow fields produced in the NWP models and GCMs are primarily geostrophic, modelers prefer the accuracy of the quasi-static primitive equations. However, quasi-geostrophic theory still plays an important role among theoreticians, whose primary goal is physical understanding rather than accuracy.

Perhaps well into the 21st century humans will possess computing devices 10^9 times as powerful as those we have today. Most NWP models and GCMs may then be based on the exact, nonhydrostatic primitive equations with far fewer assumptions on moist physical processes, and with resolutions of 100 m over the whole globe. Cumulus parameterization as we know it will not be needed, and myriads of individual clouds will be explicitly simulated. The frontier of physical parameterization will have been pushed back to cloud microphysics. But, in some dark, ivy-covered building there will be some theoreticians, bent on physical understanding, studying "simple models" that incorporate Arakawa's ideas on cumulus parameterization.

ACKNOWLEDGMENTS

I would like to take this opportunity to express a personal and enormous debt of gratitude to Akio Arakawa. I have at times felt guilty that, after Akio spent so much time trying to educate me in dynamics and general circulation modeling, and after finally getting me to a level where there was some chance to usefully contribute to his research project, I left for a job elsewhere. Perhaps this is the nature of graduate education. In any event, thank you for sharing all your profound and marvelous ideas and for all your personal encouragement.

REFERENCES

Arakawa, A. (1960). Nongeostrophic effects in the baroclinic prognostic equations. *Proceedings of the International Symposium on Numerical Weather Prediction*, Tokyo, 1960, Meteorological Society of Japan, 161–175.

Arakawa, A. (1966). Computational design for long-term numerical integration of the equations of fluid motion: Two-dimensional incompressible flow. Part I. *J. Comput. Phys.* **1**, 119–143.

Arakawa, A. (1968). Parameterization of cumulus convection *Proceedings of the WMD/IUGG Symposium on Numerical Weather Prediction*, Tokyo, 1968, Japan Meteorological Agency, IV, **8**, 1–6.

Arakawa, A., and M.-D. Cheng (1993). The Arakawa–Schubert cumulus parameterization. *In* "The Representation of Cumulus Convection in Numerical Models of the Atmosphere" (K. A. Emanuel and D. J. Raymond, eds.), pp. 123–136. American Meteorological Society.

Arakawa, A., and K.-M. Xu (1990). The macroscopic behavior of simulated cumulus convection and semi-prognostic tests of the Arakawa–Schubert cumulus parameterization. *In* "Proceedings of the Indo–US Seminar on Parameterization of Sub-Grid Scale Processes in Dynamical Models of Medium Range Prediction and Global Climate," Pune, India. IITM.

Cheng, M.-D., and A. Arakawa (1994). Effects of including convective downdrafts and a finite cumulus adjustment time in a cumulus parameterization. *In* "Tenth Conference on Numerical Weather Prediction," Portland, Oregon, July 17–22, 1994, pp. 102–104.

Moorthi, S., and M. Suarez (1992). Relaxed Arakawa–Schubert: A parameterization of moist convection for general circulation models. *Mon. Wea. Rev.* **120**, 978–1002.

Nitta, T., and S. Esbensen (1974). Heat and moisture budget analyses using BOMEX data. *Mon. Wea. Rev.* **102**, 17–28.

Pan, D.-M., and D. R. Randall (1998). A cumulus parameterization with a prognostic closure. *Quart. J. Roy. Meteor. Soc.* **124**, 949–981.

Randall, D. A., and D.-M. Pan (1993). Implementation of the Arakawa–Schubert cumulus parameterization with a prognostic closure. *In* "The Representation of Cumulus Convection in Numerical Models of the Atmosphere" (K. A. Emanuel and D. J. Raymond, eds.), pp. 137–144. American Meteorological Society.

Randall, D. A., P. Ding, and D.-M. Pan (1997a). The Arakawa–Schubert parameterization. *In* "The Physics and Parameterization of Moist Atmospheric Convection," (R. K. Smith, ed.), pp. 281–296. Kluwer Academic Publishers, Netherlands.

Randall, D. A., D.-M. Pan, P. Ding, and D. G. Cripe (1997b). Quasi-equilibrium. *In* "The Physics and Parameterization of Moist Atmospheric Convection" (R. K. Smith, ed.), pp. 359–385. Kluwer Academic Publishers, Netherlands.

Yanai, M., S. Esbensen, and J.-H. Chu (1973). Determination of bulk properties of tropical cloud clusters from large-scale heat and moisture budgets. *J. Atmos. Sci.* **30**, 611–627.

Chapter 7

On the Origin of Cumulus Parameterization for Numerical Prediction Models

Akira Kasahara

National Center for Atmospheric Research, Boulder, Colorado

I. INTRODUCTION

By the year 1972, the development of atmospheric models for weather prediction and climate simulation had progressed well along with the plan for the Global Weather Experiment which took place in 1979. The Joint Organizing Committee (JOC) of the Global Atmospheric Research Programme convened a conference in Leningrad in March 1972 to increase the understanding of the physical processes of subgrid scales in the numerical models. The title of this JOC study conference was "Parameterization of SubGrid Scale Processes." As seen from the report (JOC, 1972), the parameterization of clouds and convection was discussed, including a brief historical review.

This chapter is an essay on the early history of cumulus parameterization mainly in the 1960s to the early 1970s. The author's approach here is slightly different from usual: The origin of cumulus parameterization is

General Circulation Model Development

traced as a necessary means to perform stable time integrations of the primitive equation atmospheric models with moist physical processes. It is hoped that this unorthodox approach will provide food for thought on the future development of cumulus parameterization.

In the early days of numerical weather prediction during the 1950s, most of the prediction models were formulated using a quasi-geostrophic assumption. Because fast moving gravity-inertia waves were eliminated in quasi-geostrophic models, a relatively large time step on the order of 1 hr could be used in the time integrations. However, researchers began to notice systematic errors in the forecasts produced by quasi-geostrophic models. They found that these errors were attributable to quasi-geo-strophic approximations used in forecast models, and that the use of the original, unmodified hydrostatic (primitive equation) models would do much to correct the deficiencies (Kasahara, 1996).

The nomenclature of "primitive equation" was introduced by Charney (1955), who made initial attempts to integrate the models on an early computer. However, the formulation of models was first developed by Richardson (1922). Because inertia-gravity modes are present in the primitive equation models, their use requires extra care in handling the time integration, including the need for a shorter time step than one used in quasi-geostrophic models of comparable horizontal resolution. Otherwise, large-amplitude inertia-gravity motions may develop and overwhelm slow moving, meteorologically significant motions.

Smagorinsky (1958) and Hinkelmann (1959) demonstrated that the primitive equation models can be integrated stably as an alternative to the use of quasi-geostrophic models. Many operational primitive equation prediction models were developed by Shuman (1962) at the National Meteorological Center, U.S. Weather Bureau (now the National Center for Environmental Prediction, NOAA); by Reiser (1962) at the German Weather Service; and by Gambo (1962) at the Japan Meteorological Agency. However, these models were adiabatic, and no effects of moist convection were considered.

II. TREATMENT OF CUMULUS CONVECTION IN TROPICAL CYCLONE MODELS

In November 1960, the first international conference on numerical weather prediction was held in Tokyo. (Subsequent meetings were held in Oslo in 1962 and Moscow in 1964.) There, I presented a paper on a numerical experiment on the development of a tropical cyclone using a primitive equation model in which the release of the latent heat of condensation is explicitly treated (Kasahara, 1961, 1962). A similar attempt was also discussed by Syōno (1962) at the conference. These were the first nonlinear primitive equation model calculations to attempt to explain the formation of a tropical cyclone, despite the fact that it was already known by then, through many linear models and observational studies, that condensation heating provides the major source of energy for tropical cyclones (e.g., Yanai, 1964).

The irony of these early attempts was that, instead of getting a typhoon, grid-scale cumulus cells were developed in, and dominated over, an initial weak cyclonic circulation. I presented my reasoning for this phenomenon, which was caused by a physical process in the model, rather than a numerical instability. At the end of my talk, Jule Charney made a rather long comment, which I reproduce here from the proceedings of the conference (Kasahara, 1962), because Charney pointed out the crux of cumulus parameterization.

> You touched upon a very important problem...namely you found in your numerical experiment that, instead of getting a typhoon, you get numerous cumulus clouds and you discussed the difficulty of how one suppresses cumulus clouds.... This will probably be the central problem in the whole theory of the formation of a tropical cyclone. Why does a large-scale convective system form when the motion in the atmosphere is apparently more unstable for cumulus-cloud scale? I mention this because Ogura and I have also made a calculation which, however, was not sufficiently complete to report on here. We take an attitude that a hurricane or a typhoon and the cumulus clouds do not compete, but they cooperate. That is to say, in the tropical depression the ascending motion is organized in such a way that the cumulus clouds actually cooperate to maintain the energy of the large-scale system.... How do you handle that in the numerical prediction scheme and isn't it very difficult to deal with both small and large scales? (p. 402).

In response to Charney's question, I answered as follows, again reproducing from the proceedings.

> As I have shown in the last slide, the scale of the convection which is produced by the release of latent heat is strongly dependent upon the magnitudes of the eddy viscosity and eddy diffusivity which we choose in the model. Unfortunately the magnitudes of these quantities are not well known and these are the only ambiguous quantities in the whole equations. If you use a small ordinary magnitude for the eddy viscosity, then, as you say, you will obtain only the motions of cumulus-cloud scale. However, from the fact that we still observe the development of large-scale motions in a typhoon, I have a feeling that such small-scale cumulus motions do provide a pump of energy supply by which the large-scale motions eventually develop. In this respect, I must say that cumulus clouds are integral parts of a typhoon. One must realize, however, that our task is not to describe an individual cell of cumulus convection, but to

describe the development of a typhoon system as a whole. So my attitude is to take into account the effect of cumulus clouds in a statistical manner in the prediction model for large-scale motions. One way which I presented here is to increase the magnitudes of the eddy viscosity and eddy diffusivity in the cumulus convection area and otherwise we use ordinary magnitudes for the quantities. In order to encourage cumulus convection being an important agency to supply energy of the typhoon development, I am also going to try to incorporate in the model more physical processes such as the "entrainment" of drier outside air into the updraft and the form "resistance" operating against cumulus growth.

What had happened in the typhoon models of Kasahara and Syōno is the manifestation of gravitational instability, caused by the lapse rate of temperature being steeper than the saturation moist adiabatic lapse rate, known as *conditional instability*. The way latent heat of condensation is calculated in these models is proportional to the vertical velocity at the location where heating is added. Thus, the conditional instability is met uniformally throughout the troposphere. One way to reduce the degree of gravitational instability is to cut the direct link between condensation heating and the collocated vertical velocity. Instead of calculating condensation heating using the *in situ* vertical velocity, Charney and Eliassen (1964) proposed to calculate condensation heating in proportion to the horizonal convergence of moisture into a vertical unit column. They envisioned that the moisture convergence in the frictional boundary layer generates tall cumulus clouds, which distribute heat to the environment. This idea came from their earlier work on the effect of surface friction in quasi-geostrophic flow (Charney and Eliassen, 1949). Then, Charney and Eliassen performed a stability analysis using a two-level quasi-balanced model and demonstrated that the growth rates of perturbations are flat with respect to a wide range of the horizontal scale of perturbation from cyclones to cumulus clouds. From this analysis, they concluded that the growth of tropical cyclones can be explained by the application of this type of diabatic heating in gradient-wind balanced hurricane models. Since this type of heating formulation will yield the growth of cyclone-scale motion in a conditionally unstable environment without causing the domination of cumulus-scale motion, Charney and Eliassen (1964) proposed to call the instability mechanism associated with this particular form of heating a *conditional instability of the second kind* (CISK) to distinguish it from the conditional instability (of the first kind) attributed to cumulus convection.

Actually, Charney and Eliassen (1964) did not present any numerical integration to demonstrate the growth of a tropical cyclone using a nonlinear hurricane model. However, Ogura (1964) conducted a numerical integration with a two-level quasi-balanced hurricane model using the specification of heating in the middle of the model atmosphere in the same way as was done by Charney and Eliassen (1964); namely, in proportion to the vertical velocity at the top of the frictional boundary layer. Incidentally, Ogura (1964) based his reasoning of choosing this particular heating formulation on the diagnostic analysis of precipitation in a typhoon performed by Syōno *et al.* (1951). Since the moisture convergence in the frictional boundary layer may be expressed proportional to the relative vorticity (Charney and Eliassen, 1949; Syōno, 1951), the distribution of precipitation in a typhoon would be proportional to a positive value of the relative vorticity at the top of the frictional boundary layer. In the three panels in Fig. 1, the solid curve shows the radial distribution of surface wind in a typhoon. As the typhoon passed through a surface observation station, the tangential wind component at the top of the boundary layer relative to the typhoon center can be estimated from observed surface

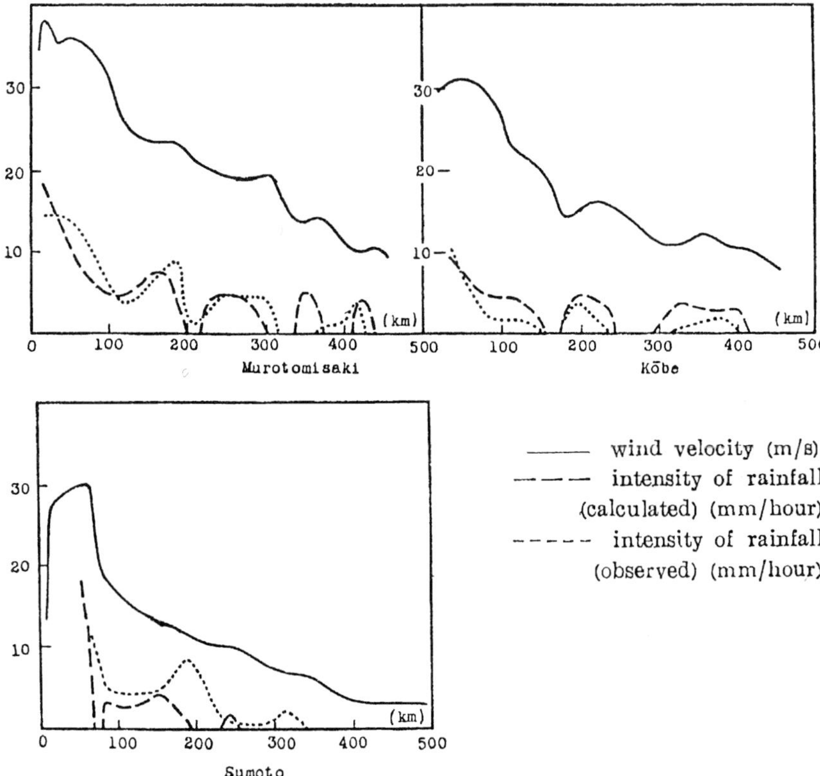

Figure 1 Distributions of surface wind (solid line), calculated (dashed line), and observed (dotted line) rainfall rates in Typhoon Jane, September 1950. (From Syōno *et al.*, 1951.)

winds by the assumption of axial symmetry. Then, the radial distribution of
the tangential wind in the typhoon was constructed from the time series of
surface wind observations at a particular station by knowing the distance
between the station and the center of the typhoon at a particular time. The
radial distribution of observed precipitation, represented by dotted lines, in
each panel was constructed in the same way applying the space and time
conversion to the observations at three different stations noted under each
abscissa. The dashed lines show the distribution of estimated precipitation
rate based on that of relative vorticity obtained from the distribution of
tangential wind. The precipitation was set to zero where the relative
vorticity was negative. In spite of many assumptions, the agreement between the observed and estimated precipitation rates was encouraging
enough to allow the parameterization of cumulus heating in a hurricane
model based on frictional mass convergence thinking. Although Ogura
(1964) was able to demonstrate the growth of cyclone-scale motion without
a contamination of cumulus-scale motions, the circulation did not approach a steady state.

At the time that the efforts of Charney, Eliassen, and Ogura were
made, independently Ooyama (1964) formulated a dynamical model for
the study of tropical cyclone development. In his two-layer quasi-balanced
cyclone model, Ooyama adopted the hypothesis that the rate of total heat
production by convective clouds in a vertical column is proportional to the
supply of water vapor into the column by the convergence of large-scale
inflow in the lowest atmospheric layer. At the first glance this hypothesis is
identical to the one adopted by Charney, Eliassen, and Ogura mentioned
earlier. However, there was a subtle difference in Ooyama's implementation of this hypothesis in his two-layer model in comparison with others: an
explicit recognition of cloud mass flux form for representation of heating
effects by convective clouds. In this connection, Ooyama introduced a
parameter, η, referred to as an "entrainment parameter." More specifically, in Ooyama's two-layer model for every unit mass of air that enters
from the boundary layer, $(\eta - 1)$ units of the lower layer air are entrained
into the cloud mass flux and η units of cloud mass flux enter into the
upper layer. Thus, the heating rate in this two-layer model can be interpreted to be proportional to η times the vertical velocity at the top of the
boundary layer. The value of η was determined from the energy balance of
the convective updraft as the ratio of the difference between the average
equivalent potential temperature of surface air and that of the lower layer
over the difference between the average equivalent potential temperature
of the upper layer and that of the lower layer.

It turned out that the parameter η played an important role in the
history of cumulus parameterization. Smith (1997a) wrote an informative

review on subtle differences in the implementation of the same hypothesis in their hurricane models by Charney, Eliassen, and Ooyama. The hypothesis is that the condensation heating is expressed proportional to the moisture flux at the top of boundary layer. And, there is that proportional factor, although the reasoning behind the introduction of the factor is very different from each other. Because Charney and Eliassen (1964) introduced the heating formulation based on this hypothesis in conjunction with the theory of CISK mentioned earlier, this particular way to represent the convective heating in numerical prediction models became known inappropriately as "CISK parameterization" with some choices of the value of parameter η. Smith (1997a) discusses some of the confusion coming from hypothetical specifications on the value of η in the ill-fated "CISK parameterization."

The numerical integration performed by Ooyama (1964) with his nonlinear two-layer cyclone model using a constant value of the parameter η did not reach a steady state, although he was able to show development of a hurricane-like vortex. In fact, his result was very similar to the finding of Ogura (1964) mentioned earlier. As reminisced by Ooyama (1997), it took him a few more years to comprehend this difficulty, and he was finally able to integrate the model to bring to a steady state as described in Ooyama (1969). The crucial step necessary to produce a steady-state cyclone was the formulation of time-dependent parameter η by considering the reduction of moist convective instability associated with the development of warm core vortex.

In those attempts to study the development of tropical cyclones with quasi-balanced hurricane models, the models consisted of only two degrees of freedom in the vertical (two layers or two levels). Therefore, it was not obvious how to specify the parameter η in the models with many degrees of freedom in the vertical. Or, to put it more generally, how we should formulate the vertical distribution of the cumulus heating function in the spirit of CISK? In this respect, Kuo (1965) made a more specific proposal that the cyclone-scale motion receives heating and moistening from cumulus cells through the mixing of air between the cumulus cells and their environment. This action tends to homogenize the differences between the respective temperature and moisture distributions throughout the conditionally unstable layer. Kuo (1974) extended the above idea further to formulate a cumulus parameterization of deep cumulus convection controlled by the convergence of moisture through the deep layer. His formulation was later improved by Anthes (1977a) by introducing a one-dimensional cloud model to replace Kuo's original idea of cloud mixing toward moist adiabat. This modified version of the cumulus parameterization has been extensively used in numerical prediction models.

Coming back to the period from the time that the idea of CISK was first introduced until Ooyama's (1969) article was published, many investigations took place to understand the role of the parameter η. Except for works of Kasahara and Syōno, earlier dynamical models used for the simulation of tropical cyclones adopted the assumption of gradient-wind balance similar in nature to the balance formulation of Charney (1962). Therefore, it remained to be answered what kind of instability would be produced in the primitive equation models if diabatic heating is specified proportional to the moisture convergence in the planetary boundary layer and how the heating profile in the vertical influences the characteristics of instability. Syōno and Yamasaki (1966) investigated this question and found different types of instability that were not present in the balanced models, depending on the vertical distribution of the diabatic heating rate. In the balanced models the manifestation of conditional instability is such that small-scale motions are discouraged in favor of producing large-scale motions, while in the primitive equation models rapid small-scale instability can dominate over slow growing large-scale instability unless the vertical distribution of diabatic heating takes a special condition. What is this special condition?

Yamasaki (1968a,b) made extensive numerical experiments using a primitive equation tropical cyclone model to find out what the relationship is between the vertical distribution of diabatic heating rate and the growth of cyclone-scale disturbances. The upshot of his study is, in simple terms, that in order to produce cyclone-scale disturbances, the diabatic heating function must take such a vertical distribution that conditional instability is reduced almost uniformly throughout the unstable layer, particularly in the upper troposphere. In fact, Yamasaki (1968c) demonstrated the growth of cyclone-scale motion in a conditionally unstable environment using a multilevel primitive equation model by adopting a heating function that is approximately proportional to the temperature difference between cumulus clouds and their environment. This is the same cumulus parameterization in essence as the one proposed by Kuo (1965, 1974). Later, applying various versions of the Kuo formulation as the cumulus parameterizations, many numerical experiments for the development of tropical cyclones were conducted by Rosenthal (1970a,b), Mathur (1974), and Anthes (1972, 1977b) using primitive equation models and Sundqvist (1970a,b) using a balanced model, with all having many degrees of freedom in the vertical. Since it is not the primary purpose of this essay to discuss the history of the study of tropical cyclones, an interested reader on this subject is referred to a monograph by Anthes (1982).

III. TREATMENT OF CUMULUS CONVECTION IN GENERAL CIRCULATION MODELS

In the mid-1960s, when research on the development of tropical cyclones was thriving, a great deal of research activity took place in the numerical simulation of atmospheric general circulation. This was stimulated by a successful experiment conducted by Phillips (1956) with a quasi-geostrophic model, but this time primitive equation models were used (Lewis, 1998). Dealing with primitive equation models having a full-blown physics package, including one for the moist physics, how did the researchers working in general circulation models (GCMs) cope with the problem of conditional instability? This question is important for understanding the role of cumulus convection in the global circulation of the atmosphere in much the same way as in the life cycle of tropical cyclones.

In an earlier work on GCM, Smagorinsky (1963) used a primitive equation model that was essentially a dry model in which the static stability was a fixed parameter. Even in Smagorinsky *et al.* (1965), the prediction of water vapor and the release of latent heat of condensation were not included explicitly in the model. Instead, the stabilizing effect of moist convection was emulated by adjusting the temperature lapse rate when it exceeded the moist adiabatic value. Therefore, the model was moist adiabatic in contrast to a dry formulation. A more satisfactory solution was proposed by Manabe *et al.* (1965) in which water vapor was treated as the prognostic variable and the condensation heating was explicitly included. When air is saturated, the temperature lapse rate is adjusted when it exceeds the moist adiabatic value (Fig. 2). The adjustment process assumes the conservation of moist entropy and the increase or decrease of the temperature is interpreted as heating or cooling due to condensation or evaporation, respectively. This procedure is referred to as *moist convective adjustment.* Even though the air is saturated, if the temperature lapse rate does not exceed the moist adiabatic value, no moist convection sets in and only stable condensation effects are calculated. If the air is not saturated, temperature adjustment is made only when the temperature lapse rate exceeds the dry adiabatic lapse rate. In that case, the temperature lapse rate is restored to the dry adiabatic lapse rate under the conservation of entropy.

The justification for adjusting the static stability of the atmosphere to account for the role of moist convection in the large-scale motions as a means to suppress conditional instability seems to be rooted in the practice

	$h<1$		$h\geq1$
$-\dfrac{\partial T}{\partial P}\leq\gamma_d$	No condensation, no convection $\delta r=0$ $\delta T'=0$ (No adjustment)	$-\dfrac{\partial T}{\partial P}\leq\gamma_m$	Large-scale condensation only $\delta r,\ \delta T$ from $\begin{cases} r+\delta r=r_s(T+\delta T,\ P) & (3.3) \\ c_p\delta T+L\delta r=0 & (3.4) \end{cases}$
$-\dfrac{\partial T}{\partial P}>\gamma_d$	Dry convection only $\delta r=0$ δT from $\begin{cases} \dfrac{\partial}{\partial P}\theta(T+\delta T,\ P)=0 & (3.1) \\ \dfrac{c_p}{g}\displaystyle\int_{P_B}^{P_T}\delta T\,dP=0 & (3.2) \end{cases}$	$-\dfrac{\partial T}{\partial P}>\gamma_m$	Moist convection and large-scale condensation $\delta r,\delta T$ from $\begin{cases} \dfrac{\partial}{\partial P}\theta_e(T+\delta T,\ r+\delta r,\ P)=0 & (3.5) \\ r+\delta r=r_s(T+\delta T,\ P) & (3.6) \\ \dfrac{1}{g}\displaystyle\int_{P_B}^{P_T}(c_p\delta T+L\delta r)dP=0 & (3.7) \end{cases}$

γ_d—dry adiabatic lapse rate.
γ_m—moist adiabatic lapse rate.
δr—adjustment of the mixing ratio of water vapor.
δT—adjustment of the temperature.
$P_T,\ P_B$—pressure at top and base of a dry or moist unstable layer containing two or more contiguous levels of the model.

θ—potential temperature.
θ_e—equivalent-potential temperature
h—relative humidity.
r_s—saturation mixing ratio.
g—acceleration of gravity.

Figure 2 Procedures of convective adjustment depending on whether the relative humidity is less than one (left column) or greater than one (right column) and whether the temperature lapse rate is less than the critical value (upper row) or greater than the critical value (lower row). The critical value is the dry adiabatic lapse rate if the environment is undersaturated and it is the moist adiabatic lapse rate in oversaturation. (From Manabe *et al.*, 1965.)

of so-called "convective adjustment" in the study of Manabe and Strickler (1964). It is well known that the lapse rate of temperature in the troposphere becomes even greater than the dry adiabatic lapse rate when only radiative processes are considered. In fact, this is why vigorous overturning of air is expected in the tropics where heating due to insolation exceeds cooling due to outgoing long-wave radiation. Thus, the simplest way to incorporate the process of air mass overturning is to adjust the temperature lapse rate when it exceeds either the dry or moist adiabatic lapse rate depending on whether the air is saturated or not. It turns out that the convective adjustment has the virtue of totally suppressing gravitational instability. Therefore, the pathological difficulty encountered in the earlier typhoon model calculations was bypassed in the GCM calculations by the application of convective adjustment.

In parallel with the development of GCM at the Geophysical Fluid Dynamics Laboratory, NOAA, many GCMs were developed about the same time by Mintz (1965) and A. Arakawa at UCLA, Leith (1965) at the Lawrence Livermore Laboratory, Kasahara and Washington (1967) at NCAR, and so on. In these GCMs, a variety of techniques were adopted in handling moist convection similar in spirit to convective adjustment. However, Mintz (1965) employed the adjustment of static stability in a similar way as Smagorinsky *et al.* (1965), since in both studies the explicit moisture forecast was not performed. Leith (1965) included the prediction of moisture, but the condensation heating rate was reduced empirically as a function of static stability. Kasahara and Washington (1967) adopted a convective adjustment scheme to control gravitational instability. In those days, the convective adjustment approach was used extensively for hemispherical and global numerical weather prediction models (e.g., Shuman and Hovermale, 1968; Miyakoda *et al.*, 1969). Thus, in the case of GCM calculations we have not had any intriguing drama to speak of in contrast to the case of tropical cyclone modeling.

Because the practice of convective adjustment to control conditional instability in the primitive equation models was in vogue already in the early 1960s, it is an interesting question to ask how much the history of the numerical modeling of tropical cyclones was altered and whether a controversy surrounding the terminology of "CISK" has ever developed (cf. Smith, 1997a). In fact, it is rather surprising to see that the application of convective adjustment as a cumulus parameterization for tropical cyclone modeling came rather late. In this connection, Kurihara (1973) developed a variation of convective adjustment that was designed to improve the original procedure of Manabe *et al.* (1965) by equilibrating the environmental temperature toward a reference state of hypothetical deep cloud.

Kurihara and Tuleya (1974) applied this adjustment scheme to a three-dimensional simulation of tropical cyclone development.

As seen from the early histories of tropical cyclone modeling and general circulation experiments just presented, there were two approaches of independent origin for attempting to stably perform the time integration of primitive equation models in a conditionally unstable environment with the explicit feedback of moisture calculation. In fact, tracing the roots of these two approaches will help to understand the concept of CISK, which is often misunderstood in a variety of ways. It is unfortunate that the "theory of CISK" often refers to the shift of instability from cumulus to cyclone scales, resulting from a particular type of diabatic heating that is proportional to the vertical motion at the top of the frictional layer. I would rather like to regard CISK as a working concept in which cumulus clouds act to liberate conditional instability and provide an energy source of latent heat for the large-scale circulations. From the standpoint of numerical modeling, I prefer to look on cumulus parameterization as a means of realizing CISK in large-scale circulation models. As Ooyama (1982) put it, "The present author views CISK in terms of the conceptual content that has grown and matured with advances in modeling work. Then, the spirit of CISK as the cooperative intensification theory is valid and alive" (p. 377).

IV. ADVENT OF ARAKAWA–SCHUBERT CUMULUS PARAMETERIZATION

In 1968, another international symposium on numerical weather prediction was held in Tokyo. Again, its proceedings serve as a valuable historical document. Many papers were presented that demonstrated the utility of primitive equation models for studying hurricane development, simulating the general circulation of the atmosphere, and even medium-range weather forecasting as referred to earlier in Section III. In addition, active discussions took place at the symposium concerning the improvement of various physical processes in the prediction models. With respect to cumulus parameterization, Arakawa (1969) presented a new cumulus scheme for the UCLA Mintz–Arakawa general circulation model. Although this scheme was soon superseded by a more elaborate formulation, known as the Arakawa–Schubert scheme and discussed later, it was the first attempt to express the vertical distributions of condensation heating and moistening rates in GCMs through the introduction of cloud types. Figure 3 shows one of the three cloud types considered by Arakawa (1969). In this

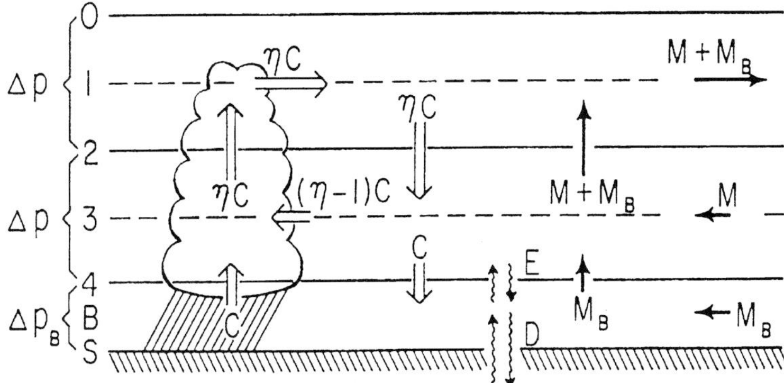

C : total upward mass flux from the boundary layer into the clouds.

$(\eta{-}1)C$: total horizontal mass flux from the surrounding air into the

clouds in layer 3.

$\eta > 1$: entrainment,

$\eta < 1$: detrainment.

ηC : total upward mass flux in the clouds at the middle-level 2.

This is also the total mass flux from the clouds into the surrounding

air in layer 1.

M_B : large-scale mass convergence in the planetary boundary layer B.

M : large-scale mass convergence in layer 3.

E : mass exchange rate, by a turbulent eddy process, between layers

B and 3.

D : mass exchange rate, by turbulent eddy process, between the

planetary boundary layer B and a thin surface layer.

Figure 3 One of the three types of clouds considered in formulating a cumulus parameterization by A. Arakawa in his effort to interpret the practice of convective adjustment under the concept of CISK. (From Arakawa, 1969.)

particular type of cloud, as reminisced by Arakawa (1997) himself, he adopted an entraining cloud model similar to the one proposed by Ooyama (1964, 1969), taking note of the entrainment parameter η. However, because Arakawa's model had three vertical levels instead of the one heating level in Ooyama's model, Arakawa designed the cumulus-induced mass circulation as seen in Fig. 3 to include the detrainment and cumulus-induced subsidence effects on the large-scale environment.

Before Arakawa's formulation, the modeling of cumulus convection had not been fully ingrained in the design of cumulus parameterization to realize the concept of CISK in primitive equation models. However, there were some earlier attempts at cumulus modeling with the intention of designing a cumulus parameterization to represent the vertical transport of heat, water vapor, and momentum by deep cumulus clouds as an alternative to the moist convective adjustment procedure. Asai and Kasahara (1967) and Kasahara and Asai (1967) proposed a model of cumulus ensemble as a collection of clouds of the same kind, which consists of ascending and descending columns as shown in Fig. 4. To evaluate the effects of an ensemble of such convective elements on the large-scale environment, one important question is how to determine the population of model clouds. It was hypothesized that the ratio between the updraft radius, a, and the downdraft radius, b, shown in Fig. 4, is determined in such a way that the vertical transport of heat is maximized. For a small updraft, the vertical heat flux is expected to be small. However, if the updraft size is too large, then the compensating downward motion acts too strongly as a brake to the updraft. Therefore, there is an optimal ratio between the sizes of updraft and downdraft that determines the cloud population. It was found that the most active cloud ensemble appears

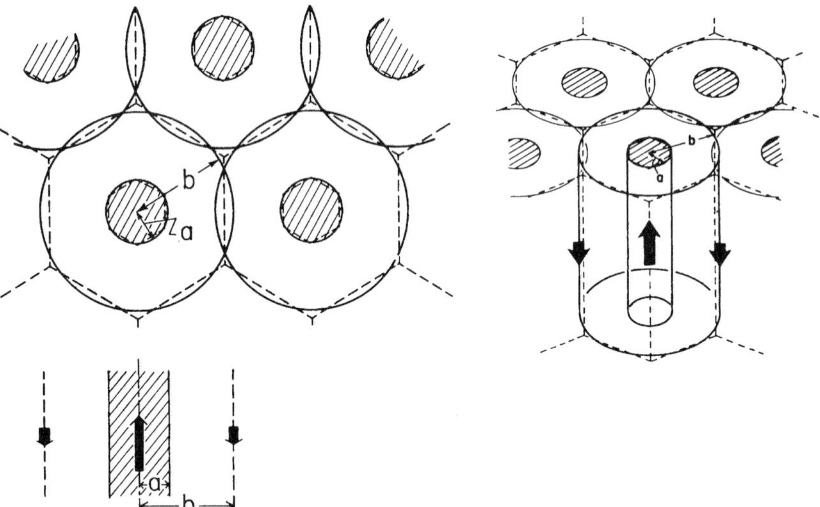

Figure 4   A model of uniform cloud ensemble in which the downdraft and the updraft play important roles in determining cloud population. A closure assumption is needed to determine the ratio $\sigma(\equiv a/b)$. (From Asai and Kasahara, 1967.)

when the cloud towers occupy several percent of a given domain. This agrees with the conventional wisdom that the coverage of cumulonimbus in the central part of a hurricane is around several percent (Malkus *et al.*, 1961), although this number could be higher depending on how narrowly the domain of the vortex is defined. This may be interpreted to mean that cumulus clouds are formed in tropical cyclones to carry heat upward with the most efficient rate.

Three factors are involved in the design of cumulus parameterization under the scenario of cooperative interaction between cumulus convection and its environment. First, one must decide what kind of cloud model is to be considered. Up to this point, it had been assumed that cumulus convection consisted of clouds of the same size. However, Ooyama (1971) attempted to devise a cumulus parameterization by considering an ensemble of clouds of different sizes dispatched from the top of the frictional layer. However, he did not elaborate on how the spectral distribution of cloud mass flux at the cloud base, called the "dispatcher function," should be determined.

Next, one must formulate the physical processes by which the ensemble of clouds acts as the source of heat and moisture in the large-scale environment. Many authors addressed this topic around 1970. The most representative work in this category seems to be that of Yanai *et al.* (1973). They have formulated the apparent heat source Q_1, which consists of radiative heating, heating/cooling due to condensation/evaporation of rain droplets, and the vertical flux divergence of sensible and latent heat due to cumulus convection. Similarly, they defined Q_2 as a measure of the apparent moisture sink, which is due to the net condensation and the vertical divergence of moisture flux by cumulus convection (Fig. 5).

Having chosen the model of cloud ensemble and how the formulation of the vertical divergence of sensible and latent heat and moisture transport by cumulus convection through such a cloud model is completed, the last

$$Q_1 \equiv \frac{\partial \bar{s}}{\partial t} + \overline{\nabla \cdot s\mathbf{V}} + \frac{\partial \bar{s}\bar{\omega}}{\partial p} = Q_R + L(c - e) - \frac{\partial}{\partial p}\,\overline{s'\omega'}$$

$$Q_2 \equiv -L\left(\frac{\partial \bar{q}}{\partial t} + \overline{\nabla \cdot q\mathbf{V}} + \frac{\partial \bar{q}\bar{\omega}}{\partial p}\right) = L(c - e) + L\frac{\partial}{\partial p}\,\overline{q'\omega'}$$

Figure 5　Definitions of Q_1 and Q_2 from Yanai *et al.* (1973). Symbols: S, dry static energy; $\mathbf{V}$, wind velocity; p, pressure; ω, p velocity; q, specific humidity; Q_R, radiative heating rate; c, condensation rate; e, evaporation rate of cloud water; L, latent heat of condensation. Also, bar and prime denote, respectively, area mean and the deviation from the area mean.

step of cumulus parameterization is to determine the cumulus mass flux at the cloud base, i.e., the dispatcher function. Diagnostically, the cumulus mass flux can be estimated if Q_1 and Q_2 are known as done, for example, by Ogura and Cho (1973) using observations of Q_1 and Q_2 by Nitta (1972). Prognostically, the cumulus mass flux must be determined *a priori* for calculations of Q_1 and Q_2. This is the art of cumulus parameterization. How can this be achieved?

With the background of these developments just addressed, Arakawa and Schubert (1974) proposed one of the most sophisticated cumulus parameterizations available today. A buoyant plume model including condensation and dynamical entrainment/detrainment processes was set up to represent the structure of the individual cloud. Then, the cumulus mass flux was expressed as the summation of cloud element mass flux with respect to all sizes. Namely, the cumulus flux is represented spectrally as an integral of cloud element mass flux as the function of a dimensionless parameter denoting the spatial scale of each cloud. One such parameter is the rate of entrainment, which determines the size of the cloud. Now, here comes the crucial question: What determines the cloud element mass flux? In other words, how does one represent the cloud element mass flux in terms of the large-scale environmental variables of temperature, moisture, etc.? More specifically, what has to be determined is the cloud element mass flux at the cloud base, since the vertical structure of each cloud element is already known from the cloud model.

Arakawa and Schubert (1974) introduced the idea of quasi-equilibrium as a closure of this problem. Before discussing how this closure works, let us write the kinetic energy equation of the cloud element in the following schematic manner:

$$\frac{dK(s)}{dt} = A(s)Mb(s) - D(s), \tag{1}$$

where s denotes a parameter representing the scale of cloud; $K(s)$ and $D(s)$ are, respectively, the kinetic energy and its dissipation rate of cloud s. Here, $Mb(s)$ denotes the cloud element mass flux at the cloud base, and $A(s)$ is the work done by each cloud element due to the buoyancy as adopted customarily in the cloud model, which represents the kinetic energy generation per unit mass flux. Therefore, we need the condition of $A(s) > 0$, namely, conditional instability to generate convection.

Equation (1) describes the temporal evolution of the kinetic energy $K(s)$. In a conditionally unstable environment in which $A(s) > 0$, the magnitude of $K(s)$ would be small initially when the cloud starts to form due to some triggering. However, $K(s)$ starts to increase very quickly as the cloud grows. In fact, $K(s)$ will increase exponentially in time unless the buoyancy term, $A(s)$, is controlled to prevent its catastrophic growth and/or the energy dissipation rate, $D(s)$, becomes sufficiently large to offset the growth of cloud. The simplest way to control this runaway growth of $K(s)$ is to make the time derivative of $K(s)$ on the left-hand side of Eq. (1) zero. This gives

$$A(s) = D(s)/Mb(s). \tag{2}$$

Because the dissipation rate, $D(s)$, is relatively small, Eq. (2) implies that $A(s)$ nearly vanishes. This is essentially what the convective adjustment method dictates. However, cumulus clouds in nature do not liberate conditional instability instantaneously. Therefore, the balance condition that $dK(s)/dt$ vanishes is not entirely desirable.

The idea of the quasi-equilibrium closure can be interpreted as a higher order balance approach to control the runaway growth of $K(s)$. Now, instead of the first derivative of $K(s)$ vanishing, let us assume that the second derivative of $K(s)$ with respect to time vanishes. This higher order scheme has the virtue of preserving relationship (1), yet controlling the fast temporal growth of $K(s)$.

Because the dissipation term $D(s)$ is generally small, and the temporal variation of $Mb(s)$ is smaller than that of $A(s)$, the condition that the second derivative of $K(s)$ with respect to time vanishes can be approximated by

$$\frac{dA(s)}{dt} = 0. \tag{3}$$

Arakawa and Schubert (1974) call $A(s)$ the "cloud work function." They express the temporal derivative of $A(s)$ as the sum of the contributions from clouds and their large-scale environment as follows.

$$\frac{dA(s)}{dt} = \left[\frac{dA(s)}{dt}\right]_c + \left[\frac{dA(s)}{dt}\right]_{ls}. \tag{4}$$

Keep in mind that terms $dA(s)/dt$ for clouds and their large-scale environment, respectively, on the right-hand side of Eq. (4) involve the temporal changes of temperature T and specific humidity q, which can be expressed through the respective prediction equations. Thus, the second term on the right-hand side of Eq. (4) represents the large-scale forcing term for cloud, s, denoted by $F(s)$.

The first term on the right-hand side of Eq. (4) is the temporal change of energy generation by clouds, which can be expressed through the use of the cloud model equations as

$$\left[\frac{dA(s)}{dt}\right]_c = \int_0^{s_{max}} G(s, s')Mb(s')\,ds', \tag{5}$$

where $G(s, s')$ denotes the interaction coefficients of $Mb(s')$, which represent the influence of cloud type s' on the temporal change of $A(s)$. Parameter s_{max} denotes the maximum cloud size in the cloud ensemble.

By combining Eqs. (3)–(5), the quasi-equilibrium closure of Eq. (3) gives

$$\int_0^{s_{max}} G(s, s')Mb(s')\,ds' = -F(s). \tag{6}$$

Thus, $Mb(s)$ is obtained as the solution of integral equation (6) for a given forcing term F and the expression of the kernel $G(s, s')$, which involves the cloud model. In practice, the solution of Eq. (6) is difficult to obtain and many simplifications to solve Eq. (6) have been proposed. One such simplification is proposed by Moorthi and Suarez (1992) who recognize that the dominant terms of $G(s, s')$ are those of self-interaction, i.e., $G(s, s)$. Therefore, by neglecting the interactions involving different types of clouds, one immediately gets

$$Mb(s) = -F(s)/[G(s, s)\Delta s], \tag{7}$$

where Δs denotes a finite cloud scale interval. Once $Mb(s)$ is determined, the sensible heat and moisture transports by cloud s and its precipitation rate can be obtained.

The view that the quasi-equilibrium closure of Arakawa and Schubert can be looked on as setting the second derivative of $K(s)$ with respect to time to be vanished is shared by Miyakoda and Sirutis (1989). They proposed application of the bounded derivative method of Kreiss (1979, 1980) to formulate the hierarchy of physical closures in order to parameterize the subgrid scale processes of cumulus convection and boundary layer turbulence. The bounded derivative method was developed to deal with the motions of multiple time scales in a physical system in a well-behaved manner. A specific application of the bounded derivative method to the primitive equations was discussed by Browning *et al.* (1980). Because fast and slow time scale motions are involved in the primitive equations, the fast time scale (noise) motions may overwhelm the slow motions of meteorological interest during the time integration of the primitive equations unless the initial conditions are suitably adjusted (Hinkelmann, 1951).

The process of adjusting the input data for the prediction models to ensure that the fast time scale motions are under control is referred to as "initialization." A breakthrough to this nagging question since the time of Richardson (1922) was made by Machenhauer (1977) and, independently, by Baer and Tribbia (1977) when they proposed so-called "nonlinear normal mode initialization" (NNMI). The basic idea of NNMI is not to entirely eliminate the fast time scale components from the input data, but to set weak fast time scale components in such a way that those time scale components do not grow undesirably in time (Leith, 1980). A connection between the idea of NNMI and the bounded derivative principle in adjusting the input data to initialize the primitive equation models was discussed by Kasahara (1982).

When multiple time scale motions are involved in one physical system such as atmospheric models, the motions of all time scales must behave well mathematically in the temporal evolution of the principal motions of our concern, namely, large-scale synoptic motions. The fast growth of grid-point cumulus clouds, despite their importance as an energy source to the large-scale motions, must be controlled. A new cumulus parameterization may be developed based on the premise that the third derivative of $K(s)$ with respect to time vanishes. Presumably, the higher the degree of the time derivatives to be bounded, the higher the degree of approximation to the "super-balanced state" as Lorenz (1980) has demonstrated in the time integration of low-order systems. The application of the higher order methods to a complex physical system is not necessarily practical. Nevertheless, understanding of what can be done helps bring us peace of mind.

V. EPILOGUE

It is not the objective of this chapter to discuss many developments in cumulus parameterization since the advent of the Arakawa and Schubert formulation. The interested reader on this topic is referred to the monograph of Emanuel and Raymond (1993), which provides excellent discussions on many aspects of cumulus parameterization schemes available today. Another useful source of information on the topic of cumulus parameterization is a recent book edited by Smith (1997b), which is a collection of lectures at an Advanced Study Institute of the North Atlantic Treaty Organization (NATO). Reference to those two books will help readers learn about the many advances that have been made during the last quarter century, not only in the design of cumulus parameterization for numerical prediction models, but also in understanding the morphology of moist atmospheric convection through the diagnostic analyses of observations from various international fields programs, such as GATE (Global Atmospheric Research Program Atlantic Tropospherical Experiment) and TOGA COARE (Tropical Ocean Global Atmosphere Coupled Ocean Atmosphere Response Experiment). However, I may not be alone in feeling that the problem of designing a suitable cumulus parametrization for climate models is far from solved and discouragingly difficult. My intent in writing this chapter is to reflect on the early history of cumulus parametrization with the hope that I can learn a lesson from the frontier story in facing up to the enormous challenges of dealing with this issue in the future modeling of weather prediction and climate simulation.

Clearly, there are two somewhat independent roots in the early history of cumulus parameterization. One is the concept of CISK as a theory of cooperative interactions between cumulus convection and its environment for cyclone-scale development. Charney first mentioned this idea in connection with the theory of tropical cyclone formation. Ooyama gave a lot of thought to how to parameterize cumulus convection in his quest to explain the mechanism of tropical cyclone development. (One should not overlook Ooyama's emphasis on the important role of air–sea interactions as an energy source of tropical cyclones, although this is not directly connected to the present subject.) Finally, Arakawa and Schubert succeeded in formulating the concept of cooperative interactions as cumulus parameterization in a closed form through the hypothesis of quasi-equilibrium of the cloud work function.

The other root is the practice of convective adjustment to stably integrate the primitive equation models in a conditionally unstable moist environment, such as in the tropics. Many variants of the convective adjustment scheme, originally proposed by Manabe, are being used successfully in the meso-scale models, as well as large-scale circulation models. One thing is common in cumulus schemes from the two roots: the role of liberating conditional instability more or less uniformly throughout the troposphere. The scheme proposed by Kuo, which has been used extensively, seems to be a blend of the two attributes of CISK and convective adjustment. The reason why it is so difficult to design an ideal cumulus parameterization is that the spatial and temporal scales of motion required to liberate conditional instability are different depending on the atmospheric phenomena of interest. Cumulus convection in nature does not liberate conditional instability instantaneously. The degree of instability left in the large-scale environment at a particular location and time dictates the subsequent development of the variety of weather disturbances.

The quest of searching for a suitable cumulus parametrization for climate models must continue. Lately, I have been interested in a modeling study to evaluate the impacts of global warming on various aspects of tropical cyclones (TCs). Tsutsui and Kasahara (1996) examined the question on how well a global climate model with a horizontal grid resolution of approximately 300 km can simulate the behaviors of TC-like disturbances in long-term climate simulations. A noteworthy finding in that study is that the T42 resolution NCAR CCM2 (Community Climate Model Version 2; see Hack *et al.*, 1993) simulated quite realistically the geographical and seasonal variations of observed TCs. However, I observed recently that the T42 resolution NCAR CCM3 does not simulate TC-like vortices very well. The CCM3 (Kiehl *et al.*, 1996) is an improved version of the NCAR Community Climate Model, which reproduces climatology far more realistically than CCM2 particularly when CCM3 is used as the atmospheric component in a coupled atmosphere–ocean–land climate system. This is well documented in many articles published in the Climate System Model special issue of the *Journal of Climate*, June 1998.

As far as I am aware, one major difference in the physics package of CCM3 from that of CCM2 is the use of a combination of Hack (1994) and Zhang and McFarlane (1995) schemes as the cumulus parameterization in CCM3 versus just the Hack scheme by itself in CCM2. Clearly, the cumulus parameterization in CCM3 works well in the simulation of mean climatology, while not producing realistic looking TC vortices; and vice versa in the case of CCM2. The question then is this: Can we design a suitable cumulus parameterization in relatively low-resolution climate models that can simulate TC-like vortices in a reasonable degree as well as reproduce a satisfactory climatology of large-scale circulations? One can argue that a higher horizontal resolution model is needed to meet such a dual goal. Clearly, testing a cumulus scheme in a high-resolution model in climate simulation mode is comupter intensive. How to resolve this difficulty will remain our challenge in the 21st century.

As computer capabilities in both speed and memory storage increase in the future, more effort will be put into developing high-spatial-resolution numerical models, including all relevant physical processes that govern the motions of all scales. As our application of numerical modeling expands from weather prediction to climate projection, careful consideration of the hydrological cycle becomes important. Thus, the prognostic treatment should be made for the calculation of liquid water and ice in the atmosphere, including cloud physics, along with the prognostic calculation of water vapor. Of course, this will not necessarily eliminate the need for subgrid scale parameterization to control an excessive growth, if any, of physical instabilities. The interpretation of the quasi-equilibrium closure of the Arakawa–Schubert cumulus parameterization from the viewpoint of the bounded derivative principle is intended to suggest an approach to controlling the catastrophic behaviors of fast time scale physical phenomena in light of well-explored solutions to initialization of the primitive equation models. A reader interested in the subject of the initialization of the primitive equation models is referred to a textbook by Daley (1991).

The inclusion of the prognostic calculation of liquid water and ice in the atmospheric model forces us to set up the initial conditions of liquid and solid water content and precipitation rate, as well as the distribution of water vapor. Unfortunately, the hydrological aspect in numerical prediction models is one of the weakest parts from the standpoint of both modeling and observation. The water vapor analysis at operational centers relies heavily on the first-guess field due to the shortage of radiosonde water vapor observations. Because different cumulus schemes are used by various operational centers and produce different precipitation rates, the analyzed moisture fields at various operational centers differ markedly depending on the choice of cumulus schemes, as well as the quality of moisture observations (Kasahara *et al.*, 1996). However, progress is being made at operational centers to improve moisture analysis through the use of four-dimensional (4-D) variational data assimilation by incorporating the Special Sensor Microwave/Imager (SSM/I) data on precipitable water and precipitation (Andersson *et al.*, 1993). As another means to measure atmospheric water vapor, a prospect is on the horizon to design an extensive observation network of precipitable water inferred from the delay of signals propagating from the global positioning system (GPS) satellites to ground-based GPS receivers (Bevis *et al.*, 1992).

In reviewing the book by G. K. Batchelor on the life and legacy of G. I. Taylor, Donnelly (1997) mentions that Taylor is quoted as saying, "... in general it seems to me that it is through particular problems which can be subjected to experimental verification or compared with natural phenomena that most advances are made" (p. 82). It is essential to improve the quality of various moisture and precipitation observations and the techniques for hydrological analyses in order to advance the state of the art in cumulus parameterization research.

ACKNOWLEDGMENTS

The National Center for Atmospheric Research (NCAR) is sponsored by the National Science Foundation. In writing this essay, I benefitted from numerous discussions with my colleagues, many of whom are the authors of the articles listed in the references. In addition, I would like to thank Rick Anthes, John Lewis, Brian Mapes, Vic Ooyama, Phil Rasch, Junichi Yano, and an anonymous reviewer who read earlier versions of this manuscript and gave me useful comments. My interest in the problem of cumulus parameterization is stimulated by my renewed study on tropical cyclones, conducted through a research collaboration between NCAR and the Central Research Institute of Electric Power Industry, Japan. The manuscript was typed by Barbara Ballard.

REFERENCES

Andersson, E., J. Pailleux, J.-N. Thèpaut, J. R. Eyre, P. McNally, G. A. Kelly, and P. Courtier (1993). Use of radiances in 3D/4D variational data assimilation. *In* "Workshop Proc. on Variational Assimilation, with Special Emphasis on Three-Dimensional Aspects," pp. 123–156. European Centre for Medium-Range Weather Forecasts, Shinfield Park, Reading, UK.

Anthes, R. A. (1972). Development of asymmetries in a three-dimensional numerical model of the tropical cyclone. *Mon. Wea. Rev.* **100,** 461–476.

Anthes, R. A. (1977a). A cumulus parameterization scheme utilizing a one-dimensional cloud model. *Mon. Wea. Rev.* **105,** 270–286.

Anthes, R. A. (1977b). Hurricane model experiments with a new cumulus parameterization scheme. *Mon. Wea. Rev.* **105,** 287–300.

Anthes, R. A. (1982). "Tropical Cyclones: Their Evolution, Structure, and Effects," *Meteor. Monog.* **19.**

Arakawa, A. (1969). Parameterization of cumulus convection. *In* "Proc. WMO/IUGG Symposium on Numerical Weather Prediction in Tokyo," Nov. 1968. pp. IV-8-1–IV-8-6. Japan Meteor. Agency.

Arakawa, A. (1997). Cumulus parameterization: An ever-challenging problem in tropical meteorology and climate modeling. *In* "Preprint Volume, 22nd Conference on Hurricanes and Tropical Meteorology," Ft. Collins, Colorado, pp. 7–12. American Meteorological Society, Boston.

Arakawa, A., and W. H. Schubert (1974). Interaction of a cumulus cloud ensemble with the large-scale environment, Part I. *J. Atmos. Sci.* **31,** 674–701.

Asai, T., and A. Kasahara (1967). A theoretical study of the compensating downward motions associated with cumulus clouds. *J. Atmos. Sci.* **24,** 487–496.

Baer, F., and J. J. Tribbia (1977). On complete filtering of gravity modes through nonlinear initialization. *Mon. Wea. Rev.* **105,** 1536–1539.

Bevis, M., S. Businger, T. A. Herring, C. Rocken, R. Anthes, and R. H. Ware (1992). GPS meteorology: Remote sensing of atmospheric water vapor using the global positioning system. *J. Geophys. Res.* **97,** D14, 15,787–15,801.

Browning, G., A. Kasahara, and H. O. Kreiss (1980). Initialization of the primitive equations by the bounded derivative method. *J. Atmos. Sci.* **37,** 1424–1436.

Charney, J. G. (1955). The use of the primitive equations in numerical weather prediction. *Tellus* **7,** 22–26.

Charney, J. G. (1962). Integration of the primitive and balance equations. *In* "Proc. Int. Symposium on Numerical Weather Prediction in Tokyo," Nov. 1960, pp. 131–152. Meteor. Soc. Japan.

Charney, J. G., and A. Eliassen (1949). A numerical method for predicting the perturbations of the middle latitudes westerlies. *Tellus* **1,** 38–54.

Charney, J. G., and A. Eliassen (1964). On the growth of the hurricane depression. *J. Atmos. Sci.* **21,** 68–75.

Daley, R. (1991). "Atmospheric Data Analysis." Cambridge Univ. Press, Cambridge, MA.

Donnelly, R. J. (1997). A book review on "The life and legacy of G. I. Taylor by G. Batchelor." *Phys. Today*, June issue, p. 82.

Emanuel, K. A., and D. J. Raymond (ed.) (1993). The representation of cumulus convection in numerical models. *Meteor. Monog.* **24**(46).

Gambo, K. (1962). The use of the primitive equations in balanced condition. *In* "Proc. International Symposium on Numerical Weather Prediction in Tokyo," Nov. 1960, pp. 121–130. Meteor. Soc. Japan.

Hack, J. J. (1994). Parameterization of moist convection in the National Center for Atmospheric Research Community Climate Model (CCM2). *J. Geophys. Res.* **99**, 5551–5568.

Hack, J. J., B. A. Boville, B. P. Briegleb, J. T. Kiehl, P. J. Rasch, and D. L. Williamson (1993). Description of the NCAR Community Climate Model (CCM2), NCAR Tech. Note NCAR/TN-336 + STR. NCAR.

Hinkelmann, K. (1951). Der Mechanismus des meteorologischen Lärmes. *Tellus* **3**, 285–296.

Hinkelmann, K. (1959). Ein numerisches Experiment mit den primitive Gleichungen. *In* "The Atmosphere and the Sea in Motion; Rossby Memorial Volume," (B. Bolin, ed.), pp. 486–500. Rockefeller Institute Press.

Joint Organizing Committee (1972). Parameterization of sub-grid scale processes, GARP Publication Series No. 8. World Meteor. Org.

Kasahara, A. (1961). A numerical experiment on the development of a tropical cyclone. *J. Meteor.* **18**, 259–282.

Kasahara, A. (1962). The development of forced convection caused by the released latent heat of condensation in a hydrostatic atmosphere. *In* "Proc. International Symposium on Numerical Weather Prediction in Tokyo," Nov. 1960, pp. 387–403. Meteor. Soc. Japan.

Kasahara, A. (1982). Nonlinear normal mode initialization and the bounded derivative method. *Rev. Geophys. Space Phys.* **20**, 385–397.

Kasahara, A. (1996). Primitive equations. *In* "Encyclopedia of Climate and Weather" (S. H. Schneider, ed.), Vol. 2, pp. 612–616. Oxford University Press, New York.

Kasahara, A., and T. Asai (1967). Effects of an ensemble of convective elements on the large-scale motions of the atmosphere. *J. Meteor. Soc. Japan* **45**, 280–291.

Kasahara, A., and W. M. Washington (1967). NCAR global general circulation model of the atmosphere. *Mon. Wea. Rev.* **95**, 389–402.

Kasahara, A., J. Tsutsui, and H. Hirakuchi (1996). Inversion methods of three cumulus parameterizations for diabatic initialization of a tropical cyclone model. *Mon. Wea. Rev.* **124**, 2304–2321.

Kiehl, J. T., J. J. Hack, G. B. Bonan, B. A. Boville, B. P. Briegleb, D. L. Williamson, and P. J. Rasch (1996). Description of the NCAR Community Climate Model (CCM3). NCAR Tech. Note NCAR/TN-420 + STR. NCAR.

Kreiss, H. O. (1979). Problems with different time scales for ordinary differential equations. *SIAM J. Num. Anal.* **16**, 980–998.

Kreiss, H. O. (1980). Problems with different time scales for partial differential equations. *Commun. Pure Appl. Math.* **33**, 399–439.

Kuo, H. L. (1965). On the formation and intensification of tropical cyclones through latent heat released by cumulus convection. *J. Atmos. Sci.* **22**, 40–63.

Kuo, H. L. (1974). Further studies of the parameterization of the influence of cumulus convection on large-scale flow. *J. Atmos. Sci.* **31**, 1232–1240.

Kurihara, Y. (1973). A scheme of moist convective adjustment. *Mon. Wea. Rev.* **101**, 547–553.

Kurihara, Y., and R. E. Tuleya (1974). Structure of a tropical cyclone developed in a three-dimensional numerical simulation model. *J. Atmos. Sci.* **31**, 893–919.

Leith, C. (1965). Numerical simulation of the earth's atmosphere. *Meth. Comput. Phys.* **4**, 1–28.

Leith, C. (1980). Nonlinear normal mode initialization and quasi-geostrophic theory. *J. Atmos. Sci.* **37**, 958–968.

Lewis, J. M. (1998). Clarifying the dynamics of the general circulation: Phillips's 1956 experiment. *Bull. Am. Meteor. Soc.* **79**, 39–60.

Lorenz, E. N. (1980). Attractor sets and quasi-geostrophic equilibrium. *J. Atmos. Sci.* **37**, 1685–1699.

Machenhauer, B. (1977). On the dynamics of gravity oscillations in a shallow water model, with applications to normal mode initialization. *Beitr. Phys. Atmos.* **50**, 253–275.

Malkus, J. S., C. Ronne, and M. Chaffee (1961). Cloud patterns in hurricane Daisy, 1958. *Tellus* **13**, 8–30.

Manabe, S., and R. F. Strickler (1964). Thermal equilibrium of the atmosphere with a convective adjustment. *J. Atmos. Sci.* **21**, 361–385.

Manabe, S., J. Smagorinsky, and R. F. Strickler (1965). Simulated climatology of a general circulation model with a hydrologic cycle. *Mon. Wea. Rev.* **93**, 769–798.

Mathur, M. B. (1974). A multiple-grid primitive equation model to simulate the development of an asymmetric hurricane (Isbell, 1964). *J. Atmos. Sci.* **31**, 371–393.

Mintz, Y. (1965). Very long-term global integration of the primitive equations of atmospheric motion. *In* "WMO-IUGG Symposium on Research and Development Aspects of Long-range Forecasting," Boulder, CO, 1965, WMO-No.162.TP.79, pp. 141–167. World Meteor. Org.

Miyakoda, K., and J. Sirutis (1989). A proposal of moist turbulence closure scheme, and the rationalization of Arakawa–Schubert cumulus parameterization. *Meteor. Atmos. Phys.* **40**, 110–122.

Miyakoda, K., J. Smagorinsky, R. F. Strickler, and G. D. Hembree (1969). Experimental extended predictions with a nine-level hemispherical model. *Mon. Wea. Rev.* **97**, 1–76.

Moorthi, S., and M. J. Suarez (1992). Relaxed Arakawa–Schubert: A parameterization of moist convection for general circulation models. *Mon. Wea. Rev.* **120**, 978–1002.

Nitta, T. (1972). Energy budget of wave disturbances over the Marshall Islands during the years of 1956 and 1958. *J. Meteor. Soc. Japan* **50**, 71–84.

Ogura, Y. (1964). Frictionally controlled, thermally driven circulations in a circular vortex with application to tropical cyclones. *J. Atmos. Sci.* **21**, 610–621.

Ogura, Y., and H.-R. Cho (1973). Diagnostic determination of cumulus cloud populations from observed large-scale variables. *J. Atmos. Sci.* **30**, 1276–1286.

Ooyama, K. (1964). A dynamical model for the study of tropical cyclone development. *Geofisica Int.* **4**, 187–198.

Ooyama, K. (1969). Numerical simulation of the life cycle of tropical cyclones. *J. Atmos. Sci.* **26**, 3–40.

Ooyama, K. (1971). A theory of parameterization of cumulus convection. *J. Meteor. Soc. Japan* **49**(Special issue), 744–756.

Ooyama, K. (1982). Conceptual evolution of the theory and modeling of the tropical cyclone. *J. Meteor. Soc. Japan* **60**, 369–380.

Ooyama, K. V. (1997). Footnotes to "conceptual evolution." *In* "Preprint Volume, 22nd Conference on Hurricanes and Tropical Meteorology," Ft. Collins, Colorado, May 1997, pp. 13–18. American Meteorological Society, Boston.

Phillips, N. A. (1956). The general circulation of the atmosphere: A numerical experiment. *Quart. J. Roy. Meteor. Soc.* **82**, 123–164.

Richardson, L. F. (1922). "Weather Prediction by Numerical Process." Cambridge University Press, Cambridge, UK.

Reiser, H. (1962). Baroclinic forecasts with the primitive equations. *In* "Proc. International Symposium on Numerical Weather Prediction in Tokyo," Nov. 1960, pp. 77–84. Meteor. Soc. Japan.

Rosenthal, S. L. (1970a). Experiments with a numerical model of tropical cyclone development: Some effects of radial resolution. *Mon. Wea. Rev.* **98,** 106–120.

Rosenthal, S. L. (1970b). A circularly symmetric primitive equation model of tropical cyclone development containing an explicit water vapor cycle. *Mon. Wea. Rev.* **98,** 643–663.

Shuman, F. G. (1962). Numerical experiments with the primitive equations. *In* "Proc. International Symposium on Numerical Weather Prediction in Tokyo," Nov. 1960, pp. 85–107. Meteor. Soc. Japan.

Shuman, F. G., and J. B. Hovermale (1968). An operational six-layer primitive equation model. *J. Appl. Meteor.* **7,** 525–547.

Smagorinsky, J. (1958). On the numerical integration of the primitive equations of motion for baroclinic flow in a closed region. *Mon. Wea. Rev.* **86,** 457–466.

Smagorinsky, J. (1963). General circulation experiments with the primitive equations. I. The basic experiment. *Mon. Wea. Rev.* **91,** 99–164.

Smagorinsky, J., S. Manabe, and J. L. Holloway, Jr. (1965). Numerical results from a nine-level general circulation model of the atmosphere. *Mon. Wea. Rev.* **93,** 727–768.

Smith, R. K. (1997a). On the theory of CISK. *Quart. J. Roy. Meteor. Soc.* **123,** 407–418.

Smith, R. K. (ed.) (1997b). "The Physics and Parameterization of Moist Atmospheric Convection." NATO ASI Series C. Math. and Phys. Sci. Vol. 506. Kluwer Academic Publishers, Boston.

Sundqvist, H. (1970a). Numerical simulation of the development of tropical cyclones with a ten-level model. Part I. *Tellus* **22,** 359–390.

Sundqvist, H. (1970b). Numerical simulation of the development of tropical cyclones with a ten-level model. Part II. *Tellus* **22,** 504–510.

Syōno, S. (1951). On the structure of atmospheric vortices. *J. Meteor.* **8,** 103–110.

Syōno, S. (1962). A numerical experiment of the formation of tropical cyclone. *In* "Proc. International Symposium on Numerical Weather Prediction in Tokyo," Nov.1960, pp. 405–418. Meteor. Soc. Japan.

Syōno, S., and M. Yamasaki (1966). Stability of symmetrical motions driven by latent heat release by cumulus convection under the existence of surface friction. *J. Meteor. Soc. Japan* **44,** 353–375.

Syōno, S., Y. Ogura, K. Gambo, and A. Kasahara (1951). On the negative vorticity in a typhoon. *J. Meteor. Soc. Japan* **29,** 397–415.

Tsutsui, J., and A. Kasahara (1996). Simulated tropical cyclones using the National Center for Atmospheric Research community climate model. *J. Geophys. Res.* **101,** D10, 15,013–15,032.

Yamasaki, M. (1968a). Numerical simulation of tropical cyclone development with the use of primitive equations. *J. Meteor. Soc. Japan* **46,** 178–201.

Yamasaki, M. (1968b). A tropical cyclone model with parameterized vertical partition of released latent heat. *J. Meteor. Soc. Japan* **46,** 202–214.

Yamasaki, M. (1968c). Detailed analysis of a tropical cyclone simulated with a 13-layer model. *Papers Met. Geophys.* **19,** 559–585.

Yanai, M. (1964). Formation of tropical cyclones. *Rev. Geophys.* **2,** 367–414.

Yanai, M., S. Esbensen, and J.-H. Chu (1973). Determination of bulk properties of tropical cloud clusters from large-scale heat and moisture budgets. *J. Atmos. Sci.* **30,** 611–627.

Zhang, G. J., and N. A. McFarlane (1995). Sensitivity of climate simulations to the parameterization of cumulus convection in the Canadian Climate Centre general circulation model. *Atmos. Ocean* **33,** 407–446.

Chapter 8

Quasi-Equilibrium Thinking

Kerry Emanuel
Program in Atmospheres, Oceans and Climate
Massachusetts Institute of Technology
Cambridge, Massachusetts

I. INTRODUCTION

Statistical equilibrium thinking is natural to us in most contexts. In fluid problems for which the scales of interest are much larger than the mean free path between molecular collisions, we are comfortable dealing with the statistics of molecules rather than their individual dynamics, so that concepts such as pressure and temperature are natural and well developed. The great conceptual simplification brought about by statistical thinking arises from explicit assumptions that the space and time scales we are interested in are large compared to those characterizing the elementary particles or processes we are averaging over.

In large-scale geophysical fluid dynamics, we have become comfortable with a variety of scaling assumptions that greatly simplify thinking about the dynamics and formulating them in numerical models. Among the first approximations we become acquainted with are the hydrostatic and anelastic approximations, which filter out sound waves. It is important to remember here that these approximations are by no means equivalent to pretending that sound waves do not exist or that they are not important; rather, we simply assume that adjustments brought about by them are so fast compared to weather systems that we may think of them as occurring infinitely fast. If we had to consider flows with speeds close to or exceeding the speed of sound, we would be forced to abandon these approximations and the special kind of thinking that goes with them. Similarly, for flows of small Rossby number, we can stop thinking about gravity waves and take it for granted that they bring about adjustments that are very fast compared to the time scale of weather systems of synoptic scale and larger. Once again, this mode of thinking should not be regarded as pretending that gravity waves do not exist; rather, we approximate their adjustment time scale as zero. The enormous simplification that this and a few other assumptions bring about is beautifully entailed in "PV thinking," as described by Hoskins *et al.* (1985).

Statistical equilibrium ideas play a crucial role in thinking about and accounting for turbulence at all scales. Almost all successful theories and parameterizations of three-dimensional turbulence rely on the idea that kinetic energy is cascaded so rapidly to small scales, where it is dissipated, that generation and dissipation are nearly in equilibrium. Even the so-called "one-and-a-half order" closure, popular in a variety of applications, allows for only small departures from this kind of equilibrium. Turbulence kinetic energy can respond with a small time lag to changes in generation and dissipation, and can be advected away from regions where it is generated.

Statistical equilibrium thinking is also the natural mode of thinking about ordinary dry convection. We regard the intensity of boundary layer convection as a statistical quantity that is directly related to the intensity of the surface heat flux. To a first approximation, we think of such convection establishing on a very short time scale a dry adiabatic lapse rate through the depth of the convecting layer. If we were asked why a circulation develops between a dry, sandy field and an adjacent irrigated pasture, we reply that the air over the pasture is cooler, owing to evaporation. We take it for granted that small-scale convection over the dry field distributes heat through the boundary layer on a short time scale. Few would state that the field-scale circulation arises from a spontaneous organization of small-scale convective elements.

In 1974, Arakawa and Schubert formally introduced their application of statistical equilibrium to wet convection, using virtually the same general idea that had met with some success in ordinary three-dimensional turbulence: the idea that generation and dissipation of turbulence kinetic energy are nearly in equilibrium. This followed more than a decade of false starts by quite a few distinguished researchers, grappling with the construction of an appropriate closure for wet convection. These failed largely because they did not regard convection as an equilibrium process, or because they assumed that water vapor, rather than energy, is the quantity in equilibrium. Among the few physically consistent representations of convection that appeared before this time was moist convective adjustment (Manabe *et al.*, 1965), which, although not based on statistical equilibrium, acts in such a way as to preserve it.

It is somewhat surprising that, almost a quarter century after the introduction of the idea of quasi-equilibrium, very little of its conceptual content has influenced the thinking of most tropical meteorologists, even while the parameterization itself is enjoying increasing use. It is still very common to hear statements to the effect that latent heating drives tropical circulations, or that such circulations arise from a cooperative interaction among cumulus clouds. In the following sections, I attempt to show why such statements are inconsistent with the notion of quasi-equilibrium and to trace the history of thinking about the interaction of cumulus convection with large-scale circulations.

II. IS "LATENT HEATING" A USEFUL CONCEPT?

We are all taught that the condensation of water vapor releases a comparatively large quantity of heat to the air in which the condensate is suspended, and we are used to thinking of this just like any other heat source, like radiation, for example. The problem with this way of thinking is that it fails to recognize and take advantage of the fact that virtually all condensation in the atmosphere is very nearly reversible, and so may be usefully incorporated into the definition of the entropy of a system consisting of dry air, water vapor, and condensed water. (Of course, the fallout of condensate is irreversible, but that is another matter. Note also that in contrast to condensation, freezing is not usually reversible because it occurs at temperatures well below 0°C.) In such a system, there is no "latent heating"; phase changes between vapor and liquid droplets hardly affect the entropy of the system.

The distinction between external heating and internal rearrangements of the terms that comprise the specific entropy is far from academic. For example, external heating applied to rotating, stratified fluid will result in a local increase in the temperature of the fluid in the vicinity of the source. But the presence of deep, precipitating cumulus convection in a rotating, stratified fluid may very well be associated with local reduction of temperature. In the first case, the correlation between heating and temperature is virtually guaranteed to be positive, while in the second case it is quite possible for the "latent heating" to be negatively correlated with temperature, resulting in a reduction of kinetic energy. Thus the "organization of convection" need not lead to the amplification of a disturbance. Despite this, the idea that certain types of tropical disturbance arise from an organization of convection persists.

We illustrate the fallacy of regarding latent heating as the cause of phenomena associated with convection by a few examples.

A. Dry Convective Turbulence

It is helpful to start out with a problem uncomplicated by the presence of moisture. One relatively simple paradigm, a version of which dates back to Prandtl (1925), consists of a shallow layer of dry soil continuously heated by a constant imposed solar radiation, underlying a fluid cooled through a finite depth by a constant imposed radiative cooling. In equilibrium, the incoming solar radiation at the top of the system matches the total outgoing radiation. But conduction of heat from the soil to the overlying fluid will destabilize the latter, resulting in convection. In statistical equilibrium, the convergence of the convective heat flux matches the radiative cooling of the fluid. This paradigm is illustrated in Fig. 1.

Now consider the entropy budget of the system. By dividing the first law of thermodynamics through by temperature, T, one obtains

$$C_{\mathrm{p}}\left(\frac{d \ln(T)}{dt}\right) - R\left(\frac{d \ln(p)}{dt}\right) = \frac{Q_{\mathrm{rad}}}{T} + \dot{s}_{\mathrm{irr}}, \tag{1}$$

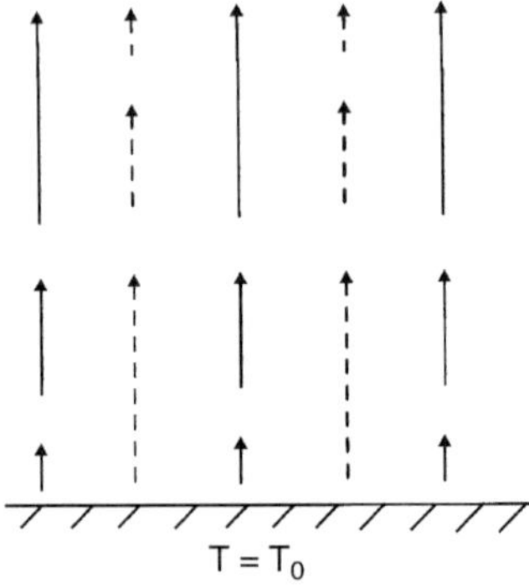

Figure 1 Radiative-convective equilibrium over dry land. Solid arrows denote long-wave radiative flux, which increases upward; dashed arrows denote turbulent convective heat flux, which decreases upward. There is no net flux divergence except at the surface, where it is balanced by absorption of solar radiation.

where C_p is the heat capacity at constant pressure, R is the gas constant for dry air, p is pressure, Q_{rad} is the radiative (and conductive) heating, and $s_{irr}^{\cdot}$ represents various irreversible entropy sources. We consider the system to be closed in mass, so that integrating Eq. (1) over the entire system and over a long enough time to average out the statistical fluctuations, we get

$$\int s_{irr}^{\cdot} = -\int \frac{Q_{rad}}{T}, \tag{2}$$

where the integral is over the entire system and time. Since, in equilibrium, the surface heating balances the net atmospheric cooling, we can express Eq. (2) as

$$\int s_{irr}^{\cdot} = F_s\left(\frac{1}{\overline{T}} - \frac{1}{T_s} \right), \tag{3}$$

where F_s is the net radiative flux at the surface, T_s is the surface temperature, and $\overline{T}$ is the average temperature at which radiative cooling occurs. Now if we assume that dissipation of kinetic energy is the dominant irreversible entropy source, then the left side of Eq. (3) is just the system integral of the dissipative heating divided by temperature. Since, in equilibrium, dissipation of kinetic energy must equal the rate of conversion of potential energy to kinetic energy, we can write Eq. (3) as

$$\left(\frac{1}{T_{diss}} \right) \int \overline{w'B'} = F_s\left(\frac{1}{\overline{T}} - \frac{1}{T_s} \right), \tag{4}$$

where $\overline{w'B'}$ is the buoyancy flux, which is also the rate of conversion of potential to kinetic energy, and T_{diss} is the mean temperature at which kinetic energy is dissipated. Expression (4) tells us what the integrated buoyancy flux is as a function of the energy input to the system and something like a thermodynamic efficiency. Given that the temperature lapse rate is not likely to be too far off the dry adiabatic lapse rate, a very good estimate can be made of the mean temperature $\overline{T}$. The mean temperature at which kinetic energy is dissipated, T_{diss}, is not as easy to estimate, but because it appears only as an absolute value, errors in its estimate will not have a serious effect on the evaluation of Eq. (4). Thus the energy-entropy method yields an appropriate scale for the buoyancy flux in the system. This scale is proportional to the radiation absorbed by the surface and the difference between the surface temperature and a mean temperature of the free atmosphere. We can think of the convection as a heat engine, converting the absorbed heating into mechanical work with an efficiency proportional to the difference between the input and output temperatures. The engine does no work on its environment; instead, the mechanical energy is dissipated and locally turned back into enthalpy.

Having described one aspect of the dry convection problem, let's apply the same methods to moist convection.

B. Moist Convective Turbulence: The Naive Approach

We use the same paradigm for moist convection, by replacing the dry soil used above with a thin layer of water. To make life simple, we assume that all of the net incoming radiation at the surface is balanced by evaporation, neglecting the sensible component of the turbulent surface enthalpy flux. We allow the resulting moist convective clouds to precipitate, so we expect to see tall cumulonimbi separated by regions of clear, subsiding air. In spite of the possibly impressive appearance of such clouds, we continue to treat the convection statistically. The general picture is illustrated in Fig. 2.

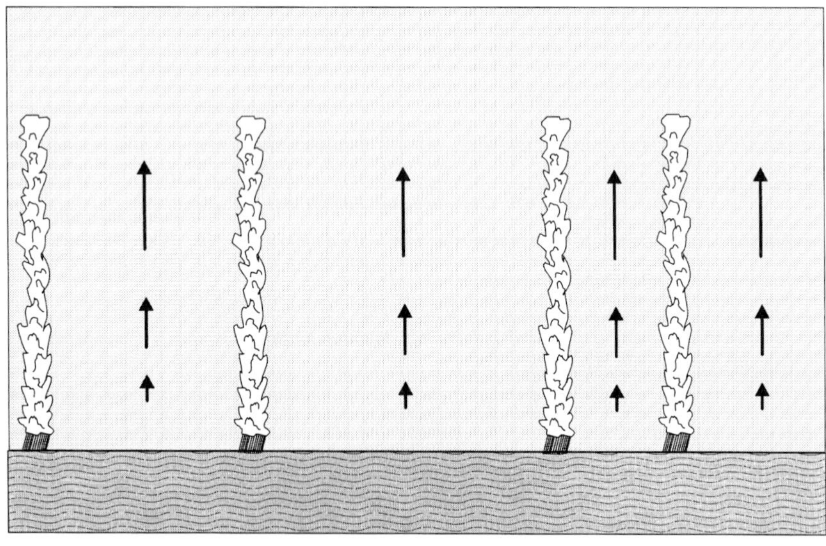

Figure 2   Radiative-convective equilibrium over a water surface. Arrows denote long-wave radiative flux.

Here we are deliberately going to engage in sloppy thermodynamics, following the habits of many large-scale dynamists when they try to do thermodynamics. In particular, we forget about the dependencies of heat capacities and gas constants on water content and do not bother to distinguish between total pressure and the partial pressure of dry air. Following the same procedure as in the previous subsection, we get, from the first law,

$$C_p\left(\frac{d\ln(T)}{dt}\right) - R\left(\frac{d\ln(p)}{dt}\right) = -\frac{L_v}{T}\left(\frac{dq}{dt}\right) + \frac{Q_{rad}}{T} + \dot{s}_{irr}, \qquad (5)$$

where L_v is the latent heat of vaporization and q is the specific humidity. The first term on the right side is the latent heating term. Once again, we integrate this over the system, conserving mass, to get

$$\int \dot{s}_{irr} = -\int\left(\frac{Q_{rad}}{T}\right) + \int\frac{L_v}{T}\left(\frac{dq}{dt}\right). \qquad (6)$$

Now we notice that, owing to the assumption that all of the absorbed solar radiation is compensated for by evaporation, the terms on the right side of Eq. (6) cancel when integrated through the thin layer of water. What we are left with is

$$\int \dot{s}_{irr} = -\int\left(\frac{Q_{cool}}{T}\right) + \int\frac{L_v}{T}\left(\frac{dq}{dt}\right)_{cloud}, \qquad (7)$$

where the remaining terms on the right are the radiative cooling of the atmosphere and the latent heating inside clouds. Inside the clouds, the latent heat release shows up as an increase of potential temperature, so that

$$-\frac{L_v}{T}\left(\frac{dq}{dt}\right) \simeq \frac{C_p}{\theta}\left(\frac{d\theta}{dt}\right),$$

where θ is the potential temperature. Outside the clouds, the radiative cooling causes a decrease in potential temperature:

$$Q_{cool} = -C_p\frac{T}{\theta}\left(\frac{d\theta}{dt}\right).$$

One can see that the two terms on the right side of Eq. (7) cancel, leaving us with no irreversible entropy production. We have gotten nowhere, except to show that radiative cooling is balanced by radiative heating. Note also that, unlike the dry problem, the surface temperature vanished and plays no role. What happened?

C. MOIST CONVECTIVE TURBULENCE: DOTTING THE i'S

Let's start over again, this time being careful with the thermodynamics. We account for the effect of water substance on heat capacities and gas constants, and we are careful to separate the total pressure into the partial pressure of dry air, p_d, and the partial pressure of water vapor (or "vapor pressure"), e. Instead of Eq. (5), we get (see Emanuel, 1994, for a derivation)

$$\left(C_{pd}(1 - q_t) + C_l q_t\right)\left(\frac{d\ln(T)}{dt}\right) - R_d(1 - q_t)\left(\frac{d\ln(p_d)}{dt}\right)$$

$$= -\frac{1}{T}\left(\frac{dL_v q}{dt}\right) + qR_v\left(\frac{d\ln(e)}{dt}\right) + \frac{Q_{rad}}{T} + \dot{s}_{irr}, \qquad (8)$$

where C_{pd} is the heat capacity at constant pressure of dry air, C_l is the heat capacity of liquid water, q_t is the total (condensed plus vapor phase) specific water content, R_d is the gas constant for dry air, and R_v is the gas constant for water vapor. Notice that, in addition to the modifications of the effective heat capacities and gas constants, there is an extra term on the right side of Eq. (8) that we neglected in Eq. (5): the part of the work done by expansion against the vapor pressure. This term does not integrate to zero through a closed system, owing to the variability of q. We can also re-express the latent heating term:

$$\frac{1}{T}\left(\frac{dL_v q}{dt}\right) = \frac{d}{dt}\left(\frac{L_v q}{T}\right) + \frac{L_v q}{T^2}\left(\frac{dT}{dt}\right). \qquad (9)$$

But, by the Clausius–Clapeyron equation (e.g., see Emanuel, 1994),

$$\frac{L_v q}{T^2}\left(\frac{dT}{dt}\right) = R_v q\left(\frac{d\ln(e^*)}{dt}\right), \qquad (10)$$

where e^* is the saturation vapor pressure. We now combine Eqs. (9) and (10), substitute the result into Eq. (8), and integrate over the system as before. In doing so, we note that, because of fallout of precipitation, q_t is not conserved following the motion of the air and this results in some additional, irreversible contributions to entropy production. Using some integrations by parts, we get

$$\int \dot{s}_{\text{irr}} = -\int \frac{Q_{\text{rad}}}{T} + R_v \ln(\mathscr{H})\left(\frac{dq}{dt}\right), \qquad (11)$$

where $\mathscr{H}$ is the relative humidity, $\equiv e/e^*$. The last term in Eq. (11) is negative definite because the vapor content can only increase by evaporation into subsaturated air; condensation always occurs with $\mathscr{H} = 1$. Therefore, it belongs on the left side of the equation, as part of the irreversible entropy production term.

What happened to the latent heating term? It canceled with a term we left out when doing things the sloppy way—the work against the vapor pressure. There is no contribution of latent heating to mechanical energy production when the thermodynamics is done properly. What we are left with is an equation identical in form to Eq. (3), except that there are more contributions to the irreversible entropy production. [A relation like that of Eq. (3) was first derived for the case of moist convection by Rennó and Ingersoll, 1996.] These include mixing of moist and dry air, evaporation of rain and surface water into subsaturated air, and frictional dissipation owing to falling rain. A complete scale analysis of these terms was performed by Emanuel and Bister (1996), who showed that mechanical dissipation still dominates, so that Eq. (4) remains approximately true. The role of moisture is to some extent hidden; its primary function is possibly to modify the mean temperature, $\overline{T}$, at which radiative cooling occurs. *In no event is it sensible to regard moist convection, in equilibrium, as being driven by "latent heat release."* Thus convective scheme closures that rely on the moisture budget are doomed to fail, because they violate causality. Convection is not caused by moisture, or "moisture convergence" any more than dry convection that happens to contain mosquitoes is caused by "mosquito convergence." In neither case do we deny that there may be a very strong *association* between the two, but it is not causal in nature.

Now one might argue that, when convection is far from being in equilibrium with large-scale processes, the concept of latent heating might be more useful. After all, the first paradigm of moist convection most of us hear about is the case of explosive, deep moist convection over middle latitude continents in spring and summer, when potential energy, stored in a conditionally unstable atmosphere with a "lid," is suddenly released by some trigger. This may be true, but in that case, the interaction with the environment is largely one way and it is not meaningful to think about parameterizing the convection as a function of large-scale variables. As put very succinctly by Arakawa and Schubert, "Unless a cumulus ensemble is in quasi-equilibrium with the large-scale processes, we cannot uniquely relate the statistical properties of the ensemble to the large-scale variables."

D. What Does Equilibrium Convection Look Like?

It is fine to imagine what moist convection in equilibrium with large-scale forcing looks like (Fig. 2), but what does it *really* look like? In the last decade, it has become possible to numerically simulate whole ensembles of convection. Figure 3 shows the distribution of low-level upward motion in a doubly periodic box of 180 km^2, using a numerical cloud model developed by the Center for the Analysis and Prediction of Storms (CAPS). The model includes representations of cloud physical and turbulent processes and is here run with a horizontal resolution of 2 km. A radiative cooling of the troposphere is imposed, and the lower surface is an ocean with fixed surface temperature. The model is run long enough for the domain-average precipitation to come into statistical equilibrium.

The convection is more or less randomly distributed, but a careful analysis (Islam *et al.*, 1993) reveals that the spacing between clouds is more nearly regular than random. This means that clouds are *less* likely to clump together than would be true if their spatial distribution were random. *There is no tendency toward spontaneous organization of clouds*, at least at these scales. (One cannot rule out the possibility of spontaneous organization at scales larger than the domain size.)

Figure 4 shows what happens, on the other hand, if we now impose a background vertical shear of the horizontal wind in the domain. (This is done by relaxing the domain horizontally averaged wind toward a prescribed value at each level.) Now we have very clear mesoscale organization of convection, with squall lines (or, more accurately, arcs) lined up across the direction of the background shear. The mechanism by which this happens was delineated by Thorpe *et al.* (1982) and Rotunno *et al.* (1988); it has to do with the interaction between the background vertical shear with the density currents created by cold, downdraft air spreading out at the surface. The spacing between the squall arcs is nearly that of the domain size, so that the domain may not be large enough to detect the true spacing such lines would have in an unbounded domain. (For different magnitudes of the shear, however, there can be several arcs within the present domain.)

One may reasonably ask whether a parameterization of moist convection should be able to simulate explicitly the actual shape of the convection; that is, to distinguish between the forms of convection in Figs. 3 and 4. The answer is no. After all, the large-scale forcing imposed in both cases

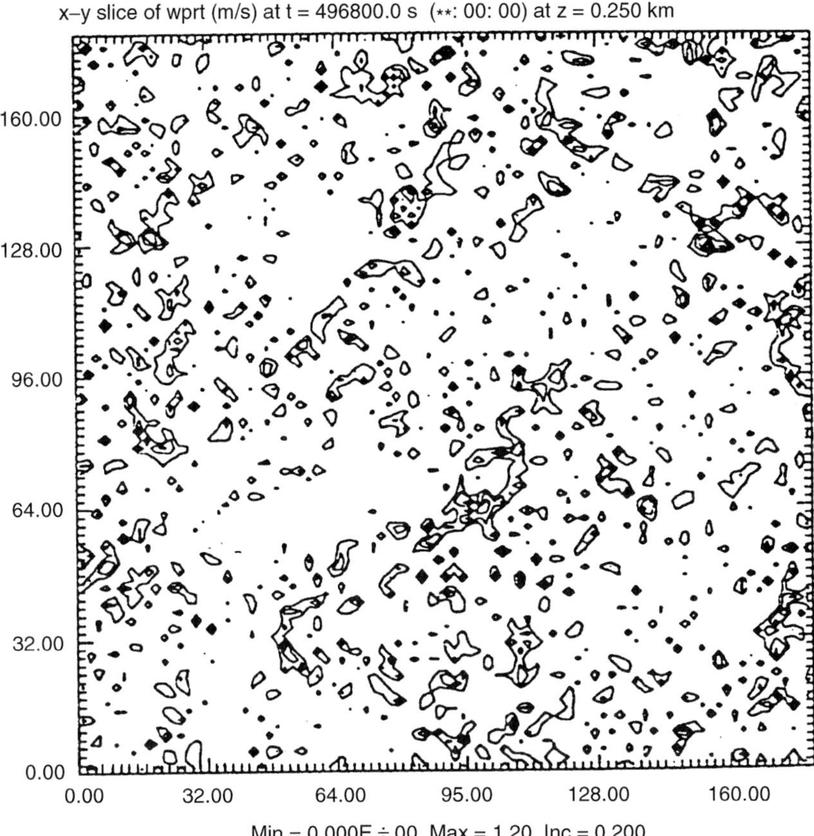

Figure 3 Distribution of upward motion at 250 m in a three-dimensional numerical simulation of radiative-convective equilibrium over a water surface. The simulation has reached statistical equilibrium at this time. (From Robe, 1996.)

is identical. (The background wind shear is not a forcing in this sense; it does not contribute to destabilizing the atmosphere to convection.) Fortunately, there is hardly any detectable difference in the equilibrium, domain-averaged vertical profiles of temperature and relative humidity between Figs. 3 and 4, so that if one is after the vertical heat and moisture fluxes, it may be permissible to neglect the background shear. The convective momentum fluxes are another matter, of course, and their parameterization remains an outstanding problem. (If the relaxation toward the background shear profile is suddenly stopped in the simulations above, the

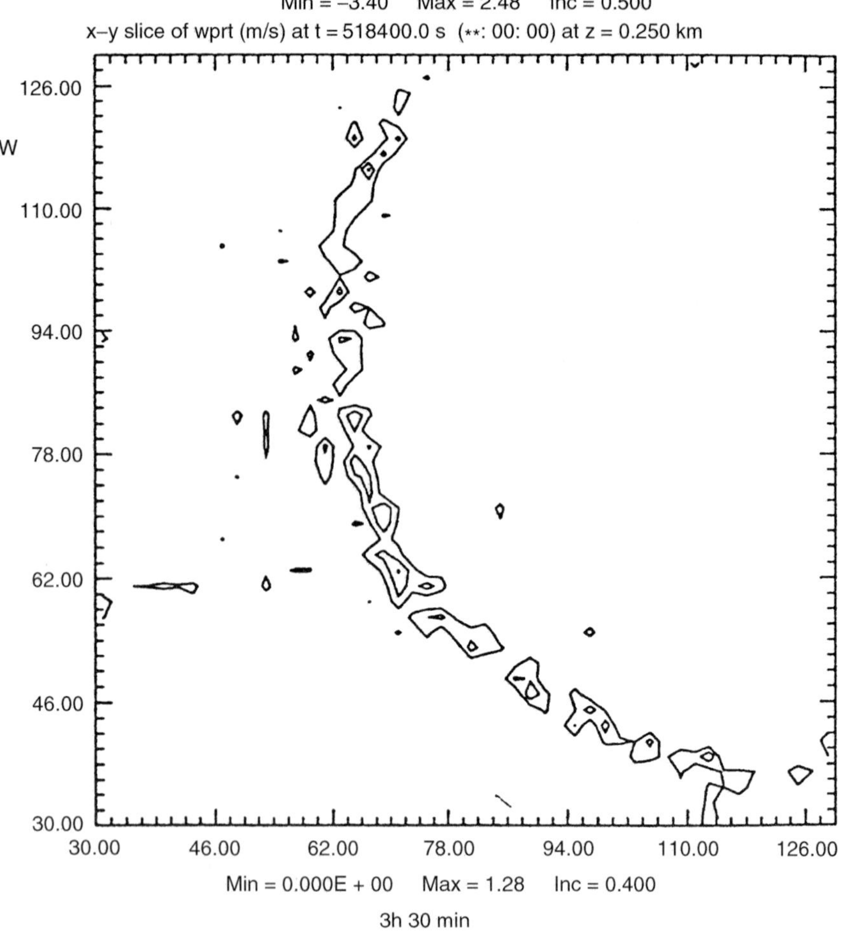

Figure 4   As in Fig. 3, but for a simulation with an imposed vertical wind shear from right to left, in the lowest 3 km. (From Robe, 1996.)

domain average shear relaxes toward zero on a surprisingly short time scale, indicating mostly down-gradient momentum transport by the convection.)

It might be possible, on the other hand, to formulate a representation of convection that regards only the active clouds as the subgrid-scale elements and that takes the mesoscale cold pools to be explicitly simulated by the model. This form of mesoscale convective parameterization would be valid if quasi-equilibrium holds for the interaction between cumulus clouds and mesoscale circulations. That is, if the clouds forming at the leading edge of the cold pool behave in such a way that the rate of destabilization of the column owing to uplift at the leading edge of cold pools is nearly balanced by convective stabilization by the small-scale cumulus elements, then this kind of mesoscale parameterization is viable. But we emphasize that in this case, the cold pools must be explicitly simulated.

E. Quasi-Equilibrium and Convective Inhibition

One peculiarity of moist convection, with no analog in dry convection, is the possibility of metastable states that are stable to small perturbations but unstable to sufficiently large ones. Textbooks almost always contain examples of metastable soundings from places like Texas, where the degree of convective inhibition can be exceptionally large, even when there is a large reservoir of convective available potential energy (CAPE). To what extent is the presence of convective inhibition (hereafter CIn) consistent with statistical equilibrium?

In numerical experiments such as those described in the previous subsection, the experimental design virtually ensures statistical equilibrium when averaged over sufficiently large space-time subdomains. How small can one make the subdomain before statistical equilibrium fails? Figure 5 shows the ratio of the standard deviation of convective rainfall to the subdomain mean, as a function of the size of the space-time subdomain, for a pure convective-radiative equilibrium experiment (Islam *et al.*, 1993) Clearly, the statistics in this case are stable down to remarkably small scales. But were the same thing done for the experiment with shear (Fig. 4), surely the statistics would be less stable and bigger subdomains would be necessary for quasi-equilibrium to be valid. A careful examination of point soundings in these experiments reveals, that, indeed, there is some CIn between active clouds in all the experiments. But it is noticeably larger in the experiments with shear. In this case, strong lifting at the leading edge of the cold pools forces convection there, but the total amount of convection over the domain is constrained by the radiative cooling. Thus the convection must be suppressed between the squall lines. The magnitude of the CIn is part of the quasi-equilibrium state; it is not imposed externally. The forecaster, trying to predict the evolution of individual clouds, might profit from looking at the CIn, but those who are trying to understand the large-scale factors that determine the mesoscale structure would be looking at part of the outcome, not part of the cause.

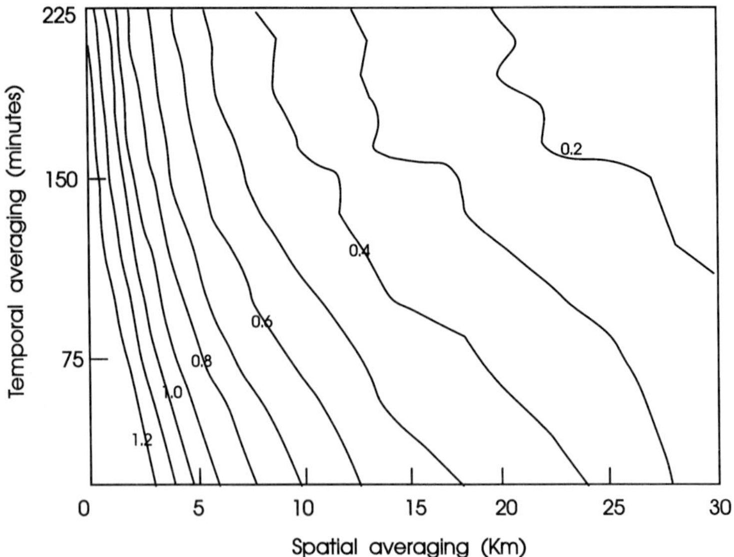

Figure 5   Ratio of the variance to the domain average of the precipitation in a three-dimensional numerical simulation of radiative-convective equilibrium over a water surface, as a function of space-time averaging. The ordinate is the length of time averaging; the abscissa is the length of averaging in space. This ratio asymptotes to $\sqrt{2}$ for short averaging intervals. (From Islam *et al.*, 1993.)

III. THE PHYSICS OF CONVECTIVE QUASI-EQUILIBRIUM

Part of the difficulty some have in accepting the quasi-equilibrium postulate may have to do with problems visualizing how it may work in nature. In the case of dry boundary layer convection, it is relatively easy to understand the process. Suppose, for example, that the rate of radiative cooling is increased in some individual atmospheric layer above the surface. At first, this layer may be expected to cool. But as soon as it does so, it is more unstable with respect to the air just below it, and less unstable with respect to the air just above it. This provides not only for an increase in the convective heat flux from the lower layer, but also for a decrease of the flux to the higher layer; both act to increase the convergence of the convective heat flux, thus warming the layer.

It is more difficult to imagine what happens in a moist convecting layer. Start with a state of pure radiative convective equilibrium and, to make life simple, specify the radiative cooling profile. Now suppose we increase the rate of cooling in some atmospheric layer above the subcloud layer. If this layer happens to be *just* above the subcloud layer, then it is not difficult to see that the convective flux from the boundary layer will increase, just as in the dry case, and there will be a compensating warming. But what happens if the extra cooling is introduced to a layer far removed from the subcloud layer? The subcloud layer simply cannot know directly about this development and there is little or no basis for thinking that there will be a compensating increase in mass flux out of the subcloud layer. Even if there were, this would entail an extra warming not only in the layer to which we added the cooling, but to all layers below that layer. The warming of these other layers, to which we did not add extra cooling, would quickly stabilize the lower atmosphere and cut off the convection.

Nature resolves this paradox in two ways, as becomes evident on examining the response of explicit ensembles to changes in imposed cooling rates. First, the mass flux can increase in the individual layer to which we add extra cooling *without* increasing the mass flux out of the boundary layer. This occurs because of entrainment. While the exact physics of entrainment into cumulus clouds is not well understood, it is becoming increasingly clear that the rate of entrainment is sensitive to the vertical gradient of the buoyancy of the clouds (Bretherton and Smolarkiewicz, 1989). Cooling an individual layer will have the effect of increasing the buoyancy of clouds rising into that layer. This increases the upward acceleration of air in the clouds and leads to greater entrainment just below the layer of extra cooling. This in turn increases the mass flux in the layer. The increased compensating subsidence outside the cloud warms the layer, opposing the initial added cooling. The physics is very different from what happens in the dry case, but the effect is the same.

The second response to the presence of a layer of extra cooling is entailed in the precipitation physics. Adding cooling to the system means that, to reach equilibrium, there must be an increase in precipitation. How this happens is complex, but it is crucial to recognize that any increase in precipitation will also, in general, increase the magnitude of any unsaturated downdrafts driven by evaporation of precipitation. This will generally occur below the layer into which extra cooling has been added. Because no cooling has been added there, the increased downdraft mass flux must be compensated by an increased updraft mass flux. One may think of it this way: The upward mass flux compensates not just the imposed radiative cooling, but also the (interactive) evaporative cooling. So there *can* be an increase in *updraft* mass flux out of the subcloud layer. This can help warm the layer to which the extra cooling has been added.

Entrainment and adjustments of the unsaturated downdraft are together very effective in compensating for changes in the imposed forcing.

To illustrate this, Fig. 6 shows the imposed radiative cooling profiles and equilibrium convective heating profiles for a variety of experiments using a single-column model with the convective scheme of Emanuel and Ziukovic-Rothman (1999). This is not explicitly a quasi-equilibrium scheme. Instead, it calculates the cloud base updraft mass flux based on an assumption of quasi-equilibrium of subcloud layer air with respect to the air *just above* the subcloud layer, as advocated by Raymond (1995). But, unlike the general quasi-equilibrium closure of Arakawa and Schubert (1974), the mass flux above cloud base is not calculated explicitly from a quasi-equilibrium assumption; rather, the rate of entrainment into clouds is allowed to respond to vertical variations of cloud buoyancy.

It is evident in Fig. 6 that even bizarre profiles of imposed radiative cooling are compensated for by the net convective heating profiles, demonstrating the efficacy of the adjustment process. Figure 7 shows that the resulting temperature profiles are all very close to a moist adiabatic profile. Thus the assumption that convection relaxes the temperature profile of a convecting layer back toward a moist adiabat is well verified in this model. Zeng, Neelin, and others discuss in Chapter 15 the profound implications that this has for understanding tropical dynamics.

IV. NONEQUILIBRIUM THINKING

Most students of meteorology are conditioned to think of convection in nonequilibrium terms, being first introduced to the concept of conditional instability through the illustration of highly metastable soundings from places like Oklahoma. Instability accumulates under some "lid" and is released suddenly when convective temperature is attained or when some mesoscale process locally removes the potential barrier to convection. This may very well be an appropriate mode of thinking about the type of convection that often results in severe thunderstorms. But it is probably inappropriate for thinking about many tropical circulation systems.

Nowhere is the disparity between equilibrium and nonequilibrium thinking more on display than in discussions about hurricanes. As reviewed very thoroughly by Yanai (1964), most of the earliest attempts to model hurricanes, beginning in the early 1960s, focused on finding a particular mode by which stored conditional instability is released. As earlier theoretical studies had predicted, conditional instability is released at the scale of individual clouds. All attempts to run numerical simulations of hurricanes as modes of release of conditional instability failed to produce a hurricane-scale vortex. Earlier theoretical work by Riehl (1950) and Kleinschmidt (1951) had shown that the warmth of the eyewall could only be

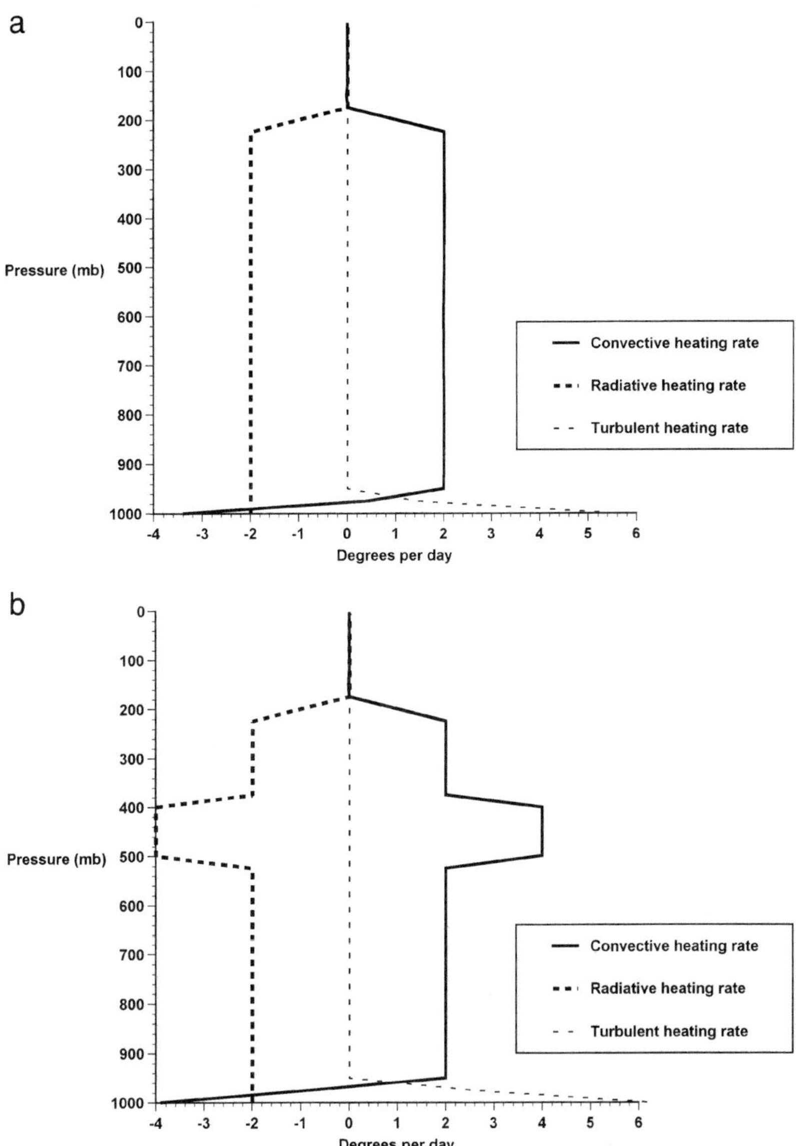

Figure 6 The heat budget of a single-column model in radiative-convective equilibrium, showing the rate of heating as a function of pressure. In each case, the solid line denotes the convective heating rate, the dashed line thc (imposed) radiative heating rate, and the thin dashed line the convergence of the dry tubulent heat flux. (a) Uniform radiative cooling in the troposphere. (b) Same as (a) but with added cooling in the 400- to 500-mb layer. (c) Same as (a) but with zero cooling in the 850- to 950-mb layer. (d) No cooling in the 500- to 950-mb layer. This shows that convection can penetrate even a deep layer of no large-scale destabilization.

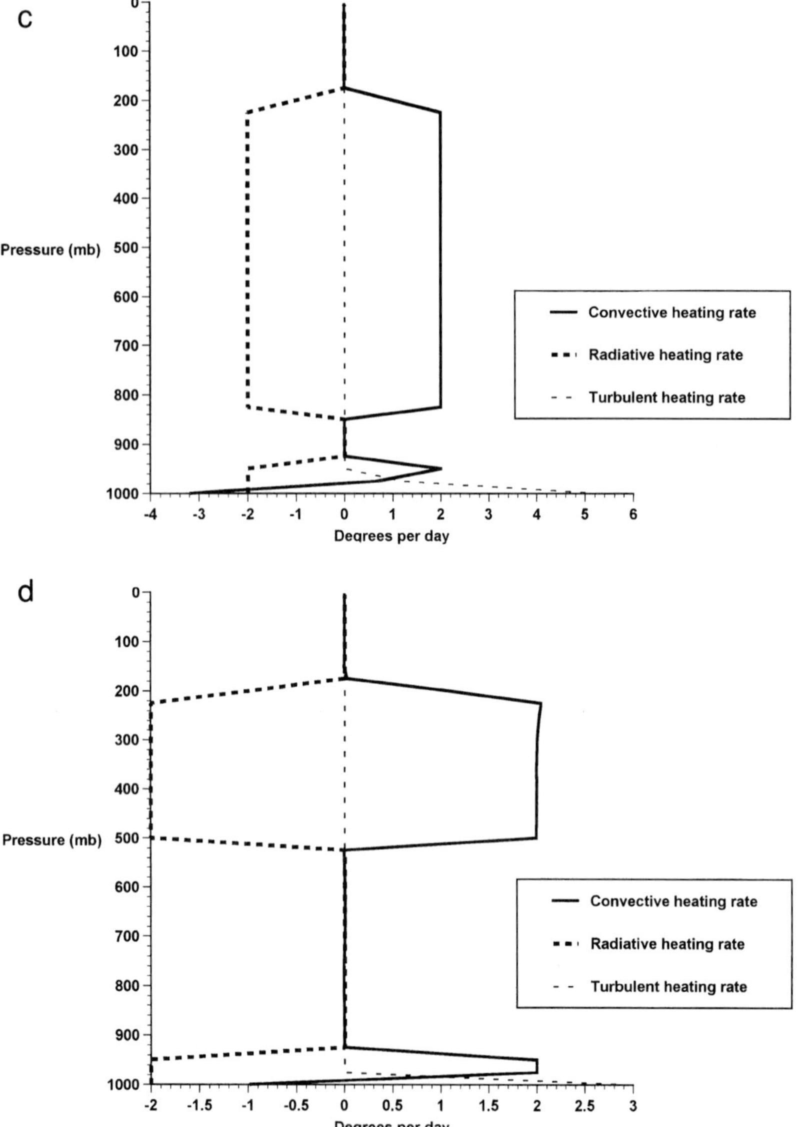

Figure 6   (*Continued*)

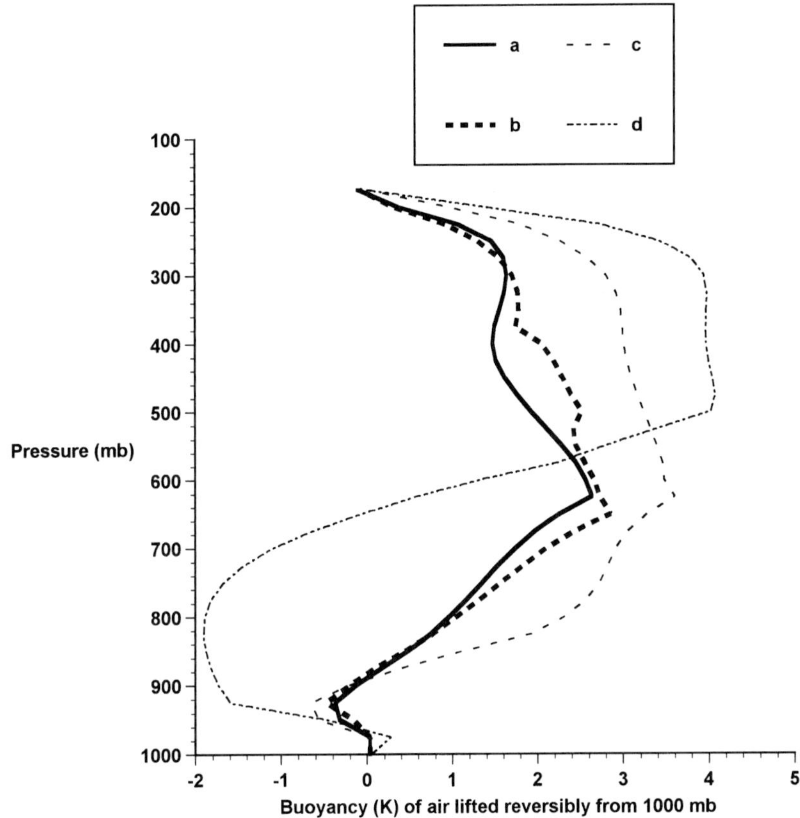

Figure 7   Departure of the ambient temperature from a reference moist adiabat for each of the experiments illustrated in Fig. 6. A positive value means that the reference adiabat is warmer than the atmosphere.

explained by the enormous enthalpy transfer from ocean to atmosphere that occurs in the high wind region of the storm. Although the principals involved in this work were undoubtedly aware of this earlier theoretical work, they evidently considered the heat transfer to be a secondary issue.

The failure of these earliest attempts at numerical simulation formed a large part of the motivation behind the development of the theory of conditional instability of the second kind (CISK) by Charney and Eliassen (1964) and Ooyama (1964). The history of the development of CISK is reviewed very nicely by Kasahara in Chapter 7 of this volume. The idea of

CISK was stated very beautifully by Charney and Eliassen (1964):

> ...we should look upon the pre-hurricane depression and the cumulus cell
> not as competing for the same energy, for in this competition the cumulus cell
> must win; rather we should consider the two as supporting one another—the
> cumulus cell by supplying the heat energy for driving the depression, and the
> depression by producing the low-level convergence of moisture into the cumu-
> lus cell.

In my view, a fatal flaw was introduced into thinking about tropical circulations by this enormously influential work. It is the idea that latent heat release can ever be an energy source for equilibrium circulations, an idea disproved earlier in Section II. This flaw was exacerbated by later work that also introduced the incorrect notion that the vertical profile of convective heating is an internal property of the convective clouds that can, to a first approximation, be specified independently of the environment.

The Charney and Eliassen work attempted to demonstrate CISK by posing a balanced model in which, as in the case of unbalanced models, the latent heat release is taken to be proportional to vertical velocity but, unlike unbalanced models, the vertical velocity was constrained to be that associated with Ekman pumping. Thus constrained, the model dutifully produced a linear instability with tropical cyclone-like characteristics, but even in this case the most rapidly growing modes were of small scale.

The difference between nonequilibrium (CISK) thinking and equilibrium thinking, in the case of a tropical cyclone, is illustrated in Fig. 8. In nonequilibrium thinking, the ambient atmosphere has a reservoir of usable potential energy for convection. The tropical cyclone is a means of releasing that instability on a large scale. In equilibrium thinking, the storm passes through an infinite sequence of convective equilibrium states, and the warmth of the eyewall is a consequence of the energy flux from the ocean. In reality, there is always some stored potential energy to balance dissipation in clouds (see Emanuel and Bister, 1996), and there is never perfect equilibrium in an evolving system. Nevertheless, as subsequent work (Emanuel, 1989) showed, approximating the evolution as passing through an infinite sequence of equilibrium states yields a realistic numerical simulation.

Five years after the CISK papers were published, Ooyama (1969) presented the first genuinely successful numerical simulation of a tropical cyclone. It captured the essential physics of the intensification process, and documented the sensitive dependence of the vortex evolution on the exchange coefficients of enthalpy and momentum at the sea surface. It confirmed the deductions of Riehl (1950) and others that surface enthalpy

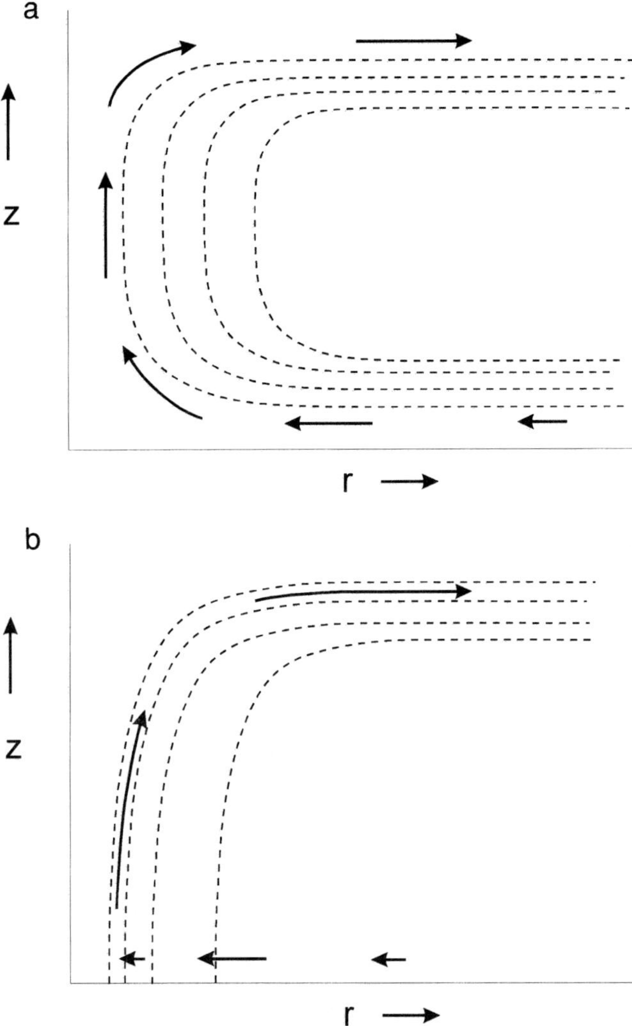

Figure 8 Two views of tropical cyclone physics. In each case, the dashed lines indicate surfaces of constant saturation θ_e. (a) The CISK view. The frictionally induced inflow allows for convection in the core. The core region is warm because the convection is there and not in the outer region. The role of the ocean is to keep the subcloud layer "stoked"; i.e., to prevent its θ_e from decreasing in the face of convective downdrafts. Intensification stops when the free troposphere saturation θ_e is the core increases to the initial value of θ_e in the boundary layer. (b) The WISHE view. The saturation θ_e is tied to the subcloud layer θ_e. The core is warm because surface fluxes have increased θ_e there. The rate of intensification is limited by the surface fluxes, not by the convection, which is very fast by comparison. Intensification stops when the frictionally induced radial advection of low θ_e air balances the surface enthalpy flux.

fluxes are essential; when they are excluded from the model very little happens.

One might have thought that Ooyama's and subsequent numerical simulations would finally lay to rest the notion that the controlling process in tropical cyclone intensification is some kind of cooperation between the cyclone circulation and cumulus clouds. That this has not happened is a testament to the attractiveness of nonequilibrium thinking and, I think, the eloquence of the Charney and Eliassen paper. The Ooyama simulation left an opening for CISK adherents: It began with a highly unstable vertical sounding and used a cumulus parameterization that could not switch on unless there was moisture convergence in the boundary layer. Together, this meant that the resting state was indeed linearly unstable to CISK, as shown by Ooyama in the same paper. But it was also clear from the nonlinear simulations that, even with the loaded gun, CISK could account for no more than a trivial initial intensification. In a later paper, Ooyama (1982) downplayed the significance of the linear instability, and Klaus and Reeder (1997) showed that Ooyama's original model, when initialized with a neutral sounding, is metastable in the same sense as the model of Emanuel (1989), which also begins with a neutral sounding. The latter model showed that a perfectly acceptable numerical simulation of a tropical cyclone results even under the extreme assumption that the model is always precisely neutral to cumulus convection.

Craig and Gray (1996), using a cloud-resolving, nonhydrostatic primitive equation model of a tropical cyclone, showed that, at all stages of intensification, the rate of development is directly proportional to the surface enthalpy exchange coefficient and inversely proportional to the drag coefficient, in direct contradiction to the predictions of CISK.

The most recent variation of nonequilibrium thinking holds that tropical cyclones cannot develop until the rotation is strong enough that the local deformation radius is of the same order as the scale of convective clouds. This notion was introduced by Ooyama (1982) and furthered by Hack and Schubert (1986). The general idea is that an isolated heat source cannot produce finite warming in a nonrotating, infinite fluid, because the warming is spread over an infinite region. Warming only commences when the deformation radius becomes appreciably small.

The problem with this idea is that it postulates a single, noninteractive cloud, whereas equilibrium thinking demands that we consider an ensemble of clouds always in equilibrium with the forcing. A large-scale circulation with a large deformation radius can be efficiently warmed by an array of clouds and there is no physical reason to postulate a single cloud. On the contrary, doing so is unphysical. In a nonrotating unbounded domain initially in radiative-convective equilibrium, a sudden increase in the surface fluxes will be accompanied by a rapid increase in the domain temperature at all levels up to the tropopause, in spite of the total absence of rotation. There are many dynamical reasons why the rate of intensification of tropical cyclones may depend on their intensity, but the ratio of the cloud scale to the local deformation radius is not one of them, as demonstrated by tropical cyclone models that constrain the free atmospheric temperature to always lie on a moist adiabat determined by the thermodynamic properties of boundary layer air.

Nonequilibrium thinkers continue to hold that tropical cyclones are powered by latent heat release in cumulus clouds, though surface energy fluxes are of course necessary to maintain the reservoir of conditional instability. This is analogous to claiming that cars are powered by their drive trains, though engines are of course necessary to keep torque on them. In both cases the awkwardness arises from a failure to separate time scales—in the case of a car, the time scale for transmission of torque through the drive train is tiny compared to the time scale determined by the power of the engine and the inertia of the car; in the case of the hurricane, the time scale for transmission of enthalpy by convection, on the order of a few hours, is small compared to the time scale determined by the rate of surface enthalpy flux and the inertia of the cyclone. No numerical simulation has ever demonstrated a critical role for convective time scales in the evolution of the vortex.

One final point to be made about nonequilibrium thinking concerns ordinary conditional instability itself. In nonequilibrium thinking, the degree of conditional instability is thought of as an external condition that determines various properties of the convection. In equilibrium thinking, by contrast, the degree of conditional instability is determined by the convection itself, together with the forcing. Quasi-equilibrium does not imply actual invariance of CAPE, any more than quasi-geostrophy implies invariance of ageostrophic velocities. In the quasi-equilibrium view, both the intensity of the convection and CAPE are determined by the forcing and, to second order, the time rate of change of the forcing. CAPE is not a predictor, though it can be an indicator.

V. EQUILIBRIUM THINKING

The underlying proposition in quasi-equilibrium thinking is that convection rapidly adjusts the temperature profile back toward a moist adiabat in a way that preserves the vertically integrated enthalpy. To a first order of approximation, convection keeps the atmospheric temperature profile on a moist adiabat tied to the subcloud layer entropy. This strongly constrains the vertical structure of the horizontal and vertical velocities as well as the temperature perturbations associated with large-scale disturbances in convecting atmospheres, as discussed by Neelin and Yu (1994) and Emanuel *et al.* (1994). It also means that, in convecting atmospheres, the problem of predicting the evolution in three dimensions of atmospheric variables reduces largely to the problem of predicting the evolution of subcloud layer entropy.

For all disturbances with time scales appreciably greater than the time scale of convective adjustment, the vertical structure of the disturbance is completely determined by the shape, in $T - p$ space, of a moist adiabat (Neelin and Yu, 1994). Moreover, the static stability felt by such disturbances is not related to the degree of conditional instability but rather to a "gross moist stability" that depends on the shape of the vertical moist static energy profile (Neelin and Yu, 1994). Calculations of the distribution and magnitude of this gross stability have been presented by Yu and Neelin (1997). The exact magnitude of this stability measure, on the other hand, depends not only on the shape of the moist static energy profile but on the relationship between static energy and moisture fluctuations. This relationship, in turn, depends on the details of cloud microphysical processes, which are poorly understood.

One of the most basic issues we may address in quasi-equilibrium thinking is what happens when an internal gravity wave passes through a background atmosphere in radiative-convective equilibrium. (Note that the equivalent question in nonequilibrium thinking is what happens when an internal wave passes through a cloudless background atmosphere that contains stored CAPE.) To assert that the convection remains close to a state of statistical equilibrium with the large scale, we must assert that the horizontal wavelength is much larger than the intercloud spacing of the background state and that the wave period turns out to be much longer than the convective adjustment time scale.

Note first that, as of this writing, the answer to the question posed above has not been obtained by direct numerical simulation with cloud-resolving models. Such models, at least in three dimensions, are not quite capable of simultaneously containing a reasonably large gravity wave and resolving individual clouds.

The first and most basic question is how the radiative-convective equilibrium atmosphere responds to large-scale ascent and descent. To begin with, we make use of the fact that, as long as all convective kinetic energy is locally dissipated, convection does not alter the mass integral of the system moist state energy, h:

$$h = C_{\mathrm{p}}T + L_{\mathrm{v}}q + gz.$$

Thus, if we neglect perturbations to the radiative heating, the equation for the vertically integrated moist static energy perturbation from the mean state is

$$\frac{\partial [h']}{\partial t} = -\left[\omega' \frac{\partial \bar{h}}{\partial \mathrm{p}} \right], \tag{12}$$

where the brackets denote an integral over the mass of the convecting layer, and the overbar signifies the background state. As discussed earlier, the vertical structure of ω' is determined by the condition that the temperature profile remains moist adiabatic. Thus, the sign of the response of h' to ascent or descent depends on the convolution of the vertical structure function of ω with $\partial \bar{h} / \partial \mathrm{p}$, as pointed out by Neelin and Yu (1994). It is straightforward to calculate this from atmospheric soundings, and this has been done (Yu and Neelin, 1997). The conclusion is that this measure of stratification is stable throughout the atmosphere, meaning that upward motion will be associated with decreasing h'.

Now if h' decreases, then either T' decreases, q' decreases, or both. In all of the observational studies of which I am aware, ascent is associated with increasing q', implying that T' must decrease with h'. Moreover, even if the extreme assumption is made that the relative humidity is invariant with large-scale vertical motion, it is easy to show that T' must have the same sign as h':

$$h' = c_\mathrm{p} T' + L_\mathrm{v} q'$$

$$= c_\mathrm{p} T' + L_\mathrm{v} \mathcal{H} \left(\frac{\partial q^*}{\partial T} \right)_\mathrm{p} T', \tag{13}$$

where $\mathcal{H}$ is the relative humidity. Because $(\partial q^* / \partial T)_\mathrm{p}$ is positive, T' and h' must have the same sign when $\mathcal{H}$ is constant. For T' to have the opposite sign as h', $\mathcal{H}$ would have to have a rather strong negative correlation with ascent, which is certainly not observed.

Thus, although the exact value of the response of temperature to ascent depends on the details of how clouds humidify the atmosphere, there is little question that ascent causes cooling, just as in a dry, stable atmosphere. The magnitude of this cooling is less, and sometimes much less, than in a nonconvecting atmosphere.

It pays to consider the same process from another angle, introduced by the author (Emanuel, 1989). It begins with the observation that the temperature of convecting atmospheres lies close to an adiabat originating in the subcloud layer. Thus, in strict statistical equilibrium as defined by Emanuel *et al.* (1994), the temperature of the troposphere is tied to the moist static energy of the subcloud layer. Now, if large-scale ascent occurs, convergence will occur in the subcloud layer, but this does *not* change the moist static energy, which is, after all, a conserved variable. But to keep the deep troposphere approximately moist adiabatic, there must be enhanced convection to counter the adiabatic cooling associated with the large-scale ascent. Enhanced convection will be associated also with enhanced downdrafts, which import low static energy into the subcloud layer. The reduction of subcloud-layer moist static energy will then be associated with a net cooling of the lower troposphere. This argument is fully equivalent to the preceding one, but makes the importance of downdrafts more explicit. Reducing the relative magnitude of the response of the convective downdraft moist static energy flux reduces the effective stratification. When this flux vanishes, as in a saturated atmosphere, so does the effective stratification.

Thus we may expect a large-scale gravity wave, propagating through a radiative-convective equilibrium background state, to propagate much more slowly than an equivalent wave propagating through a cloudless atmosphere with the same stratification.

But there is, after all, a difference. In the case of a dry atmosphere, there is no time lag between vertical displacement and buoyancy. On the other hand, it takes time for convection to respond to changes in its environment. This time lag is probably on the order of hours, but its effect on the large-scale dynamics is important. Consider again the problem of an internal gravity wave passing through a background state in radiative-convective equilibrium. As described before, upward motion will be associated with cooling, and, ordinarily, the lowest temperatures will occur $1/4$ cycle after the strongest upward motion. Now take into account the small lag in the response of the convection. Now the convection, rather than being precisely in phase with the wave ascent, will lag slightly, thus shifting slightly toward the cold phase of the disturbance. This will cause the convective heating to have a negative correlation with temperature, thereby draining perturbation potential energy from the wave. This effect is called *moist convective damping*. This was shown by the author (Emanuel, 1993) to damp linear equatorial waves of all kinds in a simple model with a quasi-equilibrium-type convective closure. Neelin and Yu (1994) showed it more generally to be true of linear disturbances in a vertically continuous atmosphere using the Betts (1986) convection scheme, and Brown and Bretherton (1995) demonstrated it in a linear model using the Emanuel (1991) convection scheme. Here I demonstrate that it applies to a quasi-linear, two-column model. The model is linear in the sense that it ignores wave–wave interaction, and retains strictly sinusoidal spatial distribution of the wave variables, but it allows for arbitrary amplitude of the wave.

Nonlinear terms in the momentum equation are ignored, but are retained in the thermodynamic equation. The linearized inviscid, hydrostatic vorticity equation in a nonrotating atmosphere can be written

$$\frac{\partial^2}{\partial z^2}\frac{\partial w}{\partial t} = \nabla_2^2 B,\tag{14}$$

where w is the vertical velocity, B is the buoyancy, and ∇_2^2 is the horizontal Laplacian operator. Assuming horizontally sinusoidal disturbances with combined horizontal wave number K, Eq. (14) becomes

$$\frac{\partial^2}{\partial z^2}\frac{\partial w}{\partial t} = -K^2 B.\tag{15}$$

This equation can be solved in a single column if K is specified. One problem that arises in doing this, however, is that to define the buoyancy, B, one must define a background state. This is easy enough for infinitesimal disturbances, but for real disturbances the background state may itself change. To avoid this difficulty, we solve Eq. (15) in a two-column model, where each column is out of phase with the other ($w_1 = -w_2$) and define B as

$$B_i = \frac{1}{2}\frac{g}{\overline{T}_v}(T_{vi} - T_{vj}),\tag{16}$$

where the subscript i denotes the column in question and the subscript j denotes the other column, while $\overline{T}_v \equiv \frac{1}{2}(T_{vi} + T_{vj})$. Each column is identical to the single-column model described in Section III, except that vertical advection terms are added to the heat and moisture equations, and Eq. (15) is also solved in each column. Moreover, to guarantee global energy conservation, heat and moisture are advected from one column to the other according to the horizontal circulation implied by mass continuity and w.

The two-column model is initialized with an arbitrary buoyancy perturbation to the radioactive convective equilibrium state. Figure 9 shows the time evolution of the vertical velocity at a particular level in one of the two columns, for wave number 1. One observes a decaying oscillation with a period of roughly 40 days. Examination of the behavior of the system for other specified horizontal wave numbers, K, shows the same general behavior, with the period decreasing and the rate of decay increasing with increasing K.

This is completely consistent with earlier results by Emanuel (1993), Neelin and Yu (1994), and Brown and Bretherton (1995). We can say with

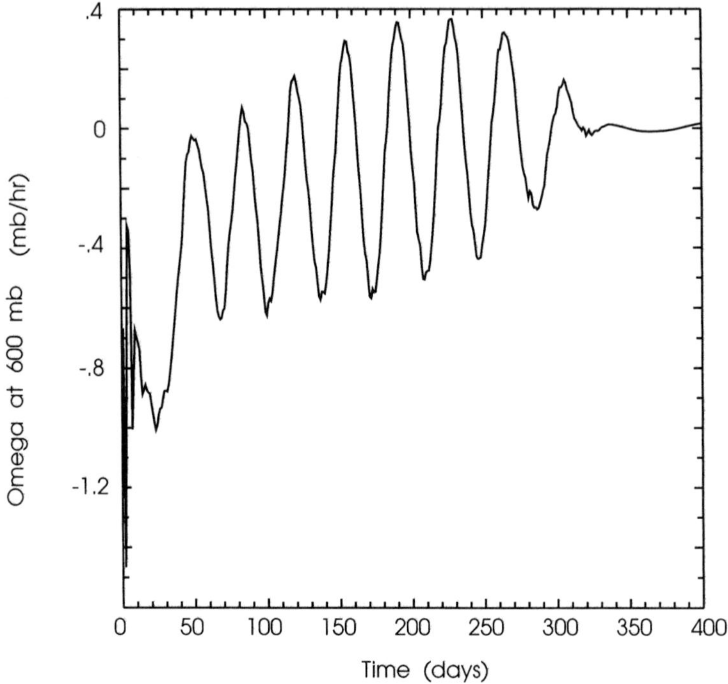

Figure 9　Variation with time of the pressure velocity (ω) at 600 mb in a two-column model that has been perturbed away from radiative-convective equilibrium.

some certainty that internal gravity waves propagating through a background state in radiative-convective equilibrium will be damped.

This finding is at odds with CISK models, dating back at least to Lindzen (1974). What is different about these models? Two key elements of CISK models allow them to produce unstable internal waves. The first is the partial or complete decoupling of convective heating from buoyancy, which occurs when the rate of heating is related to moisture convergence. This permits buoyancy to accumulate in one phase of a wave and be reduced in another and so is explicitly contrary to the quasi-equilibrium hypothesis. The second is, in the case of some models, a specification of the vertical profile of convective heating that is independent of the vertical profile of cloud buoyancy. This also allows for in-phase fluctuations of buoyancy and heating.

Other aspects of quasi-equilibrium thinking are discussed in Emanuel *et al.* (1994). The two most important conclusions that emerge from the quasi-equilibrium point of view are these:

- Large-scale disturbances in convecting atmosphere "feel" a reduced, but still positive, effective static stability.
- Such disturbances are also damped in proportion to their frequency.

If this mode of thinking is correct, then we must turn away from convection per se as an "explanation" for large-scale disturbances and think of it instead as a means of rapidly redistributing enthalpy in the vertical. Potential candidates for the sources of large-scale disturbances in the tropics include the following:

- Horizontal gradients of the radiative-convective equilibrium temperature. These are responsible for the Walker–Hadley and monsoon circulations, for example (Held and Hou, 1980; Plumb and Hou, 1992).
- Wind-induced surface heat exchange (WISHE). The feedback between wind and surface enthalpy flux is responsible for tropical cyclones and many play a role in other tropical phenomena.
- Transmission of wave energy from outside the tropics. There are now well-documented instances of this phenomenon (Kiladis, 1998).
- Dynamical instabilities. The instability of the African easterly jet in summer is the source of easterly waves, for example.

VI. SUMMARY

Until quite recently, quasi-equilibrium has been thought of primarily as a closure for convective parameterizations; its effect on the way we think about convection has been relatively slow to come about. In this paper I have reviewed and in some small ways extended Emanuel *et al.*'s (1994) exploration of the full implications of quasi-equilibrium. The main structure of quasi-equilibrium thinking emphasizes the following points:

- Latent heating is a concept that applies to the dynamics of individual clouds. In contrast, it plays *no role* in the energetics of cumulus ensembles.
- The state of radiative-convective equilibrium serves as the basic equilibrium state for quasi-equilibrium thinking in the same way that an east–west baroclinic flow serves as the basic state for quasi-geostrophic thinking about many midlatitude flows.
- Disturbances with space and time scales much larger than convective overturning time scales, and intercloud spacing characterizing the

radiative-convective equilibrium state may be considered to be in quasi-equilibrium with the convective clouds.

- Such disturbances "feel" an effective stratification that, while positive, is much less than typical dry stratifications. The stratification may also be related physically to drying of the subcloud layer by convective downdrafts. The effective stratification vanishes when the large scale becomes saturated, as happens in the core of tropical cyclones.
- Such disturbances are also damped in proportion to their frequency. This tends to filter high-frequency disturbances and to damp most nascent tropical depressions.
- Convection that is not close to being in equilibrium with explicitly simulated flows cannot be parameterized as a function of the explicitly resolved variables.

A full appreciation of the consequences of quasi-equilibrium will no doubt lead to important advances in understanding and predicting large-scale disturbances in convecting atmospheres.

REFERENCES

Arakawa, A., and W. H. Schubert (1974). Interaction of a cumulus cloud ensemble with the large-scale environment, Part I. *J. Atmos. Sci.* **31,** 674–701.

Betts, A. K. (1986). A new convective adjustment scheme. Part I: Observational and theoretical basis. *Quart. J. Roy. Meteor. Soc.* **112,** 677–691.

Bretherton, C. S., and P. K. Smolarkiewicz (1989). Gravity waves, compensating subsidence and detrainment around cumulus clouds. *J. Atmos. Sci.* **46,** 740–759.

Brown, R. G., and C. S. Bretherton (1995). Tropical wave instabilities. *J. Atmos. Sci.* **51,** 67–82.

Charney, J. G., and A. Eliassen (1964). On the growth of the hurricane depression. *J. Atmos. Sci.* **21,** 68–75.

Craig, G. C., and S. L. Gray (1996). CISK or WISHE as the mechanism for tropical cyclone intensification. *J. Atmos. Sci.* **53,** 3528–3540.

Emanuel, K. A. (1989). The finite-amplitude nature of tropical cyclogenesis. *J. Atmos. Sci.* **46,** 3431–3456.

Emanuel, K. A. (1991). A scheme for representing cumulus convection in large-scale models. *J. Atmos. Sci.* **48,** 2313–2335.

Emanuel, K. A. (1993). The effect of convective response time on WISHE modes. *J. Atmos. Sci.* **50,** 1763–1775.

Emanuel, K. A. (1994). "Atmospheric Convection." Oxford Univ. Press, New York.

Emanuel, K. A., and M. Ziukovic-Rothman (1999). Development and evaluation of a convective scheme for use in climate models. *J. Atmos. Sci.* **56,** 1766–1782.

Emanuel, K. A., and M. Bister (1996). Moist convective velocity and buoyancy scales. *J. Atmos. Sci.* **53,** 3276–3285.

Emanuel, K. A., J. D. Neelin, and C. S. Bretherton (1994). On large-scale circulations in convecting atmospheres. *Quart. J. Roy. Meteor. Soc.* **120,** 1111–1143.

Hack, J., and W. H. Schubert (1986). Nonlinear response of atmospheric vortices to heating by organized cumulus convection. *J. Atmos. Sci.* **43,** 1559–1573.

Held, I. M., and A. Y. Hou (1980). Nonlinear axially symmetric circulations in a nearly inviscid atmosphere. *J. Atmos. Sci.* **37,** 515–533.

Hoskins, B. J., M. E. McIntyre, and A. W. Robertson (1985). On the use and significance of isentropic potential vorticity maps. *Quart. J. Roy. Meteor. Soc.* **111,** 877–946.

Islam, S., R. L. Bras, and K. A. Emanuel (1993). Predictability of mesoscale rainfall in the tropics. *J. Appl. Meteor.* **32,** 297–310.

Kiladis, G. N. (1998). Observations of Rossby waves linked to convection over the eastern tropical Pacific. *J. Atmos. Sci.* **55,** 321–339.

Klaus, D., and M. J. Reeder (1997). The effects of convection and baroclinicity on the motion of tropical-cyclone-like vortices *Quart. J. Roy. Meteor. Soc.* **123,** 699–725.

Kleinschmidt, E., Jr. (1951). Gundlagen einer Theorie des tropischen Zyklonen. *Arch. Meteorologie, Geophysik Bioklimatologie,* Series A, **4,** 53–72.

Lindzen, R. S. (1974). Wave-CISK in the tropics. *J. Atmos. Sci.* **31,** 156–179.

Manabe, S., J. Smagorinsky, and R. F. Strickler (1965). Simulated climatology of a general circulation model with a hydrologic cycle. *Mon. Wea. Rev.* **93,** 769–798.

Neelin, J. D., and J. Yu (1994). Modes of tropical variability under convective adjustment and the Madden–Julian oscillation. Part I: Analytical theory. *J. Atmos. Sci.* **51,** 1876–1894.

Ooyama, K. (1964). A dynamical model for the study of tropical cyclone development. *Geofis. Int.* **4,** 187–198.

Ooyama, K. (1969). Numerical simulation of the life-cycle of tropical cyclones. *J. Atmos. Sci.* **26,** 3–40.

Ooyama, K. (1982). Conceptual evolution of the theory and modeling of the tropical cyclone. *J. Meteor. Soc. Japan* **60,** 369–379.

Plumb, R. A., and A. Y. Hou (1992). The response of a zonally symmetric atmosphere to subtropical thermal forcing: Threshold behavior. *J. Atmos. Sci.* **49,** 1790–1799.

Prandtl, L. (1925). Bericht über Untersuchungen zur ausgebildeten Turbulenz. *Zs. Angew. Math. Mech.* **5,** 136–139.

Raymond, D. J. (1995). Regulation of moist convection over the west Pacific warm pool. *J. Atmos. Sci.* **52,** 3945–3959.

Rennó, N. O., and A. P. Ingersoll (1996). Natural convection as a heat engine: A theory for CAPE. *J. Atmos. Sci.* **53,** 572–585.

Riehl, H. (1950). A model for hurricane formation. *J. Appl. Phys.* **21,** 917–925.

Robe, F. R. (1996). Sea, sun and shear: A recipe for precipitating convection, tropical rainbands and hurricane spiral arms, Ph D. thesis, Mass. Inst. Tech., Cambridge.

Robe, F. R., and K. A. Emanuel (1996). Dependence of tropical convection on radiative forcing. *J. Atmos. Sci.* **53,** 3265–3275.

Rotunno, R., J. B. Klemp, and M. L. Weisman (1988). A theory for strong, long-lived squall lines. *J. Atmos. Sci.* **45,** 463–485.

Thorpe, A. J., M. J. Miller, and M. W. Moncrieff (1982). Two-dimensional convection in non-constant shear: A model of mid-latitude squall lines. *Quart. J. Roy. Meteor. Soc.* **108,** 739–762.

Yanai, M. (1964). Formation of tropical cyclones. *Rev. Geophys.* **2,** 367–414.

Yu, J.-Y., and J. D. Neelin (1997). Estimating the gross moist stability of the tropical atmosphere. *J. Atmos. Sci.* **55,** 1354–1372.

Chapter 9

Application of Relaxed Arakawa–Schubert Cumulus Parameterization to the NCEP Climate Model: Some Sensitivity Experiments

Shrinivas Moorthi

NOAA, National Centers for Environmental Prediction
Camp Springs, Maryland

I. INTRODUCTION

Cumulus convection is a major component of the atmospheric hydrologic cycle. Accurate prediction of both heating as well as moisture change in the atmosphere due to cumulus convection is crucial to the success of numerical weather prediction, as well as climate and global change studies. The most elegant and complete theory of interaction between cumulus ensembles and the large-scale environment is presented in the pioneering work of Arakawa and Schubert (1974, hereafter AS), and has remained so for more than two decades. Nevertheless, the standard implementation of this theory in weather prediction and climate models (e.g., Lord *et al.*, 1982) is very expensive, particularly when the models are of high resolution. Consequently, the AS scheme has not achieved much wider application.

Moorthi and Suarez (1992, hereafter MS) developed the relaxed Arakawa–Schubert (RAS) scheme, which is a simpler and economical implementation of the basic ideas of AS. In RAS, MS have made several simplifications, both in the cloud model and the way in which quasi-equilibrium is achieved. The cloud model was simplified both by assuming that the normalized cloud updraft mass flux is a linear function of height and by ignoring the effects of liquid water loading and virtual temperature. Quasi-equilibrium is achieved not by simultaneously letting all cloud ensembles adjust the state, but by letting each cloud ensemble relax the state toward equilibrium. This is achieved by invoking several cloud ensembles for every time step, one by one, and letting a fraction (relaxation parameter) of the mass flux (required for full quasi-equilibrium of that ensemble) of each cloud ensemble partially modify the environment.

RAS has been quite successful in achieving its original goal of economy and simplicity, and some form of it is being used at several institutions. However, as in the original implementation of AS, RAS also suffers from excessive drying due to the lack of downdraft. A method for reducing the excessive drying was proposed by Sud and Molod (1988) in their implementation of AS in a version of the Goddard Laboratory for Atmosphere General Circulation Model (GCM). We have included a simplified version of Sud and Molod scheme for reevaporation of falling convective rain as an attempt to reduce the excessive drying in RAS. The new scheme introduces an additional tunable parameter.

This version of RAS has been incorporated into the new NCEP climate model, which is under development. We have used this opportunity to study the sensitivity of simulated climate to a few of the parameters whose values are more or less ad hoc and, thus, tunable. Here, we report the results from this exercise. In Section II, we present some details of the rain reevaporation formulation. A brief description of the new climate model that is under development at NCEP is given in Section III. In Section IV, we explore the impact of several parameter changes in a semi-prognostic test using GATE phase III data. Section V contains results from the sensitivity study with the NCEP climate model. A summary is given in Section VI.

II. MODIFICATION OF RELAXED ARAKAWA–SCHUBERT

A. REEVAPORATION OF THE FALLING CONVECTIVE PRECIPITATION

In the implementation of the Sud and Molod (1988) scheme into RAS, the reevaporation of rain from each cloud type occurs just after that cloud type has relaxed the environmental sounding toward quasi-equilibrium, but before invoking the next cloud type. In this way, the subsequent cloud type feels the effect of rain reevaporation from the previous cloud type. This differs from the approach of Sud and Molod (1988), in which reevaporation takes place after the environmental sounding is fully adjusted by all existing cloud types (i.e., standard implementation of AS scheme). In their approach, each cloud type does not recognize the moistening of the environment due to reevaporation of falling rain from both itself and other cloud types until the next time step.

Following Sud and Molod (1988), we parameterize the fraction of the precipitation that is evaporated into the environment in a layer at or below the detrainment level as (A. Molod, personal communication, 1992):

$$f = \min\left\{1, \frac{\Delta q}{P_r}\sigma\left[P_r(3600/\Delta t)(p/1000)^{0.5}\right]^{0.578}\right\}, \tag{1}$$

where Δq is the saturation moisture deficit of that layer calculated iteratively (we use three iterations), P_r is the precipitation over a time step of Δt seconds that is available for reevaporation, and p is the pressure at the middle of the layer in hecta Pascals (hPa). Here, σ is the horizontal fractional area of the grid covered by the falling convective precipitation, and is parameterized as the ratio of the amount of the mass detrained and the mass of the detrainment layer scaled by a tuning parameter R_k. Thus we take

$$\sigma = \min\{1, R_k g M_B \eta_i / \Delta p_i\}, \tag{2}$$

where M_B is the cloud base mass flux, η_i is normalized mass flux at the cloud top, Δp_i is the pressure thickness of the detrainment layer i, g is the gravitational acceleration, and $k = i, i + 1, \ldots, K$ where K is the total number of model layers. Although we have allowed R_k in Eq. (2) to be a function of height, in our experiments we have used constant values.

B. Some Additional Aspects of RAS

Here, we discuss some additional observations of the scheme presented in Moorthi and Suarez (1992). Suppose we have a vertically discrete model with K layers, the Kth layer being the boundary layer. The modification of the environment by any cloud type i, detraining at level i, can be written as

$$\left(\frac{\partial T_k}{\partial t}\right) = \alpha_i \frac{M_B(i)}{c_p} \Gamma_h(k), \qquad k = i, i+1, \ldots, K \tag{3}$$

and

$$\left(\frac{\partial q_k}{\partial t}\right) = \alpha_i \frac{M_B(i)}{c_p} [\Gamma_h(k) - \Gamma_s(k)], \qquad k = i, i+1, \ldots, K, \tag{4}$$

where T and q are the environmental temperature and specific humidity, α_i is the relaxation parameter for the ith cloud type, $M_B(i)$ is the cloud base mass flux for that cloud type, and c_p and L are the specific heat at constant pressure and latent heat of vaporization, respectively. There is no modification of the environment above the cloud top layer. The Γ_h and Γ_s values are the tendencies of environmental moist static energy h and dry static energy s per unit cloud base mass flux, and are given by

$$\Gamma_h(k) = \frac{g}{\Delta p_k} \Big[\eta_{i,k-1/2}(h_{k-1/2} - h_k) + \eta_{i,k+1/2}(h_k - h_{k+1/2})$$
$$+ \delta_i^k \eta_{i,i}(h_i^* - h_i) \Big], \qquad \text{for } k = i, i+1, \ldots, K \tag{5}$$

and

$$\Gamma_s(k) = \frac{g}{\Delta p_k} \Big[\eta_{i,k-1/2}(s_{k-1/2} - s_k) + \eta_{i,k+1/2}(s_k - s_{k+1/2})$$
$$- \delta_i^k \eta_{i,i} l_{i,i}(1 - r_i) \Big], \qquad \text{for } k = i, i+1, \ldots, K. \tag{6}$$

Here, g is the acceleration due to gravity, Δp_k is the pressure thickness of layer k, $\eta_{i,k+1/2}$ is the normalized mass flux for cloud type i at level $k + 1/2$, which is the interface between layers k and $k + 1$, δ_k^i is the Kronecker delta, $l_{i,i}$ is the specific cloud condensate at the cloud top for cloud type i, r_i is a cloud-type-dependent precipitation fraction, and the asterisk denotes saturation value. The normalized mass flux at the top of the detrainment layer and at the earth's surface are assumed to be zero. The discretization used in Eqs. (5) and (6) allows the possibility to modify the boundary layer (layer K) through cumulus subsidence, depending on the choice of $h_{K-1/2}$ and $s_{K-1/2}$ (the values of h and s at the boundary layer top). If we take $h_{K-1/2} = h_K$, and $q_{K-1/2} = q_K$, then there would be no direct modification of the boundary layer by convection. On the other hand, if $h_{K-1/2}$ and $q_{K-1/2}$ are taken as some combination of values within the boundary layer and in the layer above, then cumulus convection can affect the boundary layer directly. In AS convection is allowed to interact with the boundary layer only indirectly, by reducing the height of the boundary layer. When the boundary layer top is not a prognostic variable, modifying the boundary layer fields may be attractive. In Section IV, we show the impact of the choice of h and q at the boundary layer top through semi-prognostic tests.

We have also incorporated an additional feature into the RAS. For a level i at which $h_i^* > h_K$, even the deepest possible cloud type, i.e., the nonentraining (entrainment parameter $\lambda = 0$) cloud type, has its level of nonbuoyancy below that level. When a cloud type represented by such a level i is invoked, we allow the deepest cloud type to exist as long as its level of nonbuoyancy is within that layer, which in general happens when $h_{i+1/2}^* < h_K$. This is done by redefining the level of nonbuoyancy as the detraining level and defining appropriate variables there. The advantages of this procedure is that it reduces the possibility of sudden on/off of the deepest cloud type, which otherwise could produce noise.

III. THE NEW NCEP CLIMATE MODEL

At NCEP, a climate model provides guidance to seasonal and long-term climate prediction. The current operational climate model is based on a substantially older version of the operational medium-range forecast (MRF) model. A new initiative is under way with collaborative efforts between the Environmental Modeling Center and the Climate Prediction Center to develop a new generation climate model starting from the latest version of the operational MRF model.

The new climate model dynamics is identical to the operational model and is based on the Eulerian spectral approach. The current resolution of the climate model is T62 (triangular truncation with 62 zonal wave numbers) in the horizontal and 28 sigma (pressure normalized by surface pressure) layers in the vertical. The parameterized physical processes, identical to the current operational MRF model, include horizontal and vertical diffusion, gravity wave drag, land-surface processes, boundary layer physics, shallow convection, and large-scale precipitation. Details of the operational MRF model are available at the EMC web site: http://

www.emc.ncep.noaa.gov. A new radiation package (Hou *et al.*, 1996), which has been implemented in the operational MRF model, is also used. This package contains an improved solar radiation calculation based on the work of M.-D. Chou and collaborators at NASA/GSFC (Chou, 1992; Chou and Lee, 1996), as well as vegetation-type-dependent albedo and an enhanced diagnostic cloud prediction scheme.

For the representation of convection in the new climate model we are experimenting with an improved version of RAS (including the scheme for reevaporation of falling convective precipitation described earlier). Since January 1998, this climate model has been undergoing extensive evaluation through use of a parallel assimilation/forecast system. Other simulation experiments are also being carried out to evaluate the potential usefulness of the model for climate predictability. To study the model's ability to produce a reasonable climate, we have performed some sensitivity tests for both winter and summer.

Because of relatively thinner layers near the lower boundary of the model, in this implementation of RAS we strap the bottom three layers (~ 50 hPa thick) of the model together and consider it to be the boundary layer for RAS.[1] A time step of 20 min is used in all runs with the climate model. The cloud types detraining between sigma levels of 0.065 and 0.76 are invoked in random order. We invoke 42 random cloud types per hour. All shallower cloud types below the sigma level of 0.76 are invoked sequentially once every time step, before the deeper cloud types.

We have also added several additional features in this implementation of RAS. We allow no convection to take place when the boundary layer relative humidity is below 55%, and full convection effects when the relative humidity equals or exceeds 70%. The main reason for doing this is economy. In general, the boundary layer humidity is higher than 70% where convection occurs. Then, by not allowing the convection to take place when the relative humidity is below 55%, substantial reduction in computing time can be achieved. Between the relative humidities of 55 and 70% the cumulus effects are weighted by a factor that exponentially varies from 0 to 1 so that the transition is smooth. Additionally, convection is not allowed when the negative contribution to the total work function exceeds a certain percentage of the total. We have made this limiting value a function of local drag coefficient so that it is smaller over oceans and

---

[1] A more elegant approach would be to determine the boundary layer depth before invoking RAS at each horizontal grid point so that convection recognizes the horizontal variation of boundary layer thickness. Nevertheless, the version of RAS code used here does not have this flexibility. An advanced version of RAS under development will have this capability.

larger over rough terrain. This is a crude attempt to represent a stronger triggering mechanism over land than over water.

IV. SENSITIVITY IN SEMI-PROGNOSTIC TEST

Before examining the sensitivity of the climate model to parameters in RAS, we will first examine their impact in a semi-prognostic context. The original version of RAS underwent both semi-prognostic tests as well as single-column prognostic tests in MS. Their results did show that although the obtained cumulus heating profile was reasonable, the cumulus drying was excessive compared to the observed estimate. This result was consistent with the result of Lord (1978) for the standard implementation of AS scheme. MS also showed that in the semi-prognostic sense, the final heating and drying profiles were not very sensitive to the value of the relaxation parameter α. In a prognostic test, or in a prediction mode, however, an appropriate choice of both the relaxation parameter and the number of cloud types per time step may be needed to achieve best results. In this section we present the results from semi-prognostic test when reevaporation of falling precipitation is also included.

For this purpose, we use the same GATE phase III data employed by MS. The daily mean radiation data are from Cox and Griffith (1978) and all other data are as analyzed by Thompson *et al.* (1979). Surface latent heat flux (evaporation) and the sensible heat flux, and their vertical distribution in the boundary layer, are estimated using the boundary layer formulation of the MRF model. We use 19 layers of equal depth in the vertical between the surface and the top of the atmosphere. The lowest model layer is considered to be the boundary layer for the semi-prognostic test. Tests are performed by varying R_k, the tunable parameter in the formulation of reevaporation of falling precipitation [see Eq. (2)], from a value of 0 to 20.

Figures 1a and 1b show the time-averaged vertical profiles of cumulus heating and drying as a function of pressure. A value of $\alpha_i = 0.4$ is used for all cloud types in all experiments. In this figure, the thick dash-dot curves represent observed estimates of $(Q_1 - Q_R)/c_p$ and $-Q_2/c_p$ where Q_1 and Q_2 are the apparent heat source and apparent moisture sink (Yanai *et al.*, 1973) and Q_R is the net radiative heating. Note that there is some uncertainty in the observed profiles since the heat and moisture budgets do not balance when surface fluxes are taken into account. Therefore, for our purposes, the observed profiles should only serve as a guide. The thick solid lines in Fig. 1 are the convective heating and drying (negative moistening) rates obtained using RAS with semi-prognostic ap-

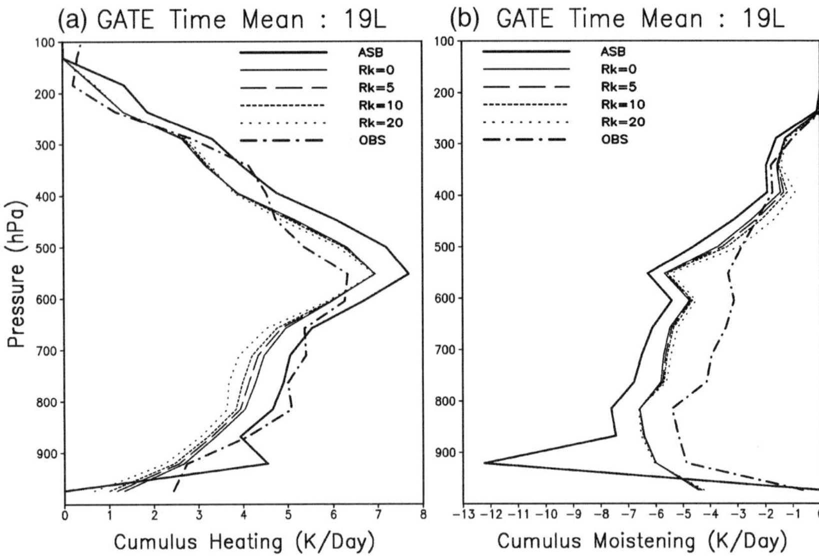

Figure 1   The time-averaged (a) cumulus heating (K) and (b) cumulus moistening (K) as a function of height obtained in the semi-prognostic test using the GATE phase III data. A value of $\alpha_i = 0.4$ is used for all cloud types. The thick dash-dot curves represent observed estimates of $(Q_1 - Q_R)/c_p$ and $-Q_2/c_p$. The thick and the thin solid lines are for convective heating and moistening rates with $h_{K-1/2} = h_K$, and $q_{K-1/2} = q_K$ and with $h_{K-1/2} = (h_K + h_{K-1})/2$ and $q_{K-1/2} = (q_K + q_{K-1})/2$, respectively, with $R_k = 0$. The long dashed, short dashed, and dotted lines correspond to the thin solid line case but with $R_k = 5$, 10, and 20, respectively.

proach, when we take $h_{K-1/2} = h_K$, and $q_{K-1/2} = q_K$ (i.e., no direct modification of the boundary layer by convection). The thin solid lines are the heating and drying rates obtained when $h_{K-1/2}$ and $q_{K-1/2}$ are taken as the average of the respective values in the boundary layer and the layer above. Both cases are with no reevaporation of falling rain (i.e., $R_k = 0$). It is obvious that in the first case (thick solid line) the cumulus drying is excessive at almost all levels below 400 hPa. The cumulus heating is comparable to the observed profile below 600 hPa, but somewhat excessive above. The predicted heating and drying rates are, of course, in exact balance when vertically integrated. The second case (thin solid line) shows dramatic reduction in the excessive drying of the first case. The heating rates also improved above 600 hPa, but somewhat cooler than the observed estimate below. Considering the uncertainty in the observed estimates, this is a significant improvement. There are three pairs of additional lines in Fig. 1 that are for the second case, with reevaporation of falling rain. The long dashed line is for $R_k = 5$, the short dashed line is for $R_k = 10$, and the dotted line is for $R_k = 20$. The reevaporation only slightly reduces the drying, at the same time reducing the heating. It also seems that a value of $R_k > 10$ may produce excessive moistening at upper levels, while producing excessive cooling. Clearly, reevaporation alone may not completely eliminate the dry bias of the Arakawa–Schubert scheme, and a downdraft parameterization may be required. Nevertheless, as shown in the next section, in the absence of downdraft, the effects of reevaporation of falling precipitation are important for global climate models. Recently, Cheng and Arakawa (1997) have shown that including a downdraft formulation in AS can drastically reduce the excessive drying, while nearly preserving the heating profiles. However, incorporation of such a downdraft is beyond the scope of this version of RAS. We are currently exploring the possibility of incorporating such a downdraft formulation in a new version of RAS (Moorthi and Suarez, personal communication, 1998).

V. SENSITIVITY EXPERIMENTS WITH THE CLIMATE MODEL

One general problem that is common to many atmospheric models is their inability to hold enough moisture in the model atmosphere. For example, some time ago in the operational model at NCEP, the assimilation of special sensor microwave imager (SSM/I) derived precipitable water was discontinued because the MRF model could not retain the initial moisture, resulting in large spindown of precipitation in the tropics. Dry bias in the tropical atmosphere over warm oceans is very common in current global models (Bony *et al.*, 1997). Bony *et al.* also point out that both the NCEP reanalysis and the NASA/DAO reanalysis are too dry over the tropical oceans when compared to satellite-based estimates. These estimates indicate that, even in the annual mean, significant areas of tropical oceans have column precipitable water in excess of 45 kg/m^2. In terms of global mean values, the annual mean column precipitable water is ~ 25 kg/m^2 with an annual cycle of amplitude ~ 4 kg/m^2 (Trenberth *et al.*, 1987; van den Dool *et al.*, 1993, 1995) with a minima around 23–24 kg/m^2 in January and maxima around 27–28 kg/m^2 in July.

We investigate the ability of the new climate model, with RAS, to retain moisture and how sensitive this is to quantities such as R_k in the reevaporation of falling precipitation and the relaxation parameter, α_i. We also investigate the sensitivity of the overall climate to changes in these parameters. Some results from these experiments are reported below. Unless stated otherwise, in all the following experiments we have taken $h_{K-1/2}$ and $q_{K-1/2}$ as a linear combination of the respective values in the boundary layer and the layer above, thus allowing the cumulus subsidence to modify the boundary layer.

A. January Case

We performed four 47-day runs starting from 15 December 1978 00 UTC and ending on 1 February 1979 00 UTC. A time-varying (daily updating) prescribed climatological sea surface temperature is used in all tests. The four runs are made with R_k values of 0, 3, 4, and 5, respectively, and a value of $\alpha_i = 0.4$ is used. When $R_k = 0$, no reevaporation of falling precipitation takes place. As R_k is increased, the area covered by the falling rain increases (subject to an upper limit of one), resulting in increased reevaporation. Because we have made only one run for each value of R_k, it may not be possible to attribute all differences in the climate to changes in these parameters. For more precise evaluation, an ensemble of such runs is necessary, but that is beyond the scope of the present study. Nevertheless, as shown below, we can still draw some conclusions from these experiments.

Figure 2a shows the evolution of global mean total precipitation (solid line), surface evaporation (dashed-line), convective precipitation (dotted line), and large-scale precipitation (dash-dot line) for the run with $R_k = 0$. The unit for all four parameters is mm/day. The rates are computed from accumulated values over a period of 12 hr. Figures 2b–d show corresponding evolution of those parameters for $R_k = 3$, 4, and 5, respectively. We infer that, after an initial spin-up period, the total precipitation and evaporation rates more or less follow each other, with a value hovering around 3.25 mm/day. The convective and large-scale precipitation rates have values of ~ 2.4 and ~ 0.8 mm/day for the case of no rain reevaporation (Fig. 2a). As seen from other figures, with rain reevaporation, the convective precipitation rate is slightly reduced with a corresponding increase in the large-scale precipitation rate.

Figure 3 shows the time evolution of the global mean precipitable water (solid line for $R_k = 0$, dashed line for $R_k = 3$, dash-dot line for $R_k = 4$, and dotted line for $R_k = 5$) for the four experiments. Note that without the reevaporation of precipitation, the atmosphere loses moisture from its initial value of ~ 23.68 kg/m^2, reaching an equilibrium value of ~ 21.6 kg/m^2. When the rain reevaporation is included, the global mean precipitable water increases from the initial value reaching an equilibrium value

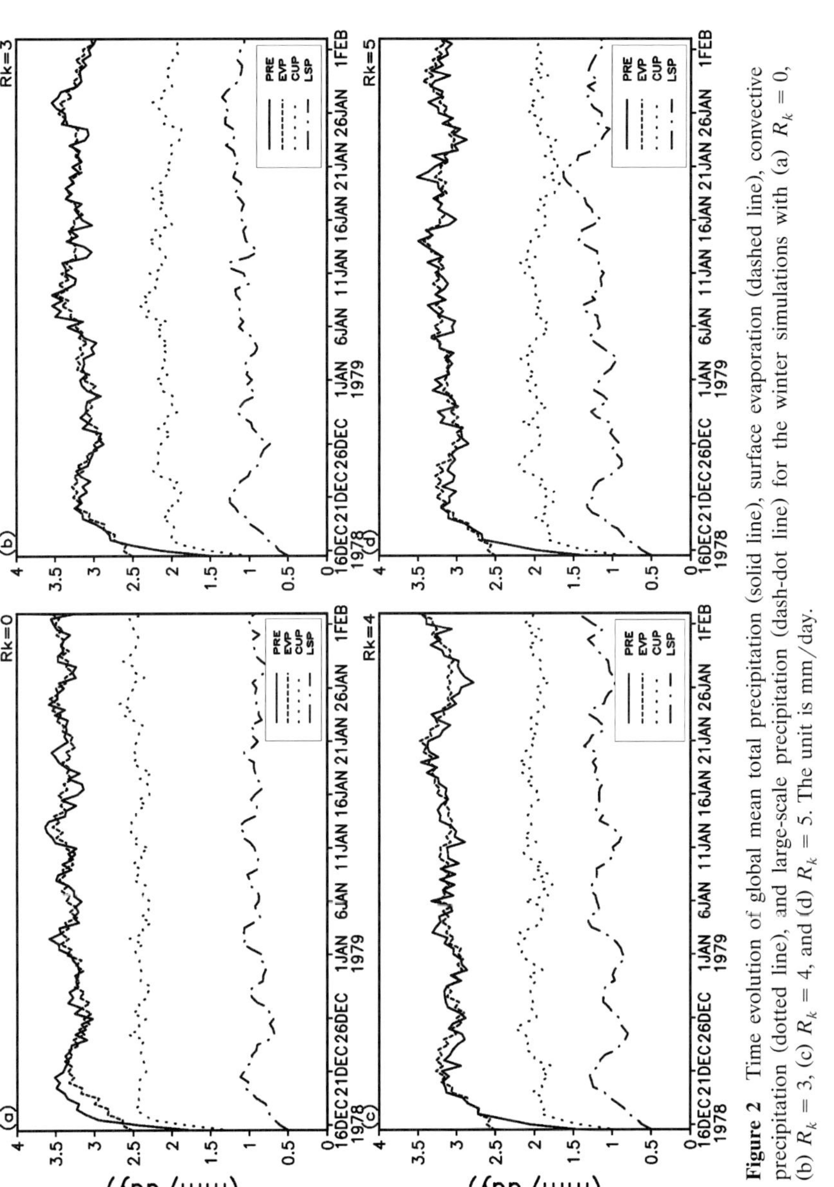

Figure 2 Time evolution of global mean total precipitation (solid line), surface evaporation (dashed line), convective precipitation (dotted line), and large-scale precipitation (dash-dot line) for the winter simulations with (a) $R_k = 0$, (b) $R_k = 3$, (c) $R_k = 4$, and (d) $R_k = 5$. The unit is mm/day.

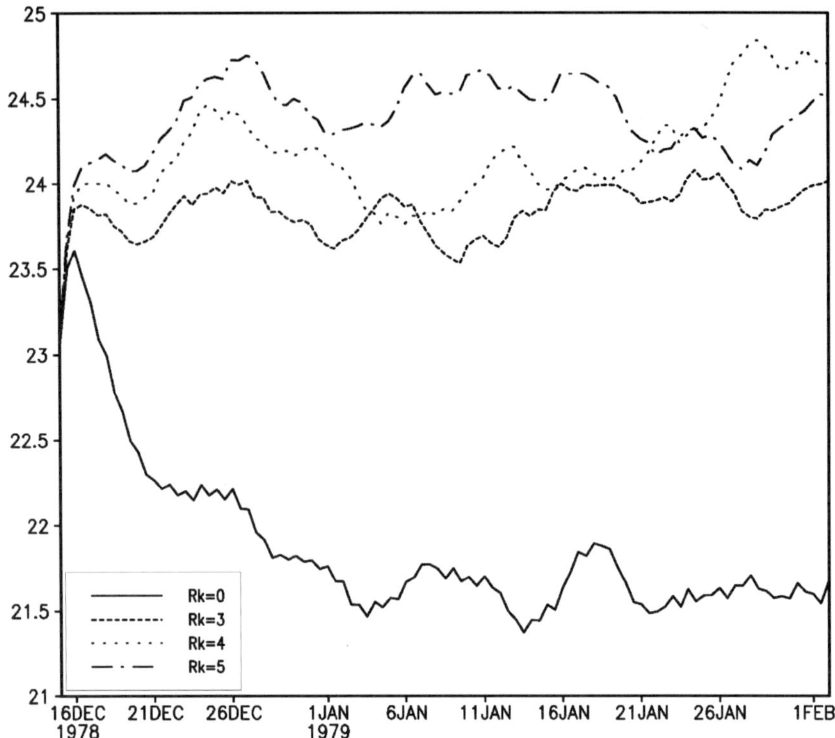

Figure 3   Time evolution of the global mean precipitable water (kg/m^2) (solid line for $R_k = 0$, dashed line for $R_k = 3$, dash-dot line for $R_k = 4$, and dotted line for $R_k = 5$) for the four winter simulations.

~ 23.8 kg/m^2 for $R_k = 3$, and higher values for $R_k = 4$ and 5. Clearly, the model atmosphere is too dry without the reevaporation of precipitation. A value of R_k in the range of 3 to 5 seems to produce reasonable global mean precipitable water for January.

Figures 4a–d show January mean total precipitable water for the four experiments with $R_k = 0$, 3, 4, and 5, respectively. The experiment with $R_k = 0$, i.e., no reevaporation of falling rain, seems to be too dry in the tropics compared to observations (Bony *et al.*, 1997) or reanalyses (Min and Schubert, 1997). Experiments with nonzero R_k considerably improve the total precipitable water with significant increase in the area over the tropical oceans having precipitable water greater than 45 kg/m^2.

Figures 5, 6, and 7 show the January mean precipitation rate, outgoing long-wave radiation at the top of the atmosphere and surface evaporation,

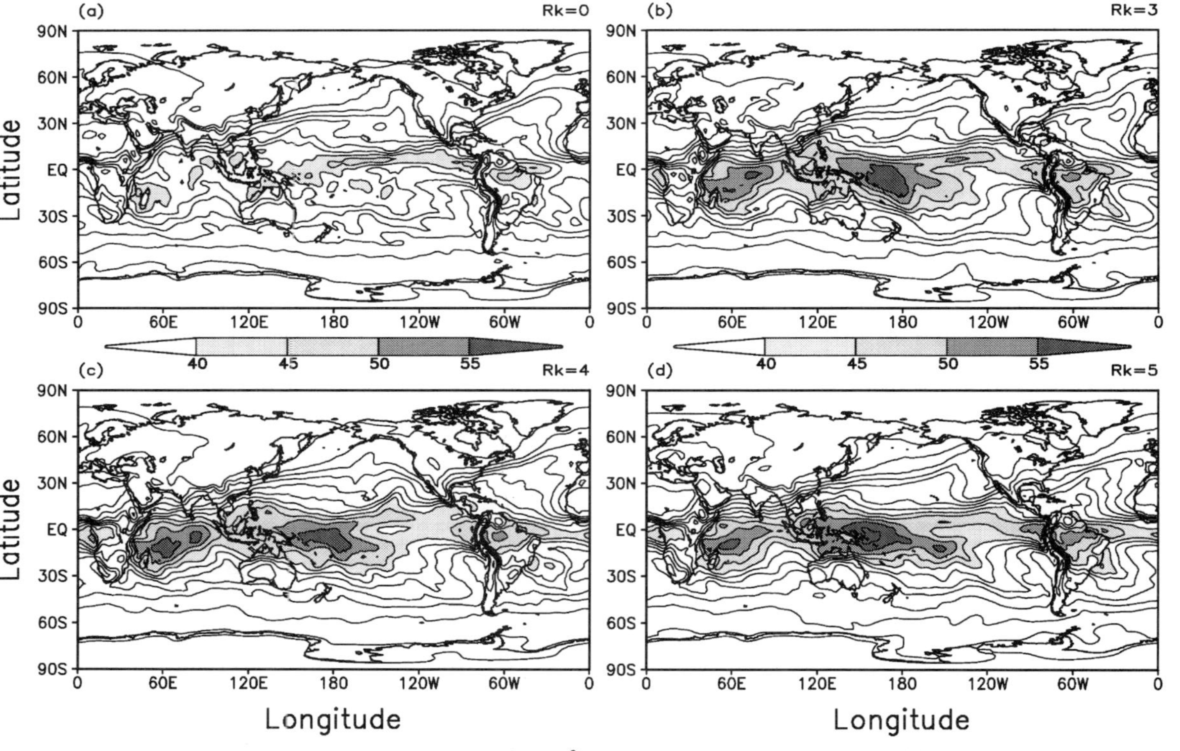

Figure 4 January mean total precipitable water (kg/m^2) for the four experiments with (a) $R_k = 0$, (b) $R_k = 3$, (c) $R_k = 4$, and (d) $R_k = 5$. The contour interval is 5 kg/m^2.

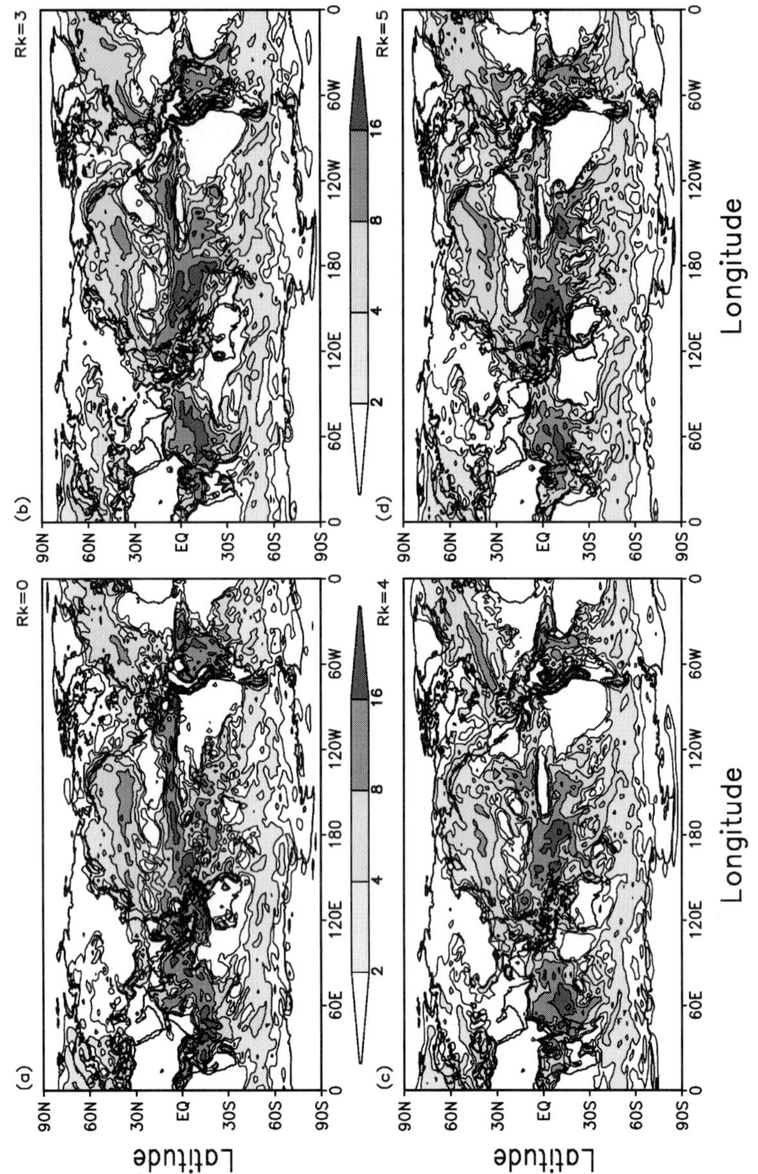

Figure 5 Same as in Fig. 4, but for January mean precipitation rate in mm/day. The contours are drawn at 1, 2, 4, 8, 16, and 32 mm/day.

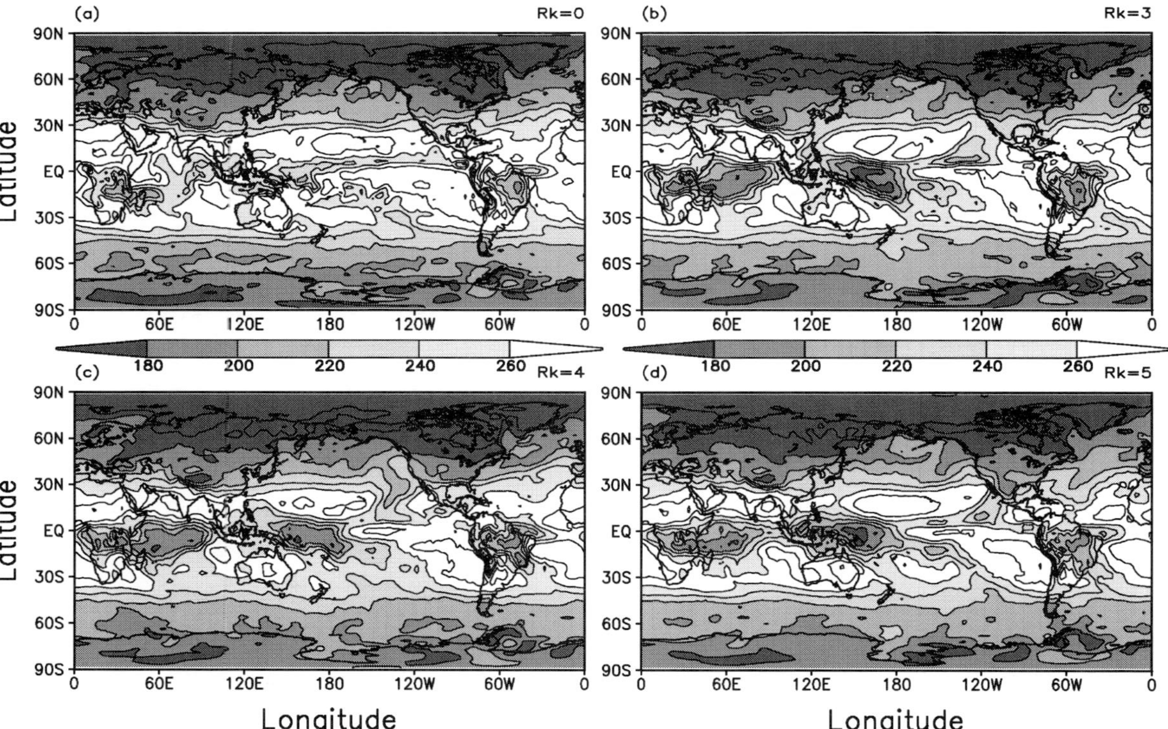

Figure 6 Same as in Fig. 4, but for January mean outgoing long-wave radiation in W/m^2. The contour interval is 20 W/m^2.

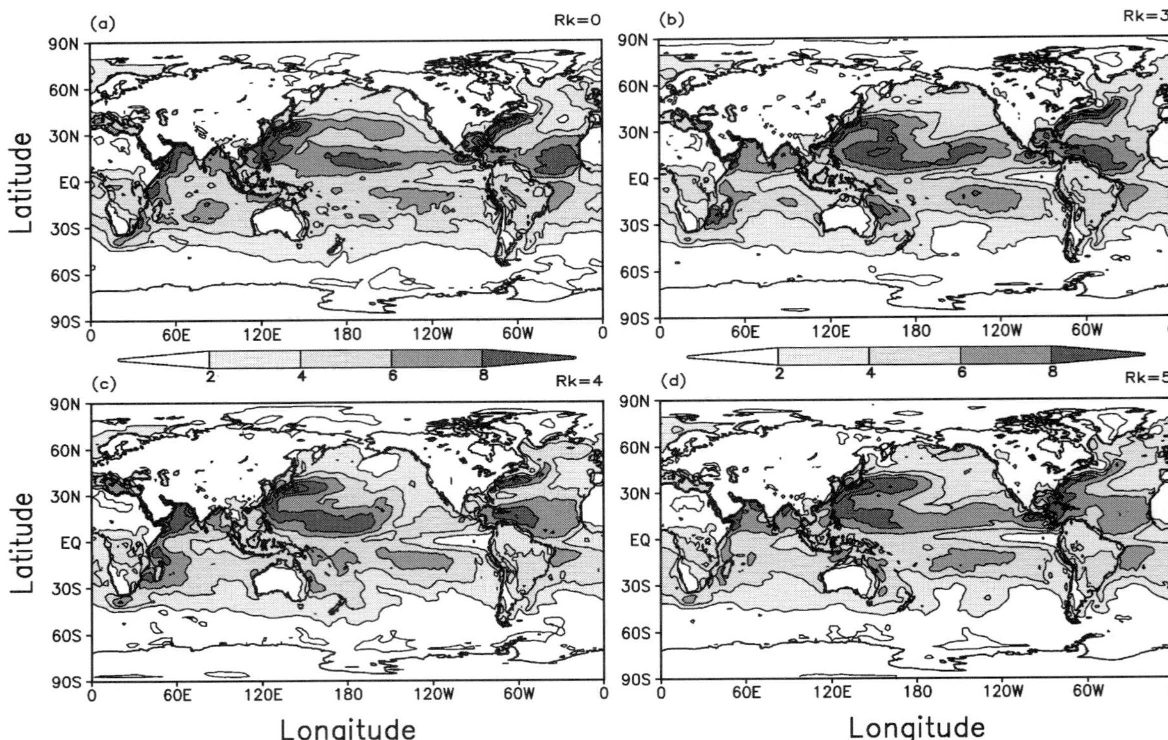

Figure 7 Same as in Fig. 4, but for January mean surface evaporation rate in mm/day. The contour interval is 2 mm/day.

respectively, again for four values of R_k. Compared to observed estimates (not shown), in general, all fields look better for nonzero values of R_k.

We have further examined the simulated dynamical fields (not shown), and, in general, they are quite similar and appear reasonable. There are some differences, but in the absence of an ensemble of runs for each R_k; it is difficult to assign a reason for them.

B. July Case

We performed four 57-day experiments starting from the NCEP/NCAR reanalysis for 5 June 1997 00 UTC to 1 August 1997 00 UTC. As before, the four runs are made with R_k values of 0, 3, 4, and 5, respectively, and a value of $\alpha_i = 0.4$ is used.

Figure 8a (similar to Fig. 2a) shows the evolution of global mean total precipitation rate (solid line), surface evaporation rate (dashed line), convective precipitation rate (dotted line), and large-scale precipitation rate (dash-dot line) for the run with $R_k = 0$. The unit for all four parameters is mm/day. Figures 8b–d show corresponding evolution of those parameters for $R_k = 3$, 4, and 5, respectively. Again, as in the January case, after an initial spin-up period, total precipitation and evaporation more or less follow each other. Both the total and convective precipitation rates are slightly higher when $R_k = 0$. However, the large-scale precipitation rate is slightly higher when $R_k > 0$. Also, the diurnal variation of the convective precipitation rate seems to be larger when no reevaporation of precipitation is included.

Figure 9 shows the time evolution of the global mean precipitable water for the four summer experiments. As expected, we note that without the reevaporation of precipitation, the atmosphere loses moisture from its initial value of ~ 25.1 kg/m^2 reaching an equilibrium value of ~ 23.75 kg/m^2. When the reevaporation is included, the global mean precipitable water increases from the initial value reaching values ~ 26.5 kg/m^2 for $R_k = 3$, and ~ 27.5 kg/m^2 for $R_k = 4$ and 5. Again, a value of R_k of about 3 to 5 seems produce reasonable global mean precipitable water for July.

Figures 10a–d show July mean total precipitable water for the four experiments with $R_k = 0$, 3, 4, and, 5, respectively. Again, the test with $R_k = 0$ seems too dry in the tropics, and a nonzero value for R_k considerably improves the total precipitable water. Figures 11, 12, and 13 show the July mean precipitation rate, outgoing long-wave radiation at the top of the atmosphere, and surface evaporation rate, respectively, for the four

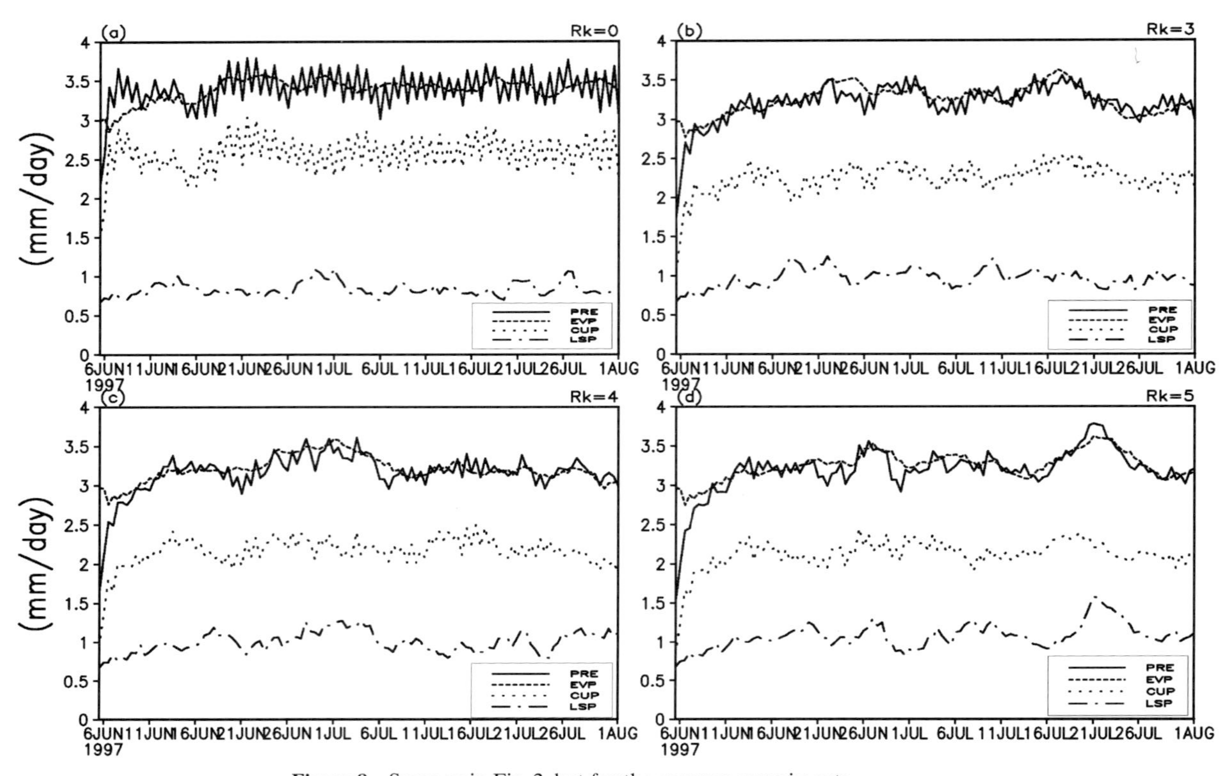

Figure 8 Same as in Fig. 2, but for the summer experiments.

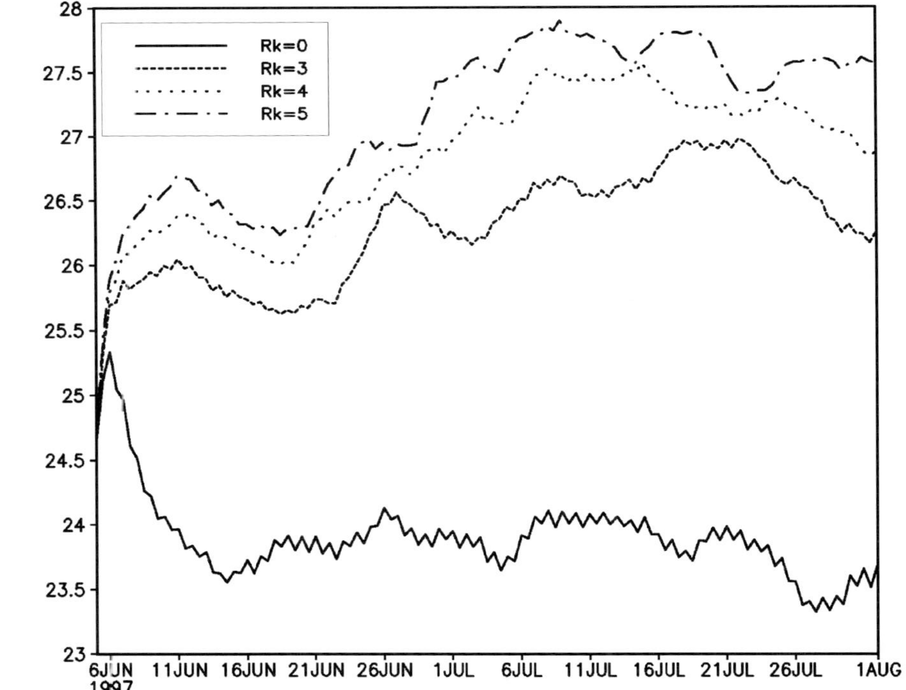

Figure 9   Same as in Fig. 3, but for the summer experiments.

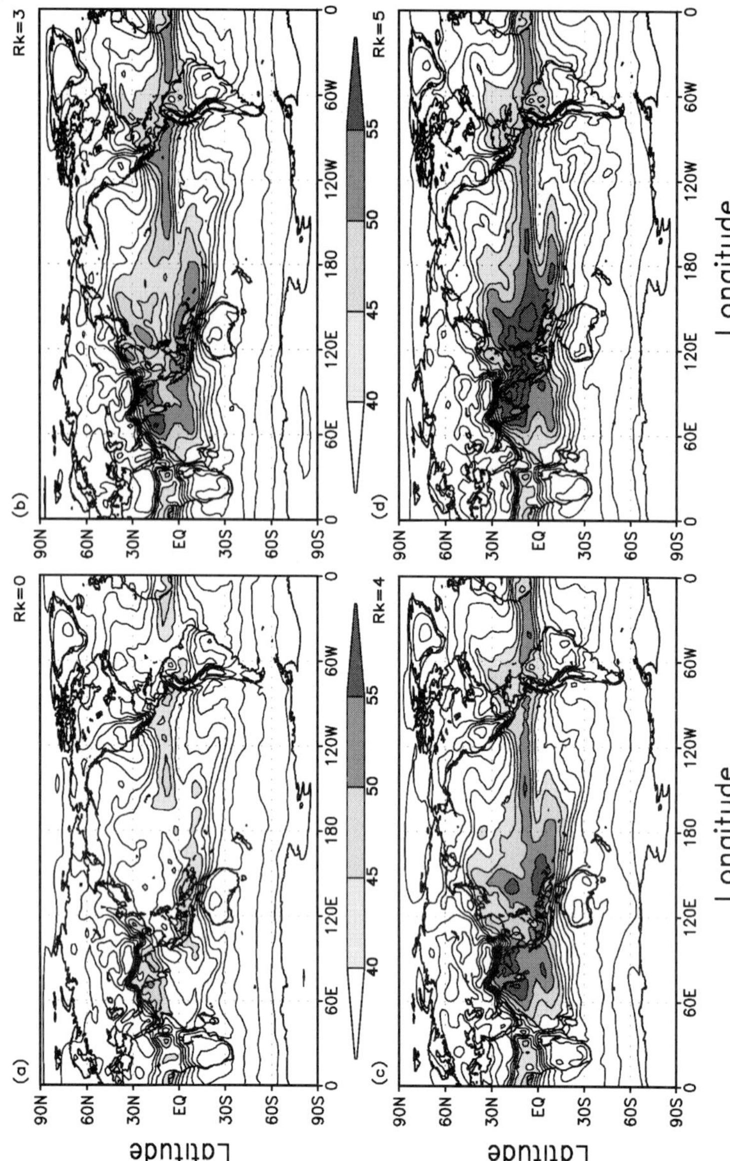

Figure 10 Same as in Fig. 4, but for July.

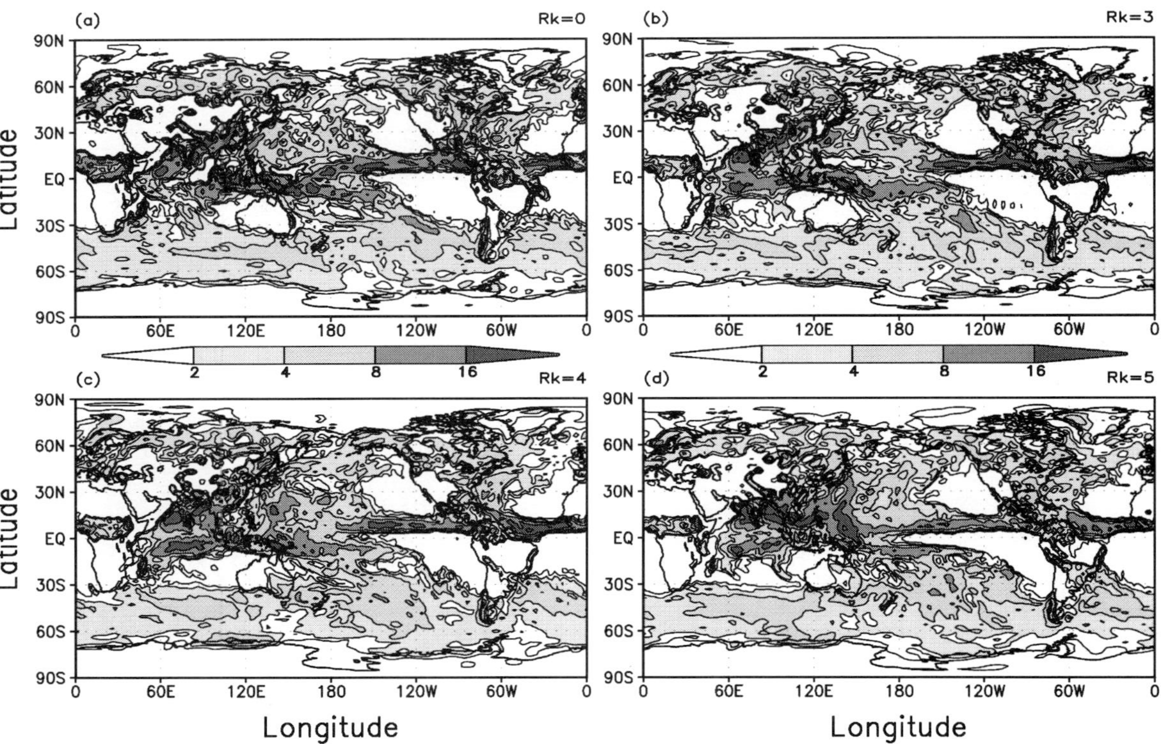

Figure 11   Same as in Fig. 5, but for July.

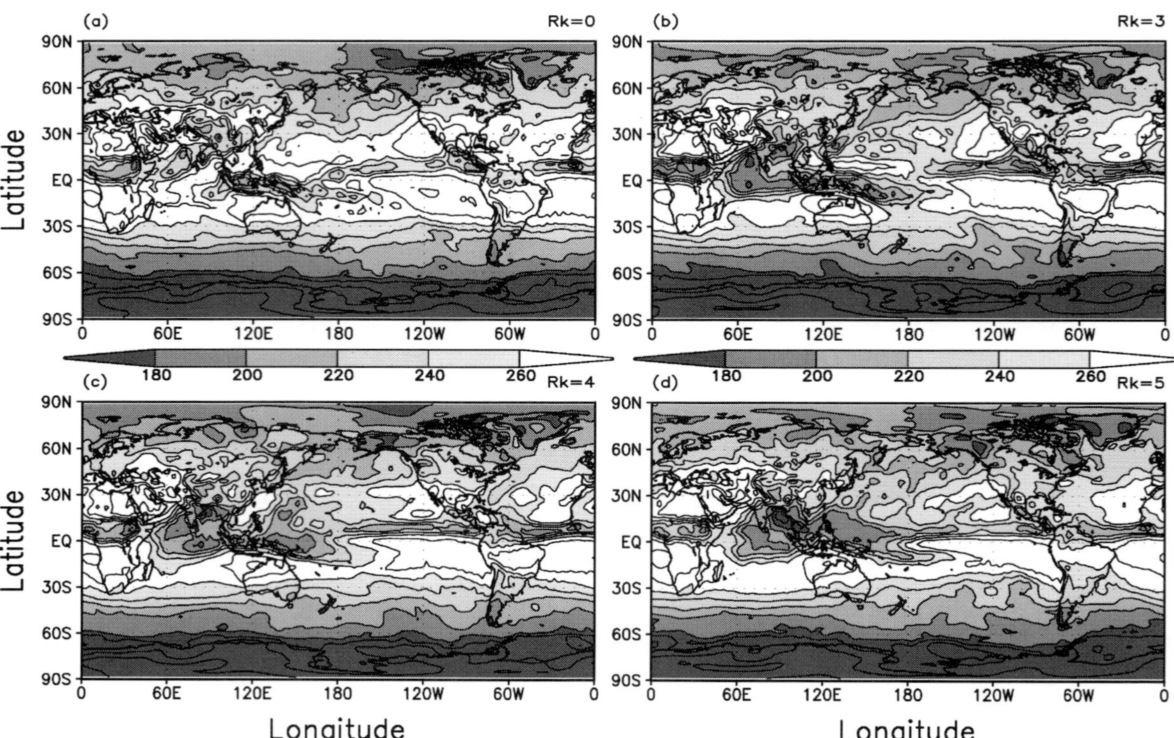

Figure 12 Same as in Fig. 6, but for July.

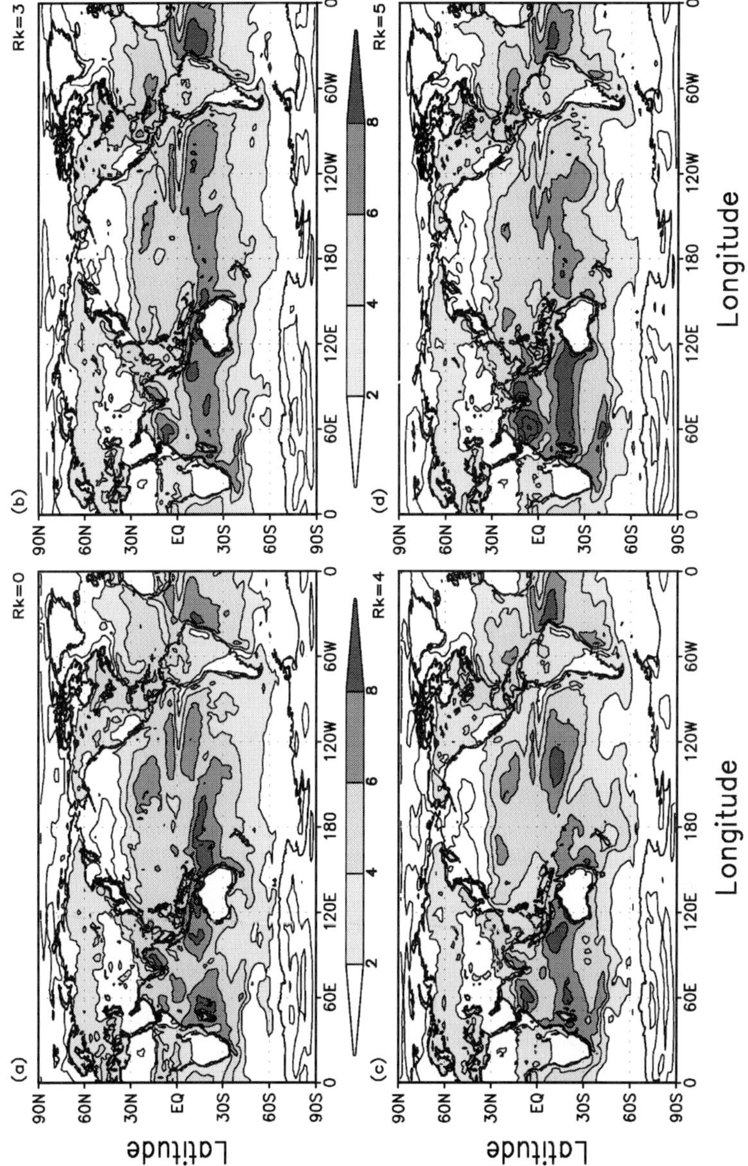

Figure 13 Same as in Fig. 7, but for July.

values of R_k. Again, the fields are more acceptable for nonzero value of R_k.

The simulated dynamical fields (not shown) are similar, with some differences, which for reasons discussed for the January case, cannot be attributed to the changes in reevaporation. However, the simulated July climate (not shown) looks quite reasonable, with a well-defined Asian summer monsoon circulation.

C. SENSITIVITY TO α_i

To explore the sensitivity of the simulation to the choice of the relaxation parameter α_i, we made two additional runs, one for the January case and one for the July case, using $R_k = 3$, $\alpha_i = 0.2$, and invoking same number of cloud types per time step as in the previous tests. The time evolution of precipitation and the total precipitable water (not shown) are similar to those presented earlier for $\alpha_i = 0.4$; however, the maximum value reached by the precipitable water is some what lower in the present case of $\alpha_i = 0.2$. Figures 14 and 15 show the January and July mean total precipitation rate, total precipitable water, outgoing long-wave radiation, and evaporation rate. Comparison of these fields with those for $\alpha_i = 0.4$ shows them to be similar, though there are some differences. If we had invoked twice the number of cloud types per time step with $\alpha_i = 0.2$, we might have obtained even greater agreement between the simulated climates with $\alpha_i = 0.2$ and $\alpha_i = 0.4$. Nevertheless, it is apparent that the simulated climate is rather insensitive to the choice of α_i, provided appropriate number of cloud types are invoked (i.e., if α_i is very small, then a larger number of cloud types may have to be invoked every time step). In practice, one should select a reasonable value of α_i so that the number of cloud types needed per time step are not too large; otherwise, the scheme becomes too expensive.

VI. SUMMARY AND CONCLUSIONS

RAS is a simple and more economical way of implementing the Arakawa–Schubert parameterization. In this paper we include a scheme for reevaporation of falling convective precipitation following the method of Sud and Molod (1988). Results from a semi-prognostic test using GATE phase III data show that RAS produces excessive drying and warming when the values of dry and moist static energies at the top of the boundary layer are assumed to be the same as those within the boundary layer.

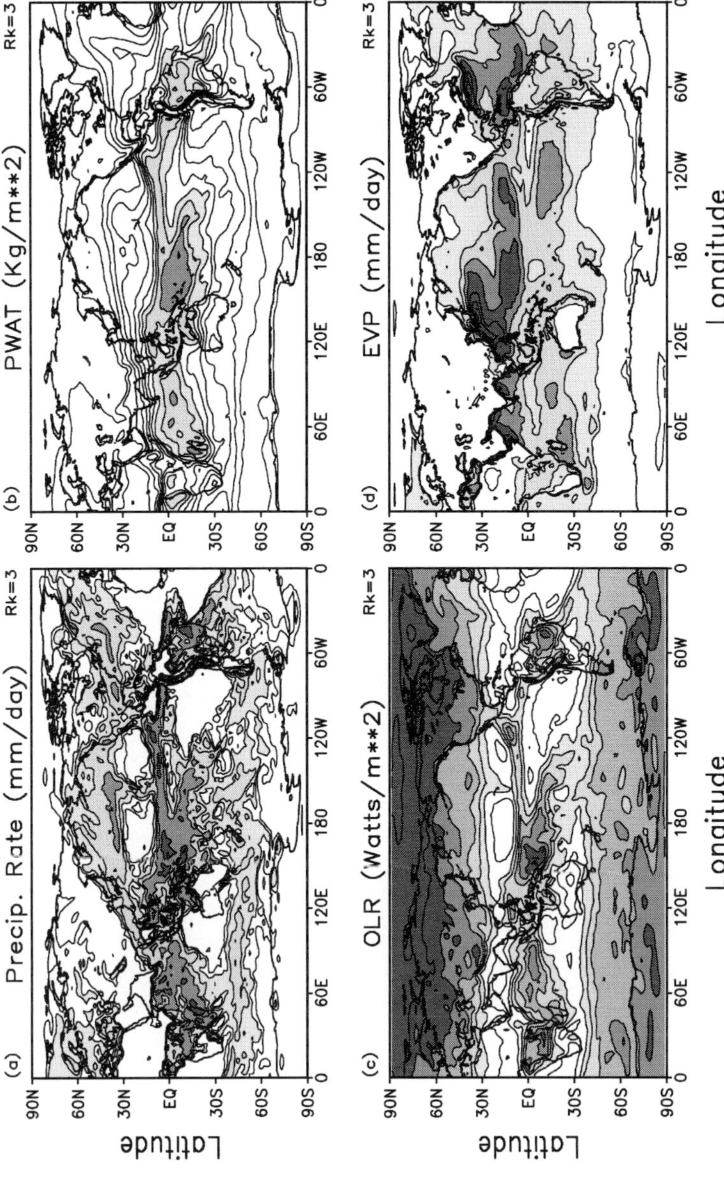

Figure 14 January mean (a) total precipitation rate (mm/day), (b) total precipitable water (kg/m^2), (c) outgoing long-wave radiation (W/m^2), and (d) evaporation rate (mm/day) for the run with $R_k = 3$ and $\alpha_i = 0.2$. Contour intervals are as in Figs. 4 to 7.

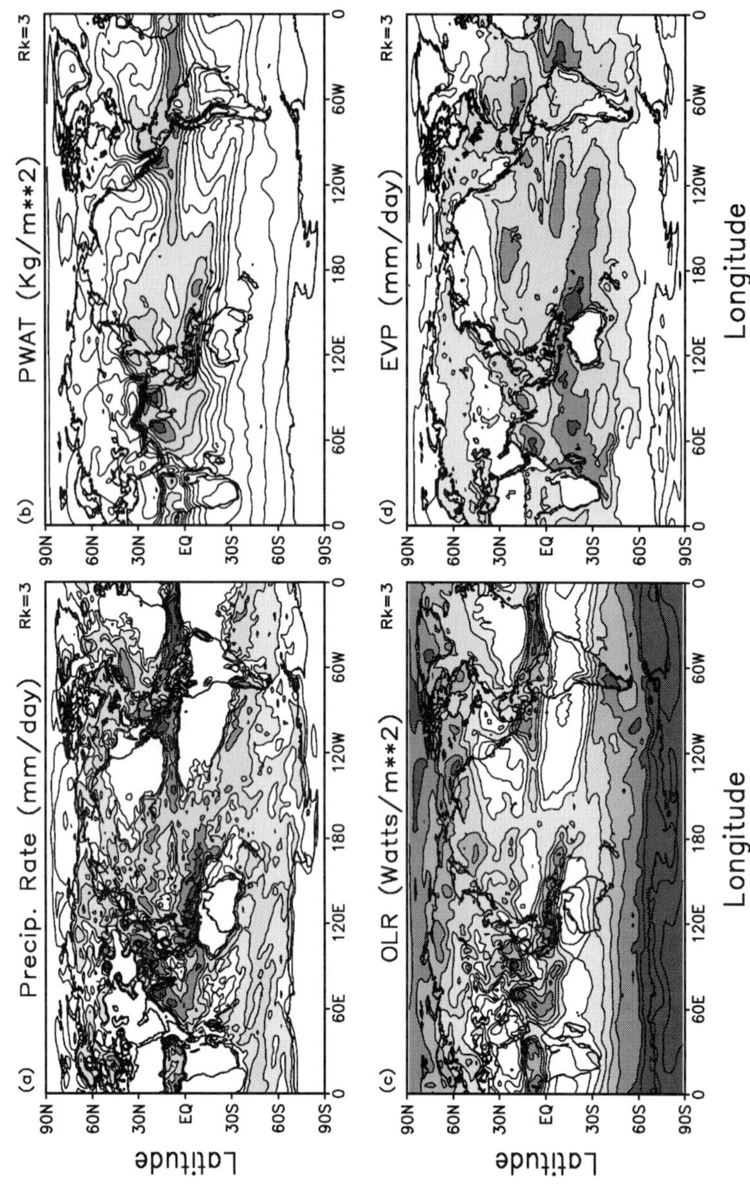

Figure 15 Same as in Fig. 14, but for July.

Computing the boundary layer top values as the average between those within the boundary layer and in the layer above, we find that the excessive drying can be substantially reduced. Reevaporation of rain, by varying the parameter R_k, did not significantly improve the excessive drying and if too strong, reevaporation could actually be harmful. This does not imply that the reevaporation is unimportant, however, in a fully nonlinear global model.

To study the impact of reevaporation in a global climate model, we used a recent version of NCEP climate model and varied the parameter R_k. We made series of 47-day integrations for January and 57-day integrations for July. We find that when reevaporation is not included, the model atmosphere indeed becomes drier, with the global mean precipitable water decreasing in time. With an increase in reevaporation, the global mean precipitable water increases to reasonably acceptable values for $R_k \sim 3$ to 5. Also the simulated climatology of precipitable water, precipitation rate, etc., is generally superior in this range of R_k. Further, the climate is not overly sensitive to the choice of the relaxation parameter in the range of 0.2 to 0.4, as long as the relaxation parameter and the number of cloud types invoked per time step are selected appropriately, i.e., the smaller the relaxation parameter, the more cloud types. In the limit, this should approach the original Arakawa–Schubert scheme, with some differences as discussed in Moorthi and Suarez (1992).

We performed an additional July experiment, where we let $h_{K-1/2} = h_K$, and $q_{K-1/2} = q_K$ (i.e., no direct modification of the boundary layer by cumulus subsidence) with $R_k = 4$ and $\alpha_i = 0.4$ for the summer case. However, the climate of this experiment seems to be very similar to those described earlier. Therefore, we did not repeat the experiment for the January case.

Currently, a version of the above climate model with $R_k = 5$ is being tested extensively in a climate parallel forecast/analysis system. The model is able to assimilate successfully the SSM/I derived precipitable water, and the forecasts from this assimilated initial state do not show any significant spin up or spin down. Further evaluation of its performance is under way. Future work involves development of an advanced version of RAS (Moorthi and Suarez, personal communication, 1998) in which several assumptions made in the original RAS are removed. Additionally, we will also test a downdraft scheme following the approach of Cheng and Arakawa (1997).

ACKNOWLEDGMENTS

I thank Dr. M. Suarez for helpful advice. I also thank Dr. S. Lord for encouragement, Mr. K. Campana for helpful review, and all those members of the global and climate modeling branches of EMC who contributed to the development and parallel testing of the new climate model, especially Drs. S. Saha, J. Schemm, and A. Kumar.

REFERENCES

Arakawa, A., and W. H. Schubert (1974). Interaction cumulus cloud ensemble with the large-scale environment. Part I. *J. Atmos. Sci.* **31**, 671–701.

Bony, S., Y. Sud, K. M. Lau, J. Susskind, and S. Saha (1997). Comparison and satellite assessment of NASA/DAO and NCEP-NCAR reanalyses over tropical ocean: Atmospheric hydrology and radiation. *J. Climate* **10**, 1441–1462.

Cheng, M. D., and A. Arakawa (1997). Inclusion of rain water budgets and convective downdrafts in the Arakawa–Schubert cumulus parameterization. *J. Atmos. Sci.* **54**, 1359–1378.

Chou, M.-D. (1992). A solar radiation model for use in climate studies. *J. Atmos. Sci.* **49**, 762–772.

Chou, M.-D., K.-T. Lee (1996). Parameterizations for the absorption of solar radiation by water vapor and ozone. *J. Atmos. Sci.* **53**, 1204–1208.

Cox, S. K., and K. T. Griffith (1978). Tropospheric radiative divergence during Phase III of GARP Atlantic Tropical Experiment (GATE), Atmos. Sci. Paper No. 291. Colorado State University, Fort Collins.

Hou, Y.-T., K. A. Campana, and S.-K. Yang (1996). Shortwave radiation calculations in the NCEP global model. International Radiation Symposium, Fairbanks, AK, Aug. 18–24, 1996.

Lord, S. J. (1978). Development and observational verification of a cumulus cloud parameterization, Ph.D dissertation. University of California, Los Angeles.

Lord, S. J., W. C. Chao, and A. Arakawa (1982). Interaction of a cumulus cloud ensemble with large-scale environment. Part IV: The discrete model. *J. Atmos. Sci.* **39**, 104–113.

Min, W., and S. D. Schubert (1997). Interannual variability and potential predictability in reanalysis products. Technical report series on global modeling and data assimilation, NASA Tech. Memo 104606, Vol. 13.

Moorthi, S., and M. J. Suarez (1992). Relaxed Arakawa–Schubert: A parameterization of moist convection for general circulation models. *Mon. Wea. Rev.* **120**, 978–1002.

Sud, Y., and A. Mood (1988). The roles of dry convection, cloud-radiation feedback processes, and the influence of recent improvements in the parameterization of convection in the GLA GCM. *Mon. Wea. Rev.* **116**, 2366–2387.

Thompson, R. M., Jr., S. W. Payne, E. E. Recker, and R. J. Reed (1979). Structure and properties of synoptic-scale wave disturbances in the intertropical convergence zone of the eastern Atlantic. *J. Atmos. Sci.* **36**, 53–72.

Trenberth, K. E., J. R. Christy, and J. G. Olson (1987). Global atmospheric mass, surface pressure and water vapor variations. *J. Geophys. Res.* **92**, 14, 815–14,826.

van den Dool, H. M., S. Saha, A. H. Oort, and W. Ebizusaki (1993). On the role of atmospheric water in the continuity equation. *In* "Proceedings of the 18th Annual Climate Diagnostic Workshop," Boulder, CO, Nov. 1–5, 1993, pp. 244–247.

van den Dool, H. M., J. Schemm, and S. Saha (1995). On the climatological annual cycle in global mean atmospheric pressure and hydrology as revealed by NCEP/NCAR reanalysis. *In* "Proceedings of the 20th Annual Climate Diagnostic Workshop," Seattle, WA, October 23–27, 1995, pp. 366–369.

Yanai, M., S. K. Esbensen, and J. H. Chu (1973). Determination of bulk properties of tropical cloud clusters from large-scale heat and moisture budgets. *J. Atmos. Sci.* **30**, 611–627.

Chapter 10

Solving Problems with GCMs: General Circulation Models and Their Role in the Climate Modeling Hierarchy

Michael Ghil and Andrew W. Robertson
Department of Atmospheric Sciences & Institute of Geophysics and Planetary Physics
University of California at Los Angeles, Los Angeles, California

I. INTRODUCTION: THE MODELING HIERARCHY

A view of climate dynamics as a modern scientific discipline first emerged about 40 years ago (Pfeffer, 1960). We understand it at the turn of the century as studying the variability of the atmosphere–ocean–cryo-sphere–biosphere–lithosphere system on time scales longer than the life span of individual weather systems and shorter than the age of our planet. When defined in these broad terms, the variability of the climate system is characterized by a power spectrum that has (1) a "warm-colored" broad-band component, with power increasing from high to low frequencies, (2) a line component associated with purely periodic forcing, annual and diurnal, and (3) a number of broad peaks that might arise from less purely periodic forcing (e.g., orbital change or solar variability), internal oscillations, or a combination of the two (Mitchell, 1976; Ghil and Childress, 1987, Chap. 11).

Understanding the climatic mechanism or mechanisms that give rise to a particular broad peak or set of peaks represents a fundamental problem of climate dynamics. The regularities are of interest in and of themselves for the order they create in our sparse and inaccurate observations; they also facilitate prediction for time intervals comparable to the periods associated with the given regularity (Ghil and Childress, 1987, Sec. 12.6; Ghil and Jiang, 1998).

The climate system is highly complex, its main subsystems have very different characteristic times, and the specific phenomena involved in each one of the climate problems defined in the preceding paragraphs are quite diverse. It is inconceivable, therefore, that a single model could successfully be used to incorporate all subsystems, capture all phenomena, and solve all problems. Hence the concept of a hierarchy of climate models, from the simple to the complex, has been developed about a quarter of a century ago (Schneider and Dickinson, 1974).

A. Atmospheric Modeling

At present, the best developed hierarchy is for atmospheric models; we summarize this hierarchy following Ghil (1995). The first rung is formed by zero-dimensional (0-D) models, where the number of dimensions, from zero to three, refers to the number of independent space variables used to describe the model domain, i.e., to physical-space dimensions. Such 0-D models essentially attempt to follow the evolution of global surface-air temperature $\bar{T}$ as a result of changes in global radiative balance (Crafoord and Källén, 1978; Ghil and Childress, 1987, Sec. 10.2):

$$c\frac{d\bar{T}}{dt} = R_i - R_o \tag{1a}$$

$$R_i = \mu Q_0\{1 - \alpha(\bar{T})\} \tag{1b}$$

$$R_o = \sigma m(\bar{T})\bar{T}^4. \tag{1c}$$

Here R_i and R_o are incoming solar radiation and outgoing terrestrial radiation, and c is the heat capacity of the global atmosphere, plus that of the global ocean or some fraction thereof, depending on the time scale of interest: One might only include in c the ocean mixed layer when interested in subannual time scales but the entire ocean when studying paleoclimate; $d\overline{T}/dt$ is the rate of change of $\overline{T}$ with time t, Q_0 is the solar radiation received at the top of the atmosphere, σ is the Stefan–Boltzmann constant, and μ is an insolation parameter, equal to unity for present-day conditions. To have a closed, self-consistent model, the planetary reflectivity or albedo α and grayness factor m have to be expressed as functions of $\overline{T}$; $m = 1$ for a perfectly black body and $0 < m < 1$ for a gray body like planet Earth.

There are two kinds of one-dimensional (1-D) atmospheric models, for which the single spatial variable is latitude or height, respectively. The former are so-called *energy-balance models* (EBMs; Budyko, 1969; Sellers, 1969), which consider the generalization of the model (1) for the evolution of surface-air temperature $T = T(x, t)$, say,

$$c(x)\frac{\partial T}{\partial t} = R_i - R_o + D. \tag{2}$$

Here the terms on the right-hand side can be functions of the meridional coordinate x (latitude, co-latitude, or sine of latitude), as well as of time t and temperature T. The horizontal heat-flux term D expresses heat exchange between latitude belts; it typically contains first and second partial derivatives of T with respect to x. Hence, the rate of change of local temperature T with respect to time also becomes a partial derivative, $\partial T/\partial t$.

The first striking results of theoretical climate dynamics were obtained in showing that slightly different forms of Eq. (2) could have two stable steady-state solutions, depending on the value of the insolation parameter μ [see Eq. (1b)] (Held and Suarez, 1974; Ghil, 1976; North *et al.*, 1981). In its simplest form, this multiplicity of stable steady states, or physically possible "climates" of our planet, can be explained in the 0-D model [Eq. (1)] by the fact that—for a fairly broad range of μ values around $\mu = 1.0$ —the curves for R_i and R_o as a function of $\overline{T}$ intersect in 3 points. One of these corresponds to the present climate (highest $\overline{T}$ value), and another one to an ice-covered planet (lowest $\overline{T}$ value); both of these are stable, while the third one (intermediate $\overline{T}$ value) is unstable. To obtain this result, it suffices to assume that $\alpha = \alpha(\overline{T})$ is a piecewise-linear function of $\overline{T}$, with high albedo at low temperature, due to the presence of snow and ice, and low albedo at high $\overline{T}$, due to their absence, while $m = m(\overline{T})$ is a smooth, increasing function of $\overline{T}$ that attempts to capture in its simplest from the "greenhouse effect" of trace gases and water vapor (Ghil and Childress, 1987, Chap. 10).

The 1-D atmospheric models in which the details of radiative equilibrium are investigated with respect to a height coordinate z (geometric height, pressure, etc.) are often called *radiative-convective models* (Manabe and Strickler, 1964; Ramanathan and Coakley, 1978; Charlock and Sellers, 1980), since convection plays a key role in vertical heat transfer. While these models preceded historically EBMs as rungs on the modeling hierarchy, it was only recently shown that they, too, can exhibit multiple equilibria (Li *et al.*, 1997; Rennó, 1997). The word (stable) *equilibrium* here and in the rest of this chapter refers simply to a (stable) steady state of the model, rather than to true thermodynamic equilibrium.

Two-dimensional (2-D) atmospheric models are also of two kinds, according to the third space coordinate that is not explicitly included. Models that resolve explicitly two horizontal coordinates, on the sphere or on a plane tangent to it, tend to emphasize the study of the dynamics of large-scale atmospheric motions (see Section II), whether they have a single layer (Charney and DeVore, 1979; Legras and Ghil, 1985) or two (Lorenz, 1963b; Reinhold and Pierrehumbert, 1982). Those that resolve explicitly a meridional coordinate and height are essentially combinations of EBMs and radiative-convective models and emphasize therewith the thermodynamic state of the system, rather than its dynamics (Saltzman and Vernekar, 1972; MacCracken and Ghan, 1988; Gallée *et al.*, 1991). Yet another class of "horizontal" 2-D models is the extension of EBMs to resolve zonal, as well as meridional surface features, in particular land−sea contrasts (Adem, 1970; North *et al.*, 1983; Chen and Ghil, 1996).

Additional types of 1-D and 2-D atmospheric models are discussed and references to these and to the types discussed above are given by Schneider and Dickinson (1974) and Ghil (1995), along with some of their main applications. Finally, to encompass and resolve the main atmospheric phenomena with respect to all three spatial coordinates, *general circulation models* (GCMs) occupy the pinnacle of the modeling hierarchy. Their genesis and the special role of successive generations of UCLA GCMs in the development and application of atmospheric GCMs to climate problems in general are covered elsewhere in this volume in great detail. Rather than dwell on this history in the present chapter, we proceed to outline, even more succinctly, the modeling hierarchies that have grown during the last quarter century in ocean and coupled ocean−atmosphere modeling.

Before doing so, it is worth noting that the results of climate simulations with GCMs, whether atmospheric or coupled, are often still interpreted in terms of the understanding gained from 0-D or 1-D EBMs. Wetherald and Manabe (1975), using a simplified sectorial GCM, confirmed the dependence of mean zonal temperature on the insolation parameter μ (the normalized "solar constant") obtained for 1-D EBMs by various authors. In fact, the sensitivity $(d\overline{T}/d\mu)\,|_{\mu=1.0}$ of global temperature $\overline{T}$ to changes in μ near the present-day climate equals about 1 K per 1% change in the insolation for both EBMs and GCMs. Many GCM studies of climate change response to increases in greenhouse trace gas concentrations actually use a linearized version of Eq. (1),

$$c\frac{dT}{dt} = -\lambda T + Q \tag{3a}$$

$$\lambda = \sum_{i=1}^{I} \lambda_i \tag{3b}$$

$$Q = \sum_{j=1}^{J} Q_j, \tag{3c}$$

for interpreting the roles of the different feedbacks λ_i, positive ($\lambda_i < 0$) or negative ($\lambda_i > 0$), and heat sources, $Q_j > 0$, or sinks, $Q_j < 0$ (e.g., Schlesinger and Mitchell, 1987; Cess *et al.*, 1989; Li and Le Treut, 1992).

B. Ocean and Coupled Modeling

The simplest 0-D ocean models are so-called *box models*, which are used to study the stability of the oceans' thermohaline circulation (Stommel, 1961; see Section IV) or biogeochemical cycles (Sarmiento and Toggweiler, 1984; Keir, 1988; Paillard *et al.*, 1993). There are 1-D models that consider the vertical structure of the upper ocean, whether the oceanic mixed layer only (Kraus and Turner, 1967; Karaca and Müller, 1989) or the entire thermocline structure.

For the oceans, 2-D models also fall into the two broad categories of "horizontal" and "vertical." Models that resolve two horizontal coordinates emphasize the study of the oceans' wind-driven circulation (Cessi and Ierley, 1995; Jiang *et al.*, 1995c; Berloff and Meacham, 1997), while those that consider a meridional section concentrate on the overturning thermohaline circulation (THC; Quon and Ghil, 1992, 1995; Thual and McWilliams, 1992). The wind-driven circulation is involved most strongly in sub- and interannual climate variability, while changes in the THC affect most strongly climate variability on the decade-to-century time scale and longer. Still, the circulation in a predominantly horizontal or vertical plane has to affect the same water masses, and 3-D ocean GCMs are thus indispensable in understanding oceanic variability (McWilliams, 1996).

Bryan and Cox's (1967) model has played a role for the development and applications of such models that resemble the one played by the UCLA GCM (e.g., Arakawa and Lamb, 1977) for atmospheric ones. A number of simplified versions of this ocean GCM (Bryan, 1986; Chen and Ghil, 1995) have been used in exploratory studies of multiple equilibria and self-sustained oscillations in the THC, in a spirit that resembles the use of a simplified atmospheric GCM by Wetherald and Manabe (1975) to complement EBM results on multiple climate equilibria.

A fairly well-developed hierarchy of coupled ocean–atmosphere models has been applied to the problem of seasonal-to-interannual variability in the tropical Pacific ocean (Neelin *et al.*, 1994; also see Section III below). Its most important rungs are, in ascending order, essentially 0-D simple models, like the delay-oscillator model of Suarez and Schopf (1988); essentially 1-D *intermediate coupled models* (Cane and Zebiak, 1985; Jin *et al.*, 1994); essentially 3-D *hybrid coupled models*, in which an ocean GCM is coupled to a much simpler, diagnostic atmospheric model (Neelin, 1990a; Barnett *et al.*, 1993); and fully coupled GCMs (Neelin *et al.*, 1992; Robertson *et al.*, 1995a,b). Recently, hybrid models of this type have also been applied to climate variability for the midlatitude (Weng and Neelin, 1998) and global (Chen and Ghil, 1996; Wang *et al.*, 1999) coupled system.

C. Dynamical Systems Theory

It has become fairly commonplace to state that the climate system contains numerous nonlinear processes and feedbacks, and that its behavior is rather irregular, but not totally random. Dynamical systems theory studies the common features of nonlinear systems of differential equations, ordinary (Smale, 1967) and partial (Constantin *et al.*, 1989). The work of Lorenz (1963a, 1964) has played a key role in establishing the relevance of this theory to studying climate variability, as well as in advancing the theory itself.

This theory can be used systematically to explore robust features of climate-system behavior on a given time scale—intraseasonal, seasonal-to-interannual, and interdecadal—as we move up and down the rungs of the modeling hierarchy for the given problem, and from the models to the relevant observations. The main features of dynamical systems theory that are important for the study of climate have been summarized by Ghil *et al.* (1991a); they involve essentially *bifurcation theory* (Guckenheimer and Holmes, 1983) and the *ergodic theory* of dynamical systems (Eckmann and Ruelle, 1985).

Bifurcation theory permits one to follow—through successive bifurcations, computed analytically or numerically—climatic behavior from the simplest kind of model solutions to the most complex, from single equilibria through multiple ones and on to periodic, chaotic, and fully turbulent solutions. Bifurcations can be computed analytically only for steady states (*fixed points* in the language of the theory) and periodic solutions (*limit cycles* in the same language), and for relatively simple models (Charney and DeVore, 1979; North *et al.*, 1981; Jin and Ghil, 1990). Transitions to more complicated behavior, quasi-periodic, chaotic, or fully turbulent, need to be investigated numerically. Furthermore, even transitions to multiple equilibria or to periodic solutions need to be computed numerically in more detailed, realistic models (Ghil, 1976; Legras and Ghil, 1985; Strong *et al.*, 1995).

The ergodic theory of dynamical systems provides statistical models for deterministically chaotic, as well as stochastically perturbed, climate evolution in space and time. This kind of nonlinear statistics permits one to evaluate systematically to which extent the behavior of fairly complex climate-model solutions matches that obtained with simpler, more easily understandable models, on the one hand, and that reflected by the existing observations, on the other (see, again, Ghil *et al.*, 1991a; Ghil, 1995, and further references therein). A number of scalar quantities are obtained by applying the theory to univariate time series, such as the leading Lyapunov exponents or the various dimensions associated with the dynamical system that presumably produced the time series in question (Guckenheimer and Holmes, 1983; Drazin and King, 1992). The leading Lyapunov exponent provides a nonlinear generalization of the linear stability of a steady state; its being positive indicates that the system it characterizes is chaotic. Of the various phase-space dimensions (Eckmann and Ruelle, 1985) used to quantify a dynamical system's number of independent degrees of freedom, the correlation dimension (Grassberger and Procaccia, 1983) became best known for its ease of computation.

In the study of climate variability across the modeling hierarchy, it is useful to apply more sophisticated numerical tools of ergodic theory than those that produce the scalar quantities above. These tools for the investigation of spatiotemporal regularities include various methods of cluster analysis for the classification of multiple weather regimes (Cheng and Wallace, 1993; Kimoto and Ghil, 1993a,b) and advanced methods for the analysis and prediction of uni- and multivariate time series (Vautard and Ghil, 1989; Plaut and Vautard, 1994; Dettinger *et al.*, 1995; Ghil and Jiang, 1998).

These ideas from dynamical systems theory are illustrated in the remainder of this chapter by applying them, in succession, to low-frequency atmospheric variability (Section II), seasonal-to-interannual climate variability (Section III), and decade-to-century variability in the THC (Section IV). In each section, the basic phenomena to be explained are presented, the main ingredients for the solution of the climate problem thus posed are derived by the use of simple models, and the rungs of the hierarchy climbed up to GCMs. In Section II these will be atmospheric GCMs, in Section III coupled GCMs, and in Section IV ocean GCMs and hybrid models. A possible road map for the further integration of GCMs into systematic climate problem solving is discussed in Section V.

II. INTRASEASONAL OSCILLATIONS: THEIR THEORY AND SIMULATION

Intraseasonal time scales range from the deterministic limit of atmospheric predictability, of about 10 days, up to a season, say, 100 days. They occupy a window of overlap between short climatic time scales and low-frequency variability intrinsic to the atmosphere. These time scales are of particular importance to extended-range weather prediction. There are two complementary ways of describing low-frequency atmospheric variability: (1) episodic, via multiple weather (Reinhold and Pierrehumbert, 1982) or flow (Legras and Ghil, 1985) regimes, and (2) oscillatory, via broad-peak, slowly modulated oscillations (Ghil *et al.*, 1991a, and references therein). We restrict ourselves here to the latter.

A. EXTRATROPICAL OSCILLATIONS: OBSERVATIONS AND THEORY

Variations in global atmospheric angular momentum (AAM) and in the length of day on intraseasonal time scales are highly correlated with each other; both quantities exhibit spectral peaks with periods near 40 and 50 days (Dickey *et al.*, 1991), among others. Essentially, the Earth–atmosphere system is closed with respect to angular momentum exchanges on this time scale, except for the well-known tidal effects of the Sun and Moon, which can be easily computed and eliminated. Once this is done, what remains is the following: When the midlatitude westerly winds pick up, or the tropical easterlies slow down, the solid Earth slows down in its rotation, and the length of day increases; hence the high positive correlation between the latter and AAM.

The latitude-frequency dependence of observed AAM variance is shown in Fig. 1, averaged over 12 years and all seasons. It is clear from the figure

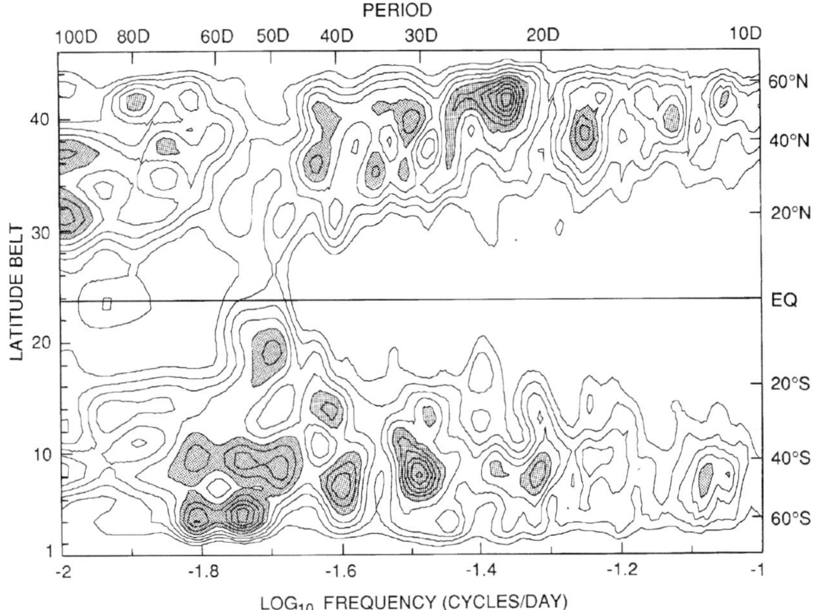

Figure 1 Latitude-frequency dependence of observed AAM variance, as shown by a contour plot of spectral power density for the 46 equal-area belts numbered from south to north. The power in each belt has been multiplied by the frequency; units are angular momentum squared $[(1/600)\text{ms}]^2$. The contour interval is 5.0, with contours starting at 20.0; values over 40.0 are shaded. (Reproduced from Dickey *et al.*, 1991, with the permission of the American Geophysical Union.)

that the 50-day peak is largely associated with AAM fluctuations in the tropics, which dominate the global AAM. The 40-day peak, however, appears to be associated primarily with variations in the strength of the midlatitude westerlies; such a peak appears both in the Northern Hemisphere (NH) and in the Southern Hemisphere. The amplitude of the 40-day oscillations in zonal winds is known, however, to be largest during boreal winter, when the winds are strongest in the NH (Weickmann *et al.*, 1985; Ghil and Mo, 1991; Strong *et al.*, 1993, 1995), and we shall thus concentrate here on the longer data sets and more detailed modeling studies for the NH.

The extent to which the tropical and NH oscillations are independent phenomena or influence each other is still the subject of active debate. The tropical oscillation was discovered by Madden and Julian (1971, 1972) in zonal winds and tropical convection over the equatorial Pacific, although its origins are still not well understood. Extratropical oscillations have been found in observed NH planetary-scale circulation anomalies with periods of 20–70 days (Branstator, 1987; Kushnir, 1987; Ghil and Mo, 1991; Plaut and Vautard, 1994). There is some evidence that the midlatitude circulation over the North Pacific is correlated to convective anomalies associated with the tropical oscillation (Weickmann *et al.*, 1985; Lau and Phillips, 1986; Higgins and Mo, 1997). On the other hand, Dickey *et al.* (1991) and Ghil and Mo (1991) found the extratropical mode to be often independent of—and sometimes to lead—the tropical one. Upper-level potential vorticity anomalies are known to propagate from the midlatitudes into the tropics, associated with northwest-to-southeast tilting troughs (Liebmann and Hartmann, 1984). They are accompanied by cold surges and can cause episodes of intense tropical convection that appear to be related to the intraseasonal oscillation in the tropics (Lau and Li, 1984; Hsu *et al.*, 1990).

Our focus here is on how a hierarchy of models can be used to formulate and test the hypothesis that the 40-day oscillation is an intrinsic mode of the NH extratropics, associated with the interaction of the jet stream with midlatitude mountain ranges. The rudiments of this hypothesis originate in the highly idealized barotropic model of Charney and DeVore (1979), which was used to study the interaction between a zonal flow and simple zonal wave-number–2 topography. Their model exhibits two stable equilibria for the same strength of the prescribed zonal forcing, which represents the strength of the pole-to-equator temperature contrast.

Figure 2a shows the model's *bifurcation diagram*, with the strength ψ_A of the zonal jet in the model's steady-state solutions plotted against the corresponding strength ψ_A^* of the forcing. The two stable equilibria—marked Z and R_-—are associated with "zonal" (higher AAM) and "blocked" (lower AAM) flow respectively, as illustrated in Fig. 2b. The near-zonal solution is close in amplitude and spatial pattern to the forcing jet and is influenced very little by the topography, while the blocked solution is strongly affected by it. In the blocked-flow solution, a ridge is located upstream of the "mountains," similar to the situation during a

---

Figure 2 Multiple equilibria of a three-mode quasi-geostrophic model with simplified forcing and topography. (a) Bifurcation diagram showing model response to changes in forcing; see text for the explanation of abscissa and ordinate. The S-shaped bifurcation curve is typical of two back-to-back *saddle-node bifurcations* that give rise to two stable solution branches (solid) separated by an unstable one (dashed). (b) Flow patterns of the zonal (upper panel) and blocked (lower panel) equilibria, corresponding to the two stable equilibria Z and R_-. (After Charney and DeVore, 1979; reproduced from Ghil and Childress, 1987, with the permission of Springer-Verlag.)

a)

ψ_A
ψ_A^*
Z
(s)
b = b''
b = b'
β_n / α_n
(b = 0)
R_+ (u)
R_-
(s)

b)

y
π
ocean
land
-2
-1
0
1
2
Z
$\pi/2$
0
x
y
π
land
ocean
R_-
$\pi/2$
0
-1/2
1/2
0
0
$\pi/2$
π
$3\pi/2$
2π
x

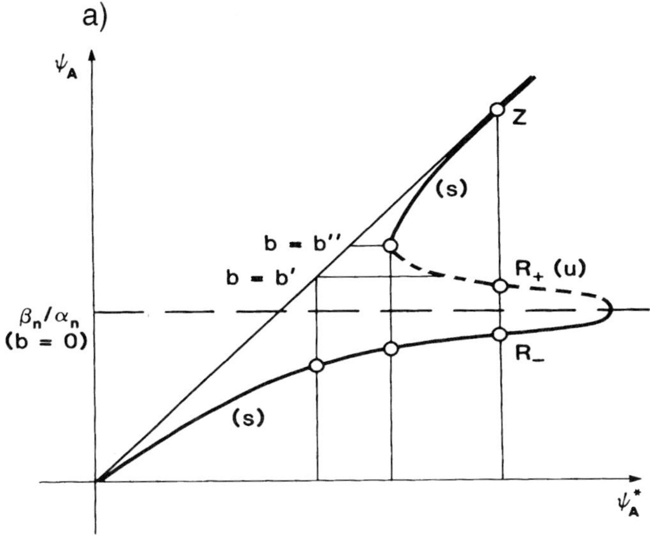

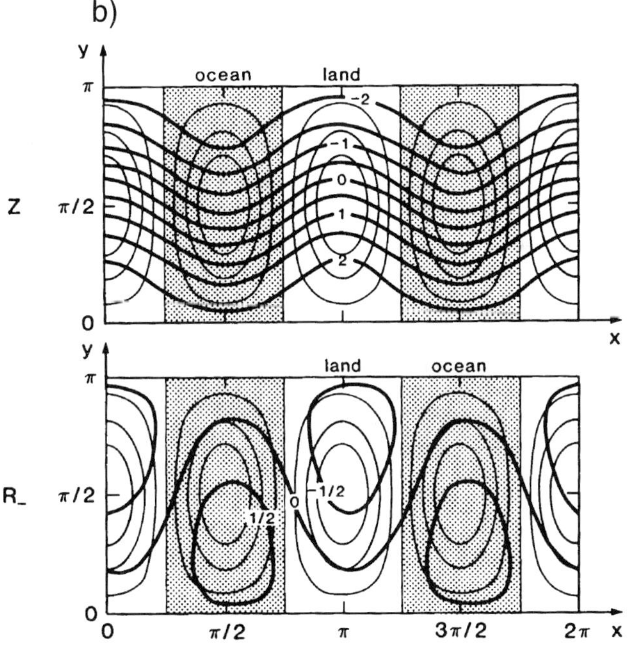

typical observed West Coast block. This configuration, with a negative zonal pressure gradient on the windward slope of the mountains, corresponds to a negative mountain torque on the atmosphere.

More complex models—both barotropic and baroclinic, with more spatial degrees of freedom than Charney and DeVore's (1979)—have been found to exhibit multiple flow patterns that are similar to those just described, for realistic values of the forcing. The crucial difference in these models is that the equilibria are no longer stable, and the system oscillates around the blocked solution or fluctuates between the zonal and blocked solutions in an irregular way (Legras and Ghil, 1985).

Jin and Ghil (1990) showed that, when a sufficiently realistic meridional structure of the solutions' zonal jet is allowed, the back-to-back saddle-node bifurcations of Fig. 2a are replaced by *Hopf bifurcation* and thus transition to finite-amplitude periodic solutions—also called limit cycles (see Section I.C)—can occur. Eigenanalyses of the unstable equilibria in a higher resolution barotropic model, as well as its time-dependent solutions, also indicate oscillatory instabilities with intraseasonal (35–50 days) and biweekly (10–15 days) time scales (Strong *et al.*, 1993). Floquet analysis of this model's limit cycles (Strong *et al.*, 1995) confirms that the 40-day oscillations that arise in it by oscillatory topographic instability are stronger in winter than in summer, like the NH observed oscillations (Knutson and Weickmann, 1987; Ghil and Mo, 1991).

B. GCM Simulations and Their Validation

Atmospheric GCMs provide a powerful tool for testing the theory of NH extratropical oscillations developed in simpler models. Marcus *et al.* (1994, 1996) made a 3-yr perpetual-January simulation with a version of the UCLA GCM that produces no self-sustained Madden–Julian oscillation in the tropics. A robust 40-day oscillation in AAM is found to arise in the model's NH extratropics when standard topography is present. Three shorter runs with no topography produced no intraseasonal oscillation, consistent with a topographic origin for the NH extratropical oscillation in the standard model. The spatial structure of the circulation anomalies associated with the model's extratropical oscillation is shown in Fig. 3, in terms of 500-mb geopotential height composites during the peak (panel a) and quadrature (panel b) phase of the AAM cycle.

The oscillation is dominated by a standing, wave-number–2 pattern, which undergoes tilted-trough vacillation. High values of AAM are associated with low 500-mb heights over the northeast Pacific and Atlantic Oceans (Fig. 3a), and vice versa. This resembles the configuration seen in

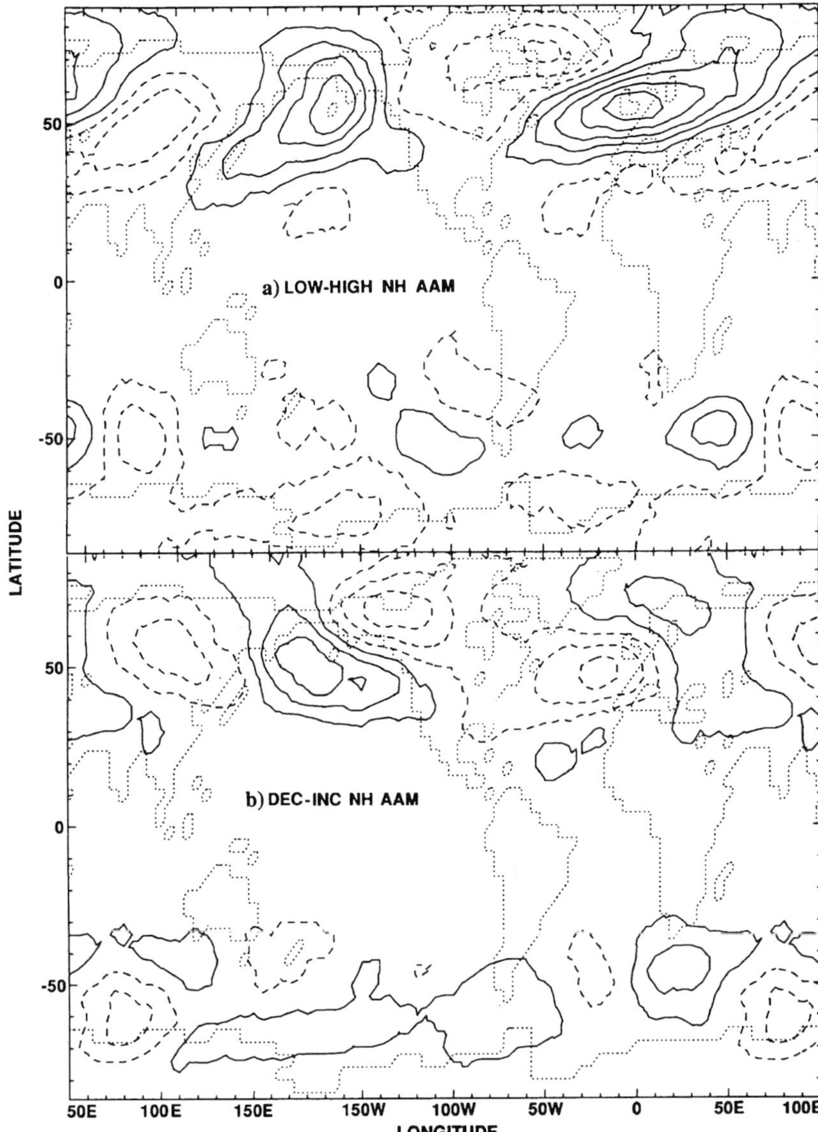

Figure 3 Composite 500-mb maps from the perpetual-January GCM experiment of Marcus *et al.* (1996). (a) For days on which the 36–60-day NH extratropical AAM anomalies exceeded 1.5 times their rms value; maps for days with a negative (positive) anomaly were added with a positive (negative) sign. (b) Constructed from maps taken 12 days earlier than those included in (a). Contour interval is 20 m, and negative contours are dashed. (Reproduced from Marcus *et al.*, 1996, with the permission of the American Meteorological Society.)

Charney and DeVore's (1979) simple model (see Fig. 2b). The GCM's NE–SW tilting phase in Fig. 3a and NW–SE tilting phase in Fig. 3b are strongly reminiscent of the extremes and intermediate phases of the 40-day oscillation that arises by Hopf bifurcation from the blocked equilibrium in the Legras and Ghil (1985) model (M. Kimoto, personal communication, 1986).

The successive phases of the 28–72-day band-passed fluctuations in 250-mb streamfunction anomalies analyzed by Weickmann *et al.* (1985; see Figs. 7 and 9a–d therein) also exhibit good agreement with the evolution of the 40-day oscillation in the work of Marcus and colleagues using the UCLA atmospheric GCM (see Ghil *et al.*, 1991b, for a video clip of the evolution of 500-mb heights, 250-mb streamfunction fields, and sea level pressures during the atmospheric GCM's 40-day oscillation). The height pattern in Fig. 3a is very similar, furthermore, to the extreme-phase patterns obtained from observed data by correlating the 10-day low-pass filtered wintertime 500-mb height fields with the sum of the mountain torques computed over the Rockies, Himalayas, and Greenland (F. Lott, A. W. Robertson and M. Ghil, in preparation, 2000).

In the GCM, the two centers of action have slightly different frequencies; this gives rise to a long-period modulation (of about 300 days) in the amplitude of the intraseasonal oscillation, similar to that observed by Penland *et al.* (1991) in globally averaged AAM time series. Global correlations with the leading empirical orthogonal functions (EOFs) of the NH extratropical 500-mb height field show NE–SW teleconnection patterns extending into the tropics, in particular into the Indian Ocean, similar to those found in observational studies (Weickmann *et al.*, 1985; Murakami, 1988). The model's zonally averaged latent heating in the tropics exhibits no intraseasonal periodicity, but a near 40-day oscillation is found in cumulus precipitation over the western Indian Ocean, suggesting an extratropical trigger of the 50-day oscillation in the tropics. Madden and Speth (1995; see Fig. 10 therein) found that (mostly extratropical) mountain torques did lead (mostly tropical) friction torques and eastward-moving convective systems during the 1987–1988 winter singled out already by Dickey *et al.* (1991; their Fig. 16).

Thus the careful analysis of perpetual-January runs with an atmospheric GCM confirms, on the one hand, the topographic origin of the NH 40-day oscillation, originally suggested by simple- and intermediate-model studies (Ghil and Childress, 1987, Sec. 6.4; Jin and Ghil, 1990). On the other hand, it provides greater realism and spatiotemporal detail, thus permitting a much better confrontation of the theory with the existing observations (Dickey *et al.*, 1991; Ghil and Mo, 1991).

III. EL NIÑO–SOUTHERN OSCILLATION, FROM THE DEVIL'S STAIRCASE TO PREDICTION

A. ENSO's Regularity and Irregularity

The El Niño–Southern Oscillation (ENSO) phenomenon dominates interannual climate variability over the tropical Pacific and influences the atmosphere globally. Figure 4 shows the power spectrum of the monthly sea-surface temperatures (SSTs) averaged over the eastern equatorial Pacific's Niño-3 area (150°W–90°W, 5°S–5°N), for the time interval 1960–1997. The spectrum is plotted in terms of the leading EOFs of a

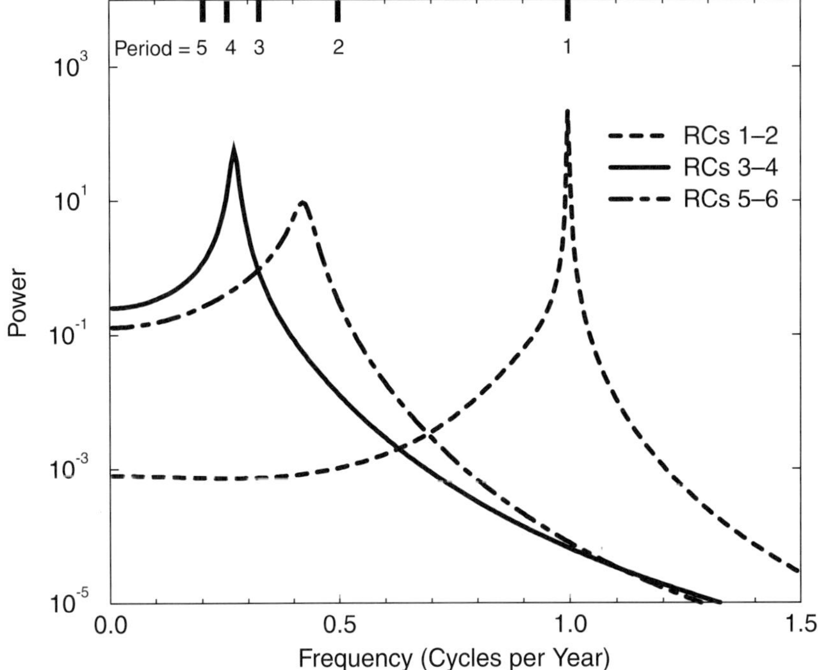

Figure 4 Power spectrum of the leading reconstructed components (RCs) of the Niño-3 SSTs for the time interval 1960–1997, using monthly data from the Climate Prediction Center of the National Centers for Environmental Prediction (NCEP). An SSA analysis with a window width of 72 months was used to derive the RCs, whose power spectra were then computed using the maximum entropy method, with 20 poles. The three curves show the annual cycle (short dashes), and the low-frequency (solid) and QB (long dashes) components of ENSO; notice logarithmic units on the ordinate.

singular-spectrum analysis (SSA; Vautard and Ghil, 1989; Vautard *et al.*, 1992). Pairs of EOFs that are in phase quadrature with each other correspond to nonlinear, anharmonic counterparts of sine and cosine pairs in standard Fourier analysis (Dettinger *et al.*, 1995; Jiang *et al.*, 1995a).

The observed SST time series contains a sharp annual cycle, together with two broader interannual peaks centered at periods of 44 and 28 months. This power spectrum provides a fine example of the distinction between the sharp lines produced by purely periodic forcing and the broader peaks resulting from internal climate variability or from the interaction of the latter with the former (see the beginning of Section I). The sharp annual peak reflects the seasonal cycle of heat influx into the tropical Pacific and the phase locking of warm events to boreal winter that gives El Niño its name. The two interannual peaks correspond to the low-frequency and quasi-biennial (QB) components of ENSO identified by a number of authors (Rasmusson *et al.*, 1990; Allen and Robertson, 1996). Jiang *et al.* (1995a) have demonstrated that major warm (El Niño) and cold (La Niña) events during the time interval 1950–1990 can be well reconstructed from ENSO's quasi-quadrennial and QB components (see Fig. 9 in Jiang *et al.*, 1995a). Together, these two components account for about 30% of the variance in the time series analyzed in Fig. 4.

In spite of the marked regularities apparent from Fig. 4 and its discussion, forecasting El Niño and La Niña events has met with mixed success, even for subannual lead times (Latif *et al.*, 1994): Year-to-year variations in forecast skill at 6–12-month lead times have been noticed by the authors of the more than a dozen models that routinely produce ENSO forecasts on a quarterly basis (see Section III.C below). Any satisfactory theory of ENSO should produce consistently reliable forecasts at such lead times. To do so, the theory must account for the observed low-frequency and QB peaks and the close relationship of these two with the annual cycle, on the one hand, as well as for the irregular occurrence of extreme events, on the other.

Much of our theoretical understanding of ENSO comes from relatively simple, essentially 0- and 1-D coupled models, consisting of a shallow-water or two-layer ocean model coupled to steady shallow-water-like atmospheric models with heavily parameterized physics (see Section I.B and references there); the more complete ones among these models are often called *intermediate coupled models* (Neelin *et al.*, 1994). In these models, ENSO-like variability has been shown to result from an oscillatory instability of the coupled ocean–atmosphere's annual-mean climatological state. Its nature has been investigated in terms of the dependence of the primary Hopf bifurcation on fundamental parameters, such as the coupling strength, oceanic adjustment time scale, and the strength of surface currents (Jin and Neelin, 1993).

The growth mechanism of ENSO is undisputed, arising from positive atmospheric feedbacks on equatorial SST anomalies via the surface wind stress, as first hypothesized by Bjerknes (1969). The cyclic nature of the unstable mode is more subtle and depends on the time scales of response within the ocean. Recently, there has been renewed interest in the thermal discharge–recharge hypothesis of Cane and Zebiak (1985) and Wyrtki (1986). Here the memory of the system resides in a disequilibrium between the wind stress and the meridional exchange of heat between equatorial and off-equatorial regions (Jin, 1997).

B. The Devil's Staircase across the Modeling Hierarchy

Regarding the relationship between El Niño and the annual cycle, Chang *et al.* (1994), Jin *et al.* (1994), and Tziperman *et al.* (1994) found that including an annual cycle in the basic state of their simple or intermediate models changed only slightly the spatial structure of the oscillatory instability. However, it caused the ENSO cycle to frequency lock to rational multiples of the annual frequency in a "Devil's Staircase." This "staircase" represents a scenario of transition to deterministic chaos, and involves two parameters: As one changes a parameter that increases the period of the intrinsic ENSO instability, frequency locking to successively longer rational multiples of the annual cycle occurs, according to the mechanism of subharmonic resonance (see Fig. 5a). As a second parameter, the coupling strength between the model's ocean and atmosphere, increases, the steps on the staircase broaden and begin to overlap (compare Fig. 5b with Fig. 5a), and the model's ENSO cycle becomes irregular due to jumps between the steps.

Moving one rung higher in the model hierarchy, hybrid coupled models, consisting of an ocean GCM coupled to a simple atmosphere, have been used for sensitivity studies in a more realistic setting. Syu *et al.* (1995) demonstrated that introducing an annual cycle caused the period of oscillation in their model to shift from 2.3 years to exactly 2 years, and thus to frequency lock with the annual cycle in the manner predicted by theory.

At the top of the model hierarchy, over a dozen coupled ocean–atmosphere GCMs have now been used to simulate the climate variability of the tropical Pacific (Neelin *et al.*, 1992; Mechoso *et al.*, 1995). These coupled GCMs have exhibited a wide range of seasonal and interannual behavior, with different amplitudes and phases of the seasonal cycle (Mechoso *et al.*, 1995), as well as different periods, amplitudes, and spatiotemporal characteristics of the interannual variability (Neelin *et al.*, 1992), and with zonally propagating or standing SST anomalies.

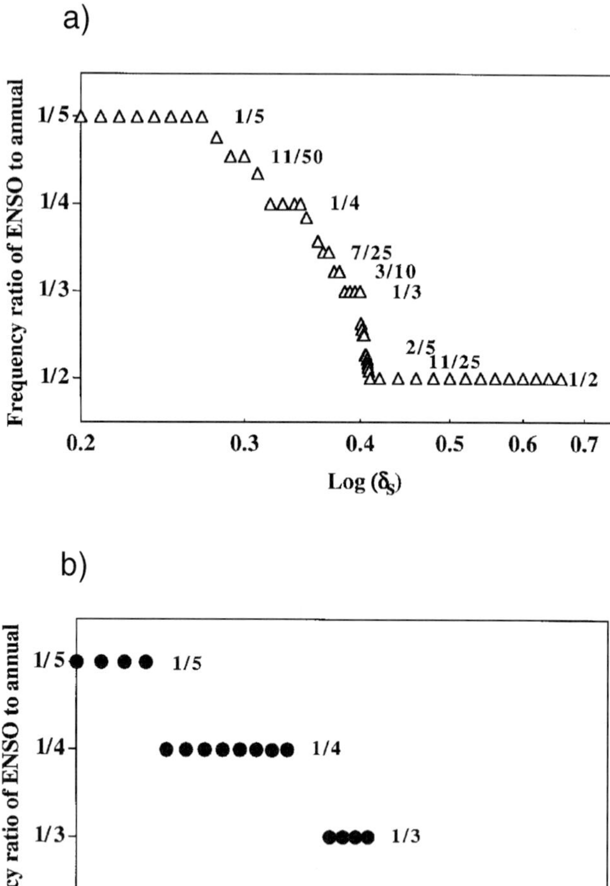

Figure 5   The Devil's Staircase in the intermediate coupled ocean–atmosphere model of Jin *et al.* (1994), plotted in terms of the frequency ratio of the model's inherent ENSO period to the annual cycle, as a function of the parameter δ_s that affects the former. The coupling parameter μ is the Hopf bifurcation parameter in the annual-average model of Jin and Neelin (1993). (a) The approximate Devil's Staircase for $\mu = \mu_0$, slightly above the primary Hopf bifurcation where the ENSO mode goes unstable; all points shown correspond to rational frequency ratios (some labeled). (b) Frequency-locked solutions for somewhat larger values of $\mu \approx 1.1\mu_0$, showing the rapid widening of the integer-period steps. (Reproduced from Jin *et al.*, 1996, with the permission of Elsevier Science B.V.)

For example, Robertson *et al.* (1995a,b) investigated the seasonal cycle and interannual variability in 45 years of simulations with the UCLA atmospheric GCM, coupled to a tropical-Pacific basin version of the Geophysical Fluid Dynamics Laboratory (GFDL) ocean GCM; the latter is a descendant of Bryan and Cox's (1967) model (see Section I.B above) and the entire coupled GCM is described in detail by Mechoso in Chapter 18 of this volume. The simulations were found to be characterized by ENSO-like QB and quasi-quadrennial modes, identified by using multichannel SSA (Plaut and Vautard, 1994) along the equator, but with weaker variability than the observed. Two simulations that differed only in details of the atmospheric GCM's surface-layer parameterizations are of particular interest, because the first (decade I) was found to be dominated by a QB oscillation (Fig. 6a), while a quasi-quadrennial period dominated the second (decade II, Fig. 6b).

A 100-year-long simulation with NASA Goddard's Aries-Poseidon coupled GCM exhibits both quasi-quadrennial and QB spectral peaks (P. Schopf and M. Suarez, personal communication, 1995) of a strength very similar to that observed in the Cooperative Ocean-Atmosphere Data Set (COADS) (Fig. 7a). These results are further confirmed by a 60-year run of NCEP's coupled GCM (Ji *et al.*, 1998; see Fig. 7b).

The complete Devil's Staircase scenario, in fact, calls for successively smaller peaks associated with the harmonics of the 4-year step, at $4/1 = 4$, $4/2 = 2$, and $4/3$ years. Both the QB and $4/3$-year $= 16$-month peak are present in observed SST data (Jiang *et al.*, 1995a; Allen and Robertson, 1996). There is a smaller and broader 18–20-month peak present in the UCLA coupled GCM in Fig. 6b, which can be interpreted as a merged version of these two peaks.

Work with intermediate coupled models suggests that the results shown in Figs. 6 and 7, together with those obtained using other coupled GCMs, can be explained in terms of each GCM simulation's location in the space spanned by a set of fundamental parameters, analogous to those that appear in the simple model of Jin and Neelin (1993). Clearly, however, the value of a simplified model parameter, such as "coupling strength," has a nontrivial expression in terms of the various physical parameterizations in the coupled GCM.

Figure 8 illustrates the spatial structure of the quasi-quadrennial mode along the equator in the UCLA coupled model, derived from multichannel SSA of a 26-year extension of the second simulation (Fig. 6b). The model's QB mode (not shown) has a structure that resembles that of Fig. 8. This common structure consists of predominantly standing oscillations in SST and zonal wind stress that peak in the central or east Pacific, accompanied by an oscillation in equatorial thermocline depth that is characterized by a

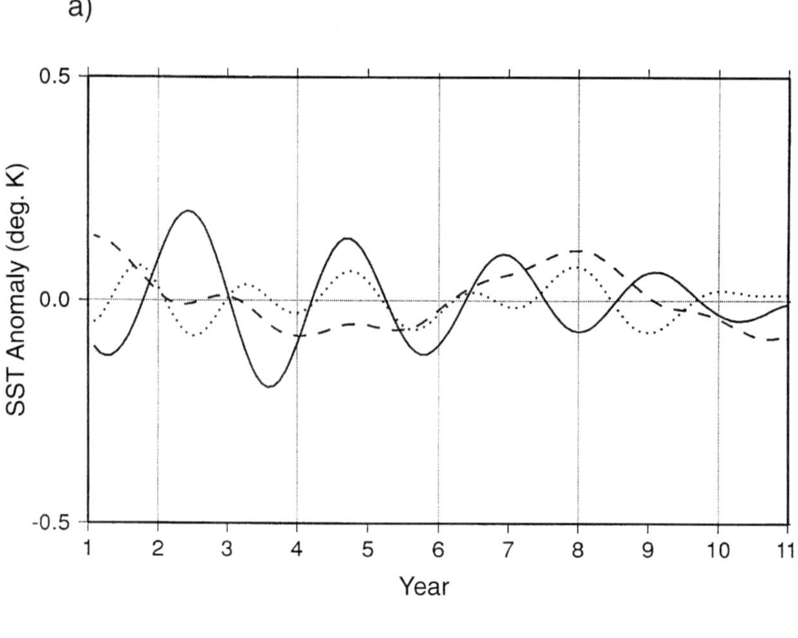

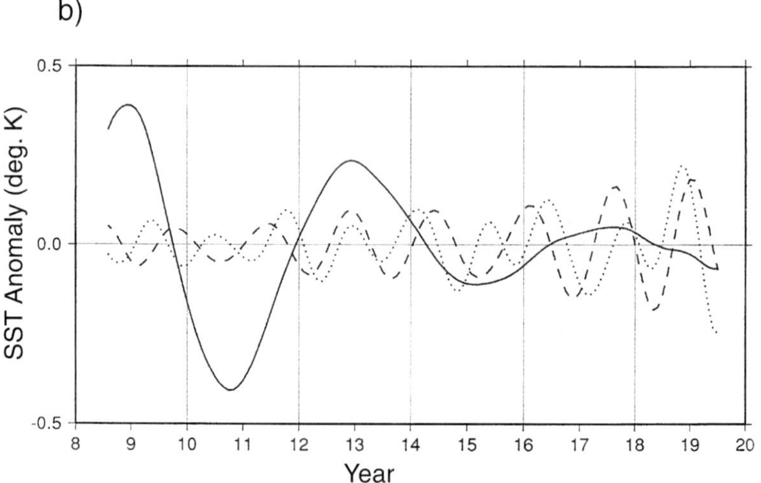

Figure 6 Equatorial SST anomaly indices from the UCLA coupled GCM, constructed from RCs of SST for (a) decade I (130°–110°W, 4°S–4°N) and (b) decade II (160°–140°W, 4°S–4°N). *Key*: Solid line, RCs 1–2; dashed line, RCs 3–4; dotted line, RCs 5–6. (Reproduced from Robertson *et al.*, 1995b, with the permission of the American Meteorological Society.)

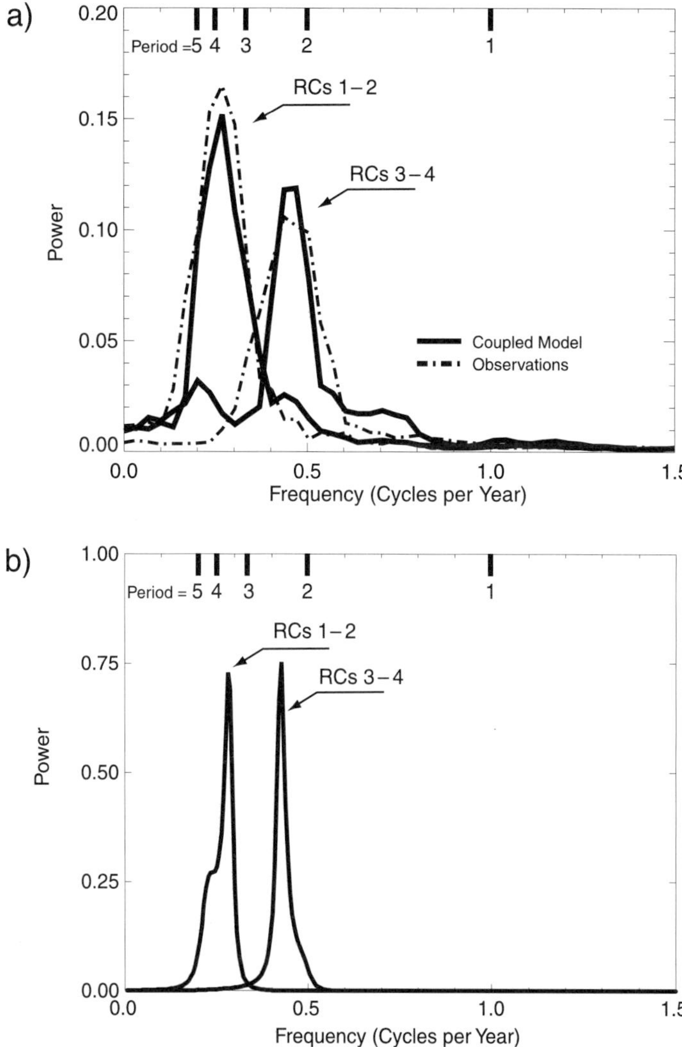

Figure 7 (a) Power spectrum of Niño-3 SST anomalies from 38 years of COADS data (1950–1987) (dash-dot line) and from a 38-year segment out of a 100-year integration with the NASA Goddard Space Flight Center's coupled GCM (solid line). Shown are the first (quasi-quadrennial) and second (QB) pair of RCs from an SSA analysis with 100 monthly lags. The good match in amplitude between the two time series, for both spectral peaks, is significant, because neither data set was normalized (P. Schopf and M. Suarez, personal communication, 1995). (b) Same as the solid line in (a) but based on a 60-year integration of NCEP's coupled GCM (Ji *et al.*, 1998); SST anomalies courtesy of M. Ji (personal communication, 1998) and spectral analysis as in Fig. 4. The monthly data in both panels were deseasonalized by subtracting the mean annual cycle, with a 3-month running average applied in (b); notice linear ordinate in both panels.

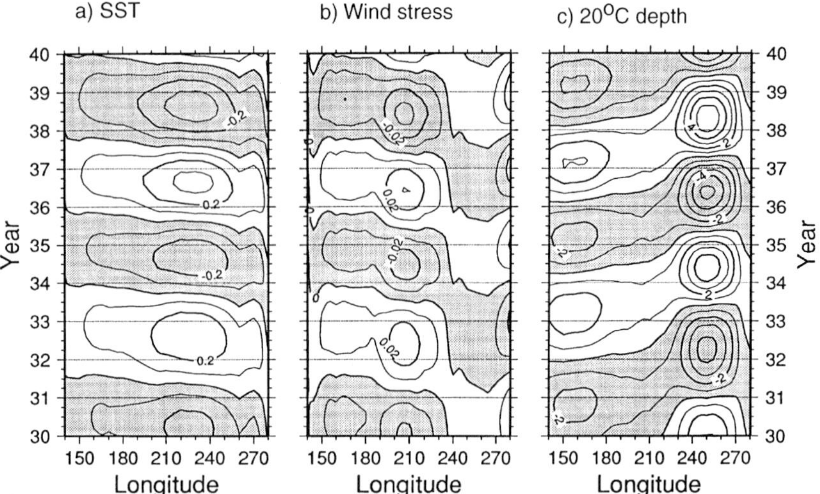

Figure 8   Hovmöller diagram along the equator, showing the structure of the quasi-quadrennial mode in the UCLA coupled GCM (compare Fig. 6b). Plotted are RCs 1–2 for the 26-year extension to decade II: (a) SST (0.05 K), (b) τ_x (0.005 dyn cm^{-2}), and (c) depth of 20°C isotherm, with greater depth shown as negative (1 m); negative anomalies are stippled. (Reproduced from Robertson *et al.*, 1995b, with the permission of the American Meteorological Society.)

phase shift of about 90° across the basin—much less than the 180° that Sverdrup balance would imply—with west leading east. These features are all characteristic of observed ENSO events (Neelin *et al.*, 1994).

Simple and intermediate models suggest the following hypotheses for producing multiple, broad spectral peaks in the interannual band: (1) the quasi-quadrennial mode might arise by *period doubling* from the QB mode (Münnich *et al.*, 1991); (2) the observed peaks may result from the linear superposition of several marginally damped modes, excited by *white-noise forcing* (Penland and Sardeshmukh, 1995); (3) *stochastic resonance* can cause intermittent jumps between the forced seasonal cycle and the lower-period internal ENSO cycle, in the presence of a certain level of noise (Stone *et al.*, 1998); (4) the interaction of the seasonal cycle and the fundamental ENSO mode can nonlinearly entrain this mode to a rational multiple of the annual frequency and produce additional peaks, according to a *Devil's Staircase* (Chang *et al.*, 1994; Jin *et al.*, 1994; Tziperman *et al.*, 1994); and (5) the quasi-quadrennial and QB peaks could represent separate oscillations, generated by *different mechanisms*, each with an independent frequency. The latter hypothesis, while not supported by explicit model results, is quite plausible when considering the differences in detailed ENSO mechanisms between models and the parameter dependence of the basic oscillation period for each model (see Neelin *et al.*, 1998, and references therein).

The results of GCM simulations, along with existing observational data, provide a means of distinguishing between these alternatives. The fact that the variance associated with the low-frequency mode is actually larger than that associated with the QB mode in the observations (cf. Jiang *et al.*, 1995a; also Fig. 4) and that the two modes have comparable variance in the GCM simulations where they are both present (Figs. 7a and 7b) essentially rules out period doubling (hypothesis 1) as an explanation of ENSO irregularity, since that scenario would imply a much smaller quasi-quadrennial mode. The similarity of the spatial structures of the quasi-quadrennial and QB modes, both in coupled GCMs (Robertson *et al.*, 1995b) and observations (Moron *et al.*, 1998), weighs against hypothesis 5 of different physics producing two independent frequencies. In the case of the UCLA coupled GCM, variations in subsurface thermocline depth were similar in all three interannual frequencies isolated, also supporting this point.

The closeness of the quasi-quadrennial and QB peaks, in observations and the GCMs cited, to integer multiples of the annual period, on the other hand, and the observed 15–16-month peak in observed data are all consistent with subharmonic frequency locking (hypothesis 4). The basic similarity between the spatial structures of the ENSO modes and the annual cycle in the UCLA coupled GCM is further evidence of an intimate relationship between ENSO variability and the annual cycle.

The low-frequency component of observed equatorial Pacific SSTs changes in period around 1960, from being near a 5-year period before, to a near-4-year period after 1960 (Moron *et al.*, 1998). This fairly abrupt change in frequency can be interpreted in terms of frequency locking to different integer multiples of the annual period, and thus to different steps on the Devil's Staircase in Fig. 5b.

The stochastic resonance hypothesis is, in a sense, intermediate between the Devil's Staircase and the stochastically forced linear-model hypothesis 2. The latter is rendered less plausible by the fact that it does not produce phase locking of individual warm and cold events to the boreal winter; it is the preference of warm events to peak in or near late December that gives El Niño its name (see, for instance, the histograms of warm events versus calendar month in Figs. 3 and 5 of Chang *et al.*, 1996). On the other hand, Jin *et al.* (1996) found that when "weather noise" is included in their model of the Devil's Staircase, the resulting irregularity still carries the signature of the subharmonic spectral peaks. It is difficult to tell in which way, if any, the stochastic-resonance model of Stone *et al.* (1998) differs from the stochastically perturbed version of the Devil's

Staircase considered by Jin *et al.* (1996) and whether a subtle difference, if it exists, can be validated by the existing observations or GCM simulations, one way or the other.

C. Regularity and Prediction

The standard application of dynamical systems theory to forecast error growth and predictability involves low-order dynamical systems that are entirely chaotic, i.e., have a purely continuous power spectrum, such as the Lorenz (1963a) system. For such systems, the leading Lyapunov exponents (e.g., Eckmann and Ruelle, 1985) are positive, and the largest one, λ_1, say, gives a rough estimate of the predictability limit T_p; T_p can be defined as the time it takes for a typical observation error e_0 to grow until it reaches an asymptotic level determined by the total energy available to the system. The rough estimate $\hat{T}_p$ is based on the error growing at the rate $\exp(\lambda_1 t)$ until it saturates at a level given by the system's climatological variance, σ_∞, say, according to

$$e_0 \exp\!\left(\lambda_1 \hat{T}_p \right) = \sigma_\infty. \tag{4}$$

The error-growth curve in forced-dissipative systems (e.g., Ghil and Childress, 1987, Sec. 5.4) actually deviates from being exponential and has an inflection point before saturation (e.g., Kalnay and Dalcher, 1987), so Eq. (4) will tend to underestimate the system's true T_p. This is compensated for by the fact that the full system's additional degrees of freedom, which are usually neglected in deriving the simplified system for which λ_1 and hence the estimate $\hat{T}_p$ in Eq. (4) is obtained, contribute additional error growth, often approximated as stochastic forcing in the simplified system. Thus Eq. (4) for the simplified, deterministic system with a few degrees of freedom—say a simple or intermediate model, in the terminology of Sections I.A and I.B—often gives a reasonable approximation for the full system, with its infinite degrees of freedom.

There is, however, an additional source of predictability in systems that have periodic or nearly periodic components, expressed as pure lines or broad peaks in their power spectrum. The fact that the days are warmer than the nights or summers than winters is useful, predictable information, independently of the change in average 24-hour or 1-year temperature over a week or a decade, as the case may be. Both examples are derived from purely periodic, externally forced phenomena. Similar considerations apply, however, to internal periodicities that might only be approximate, rather than exact and thus result in broad spectral peaks, rather than sharp lines (Ghil and Childress, 1987, Sec. 12.6). Examples of such regularities are the intraseasonal oscillations considered in Section II, the QB and low-frequency components of ENSO in this section, and the interdecadal oscillations of Section IV.

In the case of ENSO, a considerable battery of models has been used for experimental seasonal-to-interannual prediction of various indices, like the Southern Oscillation index (SOI; e.g., Keppenne and Ghil, 1992), or fields like the Niño-3 SST anomalies considered here. More than fifteen 6–12-month forecasts have been published every quarter in the *Experimental Long-Lead Forecast Bulletin* (*ELLFB*) of NOAA's Climate Prediction Center since 1992; since Fall 1997, the *ELLFB* has been published by the nonprofit Center for Ocean-Land-Atmosphere Studies and is available electronically at http://grads.iges.org/ellfb. The *ELLFB* forecasts employ both dynamically and statistically based models. The statistical models span the range from classical time-domain methods for time series prediction through neural network methods, while the dynamical ones go from intermediate coupled models all the way to fully coupled GCMs.

Table I summarizes the forecast skill of six models that have been carrying out real-time ENSO predictions for a few years' time; Zebiak and Cane's (1987) is an intermediate coupled model, Barnett *et al.*'s (1993) a hybrid coupled model, Ji *et al.*'s (1994) a fully coupled GCM, while Barnston and Ropelewski's (1992), Van den Dool's (1994), and Jiang *et al.*'s (1995b) are all statistical. The skills at 6-month lead are quite comparable, with the correlation between the field being forecast and that actually observed equal to about 0.6–0.75 for the decade 1984–1993; this level of correlation skill easily outperforms persistence and is generally considered to be useful (Barnston *et al.*, 1994).

The highest skill actually occurs in the table's last column, using a statistical model (Keppenne and Ghil, 1992; Plaut and Vautard, 1994) that is explicitly based on extracting and predicting the (nearly) periodic components of a time series. Ghil and Jiang (1998) argue that the explanation for the comparable forecast skill of dynamical and statistical models is that most of this skill—in the case of the coupled ocean–atmosphere system in the tropical Pacific—is due to the model's capturing, more or less correctly, the system's two oscillatory modes: QB and low-frequency. Still, it is quite likely that the Ji *et al.*'s (1994) coupled GCM forecast of the rapid temperature rise in the eastern tropical Pacific that occurred during late summer and early fall 1997, a few months ahead of time, is due to its modeling correctly certain features of the nonlinear dynamics that are event specific, rather than nearly periodic (see also Chapter 11).

Table I

Characteristics and 6-Month–Lead Skill of Six ENSO Forecast Models, Three of Them Dynamical and Three Statistical

Authors	Zebiak & Cane (1987)[a]		Barnett *et al.* (1993)[a]		Ji *et al.* (1994)[a]		Barnston and Ropelewski (1992)[a]		Van den Dool (1994)[a]		Jiang *et al.* (1995b)	
Model	Physical: simple coupled		Hybrid: see text		Physical: coupled GCMs		Statistical: CCA		Empirical: constructed analog		Statistical: SSA and M-SSA	
Predicted SST region (5°N–5°S)	Niño-3 90°–150°W		Central Pacific 140°–180°W		Niño-3.4 120°–170°W		Niño-3.4 120°–70°W		Niño-3.4 120°–170°W		Niño-3 90°–150°W	
Period of record	1970–93		1966–93		1984–93		1956–93		1956–93		1984–93	
Skill[b] (1982–1993)	Corr	0.62	Corr	0.65[c]	Corr	0.69[c]	Corr	0.66	Cor	0.66	Corr	0.74[d]
	RMSE	0.95	RMSE	0.97	RMSE	0.83[c]	RMSE	0.89	RMSE	0.89	RMSE	0.50[d]
	SD	1.08	SD	1.10	SD	1.00	SD	1.11	SD	1.11	SD	1.00

Reproduced from Ghil and Jiang (1998), with the permission of the American Geophysical Union.

[a] After Table 1 of Barnston *et al.* (1994).

[b] Two measures of skill are provided: (1) correlation (Corr) between prediction and validation anomaly fields (actual monthly values minus climatology) and (2) root-mean-square error (RMSE) of prediction versus validation, normalized by the variability of the validating field. SD indicates the standard deviation of the observed SSTs over the time interval for which forecast results from each model were available.

[c] See Barnston *et al.* (1994) for details.

[d] The results in this column represent retroactive real-time forecasts, like those for the Barnston and Ropelewski (1992) and Van den Dool (1994) models; the only column in the table based exclusively on actual real-time forecasts is that for the Zebiak and Cane (1987) model. The "mixed" forecast skill (i.e., based in part on retroactive real-time and in part on actual real-time forecasts) of the Jiang *et al.* (1995b) model for the time interval January 1984–December 1996 is Corr = 0.66 and RMSE = 0.53, with actual forecasts published in *ELLFB* since March 1995. The 1997–1998 warm event was predicted by this model since December 1996.

IV. INTERDECADAL OSCILLATIONS IN THE OCEANS' THERMOHALINE CIRCULATION

A. THEORY AND SIMPLE MODELS

Historically, the thermohaline circulation (THC) was first among the climate system's major processes to be studied using a very simple mathematical model and shown to possess multiple equilibria (Stommel, 1961). A sketch of the Atlantic Ocean's THC and its interactions with the atmosphere and cryosphere on long time scales is shown in Fig. 9. These interactions can lead to climate oscillations with multimillenial periods—such as the Heinrich events (Ghil, 1994, and references therein) —and are summarized in the figure's caption, following Ghil *et al.* (1987). An equally schematic view of the global THC is provided by the widely known "conveyor belt" diagram (e.g., Broecker, 1991), which does not commonly include these interactions with the atmosphere and ice.

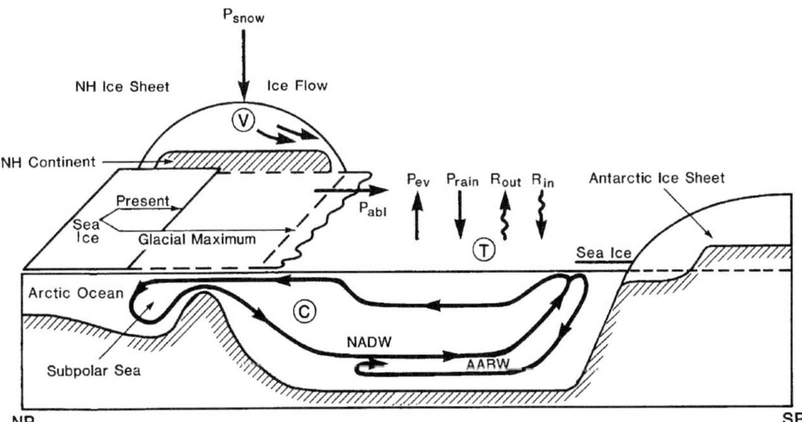

Figure 9　Diagram of an Atlantic meridional cross section from North Pole (NP) to South Pole (SP), showing mechanisms likely to affect the THC on various time scales. Changes in the radiation balance $R_{in} - R_{out}$ are due, at least in part, to changes in extent of Northern Hemisphere (NH) snow and ice cover, V, and how they affect the global temperature, T. The extent of Southern Hemisphere ice is assumed constant, to a first approximation. The change in hydrologic cycle expressed in the terms $P_{rain} - P_{evap}$ for the ocean and $P_{snow} - P_{abl}$ for the snow and ice is due to changes in ocean temperatures. Deep-water formation in the North Atlantic subpolar sea (North Atlantic deep water, NADW) is affected by changes in ice volume and extent, and regulates the intensity C of the THC; changes in Antarctic bottom water (AABW) formation are neglected in this approximation. This in turn affects the system's temperature, and is also affected by it. (Reproduced from Ghil and McWilliams, 1994, after Ghil *et al.*, 1987.)

Basically, the THC is due to denser water sinking, lighter water rising, and water-mass continuity closing the circuit through near-horizontal flow between the areas of rising and sinking. This is roughly the oceanic equivalent of the atmosphere's Hadley circulation, with two notable differences:

1. The ocean water's density ρ is a function of temperature T and salinity S, while that of the air depends on temperature and humidity.
2. Water sinks in and near fairly concentrated regions of intense convection, currently located mostly in high latitudes, and rises diffusely over the rest of the ocean; air does rise most intensely in cumulus towers, but overall the areas of net rising and sinking air in a Hadley cell are quite comparable in extent, when viewed on the synoptic and planetary scales.

The effects of temperature and salinity on the ocean water's density, $\rho = \rho(T, S)$, oppose each other: The density ρ *decreases* with increasing T and *increases* with increasing S. It is these two effects that give the *thermohaline* circulation its name, from the Greek words for T and S. In high latitudes, ρ increases as the water loses heat to the air above and, if sea ice is formed, as the water underneath is enriched in brine. In low latitudes, ρ increases due to evaporation but decreases due to heat flux into the ocean.

For the present climate, the temperature effect is commonly assumed to be stronger than the salinity effect, and ocean water is observed to sink in certain areas of the high-latitude North Atlantic and Southern Ocean—with very few and limited areas of deep-water formation elsewhere—and to rise everywhere else; thus *thermohaline, T* more important than and hence before S. During some past geological times, deep water apparently formed near the equator; such an overturning circulation of opposite sign to that prevailing today has been dubbed *halothermal, S* before T (e.g., Kennett and Stott, 1991). The quantification of the relative effects of T and S on the oceanic water masses' buoyancy in high and low latitudes is far from complete, especially for paleocirculations; the association of the latter with salinity effects that exceed the thermal ones is thus rather tentative.

Stommel (1961) considered a two-box model, with two pipes connecting the two boxes, and showed that the system of two nonlinear, coupled ordinary differential equations (ODEs) which govern the temperature and salinity differences between the two well-mixed boxes has two stable steady-state solutions, distinguished by the direction of flow in the upper and lower pipes. Stommel's paper was primarily concerned with distinct local convection regimes, and hence vertical stratifications, in the North Atlantic and Mediterranean (or Red Sea), say. Today, we mainly think of one box as representing the low latitudes and the other one the high latitudes in the global THC.

The next step in the hierarchical modeling of the THC is that of 2-D meridional-plane models (see Section I.B), in which the temperature and salinity fields are governed by coupled nonlinear partial differential equations (PDEs) with two independent space variables, latitude and depth, say. Given boundary conditions for such a model that are symmetric about the equator, as are the PDEs themselves, one expects a symmetric solution, in which water either sinks near the poles and rises everywhere else (thermohaline) or sinks near the equator and rises everywhere else (halothermal); these two symmetric solutions would correspond to the two equilibria of Stommel's (1961) box model.

In fact, Fig. 10 shows that symmetry breaking can occur, leading gradually from a symmetric two-cell circulation (Fig. 10a) to an antisymmetric one-cell circulation (approximately achieved in Fig. 10c). In between, all degrees of dominance of one cell over the other are possible, with one such intermediate state shown in Fig. 10b. A situation lying somewhere between Figs. 10b and 10c seems to resemble most closely the meridional overturning diagram of the Atlantic Ocean in Fig. 9.

This symmetry breaking can be described by a *pitchfork bifurcation* (e.g., Guckenheimer and Holmes, 1983):

$$\dot{X} = \mu - X^3. \tag{5}$$

Here X stands for the amount of asymmetry in the solution, so that $X \equiv 0$ is the symmetric branch, $\dot{X} \equiv dX/dt$ and μ is a parameter that measures the stress on the system. For $\mu < 0$ the symmetric branch is stable, while for $\mu > 0$ the two branches $X = \pm \sqrt{\mu}$ inherit its stability.

Thus, Figs. 10b and 10c both lie on a solution branch of the 2-D THC problem for which the left cell dominates: Say that North Atlantic deep water extends to the Southern Ocean's polar front, as it does in Fig. 9. According to Eq. (5), another branch exists, whose flow patterns are mirror images in the rectangular box's vertical symmetry axis (the "equatorial plane") of those in Figs. 10b and 10c. The existence of this second branch was verified numerically by Quon and Ghil (1992; their Fig. 16). Thual and McWilliams (1992) considered more complex bifurcation diagrams for a similar 2-D model and showed the equivalence of such a diagram for their 2-D model and a box-and-pipe model of sufficient complexity.

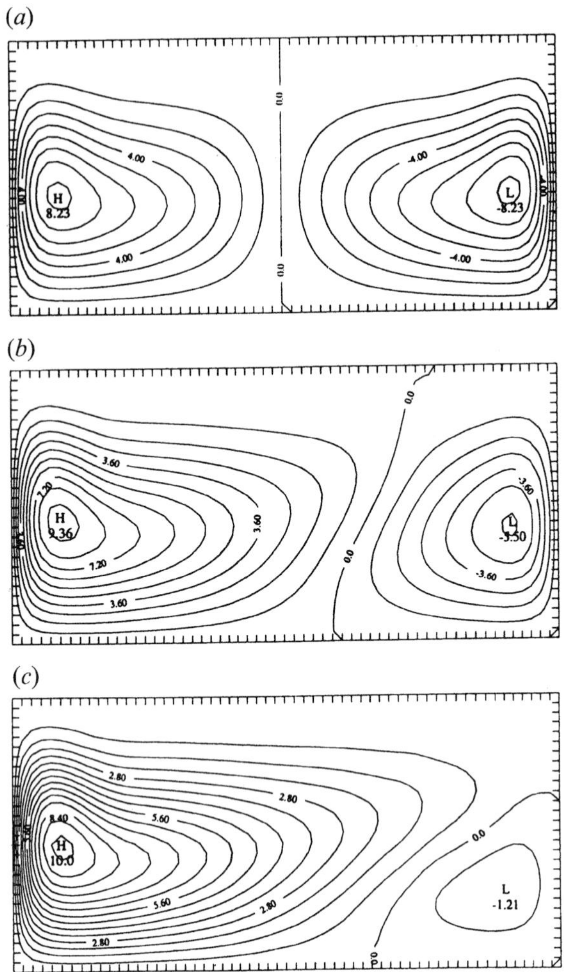

Figure 10 Streamfunction fields for a 2-D, meridional-plane THC model with so-called mixed boundary conditions: the temperature profile and salinity flux are imposed at one horizontal boundary of the rectangular box, while the other three boundaries are impermeable to heat and salt. (a) Symmetric solution for low salt-flux forcing; (b, c) increasingly asymmetric solutions as the forcing is increased. (Reproduced from Quon and Ghil, 1992, with the permission of Cambridge University Press.)

B. Bifurcation Diagrams for GCMs

Bryan (1986) was the first to document transition from a two-cell to a one-cell circulation in a simplified GCM with idealized, symmetric forcing, in agreement with the three-box scenario of Rooth (1982). Internal variability of the THC was studied simultaneously in the late 1980s and early 1990s on various rungs of the modeling hierarchy, from Boolean delay equation models (so-called "formal conceptual models": Ghil *et al.*, 1987; Darby and Mysak, 1993) through box models (Welander, 1986) and 2-D models (Quon and Ghil, 1995) to ocean GCMs. A summary of the different kinds of oscillatory variability found in the latter appears in Table II. Additional GCM references for these three types of oscillations are given by McWilliams (1996). The interaction of the (multi)millenial oscillations with variability in the surface features and processes shown in Fig. 9 is discussed by Ghil (1994).

One example of the interaction between atmospheric processes and the THC is given by Chen and Ghil (1996), who use a different kind of hybrid coupled model than that reviewed in Section III.B, to wit a (horizontally) 2-D EBM (see Section I.A) coupled to a rectangular-box version of the North Atlantic rendered by a low-resolution ocean GCM. This hybrid model's regime diagram is shown in Fig. 11a. A steady state is stable for high values of the coupling parameter λ_{ao} or of the EBM's diffusion parameter d. Interdecadal oscillations with a period of 40–50 years are self-sustained and stable for low values of these parameters.

Table II

Thermohaline Circulation Oscillations

Time scale	Phenomena	Mechanism
Interdecadal	3-D, wind-driven + thermohaline circulation	Gyre advection (Weaver *et al.*, 1991, 1993) Localized surface-density anomalies due to surface coupling (Chen and Ghil, 1995, 1996).
Centennial	Loop-type, Atlantic–Pacific circulation	Conveyor-belt advection of high-latitude density anomalies (Mikolajewicz and Maier-Reimer, 1990).
Millennial	Relaxation oscillation, with "flushes" and superimposed decadal fluctuations	Bottom-water warming, due to high-latitude freshening and its braking effect (Marotzke, 1989; Chen and Ghil, 1995)

Adapted from Ghil (1994), with the permission of Elsevier Science B.V.

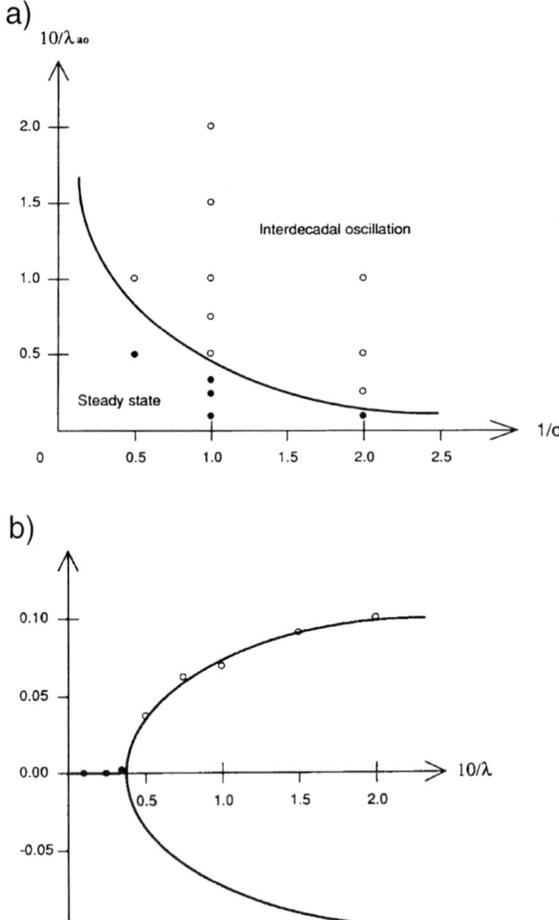

Figure 11 Dependence of THC solutions on two parameters in a hybrid coupled model (HCM); the two parameters are the atmosphere–ocean coupling coefficient λ_{ao} and the atmospheric thermal diffusion coefficient d. (a) Schematic regime diagram. The full circles stand for the model's stable steady states, the open circles for stable limit cycles, and the solid curve is the estimated neutral stability curve between the former and the latter. (b) Hopf bifurcation curve at fixed $d = 1.0$ and varying λ_{ao}; this curve was obtained by fitting a parabola to the model's numerical-simulation results, shown as full and open circles. (Reproduced from Chen and Ghil, 1996, with the permission of the American Meteorological Society.)

The self-sustained THC oscillations in question are characterized by a pair of vortices of opposite sign that grow and decay in quadrature with each other in the ocean's upper layers. Their centers follow each other anticlockwise through the northwestern quadrant of the model's rectangular domain. Both the period and the spatiotemporal characteristics of the oscillation are thus rather similar to those seen in a fully coupled GCM with realistic geometry (Delworth *et al.*, 1993).

The transition from a stable equilibrium to a stable limit cycle, via Hopf bifurcation, in Chen and Ghil's hybrid coupled model is shown in Fig. 11b. The physical characteristics of the oscillatory instability that leads to the Hopf bifurcations have been described in further detail by Colin de Verdière and Huck (1999), using both a four-box ocean–atmosphere and a number of more detailed models.

V. PERSPECTIVES

Until about two decades ago, the tools of analytical and numerical bifurcation theory could be applied only to 0-D THC models (Stommel, 1961) or 0- and 1-D climate models (Held and Suarez, 1974; Ghil, 1976; North *et al.*, 1981). We have illustrated in this review, by considering a few climate problems on different time scales, that the general theory can be combined with powerful numerical tools to study successive bifurcations across the hierarchy of climate models, all the way from 0-D global or box models (see above) to 2- and 3-D models: atmospheric (Legras and Ghil, 1985; Marcus *et al.*, 1996), oceanic (Thual and McWilliams, 1992; Quon and Ghil, 1992, 1995), and coupled (Jin *et al.*, 1994, 1996; Robertson *et al.*, 1995b; Chen and Ghil, 1996).

Each bifurcation is associated with a specific linear instability of a relatively simple climate state—oscillatory in the case of Hopf bifurcations and purely exponential in the case of saddle-node or pitchfork bifurcations —whose nonlinear saturation leads to more complicated climate behavior. Following the bifurcation tree, from one rung of the modeling hierarchy to the next, permits us, therefore, to study with increasing detail and realism the basic physical mechanisms that lead to climate variability.

Typically, the first one or two bifurcations will be captured fairly well by a lower order or otherwise very simple model of the climate problem of interest. As the model's number of degrees of freedom or otherwise its complexity increases, more complicated and realistic regimes of behavior will appear. These regimes can only be reached by additional bifurcations. The task of following bifurcations numerically off solution branches with greater and greater complexity becomes more and more challenging.

Various continuation methods (Keller, 1978; Kubicek and Marek, 1983) have been applied to follow steady-state solution branches of more and more highly resolved atmospheric (Legras and Ghil, 1985), oceanic (Speich *et al.*, 1995), and coupled (Dijkstra and Neelin, 1995) models. Projected increases in computer power should make it possible to apply such methods to currently available GCMs in the near future.

GCMs—atmospheric, oceanic, and coupled—provide climate problem solutions that have the greatest spatiotemporal detail and, one hopes, the greatest degree of realism. It is these solutions, therefore, that provide the best opportunity for evaluating our theories of climate variability—developed by climbing the lower rungs of the modeling hierarchy—against the observational evidence, such as it exists.

Such an evaluation, given the irregular character of observed climate variability, needs to be informed by the ergodic theory of dynamical systems, which can describe this irregular behavior in a consistent way. The statistical tools of the latter theory, such as singular-spectrum analysis and other advanced spectral methods, have to be applied in parallel to the GCMs' simulations and to the relevant data sets. Studying the observed and simulated climate variability with the same sophisticated tools can help pinpoint the aspects of this variability that we have understood, and can therefore predict with confidence, and those that we have not. Fortunately, there are many more of the latter, and much work remains to be done.

It is the authors' hope that the tools and points of view presented in this chapter will help to both diminish and increase the number of unsolved climate-variability problems.

ACKNOWLEDGMENTS

M. G. would like to acknowledge the warm welcome he and his ideas received at UCLA from A. A., and many informative discussions since. A. W. R. would like to thank A. A. for sharing his deep physical insights on many occasions. Both authors enjoyed the AA Fest Symposium very much and were encouraged by the quality of the other presentations to prepare their own for publication. Our coauthors and colleagues active in the three areas of climate dynamics reviewed (as shown by the list of references) are to be thanked for all we learned from them. We are especially grateful to Paul Schopf and Max Suarez for Fig. 7a, to Ming Ji for Fig. 7b, and to Alain Colin de Verdière for a preprint of his paper with Thierry Huck. Mike MacCraken, Steve Marcus, Jim McWilliams, and an anonymous reviewer read the original manuscript carefully and made constructive comments that helped improve the final version. Our work in these areas is supported by an NSF Special Creativity Award and NASA grant NAG5-317 (M.G.) and by DOE grant DE-FG03-98ER62515 (A. W. R.). Françoise J. E. Fleuriau helped with the word processing and references. This is publication 5070 of UCLA's Institute of Geophysics and Planetary Physics.

REFERENCES

Adem, J. (1970). Incorporation of advection of heat by mean winds and by ocean currents in a thermodynamic model for long-range weather prediction. *Mon. Wea. Rev.* **98,** 776–786.

Allen, M. R., and A. W. Robertson (1996). Distinguishing modulated oscillations from coloured noise in multivariate datasets. *Clim. Dyn.*, **12,** 775–784.

Arakawa, A., and V. R. Lamb (1977). Computational design of the basic dynamical processes of the UCLA general circulation model. *Methods Comput. Phys.* **17,** 173–265.

Barnett, T. P., M. Latif, N. Graham, M. Flügel, S. Pazan, and W. White (1993). ENSO and ENSO-related predictability. Part I: Prediction of equatorial Pacific sea surface temperature with a hybrid coupled ocean-atmosphere model. *J. Climate* **6,** 1545–1566.

Barnston, A. G., and C. F. Ropelewski (1992). Prediction of ENSO episodes using canonical correlation analysis. *J. Climate* **5,** 1316–1345.

Barnston, A. G., H. M. van den Dool, S. E. Zebiak, T. P. Barnett, M. Ji, D. R. Rodenhuis, M. A. Cane, A. Leetmaa, N. E. Graham, C. R. Ropelewski, V. E. Kousky, E. A. O'Lenic, and R. E. Livezey (1994). Long-lead seasonal forecasts—Where do we stand? *Bull. Am. Meteor. Soc.* **75,** 2097–2114.

Berloff, P. S., and S. P. Meacham (1997). The dynamics of an equivalent-barotropic model of the wind-driven circulation. *J. Mar. Res.* **55,** 407–451.

Bjerknes, J. (1969). Atmospheric teleconnections from the equatorial Pacific. *Mon. Wea. Rev.* **97,** 163–172.

Branstator, G. W. (1987). A striking example of the atmosphere's leading traveling pattern. *J. Atmos. Sci.* **44,** 2310–2323.

Broecker, W. S. (1991). The great ocean conveyor. *Oceanography* **4,** 79–89.

Bryan, F. O. (1986). High-latitude salinity effects and interhemispheric thermohaline circulations. *Nature* **323,** 301–304.

Bryan, K., and M. Cox (1967). A numerical investigation of the oceanic general circulation. *Tellus* **19,** 54–80.

Budyko, M. I. (1969). The effect of solar radiation variations on the climate of the Earth. *Tellus* **21,** 611–619.

Cane, M., and S. E. Zebiak (1985). A theory for El Niño and the Southern Oscillation. *Science* **228,** 1084–1087.

Cess, R. D., G. L. Potter, J. P. Blanchet, G. J. Boer, S. J. Ghan, J. T. Kiehl, H. Le Treut, Z.-X. Li, X.-Z. Liang, J. F. B. Mitchell, J.-J. Morcrette, D. A. Randall, M. R. Riches, E. Roeckner, U. Schlese, A. Slingo, K. E. Taylor, W. M. Washington, R. T. Wetherald, and I. Yagai (1989). Interpretation of cloud-climate feedbacks as produced by 14 atmospheric general circulation models, *Science* **245,** 513–551.

Cessi, P., and G. R. Ierley (1995). Symmetry-breaking multiple equilibria in quasi-geotropic, wind-driven flows. *J. Phys. Oceanogr.* **25,** 1196–1205.

Chang, P., B. Wang, T. Li, and L. Ji (1994). Interactions between the seasonal cycle and the Southern Oscillation—frequency entrainment and chaos in an intermediate coupled ocean–atmosphere model. *Geophys. Res. Lett.* **21,** 2817–2820.

Chang, P., L. Ji, H. Li, and M. Flügel (1996). Chaotic dynamics versus stochastic processes in El Niño–Southern Oscillation in coupled ocean–atmosphere models. *Physica D* **98,** 301–320.

Charlock, T. P., and W. D. Sellers (1980). Aerosol effects on climate: Calculations with time-dependent and steady-state radiative-convective model. *J. Atmos. Sci.* **38,** 1327–1341.

Charney, J. G., and J. G. DeVore (1979). Multiple flow equilibria in the atmosphere and blocking. *J. Atmos. Sci.* **36,** 1205–1216.

Chen, F., and M. Ghil (1995). Interdecadal variability of the thermohaline circulation and high-latitude surface fluxes. *J. Phys. Oceanogr.* **25,** 2547–2568.

Chen, F., and M. Ghil (1996). Interdecadal variability in a hybrid coupled ocean–atmosphere model. *J. Phys. Oceanogr.* **26,** 1561–1578.

Cheng, X., and J. M. Wallace (1993). Cluster analysis of the Northern Hemisphere wintertime 500-hPa height field: Spatial patterns. *J. Atmos.* **50,** 2674–2696.

Colin de Verdière, A., and T. Huck (1999). Baroclinic instability: An oceanic wavemaker for interdecadal variability. *J. Phys. Oceanogr.* **29,** 893–910.

Constantin, P., C. Foias, B. Nicolaenko, and R. Témam (1989). "Integral Manifolds and Inertial Manifolds for Dissipative Partial Differential Equations." Springer-Verlag, New York.

Crafoord, C., and E. Källen (1978). A note on the condition for existence of more than one steady-state solution in Budyko-Sellers type models. *J. Atmos. Sci.* **35,** 1123–1125.

Darby, M. S., and L. A. Mysak (1993). A Boolean delay equation model of an interdecadal Arctic climate cycle. *Clim. Dyn.* **8,** 241–246.

Delworth, T. S., Manabe, and R. J. Stouffer (1993). Interdecadal variations of the thermohaline circulation in a coupled ocean–atmosphere model. *J. Climate* **6,** 1993–2011.

Dettinger, M. D., M. Ghil, C. M. Strong, W. Weibel, and P. Yiou (1995). Software expedites singular-spectrum analysis of noisy time series. *EOS Trans. AGU* **76,** 12, 14, 21.

Dickey, J. O., M. Ghil, and S. L. Marcus (1991). Extratropical aspects of the 40–50 day oscillation in length-of-day and atmospheric angular momentum. *J. Geophys. Res.* **96,** 22643–22658.

Dijkstra, H. A., and J. D. Neelin (1995). On the attractors of an intermediate coupled equatorial ocean–atmosphere model. *Dyn. Atmos. Oceans* **22,** 19–48.

Drazin, P. G., and G. P. King (eds.) (1992). "Interpretation of Time Series from Nonlinear Systems" (Proc. IUTAM Symp. & NATO Adv. Res. Workshop, University of Warwick, England; *Physica D*, **58**). North-Holland, Amsterdam.

Eckmann, J.-P., and D. Ruelle (1985). Ergodic theory of chaos and strange attractors. *Rev. Mod. Phys.* **57,** 617–656 (addendum, *Rev. Mod. Phys.* **57,** 1115, 1985).

Gallée, H., J. P. van Ypersele, Th. Fichefet, C. Tricot, and A. Berger (1991). Simulation of the last glacial cycle by a coupled, sectorially averaged climate—ice-sheet model. I. The climate model. *J. Geophys. Res.* **96,** 13, 139–161.

Ghil, M. (1976). Climate stability for a Sellers-type model. *J. Atmos. Sci.* **33,** 3–20.

Ghil, M. (1994). Cryothermodynamics: The chaotic dynamics of paleoclimate. *Physica D* **77,** 130–159.

Ghil, M. (1995). Atmospheric modeling. *In* "Natural Climate Variability on Decade-to-Century Time Scales" (D. G. Martinson, K. Bryan, M. Ghil, M. D. Hall, T. R. Karl, E. S. Sarachik, S. Sorooshian, and L. D. Talley, eds.). pp. 164–168. National Academy Press, Washington, DC.

Ghil, M., and S. Childress (1987). "Topics in Geophysical Fluid Dynamics: Atmospheric Dynamics, Dynamo Theory and Climate Dynamics." Springer-Verlag, New York.

Ghil, M., and N. Jiang (1998). Recent forecast skill for the El-Niño/Southern Oscillation. *Geophysics. Res. Lett.* **25**(2), 171–174.

Ghil, M., and J. McWilliams (1994). Workshop tackles oceanic thermohaline circulation, *EOS Trans. AGU* **75,** 493, 498.

Ghil, M., and K. C. Mo (1991). Intraseasonal oscillations in the global atmosphere. Part I: Northern Hemisphere and tropics. *J. Atmos. Sci.* **48,** 752–779.

Ghil, M., and R. Vautard (1991). Interdecadal oscillations and the warming trend in global temperatures time series. *Nature* **350,** 324–327.

Ghil, M., A. Mullhaupt, and P. Pestiaux (1987). Deep water formation and Quarternary glaciations, *Clim. Dyn.* **2**, 1–10.

Ghil, M., M. Kimoto, and J. D. Neelin (1991a). Nonlinear dynamics and predictability in the atmospheric sciences. *Rev. Geophys.* **29**, *Suppl.*, 46–55.

Ghil, M., S. L., Marcus, J. O. Dickey, and C. L. Keppenne (1991b). "AAM the Movie." NTSC videocassette AVC-91-063, Caltech/NASA Jet Propulsion Laboratory, Pasadena, CA 91109. (Available also from M. Ghil upon request.)

Grassberger, P., and I. Procaccia (1983). Characterization of strange attractors. *Phys. Rev. Lett.* **50**, 346–349.

Guckenheimer, J., and P. Holmes (1983). "Nonlinear Oscillations, Dynamical Systems and Bifurcations of Vector Fields." Springer-Verlag, New York.

Held, I. M., and M. J. Suarez (1974). Simple albedo feedback models of the icecaps. *Tellus* **36**, 613–628.

Higgins, R. W., and K. C. Mo (1997). Persistent North Pacific anomalies and the tropical intraseasonal oscillation. *J. Climate* **10**, 223–244.

Hsu, H. H., B. J. Hoskins, and F.-F. Jin (1990). The 1985–86 intra-seasonal oscillation and the role of topographic instabilities. *J. Atmos. Sci.* **47**, 823–839.

Ji, M., A. Kumar, and A. Leetmaa (1994). An experimental coupled forecast system at the National Meteorological Center: Some early results, *Tellus* **46A**, 398–418.

Ji, M., D. W. Behringer, and A. Leetmaa (1998). An improved coupled model for ENSO prediction and implications for ocean initialization, Part II: The coupled model, *Mon. Wea. Rev.* **126**, 1022–1034.

Jiang, N., J. D. Neelin, and M. Ghil (1995a). Quasi-quadrennial and quasi-biennial variability in the equatorial Pacific. *Clim. Dyn.* **12**, 101–112.

Jiang, N., M. Ghil, and J. D. Neelin (1995b). Forecasts of equatorial Pacific SST anomalies by an autoregressive process using similar spectrum analysis. *In* "Experimental Long-Lead Forecast Bulletin (ELLFB)." National Meteorological Center, NOAA, U.S. Department of Commerce, **4**(1), 24–27.

Jiang, S., F.-F. Jin, and M. Ghil (1995c). Multiple equilibria, periodic, and aperiodic solutions in a wind-driven, doublee-gyre, shallow-water model, *J. Phys. Oceanogr.* **25**, 764–786.

Jin, F.-F. (1997). An equatorial ocean recharge paradigm for ENSO, Part I: Conceptual model. *J. Atmos. Sci.* **54**, 811–829.

Jin, F.-F., and M. Ghil (1990). Intraseasonal oscillations in the extratropics: Hopf bifurcation and topographic instabilities. *J. Atmos. Sci.* **47**, 3007–3022.

Jin, F.-F., and J. D. Neelin (1993). Modes of interannual tropical ocean–atmosphere interaction—a unified view. Part III: Analytical results in fully-coupled cases. *J. Atmos. Sci.* **50**, 3523–3540.

Jin, F.-F. J. D. Neelin, and M. Ghil (1994). El Niño on the Devil's staircase: Annual subharmonic steps to chaos. *Science* **264**, 70–72.

Jin, F.-F., J. D. Neelin, and M. Ghil (1996). El Niño/Southern Oscillation and the annual cycle: Subharmonic frequency-locking and aperiodicity. *Physica D* **98**, 442–465.

Kalnay, E., and A. Dalcher (1987). Forecasting forecast skill. *Mon. Wea. Rev.* **115**, 349–356.

Karaca, M., and D. Müller (1989). Simulation of sea surface temperatures with the surface heat fluxes from an atmospheric circulation model. *Tellus* **41A**, 32–47.

Keir, R. S. (1988). On the late pleistocene ocean geochemistry and circulation. *Paleoceanography* **3**, 413–446.

Keller, H. B. (1978). Global homotopies and Newton methods. *In* "Nonlinear Analysis" (C. de Boor and G. H. Golub, eds.), pp. 73–94. Academic Press, San Diego.

Kennett, R. P., and L. D. Stott (1991). Abrupt deep-sea warming, paleoceanographic changes and benthic extinctions at the end of the Palaeocene. *Nature* **353**, 225–229.

Keppenne, C. L., and M. Ghil (1992). Adaptive filtering and prediction of the Southern Oscillation index, *J. Geophys. Res.* **97**, 20449–20454.

Kimoto, M., and M. Ghil (1993a). Multiple flow regimes in the Northern Hemisphere winter. Part I: Methodology and hemispheric regimes. *J. Atmos. Sci.* **50**, 2625–2643.

Kimoto, M., and M. Ghil (1993b). Multiple flow regimes in the Northern Hemisphere winter. Part II: Sectorial regimes and preferred transitions. *J. Atmos. Sci.* **50**, 2645–2673.

Knutson, T. R., and K. M. Weickmann (1987). 30–60 day atmospheric oscillations: Composite life cycles of convection and circulation anomalies. *Mon. Wea. Rev.* **115**, 1407–1436.

Kraus, E., and J. Turner (1967). A one-dimensional model of the seasonal thermocline. *Tellus* **19**, 98–105.

Kubicek, M., and M. Marek (1983). "Computational Methods in Bifurcation Theory and Dissipative Structures." Springer-Verlag, New York.

Kushnir, Y. (1987). Retrograding wintertime low-frequency disturbances over the North Pacific Ocean. *J. Atmos. Sci.* **44**, 2727–2742.

Latif, M., T. P. Barnett, M. A. Cane, M. Flügel, N. E. Graham, H. von Storch, J.-S. Xu, and S. E. Zebiak (1994). A review of ENSO prediction studies. *Clim. Dyn.* **9**, 167–179.

Lau, K.-M., and M.-T. Li (1984). The monsoon of East Asia and its global associations—A survey. *Bull. Am. Meteor. Soc.*, **65**, 114–125.

Lau, K.-M., and T. J. Phillips (1986). Coherent fluctuations of extratropical geopotential height and tropical convection in intraseasonal time scales. *J. Atmos. Sci.* **43**, 1164–1181.

Legras, B., and M. Ghil (1985). Persistent anomalies, blocking and variations in atmospheric predictability. *J. Atmos. Sci.* **42**, 433–471.

Li, Z. X., and H. Le Treut (1992). Cloud-radiation feedbacks in a general circulation model and their dependence on cloud modeling assumptions. *Clim Dyn.* **7**, 133–139.

Li, Z.-X., K. Ide, H. Le Treut, and M. Ghil (1997). Atmospheric radiative equilibria in a simple column model. *Clim. Dyn.* **13**, 429–440.

Liebmann, B., and D. L. Hartmann (1984). An observational study of tropical-midlatitude interaction on intraseasonal time scales during winter. *J. Atmos. Sci.* **41**, 3333–3350.

Lorenz, E. N. (1963a). Deterministic nonperiodic flow. *J. Atmos. Sci.* **20**, 130–141.

Lorenz, E. N. (1963b). The mechanics of vacillation. *J. Atmos. Sci.* **20**, 448–464.

Lorenz, E. N. (1964). The problem of deducing the climate from the governing equations. *Tellus* **16**, 1–11.

MacCracken, M. C., and S. J. Ghan (1988). Design and use of zonally averaged models. *In* "Physically-Based Modelling and Simulation of Climate and Climatic Change" (M. E. Schlesinger, ed.), pp. 755–803. Kluwer Academic Publishers, Dordrecht.

Madden, R. A., and P. R. Julian (1971). Detection of a 40–50 day oscillation in the zonal wind in the tropical Pacific. *J. Atmos. Sci.* **28**, 702–708.

Madden, R. A., and P. R. Julian (1972). Description of global-scale circulation cells in the tropics with a 40–50 day period. *J. Atmos. Sci.* **29**, 1109–1123.

Madden, R. A., and P. Speth (1995). Estimates of atmospheric angular momentum, friction, and mountain torques during 1987–1988. *J. Atmos. Sci.* **52**, 3681–3694.

Manabe, S., and R. F. Strickler (1964). Thermal equilibrium of the atmosphere with a convective adjustment. *J. Atmos. Sci.* **21**, 361–385.

Marcus, S. L., Ghil, M., and Dickey, J. O. (1994). The extratropical 40-day oscillation in the UCLA General Circulation Model, Part I: Atmospheric angular momentum. *J. Atmos. Sci.* **51**, 1431–1466.

Marcus, S. L., Ghil, M., and Dickey, J. O. (1996). The extratropical 40-day oscillation in the UCLA General Circulation Model, Part II: Spatial structure. *J. Atmos. Sci.* **53**, 1993–2014.

Marotzke, J. (1989). Instabilities and multiple steady states of the thermohaline circulation. *In* "Ocean Circulation Models: Combining Data and Dynamics" (D. L. T. Anderson and J. Willebrand, eds.), pp. 501–511. Kluwer Academic Publishers, Dordrecht.

McWilliams, J. C. (1996). Modeling the oceanic general circulation. *Annu. Rev. Fluid Mech.* **28**, 215–48.

Mechoso, C. R., A. W. Robertson, N. Barth, M. K. Davey, P. Delecluse, P. R. Gent, S. Ineson, B. Kirtman, M. Latif, H. Le Treut, T. Nagai, J. D. Neelin, S. G. H. Philander, J. Polcher, P. S. Schopf, T. Stockdale, M. J. Suarez, L. Terray, O. Thual, and J. J. Tribbia (1995). The seasonal cycle over the tropical Pacific in coupled ocean–atmosphere general circulation models. *Mon. Wea. Rev.* **123**, 2825–2838.

Mikolajewicz, U., and E. Maier-Reimer (1990). Internal secular variability in an ocean general circulation model. *Clim. Dyn.* **4**, 145–156.

Mitchell, J. M., Jr. (1976). An overview of climatic variability and its causal mechanisms. *Quartern. Res.* **6**, 481–493.

Moron, V., Vautard, R., and M. Ghil (1998). Trends, interdecadal and interannual oscillations in global sea-surface temperatures. *Clim. Dyn.* **14**, 545–569.

Münnich, M., M. A. Cane, and S. E. Zebiak (1991). A study of self-excited oscillations in a tropical ocean–atmosphere system. Part II: Nonlinear cases. *J. Atmos. Sci.* **48**, 1238–1248.

Murakami, T. (1988). Intraseasonal atmospheric teleconnection patterns during the Northern Hemisphere winter. *J. Climate* **1**, 117–131.

Neelin, J. D. (1990a). A hybrid coupled general circulation model for El Niño studies. *J. Atmos. Sci.* **47**, 674–693.

Neelin, J. D. (1990b). The slow sea surface temperature mode and the fast-wave limit: Analytic theory for tropical interannual oscillations and experiments in a hybrid coupled model. *J. Atmos. Sci.* **48**, 584–606.

Neelin, J. D., M. Latif, M. A. F. Allaart, M. A. Cane, U. Cubasch, W. L. Gates, P. R. Gent, M. Ghil, C. Gordon, N. C. Lau, C. R. Mechoso, G. A. Meehl, J. M. Oberhuber, S. G. H. Philander, P. S. Schopf, K. R. Sperber, A. Sterl, T. Tokioka, J. Tribbia, and S. E. Zebiak (1992). Tropical air–sea interaction in general circulation models. *Clim. Dyn.* **7**, 73–104.

Neelin, J. D., M. Latif, and F.-F. Jin (1994). Dynamics of coupled ocean–atmosphere models: The tropical problem. *Annu. Rev. Fluid Mech.* **26**, 617–659.

Neelin, J. D., D. S. Battisti, A. C. Hirst, F.-F. Jin, Y. Wakata, T. Yamagata, and S. E. Zebiak (1998). ENSO theory. *J. Geophys. Res.* **103**, 14261–14290.

North, G. R., R. F. Cahalan, and J. A. Coakley, Jr. (1981). Energy balance climate models. *Rev. Geophys. Space Phys.* **19**, 91–121.

North, G. R., Mengel, J. G., and D. A. Short (1983). Simple energy balance model resolving the seasons and the continents: Application to the astronomical theory of the ice ages. *J. Geophys. Res.* **88**, 6576–6586.

Paillard, D., M. Ghil, and H. Le Treut (1993). Dissolved organic matter and the glacial–interglacial pCO_2 problem. *Global Biogeochem. Cycles* **7**, 901–914.

Penland, C., and P. D. Sardeshmukh (1995). The optimal growth of tropical sea surface temperature anomalies. *J. Climate* **8**, 1999–2024.

Penland, C., M. Ghil, and K. M. Weickmann (1991). Adaptive filtering and maximum entropy spectra, with application to changes in atmospheric angular momentum. *J. Geophys. Res.* **96**, 22659–22671.

Pfeffer, R. L. (ed.) (1960). "Dynamics of Climate. Pergamon Press, New York.

Plaut, G. R., and R. Vautard (1994). Spells of oscillations and weather regimes in the low-frequency dynamics of the Northern Hemisphere. *J. Atmos. Sci.* **51**, 210–236.

Quon, C., and M. Ghil (1992). Multiple equilibria in thermosolutal convection due to salt-flux boundary conditions. *J. Fluid Mech.* **245**, 449–483.

Quon, C., and M. Ghil (1995). Multiple equilibria and stable oscillations in thermosolutal convection at small aspect ratio. *J. Fluid Mech.* **291,** 33–56.

Ramanathan, V., and J. A. Coakley (1978). Climate modeling through radioactive convective models. *Rev. Geophys. Space Phys.* **16,** 465–489.

Rasmusson, E. M., X. Wang, and C. F. Ropelewski (1990). The biennial component of ENSO variability. *J. Marine Syst.* **1,** 71–96.

Reinhold, B. B., and R. T. Pierrehumbert (1982). Dynamics of weather regimes: Quasi-stationary waves and blocking. *Mon. Wea. Rev.* **110,** 1105–1145.

Rennó, N. O. (1997). Multiple equilibria in radiative-convective atmospheres. *Tellus* **49A,** 423–438.

Robertson, A. W., C.-C. Ma, C. R. Mechoso, and M. Ghil (1995a). Simulation of the tropical-Pacific climate with a coupled ocean–atmosphere general circulation model. Part I: The seasonal cycle. *J. Climate* **8,** 1178–1198.

Robertson, A. W., C.-C. Ma, M. Ghil, and R. C. Mechoso (1995b). Simulation of the tropical-Pacific climate with a coupled ocean–atmosphere general circulation model. Part II: Interannual variability. *J. Climate* **8,** 1199–1216.

Rooth, C. (1982). Hydrology and ocean circulation. *Progr. Oceanogr.* **11,** 131–149.

Saltzman, B., and A. D. Vernekar (1972). Global equilibrium solutions for the zonally averaged macroclimate, *J. Geophys. Res.* **77,** 3936–3945.

Sarmiento, J. L., and J. R. Toggweiler (1984). A new model for the role of the oceans in determining atmospheric pCO_2, *Nature* **308,** 621–624.

Schlesinger, M. E. (1986). Equilibrium and transient climatic warming induced by increased atmospheric CO_2. *Clim. Dyn.* **1,** 35–51.

Schlesinger, M. E., and J. B. Mitchell (1987). Climate model simulations of the equilibrium climatic response to increased carbon dioxide. *Rev. Geophys.* **25,** 760–798.

Schneider, S. H., and R. E. Dickinson (1974). Climate modeling. *Rev. Geophys. Space Phys.* **25,** 447–493.

Sellers, W. D. (1969). A climate model based on the energy balance of the earth–atmosphere system. *J. Appl. Meteor.* **8,** 392–400.

Smale, S. (1967). Differentiable dynamical systems. *Bull. Am. Math. Soc.* **73,** 747–817.

Speich, S., H. Dijkstra, and M. Ghil (1995). Successive bifurcations in a shallow-water model, applied to the wind-driven ocean circulation. *Nonlin. Proc. Geophys.* **2,** 241–268.

Stommel, H. (1961). Thermohaline convection with two stable regimes of flow. *Tellus* **13,** 224–230.

Stone, L., P. I. Saparin, A. Huppert, and C. Price (1998). El Niño chaos: The role of noise and stochastic resonance on the ENSO cycle. *Geophys. Res. Lett.* **25**(2), 175–178.

Strong, C. M., F.-F. Jin, and M. Ghil (1993). Intraseasonal variability in a barotropic model with seasonal forcing. *J. Atmos. Sci.* **50,** 2965–2986.

Strong, C. M., F.-F. Jin, and M. Ghil (1995). Intraseasonal oscillations in a barotropic model with annual cycle, and their predictability. *J. Atmos. Sci.* **52,** 2627–2642.

Suarez, M. J., and P. S. Schopf (1988). A delayed action oscillator for ENSO. *J. Atmos. Sci.* **45,** 3283–3287.

Syu, H., J. D. Neelin, and D. Gutzler (1995). Seasonal and interannual variability in a hybrid coupled GCM. *J. Climate* **8,** 2121–2143.

Thual, O., and J. C. McWilliams (1992). The catastrophe structure of thermohaline convection in a two-dimensional fluid model and a comparison with low-order box models. *Geophys. Astrophys. Fluid Dyn.* **64,** 67–95.

Tziperman, E., L. Stone, M. Cane, and H. Jarosh (1994). El Niño chaos: Overlapping of resonances between the seasonal cycle and the Pacific ocean-atmosphere oscillator. *Science* **264,** 72–74.

Van den Dool, H. M. (1994). Searching for analogues, how long must we wait? *Tellus* **46A,** 314–324.

Vautard, R., and M. Ghil (1989). Singular spectrum analysis in nonlinear dynamics, with applications to paleoclimatic time series. *Physica D* **35,** 395–424.

Vautard, R., P. Yiou, and M. Ghil (1992). Singular-spectrum analysis: A toolkit for short, noisy chaotic signals, *Physica D* **58,** 95–126.

Wang, X., P. H. Stone, and J. Marotzke (1999). Global thermohaline circulation, Part II: Sensitivity with interactive atmospheric transports, *J. Climate* **12,** 83–91.

Weaver, A. J., E. S. Sarachik, and J. Marotzke (1991). Freshwater flux forcing of decadel and interdecadal oceanic variability. *Nature* **353,** 836–838.

Weaver, A. J., J. Marotzke, P. F. Cummings, and E. S. Sarachick (1993). Stability and variability of the thermohaline circulation. *J. Phys. Oceanogr.* **23,** 39–60.

Weickmann, K. M., G. R. Lussky, and J. E. Kutzbach (1985). Intraseasonal (30–60 day) fluctuations of outgoing longwave radiation and 250 mb streamfunction during northern winter. *Mon. Wea. Rev.* **113,** 941–961.

Welander, P. (1986). Thermohaline effects in the ocean circulation and related simple models. *In* "Large-Scale Transport Processes in Oceans and Atmosphere" (J. Willebrand and D. L. T. Anderson, eds.), pp. 163–200. D. Reidel, Norwell, MA.

Weng, W., and J. D. Neelin (1998). On the role of ocean–atmosphere interaction in midlatitude interdecadal variability. *Geophys. Res. Lett.* **25**(2), 167–170.

Wetherald, R. T., and S. Manabe (1975). The effect of changing the solar constant on the climate of a general circulation model. *J. Atmos. Sci.* **32,** 2044–2059.

Wyrtki, K. (1986). Water displacements in the Pacific and the genesis of El Niño cycles. *J. Geophys. Res.* **91,** 7129–7132.

Zebiak, S. E., and M. A. Cane (1987). A model El Niño Southern Oscillation. *Mon. Wea. Rev.* **115,** 2262–2278.

Chapter 11

Prospects for Development of Medium-Range and Extended-Range Forecasts

Anthony Hollingsworth
European Centre for Medium-Range Weather Forecasts
Shinfield Park, Reading, United Kingdom

I. INTRODUCTION

The work of Prof. Arakawa and his school of distinguished graduates has had a worldwide impact over many years. Professor Arakawa's work on numerics and parameterization has contributed much to the development of the science and to improvements in weather forecasting. It is therefore a privilege to participate in this celebration and to have the opportunity to convey the appreciation and congratulations of my ECMWF colleagues, and myself, to Prof. Arakawa on the occasion of his 70th birthday.

Professors Arakawa and Mintz at UCLA, and Professor Smagorinsky at GFDL, were important benefactors of the fledgling ECMWF. In early

1975 Professor Wiin-Nielsen, the director-designate of ECMWF asked these distinguished scholars for their respective models and, in characteristic fashion, they generously agreed to provide them. Robert Sadourny and I had just joined the planning staff for ECMWF, so Sadourny was dispatched to his professor at UCLA and I to GFDL, there to pick up the model software and to make the integrations needed to verify the eventual implementations on our home computer. Both institutes were overwhelmingly generous with software, data sets, and help so that by the ratification of ECMWF's convention (November 1, 1975) both models had been successfully ported and run at ECMWF. The scientific content and software engineering of the two models were the objects of intensive study for several years and they were important to the development and validation of ECMWF's first operational model.

Medium-range and extended-range weather forecasting is at an exciting stage of development. A new generation of operational and research satellites is coming on line; four-dimensional variational assimilation has been established as a powerful and effective method to use all observations; numerical methods continue to provide improved accuracy and substantial economies; parametrization schemes are improving steadily through new approaches that jointly exploit field experiments, large-eddy simulations, and operational data assimilation; ensemble prediction systems are providing a new dimension in probabilistic forecasting; the development of simplified Kalman filters, based on singular vectors, will benefit both the assimilation systems and the ensemble prediction systems; and computer vendors are confident they can meet the requirements for computational power in an affordable manner. These developments will undoubtedly lead to further gains in medium- and extended-range forecast skills and will also contribute to the realization of the present exciting prospects for useful forecasts on seasonal and longer time scales.

The task of this report to Prof. Arakawa is to summarize the methods by which the models and assimilation systems are developed, and to make some extrapolations of where we may hope to be for his 80th birthday.

II. METHODS FOR THE DEVELOPMENT
OF FORECAST MODELS

The practical importance of weather forecasts (on short, medium, and extended ranges) for the protection of life and property, together with profound social and political concerns about environmental change, have made the development and validation of atmospheric and ocean models a key focus for meteorological and oceanographic science. There is close similarity between the models used for work on medium- and extended-range forecasting and the general circulation models (GCMs) used for climate research. Both modeling communities rely heavily for model validation on the extensive climate data sets prepared by national and international agencies under the aegis of the WMO/ICSU. However, there are few systematic methods to identify the sources of problems in long runs of a general circulation model, because almost all model errors are fully developed and fully interactive. It is much easier to diagnose errors (say, in a parametrization scheme) when they grow in an otherwise accurate series of forecasts that start from accurate analyses. The forecasting community thus has powerful tools for model validation, stemming from the data assimilation systems needed for forecasting practice, and from verifications of operational forecasts. Forecast centers have pioneered systematic methods to diagnose and resolve model problems, using operational and field experiment data. These methods have enabled the forecast centers to identify missing processes in the model, and to refine the representations of well-known processes.

The forecast centers' ability to compare every single instantaneous observational measurement with a forecast of the measured value is a powerful scientific resource. The comparison of forecast with measurement is the basis of the data assimilation process through which observations are interpreted so as to partition the measurement-minus-forecast difference into meteorological information on the one hand and observational noise on the other. Operational four-dimensional variational assimilation (4D-Var) systems represent the current state of the art in extraction of information from observations; they provide accurate analyses of all available *in situ* and satellite data, and result in excellent forecasts. Detailed studies of the differences can identify recurring model errors and recurring data errors. The accuracy of the operational assimilation systems is also a great benefit in the diagnosis of forecast problems and in the subsequent development of the forecast models. This power is further exploited in the active role played by forecast centers in supporting field experiments and in interpreting and exploiting field experiment data.

The forecast centers' research priorities are set by operational problems. Though this might seem a restriction on the range of development effort, it is in fact a great strength. Models are formulated through compromises between scientific uncertainties across a range of disciplines (radiation, turbulence, etc.), taking account of the available computer power. Models are so interactive that the solution of a given operational problem (e.g., erroneous nighttime skin temperatures over midlatitude continents in winter) requires reevaluation of all of the parameterizations in the model against the latest scientific information on radiation, cloud, boundary layer turbulence, and soil conductivity for heat and water (Gregory *et al.*, 1998b). In solving an operational problem, the steady goal is to ameliorate the problem while maintaining or improving overall model performance. Analysis of an operational problem will generally suggest a revision of the current formulations (or a new formulation) where the largest changes are made in the least reliable parts of the current model. This iterative method of successive reassessment, diagnosis, and reformulation has proven effective in forecasting practice. The results have also been useful in two other ways. First, the method has sometimes identified neglected physical processes that require careful and sometimes urgent attention in both forecast and climate models. Secondly the method has been useful in helping field experiment teams target their work on the issues of greatest benefit to modelers, with benefits for the science as a whole.

Focused diagnosis of operational problems, and participation at the cutting edge of field research, has thus enabled forecast centers to identify and remove many serious weaknesses in their models. The schematic in Fig. 1 summarizes the scientific and data resources available for the development of real-time medium- and extended-range forecasts, and documents the interplay between the different resources. The variety of resources available for the development of forecast models may be contrasted with the exiguous data resources available for direct validation of simulations of any climate other than the present climate. One can have confidence in simulated climate scenarios only if one has confidence in the physical formulations of the GCMs. A strong case could be made that each GCM should be equipped with a data assimilation system, so that one can diagnose its performance with field experiment data and in medium- and extended-range forecasts. Such diagnosis is bound to provide penetrating insights on how to improve the physical formulations of the GCMs.

The interplay of data assimilation, model development, and operational forecasting within the main forecast centers has led to an integration of scientific and technical developments, with tightly coupled, mission-oriented research teams backed by strong operations departments. This has had consequences for the engineering of the operational software. Any forecast system needs (1) timely good-quality observations, (2) a good assimilation scheme which can provide good analyses and useful flow-dependent estimates of analysis and forecast accuracy, (3) a model with efficient numerics and accurate parameterizations, (4) an effective ensemble prediction system, and (5) a powerful computer. Development in any one of these areas affects development in the others. To illustrate the interdependence of progress in the different elements of the forecast

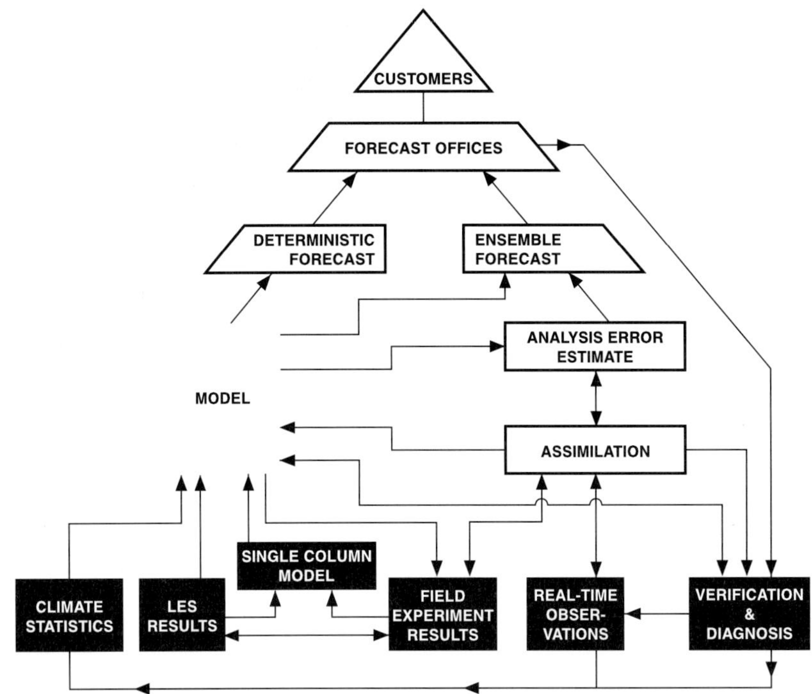

Figure 1 The scientific and data resources available for the development of real-time medium- and extended-range forecasts, and illustrates the interconnections between the uses of the different resources.

system, consider the current situation at an operational center such as ECMWF:

- The quality of the real-time observations is dependent on the assimilation system because of the importance of the assimilation system in providing long-loop monitoring of the quality of all data, and also because of the requirements for slowly varying bias corrections to certain data. In addition real-time forecast systems have contributed much to the calibration and geophysical validation of novel satellite instrumentation such as the ERS-1 radars (scatterometer, altimeter, SAR).
- The (tangent and) adjoint versions of the model's dynamics and physics are required for the four-dimensional variational assimilation (4D-Var) system, for the singular vectors used in the ensemble prediction system, and for the forthcoming simplified Kalman filter based on

those singular vectors. The latter will play a key role in cycling the error statistics in 4D-Var and in preparing improved perturbations for the ensemble prediction system.

- Development of the model's parameterizations depends on verifications of operational forecasts against accurate operational analyses, on extensive experimental assimilations of field data, on the results of large eddy simulations, on studies of extended reanalysis assimilations, and on verifications of long runs to check the climatology of the model.

- Methods for diagnosis of forecast errors depend on estimation of subtle imbalances between dynamical and physical forcing (Klinker and Sardeshmukh, 1991), and increasingly on calculations of the sensitivity of forecast errors to initial data (Rabier et al., 1996), and thus on the adjoints of the model's dynamics and physics.

- Ocean surface wave forecasts, and extended-range forecasts with coupled atmosphere–ocean models are extremely sensitive to the quality of the ocean–atmosphere fluxes, and pose important requirements for, and constraints on, atmospheric parameterization developments. Ocean data are valuable proxy data for verification of the atmospheric forecasts.

The mathematical and technical tools needed to undertake all of these tasks have been implemented in an Integrated Forecast System (IFS/Arpege) developed jointly by ECMWF and Meteo-France, which is now in its 11th year and 21st common software cycle. ECMWF uses one set of configurations of the software for medium- and extended-range forecasts, while Meteo-France uses a different set of configurations for short-range forecasting and climate research. It goes without saying that the success of the operational work and forecast research depends crucially on adequate high-performance computing resources and on powerful data handling systems.

III. DEVELOPMENT OF THE ECMWF FORECASTING SYSTEM

The scientific and technical approaches to model and assimilation development outlined above are the outgrowth of two decades of experience developing forecasting systems. ECMWF's first model and assimilation system in 1979 included many novel ideas and set new standards for medium-range forecast performance (Geleyn and Hollingsworth, 1979;

Hollingsworth *et al.*, 1980; Lorenc, 1981; Louis, 1979; Temperton and Williamson, 1981; Williamson and Temperton, 1981).

Motivated by operational forecast problems, parameterization developments since then include the first envelope orography scheme in 1983 (Wallace *et al.*, 1983); the first shallow convection scheme in 1985 (Tiedtke, 1984); two successful convection schemes, only one of which could be implemented in 1989 (Betts and Miller, 1986; Tiedtke, 1989); an advanced radiation scheme in 1989 (Morcrette, 1990, 1991); the novel 1990 treatment of ocean surface fluxes in the free convection limit (Miller *et al.*, 1992); new formulations of land surface processes including hydrological and vegetation effects in 1994 (Betts *et al.*, 1993; Beljaars *et al.*, 1995); a new parameterization of subgridscale orography in 1995 (Lott and Miller, 1997); also in 1995, a radically new cloud scheme that enforces coherence throughout the physical parameterizations (Tiedtke, 1993); the representation of soil moisture freezing in 1996 (Viterbo *et al.*, 1998); and a major reduction in 1997 of climate drift in extended-range forecasts through coupled revisions of the radiation scheme (based on new spectroscopic data), and revisions of the convection and cloud schemes based on LES studies and field measurements (Gregory *et al.*, 1998a,b). These developments were direct responses to operational forecast problems, based on the best available theories and observations.

In numerical algorithms, ECMWF was among the pioneers of the semi-implicit scheme in high-resolution operational global models in 1979; it implemented a successful global spectral model with a new vertical coordinate system in 1983 (Simmons and Burridge, 1981); it implemented a three-time-level semi-Lagrangian scheme in 1991 (Hortal and Simmons, 1991; Ritchie et al., 1995) and has produced substantial efficiency gains in the semi-Lagrangian methodology since then (Simmons and Temperton, 1996; Temperton, 1997; Hortal, 1999). These efficiency gains, together with enhancements in the Centre's computer power, have enabled the Centre to increase both horizontal and vertical resolution to provide more accurate large-scale medium-range forecasts and more detailed and useful products from the forecasts. The current horizontal resolution is $T_L 319$ ($\sim$ 60-km resolution; subscript L indicates a linear Gaussian grid) and 50 levels in the vertical. As part of these efforts, the Centre pioneered the operational use of shared-memory parallel processors in the 1980s (Dent, 1984) and then pioneered the operational use of distributed-memory parallel processors in the 1990s (Dent and Modzynski, 1996).

ECMWF has played a leading role in the development of data assimilation methods. The Centre's optimal interpolation intermittent assimilation system (Lorenc, 1981) was brought to a high level of development during the 1980s, with many innovations and refinements (Shaw *et al.*, 1987;

Lönnberg, 1988; Wergen, 1988; Unden, 1989). It provided excellent analyses of the available data, resulting in excellent forecasts; it provided the basis for powerful new methods of data monitoring and quality control (Hollingsworth *et al.*, 1986); it was a successful vehicle for the FGGE analyses (Bengtsson *et al.*, 1982) and for the 1979–1993 ERA-15 reanalyses (Gibson *et al.*, 1997); it was the basis for important studies of the global observing system; and it demonstrated the serious shortcomings of then-standard methods for using satellite data (Andersson *et al.*, 1991; Kelly *et al.*, 1991; Flobert *et al.*, 1991). It thus provided clear motivation for the development of the four-dimensional variational assimilation system (4D-Var), which entered service in November 1997 and which can make much better use of satellite data.

The phased development of the four-dimensional variational assimilation took many years, with operational implementation of a one-dimensional variational analysis (1D-Var) of TOVS radiances in 1991 (Eyre *et al.*, 1993), operational implementation of the IFS/Arpege model in 1994, operational implementation of the three-dimensional variational analysis (3D-Var) in 1996 (Courtier *et al.*, 1998; Rabier *et al.*, 1998, Andersson *et al.*, 1998), and operational implementation of the four-dimensional variational analysis (4D-Var) in 1997 (Rabier *et al.*, 1999, Mahfouf and Rabier, 1999; Klinker *et al.*, 1999). The 1997 operational implementation of 4D-Var includes a number of restrictions and simplifications that will be successively relaxed during the next few years so the system can deliver its full potential. In parallel with continued development of 4D-Var, the next major operational implementation in the development of the assimilation system will be a simplified Kalman filter (Fisher and Courtier, 1995; Fisher, 1998). In the course of these developments, the requirements of the assimilation system led to new demands on the model to provide realistic *a priori* estimates of the available observations, such as improved surface temperatures over ocean ice, improved ocean surface wind fields, more realistic tropospheric humidity structures, and improved stratospheric temperature structures. Model improvements have thus contributed to forecast skill, both directly through the forecasts and indirectly through the assimilation system.

ECMWF has played a leading role in the use of satellite data for numerical weather prediction. The Centre's feedback to the data producers contributed substantially to improvements in the quality of wind products and sounding retrievals during the last 15 years. The Centre pioneered the operational use of sounding radiances in 1992 (Eyre *et al.*, 1993), and was the first operational institute to make direct use of radiances in 3D-Var. The Centre contributed substantially to the engineering calibration of the ERS instruments; ESA's operational scatterometer algorithm was developed and validated at the Centre (Stoffelen and Anderson, 1997a,b,c). The Centre's variational assimilation system is designed to make effective use of satellite data. Among the many benefits expected from the advanced assimilation systems will be the extraction of wind information from the time sequence of satellite data on ozone and water vapor (Andersson *et al.*, 1994). Current preparations for assimilation of MSG-SEVERI radiances, ENVISAT ozone products, and METOP measurements will ensure early forecast benefits from the explanation of each of these data streams as they become available.

Since 1992, the Centre's pioneering ensemble prediction system (Buizza *et al.*, 1993; Molteni *et al.*, 1996) has provided a growing range of new products helping forecasters to deal scientifically and quantitatively with large day-to-day variations in the predictability of the atmosphere, and with the implications of these fluctuations on a wide range of weather parameters (Buizza *et al.*, 1999). Numerical efficiency gains and increased computer power have enabled the Centre to provide substantial increases in the resolution of the EPS model, with corresponding benefits for the quality of the overall EPS model and for the whole range of EPS forecast products (Buizza *et al.*, 1998).

The Centre's ocean surface wave forecasting project has consistently provided penetrating criticism of the atmospheric model and has provided equally valuable validation of successive model upgrades. Recent research demonstrated benefits for both atmospheric and wave forecasts of a direct coupling of the two models, and this was implemented in operations in 1998 (P. A. E. M. Janssen, personal communication, 1998). Work on wave assimilation has contributed much to the quality of ESA's radar-altimeter and SAR-wave algorithms (Janssen *et al.*, 1997).

ECMWF's project on experimental seasonal forecasting (Palmer *et al.*, 1990; Palmer and Anderson, 1994; Stockdale *et al.*, 1998) is helping establish the predictability of seasonal forecasts. The project has implemented a coupled atmosphere–ocean forecast system that provides experimental real-time ensemble seasonal forecasts to 6 months ahead, using the ECMWF ocean model, the HOPE ocean model from MPI-Hamburg, the BMRC-Melbourne ocean analysis, and the OASIS-CERFACS ocean–atmosphere coupler. Real-time forecasts for the 1997–1998 ENSO event were successful. Because of the exceptional nature of this event, and in response to overarching WMO requirements, the ECMWF Council decided to make a range of products from the experimental seasonal forecast project available on the ECMWF web site (http:// www.ecmwf.int).

The planned ECMWF reanalysis, ERA-40, for the period 1958–1998 together with the corresponding NCEP/NCAR reanalysis will provide the range of cases needed to assess more fully the reliability of seasonal forecasts. Progress in seasonal forecasting will continue to depend on active collaboration between data producers and the many research groups active in the area, under the aegis of WMO/ICSU.

IV. PROGRESS IN FORECASTING

To provide a statistical overview of the development of midlatitude forecast skill, Fig. 2 shows the evolution during 1980–1997 of the 500-mb Northern Hemisphere forecast skill measured by the number of days before the anomaly correlation score drops to four different levels, 85, 80, 70, and 60%. There is a sustained trend of improving skill over the period, modulated by marked interannual variability. The interannual variations in medium-range forecast skill are much larger than would be estimated from the observed interannual variations of the day 1 forecast skill by fitting a standard model for error growth, and must arise from interannual varia-tions in predictability.

Verifications of the rms position error in Northwest Pacific typhoon forecasts from three global models (ECMWF, UKMO, JMA) in 1996, as verified by JMA (N. Sato, personal communication, 1997) show that at

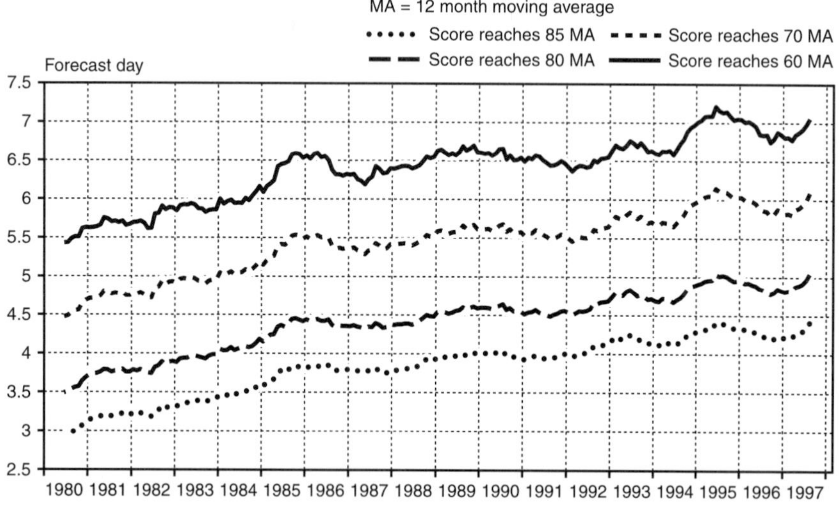

Figure 2   Evolution during 1980–1997 of the 500-mb Northern Hemisphere forecast skill measured by the number of days before the anomaly correlation score drops to four different levels, 85, 80, 70, and 60%.

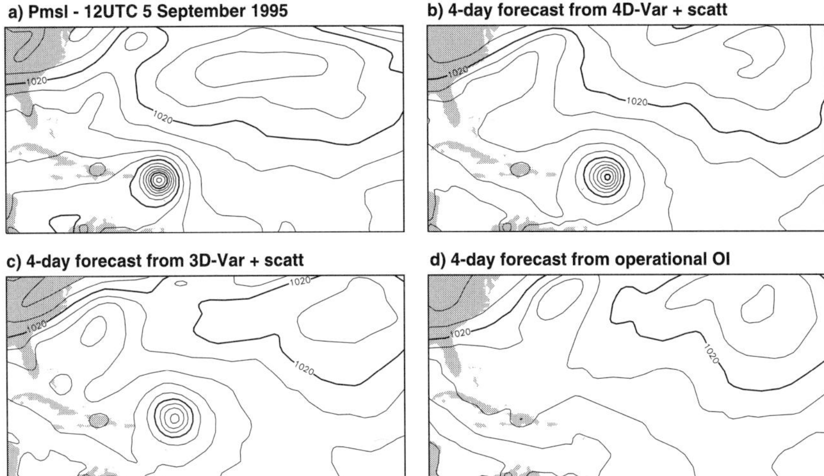

Figure 3 Verifying analysis (panel a) and a set of three 4-day forecasts for hurricane Luis in the Atlantic on September 5, 1995. The then-operational optimal interpolation system (panel d) did not use scatterometer data and produced a poor forecast. The use of 3D-Var and scatterometer data (panel c) produced a better forecast. The best forecast (panel b) used 4D-Var and scatterometer data.

$D + 3$ all three models show errors of at most 400 km. This may be compared with the typical 3-day position error of more than 600 km from more traditional methods of typhoon forecasting (Guard *et al.*, 1992).

The ERS scatterometer data are critical for the quality of ECMWF's typhoon forecasts. To illustrate the synoptic impact of assimilation developments and of new data, Fig. 3 shows the verifying analysis (panel a) and a set of three 4-day forecasts for hurricane Luis in the Atlantic on September 5, 1995. The then-operational optimal interpolation system (panel d) did not use scatterometer data and produced a poor forecast. The use of 3D-Var and scatterometer data (panel c) produced a better forecast. The best forecast (panel b) used 4D-Var and scatterometer data. The successive benefits of the more sophisticated assimilation method and the new data are quite evident (Lars Isaksen, personal communication, 1998).

V. ECMWF'S EARTH SYSTEM MODEL AND ASSIMILATION SYSTEM

Operational experience has repeatedly shown that medium-range forecast models must represent the main features of the interactions between

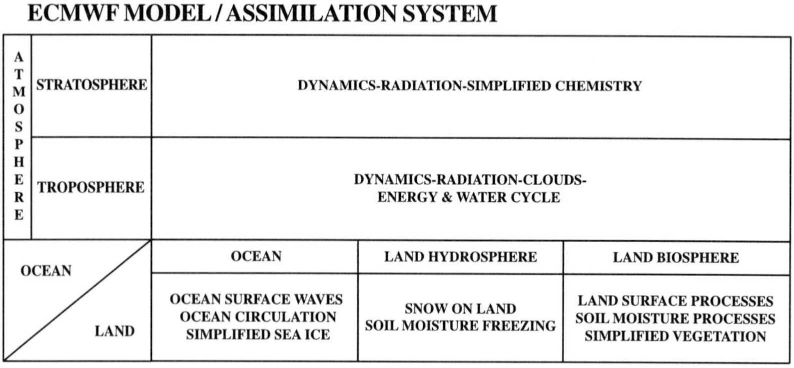

ECMWF MODEL / ASSIMILATION SYSTEM

A T M O S P H E R E	STRATOSPHERE	DYNAMICS-RADIATION-SIMPLIFIED CHEMISTRY		
	TROPOSPHERE	DYNAMICS-RADIATION-CLOUDS- ENERGY & WATER CYCLE		
OCEAN LAND		OCEAN	LAND HYDROSPHERE	LAND BIOSPHERE
		OCEAN SURFACE WAVES OCEAN CIRCULATION SIMPLIFIED SEA ICE	SNOW ON LAND SOIL MOISTURE FREEZING	LAND SURFACE PROCESSES SOIL MOISTURE PROCESSES SIMPLIFIED VEGETATION

Figure 4   Components of ECMWF's Earth system model comprising coupled modules for a coupled atmosphere–ocean general circulation model together with interacting software modules for, *inter alia*, simplified stratospheric and tropospheric chemistry; surface exchanges of energy, momentum, and gases; land surface/soil physical and (simplified) biological processes, snow, and sea ice; simplified hydrological processes; and ocean surface wave dynamics and ocean ice.

atmosphere, land, ocean, cryosphere, and biosphere, which together govern the evolution of the Earth's fluid system. Model and assimilation developments resulting from this experience, together with the requirements of seasonal forecasting, have led to a considerable elaboration of the forecast system, so that the current ECMWF system may be described as an Earth system model and assimilation system, as illustrated schematically in Fig. 4. ECMWF's Earth system model comprises the following coupled modules:

- *Atmosphere*: an atmospheric general circulation model
- *Ocean circulation*: an ocean general circulation model; ocean ice processes
- *Ocean surface waves*: ocean surface wave dynamics model
- *Land*: land biosphere module; land surface, soil, hydrological, and snow model
- *Ozone*: parametrized stratospheric ozone chemistry

Thus the modules comprise the familiar components of a coupled atmosphere–ocean general circulation model together with interacting software modules, for, *inter alia*, simplified stratospheric and tropospheric chemistry; surface exchanges of energy, momentum, and gases; land surface/soil physical and (simplified) biological processes, snow, and sea ice; simplified hydrological processes; and ocean surface wave dynamics and ocean ice. These developments in the model have enabled the Centre to make substantial improvements in the quality of its large-scale medium-range forecasts, and in the quality of its corresponding deterministic and probabilistic forecasts for local weather parameters. These developments have also contributed substantially to the progress of the ensemble seasonal forecasting project.

ECMWF's advanced four-dimensional variational data assimilation system (4D-Var) has been developed specifically to optimize the use of satellite data. By early 2000, the operational 4D-Var system will be supported by a powerful new algorithm (a simplified Kalman filter) to provide flow-dependent forecast error structures at the start of each 4D-Var cycle. The Centre's assimilation system also meets the basic requirements of an Earth fluid system assimilation system.

VI. OPPORTUNITIES FOR DEVELOPMENT OF MEDIUM-RANGE AND EXTENDED-RANGE WEATHER FORECASTS

ECMWF's long-term goal is to deliver useful weather forecasts to 8 days and beyond. In addition the Centre will contribute to the realization of a useful seasonal forecast capability. The ensemble prediction approach will play a major role in attaining these goals, and its success will depend crucially on the quality of the assimilation system and the forecast model.

Accurate and reliable medium-range weather forecasts for precipitation, wind, and temperature continue to be the Centre's most important challenges. There are known deficiencies in our ability to analyze and forecast wind, temperature, precipitation, cloud, and humidity. These arise from a lack of observations to describe the state of the atmosphere, from gaps in our scientific understanding of many detailed aspects of atmospheric behavior, and from limitations in computer power. Further progress toward the Centre's goals will require improved scientific understanding of the atmosphere and its interactions with the other main components of the Earth's fluid system (i.e., land hydrosphere and ocean), together with better techniques for modeling and assimilation of the Earth's fluid system. Continued progress will also be required in the supporting technologies of Earth observation and high performance computing.

A. Opportunities from Developments in Operational Satellites

The most critical aspect of the forecasting problem is the availability of high-quality data on all the important aspects of the atmosphere and its boundary conditions. There will be major observational opportunities to improve medium-range forecasting in the coming decade. Tables I and II lists the plans of major space agencies for operational and research satellites in the next 5–15 years. The CEOS operational satellite program, managed by EUMETSAT, NOAA, and JMA, will provide the basic space-based observations on which both operational assimilations and Earth system science assimilations will depend. Total international funding for Earth observation programs and for Earth fluid system science will be very large.

Forecast benefits are being realized from the assimilation of AMSU A/B data and from the assimilation of sounding data from the geostationary satellites. We expect to have important benefits from the ASCAT and METOP in 2003. However there is little doubt that the new generation of high vertical resolution sounders, such as IASI on METOP, present the most important opportunities to improve the accuracy of forecast initial conditions and to improve forecast models.

With a field of view of about 10 km, IASI and similar advanced sounders will provide very detailed horizontal and vertical sounding infor-

Table I

Planned Operational Satellites 1998–2010

Polar orbit	Mission	Key instruments
1997–2007	NOAA 14…	
	NOAA N′	AMSU A/B
	DMSP F14…	SSMI/T/T2
2002–2016	METOP-1	IASI, ASCAT, OMI, GRAS
2006–	NPOESS	NOAA Advanced Sounder
Geostationary orbit		
1997–2001	MESOSAT	
2001–2004	MSG	SEVIRI
1997–	GMS	
1997–2010	GOES-I-M	GOES Sounder

Table II
Planned Research Satellites 1998–2010

	Mission	Key instruments
1997	ERS-2	Scat, SAR, Alt, GOME
	TOPEX/POSEDON	Altimeter
	TRMM	Precipitation radar, TMI
1998	EOS-AM-1	Land/clouds/aerosol
1999	ADEOS-II	Seawinds scatterometer
2000	ENMSAT	MIPAS, SAR, altimeter
	TOPEX/POSEIDON follow-on	SCHIAMACHY
	EOS-PM-1	AIRS
2001	TRMM follow-on (to 55N)	
2002	EOS-Chem	
2004 +	Candidate missions	Cloud radar
		Doppler wind lidar
		Chemistry mission
		Land mission
		Gravity mission

mation on temperature and humidity. An assimilation system with comparable resolution will be required to exploit effectively the geophysical information these instruments will provide. Studies of the evolution of the potential vorticity field show that one can make substantial improvements to the assimilation of tracer data (and thus winds), and probably also to the quality of medium-range forecasts, provided one calculates potential vorticity advection very accurately, i.e., with horizontal resolution of order 10–20 km. Successful medium-range forecasts of intense small-scale phenomena will probably require resolution of this order. This level of resolution is also required to model the interactions of fine-scale dynamic and orographic structures on land and in the ocean with the other components of the Earth system; such interactions are difficult to describe or aggregate in any other way.

Currently ECMWF uses little tropospheric satellite data over land, because of the difficulties posed by the inhomogeneities of the land surface. The attrition of the land-based radiosonde network, which forms the backbone of the current Northern Hemisphere observing system, poses a major challenge. The decline in coverage of the radiosonde network requires us to exploit all possible alternative data sources. The Centre will have to address the use over land of advanced sounders such as IASI.

Given IASI's 10-km field of view, a dynamic specification of the surface radiative properties (e.g., 15 vegetation and land-surface types, with associated properties of moisture stress, bidirectional reflectance, etc.) will be required at a resolution of about 10–20 km. This, in turn, will require a capability to model as far as possible the land surface and land biosphere, and to assimilate relevant satellite information. Such a modeling capability will also be of considerable benefit for medium- and extended-range forecasts.

B. Opportunities from Developments in Research Satellites

Environmental concerns have motivated funding for research satellite missions such as ESA's Earth-Explorer program, NASA's Mission to Planet Earth program, and NASDA's Earth Observation System program. Each new satellite instrument will be supported by extensive field validation programs. Each new satellite instrument will provide the raw material for many scientific investigations of the Earth's fluid system. These space-based observation programs will provide the data needed to resolve key scientific questions in the development of both operational medium-range forecast models and science-oriented Earth fluid system models.

ECMWF's medium-range forecast activity can make substantial contributions to, and can derive substantial benefits from, the planned European initiatives in Earth fluid system science, particularly in the areas of parametrization and assimilation. A topical example is the data from the TRMM mission, which are providing unprecedented coverage of the horizontal and vertical structure of tropical rain systems. ECMWF is a partner in an EU-funded project to exploit the TRMM data, which will be an invaluable resource for parameterization and assimilation studies for many years.

Work is planned to extract wind information from ENVISAT ozone data, to improve the treatment of land processes with EOS AM-1 data, and to exploit the tropospheric data from EOS AM-1 and EOS PM-1 to improve the modeling of clouds and the hydrological cycle. Further downstream there are exciting possibilities, such as a Doppler wind lidar in orbit, and a cloud radar in orbit, both of which would contribute substantially to our forecasting and modeling ability.

Carbon dioxide is not entirely well mixed in the atmosphere; there are marked (5 ppmv) seasonal and spatial variations in CO_2 abundance. Combined use of microwave and advanced infrared sounders will make it possible to derive information on the three-dimensional distribution of CO_2 in an operational data assimilation system. Such information would improve the accuracy of the assimilation and would also permit one to calculate seasonal fluctuations in the atmospheric stock of CO_2, from which one would estimate the net sources and sinks of CO_2 at the Earth's surface. Given conventional estimates of the anthropogenic sources of CO_2, one can then estimate the natural fluctuations of CO_2 sources and sinks at the surface. Routine monitoring of the natural sources of CO_2 would be of value for many aspects of climate science, not least the validation and improvement of the land–biosphere and ocean–biosphere components of the Earth system model.

In summary, the advanced sounding capabilities of operational satellite missions in the next decade provide real opportunities to improve the initial conditions for forecasts and the forecast models. Environmental research satellites and associated research programs will also lead to much improved physical parameterizations for medium-range forecasting. Computer developments are expected to make it affordable to run global operational models with resolutions of order 10 km by the year 2010. Such resolution will be needed to assimilate fully the information provided by the advanced sounders, and will in turn enable the production of more accurate medium-range and extended-range forecasts and more detailed and accurate forecasts of local weather elements.

C. Opportunities from Developments in Data Assimilation

ECMWF's 4D-Var data assimilation system is the most advanced in the world. It has been specifically designed to handle a wide variety of satellite data. The Centre's advanced assimilation system will make important contributions to the Centre's goals. A prime contribution will arise from the operational exploitation of the full potential of the four-dimensional variational assimilation system, 4D-Var, which will be fully developed to use a longer assimilation window, and to use more refined physical parameterizations. A second contribution will come from the effective use of a wider variety of real-time assimilatable satellite data, including data from the current generation and the new generation of infrared sounders and microwave sounders in polar orbit, data from new infrared sounders and visible imagers in geostationary orbit, data from scatterometers and microwave imagers in polar orbit, cloud-track wind data, in addition to the improved and wider use of the available *in situ* and ground-based data, including that from profilers. A third contribution will come from the implementation of a simplified Kalman filter, which will improve 4D-Var assimilations through flow-dependent calculations of the variance and correlation structure of the errors of the background field at the start of the assimilation period. Further downstream the Centre will examine the possibility of using adaptive filtering to update the background and observation error statistics in real time.

These developments in data assimilation will bring with them the need for observing system experiments to reassess the role of satellite and ground-based observing systems, and to validate the performance of the assimilation system itself. The Centre is therefore very well placed to play a leading role in exploiting the new satellite data to improve modeling and forecasting.

D. OPPORTUNITIES FROM DEVELOPMENTS IN FORECAST MODELS

ECMWF's forecast model is the cornerstone of the Centre's activity in data assimilation, deterministic forecasting, ensemble forecasting, and seasonal forecasting. Improvements in the forecast model therefore benefit all aspects of the performance of the Centre's forecasting system. Parameterization developments will flow from the Centre's diagnostic and modeling work, and from collaborations with many external groups. Increases in algorithmic efficiency and computer power will enable the Centre to make important upgrades in the horizontal and vertical resolution of the model. These will be essential for effective assimilation of new satellite data and will also provide better large-scale and local medium-range and extended-range forecasts.

E. OPPORTUNITIES FROM DEVELOPMENTS IN PHYSICAL PARAMETERIZATIONS

The Centre's model provides an integrated and increasingly more unified representation of the atmospheric hydrological and energy cycles. Strong emphasis will be placed on the coupling between schemes rather than on the development of schemes individually. The cloud scheme is an example of unification as it couples clouds directly with adiabatic and diabatic processes. Much effort will be devoted to model validation, through regular monitoring of forecast errors of all types and through specific comparisons with process data from field experiments and satellites. There is a clear need to improve the treatment of soil and surface processes, to improve forward radiative modeling for assimilation purposes, to improve medium-range forecasts, and to enhance the prospects for seasonal prediction.

Increases in horizontal and vertical resolution provide an effective way of simplifying the parameterization problem: As more of the critical subgridscale processes are resolved, the complexity of parameterizing the unresolved processes is decreased. However, the problem of double-counting, when processes are partly resolved and partly parameterized, remains. There will be continuing development of linearized and adjoint versions of the physical parameterizations for use in 4D-Var, singular-vector calculations, and sensitivity studies. Later in the planning period the possibility of using variational methods to refine the parametrization schemes will be investigated.

F. Opportunities from Developments in Numerical Methods

Developments in numerical techniques have delivered substantial economies in computing cost during the last 5–10 years. In the near future there will be further gains in accuracy, through refinements of the vertical differencing scheme and of the time-stepping algorithm.

The current spectral technique is efficient at resolutions up to $T_L 639$ but may become less efficient than other numerical techniques at significantly higher resolutions. The general numerical formulation of the current model is thus likely to be efficient for the next 5 years. During this period there will be a critical review of alternative numerical methods, including different formulations of the basic equations (e.g., potential vorticity conserving schemes, hydrostatic versus nonhydrostatic schemes), different horizontal and vertical discretizations, different coordinate systems, and different time integration schemes. In the light of this review and in the light of resolution requirements for assimilation and forecasting, work will begin on efficient numerical formulations to meet the growing requirements.

G. Opportunities from Increases in Vertical and Horizontal Resolution

Model resolution is critical for the success of short-range forecasting. The Centre's winter 6-day forecasts for Europe since 1990 have been about as accurate as the typical winter 2-day forecasts made in 1970. This achievement was only possible because the Centre's 2-day forecasts in 1990 were far better than the 2-day forecasts in 1970. Much of the improvement in the 2-day forecast stems from improvements in resolution and from the substantial attendant benefits in assimilation and parametrization.

The limiting factor for medium-range predictability is the rapid growth through instability processes of small errors in the initial data or rapid growth of small model errors. Studies of the evolution of the potential vorticity field (Dritschel *et al.*, 1999) suggest that one can make substantial improvements both in data assimilation and in the quality of medium-range forecasts, provided one calculates potential vorticity advection very accurately. With current numerical schemes, this requires horizontal resolution of order 10–20 km. Successful medium-range forecasts of intense small-scale phenomena will probably require resolution of this order.

As an example of the importance of resolution in such cases, we consider the explosive development of the devastating Iberian storm of November 6, 1997. Figure 5 (see color insert) shows the operational analyses of 10-m wind and mean-sea-level pressure (Fig. 5a) and potential vorticity on the 305 K isentrope (Fig. 5d) over the Iberian Peninsula at 0000 UTC on November 6, 1997, when the rapidly developing storm caused serious loss of life, with 31 fatalities, and extensive flooding. The corresponding operational 60-h and forecasts for the event are shown in Figs. 5b and Fig. 5e. Although they were aong the most successful of any operational forecasts at this range, the $T213$ forecasts clearly underestimate the intensity of the storm. Experimental 60-h $T_L 639$ forecasts, shown in Figs. 5c and Fig. 5f are far more successful in forecasting the position and intensity of the storm. Clearly, if one is to forecast such events 3–5 days ahead of time, one needs to have the highest resolution possible for the assimilation system and for the forecast model.

The same order of resolution (10–20 km) is also required to model the interactions of fine-scale dynamic and orographic structures on land and in the ocean with the other components of the Earth system; such interactions are difficult to describe or aggregate in any other way. This resolution is also required to assimilate fully the information provided by the advanced sounders, and will in turn enable the production of more accurate medium-range forecasts and more detailed and accurate forecasts of local weather elements.

The inescapable conclusion is that, if resolution is important for short-range forecasting, it is even more important for medium-range forecasts, at least for the forecast range for which synoptic forecasts are useful (about two to three eddy turnover times).

The vertical resolution and extent of the ECMWF model in the stratosphere has recently been improved substantially [from 31 levels to 50 levels, with the top level moved upward from 10 hPa (30 km) to 10 Pa (65 km)], particular aims being a better direct assimilation of satellite radiance measurements including those from ozone channels and a better treatment of ultra-long waves. A further increase in vertical resolution is expected later in 1999 to enable an improved description of boundary-layer turbulence, clouds, and shallow convection, and facilitate a better unification of the components of the model physics.

Attention will then focus on assessing the value of enhanced horizontal resolution for both deterministic and stochastic forecasting, with a view to increasing the resolution of the assimilating model and deterministic forecast model to $T_L 511$ resolution (~ 40-km resolution), and increasing the resolution of the inner loops of 4D-Var, and the resolution of the ensemble forecast model, to $T_L 255$ by 2001.

Assessment of the value of increasing the resolution of the operational assimilation system and the deterministic forecast still further to match the resolution of IASI is expected to motivate a further upgrade in the resolution of the operational systems by mid-decade.

H. Opportunities from Development of Diagnostics

As noted earlier, ECMWF has pioneered systematic methods to diagnose and resolve model problems, using operational and field experiment data. These methods have enabled the Centre to identify missing processes in the model and to refine the representations of well-known processes. The Centre's ability to identify key scientific problems has helped to focus the efforts of international networks of field experimenters and of very high-resolution modelers on parameterization issues of most significance to large-scale modelers. Development of the Centre's model will continue to benefit from these approaches.

I. Opportunities from Developments in the Ensemble Prediction System

The capability of the EPS to produce reliable probabilistic forecasts of weather elements and predictions of the likelihood of extreme events has been enhanced by recent increases in model resolution and ensemble size, and by parametrization improvements. Research and experimentation will be directed to evaluating options for continued configuration improvements. Further increases in resolution and ensemble size, and extension of the forecast range of the EPS to 15–20 days, will be assessed. Given the role of the intraseasonal (30- to 60-day) oscillation in midlatitude blocking, and given the growing evidence that atmosphere–ocean interaction plays a key role in the propagation of the oscillation, the value of a coupled atmosphere–ocean model will be assessed for extension of the range of the EPS. The value of multianalysis and multimodel ensembles will be assessed, as will the value of stochastic perturbations of the physical parameterizations. The performance of the operational system will be kept under review and the value of new products (including probabilistic ocean wave forecasts) will be examined. The possibilities for improving the initial perturbations will be assessed. Particular emphasis will be placed on inclusion of physical processes with a view to developing the tropical aspects of the EPS, on targeting perturbations on the early medium-range forecast over Europe, and on including estimates of the analysis-error covariances in the singular-vector calculation. The latter will be intimately linked with the operational implementation of the simplified Kalman filter.

J. Opportunities from Development of Seasonal Forecasting

Promising results have been obtained to date from studies of seasonal prediction. Continued effort in this area will seek to confirm the results of the initial studies by an extended study of the skill with which past events can be reproduced. These will be diagnosed extensively and improved (largely through collaborative projects) so as to provide a well-founded operational system.

The provision of a reliable operational seasonal forecast capability is an important new venture of considerable social, political, and economic significance. Real-time experimental seasonal forecasts will be produced and assessed on a regular basis. It will be essential to confirm the Centre's striking initial results on the extensive set of cases to be provided by the 40-year reanalysis. Development of the ocean circulation model and of the atmosphere–ocean coupler will rely on collaborative work with the parent institutes. The experience of the IFS/Arpege system will be invaluable in developing an improved ocean data assimilation system in collaboration with interested bodies in Europe. The production of seasonal forecasts will probably benefit from a multimodel approach involving forecasts produced by several institutes in a collaborative European venture.

Although most of the demonstrated seasonal predictability stems from the El Niño phenomenon, there is much to be gained in the first one or two seasons from better treatment of land processes.

To improve medium-range and seasonal forecasts of temperature, rainfall, and near-surface parameters and to provide useful forecasts of hydrological conditions, vegetative moisture stress, and perhaps crop yields, there will be a sustained effort to advance the science base of atmosphere–land interactions (including soil moisture, hydrology, biosphere, and snow processes) and land surface models. Research on land surface processes will advance rapidly in 1999–2003 due to the availability of new satellite data, together with results from the major GEWEX continental scale experiments (BALTEX, GCIP, LBA, GAME, MAGS). The Centre's work on land processes will also be essential to improve the assimilation of operational satellite data over land.

One of the intriguing aspects of work on tropical aspects of wave forecasting and seasonal forecasting is the large sensitivity of ocean wave and ocean circulation models to the tropical surface wind field, and the large sensitivity of the latter to many aspects of the parameterization scheme, and even the assimilation scheme. One has a strong feeling that we need deeper insight into the balance of forces and feedbacks that maintains the present climate of the tropical wind field. The demand for better ocean and seasonal forecasts, together with the information provided by ocean satellite and *in situ* data, will undoubtedly stimulate the deepening of our understanding in this area.

K. Opportunities from Developments in Reanalysis

An important element in the development of a seasonal forecasting capability is the assessment and improvement of the ocean–atmosphere fluxes of momentum heat and moisture. Evaluation of the performance on seasonal time scales of both the assimilation system and the physical parametrization schemes shows that there is a requirement for further reanalysis of the data available since 1979. In addition the need to extend the range of cases on which one can test the seasonal forecast system leads to a requirement to reanalyze data prior to 1979. Consequently, the Centre is planning a reanalysis of the 40-year period 1958–1997, to be completed by 2002.

Such a reanalysis of the period 1958–1997 will provide an invaluable database not only for the experimental seasonal forecast project, but also for assimilation and data impact studies, for predictability studies, and for a wide range of meteorological and climate research.

VII. A FORWARD LOOK

ECMWF's prime long-term goal is to improve and deliver operational medium-range weather forecasts over the range from 3 to 10 days and beyond. ECMWF has recently been charged with a complementary long-term goal to establish and deliver a reliable operational seasonal forecasting capability. Ensemble prediction will play a major role in attaining both goals. Overall success will depend crucially on new and improved satellite observations, on improvements in the data assimilation system, and on improvements in the forecast model.

The new satellite data essential to achieve these goals will be provided over the next decade through heavy European, Japanese, and particularly U.S. investments in operational and research missions. In readiness to exploit the new satellite data, the Centre is completing development and operational implementation of a comprehensive Earth system forecast facility comprising an Earth system model and an advanced four-dimensional data assimilation facility.

The Earth system model comprises an atmospheric general circulation model coupled with an ocean general circulation model, together with interacting software modules for, *interalia*, atmosphere–ocean exchanges of energy moisture and momentum, ocean ice processes, and ocean surface wave dynamics; atmosphere–land exchanges of energy moisture and momentum, land surface, and soil physical and biological processes; hydrological and snow processes; and stratospheric ozone chemistry. Some modules of the Earth system model are quite sophisticated, but the science for other modules is at an early stage of development.

The Centre's four-dimensional variational data assimilation system (4D-Var) is the most sophisticated data assimilation system in operational use, and has been developed specifically to optimize the use of satellite data. By 2000, the operational 4D-Var system will be supported by a powerful new algorithm (a simplified Kalman filter) to provide flow-dependent forecast error structures at the start of the 4D-Var cycle.

Advances in computer technology during the next 10 years will make it possible to run such a system at a resolution necessary to extract all useful geophysical information from advanced sounders such as IASI (field of view of 10–15 km) and to provide much improved forecasts through accurate handling of the potential vorticity cascade, through better parameterizations, and through better handling of land boundary conditions.

We therefore expect that for Prof. Arakawa's 80th birthday, there will have been very substantial progress in medium- and extended-range forecasting, with forecast centers providing a broad range of new and high-quality products. This progress will depend on continued partnership between universities, field experimenters, operational centers, and the labs charged with GCM experimentation.

ACKNOWLEDGMENT

I am grateful to E. Klinker and P. Viterbo for the case study illustrated in Fig. 5.

REFERENCES

Andersson, E., A. Hollingsworth, G. Kelly, P. Lonnberg, J. Pailleux, and Z. Zhang (1991). Global observing system experiments on operational statistical retrievals of satellite sounding data. *Mon. Wea. Rev.* **119**, 1851–1864.

Andersson, E., J. N. Thepaut, J. R. Eyre, A. P. McNally, G. Kelly, P. Courtier, and J. Pailleaux (1994). Use of cloud cleared radiances in three/four dimensional variational data assimilation. *Quart. J. Roy. Meteor. Soc.* **120**, 627–653.

Andersson, E., J. Haseler, P. Undén, P. Courtier, G. Kelly, D. Vasiljevic, C. Brankovic, C. Cardinali, C. Gaffard, A. Hollingsworth, C. Jakob, P. Janssen, E. Klinker, A. Lanzinger, M. Miller, F. Rabier, A. Simmons, B. Strauss, J-N. Thépaut, and P. Viterbo (1998). The ECMWF implementation of three dimensional variational assimilation (3D-Var). Part III: Experimental results. *Quart. J. Roy. Meteor. Soc.* **124**, 1831–1860.

Beljaars, A. C. M., P. Viterbo, M. J. Miller, and A. K. Betts (1995). The anomalous rainfall over the USA during July 1993: Sensitivity to land surface parametrization and soil moisture anomalies. *Mon. Wea. Rev.*, **124**, 362–383.

Bengtsson, L., M. Kanamitsu, P. Kallberg, and S. Uppala (1982). FGGE 4-dimensional data assimilation at ECMWF. *Bull. Am. Meteor. Soc.* **63**, 29–43.

Betts, A. K., and M. J. Miller (1986). A new convective adjustment scheme. Part II: Single column tests using GATE wave, BOMEX, ATEX and arctic air-mass data sets. *Quart. J. Roy. Meteor. Soc.* **112**, 693–709.

Betts, A. K., J. H. Ball, and A. C. M. Beljaars (1993). Comparison between the land surface response of the ECMWF model and the FIFE—1987 data. *Quart. J. Roy. Meteor. Soc.* **119**, 975–1001.

Buizza, R., J. Tribbia, F. Molteni, and T. N. Palmer (1993). Computation of optimal unstable structures for a numerical weather prediction model. *Tellus* **45A**, 388–407.

Buizza, R., T. Petroliagis, T. Palmer, J. Barkmeijer, M. Hamrud, A. Hollingsworth, A. Simmons, and N. Wedi (1998). Impact of model resolution and ensemble size on the performance of an Ensemble Prediction System. *Quart. J. Roy. Meteor. Soc.* **124**, 1935–1960.

Buizza, R., A. Hollingsworth, F. Lalaurette, and A. Ghelli (1999). Probability precipitation prediction using the ECMWF Ensemble Prediction System. *Wea. Forecasting* **14**, 168–189.

Courtier, P., E. Andersson, W. Heckley, J. Pailleux, D. Vasiljevic, M. Hamrud, A. Hollingsworth, F. Rabier, and M. Fisher (1998). The ECMWF implementation of three dimensional variational assimilation (3D-Var). Part I: Formulation. *Quart. J. Roy. Meteor. Soc.* **124**, 1783–1808.

Dent, D. (1984). The multitasking spectral model at ECMWF. *In* "Multiprocessing in Meteorological Models" (G.-R. Hoffman and D. F. Snelling, eds.), pp. 203–214. Springer-Verlag, Berlin.

Dent, D., and G. Modzynski (1996). ECMWF forecast model on a distributed memory platform. *In* "Workshop on the Use of Parallel Processors in Meteorology, Proceedings of the Seventh Workshop," December 1996.

Dritschel, D. G., L. M. Polvani, and A. R. Mohebalhojeh (1999). The contour-advective semi-Lagrangian algorithm for the shallow water equations. *Mon. Wea. Rev.* **127,** 1551–1565.

Eyre, J. R., G. A. Kelly, A. P. McNally, E. Andersson, and A. Persson (1993). Assimilation of TOVS radiance information through one-dimensional variation analysis. *Quart. J. Roy. Meteor. Soc.* **119,** 1427–1463.

Fisher, M. (1998). Development of a simplified Kalman Filter, ECMWF Tech. Memo 260. Available from ECMWF.

Fisher, M., and P. Courtier (1995). Estimating the covariance matrices of analysis and forecast error in variational data assimilation, ECMWF Tech. Memo 220. Available from ECMWF.

Flobert, J.-F., E. Andersson, A. Chedin, A. Hollingsworth, G. Kelly, J. Pailleux, and N. A. Scott (1991). Data assimilation and forecast experiments using improved initialization inversion methods for satellite soundings. *Mon. Wea. Rev.* **119,** 1881–1914.

Geleyn, J. F., and A. Hollingsworth (1979). An economical analytical method for the computation of the interaction between scattering and line absorption of radiation. *Beitr. Phys. Atmos.* **52,** 1–16.

Gibson, J. K., P. Kallberg, S. Uppala, A. Hernandez, A. Nomura, and E. Serrano (1997). ECMWF ReAnalysis 1979–1993: ERA Description, ERA Project Report Series 1. Available from ECMWF.

Gregory, D., A. C. Bushell, and A. Brown (1998a). The interaction of convective and turbulent fluxes in General Circulation Models. *In* "Global Energy and Water Cycles" (K. A. Browning and R. J. Gurney, ed.). Cambridge University Press, Cambridge, MA.

Gregory, D., J.-J. Morcrette, C. Jakob, and A. Beljaars (1998b). Introduction of revised radiation, convection, cloud and vertical diffusion schemes into Cy18r3 of the ECMWF integrated forecast system, ECMWF Tech. Memo 254. Available from ECMWF.

Guard, C. P., L. E. Carr, F. H. Wells, R. A. Jeffries, N. D. Gural, and D. K. Edson (1992). Joint Typhoon Warning Centre and the challenges of multi-basin tropical cyclone forecasting. *Wea. Forecasting* **7,** 328–352.

Hollingsworth, A., K. Arpe, M. Tiedtke, M. Capaldo, and H. Savijarvi (1980). The performance of a medium range forecast model in winter—impact of physical parameterizations. *Mon. Wea. Rev.* **108,** 1736–1773.

Hollingsworth, A., D. B. Shaw, P. Lonnberg, L. Illari, K. Arpe, and A. J. Simmons (1986). Monitoring of observation and analysis quality by a data assimilation system. *Mon. Wea. Rev.* **114,** 861–879.

Hortal, M. (1999). The development and testing of a new two-time-level semi-Lagrangian scheme (SETTLS) in the ECMWF forecast model, ECMWF Tech. Memo 280. Available from ECMWF.

Hortal, M., and A. J. Simmons (1991). Use of reduced Gaussian grids in spectral models. *Mon. Wea. Rev.* **119,** 1057–1074.

Janssen, P. A. E. M., B. Hansen, and J.-R. Bidlot (1997). Verification of the ECMWF Wave Forecasting System against buoy and altimeter data. *Wea. Forecasting* **12,** 763–784.

Kelly, G., E. Andersson, A. Hollingsworth, P. Lonnberg, J. Pailleux, and Z. Zhang (1991). Quality control of operational physical retrievals of satellite sounding data. *Mon. Wea. Rev.* **119,** 1866–1880.

Klinker, E., and P. D. Sardeshmukh (1991). The diagnosis of mechanical dissipation in the atmosphere from large-scale balance requirements. *J. Atmos. Sci.* **49,** 608–627.

Klinker, E., F. Rabier, G. Kelly, and J.-F. Mahfouf (1999). The ECMWF operational implementation of four dimensional variational assimilation, Part III: Experimental results and diagnostics with operational configuration, ECMWF Tech. Memo 273. Available from ECMWF.

Lönnberg, P. (1988). Developments in the ECMWF analysis system. *In* "ECMWF Seminar Proceedings—Data Assimilation and the Use of Satellite Data, Vol. I, pp. 75–119. Available from ECMWF.

Lorenc, A. C. (1981). A global three-dimensional multivariate statistical interpretation scheme. *Mon. Wea. Rev.* **109,** 701–721.

Lott, F., and M. J. Miller (1997). A new sub-grid scale orographic drag parametrization: Its formulation and testing. *Quart. J. Roy. Meteor. Soc.* **123,** 101–127.

Louis, J.-F. (1979). A parametric model of vertical eddy fluxes in the atmosphere. *Boundary-layer Meteor.* **17,** 187–202.

Mahfouf, J.-F., and F. Rabier (1999). The ECMWF operational implementation of four dimensional variational assimilation, Part II: Experimental results with improved physics, ECMWF Tech. Memo 272. Available from ECMWF.

Miller, M. J., A. C. M. Beljaars, and T. N. Palmer (1992). The sensitivity of the ECMWF model to the parametrization of evaporation from tropical oceans. *J. Climate,* **5,** 418–434.

Molteni, F., R. Buizza, T. N. Palmer, and T. Petroliagis (1996). The ECMWF Ensemble Prediction System: Methodology and validation. *Quart. J. Roy. Meteor. Soc.* **122,** 73–120.

Morcrette, J.-J. (1990). Impact of changes to the radiation transfer parameterizations plus cloud optical properties in the ECMWF model. *Mon. Wea. Rev.* **118,** 847–873.

Morcrette, J.-J. (1991). Radiation and cloud radiative properties in the European Centre For Medium-Range Weather Forecasts forecasting system. *J. Geophys. Res.* **96,** 9121–9132.

Palmer, T. N., and D. L. T. Anderson (1994). The prospects for seasonal forecasting. *Quart. J. Roy. Meteor. Soc.* **120,** 755–793.

Palmer, T. N., C. Brankovic, F. Molteni, S. Tibaldi, L. Ferranti, A. Hollingsworth, U. Cubasch, and E. Klinker (1990). The ECMWF programme on extended-range prediction. *Bull. Am. Meteor. Soc.* **71,** 1317–1330.

Rabier, F., E. Klinker, P. Courtier, and A. Hollingsworth (1996). Sensitivity of two-day forecast errors over the Northern hemisphere to initial conditions. *Quart. J. Roy. Meteor. Soc.* **122,** 121–150.

Rabier, F., A. McNally, E. Andersson, P. Courtier, P. Undén, J. Eyre, A. Hollingsworth, and F. Bouttier, 1998: The ECMWF implementation of three dimensional variational assimilation (3D-Var). Part II: Structure functions. *Quart. J. Roy. Meteor. Soc.* **124,** 1809–1830.

Rabier, F., H. Jarvinen, E. Klinker, J.-F. Mahfouf, and A. Simmons (1999). The ECMWF operational implementation of four dimensional variational assimilation, Part I: Experimental results with simplified physics, ECMWF Tech. Memo 271. Available from ECMWF.

Ritchie, H., C. Temperton, A. Simmons, M. Hortal, T. Davies, D. Dent, and M. Hamrud (1995). Implementation of the Semi-Lagrangian method in a high resolution version of the ECMWF forecast model. *Mon. Wea. Rev.* **123,** 489–514.

Shaw, D. B., P. Lönnberg, A. Hollingsworth, and P. Undén (1987). Data assimilation: The 1984/85 revision of the ECMWF mass and wind analysis. *Quart. J. Roy. Meteor. Soc.* **113,** 533–566.

Simmons, A. J., and D. M. Burridge (1981). An energy and angular momentum conserving vertical finite-difference scheme and hybrid vertical coordinates. *Mon. Wea. Rev.* **109,** 758–766.

Simmons, A. J., and C. Temperton (1996). Stability of a two-time-level semi-implicit integration scheme for gravity-wave motion. *Mon. Wea. Rev.* **125,** 600–615.

Stockdale, T. N., D. L. T. Anderson, J. O. S. Alves, and M. A. Balmaseda (1998). Global seasonal rainfall forecasts using a coupled ocean–atmosphere model. *Nature* **392**(6674), 370–373.

Stoffelen, A., and D. Anderson (1997a). Scatterometer data interpretation: Measurement space and inversion. *J. Atmos. Ocean. Tech.* **14**, 1298–1313.

Stoffelen, A., and D. Anderson (1997b). Scatterometer data interpretation: Estimation and validation of the transfer function CMOD-4. *J. Geophys. Res. Oceans* **102**, 5767–5780.

Stoffelen, A., and D. Anderson (1997c). Ambiguity removal and assimilation of scatterometer data. *Quart. J. Roy. Meteor. Soc.* **123**, 491–518.

Temperton, C. (1997). Treatment of the Coriolis terms in semi-Lagrangian spectral models. *In* "Numerical Methods in Atmospheric and Oceanic Modeling. The André Robert Memorial Volume" (C. Lin, R. Laprise and H. Ritchie, eds.), pp. 293–302. Canadian Meteor. and Ocean. Soc., Ottawa, Canada.

Temperton, C., and D. L. Williamson (1981). Normal mode initialization for a multi-level gridpoint model. Part I: Linear aspects. *Mon. Wea. Rev.* **109**, 729–743.

Tiedtke, M. (1984). The effect of penetrative cumulus convection on the large scale flow in the general circulation model. *Beitr. Physics Atmos.* **57**, 216–239.

Tiedtke, M. (1989). A comprehensive mass flux scheme for cumulus parametrization in large-scale models. *Mon. Wea. Rev.* **117**, 1779–1800.

Tiedtke, M. (1993). Representation of clouds in large-scale models. *Mon. Wea. Rev.* **121**, 3040–3061.

Undén, P. (1989). Tropical data assimilation and analysis of divergence. *Mon. Wea. Rev.* **117**, 2495–2517.

Viterbo, P., A. Beljaars, J.-F. Mahfouf, and J. Teixeira (1998). The representation of soil moisture freezing and its impact on the stable boundary layer, ECMWF Tech Memo 255. Available from ECMWF.

Wallace, J. M., S. Tibaldi, and A. J. Simmons (1983). Reduction of systematic forecast errors in the ECMWF model through the introduction of an envelope orography. *Quart. J. Roy. Meteor. Soc.* **109**, 683–717.

Wergen, W. (1988). The diabatic ECMWF normal mode initialization scheme. *Beitr. Physics Atmosph.* **61**, 274–304.

Williamson, D. L., and C. Temperton (1981). Normal mode initialization for a multi-level gridpoint model. Part II: Nonlinear aspects. *Mon. Wea. Rev.* **109**, 744–757.

Chapter 12

Climate Services at the Japan Meteorological Agency Using a General Circulation Model: Dynamical One-Month Prediction

Tatsushi Tokioka

Japan Meteorological Agency, 1-3-4 Otemachi, Chiyoda-ku, Tokyo, Japan

I. INTRODUCTION

Many national meteorological centers/agencies are using or are considering using dynamical models in the climate services not only for data assimilation of climate systems, but also for short-term climate predictions up to an interannual time scale ahead. Interaction of atmosphere with the underlying surface, which has a longer relaxation time than the atmosphere, and predictability of surface variability to some extent are the scientific bases for short-term climate predictions. Recent efforts of national meteorological centers/agencies in this direction are summarized in the Long-Range Forecasting Progress Report for 1995/1996 (WMO, 1997). A coupled atmosphere ocean general circulation model (CGCM) has been used to predict El Niño phenomena fairly successfully up to a year ahead. Experimental results including those from many research groups have been published periodically as the Experimental Long-Lead Forecast Bulletin by the Climate Prediction Center (CPC) of the NCEP until December 1997, and by the Center for Ocean-Land-Atmosphere Studies (COLA) thereafter.

A predictability study on seasonal forecast has also been organized internationally as the Seasonal Prediction Model Intercomparison Project (SMIP) by CLIVAR NEG1 (Numerical Experimentation Group 1 of Climate Variability and Predictability Study Project). SMIP has proposed studies on predictability of atmosphere when sea surface temperature (SST) is perfectly predicted with the use of atmospheric general circulation models (AGCMs). The results so far show dependence of predictability on seasons and areas, with relatively high predictability in northern winter and spring when El Niño is prevailing and in the area where influence of El Niño extends (for example, see WMO, 1997; Bengtsson *et al.*, 1996; Palmer and Anderson, 1994).

At the Japan Meteorological Agency (JMA), both experimental prediction of El Niño phenomena with a CGCM and a predictability study of 3-month predictions with an AGCM are continuing and show promising results. Besides them, 1-month prediction has been operationally continued every week since March 1996 with the use of an AGCM developed at the JMA as the shortest range seasonal prediction. One month is the time range influenced not only by initial conditions but also by lower boundary conditions, such as SST anomaly, anomalous snow coverage, soil moisture, and so on. Interactions between the atmosphere and the lower boundary must occur even in this time range. However, currently such interactions are neglected over the ocean, i.e., SST is prescribed with the assumption that SST anomaly is constant throughout the prediction interval.

In this chapter, I briefly introduce JMA's dynamical 1-month prediction services and give information about the current skill and how JMA's AGCM is applicable to such a purpose. Experience obtained through such operational services is useful for the improvement of AGCMs as operational weather forecast has been.

II. PROCEDURE OF ONE-MONTH PREDICTION

A. Outline of the Model

The resolution of AGCM adopted is T63 with 30 vertical levels (T63L30). The top of the model is located at 1 hPa. SST anomaly is fixed to the initial analysis during the time integration and is added to the climatological SST. Ground surface conditions are calculated with a ground thermodynamic model combined with a simplified biosphere model (SiB) developed by Sellers *et al.* (1986), coded by Sato *et al.* (1989). The Arakawa–Schubert model (Arakawa and Schubert, 1974), as implemented by Randall and Pan (1993), is adopted to parameterize penetrative cumulus convection. Radiation is based on Lacis and Hansen (1974) with modifications. The model includes most other physical processes, such as boundary layer turbulence (Louis *et al.*, 1982), gravity wave drag (Iwasaki *et al.*, 1989), sub-grid-scale diffusion, and clouds.

Performance of the previous version of this model, in which the prognostic Arakawa–Schubert scheme is replaced by the Kuo scheme, is reported by Sugi *et al.* (1990) and Gadgil *et al.* (1997). Comparisons between the present and the previous versions were made by Kar and Sugi (private communication, 1996). They show that precipitation of the current version in the area and to the east of the Philippines both in summer and winter and over Brazil in boreal winter has increased substantially so that model precipitation agrees better with the observed analysis than that in the previous version.

B. Ensemble Prediction of Time-Averaged Fields

Current observation includes finite errors that might grow substantially in the course of time integration up to 1 month. Even when both observation and observational analysis are perfect, the model used for prediction is not perfect. Therefore, the predicted field departs from the real one sooner or later, and the difference eventually grows with time. The ensemble method is adopted to minimize the random part of these errors in the predicted fields statistically. One month well exceeds the deterministic range of prediction due to the chaotic nature of atmospheric flows. Therefore, a running mean for 7 days at least is applied to the model products, along with ensemble averaging, to smooth out such unpredictable parts in a simple way.

Ten ensembles are incorporated and calculated as follows. Five members are integrated from the observed analysis 2 days before the date of forecast. One member is started from the analysis itself without any initial perturbations, and four other members with perturbations determined by a singular vector (optimum mode) method (see Molteni and Palmer, 1993; Mureau *et al.*, 1993) modified by Tsuyuki (private communication, 1993). Another five members are prepared in the same way as just stated, but from the analysis 1 day before the date of forecast. Therefore, our method is a singular vector method combined with the lagged averaged forecast method (see Hoffman and Kalney, 1983).

The amplitude of perturbation is determined so that it is comparable to the probable observational errors. Currently, the horizontally averaged root mean square amplitude of perturbation wind at 200 hPa is adjusted to 2 m/s in winter and 1 m/s in summer with linear interpolation between.

C. PROBABILISTIC PREDICTION

From the ensemble prediction stated in Section II.B, we have information about ensemble average and deviation of each member from the average, from which we can derive some probabilistic information about forecast variables. Part of the final products disseminated as forecast are probabilities of occurrence of three categories—below normal, normal, and above normal—of monthly mean surface temperature, precipitation, and sunshine hours, for four prediction areas that cover Japan (Fig. 1). Three categories are defined in such a way as to have probabilities of occurrence of 30, 40, and 30%, respectively, for below normal, normal, and above normal categories based on observed data from 1961 to 1990. Currently, probabilities are rounded to numbers of 10 figures.

D. CORRECTION OF SYSTEMATIC MODEL BIAS

The current model has non-negligible systematic model biases. Model biases in the monthly mean fields are subtracted from the predicted fields currently. This process improves prediction.

III. SKILL OF ONE-MONTH PREDICTION

A. EXAMPLE OF ENSEMBLE PREDICTION

Figure 2 shows one example of predicting an 850-hPa temperature anomaly in eastern Japan (see Fig. 1) from the initial date of February 6, 1997. Predicted value is objectively determined with the use of statistical relations between the target quantity and the neighboring grid point values based on the perfect prognosis method. Even single prediction is quite

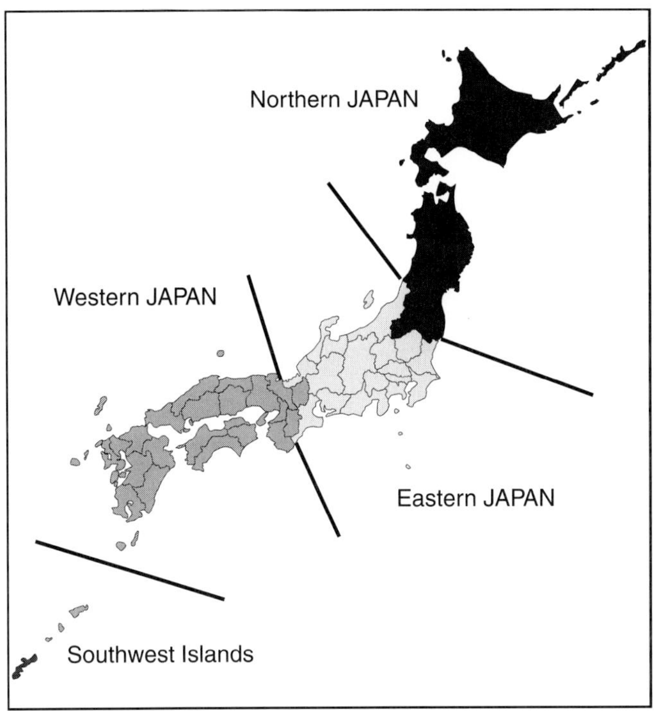

Figure 1 Forecast area division of the JMA's seasonal forecast services. One-month forecasts are released for these four areas every week on surface temperature, precipitation and sunshine hours.

good in the first half period in this case. In the latter half, the difference between members grows with time, but the ensemble average surely gives better positive predictable skill than single prediction. Figure 2 demonstrates graphically how ensemble averaging is effective for long-term predictions.

Figure 3 is an example of predicted monthly mean surface temperature anomaly (lightly shaded line) and the corresponding observed value (dark line) for northern Japan (see Fig. 1) since the beginning of operational prediction in March 1996. The agreement between the prediction and the observation is fairly good. However, in the southwest island, where the climate is of subtropical nature, the agreement is less satisfactory than that in northern Japan, which is located in the midlatitude baroclinic zone.

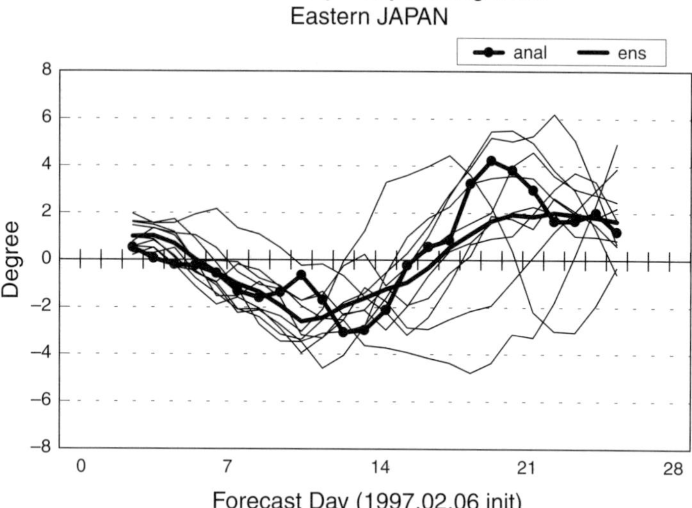

Figure 2 Example of predicting an 850-hPa temperature anomaly with 7-day running mean for eastern Japan (see Fig. 1) from the initial date of February 6, 1997. The abscissa is day from the initial and the ordinate is the 850-hPa temperature anomaly. The thin line is the anomaly of each run, the lightly shaded thick line the ensemble average of 10 members, and the thick solid line the analysis.

B. Meaning of Time Integration of the Latter Half Period of a Month

The anomaly correlation coefficient (ACC) of daily mean field between observed analysis and prediction falls rapidly with time. The ACC of daily mean 500-hPa geopotential height, for example, is less than 0.4 in the latter half of a month. It is natural, therefore, to raise the question of whether the time integration for that period is really meaningful or not for estimating monthly mean anomalies. Takano *et al.* (2000) gave an answer to this question. Figure 4a shows Northern Hemisphere (poleward of 20°N; NH) ACCs of monthly mean 500-hPa geopotential height of prediction and analysis. The white column shows ACC where full 28-day prediction is used to predict monthly mean fields, whereas the lightly and heavily shaded columns show ACCs where only the first 7-day and the 14-day predictions, respectively, are used for that purpose. Figure 4a shows that ensemble mean prediction of full 28-day integration is better than any other cases on a yearly basis and that the time integration of the latter half

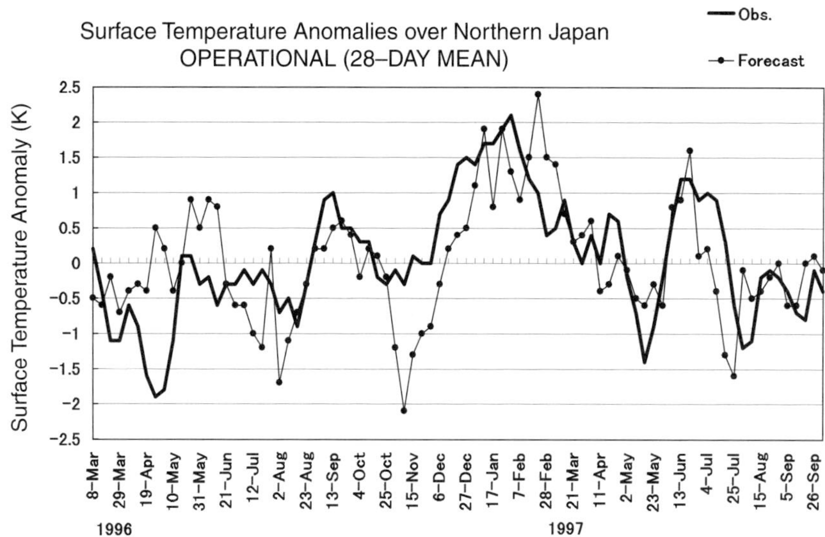

Figure 3 Example of predicted monthly mean surface temperature anomaly (lightly shaded line) and the corresponding analysis (dark line) for northern Japan (see Fig. 1) since the beginning of operational prediction. The abscissa is the initial date of prediction.

month is meaningful for 1-month predictions. Autumn is the only exceptional season when the full 28-day prediction does not necessarily give better results than the other two.

C. EFFECT OF ENSEMBLE AVERAGING

Figure 4b shows the same data as Fig. 4a but for a single prediction (Takano *et al.*, 2000). This shows that the first 7-day integration gives the best ACC among the three cases, but the value is almost the same as that of the ensemble prediction. This is because departure of each member from the ensemble average is still not large in the first 7-day integration (see Fig. 2). On the other hand, full 28-day integration gives the smallest ACC in single prediction due to the growth of random errors with time in a single prediction. We may conclude that ensemble averaging in full 28-day integration is the most effective among the three through an effective cancellation of errors in ensemble averaging (Takano *et al.*, 2000).

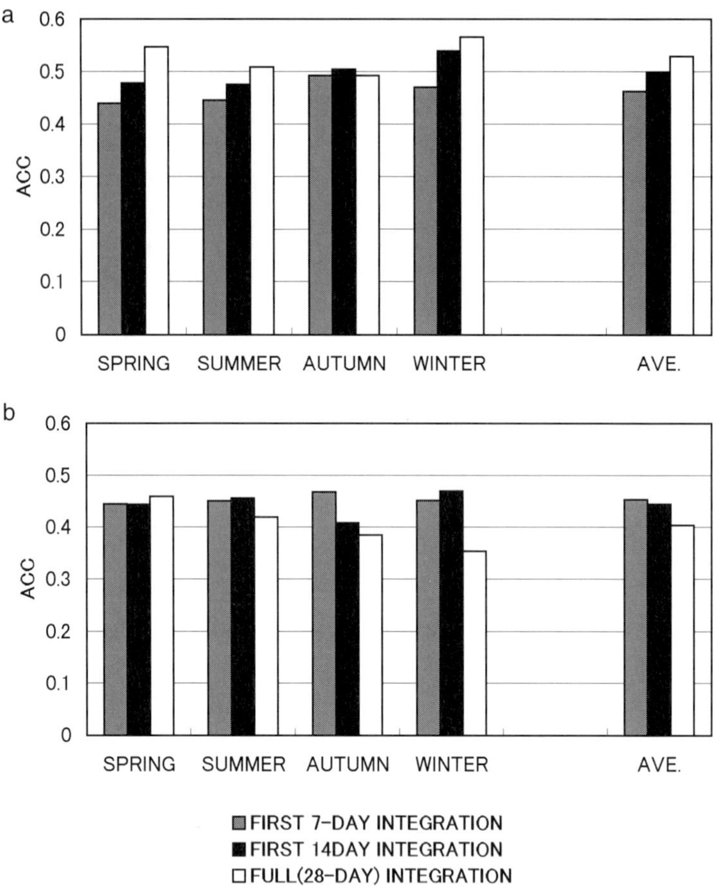

Figure 4   ACC of 500-hPa geopotential height in the Northern Hemisphere (20°N ~ North Pole, NH) between prediction and analysis for each season. White column is ACC where full 28-day prediction is used to estimate monthly anomalies, whereas the lightly and heavily shaded ones are those where the first 7-day and 14-day predictions, respectively, are used. (a) ACC when ensemble average of 10 runs is used. (b) ACC when single run is used. (From Takano *et al.*, 2000.)

D. Ensemble Size

Figure 5 shows how NH ACCs change with ensemble size and season. With the exception of autumn, ACCs increase with the increase of ensemble size up to 10 (Takano *et al.*, 2000). Better ACCs are expected for further increases in size, especially in winter and spring. Note that increas-

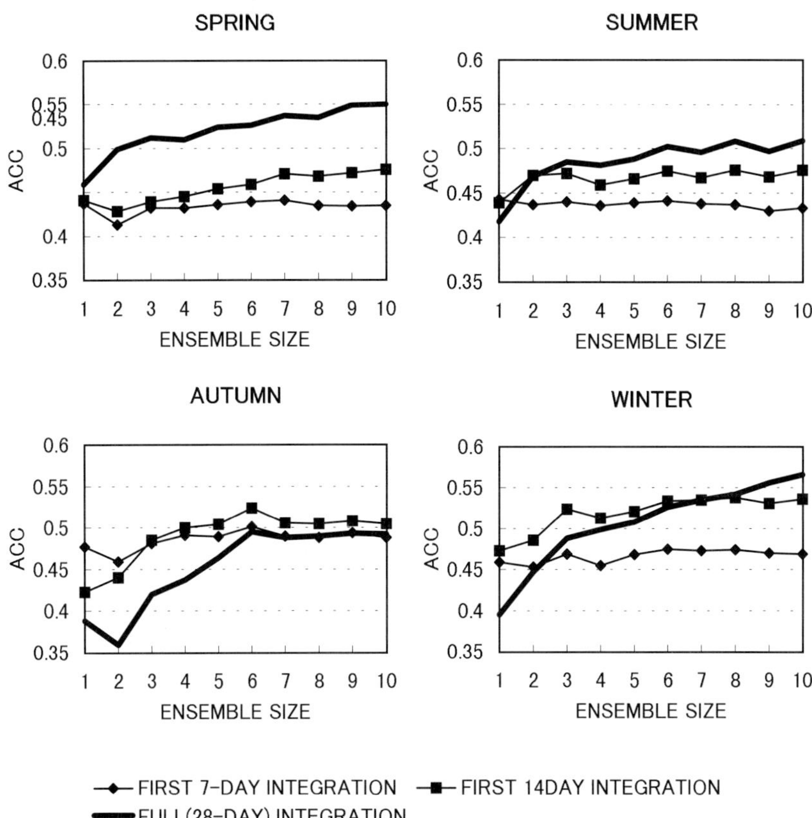

Figure 5 Dependence of NH ACC on the ensemble size for each season. ACC is calculated for monthly mean 500-hPa geopotential height. (From Takano *et al.*, 2000.)

ing trend of ACC with size in autumn has leveled off at 6 already and that ACC based on the first 14-day integration is larger than that of full 28-day integration (see Fig. 4a also). The reason for this is not clear. This might not be indifferent to the current low prediction skill in autumn, which is shown in Section III.E.

E. ACC OF GEOPOTENTIAL HEIGHT AT 500 HPA

Figure 6 shows monthly mean NH ACC of 500-hPa geopotential height between ensemble mean prediction and analysis since the beginning of operational prediction. Thick dark line shows running averaged ACC over

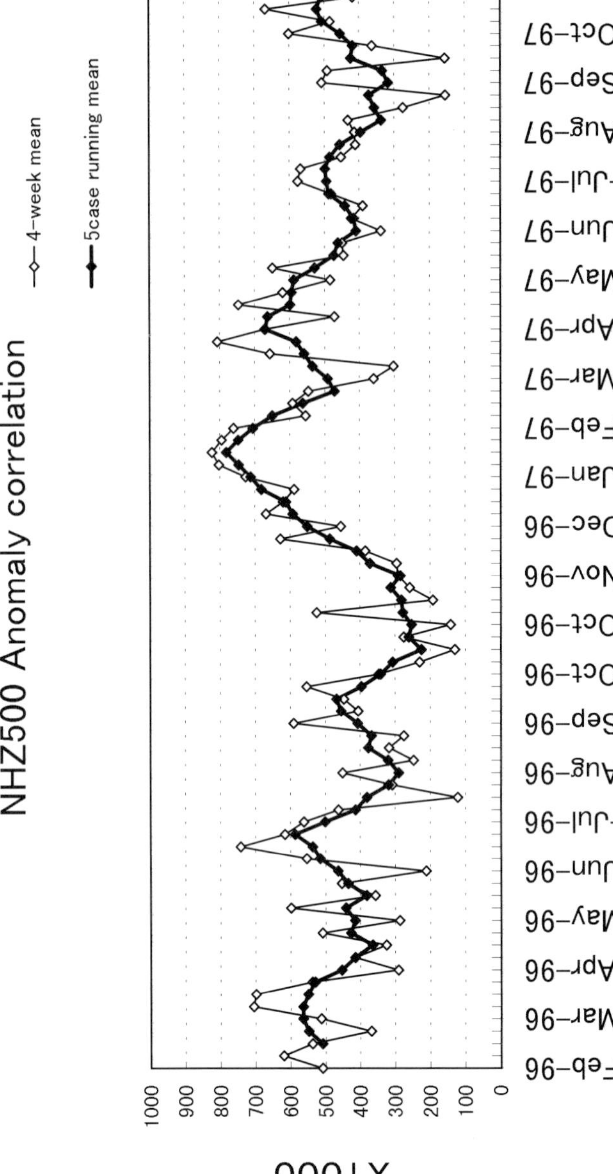

Figure 6 NH ACC of monthly mean 500-hPa geopotential height between ensemble mean prediction and analysis since the beginning of operational prediction. Abscissa is the prediction date. Thick line is running mean ACC over five consecutive predictions. (From Takano *et al.*, 2000.)

five consecutive predictions. Although the data cover only a year and a half, we can recognize seasonal dependence of ACC clearly. It is relatively high in winter and spring, while it is low in summer and autumn. The highest value 0.87 is attained in the prediction from November 21, 1997. On the other hand, low values around 0.2 appear several times in summer and autumn when overall skill of prediction is low.

Another feature noted in Fig. 6 is relatively short period fluctuations of ACC from 2 to 4 weeks. Part of these fluctuations is attributed to the relative time interval between the date of forecast and the timing of blocking development. This process is confirmed in several cases, but has not been surveyed in all cases throughout the entire operational period yet. Currently, JMA's AGCM is capable of predicting a blocking if prediction is made from the date less than 3–7 days prior to the occurrence of blocking. If a model failed to predict a blocking at a relatively early stage of the forecast period, the ACC of 500-hPa height tends to be low. On the other hand, if a blocking were either predicted successfully or absent in the prediction period, the ACC could be high.

Figure 7 is similar to Fig. 6, but for NH ACCs calculated for the first week, the second week, and the third plus fourth weeks (the latter half period of a month). ACC of the first week is around 0.8 and is higher than those of the second week and the latter half period, as expected. The time variation of the first week ACC is less than those of the other two. If we compare Figs. 6 and 7, the variation of the running mean ACC for the latter half period in Fig. 7 closely follows that in Fig. 6. In other words, the ACC of the monthly mean 500-hPa geopotential height is highly correlated with that of the latter half period. The correlation coefficient between them is 0.76, whereas the correlation with that of the first half period of the month is 0.59 (Takano *et al.*, 2000).

F. RELATIONSHIP BETWEEN ACC AND SPREAD

Spread is defined as the standard deviation of 10 members from the ensemble average normalized by the climatological standard deviation for monthly mean 500-hPa height. When spread is large, the concentrated appearance of members into one of three categories may not be likely to occur, i.e., high probability may not be assigned to one particular category in such a case. If this is true, a large/small spread may go with a low/high ACC of monthly mean 500-hPa height. Operational forecasts so far show that the relationship between NH ACC for the monthly mean 500-hPa height and spread is seasonally dependent and that it is relatively good in

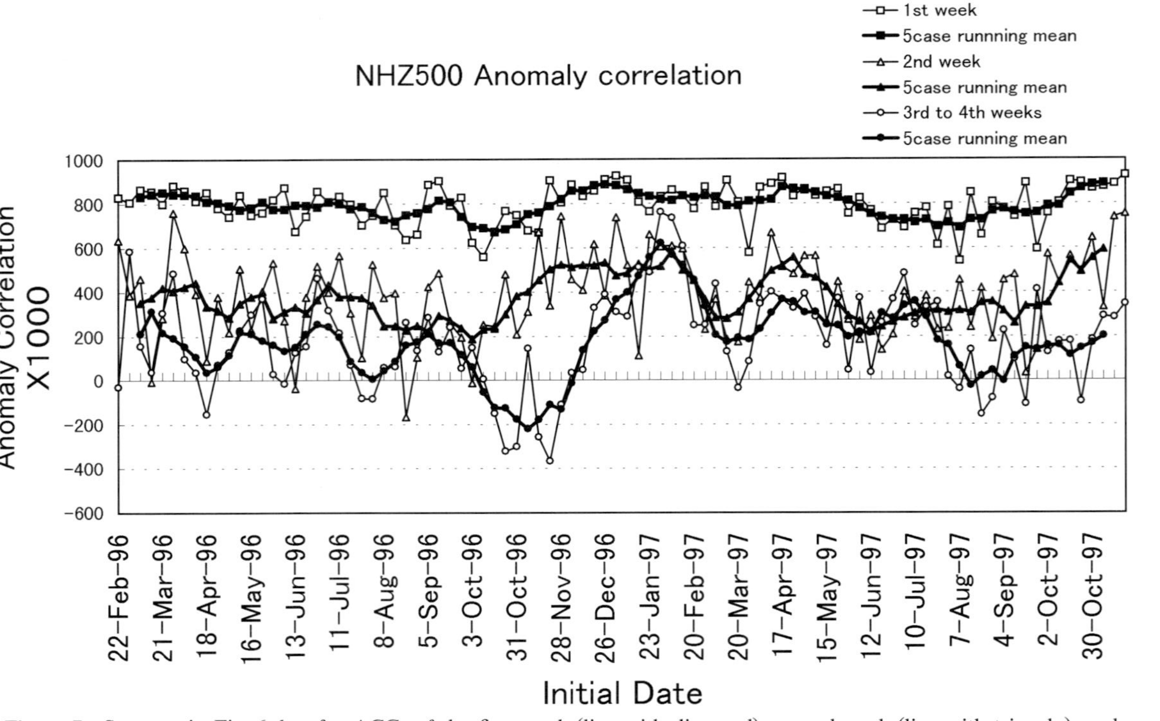

Figure 7   Same as in Fig. 6, but for ACCs of the first week (line with diamond), second week (line with triangle), and third plus fourth weeks (line with asterisk). Thick lines are running mean ACCs over five consecutive predictions. (From Takano *et al.*, 2000.)

northern winter. However, the ACC between them is about 0.5 and is not so surprisingly high.

G. SKILL OF FORECAST

One way of quantifying skill of forecast is to calculate the rate of actual occurrence of the predicted highest probability category (rate of agreement). If we predict randomly, the rate of agreement is expected to be 34% as we have categorized it, as explained in Section II.C. If we continue to predict the "normal" category, i.e., climatological prediction, the agreement rate is expected to be 40% in the long run. The rate of agreement for monthly mean surface temperature since the beginning of the forecast is about 47%, and exceeds those of the random and the climatological predictions. However, the rates of agreement for monthly mean precipitation and sunshine hours are 41 and 42%, respectively.

Verification of the predicted probability of each category in the 1-month forecast is not easy. One way of assessing this is a statistical method. Figure 8 shows such results based on operational forecasts over a year and a half, where the abscissa is predicted probability, the numbers at the top of columns are total numbers of predicting the probability, and the ordinate is the ratio of observed occurrence. This is called a *reliability diagram*. If the ratio of observed occurrence is equal to the predicted probability, i.e., the columns line up on the auxiliary line in the figure, we

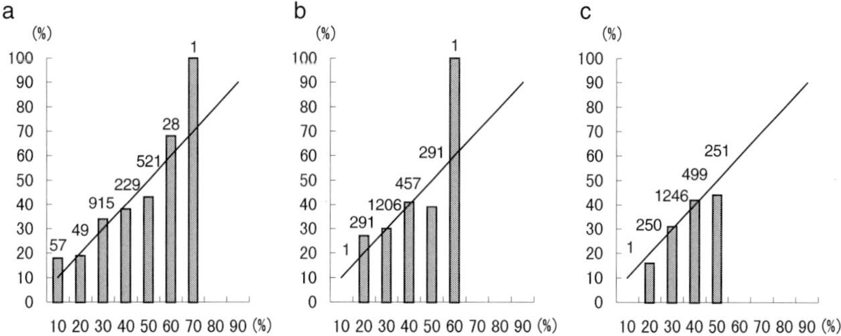

Figure 8 Reliability diagrams based on operational forecasts since March 1996. Abscissa is predicted probability, number at the top of column is total number of prediction of the probability, and ordinate is the ratio of observed occurrence. (a) Surface temperature, (b) precipitation, and (c) sunshine hours. Auxiliary line is the line when predicted probability is equal to the observed occurrence rate.

may say the predicted probability of each category is reasonable and reliable on a statistical basis. Figures 8a, 8b, and 8c are reliability diagrams for surface temperature, precipitation and sunshine hours, respectively. Results are promising. In particular, we might say the probability assigned for surface temperature is quite reasonable.

IV. FUTURE IMPROVEMENTS

JMA's current AGCM has systematic model biases. They are subtracted from the model output in the current 1-month prediction. Although this process substantially improves the prediction of monthly mean anomaly fields, it goes without saying that the reduction of model biases through the improvement of model physics and model dynamics including resolution, treatment of upper boundary conditions and others would be the way to head. Iwasaki and Kitagawa (1998) recognized not small systematic errors in the surface short-wave radiation in the present AGCM and identified them due to unsatisfactory treatment of clouds and complete neglect of aerosol effect on solar radiation. They improved those processes, studied their impacts in June, and showed that errors found over the Tibetan plateau diminish substantially. Iwasaki and Kitagawa have shown that the improvement spreads further over the globe through dynamical linkages.

The current prediction system has another unsatisfactory aspect, i.e., the initial condition of the land surface is always the climatological condition. This certainly must have non-negligible negative impacts on the model's skill of prediction. This is to be improved as soon as the data assimilation system of the land surface including snow is ready for operational use.

As was shown in Fig. 6, in some cases, monthly mean geopotential height anomaly already has a surprisingly high correlation (above 0.8 for example) with the observed counterpart. This is very encouraging for the future of dynamical 1-month prediction. On the other hand, there are several cases where the correlation is less than 0.2. Statistical study shows that when the ACC of the monthly mean field is low, the ACC of the latter half period of the month is generally low. This leads to a reasonable but plain conclusion that we have to improve model performance especially in the latter half period of the month. To this end, air–sea interaction, which is completely neglected currently, might also be important in addition to the points stated already.

We noted relatively rapid fluctuations in the ACC for predicted 500-hPa height from 2 to 4 weeks. Although full mechanisms of the fluctuations have not been identified yet, a part of them is related to the time interval between the date of forecast and the timing of occurrence of blocking. To prevent low ACC cases of this kind, we have to improve a model so that it has better predictability skills for blockings. Horizontal resolution and parameterization of sub-grid-scale topography might have to be improved for this purpose.

The prediction for summer 1997 in Japan was influenced by passages of several typhoons. In particular, the climate in the latter half of August around Japan was hot and humid due to a warm and humid air mass brought by the passage of typhoon 13 (Winnie). The current AGCM with a resolution of T63 is unable to simulate a typhoon well in its birth, movement, and growth, although T103 version has some skill as demonstrated by Sugi and Sato (1995). Even one intensive typhoon that comes close to Japan influences monthly mean fields. It would be necessary for us to increase the model's horizontal resolution, at least, to improve 1-month forecasts in such situations.

The ensemble size is currently 10. This certainly would have to be increased as discussed in Section III.D. The relationship between the spread and the NHACC for the 500-hPa height might be improved with an increase in the ensemble size, as the occurrence of blocking, for example, might be captured well probabilistically.

We have accumulated operational prediction data over a year and a half. Some information is routinely extracted from that data; however, detailed case studies have not been started yet. Such efforts, especially for cases where we had poor model performance, are indispensable to an understanding of the causes of them and to identify points to be improved in the current AGCM. Input from the predictability study of seasonal time scale with an AGCM and with observed SST and from El Niño experimental prediction with a CGCM, both of which are currently being carried out at the JMA, is also useful to recognize unsatisfactory aspects of models in general sense.

To improve 1-month forecast services, another important point is to improve methods of extracting useful information from the model products. I do not go into detail here about this problem, but merely mention that there is a lot of room for improvement in current guidance programs and that we could improve our services substantially if we could devise a new method of differentiating cases of high ACC from those of low ACC (see Figs. 6 and 7) at the stage of prediction. Spread of ensembles is used currently, but this is not always a good measure for that purpose. The analysis of operational forecast data accumulated so far is expected to provide important clues for finding new alternative methods.

ACKNOWLEDGMENTS

The author acknowledges Dr. K. Takano, Mr. K. Yagi, and Ms. C. Kobayashi of the Climate Prediction Division of the JMA and Mr. M. Sugi of the Meteorological Research Institute for providing unpublished information and figures.

REFERENCES

Arakawa, A., and W. H. Schubert (1974). Interaction of a cumulus cloud ensemble with the large-scale environment. Part I. *J. Atmos. Sci.* **31,** 674–701.

Bengtsson, L., K. Arpe, E. Roeckner, and U. Schulzweida (1996). Climate predictability experiments with a general circulation model. *Clim. Dyn.* **12,** 261–278.

Gadgil, S., S. Sajani, and Participating Modelling Groups of AMIP (1997). Monsoon precipitation in the AMIP runs. Centre for Atmospheric and Oceanic Sciences, Indian Institute of Science, Bangalore, India.

Hoffman, P., and E. Kalney (1983). Lagged average forecasting, an alternative to Monte Carlo forecasting. *Tellus* **35A,** 100–118.

Iwasaki, T., and H. Kitagawa (1998). A possible link of aerosol and cloud radiation to Asian summer monsoon and its implication in long-range numerical weather prediction. *J. Meteor. Soc. Japan,* **76,** 965–982.

Iwasaki, T., S. Yamada, and K. Tada (1989). A parameterization scheme of orographic gravity wave drag with the different vertical partitioning, Part I: Impact on medium range forecasts. *J. Meteor. Soc. Japan* **67,** 11–27.

Lacis, A. A., and J. E. Hansen (1974). A parameterization for the absorption of solar radiation in the earth's atmosphere. *J. Atmos. Sci.* **31,** 118–133.

Louis, J., M. Tiedtke, and J.-F. Geleyn (1982). A short history of PBL parameterization at ECMWF. *In* "Workshop on Planetary Boundary Layer Parameterization," pp. 59–80. ECMWF.

Molteni, F., and T. N. Palmer (1993). Predictability and finite-time instability of the northern winter circulation. *Quart. J. Roy. Meteor. Soc.* **119,** 269–298.

Mureau, R., F. Molteni, and T. N. Palmer (1993). Ensemble prediction using dynamically conditioned perturbations. *Quart. J. Roy. Meteor. Soc.* **119,** 299–322.

Palmer, T. N., and D. L. T. Anderson (1994). The prospects for seasonal forecasting—A review paper. *Quart. J. Roy. Meteor. Soc.* **120,** 755–793.

Randall, D., and D.-M. Pan (1993). Implementation of the Arakawa–Schubert cumulus parameterization with a prognostic closure. *In* "The Representation of Cumulus Convection in Numerical Models" (K. A. Emanuel and D. J. Raymond, eds.), Meteorological Monographs, Vol. 24, pp. 137–144. American Met. Soc.

Sato, N., P. J. Sellers, D. Randall, E. Schneider, J. Shukla, J. Kinter, Y.-T. Hou, and E. Albertazzi (1989). Effects of implementing the simple biosphere model in a general circulation model. *J. Atmos. Sci.* **46,** 2757–2782.

Sellers, P. J., Y. Mintz, Y. C. Sud, and A. Dalcher (1986). A simplified biosphere model(SiB) for use within general circulation models. *J. Atmos. Sci.* **43,** 505–531.

Sugi, M., and N. Sato (1995). A ten year integration of the T106 JMA Global Model. *In* "Proceedings of the First International AMIP Scientific Conference," Monterey, California, May 15–19, 1995. *WMO / TD* No. 732, pp. 505–510.

Sugi, M., K. Kuma, K. Tada, K. Tamiya, N. Hasegawa, T. Iwasaki, S. Yamada, and T. Kitade (1990). Description and performance of the JMA operational global spectral model (JMA-GSM89). *Geophys. Mag.*, **43,** 105–130.

Takano, K., K. Yoshimatsu, C. Kobayashi, and S. Maeda (2000). On the forecast skill of ensemble one-month forecast. *J. Meteor. Soc. Japan* **77,** in press.

WMO (1997). Long-range forecasting progress report for 1995/1996, WMO Technical Document No. 800, LRFP report series No. 3.

Chapter 13

Numerical Methods: The Arakawa Approach, Horizontal Grid, Global, and Limited-Area Modeling

Fedor Mesinger

NOAA Environmental Modeling Center / UCAR Visiting Scientist Program
Camp Springs, Maryland

I. INTRODUCTION: THE ARAKAWA APPROACH IN NUMERICAL METHODS

It is perhaps a remarkable characteristic of atmospheric numerical modeling that in spite of the steady progress during the past more than four decades the diversity of points of view on what are the most promising

General Circulation Model Development

principles to follow shows little sign of diminishing. Within these points of view, I find it fitting to refer to the Arakawa approach in numerical modeling as the one in which attention is focused on the realism of the physical properties of the discrete system within given computational resources. In other words, with the Arakawa approach one is not relying on these properties to automatically become satisfactory as the resolution is increasing, merely as a result of the observation of basic requirements of computational mathematics. Instead, one is striving to achieve properties deemed desirable with the resolution at hand. This is achieved by consideration of the physical properties of the finite difference analog of the continuous equations.

With this formulation, there is clearly some room left for searching as to what exactly are the physical properties to which attention is best paid, and to what should be the priorities among various possibilities. Historically, the incentive for the approach came from Norman Phillips's (1959) discovery of the mechanism of nonlinear instability as consisting of a systematic distortion of the energy spectrum of two-dimensional nondivergent flow. A straightforward remedy used by Phillips was one of Fourier filtering aimed at preventing the fatal accumulation of energy in shortest scales. Akio Arakawa, however, realized that the maintenance of the difference analogs of domain-averaged kinetic energy and enstrophy guarantees no change in the average wave number, thus preventing nonlinear instability with no damping in the terms addressed; and demonstrated a way to achieve this with his famous (Arakawa, 1966) horizontal advection scheme. (For additional historic comments see, e.g., Lilly, 1997.) The Arakawa advection scheme and subsequent numerous conservation considerations as discussed in Arakawa and Lamb (1977, hereafter AL), for example, have established the maintenance of the difference analogs of chosen integral constraints of the continuous atmosphere as the hallmark of the approach. Yet, more generally, emphasis was placed by Arakawa, and by others, on reproducing numerous other properties of physical importance of the fluid dynamical system addressed. Dispersion and phase speed properties, avoidance of computational modes, and avoidance of false instabilities are the typical examples, as succinctly summarized in Section 7 of a recent review paper by Arakawa (1997) or, more extensively, in Arakawa (1988).

In striving to achieve goals of this type, no advantage tends to be obtained from increasing the order of the accuracy of the scheme. For example, as gently stated by Arakawa (1997) in summarizing the problem of the computational mode, "The concept of the order of accuracy . . . based on the Taylor expansion...is not relevant for the existence or non-existence of a computational mode." Similarly, Mesinger (1982; see also Mesinger and Janjić, 1985) demonstrated that an increase in resolution that entails an increase in the formal Taylor series accuracy does not necessarily help in achieving a physically desirable result and can even result in an increase of the actual error.

Underlying the Arakawa approach is the determination to understand the reason of a numerical problem—including those at the shortest represented scales—and try to address its cause as opposed to using artificial diffusion or filtering to alleviate its consequences and presumably lose some of the real information in the process. Yet, a different emphasis, or different views on what may be the best road to follow, are not hard to find among leading atmospheric modelers. For example, in a recent paper by Pielke *et al.* (1997) citing also supporting sources, one reads that "such short waves [wavelengths less than $4\Delta x$] are inadequately resolved on a computation grid and even in the linearized equations are poorly represented in terms of amplitude and/or phase. For these reasons, and because they are expected to cascade to even smaller scales anyway, it is desirable to remove these waves." In yet another recent paper (Gustafsson and McDonald, 1996), one reads that "Unwanted noise is generated in numerical weather prediction models, by the orography, by the boundaries, by the 'physics,' or even sometimes by the dynamics. The spectral approach provides two useful filters for attacking this problem *at no computational cost.* ...It was now necessary to write and test new filters for the gridpoint model if it was to continue to compete with the spectral model."

I will return to some of these issues in more detail later. For examples of physical properties that have been and can be considered in the Arakawa style I will start with a retrospective of the horizontal grid topic. This will permit me to review and also present some recent developments in this area. I then proceed with an exposition on the experience from the operational running of the Eta model at the U.S. National Centers for Environmental Prediction (NCEP), to the extent that it can be viewed as a contribution to the issues raised. A number of other global and limited-area modeling topics, having to do with the pole problem, the viability of the limited-area modeling approach, and the resolution versus domain size trade-off, are also discussed. Use will again be made of the Eta model results where appropriate. I conclude by illustrating the remarkable progress that has been accomplished in the atmospheric numerical modeling field during the past decade or so and by commenting on thrusts taking place or expected.

II. THE HORIZONTAL GRID: RETROSPECTIVE

Choice of the horizontal grid could well be considered the central point of the Arakawa-style considerations because numerous conservation and other issues and trade-offs are related to one choice or another. It is also the first problem that received attention at the beginning of the "primitive equation age" of atmospheric modeling in the late 1960s.

In a primitive equations framework, AL have argued that there are two main computational problems in the simulation of atmospheric motions: simulation of the geostrophic adjustment and simulation of the slowly changing quasi-geostrophic motion after it has been established by the geostrophic adjustment. As to the former, Winninghoff (1968) and AL have analyzed the dispersion properties of the simplest centered approximations to the shallow-water equations on square horizontal grids. Their results have subsequently been summarized and discussed at a number of places (e.g., Janjić and Mesinger, 1984), most recently probably by Randall (1994), and so are only briefly stated here. The desirable property of the relative frequency monotonically increasing with wave number is achieved only for some of the grids and for some values of λ/d, λ being the radius of deformation, $(gH)^{1/2}/f$, with symbols here having their usual meaning and d being the grid distance. The results for the nonstaggered grid, A, and the fully staggered grid, D, having u and v located so that they represent velocity components tangential to h-point grid boxes, turn out to be rather unfavorable. The fully staggered grid, C, having u and v located such that they represent components normal to h-point boxes, and the semi-staggered grids, B and E, having the two velocity components at the same points, look much better. To quote Randall (1994), "the C grid does well with deep, external modes but has serious problems with high internal modes, whereas the B grid has moderate problems with all modes."

Irrespective of how much one or the other type of modes is present in the real atmosphere, the problem of deep external modes seems quite important in primitive equation models. With "physics" performed in individual gridboxes of the model, note that 29% of the heat resulting from a physics call will instantaneously be converted into the gravitational potential energy of the column. If, in a thought experiment, we consider a single column having received heat in this way, it will as a result have an excess of potential energy relative to surrounding columns. This will generate outward-directed pressure gradient forces, which will initiate a geostrophic adjustment process. How satisfactorily a process is handled that is initiated by this fraction of heat supplied by a physics call should certainly be a subject of concern.

With B Grid, as pointed out by Arakawa (1972), it is the averaging of the pressure gradient force that causes trouble. With the E Grid representing the B Grid rotated by 45 degrees, the problem is the same except that the averaging is not explicitly performed; shallow-water pure gravity-inertia wave solutions on the two grids are in fact identical (Mesinger and Arakawa, 1976). The two grids will therefore be referred to as the B/E Grid hereafter when statements are made that are applicable to both of them.

The propagation of the pure shallow-water gravity wave is the source of the geostrophic adjustment difficulties of grids other than C. Consider the B/E Grid: It can be considered to consist of two C subgrids, shifted by the B/E Grid grid distance relative to each other. If now a single h point is perturbed, a pure gravity wave that is excited will propagate only along points of a single C subgrid to which the perturbed point belongs. Points of the other C subgrid, which include the four h points nearest to the perturbed point, will not be affected. This is the lattice separation problem of the B/E Grid. In a more complete system the four h points nearest to the perturbed point will be affected, but only through much slower Coriolis and advection terms (Mesinger, 1973). The situation with the A and D Grids, in this sense, is still worse, and is not considered further here.

Lattice separation is a problem of space differencing. This can also in a formal way be demonstrated by considering the phase speed of a pure gravity wave, with the time derivative left in the differential form and space derivatives replaced by simplest centered analogs. For the E Grid, one obtains (e.g., Mesinger and Arakawa, 1976),

$$\frac{c_*}{\sqrt{gH}} = \sqrt{\frac{\sin^2 X + \sin^2 Y}{X^2 + Y^2}} \, . \tag{1}$$

Here c_* is the finite-difference phase speed, $X \equiv kd/\sqrt{2}$, $Y \equiv ld/\sqrt{2}$, with k, d, and the remaining symbols having their usual meaning. The contour plot of Eq. (1) is shown in Fig. 1; because of symmetry only one-half of the admissible wave number domain is displayed. The relative phase speed is seen to reduce to zero for the highest admissible wave number of the domain displayed, $X = \pi$. Constant values of h on one and on the other of the C subgrids, different from each other, represent a stationary solution, no matter how different the two values might be.

To address this B/E Grid lattice separation problem, Arakawa (1972) designed a time-alternating, space-uncentered scheme, which he had used in combination with Matsuno time differencing. A method of arriving at space-time centered, second-order accurate schemes was subsequently proposed by Mesinger (1973; see also Mesinger and Arakawa, 1976). The

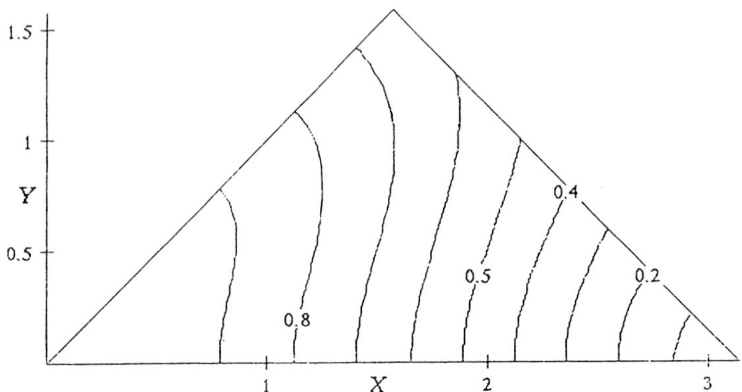

Figure 1   Relative phase speed of gravity wave with simplest centered space differencing, Eq. (1), on the Arakawa E Grid. For reasons of symmetry, only a half of the admissible wave number domain is shown.

method results in modifications of the divergence term of the continuity equation. Specifics of the modification depend on the choice of the time-differencing scheme, but will entail interaction between neighboring height points via the pure gravity wave terms, thus significantly improving on the lattice separation problem. If, for example, the forward–backward time scheme is used, with the momentum equation integrated forward,

$$u^{n+1} = u^n - g\Delta t\,\delta_x h^n, \qquad v^{n+1} = v^n - g\Delta t\,\delta_y h^n, \tag{2}$$

instead of

$$h^{n+1} = h^n - H\Delta t\left[(\delta_x u + \delta_y v) - g\Delta t\nabla_+^2 h\right]^n, \tag{3}$$

the method results in the continuity equation (Mesinger, 1974):

$$h^{n+1} = h^n - H\Delta t\left[(\delta_x u + \delta_y v) - g\Delta t\left(\frac{3}{4}\nabla_+^2 h + \frac{1}{4}\nabla_\times^2 h\right)\right]^n. \tag{4}$$

Here again the E Grid is used, n is the time index, and substitutions have been made from Eq. (2) into Eqs. (3) and (4) so as to have on their right sides values at the time level n only; the "plus" and the "cross" subscripts depict the geometry of the h values used in five-point analogs to the Laplacian; other symbols have their standard meaning. The original system, Eqs. (2) and (3), involves no communication between the two C subgrids of the E Grid. In contrast, in the system of Eqs. (2) and (4), this communication is achieved via the cross Laplacian term of Eq. (4).

For visualization of the impact of this difference, consider what happens following an increase in height at a single height point, within one forward–backward time step. With the system of Eqs. (2) and (3), height values at second nearest neighbors increase, as depicted by the plus Laplacian of Eq. (3); while the four nearest h-point neighbors undergo no change. When Eq. (3) is replaced by Eq. (4), height values at all eight neighbors increase, a much more satisfactory situation. Still, the h values at the four nearest neighbors, belonging to the C subgrid which is not the one of the perturbed point, will undergo an increase that is only two-thirds of that which occurs at the second nearest neighbors. Thus, although the improvement due to the modification is considerable, it has not completely removed the problem. Besides, the modification also results in some damping of the shortest gravity waves (e.g., Mesinger, 1974). Returning to the positive side, one can take additional comfort in the facts that the scheme remains neutral for waves for which the wave numbers in the x and y directions are the same, that the modification has no impact when the plus and the cross Laplacians are equal, and that there is no penalty in terms of the CFL stability condition of the scheme.

There are understandably numerous other considerations to be made in assessing the attractiveness of the C versus the B/E Grid. Regarding the "slowly changing quasi-geostrophic motion" the highest priority of course has been accorded to the horizontal advection scheme resulting in the Arakawa–Lamb (1981) scheme for the C Grid and in the Janjić (1984) scheme for the B/E Grid. Both schemes reduce to the Arakawa (1966) scheme in the case of horizontal nondivergent flow, and accordingly have their analogs of the famous Fjørtoft–Charney energy scale (e.g., Mesinger and Arakawa, 1976, Fig. 7.1). Energy scale analogs are different, however, with the Janjić scheme analog having the two-grid-interval wave extend to infinity so that the cascade into the shortest wave is not possible (Janjić and Mesinger, 1984, Fig. 3.12). This results in an enhanced constraint on the energy cascade into the smallest scales. Still other differences between the two schemes are their conservation properties, which are additional to the three classical ones of the Arakawa (1966) scheme with the Arakawa–Lamb scheme conserving potential enstrophy, and the Janjić scheme conserving momentum. Thus, with the Janjić scheme, the Hollingsworth–Källberg noncancellation instability (Hollingsworth *et al.*, 1983) is not a matter of concern.

Time differencing is yet another consideration. The leapfrog or the semi-implicit scheme are the choices typical of the C Grid models, and the split-explicit, forward–backward scheme of the B/E Grid models. The attractiveness of the simple two-time level split-explicit scheme, if one were to be a believer in it, is with the C Grid reduced due to a problem with the Coriolis terms.

My choice of the E Grid when writing the code that could be considered the ancestor of today's Eta model (e.g., Mesinger and Janjić, 1974) was based on two additional points. One is the simple appeal of carrying the two velocity components at the same grid points, given that it is the velocity vector which is the basic dynamical variable to be forecast, and not its individual components.

The second is the possibility of having all variables defined along a single outer boundary of a rectangular E Grid limited-area domain. This feature has enabled the design of an apparently very successful lateral boundary conditions scheme (Mesinger, 1977) along the lines of Oliger and Sundström (1978); this point is returned to later in this chapter.

III. HEXAGONAL GRIDS

With each of the square grids and centered second-order schemes experiencing a set of problems, examination of other options is obviously justified. An option considered early in the development of the primitive equation techniques has been that of the hexagonal grids. One might argue that the hexagonal grid is an attractive choice given that each grid point has a set of six nearest neighbors, with all six at the same distance from the considered point, being isotropic in that sense.

All four of the Arakawa grids have their hexagonal analogs. They are displayed in Fig. 2, using circles to denote the height points, and bars to depict the location as well as orientation of the velocity points (in the manner of Song and Tang, personal communication, 1991). In the order as displayed, they will be referred to as the HA, HD, HC, and the HB/E Grid. In their very early work Sadourny and Williamson used the first three of the H Grids, as shown here; Sadourny the D and then the C Grid, and Williamson the A Grid (e.g., Sadourny and Morel, 1969, and Williamson, 1969, and references therein). The somewhat counterintuitive HB/E Grid has been used by Thacker (e.g., 1978).

A disadvantage of the fully staggered grids, D and C, specific to their hexagonal versions is their having an excess of velocity components, three components per each height point rather than two (Sadourny, personal communication, 1981). To circumvent this disadvantage still another possibility has been pointed out by Popović *et al.* (1996): to skip every third velocity component of the HC Grid. One ends up with a grid that can be obtained by deforming a square C Grid into a hexagonal shape.

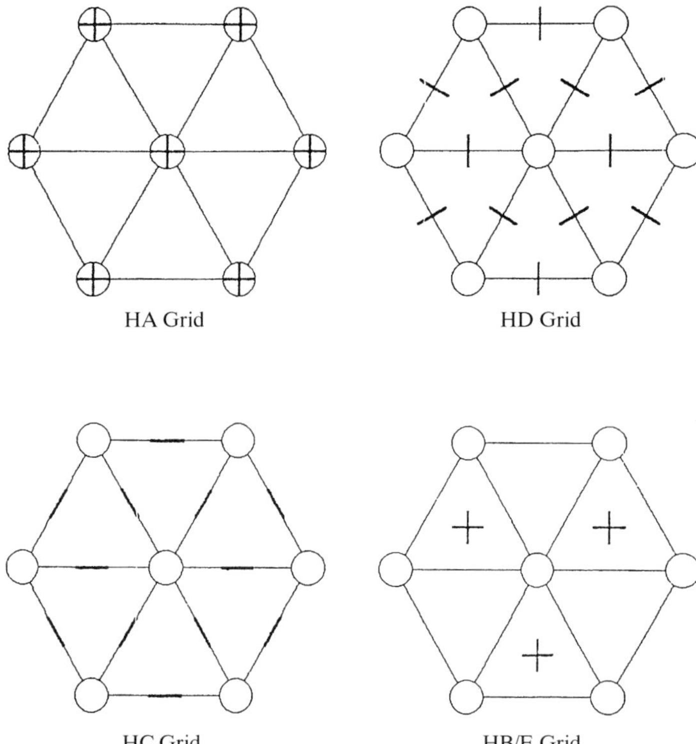

Figure 2 Hexagonal analogs of the Arakawa square horizontal grids A, D, C and B/E. Circles denote the h points, and bars denote the location as well as orientation of the velocity components.

At the time of the early work of Sadourny and Williamson little was known about the properties of the finite-difference analogs of primitive equations on various grids as summarized here and a question arises: What is the situation with the hexagonal grids regarding the issues raised? This was precisely the idea of Ničković (1994) when he recently analyzed the stability of the forward–backward scheme used on the HC Grid. He has found that the scheme is neutral provided

$$\Delta t \le \sqrt{\frac{2}{3}} \frac{d_\mathrm{h}}{\sqrt{gH}}, \tag{5}$$

where d_h is the hexagonal grid distance. For a possible comparison with the stability ranges of the square grids, one should note that

$$d_{\mathrm{h}} = \sqrt{\frac{2}{\sqrt{3}}}\, d, \tag{6}$$

where d is the grid distance of an equivalent square grid having the same number of grid points per unit area. The numerical factor on the right side of Eq. (6) is equal to about 1.075. A point emphasized by Ničković is that the hexagonal grid used on an icosahedron to construct grids for the sphere may have caused concern due to its singular points and lines, but that this would not stand in the way of using a hexagonal grid for a limited-area model.

In view of the HC Grid problem of the extra velocity components that is additional to the standard C Grid problem of the need for averaging of the Coriolis terms, properties of the HB/E Grid appear particularly intriguing. As to the forward–backward scheme, by using the simplest three-point differencing for the gravity wave terms one can demonstrate that the scheme corresponds to the centered second-order wave equation. The scheme is neutral within the same stability range as that of the HC Grid scheme [Eq. (5)]. With the time derivative kept in the differential form, the relative gravity wave speed is

$$\frac{c_*}{\sqrt{gH}} = 2\sqrt{\frac{3 - \cos X - 2\cos(X/2)\cos(\sqrt{3}\,Y/2)}{3(X^2 + Y^2)}}. \tag{7}$$

Here $X \equiv kd_{\mathrm{h}}$, $Y \equiv ld_{\mathrm{h}}$, with k and l as before being the wave numbers along the x and y axes. The admissible wave-number domain of the hexagonal grid is shown in the upper panel of Fig. 3, and the relative phase speed [Eq. (7)] in its lower panel. Because of the threefold symmetry within the positive wave-number quadrant, only one-third of the admissible domain is shown. In contrast to Fig. 1, the relative phase speed is seen never to reduce to zero; its minimum value is $(3/2)^{3/2}/\pi$, about 0.585. There is no lattice separation problem.

These attractive features of the HB/E Grid, and perhaps also of the HC Grid, call for additional attention. The geostrophic adjustment situation in the Arakawa–Winninghoff sense has been analyzed by Ničković (personal communication, 1998). The relative frequency, $|\nu|/f$, of the gravity-inertia wave Ničković obtains for the HC Grid, for the case $\lambda/d = 2$, is shown in Fig. 4a. The values seen are similar to those of the square C Grid (e.g., Arakawa and Lamb, 1977); the relative frequencies increase inside all of the admissible wave-number domain, attaining maxi-

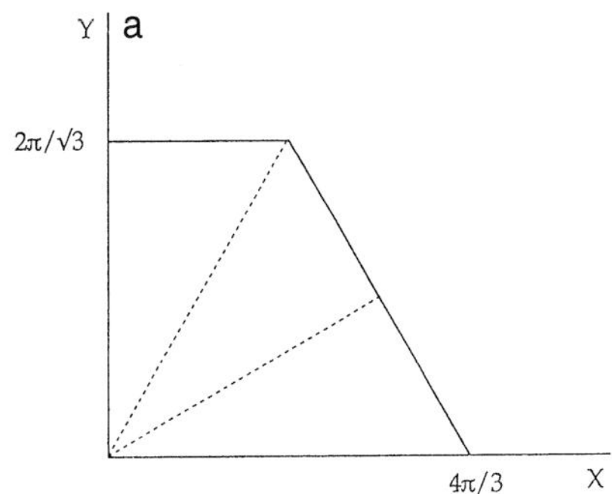

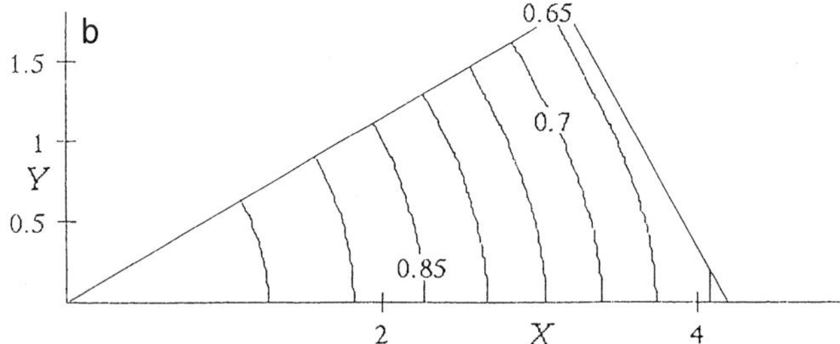

Figure 3 (a) The admissible wave-number domain of the hexagonal grid with $X \equiv kd_{\rm h}$, $Y \equiv ld_{\rm h}$. Here $d_{\rm h}$ is the grid distance of the hexagonal grid. (b) Relative phase speed of gravity wave with simplest centered space differencing [Eq. (7)] on the hexagonal B/E Grid. For reasons of symmetry, only a third of the admissible wave-number domain is shown.

mum values at its corners. In Fig. 4b, relative frequencies of the geostrophic mode are shown. They are different from zero, in striking contrast to the situation with any of the square grids.

The situation is similar with the HB/E Grid (not shown), with an additional feature that the relative frequencies of the two gravity-inertia waves are somewhat different. Once again an error in the frequency of the geostrophic mode is found.

How damaging the error in the frequency of the geostrophic mode discovered by Ničković might be is obviously an important issue. To my knowledge there are no comprehensive model integrations, in a hexagonal u, v formulation, that could be used for an attempt to answer this question. In the source-sink experiment of Ničković (1994) no detrimental effects were obvious; the result looked encouraging.

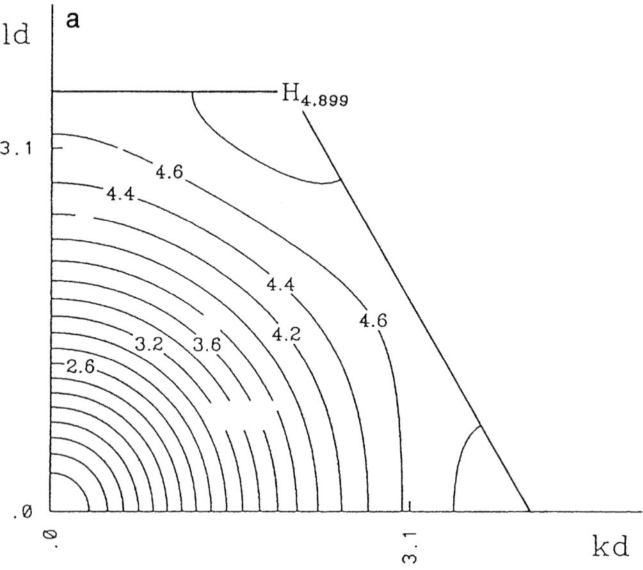

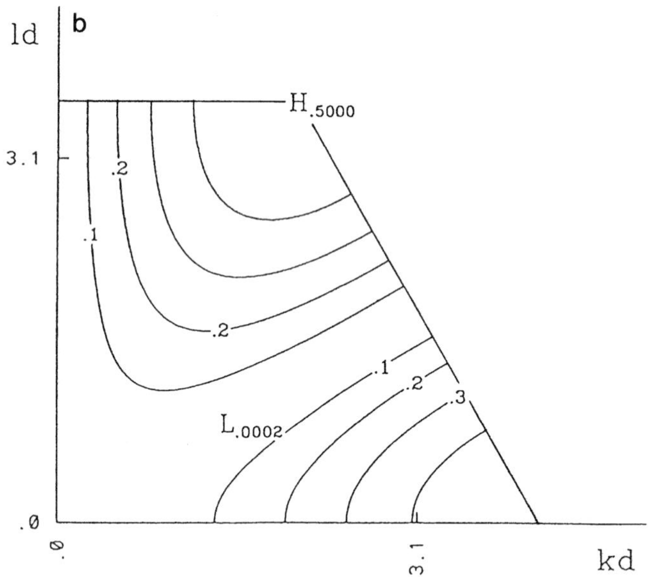

Figure 4 The relative frequency, $|\nu|/f$, on the HC Grid with simplest centered space differencing, for the case $\lambda/d = 2$. The gravity-inertia wave relative frequency is shown in the (a), and that of the geostrophic mode in (b) (Ničković, personal communication, 1998).

IV. RANDALL Z GRID AND C-GRID-LIKE B / E GRID GRAVITY WAVE SCHEMES

Excellent geostrophic adjustment properties of the unstaggered grid for the vorticity and divergence as prognostic variables ("Z" Grid) were pointed out by Randall (1994). Progress in using the vorticity/divergence formulation on a hexagonal grid, subsequent to Heikes and Randall (1995a,b), are reported elsewhere in this volume.

Still another option is to try to benefit from both the simplicity and straightforwardness of the u, v formulation and from the excellent properties of the streamfunction/velocity potential formulation for gravity-inertia waves, by switching between the two as the integration proceeds. The cost of this option in view of the need to solve for the streamfunction and for the velocity potential at each time step may appear discouraging at this time.

A radically new approach to address the lattice separation problem, however, has been advanced by Janjić (personal communication, 1992; also Janjić *et al.*, 1998). It consists of averaging the time difference in the continuity equation. If, for example, the forward–backward scheme is used with the continuity equation integrated forward, on the B Grid, and the averaging is performed over five points, we have

$$u^{n+1} = u^n - g\Delta t \delta_x (\overline{h}^y)^{n+1},$$

$$v^{n+1} = v^n - g\Delta t \delta_y (\overline{h}^x)^{n+1}, \tag{8}$$

$$\frac{1}{2}(\overline{h}^{xx} + \overline{h}^{yy})^{n+1} = \frac{1}{2}(\overline{h}^{xx} + \overline{h}^{yy})^n - H\Delta t \left(\delta_x \overline{u}^y + \delta_y \overline{v}^x \right)^n. \tag{9}$$

This scheme is referred to as the "five h-point" or FHP scheme. As shown by Janjić, the scheme is neutral for

$$\Delta t \leq \frac{d}{\sqrt{2gH}}, \tag{10}$$

which is the same as the C Grid stability condition. With the time derivative in the differential form, the relative gravity wave speed is

$$\frac{c_*}{\sqrt{gH}} = \sqrt{2\frac{\sin^2 X \cos^2 Y + \cos^2 X \sin^2 Y}{(X^2 + Y^2)(\cos^2 X + \cos^2 Y)}}. \tag{11}$$

Here $X \equiv kd/2$, $Y \equiv ld/2$. Within the admissible wave-number domain, Eq. (11) achieves its minimum value for $X = \pi/2$, $Y = \pi/2$, of about 0.65. Thus, there is no lattice separation problem.

On the downside, "deaveraging" (Janjić *et al.*, 1998) of Eq. (9) needs to be performed; this can be done by relaxation, which according to Janjić *et al.* (1998) converges "surprisingly quickly." A single-point height perturbation affects in a single time step the four nearest height points the most (a "C-Grid-like" scheme), but propagates in one time step throughout the domain. This is reminiscent of the situation with the so-called compact schemes (e.g., Leslie and Purser, 1991).

Yet another scheme can be easily designed that also employs tendency averaging to remove the B/E Grid lattice separation problem. This can be done by essentially following the idea of Janjić (1984) for his construction of the B/E Grid horizontal advection scheme. First, auxiliary C Grid velocity components are introduced in an appropriate way. For the gravity wave terms they are needed midway between the nearest height points on the B Grid so as to be defined by

$$u_C = \overline{u}^y, \qquad v_C = \overline{v}^x. \tag{12}$$

If now the forward–backward scheme is desired, one can write the scheme in terms of the C Grid velocities, and then substitute Eq. (12) to obtain a B Grid scheme. If the continuity equation is integrated forward, one obtains

$$\left(\overline{u}^y\right)^{n+1} = \left(\overline{u}^y\right)^n - g\Delta t \delta_x h^{n+1},$$

$$\left(\overline{v}^x\right)^{n+1} = \left(\overline{v}^x\right)^n - g\Delta t \delta_y h^{n+1}, \tag{13}$$

$$h^{n+1} = h^n - H\Delta t \left(\delta_x \overline{u}^y + \delta_y \overline{v}^x \right)^n. \tag{14}$$

The scheme is neutral for

$$\Delta t \leq \frac{d}{\sqrt{2gH}}, \tag{15}$$

which is once again the C Grid stability condition. The relative gravity wave speed, with the time derivative in the differential form, is

$$\frac{c_*}{\sqrt{gH}} = \sqrt{\frac{\sin^2 X + \sin^2 Y}{X^2 + Y^2}}, \tag{16}$$

with X and Y having their B Grid definitions, as in Eq. (11). Contour plots of the relative gravity wave speeds of the two schemes, Eqs. (11) and (16), are shown in Fig. 5. With the admissible domains of both X and Y being $\leq \pi/2$, the minimum value of Eq. (16) is once more seen to be about 0.65. There is no lattice separation problem. In fact, Eq. (16) can be recognized as identical to the gravity wave speed on the C Grid (e.g., Mesinger and Arakawa, 1976), which may come as no surprise given the way the scheme has been designed. Inspection of the C Grid system used to arrive at Eqs. (13) and (14) shows that a height change at a single point will in one time step propagate to the four nearest neighbors and to no other height points, as on the C Grid. A "simulated C Grid" scheme, SCG, thus seems appropriate for Eqs. (13) and (14).

How can the B Grid propagation of a single-point height perturbation take place the same as it does on the C Grid? With velocities at time step n equal zero, and heights constant except for a single grid-point value—for example, higher than the others—solution of Eq. (13) results in a wind field at the level $n + 1$ as depicted in Fig. 6. Additional to the velocities directed radially away from the perturbed point, two strips of velocity components are created as needed to have the resulting velocity divergence equal to zero at all h points except at the perturbed point and its four nearest neighbors.

Additional to the need for deaveraging, the cost for achieving a C-Grid-like propagation of single-point height perturbations is thus for both schemes a spurious wave created throughout the domain; for the FHP scheme in the height field, and for the SCG scheme in the velocity field. The constant amplitude of the spurious velocities shown in Fig. 6 may look

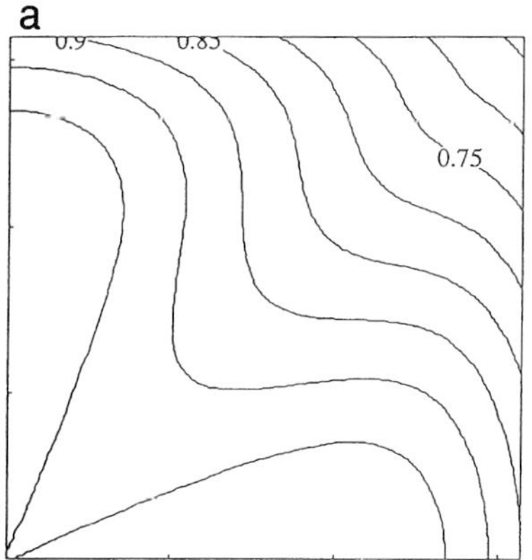

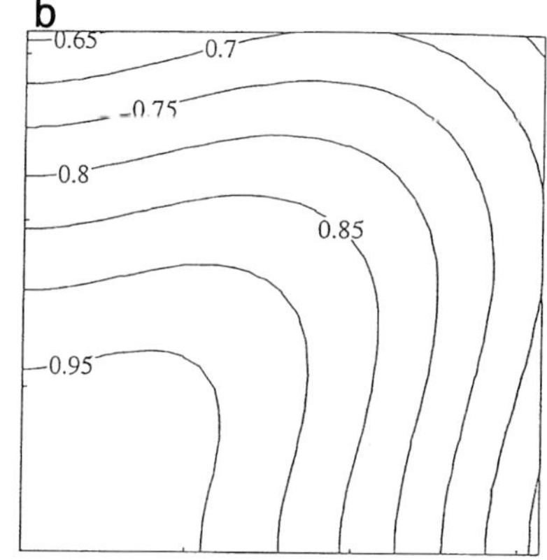

Figure 5 (a) Relative gravity wave speed of the Janjić "five h-point" scheme, Eqs. (8) and (9), and (b) of the "simulated C Grid" scheme, on the B Grid, and with time derivatives in the differential form. The coordinate axes are $X \equiv kd/2$, $Y \equiv ld/2$.

Figure 6 Solution of the B Grid "simulated C Grid" scheme, Eqs. (13) and (14), for the wind field at time level $n + 1$, following an initial condition of the velocities at time step n equal to zero, and heights constant except for a single grid-point value, that at the center of the plot, higher than the others.

worrisome; one could take some comfort in the idea that these velocities would be invisible to the Coriolis terms if the Coriolis terms were also to be included via the two-point averaging in Eq. (13).

Source-sink experiments à la Arakawa (1972) were performed for both schemes (Gavrilov, personal communication, 1998). Both schemes gave expected results (e.g., Janjić and Mesinger, 1989) and were efficient in the sense that relaxation to solve for the h or the u, v tendencies converged quickly. Thus, no preference for one or the other of the two schemes was obvious. It was recently noted by Ničković (personal communication, 1997) that five-point averaging of the velocity component tendencies also results in a scheme with gravity wave properties that are the same as those of the FHP scheme.

A favorable feature of this class of "tendency-averaged schemes" is that they can be tested in a comprehensive split model by simply replacing the adjustment stage by a stage based on one or the other of the schemes summarized. One effort of this kind, by Janjić *et al.* (1998), was already referred to. But apart from prospects offered by specific schemes or approaches reviewed in this and in the preceding section, one purpose of the material presented was to illustrate the variety of possibilities one can explore in trying to achieve the behavior of a difference scheme that is appealing from the physical point of view. Only issues related to the choice of the horizontal grid were considered; there are of course many others. Some are touched on in the following sections, but from a different perspective, namely, that of the design and performance of a specific model.

V. THE ETA MODEL: AN ARAKAWA APPROACH STORY

The so-called Eta model is a limited-area model with the numerical formulation designed following the Arakawa principles. It has been used so far primarily for weather forecasting, so one could question the appropriateness of covering it within the symposium carrying the general circulation model (GCM) development title.

My reasons for finding this appropriate are twofold. The first is that nowadays limited-area models are increasingly used as integral parts of general circulation models for simulation of regional climate. A very successful recent Eta model example of such a use is that of Ji and Vernekar (1997). Use of the Eta nested within a GCM led to improvements in their simulation of a number of observed features of Asian monsoons, compared to results of the GCM with no Eta nest.

The second is that a forecasting model is an excellent vehicle for testing the performance of a scheme or a set of schemes. In a typical operational setting, forecasts are initialized twice daily and verified against analyses. A large body of verification statistics tends to be automatically accumulated. "Clean" experiments can be and are set up in which a model with a single change is compared against the control (e.g., Rogers *et al.*, 1996, and Mesinger *et al.*, 1997, and references therein). Also, the performances of forecasting models with different properties are regularly compared and inferences made.

I expect to be able to contribute to this class of assessments, and specifically to that of the impact of the Arakawa versus what might perhaps be called a traditional approach, by reviewing some of the results of the Eta model in the remainder of this chapter. Expectation may have been widespread that the maintenance of the integral constraints and other Arakawa-type properties of the difference system while very important in climate integrations may not be a critical requirement for short-range forecasting, and that the local accuracy in short-range predictions is therefore more or less determined by the grid size and the order of accuracy of the scheme. I find that evidence accumulated during the past decade or two shows that this expectation was not justified; in fact, short-range forecasting as I hope to demonstrate may well have resulted in the most convincing indication of the potential of the approach.

There is, of course, no unique way to design the dynamics of a model following the Arakawa principles, and some of the principles may be more rewarding than others. Moreover, as the review of the horizontal grid issues shows, trade-offs are typically encountered and the best choice is frequently not obvious. Regarding the Eta model, of the various important features of its numerical formulation, the most deserving of being noted in my opinion are the following:

- *The step-mountain ("eta") vertical coordinate* (Mesinger, 1984; see also Mesinger *et al.*, 1988). The surfaces of constant eta are approximately horizontal, thereby avoiding the cause of the notorious sigma system pressure-gradient force problem (e.g., Mesinger and Janjić, 1985, 1987). Perhaps just as importantly, to simulate horizontal motion over large-scale mountain ranges, there is no need for the model to generate vertical velocities through coordinate surfaces on one and on the other side of the mountain range.

- *The Janjić* (1984) *Arakawa horizontal momentum advection scheme.* On the model's E Grid, the scheme conserves C Grid defined enstrophy for horizontal nondivergent flow. As summarized in Section II, this results in an enhanced constraint on the energy cascade toward smaller scales. Numerous other quantities are conserved, including momentum apart from the effect of mountains.

- *Gravity-wave coupling scheme of Mesinger* (1973, 1974). Rather than the scheme of Eqs. (2) and (4), the version of the modified forward–backward scheme with the continuity equation integrated forward is used (Janjić, 1979). Integration of the continuity equation forward requires less storage than the integration of the momentum equation forward, and for pure gravity-wave terms results in the same difference analog of the wave equation.

- *Energy conservation in transformations between the kinetic and the potential energy in space differencing* (Mesinger, 1984; Mesinger *et al.*, 1988). Splitting into the adjustment and the advection step is used with the pressure advection carried within the adjustment step (Janjić *et al.*, 1995); this is a necessary, although not sufficient, condition for energy conservation also in time differencing.

- *Lateral boundary conditions prescribed or extrapolated along a single outer boundary line, followed by a "buffer" row of points of four-point averaging* (Mesinger, 1977). The four-point averaging achieves coupling of the boundary conditions of the two C subgrids. Model integration from the third row of points inward is done with no "boundary relaxation" or enhanced diffusion zone ("fairly well-posed" lateral boundary conditions according to McDonald, 1997).

Within the model's physics package some of the special features are its modified Betts–Miller—or Betts–Miller–Janjić—convection scheme (Betts, 1986; Betts and Miller, 1986; Janjić, 1994), its Mellor–Yamada level 2.5 turbulence closure (Mellor and Yamada, 1982), with improved treatment of the realizability problem (Mesinger, 1993; Janjić, 1996a), its viscous sublayer scheme over both water and land surfaces (Janjić, 1996b), and its prognostic cloud water/ice scheme (Zhao and Carr, 1997). In more recent model upgrades, increasingly comprehensive land–surface parameterizations are included (e.g., Chen *et al.*, 1997, and references therein). For radiation, the Geophysical Fluid Dynamics Laboratory (GFDL) scheme is used (Fels and Schwarzkopf, 1975; Lacis and Hansen, 1974).

Until October 1995, the model was initialized with a static "regional" optimum interpolation (ROI) analysis using the Global Data Analysis System (GDAS) first guess (Rogers *et al.*, 1995). As of 12 October 1995 until February 1998, this was replaced by a 12-hr Eta-based intermittent assimilation (EDAS; Rogers *et al.*, 1996). More information on the model's physics package and its initialization/assimilation and verification system can be found in, e.g., Janjić (1994), Black *et al.* (1993), and Rogers *et al.* (1996).

Models change. It should be stressed, however, that what I believe would generally be considered major features of a model's numerical design have not changed in the Eta's case since the mid-1980s when the minimum physics version of the eta coordinate code was put together. This includes the five features described in the summary above.

The Eta model was operationally implemented at the then National Meteorological Center (NMC) on 9 June 1993, as the so called NMC early run. The term "early" refers to an early data cutoff, of 1:15 hr, aimed at providing guidance as quickly as possible. The name "early Eta" came into widespread use after the implementation of a later run of the Eta, at higher resolution, the so-called "meso Eta," in 1995.

For a regional model to be implemented at an operational center already running an operational regional model, as the NMC was at the time, the candidate model clearly needs to demonstrate superior performance—or at least obvious potential. Given the existence at NMC then as now also of an operational global model, this automatically implies an advantage of some kind as well over the NMC's global model product as available at the forecast time of the regional model. Namely, without such an advantage of the regional over the global model, running a separate regional model would be hard to justify.

The two models against which the Eta is thus naturally compared are the so-called Nested Grid Model (NGM) and the Medium-Range Forecasting (MRF) or Aviation (Avn) model. The NGM, or Regional Analysis and Forecasting System (RAFS) when referring to the entire forecast system containing the model, is a sigma coordinate gridpoint model, with an approximately 80-km inner grid nested inside its own coarser outer grid. Both grids have 16 layers in the vertical. It is initialized with a 12-hr NGM-based intermittent assimilation using ROI analysis, with a 2:00-hr data cutoff (DiMego, 1988). No change in the model nor in its analysis system have been made since August 1991 (DiMego *et al.*, 1992). The model, however, continues to be run twice daily, off 0000 and 1200 UTC data, 48 hr ahead.

The Avn/MRF model is a global spectral sigma system model. Since August 1993 it has been run with the triangular 126 truncation (T126) and 28 layers (e.g., Kanamitsu *et al.*, 1991; Pan and Wu, 1994; Hong and Pan, 1996). The two names, Avn and MRF, refer to the same model but to different data cutoff times: Until very recently, twice daily, at 0000 and 1200 UTC, the model was run 72 hr ahead with an early data cutoff, of 2:45 hr, under the name Aviation model; at 0000 UTC the Avn run is followed by the "MRF model" run with a later data cutoff, of 6:00 hr. The Avn forecasts are used for the Eta boundary conditions; however, since the Eta runs first, the 12-hr-old Avn run has been used. This changed in February 1998 as a result of the implementation of four runs per day of the Avn model.

Of the two models, comparison against the NGM was clearly the more relevant one and therefore in the early experimental stage of the Eta model care was taken to run an Eta configuration with horizontal and vertical resolution, as well as the use of computer resources, same or comparable to those of the NGM. The characteristics of the Eta forecasts which in these early tests perhaps particularly stood out were the realism of its various forecast—often smaller scale—synoptic features, such as that of multiple centers and of the placement and of the depth of surface lows (e.g., Black, 1988, Figs. 10 and 13; Mesinger and Black, 1989, Figs. 11–19; Black and Mesinger, 1989, Figs. 4 and 5). Other verification efforts of the time were addressing the mean geopotential height errors (Black and Janjić, 1988, Fig. 6; Black and Mesinger, 1989, Fig. 2) and precipitation forecasts (Mesinger and Black, 1989; Mesinger *et al.*, 1990).

Of these early tests perhaps that of the comparison of mean height errors (Black and Janjić, 1988, Fig. 6) should be particularly recalled as it was done when the Eta physics package was just about put together, in 1987, while the NGM's package was in a more mature stage (e.g., Tuccillo and Phillips, 1986). The comparison was done for a sample of 13 forecasts done by each of three models, the NGM, the Eta, and the Eta run using the sigma coordinate, with the models using the same radiation scheme. The NGM showed a steady growth of negative errors, reaching errors of more than -60 m above 150 mb at 48 hr. The Eta errors reached their greatest magnitude at about 12 hr, with values of the order of -20 m, and afterward changed very little. The errors of the Eta run in its sigma mode were considerably greater, with values of below -30 m over most of the troposphere after 12 hr, and even some below -40 m at 48 hr.

Following the early experimental stage as well as through its now more than 5 years of operational running, precipitation scores were perhaps the main guidance in assessing the overall Eta performance and in deciding on model changes. Note that this attention to the skill of precipitation forecasts was not a specialty of the Eta as "improved precipitation forecasting" was considered to be "a basic goal" already of the NGM project at its early stage (Hoke *et al.*, 1985). The precipitation analysis system of the NCEP's Environmental Modeling Center (EMC) used for that purpose is based on data provided by the National Weather Services's River Forecast Centers (RFCs); these consist of reports of accumulated precipitation for each 24-hr period ending at 1200 UTC. The analysis covers the area of the contiguous United States with reports from about 10,000 RFC rain gauge stations. In areas of poor coverage, RFC data are augmented by radar precipitation estimates if rain gauge data are available to calibrate the radar data. Data are analyzed to the gridboxes of the verification grid by simple gridbox averaging. With verification grid size on the order of 80 km, about 10 reports are available per verification box.

Three-model scores, for the Eta, the Avn/MRF model, and the NGM, for three verification periods, 00–24, 12–36 and 24–48 hr, are available and archived beginning with September 1993. Because relative model performance is to some extent season dependent, it is necessary to look at full 12-month samples, or multiples of 12 months, if the seasonal model performance is not to have an impact on the result. Accordingly, in Fig. 7, equitable threat and bias scores for the three models and for the first 24 months of the available scores, September 1993–August 1995, are shown. Recall that equitable threat score is the standard threat score corrected for skill of a random forecast (e.g., Mesinger, 1996b). The motivation for displaying the result for a 24-month sample is that during that time the resolution of the Eta model remained unchanged, at approximately 80 km in the horizontal and 38 layers in the vertical.

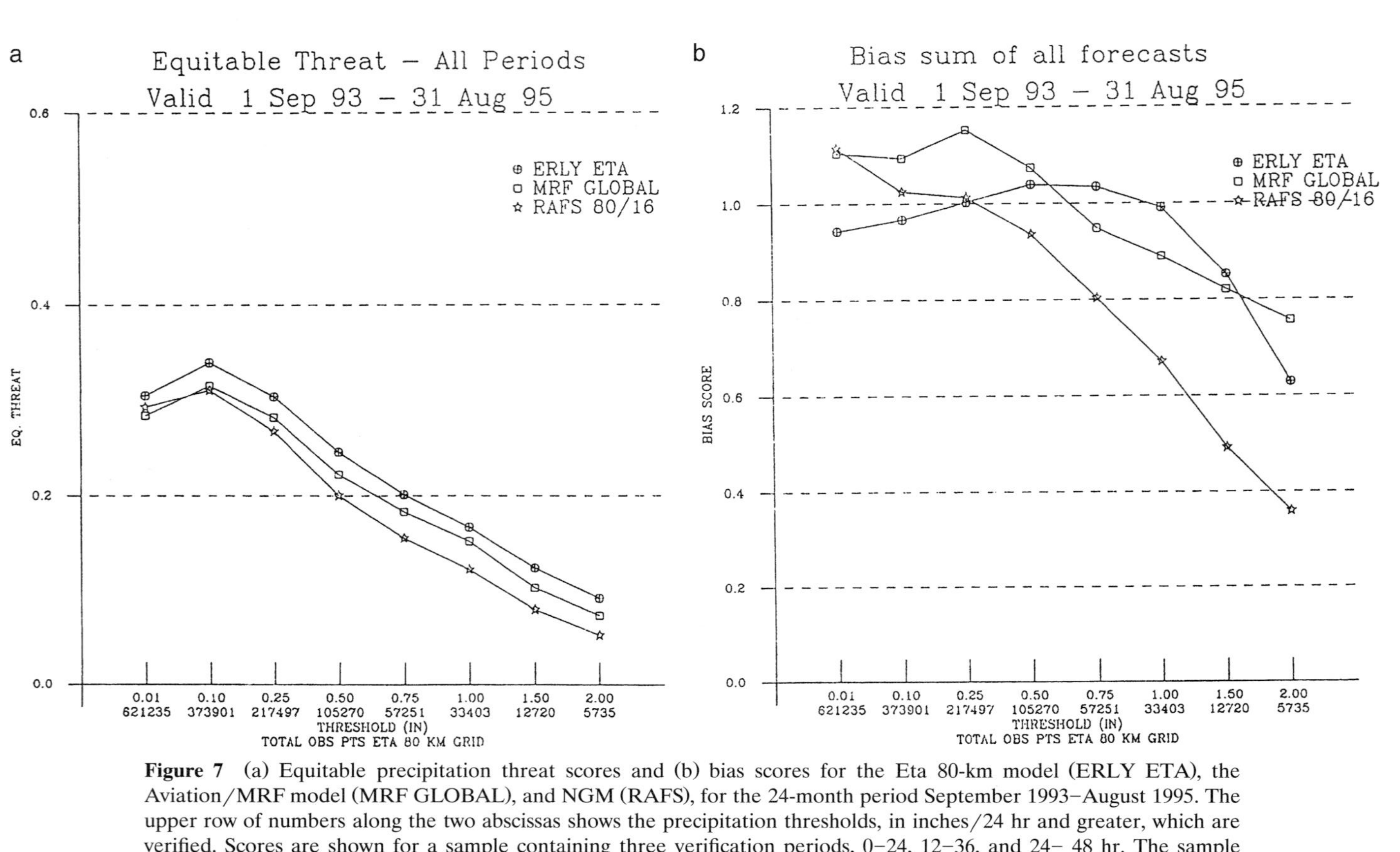

Figure 7 (a) Equitable precipitation threat scores and (b) bias scores for the Eta 80-km model (ERLY ETA), the Aviation/MRF model (MRF GLOBAL), and NGM (RAFS), for the 24-month period September 1993–August 1995. The upper row of numbers along the two abscissas shows the precipitation thresholds, in inches/24 hr and greater, which are verified. Scores are shown for a sample containing three verification periods, 0–24, 12–36, and 24– 48 hr. The sample

There are two points I wish to make based on the results shown in Fig. 7. First, note that for all the precipitation categories monitored the Eta threat scores are higher than those of its "driver" Avn/MRF model. This happens in spite of the Eta handicaps of using 12-hr "old" Avn boundary conditions, and having a shorter data cutoff, so that the Eta forecasts initialized at a given time are available before those of the global model. The Eta results thus confirm the validity of the regional limited-area modeling approach, showing that in spite of the listed handicaps, increased forecast accuracy was indeed achieved using a limited-area model. This, of course, refers to the accuracy measure chosen, that of the precipitation scores.

For the second point, I wish to emphasize that the NGM employs fourth-order accuracy schemes, along with a periodic application of a fourth-order Shapiro filter (Juang and Hoke, 1992). Its resolution and its overall use of computer resources during the period shown in the figure were comparable to those of the Eta. The average grid distance of the Eta during the time considered was in fact about 7 km greater than that of the NGM over the contiguous United States, where the verification is performed. Its vertical resolution was considerably greater, 38 layers compared to 16 levels of the NGM; but an extensive test performed in 1992 (Mesinger *et al.*, 1997, Fig. 2), showed only a very minor benefit from this higher vertical resolution of the Eta. The Eta schemes are typically of the second-order accuracy, and none are higher than the second. Yet, the Eta displays a very considerable advantage over the NGM, across all of the categories monitored.

There are of course many differences between the two models, which in one way or another contribute to the difference in precipitation scores. Different convection schemes may come to mind as the prime suspect. In this connection one should be reminded of tests made at the end of the 1980s with the then Eta model version of the Betts–Miller scheme, aimed at implementing the Eta scheme in the NGM should that prove to be beneficial. Although improvements were seen for medium and particularly for heavier precipitation, threat scores at the lower categories became worse. Thus, the overall improvement was questionable and certainly not of the magnitude as to make the NGM's scores competitive with those of the Eta (Mesinger *et al.*, 1990, Fig. 4; Plummer *et al.*, 1989). Eventually the scheme was not implemented.

Although the remaining components of the Eta's physics package of the period considered for the most part can be considered more advanced than those of the NGM, of the various sensitivity tests performed none has demonstrated impacts that would suggest physics plays a dominant role in the Eta versus NGM differences in forecast skill shown in Fig. 7. Regarding the initialization/assimilation systems of the two models, if anything, that of the NGM would be considered more advanced than that of the Eta prior to the implementation of EDAS in October 1995.

Thus, it would seem that we are left with the difference in approaches to the numerical design of the two models as the prime candidate for the leading contribution to the advantage in skill demonstrated by the Eta over the NGM during the period considered. This, of course, is not a "clean" experiment, and does not represent a proof of the role of the numerical design as suggested; other interpretations are possible. Yet, I believe the considerable evidence at hand does strongly point in that direction. A lot of care and code checking has gone into the NGM and the likelihood of, for example, a major code error is extremely remote. Discontinuation of the development of a model in an operational center of course comes as a result of a complex set of circumstances; but among those the general impression of prospects for further model improvement in a cost/beneficial sense cannot but be a matter of the highest concern. Note that from that point of view, numerical design and parameterizations are not on an equal footing, as there is no reason to look at parameterizations of one model as being less amenable to improvements than those of another. Synoptic-type and statistical verifications of the impact of the eta versus sigma coordinate, summarized to some extent above and to be returned to in Section VIII, support the idea of the major role of the numerical design in the differences in the model performance considered. Thus, I find that a strong indication is at hand pointing to the Eta versus NGM difference displayed in Fig. 7 being indeed largely an illustration of the advantage of the Arakawa over the "conventional" high-Taylor-series-accuracy, filtering-of-small-scales approach, for the comprehensive ("full-physics") atmospheric models of today. The qualification used here is motivated by the point already made in Section II, of the forcing at individual model gridboxes by the physics packages in use. Such forcing is inconsistent with the high-Taylor-series-accuracy concept, but is not in conflict with the fluid-dynamical considerations of the Arakawa approach to the design of numerical schemes, as outlined in the first two sections of this chapter.

VI. GLOBAL MODELING: THE POLE PROBLEM

A review paper with topics as covered so far would do no justice to the field without a reference to the pole problem of the Arakawa-like approach. Fourier filtering with the latitude–longitude grid is not only obviously wasteful in terms of the excessive number of grid points carried in polar regions, but is also in conflict with the basic premise of the Arakawa approach of doing no artificial filtering at small scales at which the presumably important physical parameterizations are performed.

The purpose of this section is to emphasize the apparently very good prospects of constructing well-behaved global finite-difference models using the expanded cube approach, free of the two problems just mentioned. Pioneered by Sadourny (1972) again at a very early time, the idea has been reinvigorated recently by Rančić *et al.* (1996). Two different approaches they used for the shallow-water integrations to handle the line singularities of the expanded cube, both employing the Arakawa-type B/E Grid Janjić (1977) momentum advection scheme, converged to a visually indistinguishable solution as the resolution was increased. The choice between the two approaches however was not clear, since the solution converging substantially faster, the one using a conformal grid, had a considerably less homogeneous distribution of points.

A still more recent extension of this work (Purser and Rančić, 1998) points a way to strike a balance between the two desirable features and relax the requirement of conformality to achieve a greater homogeneity, as might be found most cost effective for the task at hand.

VII. THE ETA MODEL: THE NEXT 24 MONTHS AND THE LIMITED-AREA MODELING CONCEPT

With no proof of this being impossible, it is generally expected that through increased computing power and research or developmental work, the skill of the operational prediction models should continue to improve, at present and at least for a while into the future. Indeed, regarding the past several years, various upgrades of the two "live" NCEP models/systems, the Eta and the Avn/MRF, have been taking place on a relatively regular basis during the time of and following the period of Fig. 7. For a description of some of these upgrades see, e.g., Rogers *et al.* (1996), Chen *et al.* (1997), and Hong and Pan (1996). Specifically, the Eta upgrade of 12 October 1995 included an increase in horizontal resolution to about 48 km; for the impact of this upgrade during a 6-month test period, see Mesinger (1996b).

For an assessment of this hoped-for improvement resulting from some of the implementations within the two systems, in Fig. 8 the equitable threat and bias scores for the 24-month period following that of Fig. 7 are shown. One should note that the Eta 48-km forecasts are for verification remapped to the previously used 80-km grid, in order not to penalize the

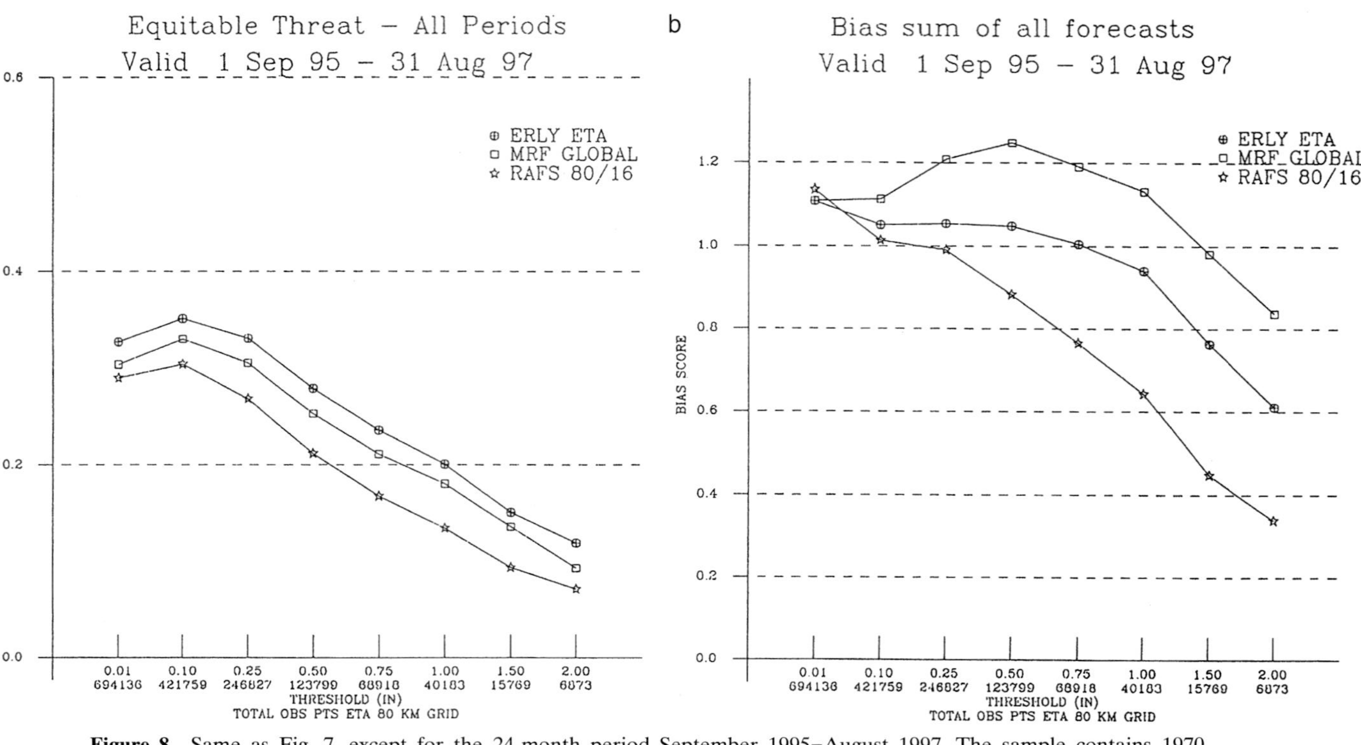

Figure 8 Same as Fig. 7, except for the 24-month period September 1995–August 1997. The sample contains 1970 verifications by each of the three models.

higher resolution model by the requirement to reproduce the increased noisiness of the 48-km box averages of the observed precipitation. Considerable improvement is indeed visible in the two live model threat scores relative to those of the frozen NGM. The scores of the NGM have of course also changed some, reflecting changes in the data entering the assimilation systems, and weather/climate variations between the two 24-month periods. Some weather events tend to result in higher scores than others, and in particular at the heaviest rain categories the impact of the frequency of occurrence in such score-friendly events is noticeable. For example, more 80-km verification boxes with amounts of 2 in. and greater per verification in the second 24 months than in the first, about 3.5 versus 3.2, is seen to be associated with increased threat not only of the two live models but of the NGM as well.

Once again, in Fig. 8 the Eta scores are significantly higher than those of its boundary condition driver Avn/MRF model. Compared to the preceding 24-month period, the difference between the two has in fact increased for most of the categories.

One might wonder: How does this advantage of the Eta depend on the forecast period, given that its lateral boundary conditions are 12-hr old? One could expect that as the forecast progresses the older and thus less accurate lateral boundary information has more and more of an impact on the contiguous U.S. area, where the verification is performed, so that the Eta skill eventually starts lagging behind that of the Avn/MRF forecast of the same initial time.

For an assessment of the situation in this respect, in Fig. 9 threat scores for the same 24-month period of the 12–36 hr forecasts (Fig. 9a) and of the 24–48 hr forecasts are shown. Inspection of the plots displayed reveals no obvious reduction in the advantage of the Eta over the Avn/MRF as the forecast period is increased from 12–36 to 24–48 hr. In fact, at several of the categories in Fig. 9 the difference in threat scores between the Eta and the Avn/MRF is at 48 hr greater than at 36 hr. Clearly, the validity of the limited-area modeling concept, with the setup and models used here, is at 48 hr not yet exhausted and a longer Eta run, provided the resources were available, would be justified.

This considerable advantage of the Eta over its driver model and in particular the resistance it displays to degradation of skill resulting from the advection of the lower accuracy (12-hr "old") boundary condition, and to contamination by the advection of the "lateral-boundary-condition-related difficulties" (Côté *et al.*, 1998) into the domain of interest I find worthy of attention. Arguments have been raised at several places regarding the relative merits of the limited-area versus the global variable-resolution strategy, in particular very recently by Côté *et al.* They summarize a

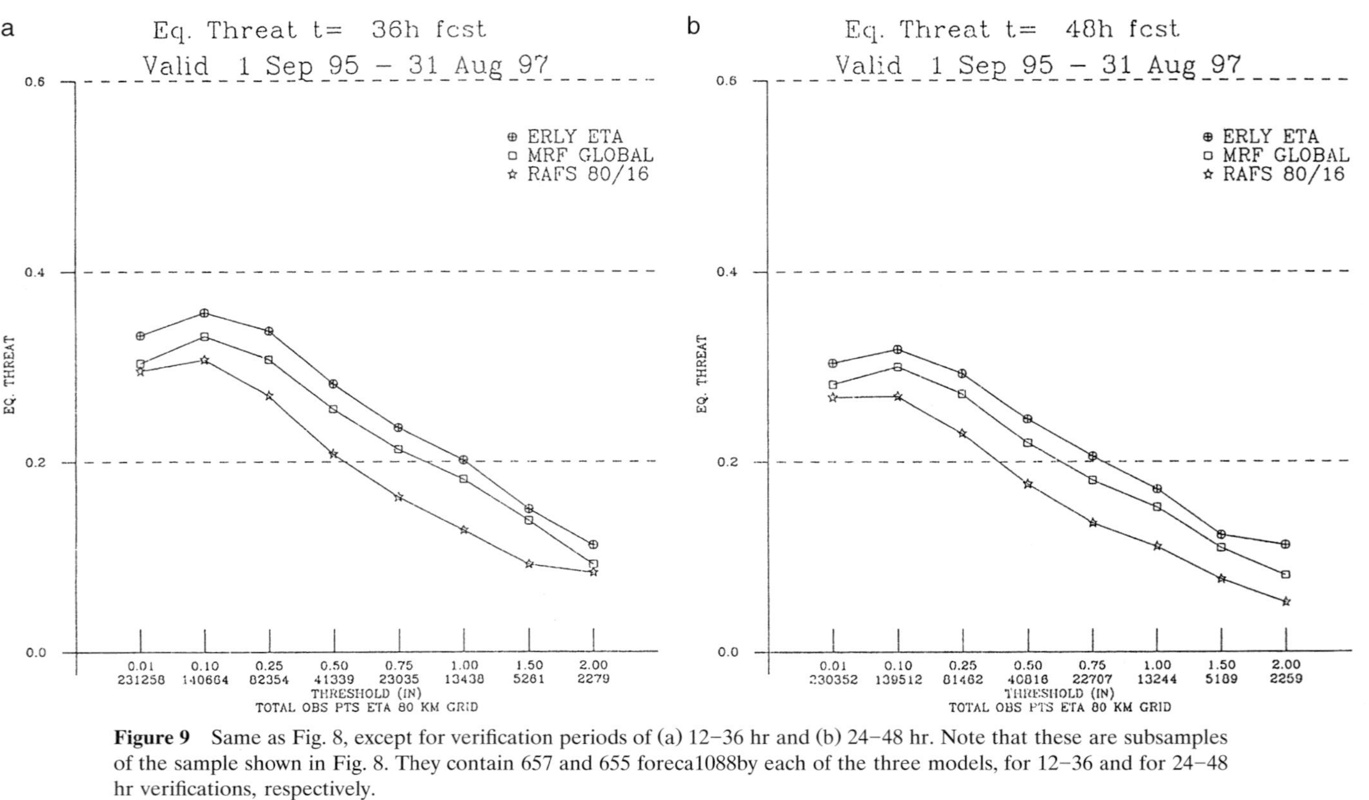

Figure 9 Same as Fig. 8, except for verification periods of (a) 12–36 hr and (b) 24–48 hr. Note that these are subsamples of the sample shown in Fig. 8. They contain 657 and 655 forecasts by each of the three models, for 12–36 and for 24–48 hr verifications, respectively.

considerable number of papers, 10 to be precise, by stating that they "all indicate that lateral-boundary-condition error can, depending upon the meteorological situation, importantly contribute to the total error." They conclude by recommending that "more needs to be done to validate the methodologies employed by today's mesoscale models."

But especially in an operational setting, I find it difficult to imagine a more appropriate type of validation than the one presented here, of demonstrating the advantage the limited-area model achieves over its driver model. Note, in particular, that this was done in spite of the imposed operational requirements that the limited-area forecasts be available *before* the driver model forecasts of the same initial time, and by using *less data*. It seems to me that this is the most rigorous "acid test," to adopt the term from Côté *et al.* (1998) and Yakimiw and Robert (1990) that "any successful limited-area model should meet," because this is the purpose for which the limited-area model has been created. This test may be contrasted with the one of the two cited papers that "the solution obtained over a limited area should well match that of an equivalent-resolution model integrated over a much-larger domain." While demonstrating the latter is an impressive computational task and of obvious interest, it is hardly one having much practical meaning since were the integration of an equivalent-resolution model over a much larger domain feasible, there would be no need to run a limited-area model in the first place. Of course, this is not meant to say that problems of the limited-area modeling are not most deserving of study. They, in fact, will continue to be addressed in the next section as well.

VIII. THE ETA COORDINATE AND THE RESOLUTION VERSUS DOMAIN SIZE TRADE-OFF

With the focus on numerical design, the question arises of whether there are any specific features of the Eta numerics or setup that could be identified as making a notable contribution to its advantage as discussed in the preceding sections.

One feature on which clean tests have been made is that of the impact of the eta coordinate. They have been done using the switch of the model which permits the same code to be run as the eta and also as a sigma system model. The original test of this kind performed with a dry/minimum physics model revealed considerable noise when running the model using the sigma coordinate (Mesinger *et al.*, 1988). This was interpreted as coming from sigma system errors. This was followed by the mean height error experiment for a sample of 13 forecasts, summarized already in Section V. In a still later study three cases were run, and a sample of nine consecutive forecasts (Mesinger and Black, 1992). This was followed by yet another study in which one case and a sample of 16 consecutive forecasts were run (Mesinger *et al.*, 1997). In both of the latter two samples, the eta version of the model resulted in higher threat scores for all precipitation categories monitored. For more confidence in the model's sigma run, note that in the second of these two samples the two other operational NCEP models were also included, with the Eta model run as sigma winning convincingly all of the categories over the NGM, and winning by a wide margin most of the categories over the Avn/MRF model.

Results of three of the four individual cases mentioned above offered perhaps still more compelling evidence in favor of the eta coordinate, in the sense that the sigma runs of the Eta reproduced to a substantial degree errors of the two NCEP sigma system models, absent or for the most part absent in the Eta. Two of these errors are well documented as highly typical of the NCEP operational models: too slow southward propagation of cold surges east of the Rockies (Sullivan *et al.*, 1993; Mesinger, 1996a), and placement of the lows as they form in the lee of the Rockies north of their analyzed positions (Mesinger *et al.*, 1996).

Regarding the latter, the error statistics summarized in Mesinger *et al.* (1996) give perhaps the most convincing evidence of the pervasiveness of the error. An area east of the Continental Divide was defined and rules set up for identification of lows within this area and within a sample consisting of 101 consecutive, 12-hr apart, 48-hr forecasts by each of the three NCEP operational models. Of 15 lee lows identified, the Avn/MRF model, for example, had placed all 15 north of their observed positions. The Eta, in the 101-forecast sample displaying little error of this kind, had reproduced the error when switched to sigma in one of the cases shown in Mesinger and Black (1992). Two more cases revealing the eta/sigma behavior of this type, of other model errors being largely or to a considerable extent reproduced when the model is switched to sigma, are the arctic surge case of Mesinger and Black (1992) and the midtropospheric cutoff case of Mesinger *et al.* (1997).

Another Eta numerical design feature on which a considerable amount of statistics has been obtained is resolution. With the 80-km Eta, a test on the impact of the increase in vertical resolution from 17 to 38 layers has been made, running a sample of 148 forecasts (Mesinger *et al.*, 1997). Three tests on the impact of the increase in horizontal resolution were made at various times (Black, 1994; Rogers *et al.*, 1996; Mesinger *et al.*, 1997), all from 80 to about 40 km, with 38 layers in the vertical. All of these tests have demonstrated a clear improvement resulting from increased resolution, with the improvement from doubling the horizontal resolution being substantially greater than that from the mentioned increase in the vertical resolution.

These results as well as evidence of numerous cases of improved simulations of orographically forced small-scale circulations (e.g., Black, 1994), along with practical considerations, have led to operational implementation in 1995 of a still higher resolution version of the Eta. It was run at about 29-km horizontal resolution, and 50 layers in the vertical. I will refer to it as the "29-km Eta"; the name "meso Eta" has also been used. The 29-km Eta was run until June 1998, when it was replaced by the "off-time" runs of the Eta. The operational setup of the 29-km Eta differed from the "early Eta" in more ways than the resolution; there were altogether five differences between the two, as follows:

1. 29 km/50 layer resolution versus 48 km/38 layer resolution of the early Eta.
2. 3:25-hr data cutoff and use of this late cutoff for initializations at 0300 and 1500 UTC versus the only 1:15-hr cutoff of the early Eta.
3. "Current" versus 12-hr-old Avn lateral boundary conditions.
4. A 3-hr "mini" data assimilation versus the 12-hr assimilation of the early Eta.
5. Smaller domain size. The 48-km Eta domain was 106 × 80 deg, while the 29-km domain was 70 × 50 deg of rotated longitude × latitude, respectively. Thus, the 29-km domain was by a factor of about 2.5 smaller than that of the 48-km Eta. The two domains are shown in Fig. 10.

The question naturally arises as to the impact of the differences between the two model setups on the model performance. Of the five differences listed, note that the first three would be expected to favor the 29-km model, and the last two the 48-km one. It would perhaps be generally expected that the first three should have by far a more dominant impact. Indeed, as stated, there are well-documented examples of benefits the 29-km Eta achieves, some of them clearly due to its better depiction of the local topography (e.g., Black, 1994; Schneider *et al.*, 1996). Precipitation scores of the early period of the running of the model have appeared to support this expectation (Schneider *et al.*, 1996).

With more than 2 years of scores available, Fig. 11 shows threat and bias score plots of the two models for the 24-month period 16 October 1995–15 October 1997, along with those of the Avn/MRF model and of the NGM. The choice of mid–October 1995 for the starting time of this sample is made because of the already referred to upgrade of the early Eta at that time. The sample contains 1245 forecasts by each of the four

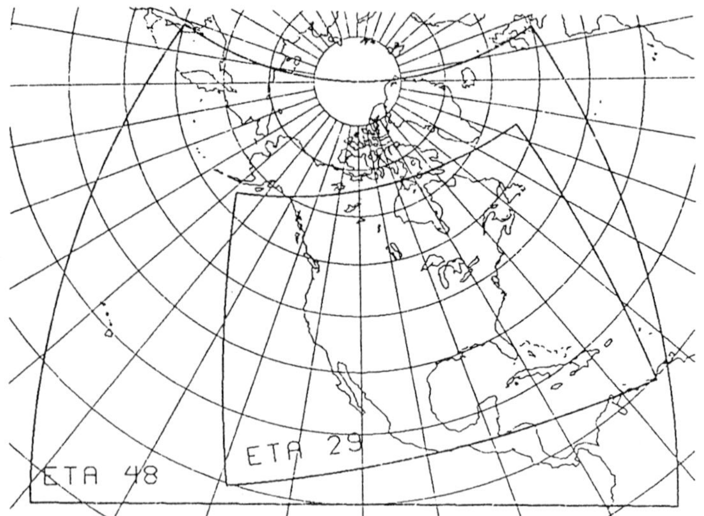

Figure 10   The domains of the Eta 48-km and of the Eta 29-km model.

models; 618 of them verifying at 24 hr and 627 verifying at 36 hr. Note that the 29-km model was run only out to 36 hr, or more precisely 33 hr, so that these two verification periods only are available for all four of the models.

Inspection of the threat score plots displayed shows that the two Eta models exhibit a very similar performance. The 29-km model is winning the two lowest categories, but it is losing the 1.5-in. category; the remaining categories are about a tie.

It would seem important to understand the reason for this relatively successful performance of the 48-km model. The EMC precipitation forecast archiving system enables examination of scores for specific forecast and time periods. Given that the influence of the model's western boundary information should be felt more at 36 hr than at 24 hr, and that it could be expected to have more impact during winter than during the summer half of the year in view of stronger westerlies in winter, one might hope to detect some kind of a signal by subdividing the sample into 24- and 36-hr forecasts and/or into "winter" and "summer" periods. Various subdivisions of this kind have been done and no clear signal was detected.

The relatively successful performance of the 48-km model thus remains somewhat of a puzzle. Recall, as referred to, that a clear benefit was obtained in clean resolution-only experiments when increasing the Eta resolution from 80 to 40 km, in three separate test periods.

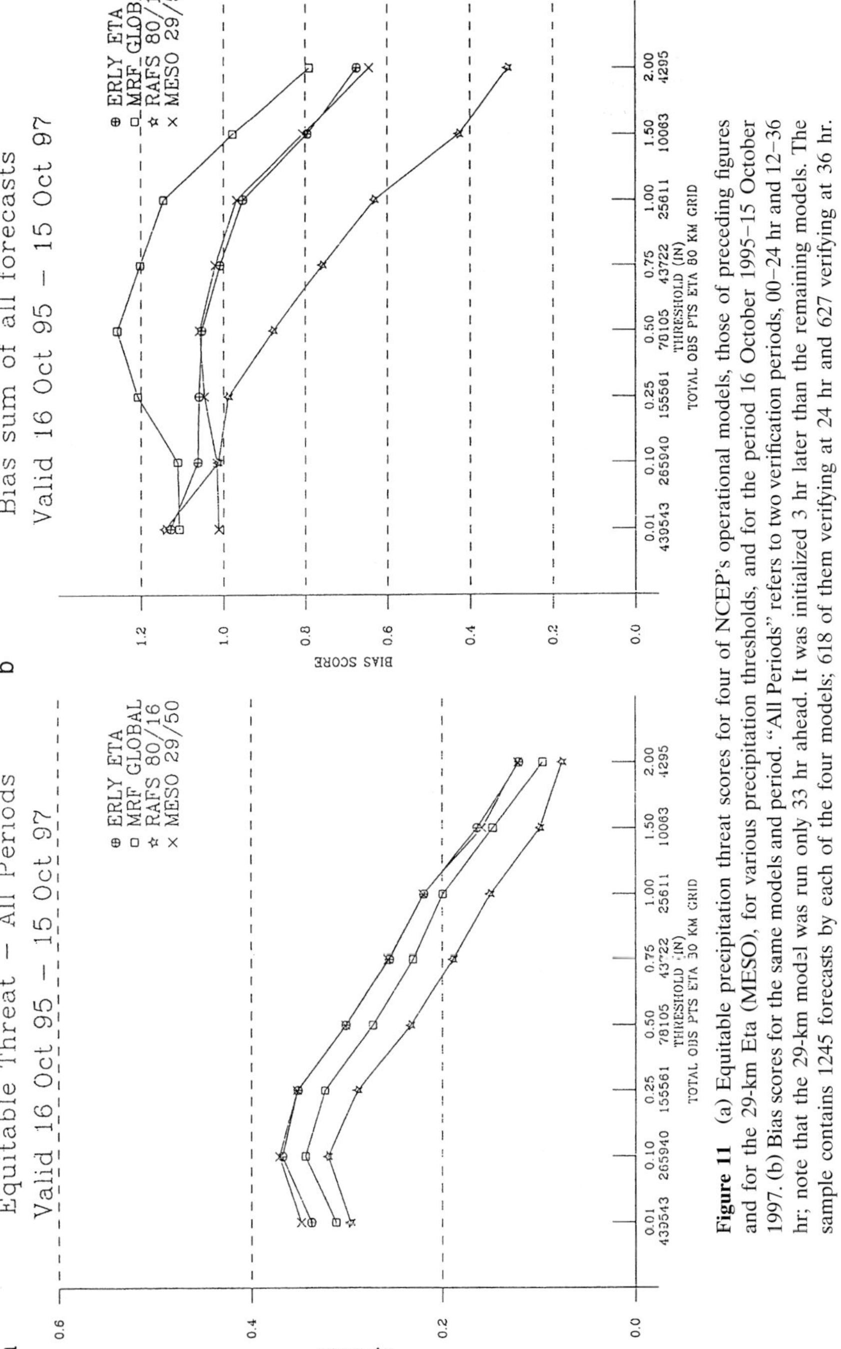

Figure 11 (a) Equitable precipitation threat scores for four of NCEP's operational models, those of preceding figures and for the 29-km Eta (MESO), for various precipitation thresholds, and for the period 16 October 1995–15 October 1997. (b) Bias scores for the same models and period. "All Periods" refers to two verification periods, 00–24 hr and 12–36 hr; note that the 29-km model was run only 33 hr ahead. It was initialized 3 hr later than the remaining models. The sample contains 1245 forecasts by each of the four models; 618 of them verifying at 24 hr and 627 verifying at 36 hr.

A possible explanation is that performance of some of the model's parameterization schemes deteriorates when the resolution is increased beyond 40–50 km. This, of course, is a subject of attention but no obvious candidate problem has been identified.

Note, however, that for a given accuracy of the boundary conditions, and for a limited-area model which for whatever reason intrinsically has more skill than its driver model, the domain size could help the performance of the limited-area model by allowing it to develop larger scales of motion, consistent with the model and also more realistic than those of the driver model.

The evidence at hand, along with the lack of credible alternatives in view of the model differences and possibilities listed, thus suggests that the much larger domain size of the 48-km model is a significant and most likely the main contributor to its successful performance relative to that of the 29-km model.

IX. HURRICANE TRACKS

Still another indication of the benefits from emphasizing a large model domain of uniform resolution, and an Arakawa-style design, may be the Eta performance in forecasting tracks of major 1996 Atlantic hurricanes in comparison with that of models employing two competing concepts (Mesinger, 1998b). Successful Eta performance in forecasting tracks of various hurricanes, tropical storms, or Pacific tropical cyclones was already noted early in the developmental work on the model (e.g., Kerr, 1990, and Lazić, 1990, among others). Thus, Kerr (1990) refers to the experimental Eta's "uncommon hurricane prediction skill during the past season." More recent examples, of the Eta skill with the tracks of the 1995 hurricanes Allison and Felix, are described in Mesinger (1996b).

The statistics put together in Mesinger (1998b) seem worthy of note in connection with the issues raised in the preceding sections. For the two most intense landfalling Atlantic hurricanes of the 1996 season, Bertha and Fran, 48-hr forecast position errors were compiled for the Eta 48-km and for three other NCEP operational or experimental models: the global (Avn) model, the regional spectral 50-km model (RSM, Juang *et al.*, 1997), and the GFDL hurricane model (GHM, Kurihara *et al.*, 1998). For Fran, errors of the NGM were also included in the comparison. NGM errors were not available for Bertha. For each storm, errors were compiled for six consecutive forecasts at 12-hr intervals, the latest six forecasts having the hurricanes centered still over water at the initial time. In other words, of forecasts at 12-hr intervals, in both cases the last six forecasts initialized before landfall were included. The errors for both hurricanes are reproduced in Table I. The median and the average errors, for all 12 forecasts considered as one sample, are also displayed.

It should be stressed that what we are looking at are not merely differences among models, but differences among forecasting systems including the data assimilation systems. In addition, the lateral boundary condition impact is involved in case of the Eta and the GHM, the Eta as stated before using the "old" and the GHM the "current" Avn boundary conditions. Note, in particular, that the Eta forecasts are available at approximately 2.5 hr after the synoptic times of 0000 and 1200 UTC, as opposed to about 5 hr in the case of the GHM.

Given the numbers of the table, an additional simple statistic is obtained by looking at how many times a model was the most accurate of the five. It is seen that the GHM and the Eta were each the most accurate three times, and have in addition shared the "first place" once; the RSM and the NGM were each the most accurate twice, and the Avn once. It is instructive to note that in a sample of only 12 forecasts, and five models of very different design including one at the time frozen for about 5 years, each of the models was best in at least one of the forecasts.

Regarding the overall performance, what is of interest here is the difference in approaches between the Eta, the RSM, and the GHM. The

Table I

NGM, Avn, RSM 50-km, Eta 48-km, and GHM 48-hr Errors[a]

Verification time	NGM	Avn	RSM50	Eta48	GHM
0000 UTC 12 July	—	425	425	625	325
1200 UTC 12 July	—	900	750	300	250
0000 UTC 13 July	—	200	225	75	75
1200 UTC 13 July	—	175	100	125	150
0000 UTC 14 July	—	350	175	300	225
1200 UTC 14 July	—	550	475	475	450
1200 UTC 5 September	275	325	325	100	150
0000 UTC 6 September	350	525	500	100	250
1200 UTC 6 September	325	350	325	125	200
0000 UTC 7 September	75	575	500	200	150
1200 UTC 7 September	75	425	350	125	100
0000 UTC 8 September	100	75	150	125	200
Median error	—	387.5	337.5	125	200
Average error	—	405	360	225	210

[a] In kilometers in forecasting the position of Hurricane Bertha (July; NGM errors not available), and Hurricane Fran (September).

Eta approach has been summarized already. The RSM is a nested spectral model, employing a perturbation approach to achieve higher resolution over the region of interest. GHM is a multiply nested grid point model, in 1996 run with two movable nests of 37- and 18.5-km resolutions. The resolution of its outside grid is 110 km. It is designed on a nonstaggered grid. As a dedicated hurricane model, it has an advanced vortex initialization scheme.

While the RSM is seen to result in some improvement over the Avn model, the improvement is not nearly of the magnitude to make it competitive with the Eta. Note that a comparison of the threat scores of the two models for 1996 (Mesinger, 1998a, Fig. 3) shows a significant advantage of the Eta, across all of the thresholds monitored.

Regarding the Eta versus the GHM, the relatively small size of the sample limits the significance of the results shown. Still, the fact that the GHM and the Eta are of a comparable accuracy in this sample raises questions relevant to modeling in general and, in particular, to limited-area modeling. It suggests that there are components in the Eta system which compensate for the Eta disadvantages of using less data, using 12-hr-old lateral boundary conditions, having less horizontal resolution in the domain of the hurricane, and having no vortex initialization scheme—if all of these indeed are advantages of the GHM compared to the Eta. To the extent they are, once again the uniform horizontal resolution over a very large model domain (and higher than the GHM coarse grid resolution) and also the staggered horizontal grid seem to me to be the prime candidates that should come to mind.

The desirability of a large "pristine uniform-resolution" area, to adopt once more a term from Côté *et al.* (1998), is of course in conflict with the also very desirable increase in resolution. The experience summarized here I believe strongly points toward a careful cost/benefit analysis of the two, as opposed to succumbing to the intense lure of the high-visibility value of the resolution kilometer number. Note that in the framework of the 48- and the 29-km Eta versions discussed earlier, assuming a proportionate increase in the vertical resolution and various economies as practiced, a 1-km increase in horizontal resolution is in terms of computer resources demand equivalent to about a 10% increase in the domain size. Thus, domain size is not expensive compared to the benefits that it appears to offer.

X. THE PROGRESS ACHIEVED

As background for a look to follow into the gain in the time validity of the precipitation forecasts, Fig. 12a shows equitable threat scores for

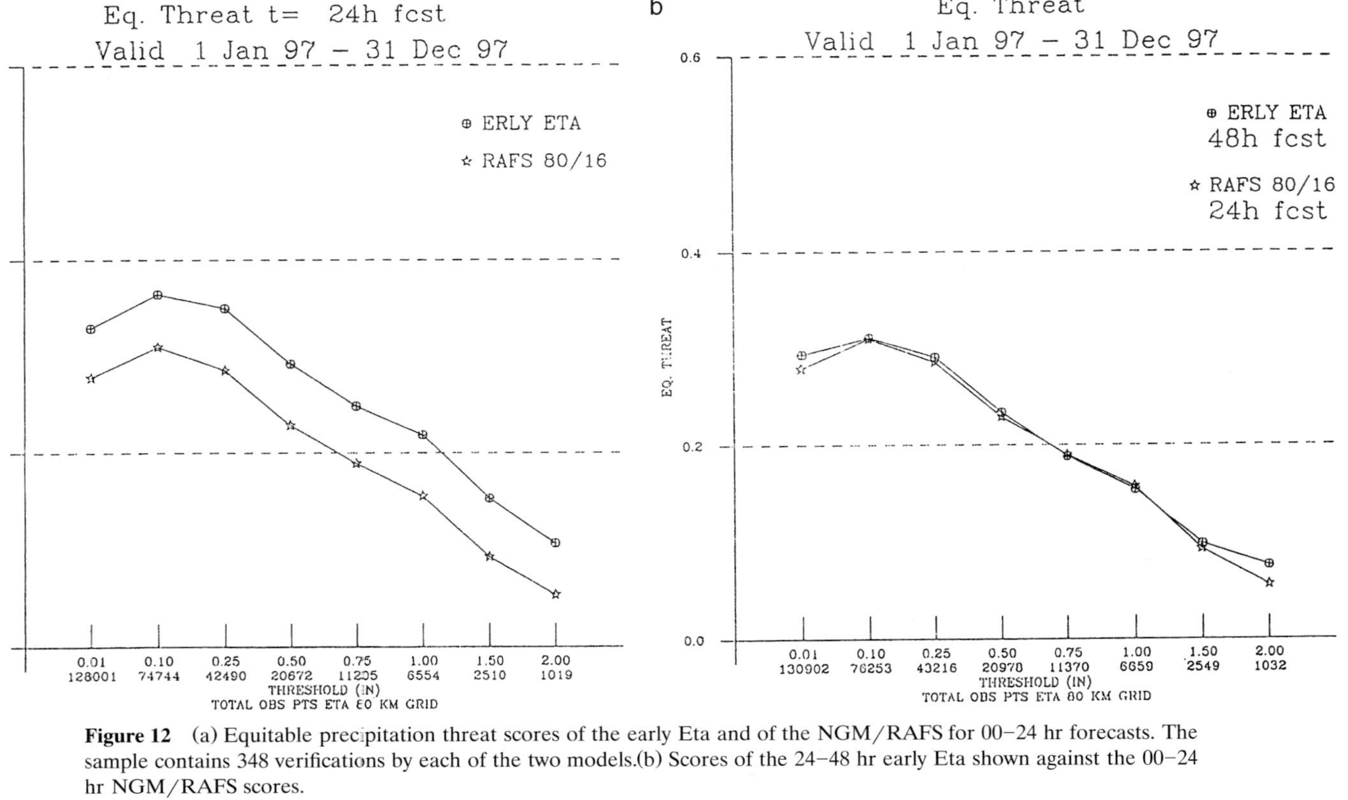

Figure 12 (a) Equitable precipitation threat scores of the early Eta and of the NGM/RAFS for 00–24 hr forecasts. The sample contains 348 verifications by each of the two models.(b) Scores of the 24–48 hr early Eta shown against the 00–24 hr NGM/RAFS scores.

00–24 hr forecasts of the early Eta and of the NGM/RAFS for 1997. Considerable gain in the accuracy of 24-hr forecasts is seen achieved by the Eta over the NGM, going from about 20% in terms of the threat scores for low precipitation categories to about a factor of 2 at the highest category monitored of 2 in./24 hr and greater.

In Fig. 12b, the 24-hr NGM threat score plot is reproduced along with that of the early Eta 24–48 hr forecasts. At the two ends the Eta scores are seen to be visibly higher, whereas for the intermediate categories the scores are very similar with a slight advantage overall in favor of the Eta as well.

Given that as stated (DiMego *et al.*, 1992) the NGM has been frozen since August 1991, Fig. 12b thus demonstrates that in terms of the official NMC/NCEP model-produced quantitative precipitation forecasts (QPF) in 6 years a gain in forecast validity of 24 hr has been achieved. Put differently, for given accuracy, in 6 years the time validity of the NMC/NCEP model-produced QPF forecasts has doubled.

One should note, however, that this increase in the NMC/NCEP model QPF accuracy is the result of a cumulative impact of at least two and very likely several effects, only one of which is the Eta model/system improvement during the 6 years. One major remaining factor is the increased resolution included as one change of the upgrade of October 1995, made possible by the increased computing resources available. A very likely other factor is one of the Eta QPFs being superior to those of the NGM already at the time the NGM was frozen. Note, for example the Eta's substantially higher threat scores across all of the categories of a sample of 58 forecasts of summer 1989, shown in the lower panel of Fig. 4 of Mesinger *et al.* (1990), with the two models using the same convection scheme. In that experiment, the same convection scheme had resulted in a similar bias pattern for the two models, which reduced what would appear to have been a very large impact of the bias on standard two-model threat scores computed at the time (see the upper panel of the same figure).

XI. EXAMPLE OF A SUCCESSFUL FORECAST

To complement various statistics summarized so far, in Fig. 13 a recent example of a successful precipitation forecast is shown. The case is for southern California heavy rains of early December 1997, affecting the symposium venue area as part of a pattern widely blamed on the intense El Niño of 1997/1998. In Fig. 13a, the 48-km Eta forecast of 24–48 hr accumulated precipitation, in inches, is shown. The amounts plotted are the 80-km box values as remapped for verification. In Fig. 13b, the

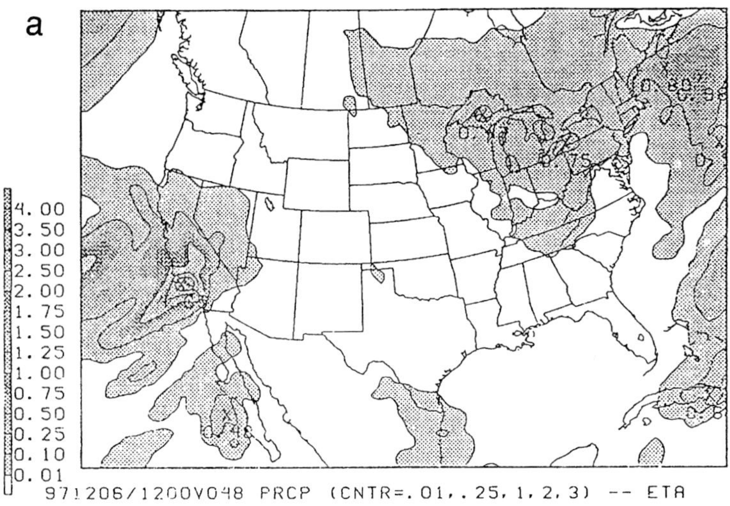

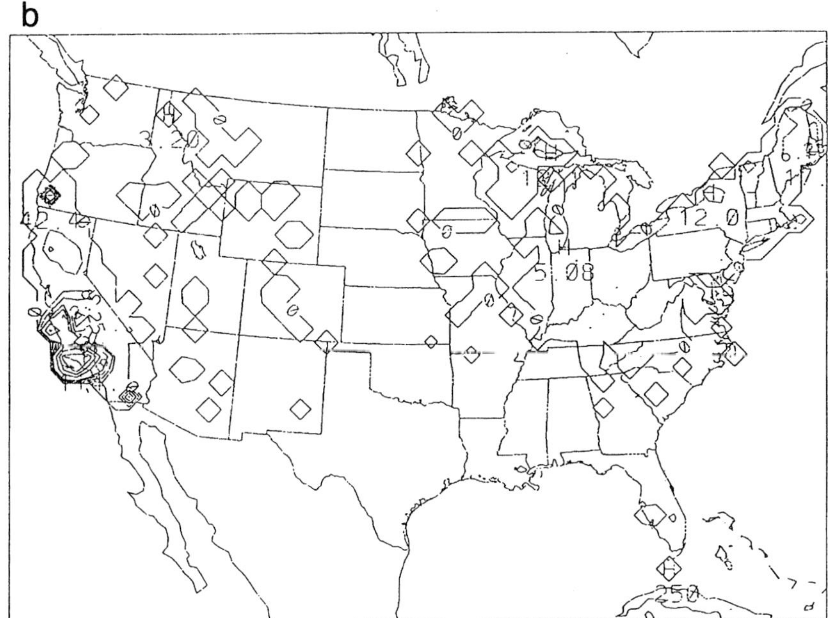

Figure 13 (a) The 48-km Eta forecast of 24–48 hr accumulated precipitation, in inches, verifying 1200 UTC 6 December 1997. (b) EMC verification analysis of 24-hr accumulated precipitation, in millimeters, verifying at the same time. Both are values valid for 80-km boxes, remapped for the forecast, and averaged for the observed precipitation.

observed precipitation in mm/24 hr is shown, also averaged over 80-km boxes.

The forecast maximum of 4.65 in./24 hr is very close to the observed box maximum of 114 mm/24 hr. The location of the forecast maximum is also well predicted, as was the one of the preceding 00–24 hr forecast period over Florida (not shown). The verification analysis of the two 24-hr periods resulted in 2 "points" (80-km boxes) of precipitation of 2 in. and greater during the first 24-hr period, over Florida, and 6 points the second, over southern California. The model had forecast 1 point of 2 in. and greater the first period, and 3 the second. All four forecast points verified. These were four "hits" out of a total of 114 that resulted in the Eta 24–48 hr scores' advantage over the NGM's 00–24 hr ones at 2 in. and greater that is seen in Fig. 12b.

XII. CONCLUSION

The intention of the leading sections of this chapter was to summarize some of the issues of the Arakawa approach in numerical methods, as an illustration of the variety of considerations that are made and opportunities that present themselves. This has also enabled me to review some of the recent ideas and results in addressing the horizontal grid problems. Propagation of gravity waves forced at individual grid points on a semi-staggered B/E Grid was the central theme of this review.

The sections that followed for the most part contained a synopsis of the results of the Eta model, viewed in comparison with other models of NMC (now NCEP). The Eta model's early success in its mission of advancing the short-range forecasting skill over the contiguous United States I find strongly supportive of the benefits obtainable through an Arakawa-style numerical formulation used by the model, as opposed to a competing high-formal-accuracy, filtering-of-small-scales approach of the NGM.

These as well as some of the more recent Eta model results summarized in later sections in addition seem to offer guidance regarding a number of central issues of the limited-area modeling strategy. One is that of the trade-off between the model resolution and domain size with indications of a considerable advantage the 48-km Eta is enjoying as a result of its uncommonly large "pristine uniform-resolution" (Côté *et al.*, 1998) domain. Another is the related issue of the advection of lower accuracy lateral boundary information with the Eta displaying a perhaps surprising degree of ability to overcome the contamination by the 12-hr-old Avn lateral boundary data.

Large-sample comparisons of the typically second-order accurate Eta against the fourth-order NGM of the early period, and against the RSM with its spectral "infinite"-order accuracy in the horizontal at more recent times, are favorable to the Eta and bring once again into focus the issue of the numerical design priorities. While it has been stressed that the concept of the order of accuracy based on Taylor expansion "is not relevant"—to adopt the expression of Arakawa (1997)—regarding various issues addressed by the Arakawa approach, one might still wonder about reasons as to why the benefits of the higher formal accuracy which are typically very visible in simple test problems seem to be absent to such a degree in complex model integrations. Note, for example, the statement of Cullen *et al.* (1997) that "use of the fourth order Heun [advection] scheme is essential to obtain results of this quality in the test problem. However, the sensitivity of the complete model to the choice between second and fourth order schemes at forecast resolutions (grid lengths less than 100 km) has been slight."

The forcing at individual model grid points done by physics packages of "complete" models is the obvious prime candidate answer. Recall that in physics the grid point values are treated as box averages. This means that box boundaries are treated as discontinuities. Of numerical schemes that have seen some application in meteorology perhaps the only one that treats grid point values as box averages allowing for possible discontinuities at box boundaries is the piecewise polynomial scheme, studied at some length by Carpenter *et al.* (1990). Enthusiasm seems lacking for a wider application of this approach, probably because of the cost involved. Note, however, that various conservation constraints in the Arakawa style are not in conflict with the box-average interpretation of the grid point values, because this is precisely how the grid point values are treated in the difference analogs that are maintained.

Expectations are perhaps universal that further increases in resolution will result in significant further increases in weather prediction and climate simulation skill. This is based not only on the fact that smaller scale terrain and other features will thereby be taken into account but also on encouraging results of actual integrations with higher resolution model versions. Regarding the Eta, note for example the reports on the Eta 10-km effort by Staudenmaier and Mittelstadt (1998) and by Black *et al.* (1998). Many other examples could be cited, including also some from nonhydrostatic models run in a weather prediction mode.

Yet, I do not find that we should expect that the impact of the differences in numerical design should automatically diminish and eventually disappear as the resolution increases. Note that the choices, if anything, are getting even more numerous due to the trend toward relaxation or removal of the hydrostatic assumption, and due to increasing computing power at our disposal. It is my expectation that, as a result, the rewards from judicious choices in the numerical design area, made with due regard to dynamical properties of the atmosphere/ocean, and physical parameterizations used, will only grow bigger.

ACKNOWLEDGMENTS

Regarding the part of this paper summarizing the performance of the Eta model, it is a pleasure to acknowledge that the development and implementation of a forecasting system, of which a model is only a part, results from the dedicated work of many people. Only some of the contributors are mentioned in various references. The EMC's precipitation analysis and verification system was originally established by John Ward, with a contribution of the grid-to-grid remapping system by the present author; and was subsequently further developed and has been maintained now for many years by Mike Baldwin.

The model's graphical archive resulting in plots such as the one of Fig. 13 and leading to numerous essential conclusions regarding model performance was established and is maintained by Keith Brill. Tom Black read the manuscript and provided many useful suggestions, which have improved the quality of the text. Jim Purser noted an error in the draft version of Fig. 3, upper panel and, along with Henry Juang, provided other useful comments as part of the NCEP internal review of the manuscript. Additional very helpful comments came from an anonymous outside reviewer.

On a more general level, the Eta model research reported here would not have been possible without recourse to the extraordinary research environment of the Environmental Modeling Center and of other unique facilities at the U.S. National Centers for Environmental Prediction, Camp Springs, Maryland. In this sense and otherwise, support of Drs. Eugenia Kalnay, director of the center for most part of the period covered, Geoff DiMego, chief of its regional and mesoscale modeling branch, and Ron McPherson, NCEP's director, were vital.

REFERENCES

Arakawa, A. (1966). Computational design for long-term numerical integration of equations of fluid motion: Two dimensional incompressible flow. Part I. *J. Comput. Phys.* **1,** 119–143.

Arakawa, A. (1972). Design of the UCLA general circulation model. *In* "Numerical Simulation of Weather and Climate," Tech. Rep. 7. Department of Meteorology, University of California, Los Angeles.

Arakawa, A. (1988). Finite-difference methods in climate modeling. *In* "Physically-Based Modelling and Simulation of Climate and Climatic Change" (M. E. Schlesinger, ed.), Part I, pp. 79–168. Kluwer, Dordrecht.

Arakawa, A. (1997). Adjustment mechanisms in atmospheric models. *J. Meteor. Soc. Japan* **75,** No. 1B, 155–179.

Arakawa, A., and V. R. Lamb (1977). Computational design of the basic dynamical processes of the UCLA general circulation model. *In* "Methods in Computational Physics" (J. Chang, ed.), Vol. 17, pp. 173–265. Academic Press, New York.

Arakawa, A., and V. R. Lamb (1981). A potential enstrophy and energy conserving scheme for the shallow water equations. *Mon. Wea. Rev.* **109**, 18–36.

Betts, A. K. (1986). A new convective adjustment scheme. Part I: Observational and theoretical basis. *Quart. J. Roy. Meteor. Soc.* **112**, 693–709.

Betts, A. K., and M. J. Miller (1986). A new convective adjustment scheme. Part II: Single column tests using GATE wave, BOMEX and arctic air-mass data sets. *Quart. J. Roy. Meteor. Soc.* **112**, 693–709.

Black, T. L. (1988). The step-mountain eta coordinate regional model: A documentation. National Oceanic and Atmospheric Administration/National Weather Service. (Available from NOAA Environmental Modeling Center, Room 207, 5200 Auth Road, Camp Springs, MD 20746.)

Black, T. L. (1994). The new NMC mesoscale Eta model: Description and forecast examples. *Wea. Forecasting* **9**, 265–278.

Black, T. L., and F. Mesinger (1989). Forecast performance of NMC's eta coordinate regional model. *In* "12th Conf. on Weather Analysis and Forecasting," Monterey, CA, pp. 42–43. Amer. Meteor. Soc.

Black, T. L., and Z. I. Janjić (1988). Preliminary forecast results from a step-mountain eta coordinate regional model. *In* "Preprints, 8th Conf. on Numerical Weather Prediction," Baltimore, MD, pp. 442–447. Amer. Meteor. Soc.

Black, T., D. Deaven, and G. DiMego (1993). The step-mountain eta coordinate model: 80 km 'early' version and objective verifications, NWS Technical Procedures Bull. 412. National Oceanic and Atmospheric Administration/National Weather Service. (Available from National Weather Service, Office of Meteorology, 1325 East-West Highway, Silver Spring, MD 20910.)

Black, T., M. Baldwin, G. DiMego, and E. Rogers (1998). Results from daily forecasts of the NCEP Eta-10 model over the western United States. *In* "Preprints, 12th Conf. on Numerical Weather Prediction," Phoenix, AZ, pp. 246–247. Amer. Meteor. Soc.

Carpenter, R. L., Jr., K. K. Droegemeier, P. W. Woodward, and C. E. Hane, (1990). Application of the piecewise parabolic method (PPM) to meteorological modeling. *Mon. Wea. Rev.* **118**, 586–612.

Chen, F., Z. Janjić, and K. Mitchell (1997). Impact of atmospheric surface-layer parameterizations in the new land-surface scheme of the NCEP mesoscale Eta model. *Bound.-Layer Meteor.* **85**, 391–421.

Côté, J., S. Gravel, A. Méthot, A. Patoine, M. Roch, and A. Staniforth, (1998). The operational CMC-MRB Global Environmental Multiscale (GEM) model. Part I: Design considerations and formulation. *Mon. Wea. Rev.* **126**, 1373–1395.

Cullen, M. J. P., T. Davies, M. H. Mawson, J. A. James, and S. C. Coulter, (1997). An overview of numerical methods for the next generation U.K. NWP and climate model. *In* "Numerical Methods in Atmospheric and Oceanic Modelling, The André J. Robert Memorial Volume" (C. Lin, R. Laprise, and H. Ritchie, eds.), pp. 425–444. Canadian Meteorological and Oceanographic Society/NRC Research Press, Ottawa.

DiMego, G. J. (1988). The National Meteorological Center regional analysis system. *Mon. Wea. Rev.* **116**, 977–1000.

DiMego, G. J., K. E. Mitchell, R. A. Petersen, J. E. Hoke, J. P. Gerrity, J. J. Tuccillo, R. L. Wobus, and H.-M. H. Juang (1992). Changes to NMC's regional analysis and forecast system. *Wea. Forecasting* **7**, 185–198.

Fels, S. B., and M. D. Schwarzkopf (1975). The simplified exchange approximation: A new method for radiative transfer calculations. *J. Atmos. Sci.* **32**, 1475–1488.

Gustafsson, N., and A. McDonald (1996). A comparison of the HIRLAM gridpoint and spectral semi-Lagrangian models. *Mon. Wea. Rev.* **124**, 2008–2022.

Heikes, R., and D. A. Randall (1995a). Numerical integration of the shallow-water equations on a twisted icosahedral grid. Part I: Basic design and results of tests. *Mon. Wea. Rev.* **123,** 1862–1880.

Heikes, R., and D. A. Randall (1995b). Numerical integration of the shallow-water equations on a twisted icosahedral grid. Part II: A detailed description of the grid and an analysis of numerical accuracy. *Mon. Wea. Rev.* **123,** 1881–1887.

Hoke, J. E., N. A. Phillips, G. J. DiMego, and D. G. Deaven (1985). NMC's Regional Analysis and Forecast System—results from the first year of daily, real-time forecasting. *In* "Preprints, 7th Conf. on Numerical Weather Prediction," Montreal, pp. 444–451. Amer. Meteor. Soc.

Hollingsworth, A., P. Källberg, V. Renner, and D. M. Burridge (1983). An internal symmetric computational instability. *Quart. J. Roy. Meteor. Soc.* **109,** 417–428.

Hong, S.-Y., and H.-L. Pan (1996). Nonlocal boundary layer vertical diffusion in a medium-range forecast model. *Mon. Wea. Rev.* **124,** 2322–2339.

Janjić, Z. I. (1977). Pressure gradient force and advection scheme used for forecasting with steep and small scale topography. *Contrib. Atmos. Phys.* **50,** 186–199.

Janjić, Z. I. (1979). Forward–backward scheme modified to prevent two-grid-interval noise and its application in sigma coordinate models. *Contrib. Atmos. Phys.* **52,** 69–84.

Janjić, Z. I. (1984). Nonlinear advection schemes and energy cascade on semi-staggered grids. *Mon. Wea. Rev.* **112,** 1234–1245.

Janjić, Z. I. (1994). The step-mountain eta coordinate model: Further developments of the convection, viscous sublayer, and turbulence closure schemes. *Mon. Wea. Rev.* **122,** 927–945.

Janjić, Z. I. (1996a). The Mellor–Yamada level 2.5 scheme in the NCEP Eta model. *In* "Preprints, 11th Conf. on Numerical Weather Prediction," Norfolk, VA, pp. 333–334. Amer. Meteor. Soc.

Janjić, Z. I. (1996b). The surface layer in the NCEP Eta model. *In* "Preprints, 11th Conf. on Numerical Weather Prediction," Norfolk, VA, pp. 354–355. Amer. Meteor. Soc.

Janjić, Z. I., and F. Mesinger (1984). Finite difference methods for the shallow water equations on various horizontal grids. *In* "Numerical Methods for Weather Prediction, Seminar 1983," pp. 29–101. ECMWF, Shinfield Park, Reading, U.K.

Janjić, Z. I., and F. Mesinger (1989). Response to small-scale forcing on two staggered grids used in finite-difference models of the atmosphere. *Quart. J. Roy. Meteor. Soc.* **115,** 1167–1176.

Janjić, Z. I., F. Mesinger, and T. L. Black (1995). The pressure advection term and additive splitting in split-explicit models. *Quart. J. Roy. Meteor. Soc.* **121,** 953–957.

Janjić, Z. I., T. L. Black, and G. DiMego (1998). Contributions toward development of a future community mesoscale model (WRF). *In* "Preprints, 16th Conf. on Weather Analysis and Forecasting," Phoenix, AZ, pp. 258–261. Amer. Meteor. Soc.

Ji, Y., and A. D. Vernekar (1997). Simulation of the Asian summer monsoons of 1987 and 1988 with a regional model nested in a global GCM. *J. Climate* **10,** 1965–1979.

Juang, H.-M. H., and J. E. Hoke (1992). Application of fourth-order finite differencing to the NMC Nested Grid Model. *Mon. Wea. Rev.* **120,** 1767–1782.

Juang, H.-M. H., S.-Y. Hong, and M. Kanamitsu (1997). The NCEP Regional Spectral Model: An update. *Bull. Amer. Meteor. Soc.* **78,** 2125–2143.

Kanamitsu, M., J. C. Alpert, K. A. Campana, P. M. Caplan, D. G. Deaven, M. Iredell, B. Katz, H.-L. Pan, J. Sela, and G. H. White (1991). Recent changes implemented into the global forecast system at NMC. *Wea. Forecasting* **6,** 425–435.

Kerr, R. A. (1990). Hurricane forecasting shows promise. *Science* **247,** 917.

Kurihara, Y., R. E. Tuleya, and M. A. Bender (1998). The GFDL hurricane prediction system and its performance in the 1995 hurricane season. *Mon. Wea. Rev.* **126**, 1306–1322.

Lacis, A. A., and J. E. Hansen (1974). A parameterization of the absorption of solar radiation in the earth's atmosphere. *J. Atmos. Sci.* **31**, 118–133.

Lazić L. (1990). Forecasts of AMEX tropical cyclones with a step-mountain model. *Austr. Meteor. Mag.* **38**, 207–216.

Leslie, L. M., and R. J. Purser (1991). High-order numerics in an unstaggered three-dimensional time-split semi-Lagrangian forecast model. *Mon. Wea. Rev.* **119**, 1612–1623.

Lilly, D. K. (1997). Introduction to "Computational design for long-term numerical integration of the equations of fluid motion: Two-dimensional incompressible flow. Part I." *J. Comput. Phys.* **135**, 101–102.

McDonald, A. (1997). Lateral boundary conditions for operational regional forecast models; a review, HIRLAM Tech. Rep. 32. (Available from A. McDonald, Irish Meteorological Service, Glasnevin Hill, Dublin 9, Ireland.)

Mellor, G. L., and T. Yamada (1982). Development of a turbulence closure model for geophysical fluid problems. *Rev. Geophys. Space Phys.* **20**, 851–875.

Mesinger, F. (1973). A method for construction of second-order accuracy difference schemes permitting no false two-grid-interval wave in the height field. *Tellus* **25**, 444–458.

Mesinger, F. (1974). An economical explicit scheme which inherently prevents the false two-grid-interval wave in the forecast fields. *In* "Proc. Symp. Difference and Spectral Methods for Atmosphere and Ocean Dynamics Problems," Part II, pp. 18–34. Academy of Sciences, Novosibirsk.

Mesinger, F. (1977). Forward–backward scheme, and its use in a limited area model. *Contrib. Atmos. Phys.* **50**, 200–210.

Mesinger, F. (1982). On the convergence and error problems of the calculation of the pressure gradient force in sigma coordinate models. *Geophys. Astrophys. Fluid. Dyn.* **19**, 105–117.

Mesinger, F. (1984). A blocking technique for representation of mountains in atmospheric models. *Riv. Meteor. Aeronautica* **44**, 195–202.

Mesinger, F. (1993). Forecasting upper tropospheric turbulence within the framework of the Mellor–Yamada 2.5 closure, Res. Activities Atmos. Oceanic Modelling, Rep. 18, 4.28–4.29. WMO, Geneva.

Mesinger, F. (1996a). Forecasting cold surges east of the Rocky Mountains. *In* "Preprints, 11th Conf. on Numerical Weather Prediction," Norfolk, VA, pp. 68–69. Amer. Meteor. Soc.

Mesinger, F. (1996b). Improvements in quantitative precipitation forecasts with the Eta regional model at the National Centers for Environmental Prediction. The 48-km upgrade. *Bull. Amer. Meteor. Soc.* **77**, 2637–2649; Corrigendum, **78**, 506.

Mesinger, F. (1998a). Quantitative precipitation forecasts of the "early" Eta model: An update. *In* "Preprints, 16th Conf. on Weather Analysis and Forecasting," Phoenix, AZ, pp. 184–186. Amer. Meteor. Soc.

Mesinger, F. (1998b). Performance of the 48-km Eta in forecasting tracks of major landfalling Atlantic hurricanes of the 1966 season, Bertha and Fran. *In* "Preprints, 16th Conf. on Weather Analysis and Forecasting, Symp. on the Research Foci of the U.S. Weather Research Program," Phoenix, AZ, pp. 526–528. Amer. Meteor. Soc.

Mesinger, F., and A. Arakawa (1976). Numerical methods used in atmospheric models, GARP Publ. Ser. 17, Vol. I. (Available from World Meteorological Organization, Case Postale No. 5, CH-1211 Geneva 20, Switzerland.)

Mesinger, F., and T. L. Black (1989). Verification tests of the Eta model, October–November 1988, NOAA/NWS National Meteorological Center, Office Note No. 355. (Available from National Weather Service, Office of Meteorology, 1325 East-West Highway, Silver Spring, MD 20910.)

Mesinger, F., and T. L. Black (1992). On the impact on forecast accuracy of the step-mountain (eta) vs. sigma coordinate. *Meteor. Atmos. Phys.* **50,** 47–60.

Mesinger, F., and Z. I. Janjić (1974). Noise due to time-dependent boundary conditions in limited area models, GARP Programme on Numerical Experimentation, Rep. 4, pp. 31–32. WMO, Geneva.

Mesinger, F., and Z. I. Janjić (1985). Problems and numerical methods of the incorporation of mountains in atmospheric models. *In* "Large-scale Computations in Fluid Mechanics," (B. E. Engquist, S. Osher, and R. C. J. Somerville, eds.), Part 2, *Lect. Appl. Math.* **22,** pp. 81–120. Amer. Math. Soc., Providence, RI.

Mesinger, F., and Z. I. Janjić (1987). Numerical techniques for the representation of mountains. *In* "Observation, Theory and Modelling of Orographic Effects," Seminar/workshop 1986, pp. 29–80. ECMWF, Shinfield Park, Reading, UK.

Mesinger, F., Z. I. Janjić, S. Ničković, D. Gavrilov, and D. G. Deaven (1988). The step-mountain coordinate: Model description and performance for cases of Alpine lee cyclogenesis and for a case of an Appalachian redevelopment. *Mon. Wea. Rev.* **116,** 1493–1518.

Mesinger, F., T. L. Black, D. W. Plummer, and J. H. Ward (1990). Eta model precipitation forecasts for a period including tropical storm Allison. *Wea. Forecasting* **5,** 483–493.

Mesinger, F., R. L. Wobus, and M. E. Baldwin (1996). Parameterization of form drag in the Eta model at the National Centers for Environmental Prediction. *In* "Preprints, 11th Conf. on Numerical Weather Prediction," Norfolk, VA, pp. 324–326. Amer. Meteor. Soc.

Mesinger, F., T. L. Black, and M. E. Baldwin (1997). Impact of resolution and of the eta coordinate on skill of the Eta model precipitation forecasts. *In* "Numerical Methods in Atmospheric and Oceanic Modelling, The André J. Robert Memorial Volume" (C. Lin, R. Laprise, and H. Ritchie, eds.), pp. 399–423. Canadian Meteorological and Oceanographic Society/NRC Research Press, Ottawa.

Ničković, S. (1994). On the use of hexagonal grids for simulation of atmospheric processes. *Contrib. Atmos. Phys.* **67,** 103–107.

Oliger, J., and A. Sundström (1978). Theoretical and practical aspects of some initial boundary value problems in fluid dynamics. *SIAM J. Appl. Math.* **35,** 419–446.

Pan, H.-L., and W.-S. Wu (1994). Implementing a mass-flux convective parameterization package for the NMC Medium-Range Forecast model. *In* "Preprints, 10th Conf. on Numerical Weather Prediction," Portland, OR, pp. 96–98. Amer. Meteor. Soc.

Phillips, N. A. (1959). An example of non-linear computational instability. *In* "The Atmosphere and the Sea in Motion (The Rossby Memorial Volume)," pp. 501–504. Rockefeller Institute Press, New York.

Pielke, R. A., M. E. Nicholls, R. L. Walko, T. A. Nygaard, and X. Zeng (1997). Several unresolved issues in numerical modelling of geophysical flows. *In* "Numerical Methods in Atmospheric and Oceanic Modelling, The André J. Robert Memorial Volume" (C. Lin, R. Laprise, and H. Ritchie, eds.), pp. 557–581. Canadian Meteorological and Oceanographic Society/NRC Research Press, Ottawa.

Plummer, D. W., T. L. Black, N. A. Phillips, and J. E. Hoke (1989). Tests of the Betts–Miller convective parameterization in the Nested Grid Model. *In* "Research Highlights of the NMC Development Division: 1987–1988," pp. 23–25. NOAA/NWS, Camp Springs. (Available from the NOAA Environmental Modeling Center, 5200 Auth Road, Camp Springs, MD 20746.)

Popović, J. M., S. Ničković, and M. B. Gavrilov (1996). Frequency of quasi-geostrophic modes on hexagonal grids. *Meteor. Atmos. Phys.* **58,** 41–49.

Purser, R. J., and M. Rančić (1998). Smooth quasi-homogeneous gridding of the sphere. *Quart. J. Roy. Meteor. Soc.* **124,** 637–647.

Rančić, M., R. J. Purser, and F. Mesinger (1996). A global shallow-water model using an expanded spherical cube: Gnomonic versus conformal coordinates. *Quart. J. Roy. Meteor. Soc.* **122,** 959–982.

Randall, D. A. (1994). Geostrophic adjustment and the finite-difference shallow-water equations. *Mon. Wea. Rev.* **122,** 1371–1377.

Rogers, E., D. G. Deaven, and G. J. DiMego (1995). The regional analysis system for the operational "early" Eta model. *Wea. Forecasting* **10,** 810–825.

Rogers, E., T. L. Black, D. G. Deaven, G. J. DiMego, Q. Zhao, M. Baldwin, N. W. Junker, and Y. Lin (1996). Changes to the operational "early" Eta analysis/forecast system at the National Centers for Environmental Prediction. *Wea. Forecasting* **11,** 319–413.

Sadourny, R. (1972). Conservative finite-differencing approximations of the primitive equations on quasi-uniform spherical grids. *Mon. Wea. Rev.* **100,** 136–144.

Sadourny, R., and P. Morel (1969). A finite-difference approximation of the primitive equations for a hexagonal grid on a plane. *Mon. Wea. Rev.* **97,** 439–445.

Schneider, R. S., and associates (1996). The performance of the 29 km meso Eta model in support of forecasting at the Hydrometeorological Prediction Center. *In* "Preprints, 11th Conf. on Numerical Weather Prediction," Norfolk, VA, pp. J111–J114. Amer. Meteor. Soc.

Staudenmaier, M., Jr., and J. Mittelstadt (1998). Results of the Western Region evaluation of the Eta-10 Model. *In* "Preprints, 12th Conf. on Numerical Weather Prediction," Phoenix, AZ, pp. 131–134. Amer. Meteor. Soc.

Sullivan, B. E., B. D. Terry, and N. W. Junker (1993). The use of dynamical model guidance in short-range forecasting. *In* "Preprints, 13th Conf. on Weather Analysis and Forecasting," Vienna, VA, pp. 35–43. *Am. Meteor. Soc.*

Thacker, W. C. (1978). Comparison of finite-element and finite-difference schemes. Part II: Two-dimensional gravity-wave motion. *J. Phys. Oceanogr.* **8,** 680–689.

Tuccillo, J. J., and N. A. Phillips (1986). Modeling of physical processes in the Nested Grid Model, NWS Technical Procedures Bull. 363. National Oceanic and Atmospheric Administration/National Weather Service. (Available from National Weather Service, Office of Meteorology, 1325 East-West Highway, Silver Spring, MD 20910.)

Williamson, D. (1969). Integration of the barotropic vorticity equation on a spherical geodesic grid. *Mon. Wea. Rev.* **97,** 512–520.

Winninghoff, F. J. (1968). On the adjustment toward a geostrophic balance in a simple primitive equation model with application to the problems of initialization and objective analysis, Ph.D. Thesis. Dept. Meteor. Univ. California, Los Angeles.

Yakimiw, E., and A. Robert (1990). Validation experiments for a nested grid-point regional forecast model. *Atmos.-Ocean* **28,** 466–472.

Zhao, Q., and F. H. Carr (1997). A prognostic cloud scheme for operational NWP models. *Mon. Wea. Rev.* **125,** 1931–1953.

Chapter 14

Formulation of Oceanic General Circulation Models

James C. McWilliams

Department of Atmospheric Sciences and Institute of Geophysics and Planetary Physics, University of California, Los Angeles

I. INTRODUCTION

The practice of oceanic numerical modeling is growing rapidly. Among the reasons for this are a widespread realization that model solutions can, either now or at least in the near future, be skillful in mimicking observed oceanic features; an understanding of the limitations of the alternative and more traditional scientific methodologies of making measurements in the oceans and developing analytic theories for highly nonlinear dynamical systems; an appreciation of the importance of the oceans in the socially compelling problems of anthropogenic changes in climate and the environment; and an exploitation of the steady increase in computing power that makes meaningfully comprehensive oceanic calculations ever more feasible.

General Circulation Model Development

The oceanic general circulation is defined as the currents on horizontal space and time scales larger than the mesoscale (of order 100 km and 3 months), the associated pressure, density, temperature, and salinity fields, plus all other elements involved in establishing the dynamical balances for these fields. The latter includes the forcing fields, the domain geometry, and the transport contributions by currents on meso- and microscales. In some contexts the general circulation also includes the biogeochemical processes associated with other material property fields such as nitrate or Freon. The term *tracers* denotes the scalar variables moving with fluid parcels, including temperature and salinity, T and S, respectively, which influence dynamics through the density, ρ.

This chapter is an overview of the formulation of oceanic general circulation models (OGCMs), with less of the history of their usage than is customary in review articles (e.g., Haidvogel and Bryan, 1992; McWilliams, 1996). OGCMs are defined as numerical models that include all of the major influences on the general circulation, constrained by limitations in both our knowledge of how to formulate the model and the available computing power. Of course, this is a somewhat loose definition, but I take it to imply that OGCMs necessarily have wind, heat, and water forcing, geographically correct domain geometry, and the equation of state for seawater—all with their inescapable approximations.

Why are OGCMs useful? It is obvious that they would be so for the study of ocean currents, but they have many other important applications as well: dynamical coupling with the atmosphere, sea ice, and land runoff that in reality jointly determine the oceanic boundary fluxes (e.g., El Niño); transport of biogeochemical materials; interpretation of the paleoclimate record; climate prediction for both natural variability and anthropogenic changes; assimilation of sparse measurements to provide dynamically consistent interpolations and syntheses; pollution dispersal; and fisheries and other biospheric management.

The scientific practices of using OGCMs include a continuing development process to improve their accuracy, a testing process against observations to assess their accuracy, and conceptual inquiries about how different physical influences combine and compete to produce the general circulation. A particularly instructive way to use an OGCM, once it has captured some particular behavior of interest, is to systematically reduce its component processes to distill the essential cause. Particularly interesting uses of OGCMs are to explore hypothetical realities, such as the ocean circulation of ancient earth epocs, the consequences of deliberate biological fertilization of the oceans, or future climates under alternative planetary management strategies.

II. DYNAMICS

Broadly summarized, the historical path of OGCM usage is first to model only the largest scales of motion with an excessively coarse numerical grid and a concomitant excessively linear and diffusive dynamics, where the "diffusion" represents the effects of, or parameterizes, the unresolved (i.e., sub-grid-scale, SGS) currents. Subsequently, more ambitious problems are posed that refine the grid; reduce the SGS diffusivities and thereby increase the nonlinearity of the dynamics; and examine both atmospherically forced fluctuations and intrinsic variability arising from instabilities of large-scale currents. The most energetic forms of the intrinsic variability are mesoscale eddies (and even larger scale, lower frequency fluctuations) that are generated by instability of the wind-driven gyres, and transient, three-dimensional overturning cells that are generated by instability of the thermohaline circulations. An increasingly broad range of scales of motion is thereby incorporated into the OGCM solutions, thus decreasing the scope of the requisite SGS parameterizations. Nevertheless, it is inconceivable that a numerical model could incorporate all excited degrees of freedom, which are of $\mathcal{O}(10^{40})$, if 1 mm and 1 sec are taken as the minimum scales and the whole ocean and 10^4 year are taken as the maximum scales. Therefore, SGS parameterizations will always be essential elements of an OGCM. The present frontiers for OGCM solutions lie both in fluctuations of decadal or longer periods, coupled with the atmosphere, and in the turbulent behavior of mesoscale currents and eddies.

The fundamental fluid dynamics of oceanic circulation is the Navier–Stokes equations for the rotating earth and a compressible liquid, seawater, comprised of water plus a suite of dissolved salts that occur in nearly constant ratio but variable amount (the salinity S), with an empirically determined equation of state. However, since the target of OGCMs is large-scale, low-frequency currents, we can choose among several levels of dynamical approximation (see Fig. 1):

- The Boussinesq equations neglect variations of density ρ everywhere in the momentum and mass balances except for the gravitational force, while retaining the full effects of compressibility in the equation of state. This excludes acoustic modes. This is a safe approximation for ocean currents because $\delta\rho/\rho \ll 1$.
- The primitive equations further make the hydrostatic approximation in the vertical momentum balance:

$$\frac{\partial P}{\partial z} = -g\rho, \tag{1}$$

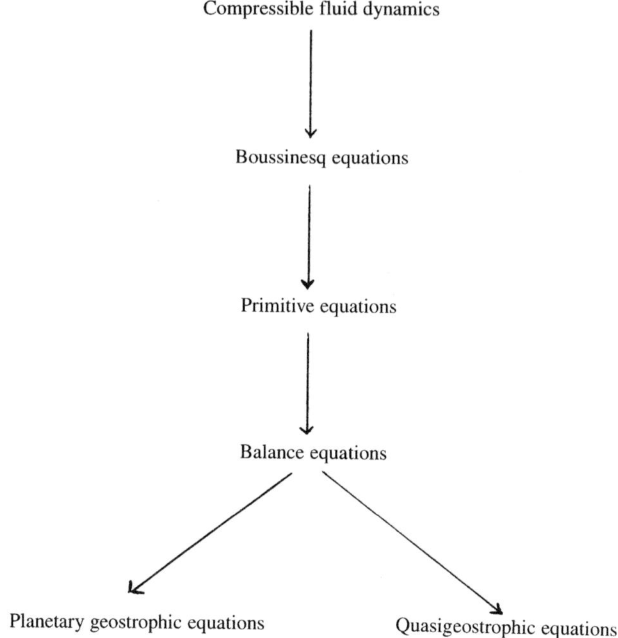

Figure 1   A hierarchy of dynamical approximations.

where P is the pressure, ρ is the density, z is the local vertical coordinate
(parallel to the gravitational force), and g is the gravitational acceleration.
This excludes convection, which requires vertical accelerations, and it
distorts high-frequency gravity waves. This is a safe approximation as long
as $H/L \ll 1$, where H and L are vertical and horizontal length scales,
respectively, characteristic of the solutions. This is true for all OGCM
solutions to date, though just barely so. The same condition of small aspect
ratio justifies the neglect of the earth's rotation vector $\mathbf{\Omega}_e$ except for its
local vertical projection, i.e., the Coriolis frequency, $f = 2\mathbf{\Omega}_e \cdot \hat{\mathbf{z}}$, where $\hat{\mathbf{z}}$
denotes the local unit vertical vector.

  • The balance equations further make approximations to the horizontal
momentum equations that are consistent with gradient-wind balance:

$$\frac{1}{\rho_0}\nabla_h^2 P = \hat{\mathbf{z}} \cdot \mathbf{\nabla}_h \times f\mathbf{u}_h + 2\frac{\partial(u,v)}{\partial(x,y)}, \tag{2}$$

where ρ_0 is the mean density, boldface denotes a vector, the subscript h
denotes horizontal component, $\mathbf{u}_h = (u,v)$, and $\mathbf{x}_h = (x,y)$. This is an

approximate balance of force divergences among the pressure gradient, Coriolis, and centrifugal forces that is usually a quite accurate approximation for general circulation currents (Gent and McWilliams, 1983). This model excludes gravity and inertial waves. There are a variety of particular formulations of balanced models, but among those whose balance relation asymptotically coincides with Eq. (2) for small Ro, the differences may not be quantitatively significant.

• The quasigeostrophic and planetary geostrophic equations are leading-order asymptotic approximations for small Rossby number, $Ro = V/fL \ll 1$ (where V is a characteristic horizontal velocity). They satisfy geostrophic balance:

$$\frac{1}{\rho_0}\boldsymbol{\nabla}_h P = -\hat{\mathbf{z}} \times f\mathbf{u}_h. \tag{3}$$

The two approximations differ by whether L/R_d is assumed to be order one or large, respectively, where $R_d \sim NH/f$ is the baroclinic deformation radius and N is the buoyancy frequency for the vertical density stratification. The size of R_d is typically in the range of 10–100 km, with smaller values at higher latitudes. Each of these models is a subset of the balance equations. They suffer from having nonuniform validity in L/R_d and Ro; from inaccuracy of Eq. (3) near the equator where $f = 0$; and, for the quasigeostrophic model, from the neglect of horizontal and temporal variations in $N(z)$. Nevertheless, these two maximally simplified models are accurate for many particular phenomena in the general circulation and therefore are useful simple models for the theory of ocean currents and, hence, for the interpretation of OGCM solutions.

When faced with this range of approximations, the early general circulation modelers chose the primitive equations because they were confident of the underlying assumptions, and the necessary numerical solution methods were simpler than those required for the Boussinesq or balance equations. From a modern dynamical perspective, the alternative choices of either the Boussinesq or balance equations could be defended, but their respective advantages—allowing solutions with H/L not small or having both an intrinsically smoother evolutionary rate (hence permitting a large step size, Δt) and a reduced dynamical phase space within which to interpret the solution—are unlikely to be enormous in most OGCM applications. Another basis for distinguishing among these three models has been their permissible time step size under the constraint of computational stability for explicit time-integration methods. The permissible step size is larger the greater the degree of physical approximation because of the implied exclusion of faster modes of behavior. This distinction would disappear if unconditionally stable, implicit time-integration methods were used.

The common basis for an OGCM is the primitive equations for a thin fluid layer in spherical coordinates (i.e., replacing the radial coordinate $r = z + a$ by the mean radius of the earth a wherever it appears as an algebraic factor):

$$\frac{Du}{Dt} - \left(f + \frac{u \tan \phi}{a}\right) v = -\frac{1}{a \cos \phi} \frac{\partial P}{\partial \lambda} + \text{SGS}$$

$$\frac{Dv}{Dt} + \left(f + \frac{u \tan \phi}{a}\right) u = -\frac{1}{a} \frac{\partial P}{\partial \phi} + \text{SGS}$$

$$\frac{\partial P}{\partial z} + \frac{g \rho}{\rho_0} = 0$$

$$\frac{1}{a \cos \phi} \left(\frac{\partial u}{\partial \lambda} + \frac{\partial v \cos \phi}{\partial \phi}\right) + \frac{\partial w}{\partial z} = 0$$

$$\frac{D(T, S)}{Dt} = \text{SGS}$$

$$\rho = \rho[T, S, P]. \tag{4}$$

Here (λ, ϕ, z) are longitude, latitude, and height, respectively; (u, v, w) are the associated velocities; T is the potential temperature (i.e., invariant under adiabatic compression); S is the salinity; and D/Dt is the substantial time derivative,

$$\frac{D}{Dt} = \frac{\partial}{\partial t} + \frac{u}{a \cos \phi} \frac{\partial}{\partial \lambda} + \frac{v}{a} \frac{\partial}{\partial \phi} + w \frac{\partial}{\partial z}. \tag{5}$$

Where SGS appears in Eq. (4), there are important, small-scale, nonconservative processes. These provide spatial transports and pathways to the dissipation that actually occurs by molecular kinetic processes on scales of $\mathcal{O}(1)$ mm. Implicitly these equations are the result of a low-pass filter at the space-time resolution of the numerical model, with SGS denoting the contributions from the unresolved scales. These equations can be augmented by additional tracer equations similar to those for (T, S), adding whichever chemical reactions are required.

The boundary conditions for an OGCM are comprised of the appropriate kinematic conditions on the normal component of velocity and flux rules for the momentum and tracers. These conditions are particularly simple at solid boundaries, where the normal velocity is zero and tracer fluxes are usually neglected. A bottom stress law is also required there, but this is part of the SGS parameterization (Section VII.E). Because OGCM domains usually have sides of finite depth as part of their spatial discretization (Section V)—unlike the true shoreline in most places—the usual practice there is to set the horizontal tangent velocity to zero (i.e., no-slip), although alternative choices of free-slip (no stress) and partial-slip (stress proportional to slip) are sometimes used. These artificial sidewall conditions exert significant influence on OGCM solutions with fine resolution and strong boundary currents. Thus, they are a very problematic aspect of the model formulation.

At the upper free surface, $z = \eta(\lambda, \phi, t)$, the kinematic condition is $w = D\eta/Dt$. Current practice is divided between the full use of this condition and the rigid-lid approximation, wherein $w = 0$ at $z = 0$ and the associated sea level is calculated diagnostically from the model's surface pressure, $\eta = P(\lambda, \phi, 0, t)/g\rho_o$. This approximation is based on $\eta/H \ll 1$ and an assumption of slow evolution compared with surface gravity waves (with a long-wave speed of $\sqrt{gH} \sim 200$ m s^{-1} with $H/L \ll 1$) and long barotropic Rossby waves (with speed $\beta gH/f^2 \sim 100$ m s^{-1}, where $\beta = 1/a$ $\partial f/\partial\phi$). This is an accurate approximation for most types of currents, excluding the tides and surface gravity waves and distorting modestly the response to synoptic weather events. So this choice is often made more for computational convenience or efficiency than for dynamical content. There are also surface momentum and tracer flux conditions (Section III).

When the domain is less than global or full depth, open boundary conditions are required at an internal fluid boundary. There is no fundamentally correct basis for specifying such conditions. However, various rules have been devised that sometimes suffice for the target phenomena of the calculation (e.g., Barnier *et al.*, 1998). These often include specified inflow, outward wave radiation, restoration of tracers toward their climatological values, and enhanced damping in the neighborhood of the internal boundary (a.k.a., a sponge layer). This is another problematic aspect of OGCM formulation.

III. FORCING

The primary forcing of a full-depth, global OGCM is by surface fluxes of momentum (stress), T (heat), S (water), and other material properties, while side (e.g., rivers) and bottom tracer fluxes may provide secondary forcing but usually are neglected.

The surface stress is due to the drag by the overlying wind. It is calculated from an empirical wind climatology using bulk regression formulas for stress. This climatology is readily available in several forms, and it is now being systematically improved through satellite wind observations and climatological reanalyses at operational weather forecast centers (e.g., Kalnay *et al.*, 1996). In polar regions the stress transmission may be mediated by sea ice. Sea ice changes the drag coefficient in the bulk regression formula for surface stress, and it can inhibit stress transmission to the ocean if ice jams develop. An ice model may be needed to adequately incorporate these effects.

The heat and water fluxes are more problematic, since there is no comparably good climatology for them. Bulk regression formulas can be used together with atmospheric surface climatologies for some locally determined components (e.g., sensible and latent heat, evaporation), but other components are nonlocally determined (e.g., precipitation and radiation). Again the presence of sea ice modifies the fluxes, both by blocking air–sea material exchanges and through freezing and melting. The historically most common practice has been to replace the uncertain flux boundary conditions with restoring terms of the form $1/\tau \cdot (T_{\text{clim}} - T)$ in the temperature tendency equation for the uppermost model grid level (ditto for S), where T_{clim} is the observed sea surface temperature and τ is a specified relaxation time (usually on the order of many days). This term can then be diagnostically interpreted as a surface heat flux divided by the grid thickness Δz. This has the seeming virtue of giving model solutions whose T and S fields are close to the observations (but note that the agreement cannot be exact or the implied surface flux would be zero). Yet OGCM experience shows that the implied fluxes obtained by this method are not physically plausible because of too much small-scale variation and probably even some large-scale bias. These defects are especially severe in S. The physical error in using restoring conditions is that they imply excessively strong local atmospheric feedbacks in which any tendency of the ocean to depart from T_{clim} elicits an atmospheric response that supplies compensating fluxes. On the other hand, choosing the opposite extreme, specified flux, has the dual problems of uncertainty in what to specify and the implied absence of any feedback. The lack of feedback allows the ocean solution to drift far away from climatology due to errors in the model and the fluxes.

The reality of atmospheric feedbacks is between these extremes. The flux components that are locally determined do have a greater degree of negative feedback than do the nonlocally determined ones. An approach that is preferable to either restoring or specified flux is a mixture of specifying certain components and calculating others with negative feedback forms, using an atmospheric climatology from reanalyses as the underlying database (e.g., Barnier *et al.*, 1995; Large *et al.*, 1997). In most aspects the accuracy in the climatology of atmospheric state variables (e.g., surface air temperature and cloudiness) is better than for the air–sea fluxes. Thus, the former may provide a better empirical basis for specifying surface fluxes than the latter. This approach is a complicated one with many arguable steps, but its resulting OGCM solutions have been found to be better than with the alternatives. Ultimately, of course, the fluxes should be self-consistently determined with sea–ice and atmospheric general circulation models.

IV. INITIAL CONDITIONS AND EQUILIBRIUM

The state of the oceanic general circulation is not observed in anywhere near the detail required for a complete initialization of an OGCM, nor is it likely to be any time soon. A better global observing system would, however, greatly improve the quality of approximate initial conditions. I believe there is also some utility in retrospective spin-ups using the atmospheric climatology for forcing and data assimilation methods to provide oceanic constraints. This has not yet been done in a substantial way, and there are interesting open questions about the ways in which the ocean is sufficiently predictable for this approach to yield a unique answer. Some aspects of OGCM solutions, such as wind-driven Rossby waves and upper-ocean thermal fluctuations, do seem to be largely predictable from surface fluxes, but the modes of intrinsic variability are likely to be much less so.

Typical OGCM initial conditions are climatological T and S fields (e.g., as in Levitus *et al.* 1994) and zero motion. From such a state there is a geostrophic adjustment to the tracer fields within days, followed by boundary, Kelvin, and Rossby wave adjustments to the wind forcing within a few years that leave behind currents that at least grossly resemble the long-time equilibrium state. The true equilibrium state occurs only after thousands of years when advection and SGS transport have redistributed the T and S fields consistent with the OGCM problem as posed (see Suginohara and Fukasawa, 1988, and Danabasoglu *et al.*, 1996, for analyses of spin-up). Unless approximately correct tracer fields are given as initial conditions, an OGCM solution will differ greatly from its equilibrium state after an integration of only several years. Scientists who wish to avoid the costly computations to reach full equilibrium do bear the burden of demonstrating that shorter integration times do not excessively bias their solutions. At present it is computationally infeasible to integrate a global OGCM to equilibrium with mesoscale resolution.

Another obvious approach to initialization is by bootstrapping (i.e., using one equilibrium OGCM solution as an initial condition for a differently posed problem). We might expect this method to be helpful in shortening the approach to equilibrium if the family of solutions were all sufficiently close to each other. The common experience to date, however, is that this method is rarely cheaper than restarting from a stratified state of rest when the goal is to closely approach an equilibrium state.

V. NUMERICAL METHODS

The computational algorithms that have been used for OGCMs have mostly been rather simple ones, as presented in Bryan (1969) and by others with minor variations. They are a finite-difference discretization of Eqs. (4) and (5), using centered, nearest neighbor differences that are second-order accurate in the grid spacing for uniform grids [the usual choice in (λ, ϕ)] and formally first-order for the (usually weakly) nonuniform vertical grids with finer resolution in the more stably stratified upper ocean. The grids are staggered in their distribution of the dependent variables, using one of several alternative schemes, with the B and C schemes most commonly used. The spatial difference operators are integrally conservative for tracer content and variance and for kinetic energy. Near the poles extra smoothing is required if the grid spacing becomes very small, as it does on a uniform longitude–latitude grid.

The time stepping is by a mixture of first- and second-order accurate procedures, with time splitting often used to solve separately for the SGS vertical mixing by an implicit algorithm that is computationally stable for arbitrarily large SGS vertical diffusivity and for everything else by an explicit algorithm. If the full free-surface kinematic condition is used, then it too is handled though a time-splitting procedure. In these situations the time step size Δt is limited by CFL stability conditions for advection, internal gravity and barotropic Rossby wave propagation, and SGS lateral transport. Over a rather wide range of spatial grid size, $\Delta t = \mathcal{O}(1)$ hr. Therefore, integrating for $\mathcal{O}(10^3)$ years to approach equilibrium requires $\mathcal{O}(10^7)$ time steps or more.

The coarsest spatial grids used in global OGCMs have $\mathcal{O}(10^5)$ grid points, corresponding to a horizontal spacing of hundreds of kilometers and a vertical spacing of hundreds of meters. To be adequate for mesoscale eddies, the grid resolution must have horizontal spacing appreciably finer than the internal deformation radius, R_d. This requires a scale of $\mathcal{O}(10)$ km (also see Section VIII), which increases the size of a global grid to at least $\mathcal{O}(10^8)$ points. At present it is infeasible to combine full mesoscale resolution with full equilibrium in a single global calculation, although it is likely to become so within the next decade with faster computers. Because this computer power probably will be achieved only with substantial parallelism, considerable incentive exists to create suitably structured OGCM codes (e.g., Dukowicz *et al.*, 1993; Bleck *et al.*, 1995; Marshall *et al.*, 1997).

A shortcut for reaching equilibrium can often be taken for coarse-resolution OGCMs by acceleration. This is a time-stepping technique in Eq. (4) with a larger Δt for (T, S) than for (u, v), and a further increase in Δt for the deep tracers (Bryan, 1984). The rationale is that the CFL constraint is usually most severe on (u, v) through the SGS horizontal momentum diffusion and the fastest waves, whereas the constraint for tracers is set by their slower SGS lateral transport and advection (which decrease with depth). Because deep tracers are the slowest solution components to approach their equilibrium values, this acceleration technique can reduce the integration time by a factor of $\mathcal{O}(100)$. If the OGCM approaches a steady solution under steady forcing, then this distortion of the transient dynamics causes only a transient error. Danabasoglu *et al.* (1996) show that, even with periodic seasonal forcing, acceleration leads to a valid equilibrium solution if it is followed by a synchronous integration, with uniform Δt, for $\mathcal{O}(10)$ years to allow the seasonal cycle to equilibrate. No doubt this particular technique has a limited range of validity in model resolution and solution transience, but more robust acceleration methods are well worth seeking because of the high computational cost of calculating OGCM equilibria.

Because far more sophisticated numerical methods than those described above are now well established in other computational contexts, a spate of numerical developments for OGCMs is likely to occur soon. Some of the currently interesting areas of development are higher order spatial and temporal discretizations (e.g., Haidvogel *et al.*, 1995); advection operators not so prone to creating erroneous extrema (e.g., Leonard, 1979; Shchepetkin and McWilliams, 1998); fuller temporal implicitness to escape physically unimportant computational stability limits (e.g., Yavneh and McWilliams, 1995); more efficient procedures for solving for a reference-level $P(\lambda, \psi, t)$ in the vertical integration of the hydrostatic relation, via the barotropic streamfunction, rigid-lid surface pressure (Pinardi *et al.*,

1995), or free-surface elevation (e.g., Dukowicz and Smith, 1994); and local grid refinement or nesting to allow a regional focus with less physical uncertainty and numerical error in the open-ocean side boundary conditions.

The governing equations [Eqs. (4) and (5)] are physically equivalent in any coordinate transformation, but uniform grids in different coordinate systems imply different discretizations of the equations. One useful application is to achieve a more uniform tiling of the sphere in the horizontal coordinates than in a regular latitude–longitude grid. This improvement is most important near the north pole where $\Delta\lambda \to 0$ (e.g., Ebby and Holloway, 1994a; Smith *et al.*, 1995). Another potentially useful application is through transformation of the vertical height coordinate z to either a potential-density (e.g., Bleck and Chassignet, 1994; Lunkeit *et al.*, 1996) or a topography-following (i.e., σ) coordinate (e.g., Ezer and Mellor, 1994; Song and Haidvogel, 1994; Barnier *et al.*, 1998). Advantages of a potential-density coordinate are a natural concentration of resolution in the pycnocline where vertical gradients are often largest; the simplification of the conservative nonlinear terms in Eqs. (4) and (5) to involve only the horizontal velocity; and a relative ease in ensuring integral conservation properties for the discretized, isopycnally oriented SGS operator forms (although this has also been achieved satisfactorily with other coordinates; see Section VII.B). Disadvantages are an uncontrollable sparseness of resolution in weakly stratified regions, such as in boundary layers and convective sites; errors in implementing the equation of state associated with its nonlinear form of compressibility, such that globally defined density or potential density surfaces are not uniformly accurate in representing the static stability (McDougall, 1987); and a greater complexity in specifying the discrete diapycnal SGS operator forms. Advantages of a topography-following coordinate are a more accurate representation of the bottom kinematic boundary condition and bottom boundary-layer SGS mixing. A well-known disadvantage is the error in the horizontal pressure gradient force where the topography is especially steep (though there may be similar, albeit less obvious, errors in other coordinate systems). Another potential disadvantage is the vanishing grid spacing as the depth becomes very shallow at the ocean margins, with a possibly severe constraint on the time step size for computational stability. In my opinion no vertical coordinate choice is clearly the superior one. Probably the more important choices are for the discrete operators that arise within a given coordinate system. Arakawa (1988) gives an interesting perspective on these different vertical coordinates and discretizations in atmospheric GCMs.

VI. DOMAIN GEOMETRY

The shape of an ocean basin is defined by its bottom topography, including its intersection with the top surface, the coastline. Topography has a profound influence on the direction of currents, especially near the bottom. The reason for this is the approximate material conservation of potential vorticity, whereby a change in the thickness h between a deep interior isopycnal surface and the bottom implies a compensating change in the fluid vorticity that acts to turn the fluid trajectory toward a path of constant f/h.

It is far less certain, however, when and where topography fundamentally controls the existence and strength of the large-scale currents. The latter are largely accelerated by internal pressure gradient forces, which are closely related to (T, S) distributions through hydrostatic balance and the equation of state. Momentum balance against this acceleration can be provided either internally by large-scale advection and SGS transport or by topographic effects. The important topographic effects are form stress (i.e., integrated horizontal pressure force on the bottom), boundary-layer drag, and hydraulic control (i.e., throughflow limited by a critical Froude number, where the velocity equals the gravity wave speed). This competition can be posed conceptually as a comparison of model solutions with simpler, smoother topography and ones with rougher, more complex topography. While the desire for geographical realism favors the latter, achievement of computational accuracy favors the former (i.e., accurate numerical solutions require smoothness near the grid scale), and the choice of numerical algorithms for topography may be quite important in determining the balance point between these competing goals. There is no satisfactory resolution of these issues at present, and the common practice is one of trial and error. Increasing the grid resolution allows the incorporation of finer topographic features. This diminishes the scope of the ambiguity in specifying topographic smoothness, although the true topography in the oceans remains rough on all conceivable OGCM scales.

Some aspects of topographic influence must be represented in the SGS parameterizations: the bottom boundary layer, coastal shoaling, small-scale roughness, and, probably, hydraulic control. (I am not aware of any OGCM solutions that achieve this state in their resolved dynamics, although the governing equations admit the possibility.) But at least part of the effects of form stress and impedance by constricting straits are resolved dynamical processes.

Topographic form stress is of central importance in maintaining the depth-integrated transport of the Antarctic circumpolar current against the surface wind stress. This was first suggested by Munk and Palmen (1951), and it has since been demonstrated in many OGCM solutions (e.g., Ivchenko and Stevens, 1995). Treguier and McWilliams (1990) found that most of the form stress in this current is associated with topographic features of large spatial scale, and this, of course, is favorable to its modelability. The role form stress plays in other large-scale currents is as yet less certain, but it probably is important near midlatitude western boundary currents (e.g., Hurlburt *et al.*, 1995; Sakamoto and Yamagata, 1996; Kagimoto and Yamagata, 1997).

Constricting straits can limit throughflow, either by boundary stress or hydraulic control (e.g., Pratt, 1990). In OGCM solutions a current through a strait whose width is too close to the grid scale will have primarily viscous dynamics. A small width probably exaggerates the boundary stress and precludes the inertially controlled dynamics of hydraulic control. This presents a modeler with the temptation to artificially widen or deepen the channel to compensate for these biases, but this is an essentially arbitrary choice. The two most important straits in the global ocean are the Drake Passage and the Indonesian Archipelago, both of which have large throughflows; demonstrations of the extreme consequences of blocking them entirely are given in Ishikawa *et al.* (1994) and Hirst and Godfrey (1993), respectively. Coarsely resolved OGCM solutions have about the same transport as is observed when these straits are artificially widened (e.g., Danabasoglu and McWilliams, 1995). Straits occur in all sizes, of course, and some will remain part of the SGS specification at all foreseeable resolutions. Probably the most important of these is the Strait of Gibraltar, whose outflow of salty Mediterranean seawater has significance in the S budget of at least the North Atlantic Ocean; Armi and Farmer (1988) argue for its hydraulic control. There are analogous difficulties with flow over sills and ridges whose size is too close to the grid scale. A notorious example is the flow of deep seawater from the Greenland Sea into the Atlantic across the Icelandic Ridge (e.g., Roberts *et al.*, 1995).

VII. PARAMETERIZATIONS

The art of parameterization is to augment the resolved dynamics in Eq. (4) with mathematical operators that accomplish the necessary physical effects by unresolved SGS processes, all of which involve turbulence and thus do not have complete analytic theories to make use of. I believe it is impossible to make a parameterization rule that is entirely correct, in the sense that it will yield the same answer as in a model that fully resolves the process. Therefore, a hypothesis must be declared about what the minimum necessary effects are. Then, after finding a suitable operator that satisfies this hypothesis, any free parameters are chosen to either achieve some quantitative effect in an OGCM solution or match independent data about the process—both, if possible. The most common parameterization hypothesis about turbulent processes is that they mix material properties, hence the most common operator form is eddy diffusion (e.g., by spatial Laplacians) with an eddy diffusivity as the free parameter.

We should be humble about our ability to represent turbulent processes. So, a guideline for choosing a good parameterization is to keep it as simple as possible, with as few free parameters as possible, consistent with achieving both the hypothesized effect and a significant impact on the solution. Nevertheless, for an OGCM there are many needed effects associated with many different SGS processes, hence many different parameterization forms are involved.

Material transport in the oceans is strongly constrained by the isosurfaces of entropy, which are locally tangent to isodensity surfaces (Fig. 2). (Momentum transport is less constrained because of the pressure-gradient

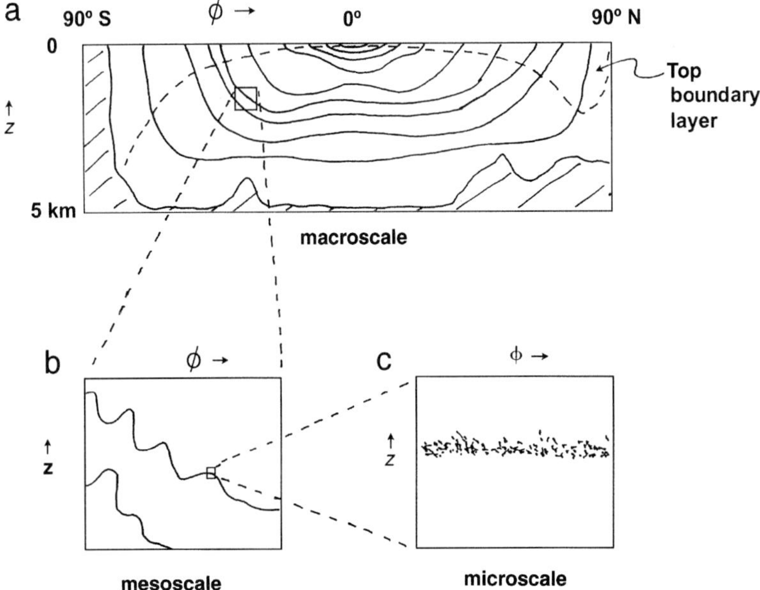

Figure 2 (a) Macroscale, (b) mesoscale, and (c) microscale views of oceanic isopycnal surfaces in a meridional plane. Material fluxes across these surfaces are much weaker than along them, except in the surface and bottom boundary layers and perhaps above regions of rough topography as well.

force.) In a stably stratified region, gravitational work is required to move matter across surfaces of neutral buoyancy (i.e., isopycnal surfaces). Hence, transport along these surfaces is much more efficient than across them, with an observed ratio of cross-isopycnal (i.e., diapycnal) to isopycnal eddy diffusivities of about 10^{-8} in the oceanic thermocline. The principal agents of turbulent isopycnal tracer flux are mesoscale eddies, which crinkle the basin-scale isopycnal surfaces without disrupting their stable vertical ordering. The agents of diapycnal tracer flux are microscale motions that can cause overturning and fragment the surfaces. Within and near the surface and bottom boundary layers, where the density stratification is weak, this constraint is lifted and microscale transports are much more efficient than in the interior.

A. Lateral Momentum Transport

For the general circulation, the dominant mechanism for lateral momentum transport is by mesoscale eddies acting through the horizontal Reynolds stress, $\overline{\mathbf{u}'_h \mathbf{u}'_h}$. This process can either be wholly parameterized or be partially resolved in an OGCM. The common parameterization is horizontal eddy diffusion, with an eddy viscosity $\nu_h \geq 0$, although hyperdiffusion (i.e., an iterated Laplacian operator) is sometimes used to strengthen a marginally resolved instability process such as gulf stream meandering. The value of ν_h depends on the grid size. It must be large enough to resolve the viscous boundary layers of width $\sim (\nu_h / \beta)^{1/3}$ and to suppress nonlinear computational instability on the grid scale (i.e., a grid Reynolds number, $\mathrm{Re} = V \Delta s / \nu_h$ cannot be large). Fundamental symmetry principles about the nature of a stress divergence imply that the operator must have additional terms beyond a simple Laplacian operator in a spherical domain (e.g., Wajsowicz, 1993). A similar consideration applies to spatially nonuniform or anisotropic eddy viscosities that are sometimes used when the grid spacing has the same attributes.

There is a well-known counterexample for this parameterization, which prompted Victor Starr to talk about "negative eddy viscosity." It is the counter-gradient eddy momentum flux that occurs in the core of a broad (i.e., with $L/R_d \gg 1$), baroclinically unstable current, such as the Antarctic circumpolar current (e.g., McWilliams and Chow, 1981). This has led some to suggest replacing lateral eddy viscosity with potential vorticity diffusion (see Section VII.B), although this has not yet been implemented in an OGCM as far as I know. Furthermore, because the momentum flux need not be related to the local large-scale flow, as with Rossby wave propagation through a region, an eddy diffusion form may not always be valid. Analyses of eddy-resolving OGCMs indicate that the resolved horizontal Reynolds stress divergence patterns are locally much different from the eddy viscosity parameterization, especially near midlatitude western boundary currents where they are largest. With fine enough grid resolution and small enough ν_h in an OGCM, these confounding behaviors can become part of the resolved dynamics, and an eddy viscosity parameterization may suffice.

B. Isopycnal Material Transport

The traditional parameterization for lateral tracer transport is horizontal eddy diffusion, but this violates the constraint of tracers mostly staying on isopycnal surfaces, with $\overline{\mathbf{u}c'}$ nearly perpendicular to $\nabla\bar{\rho}$ (here c is any tracer concentration). A parameterization by Gent and McWilliams (1990) that is isopycnally oriented and integrally adiabatic (i.e., without interior sources or sinks of any material property that alters its inventory on isopycnal surfaces) has had quite beneficial effects on OGCM tracer distributions and fluxes (Danabasoglu *et al.*, 1994; Boening *et al.*, 1995). In addition to tracer diffusion along isopycnals (Redi, 1982), there is an incompressible eddy-induced transport velocity. Its horizontal component is defined by

$$\mathbf{u}_h^* \equiv \left(\frac{\partial\bar{z}}{\partial\rho}\right)^{-1}\overline{\frac{\partial z'}{\partial\rho}\mathbf{u}_h'}, \tag{6}$$

where z is height of an isopycnal surface and the overbar is an average over the mesoscale on that surface; $\mathbf{u}^*$ combines with the large-scale $\mathbf{u}$ as the large-scale Lagrangian velocity that advects the large-scale tracers (Gent *et al.*, 1995); $\mathbf{u}^*$ also advects the isopycnal surface itself, thereby causing vertical transport of momentum, as in isopycnal form stress, and depletion of available potential energy, as in baroclinic instability. In present OGCM implementations, both $\mathbf{u}^*$ and the isopycnal mixing are represented as eddy diffusion of isopycnal layer thickness and tracers. The originally suggested parameterization form for the eddy-induced velocity is

$$\mathbf{u}_h^* = -\frac{\partial}{\partial z}[\kappa_i\mathbf{L}_h], \qquad w^* = \nabla_h\cdot[\kappa_i\mathbf{L}_h], \tag{7}$$

although variant forms have since been suggested by others. Here $\mathbf{L}_h = -\nabla_h\rho/(\partial\rho/\partial z)$ is the slope of the isopycnal surface. The eddy diffusivity, $\kappa_i \sim 10^3$ m^2 s^{-1}, can be identified with the rate of dispersion of neutrally buoyant floats in the ocean. So far, κ_i has been used in coarse resolution OGCMs with essentially no grid-size dependence in its value. Even in fine-resolution calculations that partially resolve the mesoscale eddies, there is benefit to using this parameterization with a smaller κ_i, if only to avoid false diapycnal flux (Roberts and Marshall, 1998). Because mesoscale variability is strongly inhomogeneous in the ocean, it seems likely that κ_i should also be variable. Visbeck *et al.* (1997) suggest a flow-dependent rule for κ_i based on linear baroclinic instability theory.

If one assumes a simplified $\rho[T, S, P]$ and neglects SGS terms in Eq. (4), then the hydrostatic form of Ertel potential vorticity, i.e.,

$$
q = \left(f(\phi) + \frac{1}{a \cos \phi} \left[\frac{\partial v}{\partial \lambda} - \frac{\partial (u \cos \phi)}{\partial \phi} \right] \right) \frac{\partial \rho}{\partial z}
$$
$$
- \frac{1}{a \cos \phi} \frac{\partial u}{\partial z} \frac{\partial \rho}{\partial \lambda} + \frac{1}{a} \frac{\partial v}{\partial z} \frac{\partial \rho}{\partial \phi}, \tag{8}
$$

is conserved along parcel trajectories, which lie on isopycnal surfaces. (This is also true in the other models in Fig. 1 with alternative definitions for q.) If mesoscale eddies are strong enough, they sometimes can mix, or homogenize, large-scale q (Marshall, 1981; McWilliams and Chow, 1981; Rhines and Young, 1982), although they do not always do so. A large-scale approximation to q is $f\rho_z$, and there are some regions in the ocean where this latter quantity is observed to be nearly uniform along isopycnal surfaces. Also, at least in zonally symmetric channel flows, eddy potential vorticity fluxes may be more uniformly down-gradient than eddy thickness fluxes (Treguier *et al.*, 1997). This suggests that isopycnal potential vorticity mixing is a possible parameterization hypothesis. It would be expressed as a composite of the SGS terms in the momentum, T, and S equations in Eq. (4), and as such it must be reconciled with them in a way that does not involve double counting of processes. An eddy diffusion of potential vorticity should avoid the fallacy of generating a large-scale flow from a resting stratified state, which suggests that the diffusivity has to be flow dependent. For those who believe that a resting state is physically unrealizable because nature always contains many sources that excite currents (e.g., Holloway, 1992), this consideration may be considered irrelevant. Because the usual methods for solving Eqs. (4) and (5) do not involve an equation for q explicitly, a different method would be required. There are thus significant challenges to be overcome in implementing a q-mixing parameterization in an OGCM, and potential vorticity mixing may prove to be more useful as a theoretical concept than as a practical parameterization form.

C. SURFACE BOUNDARY LAYER AND SURFACE GRAVITY WAVES

In the surface planetary boundary layer (PBL), small-scale turbulence is relatively strong because of shear and buoyancy instabilities resulting from the surface fluxes, hence $\overline{w'\mathbf{u}'_h}$ and $\overline{w'c'}$ are much larger than in the interior. In the default SGS form, the PBL is implicitly contained within the top grid cell of an OGCM since there are no elevated diffusivities in the grid interior. An essential extension beyond the default form is to increase the vertical SGS fluxes where the density stratification is gravitationally unstable (i.e., with a positive vertical gradient in potential density), mimicking the convection that cannot be dynamically resolved in Eq. (4). This is often done either by convective adjustment (a minimal vertical rearrangement of tracers until stable stratification is achieved) or by a substantial enhancement of the vertical diffusivities in the unstable region. Insofar as the convection arises from an upward surface buoyancy flux—as it usually does—then the PBL extends from the surface throughout the depth interval of adjustment or enhanced mixing. In this situation an explicit boundary layer model for the PBL transport is preferable to a local mixing rule.

The PBL model types most commonly used in OGCMs are either bulk mixed-layer models (e.g., Kraus and Turner, 1967) or single-point, moment-closure turbulence models (e.g., Mellor and Yamada, 1982). Large *et al.* (1994) reviews the usage of oceanic PBL models and proposes a K-profile parameterization (KPP). KPP does not impose such a strong constraint on the boundary-layer profiles of large-scale variables as does a mixed-layer model, and it escapes the dubious spatial locality assumption of moment-closure models. It also includes counter-gradient buoyancy and fluxes in convective situations. The KPP hypotheses are (1) the boundary layer thickness h maintains a bulk Richardson number, $\mathrm{Ri}_b = -gh(\Delta\rho)/(\Delta\mathbf{u})^2$ (where Δ indicates the difference across the boundary layer), at a critical $\mathscr{O}(1)$ value, and (2) the diffusivity profile smoothly joins with the Monin–Obukhov form near the surface and with the interior form at the outer edge of the boundary layer. In the few assessments of its performance in OGCMs to date, KPP does well (Large *et al.*, 1997; Li *et al.*, 1998; Large and Gent, 1998). It is important that PBL vertical mixing not occur simultaneously with isopycnal tracer transport (which presumes a strongly stable stratification) in an OGCM, or else the exchange between the boundary layer and the interior will be too rapid. Accordingly, a rule is used that turns off κ_i where $z \geq -h$ or $\mathbf{\nabla}_h\rho\,|/\rho_z$ is too large.

As yet no OGCM includes several SGS effects of surface gravity waves that probably are important for large-scale flows under some conditions.

One effect is enhanced mixing by breaking waves near the surface and by wave-driven Langmuir cells throughout the PBL. Another effect is that the wave-averaged dynamics contain extra terms in both the horizontal momentum and tracer equations in Eq. (4). These terms are proportional to the mean Lagrangian velocity of the waves, the Stokes drift, and their effect is somewhat analogous to the mesoscale-eddy $\mathbf{u}^*$ term discussed above. Furthermore, averaging over the waves yields extra terms in the surface kinematic, and pressure boundary conditions are related to the divergence of the Stokes-drift transport and the vertical velocity variance of the waves, respectively. See McWilliams and Restrepo (1998).

D. Interior Vertical or Diapycnal Mixing

The default SGS parameterizations for diapycnal momentum and tracer transports in the oceanic interior (i.e., $\overline{\mathbf{u}'\mathbf{u}'_h} \cdot \nabla\rho$ and $\overline{\mathbf{u}'c'} \cdot \nabla\rho$ are eddy diffusion, with uniform $\nu_v \sim 10^{-4}$ m^2 s^{-1} and $\kappa_v \sim 10^{-5}$ m^2 s^{-1}. Not uncommonly, these values are made to increase with depth in the deep ocean, which has some observational support (Polzan *et al.*, 1997). Such a tracer diffusivity is inherently nonadiabatic, in the sense defined above, and it implies a breaking of the isopycnal surfaces (Fig. 2). These are usually represented as a vertical flux, since this direction is nearly perpendicular to isopycnal surfaces in most places. [Isopycnal slopes are typically $\mathcal{O}(10^{-4})$.] For a given, nonuniform vertical grid, a sufficiently large vertical eddy diffusivity is needed to dominate the implicit numerical diffusion arising from discretization errors (Yin and Fung, 1991; but see a response by Treguier *et al.*, 1996, challenging this interpretation). Therefore, larger (ν_v, κ_v) values are often chosen for coarser vertical grids. There is a continuing controversy over whether the global oceanic circulation is consistent with the small diffusivity values above, or whether "missing mixing" processes must be found. My opinion is that the case for missing diapycnal mixing has not yet been conclusively made for processes in the upper ocean since OGCM solutions with comparably small κ_v values do not have clear discrepancies with observations (McWilliams *et al.*, 1996). OGCM solutions probably do require larger κ_v values in the abyssal ocean to match observed tracer distributions, although this has not yet been clearly demonstrated.

Various small-scale processes, in addition to convection, make the interior diabatic mixing spatially and temporally nonuniform and sometimes elevated above the default values. Most notable among these processes are stratified shear instability, breaking internal waves, and double diffusion caused by the larger molecular diffusivity of T than of S (see Large *et al.*, 1994, for a review of the SGS parameterization forms). Pacanowski and Philander (1981) show that shear instability influences the equatorial undercurrent and thermocline profiles in OGCM solutions using a parameterization where (ν_v, κ_v) vary inversely with the gradient Richardson number, $\mathrm{Ri}_g = N^2/(\partial\mathbf{u}/\partial z)^2$. Demonstrations of potentially substantial consequences on the abyssal tracer distributions are given in Cummins *et al.* (1990) and Cummins (1991) for a κ_v varying inversely with N, and in Gargett and Holloway (1992) for double diffusion with unequal κ_v values for T and S. Zhang *et al.* (1998) show moderate sensitivities of the meridional overturning circulation and heat flux to the double diffusion parameterization. The nonlinearity in $\rho[T, S, P]$ also allows for locally enhanced vertical transport through thermobaric and cabelling instabilities. The effects of $\rho[T, S, P]$ nonlinearity are present in principle in OGCM solutions that have enhanced vertical diffusivity in the presence of convective instability, but their significance has not been broadly assessed.

E. Bottom Boundary Layer and Gravity Currents

There is an analogous PBL at the ocean bottom, driven principally by boundary stress without any significant buoyancy flux. A primary effect of the bottom PBL is to provide a drag force on the adjacent flow, as in a simple Ekman layer. Usually in OGCMs this is represented as a boundary stress proportional to $\mathbf{u}$ or $|\mathbf{u}|\mathbf{u}$, acting only in the deepest grid cell. Obviously, an explicit PBL model, of the types discussed above, could be used at the bottom. Armi (1978) and Garrett (1991) argue for the possible significance of tracer transport parallel to the boundary by PBL turbulence. This implies a diapycnal flux for any isopycnal surfaces intersecting the bottom, which locally diminishes the impedance to diapycnal exchanges (Fig. 2). This effect has not yet been examined in an OGCM, although Marotzke (1997) shows plausible solutions in an simple domain geometry while limiting κ_v to be nonzero only near the sides.

Gravity currents are downslope, bottom flows with a density greater than their surroundings at the same level. This makes gravity currents doubly turbulent, due both to the stress against the bottom and to gravitational instability compared to their downstream environment. Therefore, they efficiently entrain and mix with the ambient fluid as they flow. Gravity currents occur where convection over a slope penetrates to the bottom or dense water overflows a continental shelf or a sill, and they will continue either until entrainment quenches the density anomaly, geostrophic adjustment turns the flow to be across the slope (though bottom stress keeps the flow turbulent and allows a continuing, albeit slower, downslope progression), or the bottom of the slope is reached. Several sites in the global ocean have significant water mass flows in gravity currents. These include the Mediterranean outflow, the Denmark Straits overflow, and outflows from the Antarctic coastal shelves. The usual practice in OGCMs is to ignore this process, and this is probably unwise. Beckmann and Doescher (1997) propose a particular OGCM parameterization that combines the kinematic condition with a stress-driven boundary layer but neglects entrainment, and Gnanadesikan (1998) makes a somewhat more general proposal. Price and Beringer (1994) present a gravity current model that includes entrainment that perhaps could be adapted as an OGCM parameterization.

F. Topographic Effects

The common practice with OGCMs is to include only the effects of topography that are resolved on the model grid, as discussed in Section VI. Here we consider some additional SGS effects that require parameterization.

Small-scale topography influences large-scale circulation. The most direct possibility is through a turbulent rectification process wherein mesoscale eddies above topography cause a form stress that drives large-scale currents along large-scale contours of f/h, where h is the effective vertical scale above the bottom of the affected currents (Bretherton and Karweit, 1975; Bretherton and Haidvogel, 1976; Salmon *et al.*, 1976). Holloway (1992) argues that this has a significant effect on continental slopes by driving barotropic, cyclonic rim currents (i.e., having the same rotational sense around the basin as f), and Merryfield and Holloway (1998) examine the modifications due to stable stratification in a quasi-geostrophic model. Ebby and Holloway (1994b), using a parameterization for this rectification, and Hurlburt *et al.* (1995), with very fine mesoscale resolution, show examples of this possibility in OGCM solutions. I believe that SGS form stress is a qualitatively important effect to include in OGCMs, although its strength, its vertical structure, and its relationship to the resolved form stresses need further clarification. A less direct effect of topographic roughness is its tendency to make the vertical structure of mesoscale eddies more surface intensified, with implications for their transport properties in the general circulation (Rhines, 1977; Owens and Bretherton, 1978; Boening, 1989). Large-scale currents over rough topography can generate gravity waves, especially stationary lee waves, that propagate vertically into the interior, depositing their momentum and enhancing diapycnal mixing where they break and dissipate. This is an important mechanism for the atmospheric general circulation, and recent observations indicate the occurrence of the process in the oceans as well (Polzan *et al.*, 1997). Finally, topography on the scale of the bottom boundary layer generates local secondary circulations that may substantially alter the vertical momentum flux and diapycnal mixing rate. No SGS proposal has yet been made to account for this effect.

G. RIVERS AND MARGINAL SEAS

The common practice for OGCMs is to neglect river inflows and SGS exchanges with marginal seas. Restoring terms, if used, may indirectly contribute some of these otherwise missing effects. If restoring terms are not used for the surface fluxes, then lateral tracer restoring conditions could be used in places with rivers or significant marginal sea exchanges, in the form of an open boundary condition.

It is better physics to have flux boundary conditions for river inflow, with the inflow specified either from data or a land hydrology model. Then the river inflow appears as just another component of the surface water flux for that location. The exchanges with SGS marginal seas or straits (e.g., Bering and Gibraltar) could be specified as lateral flux conditions, determined either from data or from coupling with a local model of their circulation.

VIII. SPATIAL RESOLUTION

From the perspective of a computational mathematician, there is nothing mysterious about the influence of the spatial grid spacing in a numerical calculation. In a given problem, once the resolution is fine enough for the solution to be smooth on the grid scale, the solution converges with increasing resolution according to the order of the discretization method. For oceanographers the influence of resolution in OGCMs often has been a matter of confusion, and public discussions have often generated more heat than light. There are at least three reasons why the oceanographic issues go beyond the computational mathematical dictum:

1. It is common practice to change the problem with the resolution, both by decreasing at least some of the eddy diffusivities (the lateral viscosity, in particular) to its computational stability limit and by adding finer scales to the forcing and domain geometry.

2. Qualitative changes in the solution behavior occur with resolution, due to changing the eddy diffusivity, as instability thresholds are passed. The most prominent change is the emergence of mesoscale eddies, but this transition actually involves a suite of thresholds for the many different currents in the general circulation.

3. The general circulation has many measures, and no complete understanding yet exists for how different features differently depend on resolution and each other.

There are remarkably few published studies of the influence of resolution in OGCMs, and more certainly would be worthwhile. The evidence in hand suggests that it is relatively unimportant for those solution properties, such as tracer fluxes and water mass distributions, that are simulated reasonably well in OGCMs without resolved mesoscale instabilities (i.e., with horizontal grid spacings $dx > 50$ km); see Covey (1995). In such models, the boundary currents typically are too weak and broad but are not as incorrect in their transport, the eddy-driven western recirculation gyres are missing, and the equatorial currents are too weak and broad except when the meridional resolution is locally refined to a scale smaller than 50 km.

Once the grid is fine enough for eddies to arise, the resolution seems to be extremely important, up to an as yet poorly determined threshold at least as small as 10 km, in order to calculate mesoscale eddies and intense, narrow currents credibly (i.e., with qualitative similarity to observations in eddy energy level and current location). For eddy-resolving models, resolution convergence has not yet been demonstrated and remains an important open issue. Several recent OGCM solutions for the Atlantic with especially high vertical and horizontal resolution do appear to have achieved several improvements, lacking in eddy-containing solutions at lower resolution, in their correspondences with observations. These improved quantities include surface height variance, gulf stream separation site and offshore path, and meridional heat flux (Chao *et al.*, 1996; Smith *et al.*, 1998; Chassignet *et al.*, 1999). On the other hand, some features have not yet been seen in OGCM solutions that are present in idealized models at very fine resolution. Examples are the instability of the western boundary current before its separation site (Berloff and McWilliams, 1998a), an intrinsic low-frequency variability on decadel periods (Berloff and McWilliams, 1998b), and the emergence of an abundant population of mesoscale and sub-mesoscale coherent vortices (Siegel *et al.*, 1998). No doubt other such phenomena are yet to be discovered as well.

The computational cost scales with horizontal resolution roughly as $1/dx^3$, assuming that the vertical resolution, duration of integration, and domain size are not varied. This implies roughly a thousand-fold disparity in computation costs for any given problem configuration between OGCMs that resolve the eddies and those that do not. Computer limitations, therefore, cause OGCM usage to be split between these two types of configurations. At present, eddy-resolving models can be used well only for intervals as long as decades and domains as large as basins, whereas coarse-resolution models are also suitable for centennial and millennial fluctuations and the approach to equilibrium in global domains. Although growing computer power will narrow this division, it will be at least a decade, and perhaps much longer, before it disappears and everyone will prefer an eddy-resolving model. The fact that OGCMs without eddies and with sensible parameterizations can do reasonably well in calculating the large-scale thermohaline circulations, heat and water fluxes, and water mass distributions remains somewhat mysterious: Why aren't the complex details of mesoscale eddies and narrow currents more consequential? So, perhaps this result must be accepted only provisionally. Nevertheless, it does suggest that there is some, as yet poorly explained, type of dynamical decoupling between these large-scale phenomena and the mesoscale eddies, strong currents, and other small-scale phenomena.

IX. ROLE OF THE OCEAN IN CLIMATE SYSTEM MODELS

The direct roles played by the ocean in maintaining the earth's climate are the storage and geographical transport of heat and freshwater. Indirectly, of course, the whole of the oceanic general circulation is involved in these processes. An indication of how well an OGCM with climatological forcing can play these roles comes from a recent solution by Gent *et al.* (1998). The model resolution is 2.4° in longitude, 1.2–2.4° in latitude (finer in tropical and polar regions), and 45 levels in the vertical with a vertical spacing expanding with depth from 12.5 to 250 m. Its surface forcing is a mean annual cycle with a combination of specified fluxes and feedback/restoring relations as described in Large *et al.* (1997). The model uses the isopycnal tracer transport parameterization of Gent and McWilliams (1990) and the KPP boundary layer and interior vertical diffusivity parameterizations of Large *et al.* (1994). The calculation is carried to equilibrium using the acceleration technique, with a final 17 years of synchronous integration. Time- and horizontal-mean vertical profiles of $T(z)$ and $S(z)$ are given in Fig. 3, both from the model solution and from a hydrographic climatology. The overall shapes of the profiles match fairly well, although

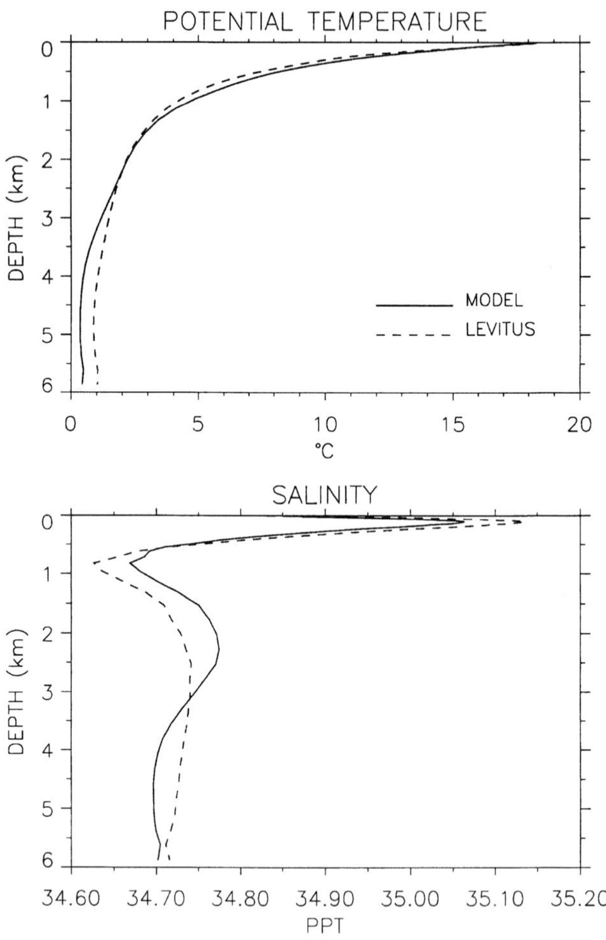

Figure 3   Time- and horizontal-mean profiles of T and S from a coarse-resolution OGCM (Gent *et al.*, 1998) and from the observational climatology of Levitus (1982) and Levitus *et al.* (1994).

the upper ocean extrema in S are somewhat smaller in the model solution. Time-mean, longitude-integrated meridional transports of heat and fresh-water are shown in Figs. 4 and 5, both from the model solution and from several empirical estimates. The heat transport is poleward in both hemi-spheres, with its peaks near $\pm 20°$. In the Southern Hemisphere, the heat transport becomes quite small across the Antarctic circumpolar current, due to the canceling effect there of the meridional overturning by the

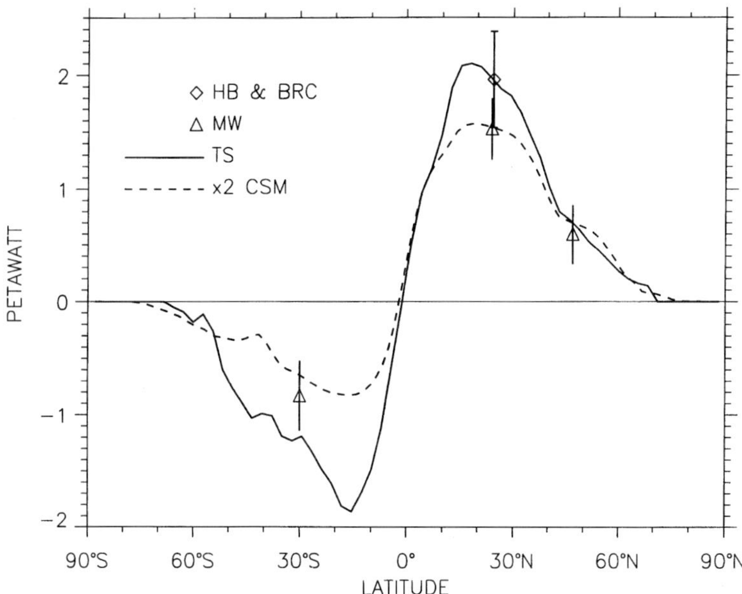

Figure 4 Time-mean northward heat transport from a coarse-resolution OGCM (Gent *et al.*, 1998) [CSM] and from empirical estimates (and their uncertainty) by Hall and Bryden (1982) [HB], Bryden *et al.* (1991) [BRC], Trenberth and Solomon (1994) [TS], and MacDonald and Wunsch (1996) [MW].

eddy-induced circulation. The model heat transport is within the uncertainties of the empirical estimates. Although the general meriodional profiles of the model and empirical water transport agree fairly well, estimating the uncertainties is more problematic than for heat. We conclude that current OGCMs are capable of playing their assigned roles in climate dynamics, at least in a time average sense, given adequate surface forcing.

The scientific problems of ocean circulation and climate have a fundamental mutual dependency. Ocean models have an excessive uncertainty in their surface boundary conditions as specified from observations (Section III), and the equilibrium climate of the earth strongly depends on the ocean's influences. A more subtle dependency is that fluctuations in either the atmosphere or ocean may elicit a response in the other medium that importantly alters the evolution by feedbacks. ENSO is the best known example of an interannual coupled mode of climate variability. Both the ocean and atmosphere exhibit many other forms of variability than ENSO, and we do not yet know very well how many of these modes are significantly coupled. The paleoclimate record shows many episodes of rapid

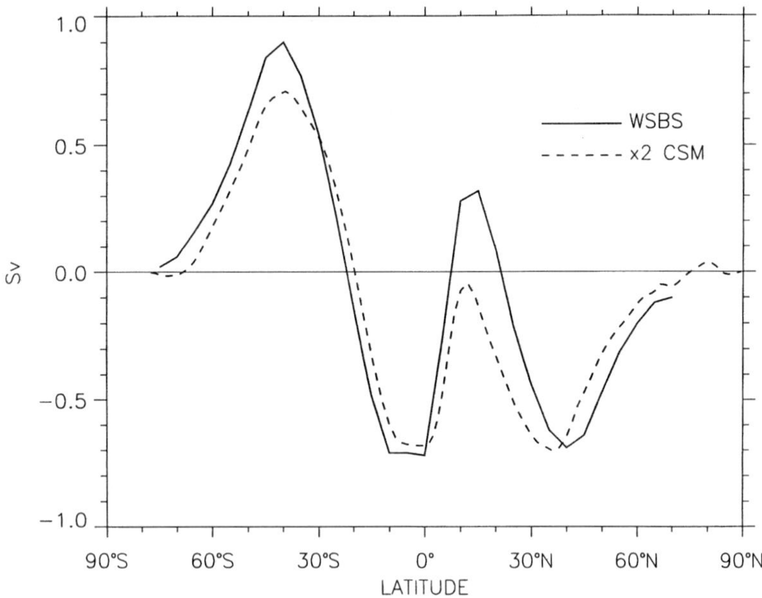

Figure 5   Time-mean northward water transport from a coarse-resolution OGCM (Gent *et al.*, 1998) [CSM] and from empirical estimates by Wijffels *et al.* (1992) [WSBS].

surface temperature changes of $\mathscr{O}(10)$ K with a probably global extent, most recently during the Alerod–Younger Dryas fluctuations at the end of the last ice age about 12,000 years ago (e.g., Dansgaard *et al.*, 1984). Within the modern era, ocean surface temperatures show substantial decadal variability in the regional spatial patterns of atmospheric "teleconnection" modes such as the North Atlantic oscillation (e.g., Deser and Blackmon, 1993). My view is that both of these phenomena are related to the delicacy and intrinsic variability of the oceanic thermohaline circulation (Saravanan and McWilliams, 1995, 1997). In addition, radiative forcing from anthropogenic greenhouse gas emissions is forcing a new coupled behavior in the atmosphere and ocean, which may substantially involve the thermohaline circulation because of the associated changes in planetary-scale surface buoyancy flux (Manabe and Stouffer, 1993; Saravanan and McWilliams, 1995).

A coupled climate model includes a global OGCM, together with models for sea ice, the land surface, and the atmosphere. This type of model has a long history of usage (e.g., Manabe and Bryan, 1969). It has been the principal means for forecasting global warming due to increasing greenhouse gases, and therefore it will undoubtedly become an even more important tool as mankind begins to practice planetary environmental management. One of the least convincing aspects of coupled models has been the prevalent tendency for climate drift away from the actual climate. To avoid this, the common practice has been to impose surface flux corrections or flux adjustments as an artificial stabilization technique. This consists of computing equilibrium states for the uncoupled oceanic and atmospheric models [e.g., using surface (S, T) restoring terms and specified sea surface T, respectively], computing the associated heat and water air–sea fluxes for each model, and then imposing the difference between the models' fluxes as spatially and seasonally varying added source terms to one of the models, usually the ocean. This does reduce climate drift in the subsequent coupled solution, and different modeling groups have found differing degrees of drift suppression by this technique.

An obvious goal for coupled modeling, therefore, is to eliminate flux corrections while avoiding excessive climate drift. A notable advance toward this goal is the recent solution of the NSF/NCAR Climate System Model. Without flux corrections (other than a global air–seawater balance constraint to compensate for the neglect of river inflow to the ocean), the average surface temperature shows no significant drift over an integration of 300 years (Fig. 6; Boville and Gent, 1998). The reasons behind this success appear primarily to be advances in SGS parameterization (clouds and radiation in the atmospheric model, the surface PBL in both models, and isopycnal tracer transport in the ocean model), as well as having an initial condition with an equilibrium, uncoupled ocean solution forced by reasonably accurate surface fluxes. This apparent lack of drift in surface temperature would not continue indefinitely in this model; e.g., there is appreciable drift in the oceanic S field associated with the lack of river inflow and excessive water cycling in the sea–ice model. As always, there are many properties of this coupled solution that are unrealistic and warrant further attention.

Achieving skillful coupled models is an important and difficult scientific challenge. So many physical processes must be included for a physically consistent calculation of climate, and there are probably many failure modes that lead to excessive drift (many of which may not yet have been discovered). Of course, some blemishes in oceanic behavior will occur because of failures in the other component models in a coupled solution. Two examples of known problems of this type are an incorrect location of the sea–ice boundary and excessive surface solar radiation due to underestimation in the atmospheric model of stratus cloud amounts or cloud absorption (i.e., the current controversy over the "anomalous absorption" in nature compared to standard radiative transfer calculations; see Cess *et al.*, 1995).

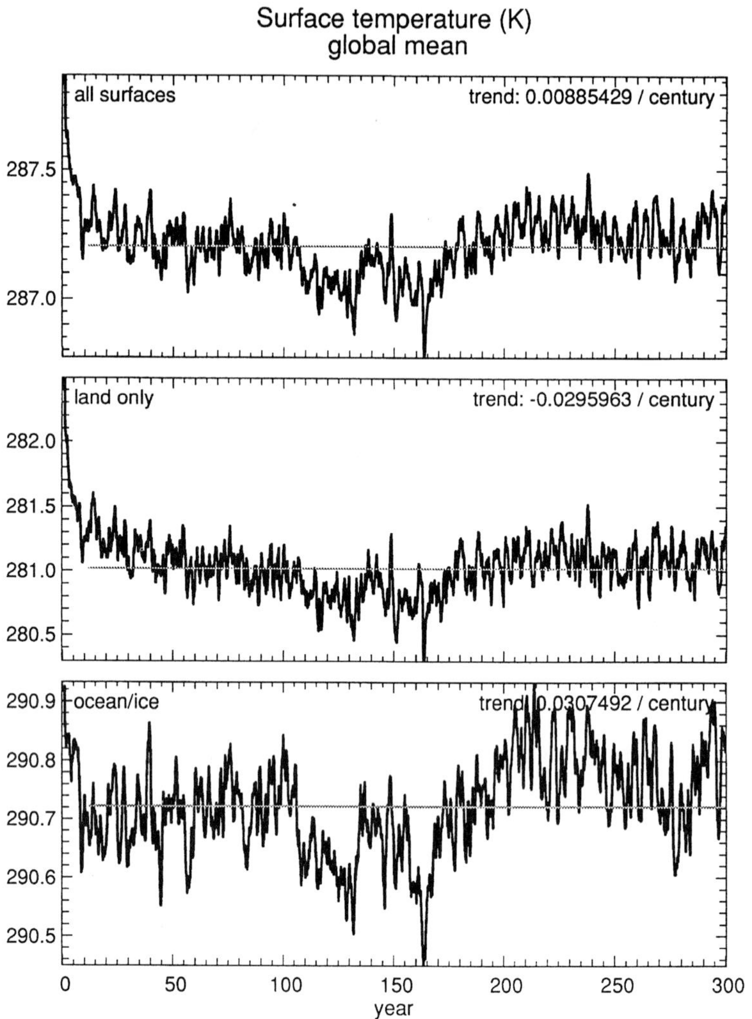

Figure 6 Time series of spatially averaged surface temperature in a 300-year integration of the NCAR Climate System Model (Boville and Gent, 1998).

OGCMs must avoid their own contributions to coupled failure modes. The use of surface restoring terms for T and S in ocean models precludes many of these failure modes. To make progress, then, ocean models must evolve away from using this procedure for the buoyancy forcing. This poses a new requirement that an OGCM yield a correct surface T (necessary for having correct air–sea fluxes) as a result of the forcing it receives and its own intrinsic variability. In principle, meeting this requirement involves all aspects of the OGCM, but there is probably particular importance to the surface PBL, marginal seas and river inflows, and interactions with sea–ice.

For the reasons discussed in Section VIII, OGCMs with coarse spatial grids will be most commonly used in coupled calculations, although some exploratory calculations with eddy-resolving models no doubt will be made. Yet the prospect of either not resolving eddies or doing so inaccurately is unpalatable. This gives us added incentive to develop more robust methods for accelerating OGCM solutions (Section IV) and more skillful parameterizations for marginally resolved eddying solutions, like those used in large eddy simulations of small-scale turbulence.

X. CONCLUSION

Models of the oceanic general circulation are currently in a phase of rapid development and expanding utilization. Within the range of legitimate choices for model formulation and boundary conditions, surveyed above, present model solutions do seem to encompass the major features of the circulation as observed. In the coming decades the primary challenges are to extend and refine the modeling choices and to expand the observational system to reduce the empirical uncertainties for testing the models. The goal, of course, is to end up by achieving a useful degree of consistency between OGCM solution behaviors and nature. The biggest surprises along the way are likely to come in weaning the ocean model from specified, or at least strongly constrained, surface boundary conditions, as it assumes its proper, more fundamental role in modeling the Earth's climate system.

ACKNOWLEDGMENTS

I wish to express my admiration for the rigor and creativity that Prof. Akio Arakawa brings to the scientific enterprise of formulating GCMs. His influence on oceanic modeling is significant and will be long lasting. I thank Profs. David Randall and Roger Wakimoto for organizing the symposium honoring Prof. Arakawa and this book. Research support was provided by the National Air and Space Agency (NAG 5-3982) and the National Science Foundation (OCE-9633681).

REFERENCES

Arakawa, A. (1988). Finite-difference methods in climate modeling. *In* "Physically based modeling and simulation of climate and climate change," Part I (M. E. Schlesinger, ed.), pp. 79–168. Kluwer, Dordrecht.

Armi, L. (1978). Some evidence for boundary mixing in the deep ocean. *J. Geophys. Res.* **83,** 1971–1979.

Armi, L., and D. M. Farmer (1988). The flow of Mediterranean and Atlantic water through the strait of Gibraltar. *Prog. Oceanogr.* **21,** 1–106.

Barnier, B., L. Siefridt, and P. Marchesiello (1995). Thermal forcing for a global ocean circulation model using a three-year climatology of ECMWF analyses. *J. Mar. Syst.* **6,** 363–380.

Barnier, B., P. Marchesiello, A. P. de Miranda, J. M. Molines, and M. Coulibaly (1998). A sigma-coordinate, primitive equation model for studying the circulation in the South Atlantic. Part I: Model configuration with error estimates. *Deep-Sea Res.*, in press.

Beckmann, A., and R. Doescher (1997). A method of improved representation of dense water spreading over topography in geopotential-coordinate models. *J. Phys. Oceanogr.* **27,** 581–591.

Berloff, P., and J. C. McWilliams (1999a). The quasigeostrophic dynamics of western boundary currents. *J. Phys. Oceanogr.* **29,** 2607–2634.

Berloff, P., and J. C. McWilliams (1999b). Low-frequency intrinsic variability in wind-driven gyres. *J. Phys. Oceanogr.* **29,** 1925–1949.

Bleck, R., and E. Chassignet (1994). Simulating the oceanic circulation with isopycnic-coordinate models. *In* "The Oceans: Physical-Chemical Dynamics and Human Impact" (S. K. Majumdar and E. W. Miller, eds.), pp. 17–39. Penn. Acad. Sci.

Bleck, R., S. Dean, M. O'Keefe, and A. Sawdey (1995). A comparison of data-parallel and message-passing versions of the Miami isopycnic coordinate ocean model. Submitted for publication.

Boening, C. W. (1989). Influences of a rough bottom topography on flow kinematics in an eddy resolving model. *J. Phys. Oceanogr.* **19,** 77–97.

Boening, C. W., W. R. Holland, F. O. Bryan, G. Danabasoglu, and J. C. McWilliams (1995). An overlooked problem in model simulations of the thermohaline circulation and heat transport in the Atlantic Ocean. *J. Climate* **8,** 515–523.

Boville, B. A., and P. R. Gent (1998). The NCAR Climate System Model, Version One. *J. Climate* **11,** 1115–1130.

Bretherton, F., and D. B. Haidvogel (1976). Two-dimensional turbulence above topography. *J. Fluid Mech.* **78,** 129–154.

Bretherton, F., and M. Karweit (1975). Mid-ocean mesoscale modeling. *In* "Numerical Models of Ocean Circulation" (R. O. Reid, ed.), pp. 237–249. National Academy Press, Washington, DC.

Bryan, K. (1969). A numerical method for the study of the circulation of the world ocean. *J. Comput. Phys.* **4,** 347–376.

Bryan, K. (1984). Accelerating the convergence to equilibrium of ocean-climate models. *J. Phys. Oceanogr.* **14,** 666–683.

Bryden, H. L., D. H. Roemmich, and J. A. Church (1991). Ocean heat transport across 24°N in the Pacific. *Deep-Sea Res.* **38**, 297–324.

Cess, R. D., and 19 co-authors (1995). Absorption of solar radiation by clouds: Observations versus models. *Science* **267**, 496–499.

Chao, Yi, A. Gangopadhyay, F. O. Bryan, and W. R. Holland (1996). Modeling the Gulf Stream system: How far from reality? *Geophys. Res. Lett.* **23**, 3155–3158.

Chassignet, E. P., Z. Garraffo, and A. Paiva (1999). Fine-mesh (1/12 degree) modeling of the North Atlantic: The spin-up phase. *J. Marine Sys.* **21**, 307–320.

Covey, C. (1995). Global ocean circulation and equator-pole heat transport as a function of ocean GCM resolution. *J. Climate* **11**, 425–437.

Cummins, P. F. (1991). The deep water stratification of ocean general circulation models. *Atmos. Ocean* **29**, 563–575.

Cummins, P. F., G. Holloway, and A. E. Gargett (1990). Sensitivity of the GFDL ocean general circulation model to a parameterization of vertical diffusion. *J. Phys. Oceanogr.* **20**, 817–830.

Danabasoglu, G., and J. C. McWilliams (1995). Sensitivity of the global ocean circulation to parameterizations of mesoscale tracer transports. *J. Climate* **8**, 2967–2987.

Danabasoglu, G., J. C. McWilliams, and P. R. Gent (1994). The role of mesoscale tracer transports in the global ocean circulation. *Science* **264**, 1123–1126.

Danabasoglu, G., J. C. McWilliams, and W. G. Large (1996). Approach to equilibrium in global ocean models. *J. Climate* **9**, 1092–1110.

Dansgaard, W., S. J. Johnson, H. B. Clausen, D. Dahl-Jensen, N. Gundestrup, C. U. Hammer, and H. Oeschger (1984). North Atlantic climatic oscillations revealed by deep Greenland ice cores. *In* "Climate Processes and Climate Sensitivity" (J. E. Hansen and T. Takahashi, eds.), pp. 288–298. Geophysical Monographs, vol. 29. AGU Press.

Deser, C., and M. Blackmon (1993). Surface climate variations over the North Atlantic Ocean during winter: 1900–1989. *J. Climate* **6**, 1743–1753.

Dukowicz, J. K., and R. D. Smith (1994). Implicit free-surface method for the Bryan–Cox–Semtner ocean model. *J. Geophys. Res.* **99**, 7991–8014.

Dukowicz, J. K., R. D. Smith, and R. C. Malone (1993). A reformulation and implementation of the Bryan–Cox–Semtner ocean model on the Connection Machine. *J. Atmos. Ocean. Tech.* **10**, 195–208.

Ebby, M., and G. Holloway (1994a). Grid transform for incorporating the Arctic in a global ocean model. *Clim. Dyn.* **10**, 241–247.

Ebby, M., and G. Holloway (1994b). Sensitivity of a large-scale ocean model to a parameterization of topographic stress. *J. Phys. Oceanogr.* **24**, 2577–2588.

Ezer, T., and G. L. Mellor (1994). Diagnostic and prognostic calculations of the North Atlantic circulation and sea level using a sigma coordinate ocean model. *J. Geophys. Res.* **99**, 14159–14171.

Gargett, A. E., and G. Holloway (1992). Sensitivity of the GFDL ocean model to different diffusivities for heat and salt. *J. Phys. Oceanogr.* **22**, 1158–1177.

Garrett, C. (1991). Marginal mixing theories. *Atmos.-Ocean* **29**, 313–339.

Gent, P. R., and J. C. McWilliams (1983). Regimes of validity for balanced models. *Dyn. Atmos. Oceans* **7**, 167–183.

Gent, P. R., and J. C. McWilliams (1990). Isopycnal mixing in ocean circulation models. *J. Phys. Oceanogr.* **20**, 150–155.

Gent, P. R., J. Willebrand, T. J. McDougall, and J. C. McWilliams (1995). Parameterizing eddy-induced tracer transports in ocean circulation models. *J. Phys. Oceanogr.* **25**, 463–474.

Gent, P. R., F. O. Bryan, G. Danabasoglu, S. C. Doney, W. R. Holland, W. G. Large, and J. C. McWilliams (1998). The NCAR Climate System Model global ocean component. *J. Climate* **11**, 1287–1306.

Gnanadesikan, A. (1998). Representing the bottom boundary layer in the GFDL ocean model: Model framework, dynamical impacts, and parameter sensitivity. Submitted for publication.

Haidvogel, D. B., and F. O. Bryan (1992). Ocean general circulation modeling. *In* "Climate System Modeling" (K. Trenberth, ed.), pp. 371–412. Oxford Press.

Haidvogel, D. B., E. Curchitser, M. Iskandarani, R. Hughes, and M. Taylor (1995). Global modeling of the ocean and atmosphere using the spectral element method. *Atmos. Ocean*, in press.

Hall, M. M., and H. L. Bryden (1982). Direct estimates and mechanisms of ocean heat transport. *Deep-Sea Res.* **29**, 339–359.

Hirst, A. C., and J. S. Godfrey (1993). The role of Indonesian throughflow in a global ocean GCM. *J. Phys. Oceanogr.* **23**, 1057–1086.

Holloway, G. (1992). Representing topographic stress for large-scale ocean models. *J. Phys. Oceanogr.* **22**, 1033–1046.

Hurlburt, H. E., P. J. Hogan, E. J. Metzger, W. J. Schmitz, and A. J. Wallcraft (1995). Dynamics of eddy-resolving models of the Pacific Ocean and the Sea of Japan. *In* "Workshop on Numerical Prediction of Oceanographic Conditions," pp. 51–57. Japan Meteorological Agency, Tokyo.

Ishikawa, I., Y. Yamanaka, and N. Suginohara (1994). Effects of presence of a circumpolar region on buoyancy-driven circulation. *J. Oceanogr.* **50**, 247–263.

Ivchenko, V. O., and D. P. Stevens (1995). The zonal momentum balance in a realistic eddy resolving general circulation model of the Southern Ocean. *J. Phys. Oceanogr.* **26**, 753–774.

Kagimoto, T., and T. Yamagata (1997). Seasonal transport variations of the Kuroshio: An OGCM simulation. *J. Marine Res.*, in press.

Kalnay, E., and 22 co-authors (1996). The NCEP/NCAR 40-year reanalysis project. *BAMS* **77**, 437–471.

Kraus, E. B., and S. Turner (1967). A one-dimensional model of the seasonal thermocline. II: The general theory and its consequences. *Tellus* **19**, 98–106.

Large, W. G., and P. R. Gent (1998). Validation of vertical mixing in an equatorial ocean model using large-eddy simulations and observations. *J. Phys. Oceanogr.*, in press.

Large, W. G., J. C. McWilliams, and S. C. Doney (1994). Oceanic vertical mixing: A review and a model with a non-local K-profile boundary layer parameterization. *Rev. Geophys.* **32**, 363–403.

Large, W. G., G. Danabasoglu, S. Doney, and J. C. McWilliams (1997). Sensitivity to surface forcing and boundary-layer mixing in a global ocean model: Annual-mean climatology. *J. Phys. Oceanogr.* **27**, 2418–2447.

Leonard, B. P. (1979). A stable and accurate convective modelling procedure based on quadratic upstream interpolation. *Comp. Meth. Appl. Mech. Eng.* **19**, 59–98.

Levitus, S. (1982). "Climatological Atlas of the World Ocean," NOAA Prof. Pap. 13. U.S. Gov. Print. Off.

Levitus, S., R. Burgett, and T. P. Boyer (1994). "World Ocean Atlas 1994: Vol. 3, Salinity." *NOAA Atlas NESDES* **3**, U.S. Gov. Print. Off.

Li, X., Y. Chao, J. C. McWilliams, and L.-L. Fu (1998). Sensitivity to vertical mixing in a Pacific Ocean General Circulation Model. Submitted for publication.

Lunkeit, F., R. Sausen, and J. M. Oberhuber (1996). Climate simulations with a global coupled atmosphere-ocean model ECHAM2/OPYC. Part I: present-day climate and ENSO events. *Clim. Dyn.* **12**, 195–212.

MacDonald, A. M., and C. Wunsch (1996). An estimate of global ocean circulation and heat fluxes. *Nature* **382**, 436–439.

Manabe, S., and K. Bryan (1969). Climate calculations with a combined ocean-atmosphere model. *J. Atmos. Sci.* **26,** 786–789.

Manabe, S., and R. J. Stouffer (1993). Century-scale effects of increased atmospheric CO_2 on the ocean–atmosphere system. *Nature* **364,** 215–218.

Marotzke, J. (1997). Boundary mixing and the dynamics of three-dimensional thermohaline circulations. *J. Phys. Oceanogr.* **27,** 1713–1728.

Marshall, J. (1981). On the parameterization of geostrophic eddies in the ocean. *J. Phys. Oceanogr.* **11,** 257–271.

Marshall, J., C. Hill, L. Perelman, and A. Adcroft (1997). Hydrostatic, quasi-hydrostatic, and nonhydrostatic ocean modeling. *J. Geophys. Res.* **102,** 5733–5752.

McDougall, T. J. (1987). Neutral surfaces. *J. Phys. Oceanogr.* **17,** 1950–1964.

McWilliams, J. C. (1996). Modeling the oceanic general circulation. *Ann. Rev. Fluid Mech.* **28,** 1–34.

McWilliams, J. C., and J. H. S. Chow (1981). Equilibrium geostrophic turbulence: I. A reference solution in a β-plane channel. *J. Phys. Oceanogr.* **11,** 921–949.

McWilliams, J. C., and J. M. Restrepo (1998). The wave-driven ocean circulation. *J. Phys. Oceanogr.*, in press.

McWilliams, J. C., N. J. Norton, P. R. Gent, and D. B. Haidvogel (1990). A linear balance model of wind-driven, mid-latitude ocean circulation. *J. Phys. Oceanogr.* **20,** 1349–1378.

McWilliams, J. C., G. Danabasoglu, and P. R. Gent (1996). Tracer budgets in the Warm Water Sphere. *Tellus* **48A,** 179–192.

Mellor, G., and T. Yamada (1982). Development of a turbulence closure model for geophysical fluid problems. *Rev. Geophys. Space Phys.* **20,** 851–875.

Merryfield, W. J., and G. Holloway (1999). Eddy fluxes and topography in stratified quasi-geostrophic models. *J. Fluid Mech.* **380,** 59–80.

Munk, W. H., and E. Palmen (1951). Note on the dynamics of the Antarctic circumpolar current. *Tellus* **3,** 53–55.

Owens, W. B., and F. P. Bretherton (1978). A numerical study of mid-ocean mesoscale eddies. *Deep-Sea Res.* **25,** 1–14.

Pacanowski, R., and G. Philander (1981). Parameterization of vertical mixing in numerical models of tropical oceans. *J. Geophys. Res.* **11,** 1443–1451.

Pinardi, N., A. Rosati, and R. C. Pacanowski (1995). The sea surface pressure formulation of rigid lid models. Implications for altimetric data assimilation studies. *Jour. Marine Syst.* **6,** 109–120.

Polzan, K. L., J. M. Toole, J. R. Ledwell, and R. W. Schmitt (1997). Spatial variability of turbulent mixing in the abyssal ocean. *Science* **276,** 93–96.

Pratt, L. J., ed. (1990). "The Physical Oceanography of Sea Straits." Kluwer, Dordrecht.

Price, J. F., and M. O. Beringer (1994). Outflows and deep water production by marginal seas. *Prog. Ocean.* **33,** 161–200.

Redi, M. H. (1982). Oceanic isopycnal mixing by coordinate rotation. *J. Phys. Oceanogr.* **12,** 1154–1158.

Rhines, P. B. (1977). The dynamics of unsteady currents. *In* "The Sea" (E. Goldberg, ed.), vol. VI, pp. 189–318. Wiley, New York.

Rhines, P. B., and W. R. Young (1982). Homogenization of potential vorticity in planetary gyres. *J. Fluid Mech.* **122,** 342–367.

Roberts, M., and D. Marshall (1999). Do we require adiabatic dissipation schemes in eddy-resolving ocean models? Submitted for publication.

Roberts, M. J., A. L. New, R. A. Wood, and R. Marsh (1995). An intercomparison of a Bryan–Cox type ocean model and an isopycnic ocean model. Part 1: The subpolar gyre and high-latitude processes. *J. Phys. Oceanogr.* **26,** 1495–1527.

Sakamoto, T., and T. Yamagata (1996). Seasonal transport variations of the wind-driven circulation in a tow-layer planetary-geostrophic model with a continental slope. *J. Mar. Res.* **54,** 261–284.

Salmon, R., G. Holloway, and M. C. Hendershott (1976). The equilibrium statistical mechanics of simple quasi-geostrophic models. *J. Fluid Mech.* **75,** 691–703.

Saravanan, R., and J. C. McWilliams (1995). Multiple equilibria, natural variability, and climate transitions in an idealized ocean–atmosphere model. *J. Climate* **8,** 2296–2323.

Saravanan, R., and J. C. McWilliams (1997). Stochasticity and spatial resonance in interdecadal climate fluctuations. *J. Climate,* in press.

Shchepetkin, A., and J. C. McWilliams (1998). Quasi-monotone advection schemes based on explicit locally adaptive dissipation. *Mon. Wea. Rev.,* in press.

Siegel, A., P. Berloff, J. C. McWilliams, J. B. Weiss, and I. Yavneh (1999). The quasi-geostrophic dynamics of wind-driven ocean gyres at very high resolution. Submitted for publication.

Smith, R. D., S. Kortas, and B. Meltz (1995). Curvilinear coordinates for global ocean models. Los Alamos Tech. Rep., LAUR-95-1146, Los Alamos National Laboratory, Los Alamos, NM.

Smith, R. D., M. E. Maltrud, M. W. Hecht, and F. O. Bryan (1999). Numerical simulation of the North Atlantic Ocean at $1/10°$. Submitted for publication.

Song, Y., and D. Haidvogel (1994). A semi-implicit ocean circulation model using a generalized topography-following coordinate system. *J. Comp. Phys.* **115,** 228–244.

Suginohara, N., and M. Fukasawa (1988). Set-up of deep circulation in multi-level numerical models. *J. Ocean Soc. Japan* **44,** 315–336.

Treguier, A. M., and J. C. McWilliams (1990). Topographic influences on wind-driven, stratified flow in a β-plane channel: An idealized model of the Antarctic Circumpolar Current. *J. Phys. Oceanogr.* **20,** 324–343.

Treguier, A. M., J. K. Dukowicz, and K. Bryan (1996). Properties of nonuniform grids used in ocean general circulation models. *J. Geophys. Res.* **101,** 20877–20881.

Treguier, A. M., I. M. Held, and V. D. Larichev (1997). On the parameterization of quasigeostrophic eddies in primitive equation ocean models. *J. Phys. Oceanogr.* **27,** 567–580.

Trenberth, K. E., and A. Solomon (1994). The global heat balance: Heat transports in the atmosphere and ocean. *Clim. Dyn.* **10,** 107–134.

Visbeck, M., J. Marshall, T. Haine, and M. Spall (1997). Specification of eddy transfer coefficients in coarse-resolution ocean circulation models. *J. Phys. Oceanogr.* **27,** 381–402.

Wajsowicz, R. C. (1993). A consistent formulation of the anisotropic stress tensor for use in models of the large-scale ocean circulation. *J. Comp. Phys.* **105,** 333–338.

Wijffels, S. E., R. W. Schmitt, H. L. Bryden, and A. Stigebrandt (1992). Transport of freshwater by the oceans. *J. Phys. Oceanogr.* **22,** 155–162.

Yavneh, I., and J. C. McWilliams (1995). Robust multigrid solution of the shallow-water balance equations. *J. Comp. Phys.* **119,** 1–25.

Yin, F. L., and I. Y. Fung (1991). On the net diffusivity in ocean general circulation models. *J. Geophys. Res.* **96,** 10773–10776.

Zhang, J., R. W. Schmitt, and R.-X. Huang (1998). Sensitivity of GFDL Modular Ocean Model to the parameterization of double-diffusive processes. *J. Phys. Oceanogr.,* in press.

Chapter 15

Climate and Variability in the First Quasi-Equilibrium Tropical Circulation Model

Ning Zeng and J. David Neelin
Department of Atmospheric Sciences and
Institute of Geophysics and Planetary Physics
University of California, Los Angeles

Chia Chou, Johnny Wei-Bing Lin, and Hui Su
Department of Atmospheric Sciences
University of California, Los Angeles

I. INTRODUCTION

Quasi-equilibrium (QE) convective closures for the relation between moist convection and large-scale dynamics originally proposed by Arakawa and Schubert (1974) assert that convective ensembles at scales smaller than the Reynolds average (sub-Reynolds scales) act to remove convective instability within the vertical column (conditional instability of the first kind). The convective motions thus tend to establish a statistical equilibrium between the variables that affect parcel buoyancy, i.e., the large-scale temperature and moisture, when an ensemble average is taken over a convective region (Arakawa, 1993; Emanuel *et al.*, 1994).

A new class of model for the tropical circulation is proposed (Neelin and Zeng, 2000, NZ hereafter; Zeng *et al.*, 2000, ZNC hereafter) that directly exploits constraints placed on the flow by deep convection, as represented by quasi-equilibrium thermodynamic closures in the convective parameterization. We refer to these as quasi-equilibrium tropical circulation models (QTCMs). In particular, the model presented in NZ uses a version of the Betts and Miller (1986, 1993) deep convection scheme, in which it is assumed that the ensemble effect of deep convection is to reduce a certain measure of convective available potential energy (CAPE). This tends to constrain the large-scale temperature profile, and thus the baroclinic pressure gradients. Analytical solutions for deep convective regions based on this were produced as part of an ongoing project, summarized in Neelin (1997, hereafter N97). The QTCM makes use of these analytical solutions as basis functions within the numerical model, an approach referred to as "tailored basis functions," because the retained vertical structures are tailored to the dominant physics of interest. This provides a way of taking "quasi-equilibrium thinking" (reviewed in Chapter 8 of this volume) into the realm of quantitative simulation.

A small hierarchy of QTCMs is anticipated, in which successively higher accuracy is obtained by retaining additional vertical structures or by embedding additional physics. In NZ the simplest QTCM is chosen that adequately simulates primary features of the tropical climatology. This is termed QTCM1 because it retains a single vertical structure for temperature and humidity. By its derivation, it is expected to give an accurate solution in deep convective regions (compared to, say, a GCM with Betts–Miller convection). Within one radius of deformation of deep convective regions, it should remain reasonably accurate, because temperature gradients are not large. At midlatitudes it is simply a highly truncated vertical representation, roughly equivalent to a two-layer model, since the Galerkin representation is tailored to tropical vertical structures.

To accompany this representation of dynamics, a radiation package is derived (after Chou and Neelin, 1996; Chou, 1997) that represents the main radiative processes at an intermediate level of complexity. This includes leading cloud–radiation interaction effects, because these are important to the general circulation and because the dynamics of convection zones is of interest. Likewise, a land–surface model is presented that includes the essentials of more complex land–surface models such as the biophysical control on evapotranspiration and surface hydrology, but retains computational and diagnostic simplicity (Zeng and Neelin, 1999). The QTCM thus occupies a niche intermediate between GCMs and simpler models. It is related to GCMs in having a step-by-step derivation from the primitive equations, a convective parameterization based on parcel buoyancy considerations, and two-stream radiative schemes with cloud interaction, while remaining computationally efficient and simple to analyze. For instance, Zeng and Neelin (1999) and NZ discuss ways in which the moist static energy budget allows more direct access to fundamental dynamics in convection zones, especially over land regions. An example of testing the impact of a process by intervening in the model to suppress it is provided in Lin *et al.* (1998), in which the impact of midlatitude disturbances on intraseasonal oscillations is tested.

We provide here a sampling of the simulation for a variety of phenomena, reviewing results that are presented in more detail in NZ, ZNC, and Lin *et al.* (2000). The model dynamical framework, the implementation of cloud prediction, short-wave and long-wave radiation schemes, and the land–surface scheme are summarized in Section II. Section III presents, in turn, the simulated climatology, sensitivity to departures from the QE parameterization, the intraseasonal oscillation, interannual variability forced by observed SST, including El Niño–southern oscillation variability, the seasonal and interannual evolution of the southeast Asian monsoon, and the Amazon water budget. Conclusions are provided in Section IV.

II. MODEL DESCRIPTION / IMPLEMENTATION

A. DYNAMICS AND CONVECTION

Approximations associated with the quasi-equilibrium assumption are used to generate analytical solutions which are then used as the first basis function in a Galerkin framework. The detailed derivations starting from the primitive equations, and interpretation of the dynamics and moist convection in the model, are discussed in NZ. The major steps includes these: (1) A vertical structure of temperature is chosen that can match the QE profile of the Betts–Miller scheme when QE applies (although QE is not assumed to always hold); (2) using the hydrostatic equation, the choice of the temperature profile gives the vertical structure of the baroclinic pressure gradients; (3) the baroclinic and barotropic components of the pressure gradient term yield corresponding vertical structures of the horizontal velocity components; (4) nonlinear advection terms and vertical eddy viscosity terms are projected onto these velocity components; (5) a vertical structure for the vertical velocity corresponding to the baroclinic component is derived using the continuity equation; and (6) an effective moist stability applicable to deep convective regions is derived based on the vertical velocity profile and the temperature and humidity structure.

The main prognostic equations of QTCM1 are for the amplitudes of the vertical structures of temperature T_1, humidity q_1, the baroclinic component of the horizontal velocity $\mathbf{v}_1$, and the barotropic component $\mathbf{v}_0$ (the barotropic vorticity ζ_0 is the actual prognostic variable):

$$\partial_t \mathbf{v}_1 + \mathscr{D}_{V1}(\mathbf{v}_0,\mathbf{v}_1) + f\mathbf{k} \times \mathbf{v}_1 = -\kappa\nabla T_1 - \epsilon_1\mathbf{v}_1 - \epsilon_{01}\mathbf{v}_0, \tag{1}$$

$$\partial_t \zeta_0 + \mathrm{curl}_z(\mathscr{D}_{V0}(\mathbf{v}_0,\mathbf{v}_1)) + \beta v_0$$
$$= -\mathrm{curl}_z(\epsilon_0\mathbf{v}_0) - \mathrm{curl}_z(\epsilon_{10}\mathbf{v}_1), \tag{2}$$

$$\hat{a}_1(\partial_t + \mathscr{D}_{T1})T_1 + M_{S1}\nabla \cdot \mathbf{v}_1$$
$$= \epsilon_c^*(q_1 - T_1)$$
$$+ (g/p_T)(-R_t^\uparrow - R_s^\downarrow + R_s^\uparrow + S_t - S_s + H), \tag{3}$$

$$\hat{b}_1(\partial_t + \mathscr{D}_{q1})q_1 - M_{q1}\nabla \cdot \mathbf{v}_1 = \epsilon_c^*(T_1 - q_1) + (g/p_T)E, \tag{4}$$

where $\mathscr{D}$ represents the projected advection and diffusion of the corresponding variables, a hat denotes vertical averaging, M_{S1} and M_{q1} are effective dry static stability and moisture stratification, ϵ denotes various effective damping/coupling coefficients, E is evaporation, H is sensible heat flux, S is short-wave flux, and R is long-wave radiative flux. Note that the projections lead to terms containing vertical averages of the profiles such as $\hat{a}_1$. The convective heating term $(q_1 - T_1)$ is a projected CAPE and ϵ_c^* is inversely proportional to the relaxation time τ_c of the Betts–Miller scheme and allows departures from QE. Outside convection zones (when CAPE is negative) $\epsilon_c^* = 0$.

An alternate prognostic equation that is very useful in analyzing model dynamics is the moist static energy equation [the sum of Eqs. (3) and (4)]:

$$\hat{a}_1(\partial_t + \mathscr{D}_{T1})T_1 + \hat{b}_1(\partial_t + \mathscr{D}_{q1})q_1 + M_1\nabla \cdot \mathbf{v}_1 = (g/p_T)F^{\mathrm{net}}. \tag{5}$$

This dominates the dynamics of convective regions, where the temperature and moisture equations alone have large canceling terms. It avoids fast, $O(\tau_c)$, time scales which go into the adjustment of CAPE. The projected moist static energy equation Eq. (5), is very similar to the vertically integrated moist static energy equation of the primitive equations. In both equations net flux into the column F^{net} balances transport by the retained degrees of freedom. Although our coding uses separate T_1 and q_1 equations, it would be possible to run the model using Eq. (5) and either Eq. (3) or a CAPE equation formed from the difference of Eq. (3)/$\hat{a}_1$ minus Eq. (4)/$\hat{b}_1$. Within convective regions, to the extent that QE ties q_1 to T_1. Eq. (5) contains all thermodynamic information necessary to solve for large scales, and this can be exploited to assist in understanding results.

The vertical profiles of these prognostic variables are derived offline. The vertical profiles a_1, B_1, V_1, and V_0, as well as the profile for vertical velocity Ω_1, are shown in Fig. 1. Note that the baroclinic profile V_1 is not exactly symmetric about midatmospheric level as assumed in simpler models (e.g., Gill, 1980), but shows more subtle structure resulting from integrating the hydrostatic equation using the retained temperature profile. Because the vertical moisture structure in moist convective regions is less constrained (see NZ), we have more freedom in choosing the actual

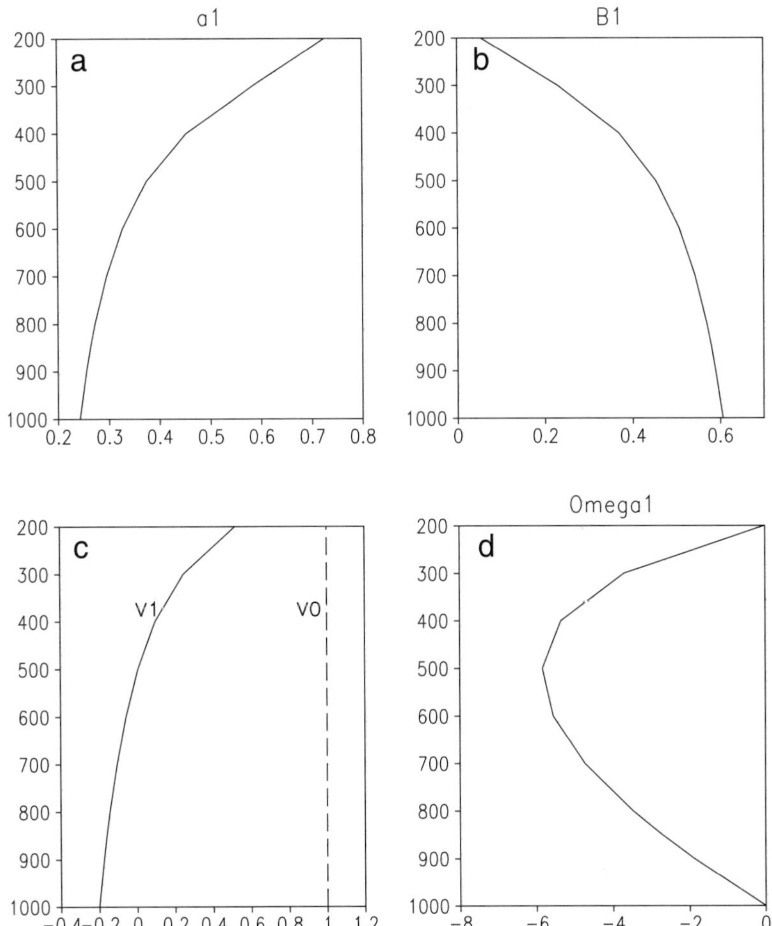

Figure 1 Vertical profiles of basis functions for (a) temperature a_1, (b) humidity B_1, (c) horizontal velocity V_1 and V_0, and (d) implied profile of vertical velocity Ω_1.

humidity profile b_1 such that it does not have to be the same as the fixed sub-saturation moist adiabatic profile B_1.

Given the vertical profiles, the total fields can be reconstructed at any vertical level as

$$T = T_r(p) + a_1(p)T_1(x, y, t), \tag{6}$$

$$q = q_r(p) + b_1(p)q_1(x, y, t), \tag{7}$$

$$\mathbf{v} = \mathbf{v}_0(x, y, t) + V_1(p)\mathbf{v}_1(x, y, t), \tag{8}$$

where T_r and q_r are the reference profiles chosen to be typical for the tropical deep convective regions, x is longitude, y is latitude, and p is pressure. Note that the model predicts the total wind while temperature and humidity are deviations from a reference profile. The reference profiles are independent of space and time, and so do not appear in differentiated terms of the equations. They are not assumed to be solutions of the equations. See NZ's Appendix A for discussion of the impact of the reference profiles.

B. Cloud Prediction and Radiation

1. Cloud Prediction

As shown in Chou (1997), four cloud types—deep cloud, cirro-cumulus/cirro-stratus (CsCc), cirrus and stratus—as classified in the International Satellite Cloud Climatology Project (ISCCP) data (Rossow and Schiffer, 1991) can capture 80% of radiation budget in the tropics. We aim to develop parameterizations for these four cloud types in the QTCM. For the QTCM, the primary concern is to simulate mean cloudiness averaged over a weekly time scale and large spatial scales, rather than to simulate cloudiness variations at every time step. We thus work with cloud cover types rather than the usual GCM level-by-level and prognostic approach. Cloud cover types have a defined vertical structure, for instance, a deep convective tower accompanied by a fixed area of anvil. A single cloud type that includes the deep and anvil clouds is found to have a good linear relationship with large-scale precipitation.

2. Long-Wave Radiation Scheme

Long-wave radiation is derived from a simplified long-wave radiation scheme (Chou and Neelin, 1996) with the Green's functions projected on the retained basis function in temperature and moisture. Because of this projection, only three components of long-wave radiative fluxes $R_t^\downarrow$, $R_s^\uparrow$, and $R_t^\uparrow$ are needed in the QTCM. The weakly nonlinear long-wave radiation scheme coded in the packages of clrad1 and clrad1-d is

$$R_t^\uparrow = R_{rt}^\uparrow + \sum_{n=0}^{N} \alpha_n \left[\epsilon_{RT_1 t}^{\uparrow n} T_1 + \epsilon_{R q_1 t}^{\uparrow n} q_1 + \epsilon_{RT_s t}^{\uparrow n} (T_s - T_{sr}) \right]$$

$$+ \sum_{n=1}^{N} \epsilon_{R\alpha_n t}^{\uparrow} (\alpha_n - \alpha_{rn}), \tag{9}$$

$$R_s^\downarrow = R_{rs}^\downarrow + \sum_{n=0}^{N} \alpha_n \left[\epsilon_{RT_1 s}^{\downarrow n} T_1 + \epsilon_{R q_1 s}^{\downarrow n} q_1 \right] + \sum_{n=1}^{N} \epsilon_{R\alpha_n s}^{\downarrow} (\alpha_n - \alpha_{rn}),$$

$$R_s^\uparrow = R_{rs}^\uparrow + \epsilon_{RT_s s}^{\uparrow} (T_s - T_{sr}),$$

where α_n and α_{rn} are cloud fraction and reference cloud fraction, respectively, for cloud type n, and $n = 0$ represents clear sky. In Eq. (9), α_{rn} can be set to zero, and then $R_{rt}^\uparrow$ and $R_{rs}^\downarrow$ are the values for clear sky. Here α_{rn} is just for consistency with the linear scheme [see NZ, Eq. (4.45)]. Coefficients in the scheme are precalculated as, for instance,

$$\epsilon_{R q_1 t}^{\uparrow n} = \int_{p_0}^{p_s} G_q^{\uparrow n} (p_t, \grave{p}) b_1(\grave{p}) d\grave{p}, \tag{10}$$

with $G_q^{\uparrow n}$ is obtained from the Harshvardhan *et al.* (1987) scheme, a full long-wave radiation scheme. This scheme is not that much more complex than a Newtonian cooling in formula but maintains the complexity of a full long-wave radiation scheme in physics.

3. Short-Wave Radiation Scheme—clrad1 and clrad1-d

Surface solar irradiance $(S_s^\downarrow)$ and net solar absorption by the atmospheric column $(S_t^\downarrow - S_t^\uparrow - S_s^\downarrow + S_s^\uparrow)$ are the primary components of short-wave radiative fluxes required in the QTCM equations during integration. The solar radiative fluxes mainly depend on solar zenith angle (θ) and surface albedo (A_s). The impact of variations of aerosol and atmospheric gases, such as ozone and CO_2, on the solar radiative fluxes are relatively small compared to the dependence on θ and A_s. Therefore, the first-order variation of the solar radiative fluxes can be approximated by simple formulas,

$$S_t^\downarrow - S_t^\uparrow - S_s^\downarrow + S_s^\uparrow = S_0 \cos \theta \sum_{n=0}^{N} \alpha_n f_{an}(\theta) g_{an}(A_s), \tag{11}$$

$$S_s^\downarrow = S_0 \cos \theta \sum_{n=0}^{N} \alpha_n f_{sn}^\downarrow(\theta) g_{sn}^\downarrow(A_s), \tag{12}$$

where S_0 is solar constant. Subscripts a indicate quantities associated with atmospheric absorption, and subscripts s indicate quantities associated with (downward) surface flux. Subscripts n indicate cloud type. To obtain the functions $f_{an}(\theta)$, $g_{an}(A_s)$, $f_{sn}^{\downarrow}(\theta)$, and $g_{sn}^{\downarrow}(A_s)$, we use the Fu and Liou (1993) solar radiation scheme and input a typical vertical profile of water vapor, temperature, CO_2, ozone, and aerosol. We then use curve fitting to approximate these two functions for conditions with clear sky and with different cloud types. This stripped-down short-wave radiation scheme captures the first-order effects of radiative processes in the Fu and Liou scheme implicitly, for instance, multiple scattering between cloud base and the surface. Surface albedo in the current QTCM is monthly climatology derived from Darnell *et al.* (1992), which is consistent with the Earth Radiation Budget Experiment (ERBE) data.

4. Short-Wave Radiation Scheme—clrad0

A simpler short-wave radiation scheme is also provided for use in QTCM1. The scheme clrad0 assumes highly simplified physics, but the assumptions are more transparent. Following Kiehl (1992) and Zeng and Neelin (1999), the scheme assumes a single layer of cloud/atmosphere with a lumped reflectivity A_c (including contributions from clouds, aerosol and atmospheric backscattering) and absorptivity a_{bs} (including contributions from water vapor and clouds). A single cloud type combines high and middle clouds and the cloud cover is proportional to the model precipitation (Section II.B.1). Given a surface albedo A_s, one can derive the fluxes at surface and top (see Zeng and Neelin, 1999, for details). For instance, the downward flux at the surface and upward flux at the top are

$$S_s^{\downarrow} = (1 - A_c)(1 - a_{bs})S_0 \cos \theta, \tag{13}$$

$$S_t^{\uparrow} = \left\{ (1 - A_c)^2 (1 - a_{bs})^2 A_s + A_c \right\} S_0 \cos \theta. \tag{14}$$

A diurnally averaged solar zenith angle dependence has been absorbed in A_c and a_{bs}.

C. Land-Surface Model

We have developed a land-surface parameterization scheme of intermediate complexity. It is much simpler than most current land-surface models (e.g., Dickinson *et al.*, 1986; Sellers *et al.*, 1996), but nevertheless models the first-order effects relevant to climate simulation including biophysical aspects, while from a diagnostic and computational point of view, it is only moderately more complicated than a bucket model (Manabe *et al.*, 1965). It is termed Simple-Land, or simply SLand. It does not attempt to resolve accurately the diurnal solar and environmental control on photosynthesis. Thus the soil moisture and seasonal variation of radiation are the main controlling factors. The most essential features for climate simulation are the low heat capacity of the land surface and specification of land albedo for the surface energy budget, and soil moisture and its consequences for the surface water budget. Sub-grid-scale variability of rainfall can significantly influence surface runoff and interception loss and, therefore, evaporation. Various analytical formulations have been proposed (e.g., Entekhabi and Eagleson, 1989). We essentially follow the same statistical approach but with choices more in line with the level of complexity of our atmospheric model.

In SLand, a single soil layer is assumed but with different depth for the energy and the water balance. For the energy balance, it essentially models the top soil layer with a typical thickness of 10 cm. The prognostic equation for ground temperature T_s is:

$$C_s \frac{\partial T_s}{\partial t} = F_s^{\mathrm{rad}} - E - H, \tag{15}$$

where C_s is the soil heat capacity, $F_s^{\mathrm{rad}} = S_s^{\downarrow} - S_s^{\uparrow} + R_s^{\downarrow} - R_s^{\uparrow}$ is downward net radiation at surface, E is the total evaporation, and H is the sensible heat flux. A small heat capacity C_s leads to a damping time scale on the order of 1 hr, so on time scales longer than a day, one has

$$F_s^{\mathrm{rad}} - E - H \approx 0. \tag{16}$$

This flux zero condition has been used explicitly by some early GCMs, and it imposes arguably the most important control on land-surface–atmosphere interaction (Zeng and Neelin 1999, NZ).

The water budget equation in a single soil layer that represents the root zone is:

$$\frac{\partial W}{\partial t} = P - E_I - R_s - E_T - R_g, \tag{17}$$

where W is the soil moisture content per unit area and P is the precipitation, E_I is the interception loss, E_T is and the evapotranspiration, R_s is the surface runoff (the fast component), and R_g is the subsurface runoff (the slow component). The soil is saturated when W equals the field capacity W_0, which is surface-type dependent. It is useful to define a relative soil wetness

$$w = W/W_0 \tag{18}$$

such that w is unity at saturation.

A main objective of SLand is to model the land–surface fluxes at large spatial and long temporal scales by statistically taking into account smaller and faster scale variations. In the above parameterizations the dependent variables are w, F_s^{rad}, and P. In the current version of the model, we use the following parameterizations: For interception loss:

$$E_I = E_I(P, F_s^{\text{rad}}). \tag{19}$$

The intercepted water is not available for surface runoff:

$$R_s = w^4(P - E_I), \tag{20}$$

and subsurface runoff is:

$$R_g = w^{2B+3}R_{g0}, \tag{21}$$

where R_{g0} is the subsurface runoff at saturation and B is the Clapp–Hornberger exponent. For evapotranspiration:

$$E_T = (r_s + r_a)^{-1}\rho_a(q_{\text{sat}}(T_s) - q_a), \tag{22}$$

where $r_a = (C_D V_s)^{-1}$ is the aerodynamic resistance, and r_s is a bulk surface resistance including stomatal/root resistance parameterized as:

$$r_s = r_{s_{\min}}w^{-\frac{1}{4}}, \tag{23}$$

where $r_{s_{\min}}$ is the minimum value of r_s occurring at no water stress ($w = 1$). The soil moisture dependences in R_s and R_g are essentially the formulations used in the Biosphere-Atmosphere Transfer Scheme (BATS, Dickinson *et al.*, 1986). The nonlinear dependence of r_s on w takes into account effects of the soil moisture uptake by deep roots under relatively dry conditions, such as what happens during the Amazon dry season. The actual form is chosen in accordance with observations and physically based parameterizations including heterogeneity effects (e.g., Entekhabi and Eagleson, 1989).

Although the scheme allows many land-surface types as long as the relevant parameter values are provided, we opt for a simple classification in the standard version with three surface types: forest, grass, and desert. The most important surface properties include surface albedo and field capacity, which play critical roles in the energy and the water balance, respectively. In the standard version of QTCM1, we use prescribed surface albedo derived from satellite observations (Darnell *et al.*, 1992), while one can easily switch to using the surface albedos linked to surface type. Sensitivity studies show discernible differences at regional scales between the two methods, but these have little impact on the global patterns. Snow hydrology is not simulated in this version, since the atmospheric model is aimed at the tropics.

D. IMPLEMENTATION

The standard version of QTCM1 includes full nonlinearities in advection, convection, and land-surface processes. Clouds associated with deep convection are predicted as one combined cloud type based on a simple precipitation/cloud-cover relationship (Section II.B). In the radiation package clrad1, the climatological monthly means of stratocumulus and cirrus clouds are prescribed from observations, while the tropical mean of middle clouds is used (constant in space and time). In the simpler package clrad0, the tropical mean of low clouds is used while high and middle clouds are predicted as one lumped cloud type according to the model precipitation. No cloud data are input for clrad0 runs. The linearized long-wave radiation scheme is used in clrad0, whereas clrad1 uses the weakly nonlinear version. Sensitivity studies show some differences between the two packages but the overall behaviors are similar. The radiation packages are used with diurnally averaged incoming solar radiation, although a diurnal cycle can be included as an option. The topographic effects on the barotropic vorticity equation are turned off in the standard version. Depending on the target phenomenon under study, one may wish to use different options, e.g., the radiation package option. The coding of the model is modularized and sufficiently transparent that this is relatively easy. An example where temperature advection is "frozen" at climatological values in order to study midlatitude–tropical interaction is given in Lin *et al.* (2000).

The surface fluxes are parameterized using the bulk transfer formulas following Deardorff (1972). These include momentum flux (stress) τ_s, evaporation E, and sensible heat flux H. Because the current version of the QTCM lacks an explicit boundary layer, the wind speed in the bulk transfer formulas is reduced by a factor η. This mimics an extrapolation from the free atmosphere into the boundary layer based on empirical relationships.

The model domain covers the whole tropics in longitude and extends to 60°N and 60°S in latitude. A sponge boundary is applied outside 45°

latitude. A horizontal resolution of 5.625° × 3.75° is used here. Selected results from the following sections have been tested at doubled resolution (in both latitude and longitude), without major changes. The prognostic equations are finite differenced on an Arakawa C Grid (e.g., Chapter 1 in this volume). The baroclinic component is similar to a shallow-water equation and is solved by applying a forward–backward scheme using the updated variables immediately. The barotropic component is solved in a vorticity/streamfunction formulation using an Adams–Bashforth scheme. The numerical CFL instability criterion limits the model time step at about 20 min, mostly due to the momentum advection associated with midlatitude baroclinic waves. On a Sun/Ultra2 workstation, it takes less than 5 min of CPU time per year of simulation.

III. MODEL RESULTS

A. CLIMATOLOGY

Results presented in this section come from two types of runs that differ only in the boundary condition of prescribed SST: (1) 1982–1998 runs driven by the observed SST of Reynolds and Smith (1994), and diurnally averaged solar radiation at top. The latter is most relevant to the land regions. Monthly mean output from these runs is used to analyze the climatology and interannual variability. Because these runs are similar to GCM runs conducted in the Atmospheric Model Intercomparison Project (AMIP) we refer to them as AMIP-like runs. (2) Seasonal runs driven by climatological SST and diurnally averaged solar radiation. Daily mean output is used in the analyses of intraseasonal oscillation. The cloud-radiation package clrad0 is used in the runs.

The presence of temperature advection and the baroclinic and barotropic wind components in velocity (similar to a two-layer model in terms of degrees of freedom) leads to baroclinic instability and generates midlatitude storms. The "tails" of storms penetrate into the tropical convergence zones and can contribute to tropical variance. These connections tend to occur at preferred locations and this effect appears to play a role in forming the climatological southern convergence zones (Fig. 2), namely, the South Pacific Convergence Zone (SPCZ), the South Atlantic Convergence Zone (SACZ), and a similar feature in the South Indian Ocean. In observations, these southeastward extensions of the convergence zones are generally quite broad, whereas the model appears to have narrower versions of these. The storms play an important role in drying the

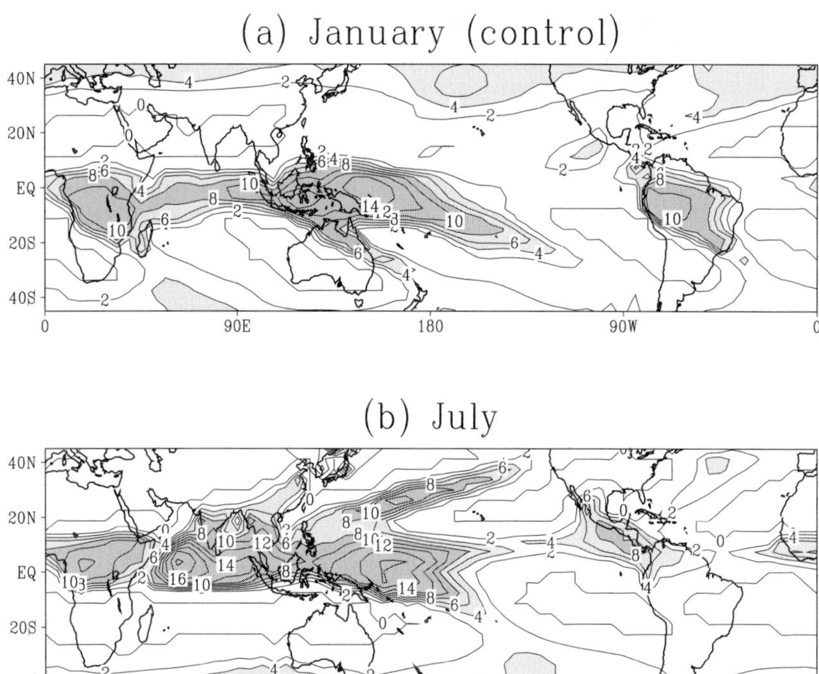

Figure 2 Model climatological monthly mean precipitation for the period 1982–1998. (a) January and (b) July. Contour interval 2 mm day^{-1}; shaded above 4 mm day^{-1}.

subtropics and feeding the moisture to midlatitudes. The storms appear to organize themselves somewhat in the North Pacific and North Atlantic, albeit not at realistic locations, likely due to the lack of topographic effects. However, the poleward edges of the subtropical dry zones are a little too wide, and the storms are too far poleward. The subtropical high-pressure regions such as in the southeastern Pacific appear to be slightly too dry, and extend too far east such that southern South America is too dry. This is also partly responsible for the narrowness of the southeastern extensions of the convergence zones discussed above. For a model of this level of complexity aimed at tropical climate simulation, we do not expect a perfect simulation of midlatitude storms. Major effects relevant to the tropics are the eddy flux of moisture, which dries the subtropics, and the eddy flux of momentum, which maintains the tropical zonal mean easterlies as discussed in NZ. The link between the subtropical and tropical convergence zones also appears relevant to teleconnections of climate anomalies. We now shift our attention to the seasonal cycle of the rainfall patterns (Fig. 2).

In January, the major convective band runs over most of the equatorial tropics, including the western Pacific warm pool region, the Amazon, Central Africa, and the Indian Ocean. However, parts of the ITCZ in the eastern Pacific and Atlantic are too weak. Sensitivity studies show that these features are sensitive to the parameterization of surface fluxes. The present version also lacks a separate explicit boundary layer, which is thought to be important in maintaining the strength of these narrow convergence zones (Wang and Li, 1993). As the sun moves northward, the equatorial convergence band also moves northward from January to July. The ITCZ over the eastern Pacific and the Atlantic strengthens. Many detailed features are also captured by the model, such as the dry region in Northeast Brazil between the ITCZ and SACZ, the wet region around Madagascar, and the mid-Pacific trough near Hawaii. Throughout a seasonal cycle, rainfall over the Amazon basin is reasonable, but rainfall over Africa appears to be too strong all year round.

In July (Fig. 2b), the African and North American monsoons are well established. The largely zonal monsoon rain in West Africa extends from the rainforest into the Sahel at about 15°N. A wet tongue extends from Central America along the Pacific coast into Mexico and the Southwest United States. The patterns and locations are quite similar to the observations (not shown). The Asian monsoon region, India, and Southeast and East Asia receive large amounts of monsoon rainfall. However, the Asian monsoon appears to be too weak, especially in contrast with the unrealistically large maximum rainfall sitting over the equatorial Indian ocean. A maximum over the Bay of Bengal is missing in the model. Another unrealistic feature is a large maximum around the dateline. A similar tendency has been noted in some runs of the UCLA AGCM (R. Mechoso, 1997, personal communication). This is found to be sensitive to the evaporation parameterization (ZNC).

Overall, the model simulates a reasonable seasonal migration of tropical and subtropical convective rainfall centers. The African and American monsoons are well simulated, but the rainfall over the Indian Ocean and Pacific Ocean is concentrated too much along the equator. Correspondingly the Asian monsoon appears somewhat weak. This suggests that the warmed continent in the model is able to shift the overall circulation, but not enough in the case of the Asian monsoon where other factors such as topography also play an important role.

We now examine various other fields. Figure 3 depicts the January climatology of evaporation, OLR (outgoing long-wave radiation), and net downward energy flux at surface. These fields are representative of the

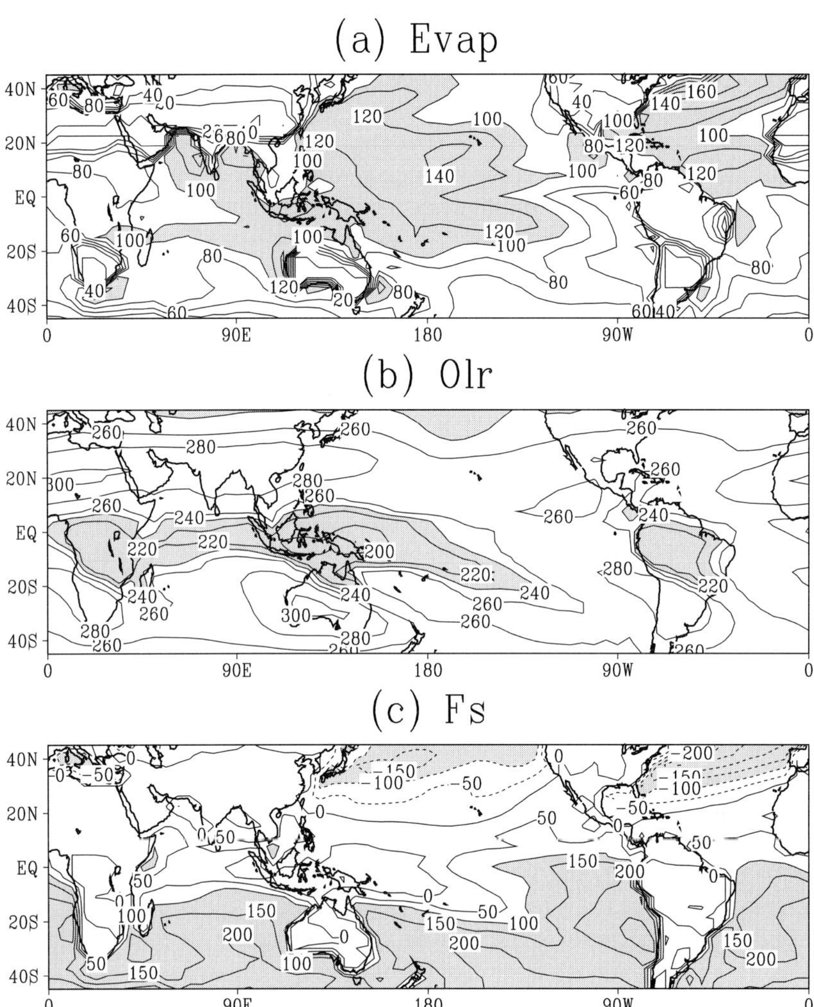

Figure 3 Model simulated January fluxes of (a) evaporation, contour interval 20 W m^{-2}, shaded above 100; (b) outgoing long-wave radiation, contour interval 20 W m^{-2}, shaded below 240; and (c) downward net surface energy flux F_s^{net}, contour interval 50 W m^{-2} with dark shading above 100 and light shading below -100.

behavior of the model's physics packages, especially surface fluxes, cloud, and radiation. Over the tropical oceans, the evaporation (Fig. 3a) pattern is determined mostly by SST and surface wind speed. The SST dependence is seen in the contrast between, say, the western Pacific warm pool and the cold tongue region. The wind speed dependence is apparent in the trade wind regions such as North and South Pacific in the subtropics. Similar to the observations, the warm pool also has a local minimum due to the weak winds there. Experiments show that the evaporation in the weak wind region depends critically on the minimum wind in the evaporation formula, which represents sub-Reynolds-scale wind variability. The overall pattern and magnitude of evaporation are similar to the observations. Over land, the evaporation tends to follow precipitation but lasts longer into the dry season due to the soil moisture memory so that the pattern of evaporation is generally smoother than that of precipitation.

The model captures the pattern of low OLR (Fig. 3b) associated with tropical convergence zones due to the long-wave trapping effects of the deep convective clouds, which are predicted according to precipitation in the model. Minimum values of near 200 W m^{-2} are seen in the warm pool region and over the Amazon. As in the observations, the OLR increases toward the dry subtropics, and decreases again at midlatitudes due to the presence of clouds and colder temperatures. The OLR is slightly high, partly due to a dry bias resulting in not enough greenhouse trapping.

The net surface flux F_s (Fig. 3c) is the sum of the net radiation, evaporation, and the sensible heat flux. It is nearly zero over land at climate scales due to the low heat capacity of land (Section II.C). Over ocean this quantity is important for coupling to ocean models. The simulated pattern is largely similar to the observations (not shown), exhibiting a gross seasonal contrast between the two hemispheres. In the tropics, this gradient is significantly modified by the circulation patterns. The SPCZ has a strong signature as can be seen by tracing, say, the 50 W m^{-2} contour line. The warm pool region has nearly zero net flux, similar to the observations. Near the eastern side of the ocean basins, F_s is overestimated because of the lack of stratus clouds. The fluxes under the storm tracks are captured, but somewhat underestimated.

Figure 4 depicts the model 850- and 300-mb winds [reconstructed using Eq. (8)]. At 850 mb, the model simulates trade winds blowing across the subtropics. The trades are especially strong in the winter Northern Hemisphere. Midlatitude westerlies develop associated with the storms. The anticyclonic motions around the subtropical highs are well simulated, especially in the Southern Hemisphere. The winds turn southeastward correctly over the Amazon Basin despite the lack of topography, suggesting that this is mostly a thermodynamical feature rather than a topo-

(a) 850 mb

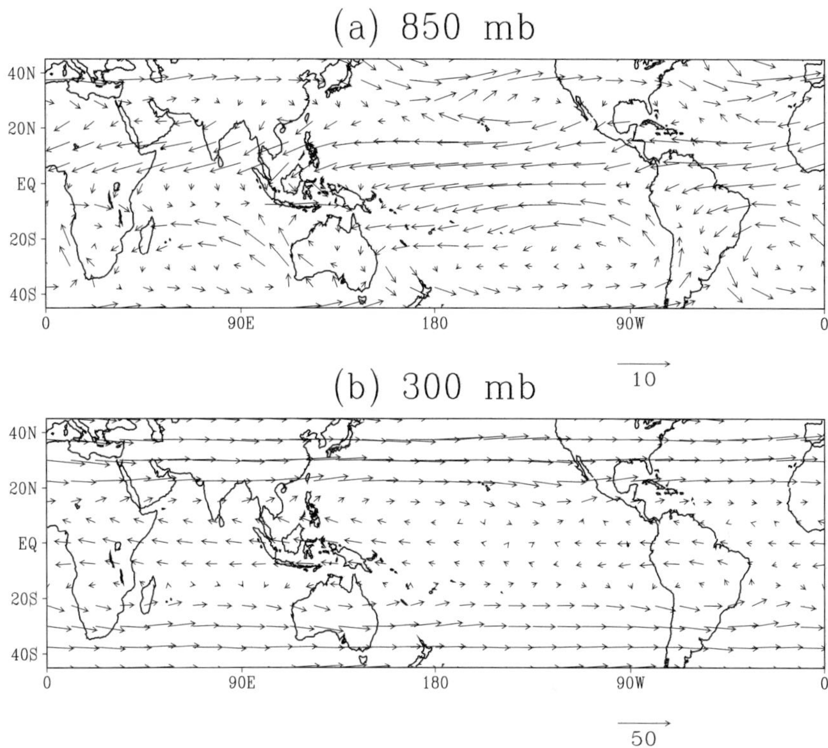

Figure 4 Model simulated January climatological winds (in m s^{-1}) at (a) 850 mb and (b) 300 mb.

graphic one. Equatorial westerlies are also seen over Africa and the Indonesian region, although they are not strong enough over the Indian Ocean. The 300-mb winds (Fig. 4b) are dominated by the subtropical westerly jets. The Northern Hemisphere jet is located at about 30°N, similar to observations, but the Southern Hemisphere jet is somewhat too close to the equator. Both jets are slightly too strong. The tropics are dominated by the easterlies with a hint of change to westerlies in the so-called westerly duct region in the Eastern Pacific, although not as strongly westerly as in the observations.

Figure 5 shows the zonal average zonal velocity. The Northern Hemisphere trades are centered at about 15°N, similar to the observations. The subtropical jets are somewhat too strong, and the easterlies extend slightly too far poleward. These correspond to a slightly too high wind shear at

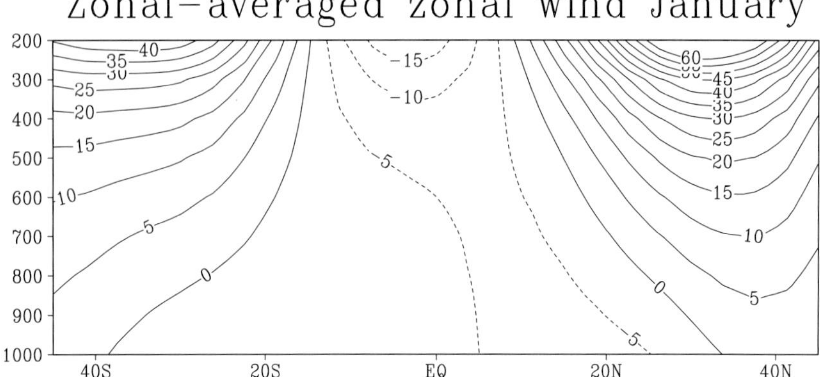

Figure 5   Model simulated January zonal averaged zonal wind (in m s^{-1}).

midlatitudes in the baroclinic mode, which is derived from a tropical moist adiabat profile by design. Overall, given the limited degrees of freedom, the model simulated wind field is quite encouraging.

The July climatology of the atmospheric temperature averaged between 500 and 200 mb [reconstructed using Eq. (6)], surface temperature, soil wetness, and 850-mb wind are shown in Fig. 6. The atmospheric temperature at 500–200 mb is chosen so that a direct comparison with the observation of Li and Yanai (1996, their Fig. 3) can be made. The most prominent feature in Fig. 6a is the warm temperature center around the Tibetan Plateau, corresponding to high pressure aloft and low pressure at low levels for the baroclinic component. Careful examination indicates that this monsoon depression is located slightly too far west of the Tibetan Plateau (which does not exist in this model run). At 850 mb, the southwesterly winds over the Indian Ocean turn around this monsoon depression and then join the trade winds from the western Pacific, flowing northward into the Eastern Asian trough (Fig. 6b). Over North America, a temperature ridge has developed associated with the North American monsoon.

B. How Much Do Departures from Quasi-Equilibrium Affect Climatology?

A model run was conducted with a convective adjustment time τ_c [related to ϵ_c^* in Eqs. (3) and (4); NZ] of 8 hr rather than 2 hr as used in the control run. The January precipitation is shown in Fig. 7. Compared to

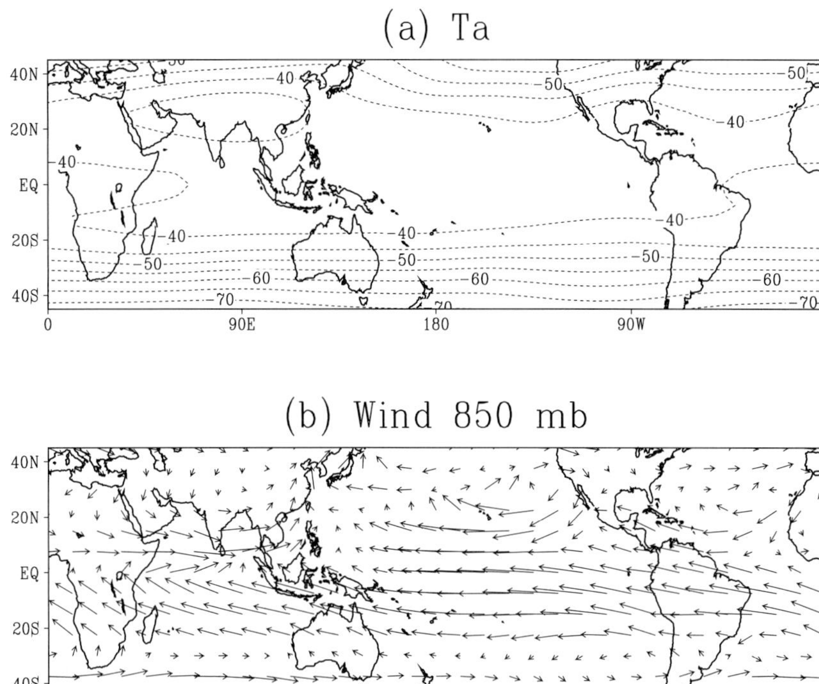

Figure 6 Model July climatology: (a) air temperature averaged between 500 and 200 mb, contour interval 5°C; (b) 850-mb winds (in m s^{-1}).

the control case (Fig. 2a), the tropical convergence zones appear to be slightly weaker, especially the eastern Pacific ITCZ. However, they are very similar to each other overall. This low sensitivity to τ_c is comforting because in the real tropical atmosphere, the adjustment time might vary significantly due to mesoscale organization and other disturbances. This insensitivity is rooted in the enthalpy constraint (latent heating comes from water vapor condensation) so that the adjustment time does not enter the moist static energy equation Eq. (5), which tends to determine the large-scale circulation patterns (NZ, Sections 5.2, 7). However, this finite relaxation time can have non-negligible impact on waves of faster time scale (N97).

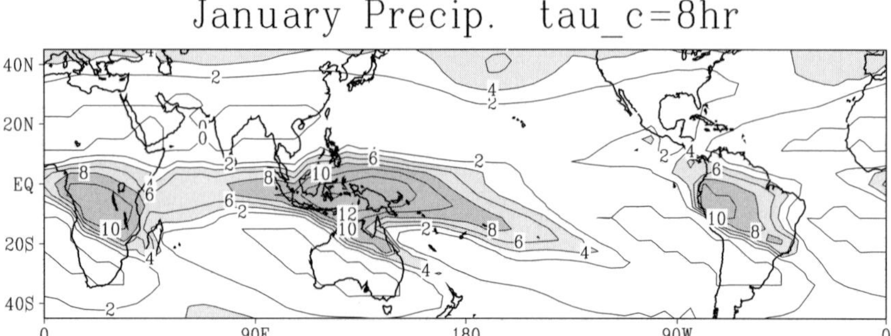

Figure 7 January climatological precipitation as in Fig. 2a, but from a run with the convective adjustment time τ_c = 8 hr (2 hr in control).

C. INTRASEASONAL OSCILLATION

The model lacks mesoscale organization in the tropics and is relatively steady compared to storm-related variations in midlatitudes. Nonetheless, it does exhibit a significant amount of intraseasonal variability. Figure 8 is a time-longitude plot showing 850-mb winds at the equator over a year. An eastward propagating Madden–Julian oscillation (MJO)-like signal is apparent, though the amplitude is somewhat weak. The amplitude and phase speed vary both seasonally and between events, and are sensitive to various parameters including the evaporation formulation. Spectral analysis (Fig. 9a) indicates a broad peak around 30 days at wave-number 1. Significant spectral peaks are also found for wave-numbers 2 and 3 (Lin *et al.*, 1998).

To study the mechanisms of the excitation and maintenance of the model intraseasonal oscillation, experiments with an earlier version of the model (beta version) were conducted to examine the effects of the evaporation-wind feedback (EWF, Emanuel, 1987; Neelin *et al.*, 1987) and midlatitude disturbances. The following cases are simulated: (1) a control run, using the standard version of QTCM1 V1.0, with evaporation-wind feedback and extratropical disturbances included (Figs. 8 and 9a are from this control run), (2) evaporation-wind feedback turned off, (3) extratropical disturbances turned off by using the mean temperature advection from a previous run, and (4) both the EWF and extratropical disturbances turned off. The corresponding spectra are shown in Fig. 9. The lack of EWF significantly reduces the amplitude of variances in spectral bands associated the model MJO (Fig. 9b). Variance remains, and is even enhanced at lower frequencies. The most dramatic impact occurs in the case where extratropical disturbances are suppressed. The peak in the 850-mb wind spectra is almost completely eliminated (Fig. 9c). The experi-

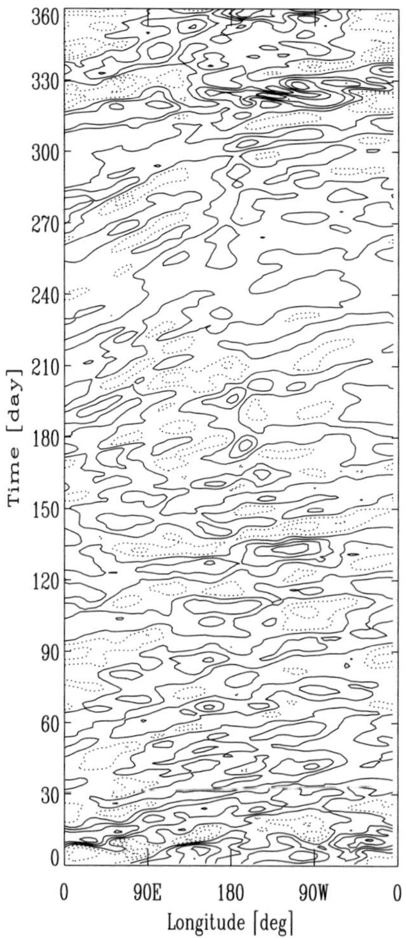

Figure 8 Time–longitude plot of equatorial (averaged from 7.5°S to 7.5°N) daily mean anomalies of 850-mb zonal wind (contour 0.5 m s^{-1}) for the control run. Negative anomalies are dotted. From January to December.

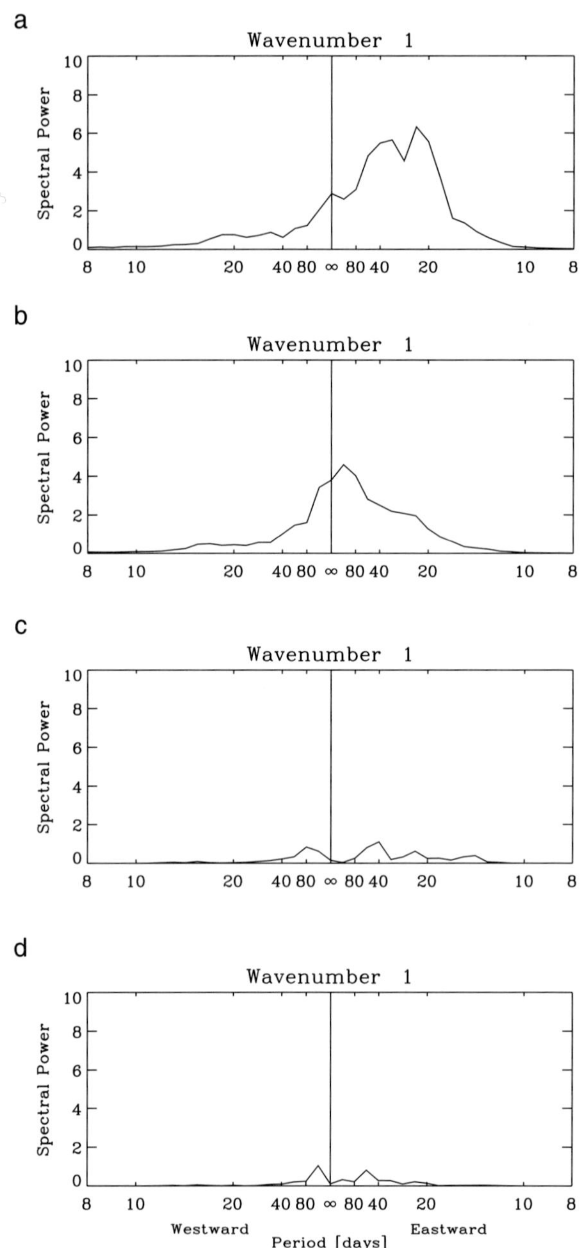

Figure 9 Power spectral density (PSD) of equatorial (average 7.5°S–7.5°N) daily mean anomalies of 850-mb zonal wind for (a) the control run, (b) evaporation-wind feedback (EWF) suppressed, (c) extratropical excitation suppressed, and (d) both EWF and extratropical excitation suppressed. Units of $(m\ s^{-1})^2$ day. Standard deviation of PSDs is 10%.

ments suggest that both the EWF and extratropical excitation act to maintain the MJO. If either of the mechanisms is removed, the MJO-like oscillation is reduced. Of the two mechanisms, the removal of extratropical excitation decreases the amplitude of the oscillation more than removing the EWF.

D. Interannual Variability

The AMIP-like runs with observed SST permit examination of the model simulated climate variability on interannual time scales. Figure 10 shows the precipitation on the equator from the model during January 1982 to March 1998. The model captures the major El Niño warm events of 1982–1983, 1986–1987, 1991–1992, and 1997–1998, which have anomalies extending all the way across the eastern Pacific, as well as the long-lasting warming events in the first half of the 1990s. The cold La Niña events of 1984, 1988–1989, 1995–1996 are also captured. During 1986–1987 and 1991–1992, the modeled rainfall anomalies in the eastern Pacific are not as strong as in observations, although the eastward extension in 1982–1983 and 1997–1998 is reasonable. The maxima around the dateline are often too large during the summer. Our sensitivity studies suggest that these are both related to deficiencies in climatology, namely, the weak ITCZ and the maximum around the dateline in summer (see discussion in Section III.A). An annual modulation of the longer warm or cold events can be seen clearly around the dateline, again an indication of the anomaly dependence on the climatology. This annual modulation is also seen in observations although it is noisier. In addition to rainfall increases directly associated with warm SST during El Niño, the model also captures several aspects of the rainfall reductions in the western Pacific/Indonesian region, and vice versa during La Niña. Over South America and the Atlantic, the correspondence in magnitude and seasonal timing of anomalies is imperfect but the model produces counterparts to several of the major anomalies, e.g., in 1984, 1987, 1988, 1991–1992, 1996, and 1997. A significant success is the longitudinal position of the ENSO anomalies, and the variation of the extent and position of these anomalies during the larger El Niños of 1982–1983 and 1997–1998. Simple models often have difficulty with these aspects, and the dynamics involves the nonlinear simulation of the convecting/nonconvecting boundary.

To see the spatial pattern of the interannual variability, we plot in Fig. 11 a composite winter precipitation difference between three El Niño years (1986–1987, 1991–1992, 1994–1995) and three La Niña years (1983–1984, 1988–1989, 1995–1996; warm events minus cold events). The main positive precipitation anomaly is centered just east of the dateline,

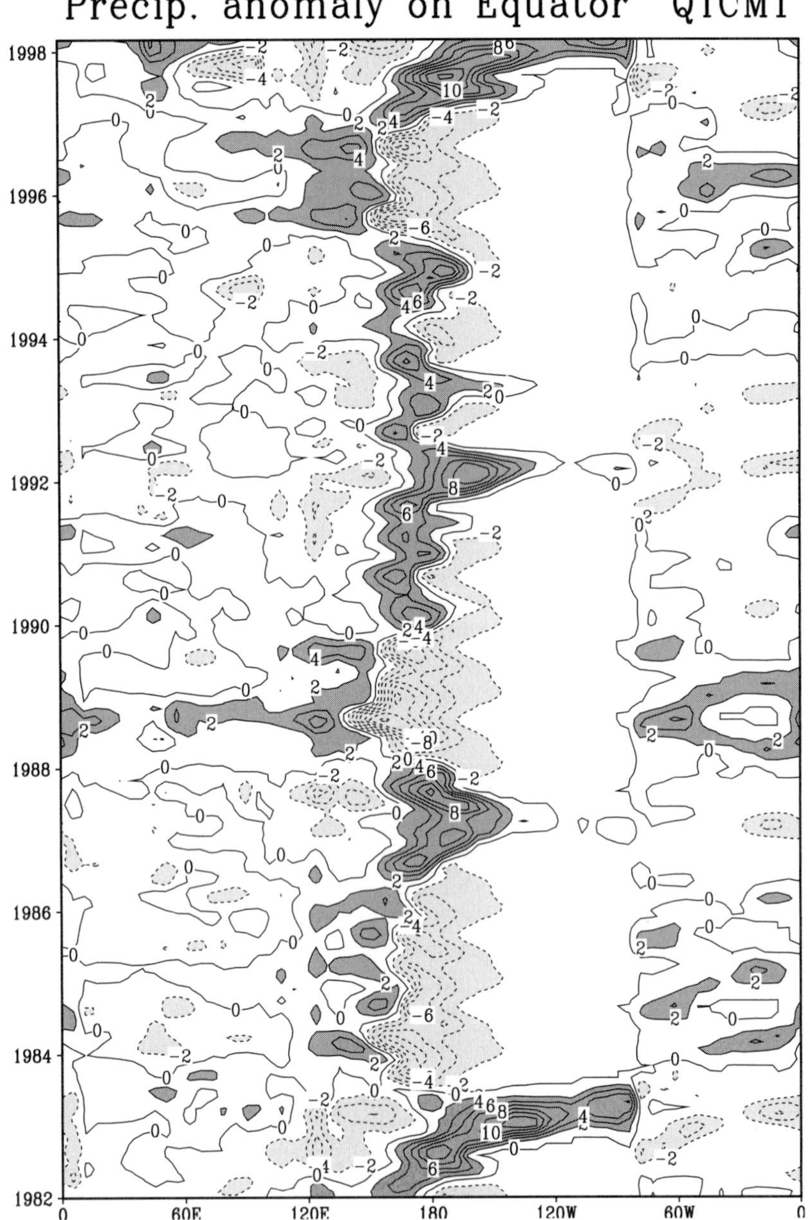

Figure 10 Model simulated monthly mean precipitation on the equator through the period 1982–1998. Contour interval 2 mm day^{-1}, with dark shading above 2 and light shading below -2.

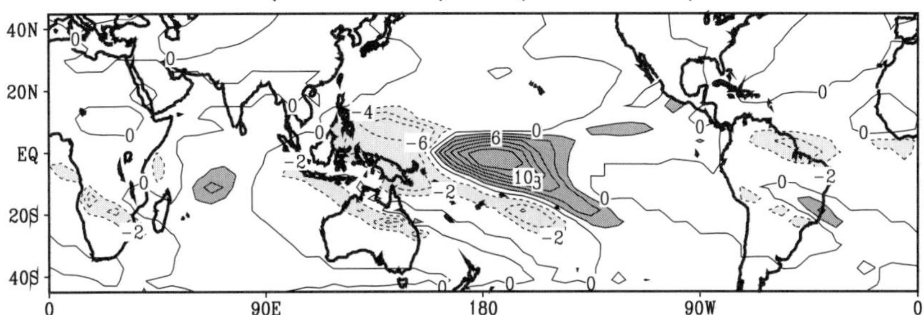

Figure 11 Winter (Dec.–Feb.) ENSO composite precipitation difference. Three warm events (1986–1987, 1991–1992, 1994–1995) minus three cold events (1983–1984, 1988–1989, 1995–1996). Contour interval 2 mm day^{-1}. Dark shading above 2 and light shading below -2 mm day^{-1}.

showing changes in the ITCZ and SPCZ. The model's positive anomaly does not extend quite as far west along the equator as the observations. Surrounding this are negative anomalies over the Indonesian region, subtropical south, and north Pacific. Further away, reduced precipitation is seen over the Amazon and northeastern Australia. In general, the model captures these observed teleconnection patterns, although some details are questionable. Interestingly, some teleconnections to subtropical latitudes are also present, including over southeastern Africa, the enhanced SACZ over southeastern Brazil, and a hint of a positive anomaly along the California coast, although the latter two are hard to see at this contour interval. These presumably involve Rossby wave dynamics (Wallace and Gutzler, 1981), but, at least in the Southern Hemisphere, involve interactions with the convective zones, so may be more complex than simple external mode Rossby waves.

The mature phase of the monsoons was discussed in Section III.A. We now wish to take a closer look at its interannual variability. The monsoons have strong interannual variability and this variability is not in general satisfactorily simulated in GCMs with interannual SST forcing (Sperber and Palmer, 1996). It is of great interest to see how the QTCM, a much simpler model but nevertheless with the main physics included, simulates this. In doing so, a practical difficulty is the lack of a universally accepted monsoon index. Here we choose the Asian monsoon wind shear index (zonal velocity at 850 mb minus 500 mb in the region 0–20°N, 40–110°E) of Webster and Yang (1992). This is a natural index for the present model because it is simply proportional to the strength of the baroclinic component (the barotropic component has constant velocity in the vertical, Fig. 1), which is driven directly by horizontal gradients in the thermodynamics.

An ensemble of 10 model runs that differ only in the initial condition are analyzed. Figure 12 depicts interannual variation of the model simulated Asian monsoon wind shear index averaged for June, July, and August of each year. Unlike most GCMs (Sperber and Palmer, 1996), the present model shows variation among the 10 ensemble runs of less than 0.5 m s^{-1} while the range of interannual variation of this index is about 5 m s^{-1}. This lack of variation among ensemble runs is likely due to omission of some processes responsible for some types of atmospheric internal variability. On the other hand, this lack of "noise" can be an advantage for analysis of the processes and mechanisms the model does represent.

Because the observations contain a significant amount of atmospheric internal variability that is not reproducible, it is not clear to what extent agreement between models forced by SST and observations is expected. The general level of agreement between the QTCM and the observations is similar to what has been found for the GCM simulations in the AMIP project (Sperber and Palmer, 1996). The results show a tendency toward negative correlation with ENSO. For instance, the warm El Niño events in the summers of 1983, 1987, and 1997 correspond to weak monsoon by the wind shear index. The relation with the cold phases is less clear. In 1989 an increase is noted in both model and observations, but this does not hold well in other La Niña years. Because the model does not simulate snow

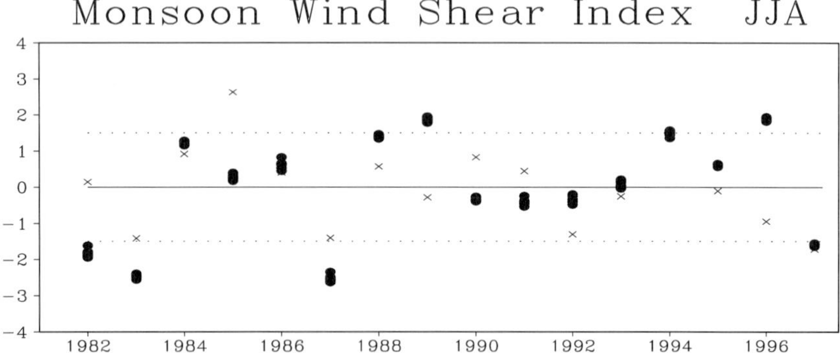

Figure 12 Interannual variation of the monsoon wind shear index (difference between the winds at 850 and 250 mb over the region 0–20°N/70–120°E) as defined by Webster and Yang (1992) for the months June–August. Filled circles are results from an ensemble of 10 model runs differing only by their initial conditions, while crosses are results from NCEP/NCAR reanalysis.

hydrology (Section II.C), this level of agreement suggests that the SST may play a more important role in the interannual variability of large-scale monsoon.

The strong hydrological cycle in the Amazon Basin, with a rainfall rate of more than 2000 mm a year, provides an excellent testbed for the land-surface model and the convective dynamics. Figure 13 shows the 12-month running means of various components of the hydrologic cycle. A negative correlation with ENSO is clearly seen in precipitation, runoff, and soil moisture content while evaporation varies only slightly for the reasons discussed in Section III.A. These trends are found also in observations (e.g., Zeng, 1999), although details of magnitude can differ. For instance, the 1988–1989 La Niña event leads to an increase of about 1 mm day^{-1} in precipitation and a corresponding increase in soil water storage. A delay

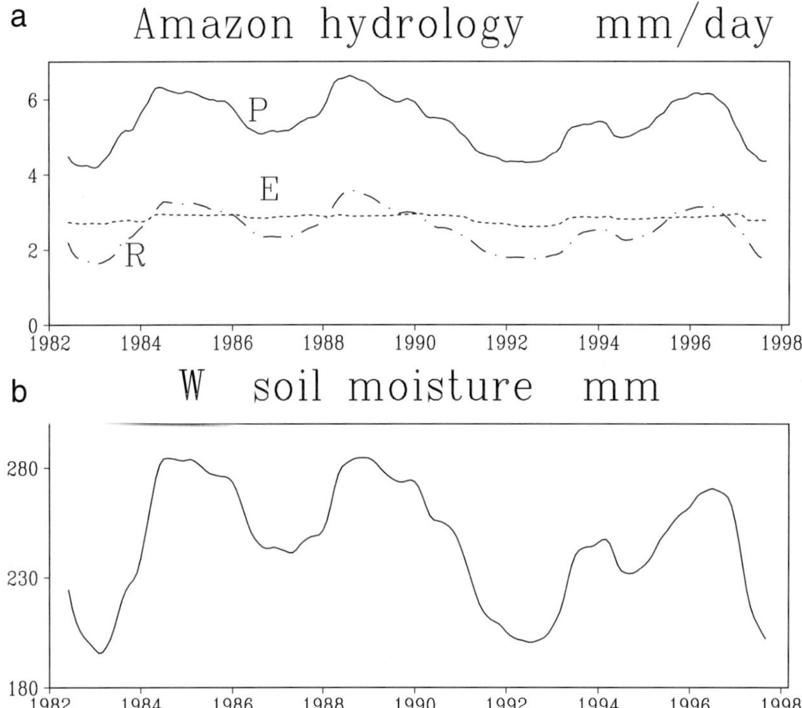

Figure 13 Interannual variations of the hydrologic cycle over the Amazon basin: (a) Precipitation P (solid line), evaporation E (dashed line), runoff R (dash-dot line), all in mm day^{-1}; (b) soil moisture content W, in mm. The monthly data are low-pass filtered by a 12-month running mean.

on interannual time scales in soil moisture and runoff from precipitation is also noticeable. This memory effect might play a role in season-to-season climate anomalies and act as a link between climate variation in the Pacific and Atlantic. The atmosphere and land-surface models developed here thus provide an efficient tool for studying this type of problem.

IV. CONCLUSION

QTCM1 aims to simulate reasonably complex tropical phenomena with a relatively streamlined tool. A unique feature is that aspects of the dominant sub-grid-scale processes, namely, moist convective dynamics, are employed in the model design. In this approach, we benefit from two strands of work pioneered by Akio Arakawa: QE convective parameterization and numerical modeling of the atmosphere. We attempt to use constraints provided by the first to produce shortcuts in the latter, at least for specific aspects of simulation of the general circulation. In terms of the boxes in Fig. 18 of Chapter 23 of this volume, we provide a tool that explicitly bootstraps solution of the "dynamical processes" by making use of properties of the "cloud processes." In doing so, the links between the two become clearer, consistent with the theoretical approach of "QE thinking" discussed in Chapter 8 of this volume. In this chapter, we present a sample of results from the first QTCM.

The tropical climatology simulated by QTCM1 appears reasonable, showing the seasonal migration of the tropical convective zones and the variation of the trades. The African and American monsoons are well represented, but the Asian monsoon is somewhat too weak. For a model with only two degrees of freedom in the vertical (a baroclinic and a barotropic component), the reconstructed wind fields are quite decent; the good accuracy despite the high truncation indicates that the effort spent on the analytical solution prior to turning to numerical methods (see NZ) was well invested. The model does a reasonable job in simulating fields such as surface evaporation, OLR, and net surface energy flux that are representative of the model physical parameterizations. These fields are of importance to coupled ocean–atmospheric modeling and are not usually simulated in simple models.

A multiyear model run driven by the observed SST anomaly from 1982 to 1998 demonstrates the model's ability to simulate the atmospheric interannual response in precipitation and other key fields. The primary ENSO rainfall anomalies are simulated near the dateline, extending further eastward during large warm events. Also simulated are the reduced rainfall regions in the tropical western Pacific/Indonesian region, and

South America. Certain subtropical teleconnections are also captured. For instance, the Amazon hydrologic cycle exhibits a correlation with ENSO. Differing teleconnection mechanisms over land and ocean regions in response to ENSO anomalies are noted. These involve feedbacks from moist convection, cloud radiative effects, and land-surface processes, as described elsewhere.

The model monsoon wind shear index as defined by Webster and Yang (1992) shows significant interannual variability and some correlation with ENSO warm events. The behavior and level of agreement with observations are similar to those of GCMs in the AMIP project. This suggests a significant role of SST variations in influencing the land–sea thermal contrast, with feedbacks from land processes.

Having originally derived the model framework with theoretical applications in mind, the quantitative simulation presented here can be viewed with some satisfaction. Although we did not expect it from a model of this complexity, the simulation of certain fields is quite comparable with that of many GCMs at similar resolution. The climatology of tropical precipitation, for instance, while imperfect with respect to observations, appears to be better than that of several GCMs of the previous generation, including the GFDL R15 model as analyzed in Lau (1985), the older UCLA GCM version (Mechoso *et al.*, 1987), and the older versions of the community climate model CCM1 and CCM2 (Hurrell *et al.*, 1993). These GCMs have proven their worth in many studies, so the comparison is encouraging. Comparison to more recent GCM versions is left to the reader. In comparing with GCMs, it is understood that the problem solved is less complex in several respects. The point made here is simply that QTCM1 can be used to reproduce certain types of results studied in GCMs but within a system that is much more accessible to theoretical analysis.

In developing the model, including the cloud–radiative parameterizations and land-surface model, we were always aware of the parallel with earlier work carried out in GCMs, including Arakawa's experience with the early two-layer UCLA GCM (see Chapter 1). Have we merely revamped a two-layer GCM in Galerkin form? Some important differences include an underlying analytical solution of the primitive equations plus deep convective parameterization that holds under certain conditions; a connection to parcel buoyancy considerations in the convective parameterization by reduction of CAPE, even in the projected system; more general vertical structures, including for cloud types, than are normally admitted in level or layer models; and formulation and parameterizations that are designed to be reducible into even simpler versions for analysis. The model also contains an underlying physical hypothesis about the dominant effects in the tropical general circulation, as reviewed in more detail in N97. The effect of deep convection in communicating surface warming through the troposphere is seen via the effect on pressure gradients of pulling temperature toward a moist convective profile. Column energy and moisture budgets, together forming the moist static energy budget, are important in setting the degree to which dynamics can flatten these pressure gradients. The reasonable success of the model in quantitative simulation of many of the major tropical climate phenomena is suggestive that QE thinking is not only a powerful concept, but can also be a basis for the development of models to simulate complex tropical climate and variability.

ACKNOWLEDGMENTS

This review is condensed from manuscript versions of Neelin and Zeng (2000), Zeng *et al.* (2000), and Lin *et al.* (2000), and was presented by the lead authors as an invited talk and poster at the Symposium on General Circulation Model Development: Past, Present, and Future, held at UCLA, January 20–22, 1998. Conversations with A. Adcroft, A. Arakawa, C. Bretherton, C. Cassou, R. Dickinson, P. Dirmeyer, K. Emanuel, R. Koster, K.-N. Liou, C. R. Mechoso, C. Perigaud, D. A. Randall, R. Seager, J. Shuttleworth, P. Webster, W. Weibel, M. Yanai, J.-Y. Yu, and S. Zebiak during the course of the project were helpful. This work was supported in part by National Science Foundation grant ATM-9521389 and National Oceanographic and Atmospheric Administration grant NA86GP0314.

REFERENCES

Arakawa, A. (1993). Closure assumptions in the cumulus parameterization problem. *In* "The Representation of Cumulus Convection in Numerical Models of the Atmosphere" (K. A. Emanuel and D. J. Raymond, eds.). American Meteorological Society.

Arakawa, A., and W. H. Schubert (1974). Interaction of a cumulus cloud ensemble with the large-scale environment, Part I. *J. Atmos. Sci.* **31**, 674–701.

Betts, A. K., and M. J. Miller (1986). A new convective adjustment scheme. Part II: Single column tests using GATE wave, BOMEX, ATEX and arctic air-mass data sets. *Quart. J. Roy. Meteor. Soc.* **112**, 693–709.

Betts, A. K., and M. J. Miller (1993). The Betts–Miller scheme. Chapter 9 in "The Representation of Cumulus Convection in Numerical Models of the Atmosphere" (K. A. Emanuel and D. J. Raymond, eds.,) Meteor. Mon. 24, No. 46, pp. 107–121. American Meteorological Society.

Chou, C. (1997). Simplified radiation and convection treatments for large-scale tropical atmospheric modeling, Ph.D. dissertation. University of California, Los Angeles.

Chou, C., and J. D. Neelin (1996). Linearization of a longwave radiation scheme for intermediate tropical atmospheric models. *J. Geophys. Res.* **101**, 15129–15145.

Darnell, W. L., W. F. Staylor, S. K. Gupta, N. A. Ritchey, and A. C. Wilber (1992). Seasonal variation of surface radiation budget derived from international satellite cloud climatology project C1 data. *J. Geophys. Res.* **97**, 15741–15760.

Deardorff, J. W. (1972). Parameterization of the planetary boundary layer for use in general circulation models. *Mon. Wea. Rev.* **100,** 93–106.

Dickinson, R. E., A. Henderson-Sellers, P. J. Kennedy, and M. Wilson (1986). Biosphere-Atmosphere Transfer Scheme (BATS) for the NCAR Community Climate Model, NCAR Tech. Note TN-275 + STR. NCAR, Boulder, CO.

Emanuel, K. A. (1987). An air–sea interaction model of intraseasonal oscillations in the tropics. *J. Atmos. Sci.* **44,** 2324–2340.

Emanuel, K. A., J. D. Neelin, and C. S. Bretherton (1994). On large-scale circulations in convecting atmospheres. *Quart. J. Roy. Meteor. Soc.* **120,** 1111–1143.

Entekhabi, D., and P. S. Eagleson (1989). Land surface hydrology parameterization for atmospheric general circulation models including subgrid scale spatial variability. *J. Climate* **2,** 816–831.

Fu, Q., and K. N. Liou (1993). Parameterization of the radiative properties of cirrus clouds. *J. Atmos. Sci.* **50,** 2008–2025.

Gill, A. E. (1980). Some simple solutions for heat induced tropical circulation. *Quart. J. Roy. Meteor. Soc.* **106,** 447–462.

Harshvardhan, R. Davies, D. A. Randall, and T. G. Corsetti (1987). A fast radiation parameterization for general circulation models. *J. Geophys. Res.* **92,** 1009–1016.

Hurrell, J. W., J. J. Hack, and D. P. Baumhefner (1993). Comparison of NCAR Community Climate Model (CCM) climates, NCAR Tech. Note TN-395 + STR. NCAR, Boulder, CO.

Kiehl, J. T. (1992). Atmospheric general circulation modeling. *In* "Climate System Modeling" (K. E. Trenberth, ed.), pp. 319–370. Cambridge University Press, Cambridge, MA.

Lau, N.-C. (1985). Modeling the seasonal dependence of the atmospheric response to observed El Niños in 1962–76. *Mon. Wea. Rev.* **113,** 1970–1996.

Li, C., and M. Yanai (1996). The onset and interannual variability of the Asian summer monsoon in relation to land–sea thermal contrast. *J. Climate* **9,** 358–375.

Lin, J. W.-B., J. D. Neelin, and N. Zeng (2000). Maintenance of tropical intraseasonal variability: Impact of evaporation-wind feedback and midlatitude storms. *J. Atmos. Sci.,* in press.

Manabe, S., J. Smagorinsky, and R. F. Strickler (1965). Simulated climatology of a general circulation model with a hydrological cycle. *Mon. Wea. Rev.* **93,** 769–798.

Mechoso, C. R., A. Kitoh, S. Moorthi, and A. Arakawa (1987). Numerical simulations of the atmospheric response to a sea surface temperature anomaly over the equatorial eastern Pacific ocean. *Mon. Wea. Rev.* **115,** 2936–2956.

Neelin, J. D. (1997). Implications of convective quasi-equilibrium for the large-scale flow. *In* "The Physics and Parameterization of Moist Atmospheric Convection" (R. K. Smith, ed.), pp. 413–446. Kluwer Academic Publishers, Dordrecht.

Neelin, J. D., and N. Zeng (2000). The first quasi-equilibrium tropical circulation model—formulation. *J. Atmos. Sci.,* in press.

Neelin, J. D., I. M. Held, and K. H. Cook (1987). Evaporation-wind feedback and low-frequency variability in the tropical atmosphere. *J. Atmos. Sci.* **44,** 2341–2348.

Reynolds, R. W., and T. M. Smith (1994). Improved global sea surface temperature analyses using optimum interpolation. *J. Climate* **7,** 929–948.

Rossow, W. B., and R. A. Schiffer (1991). ISCCP cloud data products. *Bull. Am. Meteor. Soc.* **72,** 2–20.

Sellers, P. J., and co-authors (1996). A revised land surface parameterization (SiB2) for atmospheric GCMs. Part I: Model formulation. *J. Climate* **9,** 676–705.

Sperber, K. R., and T. N. Palmer (1996). Interannual tropical rainfall variability in general circulation model simulations associated with the atmospheric model intercomparison project. *J. Climate* **9,** 2727–2750.

Wallace, J. M., and D. S. Gutzler (1981). Teleconnections in the potential height field during the Northern Hemisphere winter. *Mon. Wea. Rev.* **109**, 784–812.

Wang, B., and T. Li (1993). A simple tropical atmospheric model of relevance to short-term climate variation. *J. Atmos. Sci.* **50**, 260–284.

Webster, P. J., and S. Yang (1992). Monsoon and ENSO: Selectively interactive systems. *Quart. J. Roy. Meteor. Soc.* **118**, 877–926.

Zeng, N. (1999). Seasonal cycle and interannual variability in the Amazon hydrologic cycle. *J. Geophys. Res.* **104**, D8, 9097–9106.

Zeng, N., and J. D. Neelin (1999). A land-atmosphere interaction theory for the tropical deforestation problem. *J. Climate* **12**, 857–872.

Zeng, N., J. D. Neelin, and C. Chou (2000). The first quasi-equilibrium tropical circulation model—implementation and simulation. *J. Atmos. Sci.*, in press.

Chapter 16

Climate Simulation Studies at CCSR

Akimasa Sumi

Center for Climate System Research, University of Tokyo, Meguro-ku, Tokyo, Japan

I. INTRODUCTION

Human instinct drives us to know what will happen in the future. It is the eternal dream of meteorologists to predict tomorrow's weather and/or the coming season's weather perfectly. As physical scientists, we would like to predict the future by applying physical laws. The first attempt to predict the weather by applying physical laws was made by Richardson (1922) and many scientists have followed in his footsteps.

Weather prediction cannot be achieved purely by applying our scientific knowledge to the study of nature, because a sufficiently well-developed technology is also necessary. For this reason, weather prediction is an interdisciplinary science whose pursuit entails (1) deepening our understanding of the atmosphere, (2) development of very powerful computers, and (3) development of numerical methods.

Here I would like to emphasize the important role of computers, which are the tools by which we can obtain answers to questions suggested by nature. It is amazing that Richardson spent 4 years computing the tendencies of meteorological fields for one time level! If he could have repeated his computations, an iterative analysis of the results could have suggested the right path to the next step. This shows that the development of numerical models cannot be separated from the development of computers.

It is said that a new era of supercomputing is beginning now, because of the advent of parallel computing. However, parallel computing technology is still developing and many problems remain to be solved. For this reason, the future development of atmospheric numerical modeling will benefit from interactions with computer science.

I would also like to emphasize that both theoretical analysis and the practical implementation of numerical methods are very important for simulation studies. We should remember that we could not conduct a long-time integration before Arakawa's (1966) Jacobian was proposed.

In general, numerical weather prediction (NWP) and climate simulation are considered to be applications of simulation techniques to the study of nature. NWP is considered to be an initial value problem and climate simulation is considered to be a boundary-value problem. The purpose of NWP is to predict the future state of the atmosphere and to provide guidance to forecasters. The development of NWP has accelerated the applications of prediction to the needs of society; in fact, more reliable forecasts and warnings are much appreciated by society. Simulation, on the other hand, is considered to be a tool for making new discoveries about nature and to assess the accuracy of our understanding of nature. Both prediction and simulation are strongly related to the accuracy of the models, which can only be measured by evaluating the models against observations.

In NWP, comparisons between observed and predicted fields are considered to be a measure of success. For example, RMSE (root mean square error) and anomaly correlation are metrics that can be used for evaluating the performance of our models. At many forecast centers, day-to-day evaluations of NWP products are conducted, and these contribute to the evaluation of models and spur model development to correct the various problems that are uncovered through forecast evaluation. However, in climate simulations, comparisons between simulations and observed climate fields are considered to be inadequate at the present time. We have to establish an improved methodology by which we can evaluate our models and improve their formulations.

In Section II of this chapter, I introduce our climate simulation activities at the Center for Climate System Research (CCSR) at the University of Tokyo. Numerical experiments at CCSR are discussed in Section III. Finally, I discuss issues pertaining to model evaluation in Section IV.

II. CLIMATE SIMULATIONS AT CCSR

CCSR was established in 1991. Its objectives are the development of climate models and achieving an understanding of the dynamics of the climate system by using numerical models. When we began to discuss the implementation plan for the center, we defined the following strategies for model development:

1. Because the center belongs to the university, emphasis should be placed on understanding climate rather than simulating and predicting climate.
2. Much attention should be focused on attacking problems by using numerical models.
3. More emphasis should be placed on the interactions between modeling and remote sensing, because global modeling requires global data for many currently unobserved quantities, which may be obtained in the future through satellite remote sensing with surface validation sites. This idea is the core of our model evaluation strategy.
4. We have to educate students to focus not on the atmosphere or ocean alone, but rather to dedicate themselves to climate issues involving both the atmosphere and ocean.

In the following subsections, we briefly present our models and introduce research results that have been obtained by using these models in global warming experiments and simulations of important climate phenomena such as the quasi-biennial oscillation.

A. THE CCSR ATMOSPHERIC GENERAL CIRCULATION MODEL

Ever since CCSR was established, development of our atmospheric general circulation model (AGCM) has been conducted jointly with the National Institute for Environmental Studies (NIES). CCSR has received computer time and manpower from the NIES. The first version of our AGCM was completed in 1995 (Numaguti *et al.*, 1997). The key features of the model can be summarized as follows:

1. It is a three-dimensional hydrostatic primitive global spectral model with a sigma coordinate. Finite-difference methods are used to represent the vertical structure of the atmosphere. The standard model resolution is T21 or T42, with L20.

2. The prognostic variables of the CCSR AGCM are vorticity and divergence, temperature, surface pressure, specific humidity, soil temperature, soil moisture, snow amount, and river water storage.
3. The radiation code is based on the two-stream discrete ordinate method and the k distribution method (Nakajima and Tanaka, 1986). The radiative flux is calculated in 18 wavelength bands. Band absorptions by H_2O, CO_2, O_3, N_2O, CH_4, and 16 species of CFCs are considered. Continuum absorption by H_2O, O_2, and O_3 are included. Rayleigh scattering by gases and particle scattering and absorption by cloud and aerosol particles are considered.
4. Convection is parameterized using a simplified Arakawa–Schubert scheme. Large-scale condensation is parameterized based on the scheme of Le Treut and Li (1991). The cloud water mixing ratio is a prognostic variable.
5. Surface fluxes are based on the bulk formula (Louis, 1979). The planetary boundary layer is parameterized using a closure model developed by Mellor and Yamada (1974, 1982).
6. The effects of sub-grid-scale orographic gravity waves are included following McFarlane (1987).
7. The land-surface model is a modified bucket scheme (Manabe *et al.*, 1965).

B. The CCSR Ocean General Circulation Model

The CCSR ocean general circulation model (OGCM) has been developed by members of the CCSR staff. Its finite-difference scheme is almost the same as that of the Geophysical Fluid Dynamics Laboratory (GFDL) model (Bryan, 1969), except that a weighted upcurrent scheme is adopted for the advection of temperature and salinity (Suginohara and Aoki, 1991). The OGCM has exceptional computational efficiency.

C. An AMIP Run

Using these models, we participated in various comparison projects, e.g., AMIP (the Atmospheric Model Intercomparison Project), CMIP (the Coupled Model Intercomparison Project), and PMIP (the Paleo-climate Model Intercomparison Project). Unfortunately, we were late in sending our results, and so our results are not included in the AMIP reports. Our AMIP results are presented here, however. Figure 1a shows the zonal mean temperature and Fig. 1b the zonal wind velocity for June, July, and

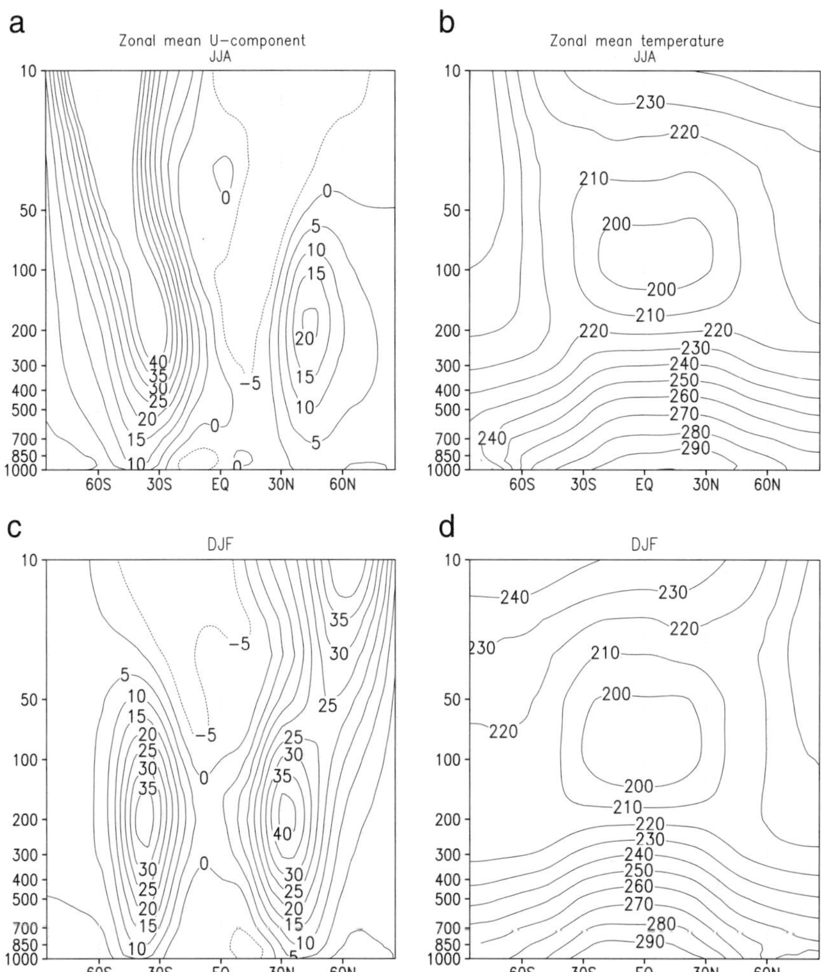

Figure 1 Height-latitude cross section for the zonal mean climatology based on the CCSR/NIES AGCM AMIP results. (a) Zonal wind ($\overline{U}$) and (b) temperature ($\overline{T}$) for summer (June, July, and August). (c) Zonal wind and (d) temperature for winter (December, January, and February). The contour interval for the winds is 5 m s^{-1}. Broken lines indicate negative values. The contour interval for the temperature is 10 K.

August. The zonal mean temperature is shown in Fig. 1c and the zonal wind velocity in Fig. 1d for December, January, and February. The corresponding figures based on NCEP data are presented in Fig. 2. Comparison shows that the performance of our model is average, although there are systematic errors. For example, a cold bias and westerly bias exist in the

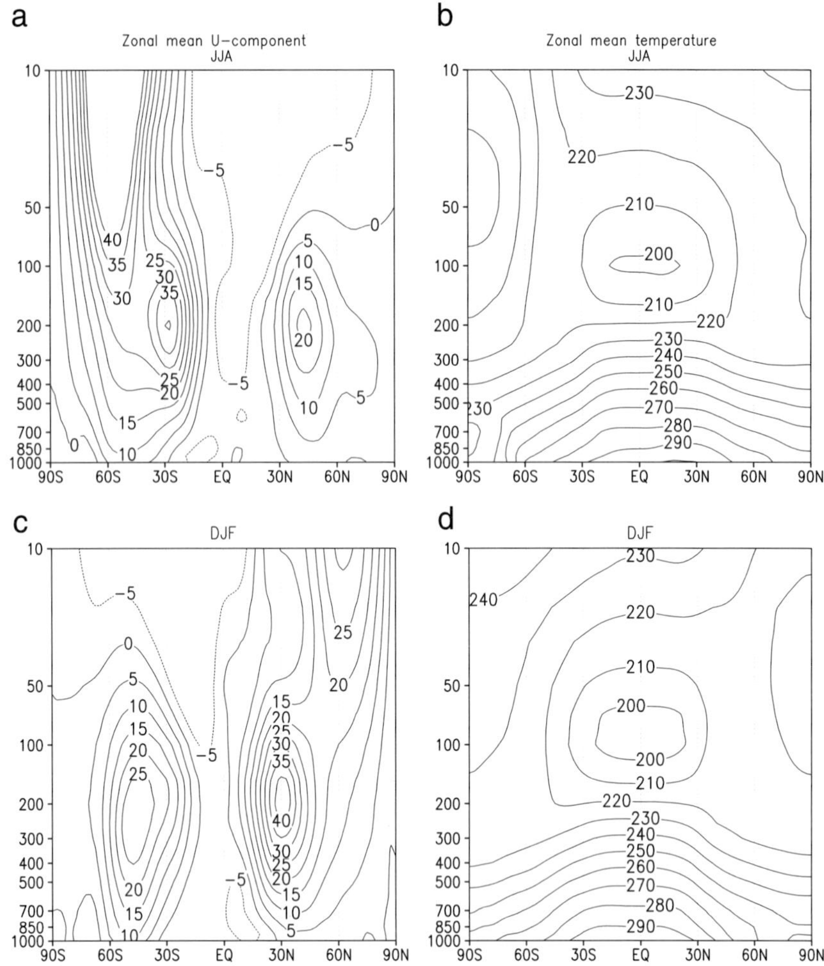

Figure 2 As in Fig. 1, but for the NCEP climatology.

polar stratosphere. In addition to their general performance, specific aspects of model performance have been investigated in AMIP. For example, a comparison of precipitation and evaporation over the continents was made by Lau *et al.* (1996). We present our results in the same format in Fig. 3. Our result is close to the observed value. This result also confirms that the performance of our model is comparable with that of other models.

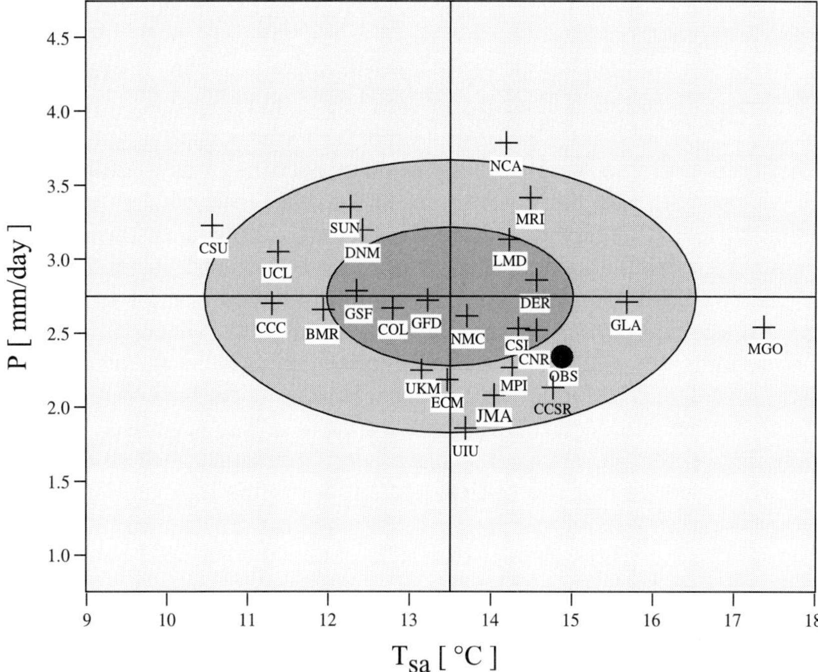

Figure 3 Mean rainfall and surface air temperature averaged over land based on the AMIP results. The CCSR result is plotted on Fig. 1 by Lau *et al.* (1996). Abbreviations: BMR, Bureau of Meteorology; CCC, Canadian Climate Center; CCSR, Center for Climate System Research; CNR, Centre National de Recherches Météorologiques; COL, COLA; CSI, Commonwealth Scientific and Industrial Research Organization; CSU, Colorado State University; DER, Dynamical Extended Range Forecasting at GFDL; DNM, Department of Numerical Mathematics of the Russian Academy of Sciences; ECM, ECMWF; GFD, GFDL; GIS, GISS; GSF, GSFC; IAP, Institute of Atmospheric Physics; JMA, Japan Meteorological Agency; LMD, Laboratoire de Météorologie Dynamique; MGO, Main Geophysical Observatory; MPI, Max-Plank-Institut; MRI, Meteorological Research Institute; NCA, NCAR; NRL, Naval Research Laboratory; OBS, Observation; SUN, State University of New York at Albany; UCL, UCLA; UGA, U.K. Universites' modeling program, UIU, University of Illinois; UKM, U.K. Meteorological Office; YON, Yonsei University, Korea.

D. Transient Experiments to Explore the Effects of Increasing CO_2

The AGCM and the OGCM have been coupled to each other, along with a thermodynamic sea-ice model and a simple river runoff model. The AGCM used in the coupled model is the T21L20 version, and the OGCM grid uses the same Gaussian grid as the AGCM (approximately $2.8° \times 2.8°$). The OGCM has 17 levels.

Using this coupled model, transient experiments on CO_2 doubling were conducted, including a control run with CO_2 fixed at present values. We used a 1% per year increase of CO_2 in order to participate in CMIP. Such participation in international comparison projects helps us to evaluate our model's performance, because the results of other research centers can be obtained and it is easy to compare our model results with them. In the transient experiment, we decided to use flux correction. This ensures that the mean SST distribution is similar to the observed one for the control run. In addition to the means, note that variabilities are well simulated. For example, the coupled model produces an ENSO-like fluctuation over the tropical Pacific Ocean and a decadal fluctuation over the northern Pacific Ocean, although their amplitudes are less than observed.

In Fig. 4, the globally averaged surface temperatures are shown for a run with a sudden increase of CO_2 and a run with the transient increase of CO_2, along with the control run. There is about a 2 K increase in the transient case, which is similar to results obtained with other models. The spatial pattern of the surface temperature increase is also similar to those

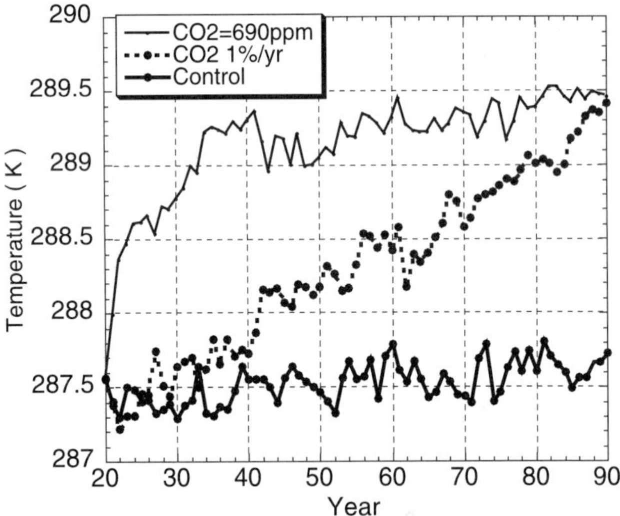

Figure 4 The time sequence of the annual average of global surface temperature based on the transient CO_2 experiments by the CCSR climate model. The thick full line is for the control case, and the dotted line is for the transient CO_2 case (1% increase). The thin full line is for the sudden CO_2 doubling case.

obtained with other models: There is more warming over the continents in the Northern Hemisphere, and less warming over the oceans and in the Southern Hemisphere.

In Fig. 5, the difference of the precipitation fields between the doubled-CO_2 case and the control case is shown. Positive differences occur in the equatorial central Pacific and the Asian monsoon region. Note that the difference over the tropical Pacific basin is similar to the ENSO pattern. To investigate the reason for this change, we calculated the difference of the surface heat budget over the tropical Pacific Ocean. Based on these results, the following scenario can be presented: (1) When CO_2 increases, the sea surface temperature (SST) increases over the tropical Pacific basin. (2) However, in the western Pacific region, the reflection of solar radiation by clouds increases and so the increase of SST in this region is not as large as in the eastern Pacific. (3) As a result, the east–west SST gradient tends to decrease, that is, an ENSO-like situation is realized. (4) The center of convection tends to shift eastward and becomes located in the central Pacific. The circulation pattern also changes. For example, the southeasterlies over southern China are enhanced. This feature has been noted in observations (Zhang *et al.*, 1996). These responses to an increase of CO_2 are similar to the results of others (Meehl and Washington, 1996). Further work is needed to determine whether or not this response is model dependent.

E. SIMULATION OF THE QBO

An example of scenario 2 given in the preceding paragraph is simulation of the QBO (quasi-biennial oscillation). In the past, climate models have been criticized because they could not simulate the QBO, even though it has been theoretically explained (Lindzen and Holton, 1968). However, the physics contained in our climate model should include that of the simple theory, so we could not believe that our model could not simulate the QBO as long as it is dynamically maintained by the Kelvin and the mixed-Rossby waves. The AGCM has been expanded to include the stratosphere in order to investigate the QBO. A simulation of a QBO-like phenomenon by using three-dimensional GCM was first conducted by Takahashi and Boville (1992), who noted that the amplitude of the mixed-Rossby wave is larger than observed. Several researchers pointed out that the contributions of gravity waves are essential to simulate the QBO, so we introduced high vertical and horizontal resolution and weak diffusion in our simulation. As a result of these efforts, we were able to simulate the QBO (Takahashi, 1996; see Fig. 6).

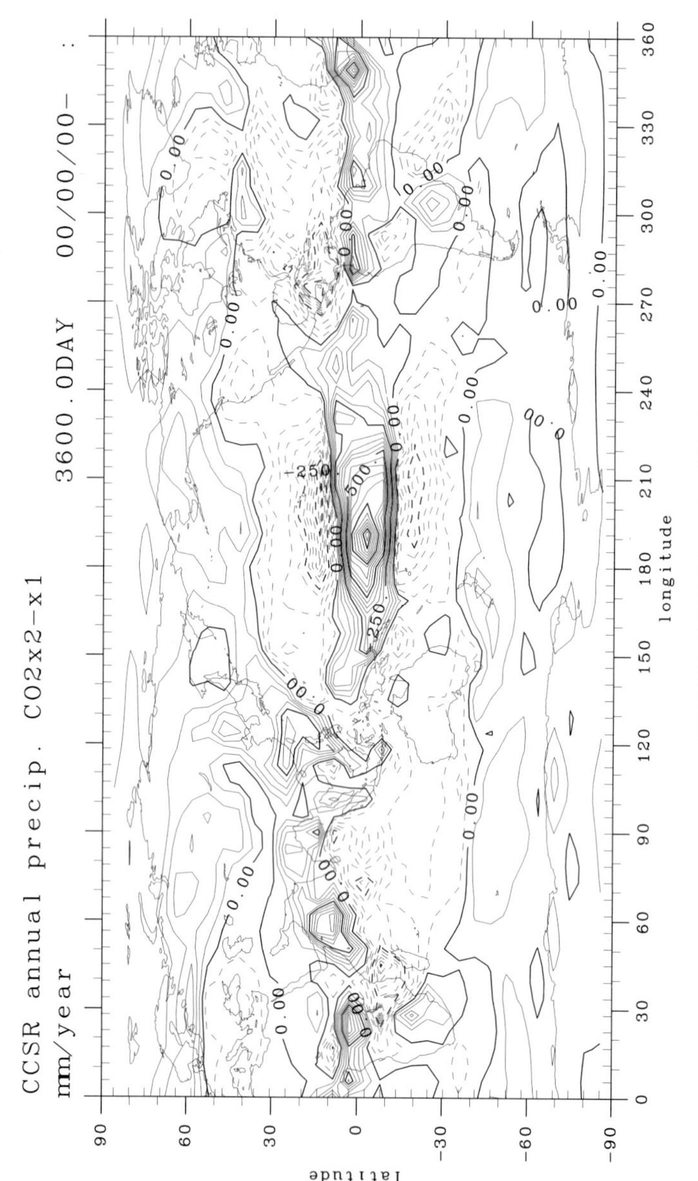

Figure 5 The difference in annual precipitation between the transient case and the control case. Units are mm year^{-1}, and contours are 50 units. Full lines are for positive values; broken lines are for negative values.

F. Use of Remote Sensing Data with Climate Models

An example of scenario 3 is the inclusion of an aerosol distribution as estimated by remote sensing. We plan to conduct global warming simulations with aerosol effects. Although the effects of aerosols on climate have been emphasized by the IPCC, many problems are associated with numerical experiments including the aerosol effects in climate models. For example, the global and size distributions of aerosols are still unknown. We have even less knowledge of the indirect effect of aerosols. The radiation parameterization in our model has been designed to compute the direct effects of aerosols. The global distribution of aerosols has recently been estimated through the use of AVHRR and OCTS data (Nakajima and Higurashi, 1997). This data set will be used in our simulations. We will conduct CO_2 doubling experiments based on the IPCC scenario.

III. CLIMATE SYSTEM DYNAMICS

We need a strategy to investigate the dynamics of the climate system, because we lack sufficient data to describe the behavior of the climate

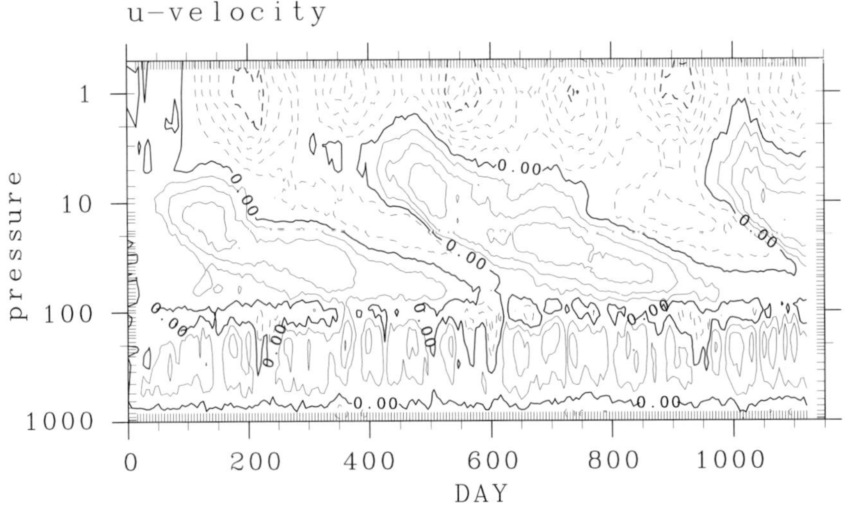

Figure 6 Height–time cross section of the zonally averaged *u*-component of simulated wind field over the equator. Units are m s^{-1}, and intervals are 6 units. Full lines are for positive values; broken lines are for negative values.

system. One way to overcome our lack of data is to recover hidden data and use paleo-data. However, note that the accuracy of paleo-data and amounts of hidden data are limited. Even though we use paleo-data, our knowledge of the behavior of the climate system is still very limited compared with our knowledge of the weather. For day-to-day weather, we have at least 50 years of reliable global data, since World War II, and 28 years of data after FGGE (the First GARP Global Experiment; GARP is the Global Atmospheric Research Program), which was conducted in 1979. For climate, we need very long time series of data, because climate works on such a long time scale.

Another way to obtain knowledge of the climate system is through numerical experiments in which sophisticated climate models are integrated for both idealized and extreme conditions. For example, the dynamical range of atmospheric circulations has been investigated by Williams (1988), who integrated a global circulation model over a wide range of parameter values. We think that this type of research is important and should be conducted together with more realistic simulations. We refer to idealized experiments like those of Williams (1978) as the computational geophysical fluid dynamics (CGFD) type approach. A simple model is used to investigate the essence of problems in the usual GFD experiments (Williams, 1978), while a full climate model that has been tuned for the present climate is used in simple boundary conditions in the CGFD experiments. Of course, our model is not perfect and there our results have limited applicability due to these imperfections. Through CGFD numerical integrations, we can try to obtain experimentally the data by which we can predict the state of the climate system for the parameter ranges.

The first CGFD runs conducted at the University of Tokyo were the "Aqua-Planet" experiments, through which it was shown that there exist typical coherent structures of convection: (1) a 30-day eastward propagating mode and (2) a double ITCZ structure (Hayashi and Sumi, 1986). After our experiment, many similar experiments were carried out and similar eastward propagating modes were simulated, but a single ITCZ was obtained in some experiments (Lau and Peng, 1987; Palmer, 1987; Swinbank *et al.*, 1988; Numaguti and Hayashi, 1991a,b; Sumi, 1992; Kuma, 1994). These experiments focus on the following issue: What is the distribution of convection for a given set of boundary conditions?

Sumi (1992) proposed a conceptional diagram summarizing the distribution of convection on the Aqua-Planet for various SST distributions (Fig. 7). He suggested that spatial organization and coherent motion can be generated by changes in the boundary conditions. These results are dependent on the model's formulation, including its cumulus parameterization

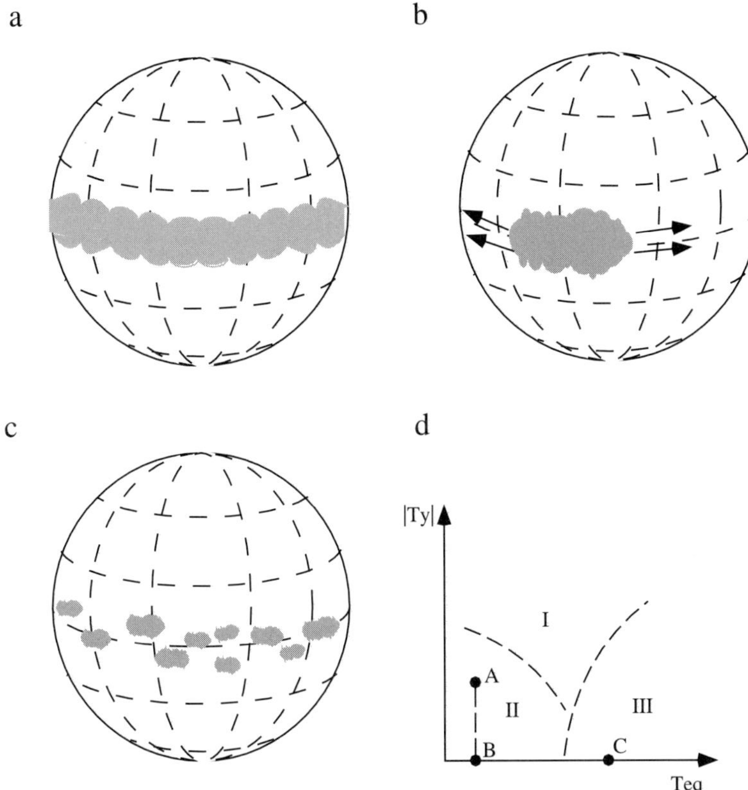

Figure 7 (a–c) Schematic diagram of various patterns of the distribution of convective activity over the Aqua-Planet and (d) a corresponding phase diagram. The ordinate of the phase diagram is the meridional gradient of SST on the equator, and the abscissa denotes the magnitude of the equatorial SST. (a) In case I, convective activity is concentrated on the equator, corresponding to region I in the phase diagram. (b) In case II, zonal asymmetry is generated in the tropics. In this case convective activities should move eastward or westward, because the time-averaged convective activity should be uniform over the entire tropical belt. (c) In case III, convective activity is occurring randomly in time and space over the sphere. A numerical experiment conducted by Hayashi and Sumi (1986) is considered to correspond to point A in the phase diagram. Numerical experiments conducted by Sumi (1992) correspond to points B and C in the phase diagram.

and its radiative parameterization. Note that the cumulus parameterization locally defines a mode of interaction between convective heating and large-scale flows, which can define a heating profile with the radiative cooling. It was pointed out by Numaguti and Hayashi (1991b) that the single ITCZ corresponds to net heating on the equator and the double

ITCZ corresponds to net cooling on the equator. These vertical heating profiles are closely related to the horizontal motion fields.

In the Aqua-Planet experiments, there is no zonal asymmetry. What will happen if we introduce zonal asymmetry in the Aqua-Planet? Sumi (1987) investigated the distribution of convective activity over the tropical ocean under the idealized land–ocean contrast.

The above experiments were conducted for given SST distributions. In a real climate system, the SST is internally determined. For this reason, further experiments are being conducted using a climate model with mixed-layer ocean. Because land–ocean contrast is a key factor of our climate system, experiments are now being conducted at CCSR using idealized land–ocean contrasts, in order to understand the basic behavior of the climate system (Abe, personal communication, 1999). In these experiments, half of the Earth is covered by ocean and the rest is assumed to be land without mountains. Three cases are investigated: case A, an east–west land–ocean contrast case; case B, a north–south land–ocean contrast case; and case C, an Aqua-Planet case (Fig. 8). The model is integrated from a resting initial state for 15 years with seasonal and diurnal cycles, and the last 10 years are used for diagnosis.

Here, we present preliminary results. In Fig. 9 the globally averaged surface heat balance and water balance are shown for three cases. The precipitation in case B is larger than in case A. This is mainly because the monsoon rainfall is enhanced by the north–south land–ocean contrast. This can also be inferred observationally. In the real world, the strongest monsoons are the Indian monsoon, the Mexican monsoon, the West Africa monsoon, and the Australian monsoon. The first three of these are associated with the north–south land–ocean contrast. This indicates that meridional land–sea contrast causes stronger convergence than the zonal

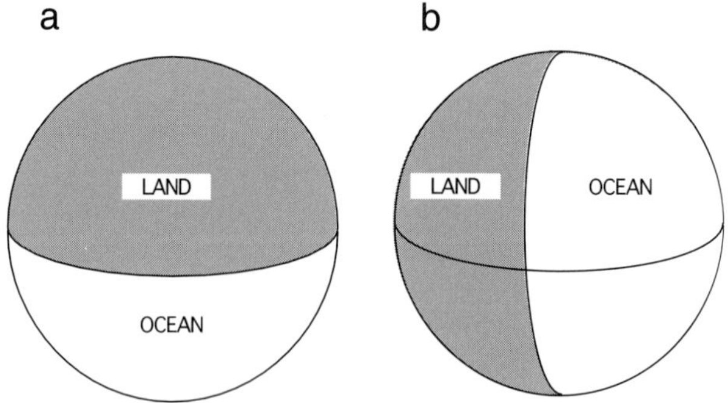

Figure 8   Land ocean configuration for (a) case A and (b) case B.

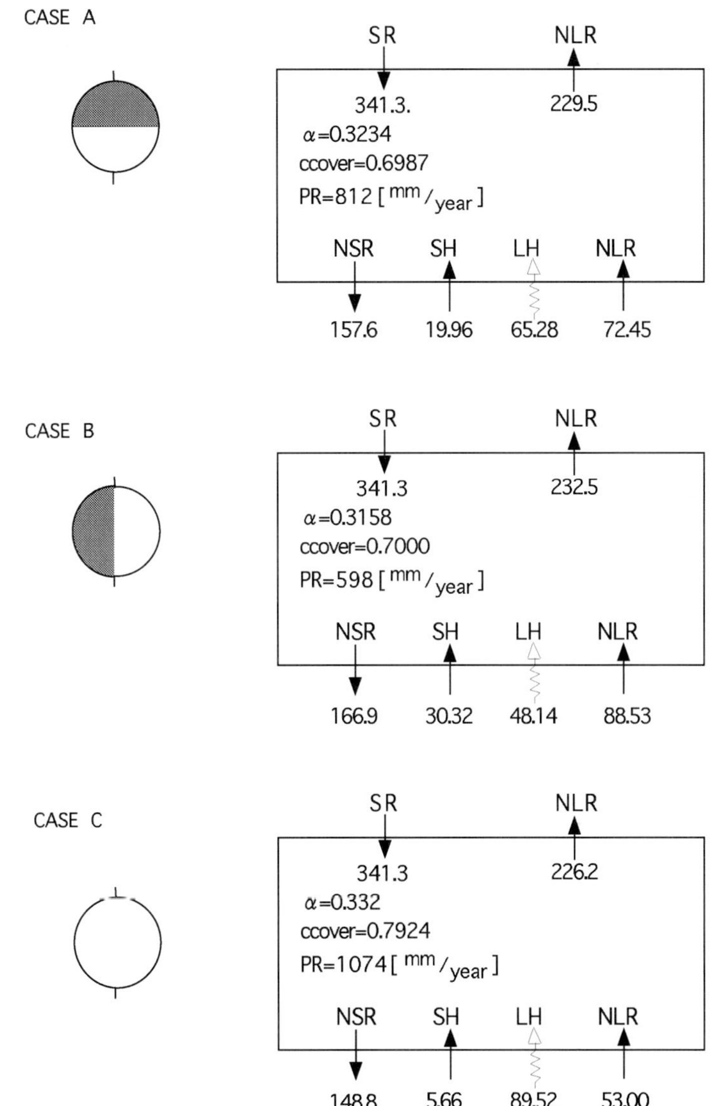

Figure 9 Results of the global averaged energy balance and water balance for case A (top), case B (middle), and Case C (bottom). SR, NLR, NSR, SH, and LH stand for short radiation, net long-wave radiation, net short-wave radiation, sensible heat flux, and latent heat flux, respectively; α and PR stand for the planetary albedo and precipitation. "Cover" denotes cloud cover. The precipitation rate is given in mm year^{-1}, and energy fluxes are in W m^{-2}.

land–sea contrast. The precipitation in case C is larger than that of case A and is almost double that of case B. This can be considered to be the maximum precipitation realized on an Earth-like planet.

In contrast, the planetary albedo is almost the same for these cases, although a small 0.9% difference is noted. This means that when the precipitation is increased, the albedo due to clouds does not increase as much as the precipitation. Interestingly, the planetary albedos are different in summer and winter for case A and case B, but when we take the annual average, they are almost the same. We have no theoretical explanation for how the planetary albedo is determined. It is strongly dependent on the mechanisms that generate clouds. It is not obvious that the same results will be obtained with other models. Whether this feature is peculiar to our model or not should be investigated further.

We have to be very careful in drawing general conclusions from these numerical experiments, but we emphasize that our experiment poses novel issues with respect to the characteristics of the climate system.

IV. HOW SHOULD WE EVALUATE OUR SIMULATIONS?

The purpose of simulation is to understand physical mechanisms by simulating the behavior of nature using known physical laws and/or empirical laws. We need a measure by which we can decide whether the results of a simulation are acceptable or not. This is the issue of model evaluation.

One measure of model performance is the RMSE (root mean square error), which is associated with the associated pattern correlation. The distance between the simulated state and the observed state is a measure of the success of the simulation, i.e., less error means a more reliable model. The RMSE is often used in numerical weather prediction, because NWP can be considered as an initial value problem and current observations suffice for detecting large-scale features.

The next question is how to evaluate climate simulations or how to evaluate the performance of a model in the simulation of climate. Usually, we use comparisons between time-averaged simulated fields and observed fields, as in NWP. We should remember that climate states consist of the mean states based on the boundary forcing and the internal dynamics. If the boundary forcing is wrong, it can lead to a systematic error. The concept of systematic error was first investigated by an intercomparison project led by the Working Group on Numerical Experimentation (WGNE;

Bengtsson and Lange, 1981). Through that project, the monthly mean and seasonal mean states were compared, and it was noted that model results look like each other, although a distance exists between the model results and the observed states. Note that there exists a strong westerly bias and a northward shift of the jet in the midlatitudes, and a warm bias in the polar lower troposphere (Sumi and Kanamitsu, 1984). This suggests that systematic errors exist, which means that our models fail to include important processes. After the completion of this research, a gravity wave drag was introduced (McFarlane, 1987). This was the first time that model performance was improved as a result of the diagnosis of model results.

Typically, we use summer and winter mean states as representative climate states, because boundary forcing is strong in summer and winter and therefore the mean states of summer and winter reveal boundary forcing effects more than internal dynamical effects. Nevertheless, model errors exist in the internal dynamics itself. Besides the stationary components, the transient aspects should be compared. The Madden–Julian oscillation is a good example.

Finally, I would like to emphasize that much attention should be paid to spring and autumn, i.e., transitional periods. For example, a frontal zone called the "Baiu front" is located ever Eastern Asia during the transition period from spring to summer, and the "Akisame front" is found in the same region during the transition from summer to autumn. These characteristic weather phenomena can be used to evaluate the performance of a model.

Transient behavior has been systematically investigated in the AMIP project (Sengupta and Boyle, 1993; Slingo *et al.*, 1992; Boyle, 1996). In earlier models, the amplitudes of the transient components were small compared to observed ones, but the current models can simulate almost the right amplitudes of the transient components. However, results are not yet satisfactory and further investigation is necessary.

V. CONCLUSION

I have summarized the ongoing activities conducted at the CCSR. Various aspects of climate system dynamics are being investigated there. We are convinced that climate simulations with a sophisticated climate model are necessary for understanding the climate system.

Climate simulation under the realistic conditions can convince us of a model's reliability. This is the real-world approach. However, climate simulations alone cannot reveal the mysteries of nature; we should strategically combine various methods for this purpose. At the same time, we can explore various characteristics of the real world by using numerical simulations.

A second approach is based on the use of simple models. A simple model can reveal the essence of the climate system and contribute to the establishment of our understanding for the climate dynamics. A hierarchy of models (zero-dimension models to three-dimensional models, simple models to a sophisticated models, by way of intermediate models) can lead to a deepening of our understanding.

A third, heuristic approach is based on CGFD. Numerical experiments using a full model under idealized conditions lead us to consider behaviors of the climate system that cannot be seen in real data.

In summary, we should study the climate system using the three approaches illustrated in Fig. 10. These three approaches are mutually related and should be systematically applied. In addition to simulation studies, we must remember the importance of observational studies. For climate studies, we cannot rely on the meteorological and climatological observations collected by governmental agencies. Satellite remote sensing is indispensable for global observations, although the calibration and evaluation of such data are always difficult and important issues. In addition to satellite data, we should use long-term historical data. It is also very important to make good use of paleo-climate data.

Finally, we mention that an important path to the understanding of climate system dynamics comes from the comparative planetary sciences.

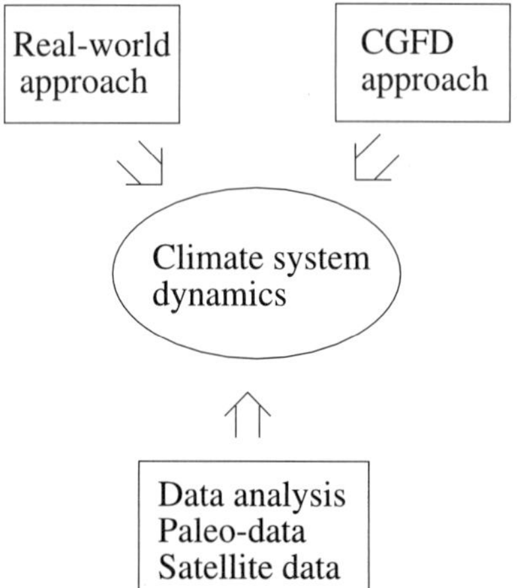

Figure 10 A schematic diagram illustrating a strategy for the study of climate system dynamics. See the text for an explanation.

Information on the climates of the other planets such as Mars and Venus can expand our framework for thinking about planetary climates.

REFERENCES

Arakawa, A. (1996). Computational design for long-term numerical integrations of the equations of atmospheric motion. *J. Comp. Phys.*, **1**, 119–143.

Bengtsson, L., and A. Lange (1981). Results of the WMO/CAS numerical weather prediction data study and intercomparison project for forecasts for the northern hemisphere in 1979–1980, PWPR Report No. 1.

Boyle, J. S. (1996). Intercomparison of low-frequency variability of the global 200hPa circulation for AMIP simulations, PCMDI Report No. 32.

Bryan, K. (1969). A numerical method for the study of the circulation of the world ocean. *J. Comp. Phys.*, **4**, 347–376.

Hayashi, Y. Y., and A. Sumi (1986). The 30–40 day oscillations simulated in an "aqua-planet" model. *J. Meteor. Soc. Japan* **64**, 451–467.

Kuma, K. (1994). The Madden and Julian oscillation and tropical disturbances in an aqua-planet version of JMA global model with T63 and T159 resolution. *J. Meteor. Soc. Japan* **72**, 147–172.

Lau, K.-M., and L. Peng (1987). A preliminary theory of the origin of the "40–50" day oscillation. *J. Atmos. Sci.* **44**, 950–972.

Lau, K.-M., J. H. Kim, and Y. Sud (1996). Intercomparison of hydrologic processes in AMIP GCMs. *Bull. Am. Meteor. Soc.* **77**, 2209–2227.

Lindzen, R., and J. R. Holton (1968). A theory of the quasi-biennial oscillation. *J. Atmos. Sci.* **25**, 1095–1167.

Le Treut, H., and Z.-X. Li (1991). Sensitivity of an atmospheric general circulation model to prescribed SST changes: feedback effects associated with the simulation of cloud optical properties. *Climate Dynamics*, **5**, 175–187.

Louis, J. (1979). A parametric model of vertical eddy fluxes in the atmosphere. *Bound.-Layer Meteor.* **17**, 187–202.

Manabe, S., J. Smagorinsky, and R. F. Strickler (1965). Simulated climatology of a general circulation model with a hydrologic cycle. *Monthly Weather Rev.* **93**, 769–798.

McFarlane, N. A. (1987). The effect of orographically excited gravity drag on the general circulation of the lower stratosphere and troposphere. *J. Atmos. Sci.* **31**, 1987.

Meehl, G. A., and W. M. Washington (1996). El Niño-like climate change in a model with increased atmospheric CO_2 concentrations. *Nature* **382**, 56–60.

Mellor, G. L., and T. Yamada (1974). A hierarchy of turbulent closure models for planetary boundary layers. *J. Atmos. Sci.* **31**, 1791–1806.

Mellor, G. L., and T. Yamada (1982). Development of a turbulent closure model for geophysical fluid problems. *Rev. Geophys. Space Phys.* **20**, 851–875.

Nakajima, T., and A. Higurashi (1997). AVHRR remote sensing of aerosol optical properties in the Persian Gulf region, summer 1991. *J. Geophys. Res.* **102**, D14, 16,935–16,946.

Nakajima, T., and M. Tanaka (1986). Matrix formulation for the transfer of solar radiation in a plain-parallel scattering atmosphere. *J. Quant. Spectrosc. Rad. Transfer* **35**, 13–21.

Numaguti, A., and Y. Y. Hayashi (1991a). Behavior of cumulus activity and the structures of circulations in an "Aqua Planet" model. Part I: The structure of the super clusters. *J. Meteor. Soc. Japan* **69**, 541–561.

Numaguti, A., and Y. Y. Hayashi (1991b). Behavior of cumulus activity and the structures of circulations in an "Aqua Planet" model. Part II: Eastward moving planetary structure and the intertropical convergence zone. *J. Meteor. Soc. Japan* **69**, 563–579.

Numaguti, A., S. Sugata, M. Takahashi, T. Nakajima, and A. Sumi (1997). Study on the climate system and mass transport by a climate model, CGER Monograph Report No. 3. (Available from CGER, NIES, Japan.)

Palmer, T. N. (1987). Modeling atmospheric low frequency variability, *Atmospheric and oceanic variability*. *Royal Meteorological Society*, 75–103.

Richardson, L. R. (1922). "Weather Prediction by Numerical Process." Cambridge University Press; reprinted by Dover, New York, 1965.

Sengupta, S. K., and J. S. Boyle (1993). Statistical intercomparison of global climate models: A common principal component approach, PCMDI Report No. 13.

Slingo, J. M., K. R. Sperber, J.-J. Morcrette, and G. L. Potter (1992). Analysis of the temporal behavior of tropical convection in the ECMWF model, PCMDI Report No. 2.

Suginohara, N., and S. Aoki (1991). Buoyancy-driven circulation as horizontal convection on beta-plane. *J. Mar. Res.* **49**, 295–320.

Sumi, A. (1987). Characteristics of simulated convective activity over a tropical ocean with zonally uniform SST surrounded by the dry continents. *J. Meteor. Soc. Japan* **65**, 853–870.

Sumi, A. (1992). Pattern formation of convective activity over the aqua-planet with globally uniform sea surface temperature (SST). *J. Meteor. Soc. Japan* **70**, 855–876.

Sumi, A., and M. Kanamitsu (1984). A study of systematic errors in a numerical weather prediction model. Part 1: General aspects of the systematic errors and their relation with the transient eddies. *J. Meteor. Soc. Japan* **62**, 234–251.

Swinbank, R., T. N. Palmer, and M. K. Davey (1988). Numerical simulations of the Madden–Julian oscillation. *J. Atmos. Sci.* **45**, 774–788.

Takahashi, M. (1996). Simulation of the stratospheric quasi-biennial oscillation using a general circulation model. *J. Geophys. Lett.* **23**, 661–664.

Takahashi, M., and B. A. Boville (1992). A three-dimensional simulation of the equatorial quasi-biennial oscillation. *J. Atmos. Sci.* **49**, 1020–1035.

Williams, G. P. (1978). Planetary circulations: 1. Barotropic representation of Jovian and terrestrial turbulence. *J. Atmos. Sci.* **35**, 1399–1426.

Williams, G. P. (1988). The dynamical range of global circulations I and II. *Clim. Dyn.* **2**, 205–260.

Zhang, R., A. Sumi, and M. Kimoto (1996). Impact of El Niño on the East Asian monsoon: A diagnostic study of the '86/87 and '91/92 events. *J. Meteor. Soc. Japan* **74**, 49–62.

Chapter 17

Global Atmospheric Modeling Using a Geodesic Grid with an Isentropic Vertical Coordinate

David A. Randall, Ross Heikes, and Todd Ringer
Department of Atmospheric Science
Colorado State University
Fort Collins, Colorado

I. INTRODUCTION

From the beginning of global atmospheric modeling, solution of the governing equations on the sphere has been a challenging problem. Today, most global atmospheric models use either the "spectral" method, based on expansions in terms of triangularly truncated spherical harmonics (Jarraud and Simmons, 1983) or finite differences implemented on regular grids in spherical (longitude–latitude) coordinates. Each of these methods has strengths and weaknesses. Finite-difference models based on latitude–longitude grids must deal with the convergence of the meridians at the poles, which demands a short time step for computational stability; this is the well known "pole problem." Filtering techniques (Arakawa and Lamb, 1977) are typically employed to allow a moderately long time step. In addition, fine zonal spatial resolution near the poles entails additional computations (e.g., for parameterized physical processes) that contribute little to the overall accuracy of the solution. On the other hand, increased zonal resolution in middle and high latitudes may be useful, particularly when the radius of deformation decreases near the poles.

The spectral method based on spherical harmonics with triangular truncation (e.g., Jarraud and Simmons, 1983) elegantly eliminates the pole problem. On the other hand, it gives poor results for the advection of strongly varying non-negative scalars such as the mixing ratios of water vapor and cloud water (e.g., Williamson and Rasch, 1994). The problem of negative water is sufficiently serious that most spectral models have now been modified to use semi-Lagrangian techniques for advection, rendering the modified models distinctly less spectral in character. Spectral computation of mass convergence and the horizontal pressure gradient force allows fast and easy implementation of semi-implicit time-differencing schemes, which permit a relatively long time step; in contrast, semi-implicit schemes are both more expensive and more complicated in finite-difference models, although they have the potential to eliminate the need for polar filtering.

Several efforts have been made to devise alternative discretizations of the sphere, to permit the use of finite-difference methods without the need for polar filtering. These include grids that skip some longitudinal grid points near the poles (Kurihara, 1965; Halem and Russell, 1973), matched and/or patched polar stereographic grids (Phillips, 1957; Browning *et al.*, 1989), and grids based on polyhedra (Sadourny *et al.*, 1968; Sadourny and Morel, 1969; Williamson, 1968, 1970; Sadourny, 1972; Thacker, 1978; Augenbaum and Peskin, 1985; Baumgardner and Frederickson, 1985; Masuda and Ohnishi, 1986; Nickovic, 1994; Popovic *et al.*, 1996; McGregor, 1996; Rancic *et al.*, 1996; Purser and Rancic, 1998; Thuburn, 1997). As recognized in early work by Arakawa and colleagues, grids based on icosahedra offer an attractive framework for simulation of the global circulation of the atmosphere; their advantages include almost uniform and quasi-isotropic resolution over the sphere. Such grids are termed *geodesic* because they resemble the geodesic domes designed by Buckminster Fuller. Geodesic grids caught the attention of Arakawa during the 1960s (Sadourny *et al.*, 1969).

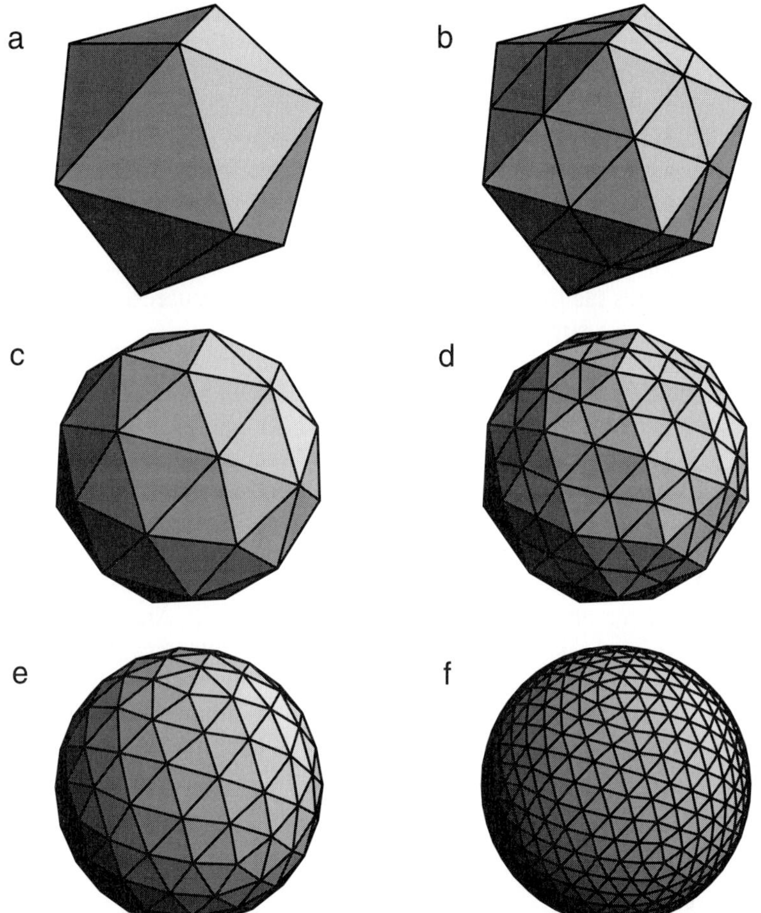

Figure 1 (a) An icosahedron inscribed in a unit sphere. (b) Bisect each edge forming four new faces. (c) Project the new vertices onto the unit sphere. The process is repeated in panels (d)–(f).

A geodesic grid is constructed by starting with an ordinary icosahedron inscribed inside a unit sphere, as shown in Fig. 1a. The icosahedron has 12 vertices. As a first step in the construction of a spherical geodesic grid, each face of the icosahedron is subdivided into four new faces by bisecting the edges. The result of this process is shown in Fig. 1b. Next, the new vertices are "popped out" onto the unit sphere, creating the polyhedron shown in Fig. 1c. This recursive process can be continued indefinitely, yielding arbitrarily fine meshes. When a finer geodesic grid is derived by subdividing a coarser one, the finer one "nests" within the coarser one. A geodesic grid of any resolution consists of exactly 12 pentagons, corresponding to the 12 vertices of the original icosahedron, and a (typically much) larger number of hexagons. Such grids are quasi-homogeneous in the sense that the area of the largest cell is only a few percent greater than the area of the smallest cell. For more details, see Heikes and Randall (1995a).

Hexagonal grids are also quasi-isotropic. As is well known, only three regular polygons tile the plane: equilateral triangles, squares, and hexagons. Figure 2 shows planar grids made up of each of these three possible polygonal elements. On the triangular grid and the square grid, some of the neighbors of a given cell lie directly across cell walls, while others lie across cell vertices. As a result, finite-difference operators constructed on these grids tend to use "wall neighbors" and "vertex neighbors" in different ways. For example, the simplest second-order finite-difference approximation to the gradient, on a square grid, uses only wall neighbors; vertex neighbors are ignored. Although it is certainly possible to construct finite-difference operators on square grids (and triangular grids) in which information from all neighboring cells is used (e.g., the Arakawa Jacobian, as discussed by Arakawa, 1966), the essential anisotropies of these grids remain, and are unavoidably manifested in the forms of the finite-difference operators. Hexagonal grids, in contrast, have the property that all neighbors of a given cell lie across cell walls; there are no vertex neighbors. As a result, finite-difference operators constructed on hexagonal grids treat all neighboring cells in the same way; in this sense, the operators are as symmetrical and isotropic as possible. The 12 pentagonal cells also have only wall neighbors; there are no vertex neighbors anywhere on the sphere.

II. THE Z GRID

Winninghoff (1968) and Arakawa and Lamb (1977) studied the simulation of geostrophic adjustment on a variety of logically rectangular staggered grids for the solution of the shallow water equations. The variables that they considered were the zonal and meridional components of the wind, and the mass. They discussed five distinct staggered grids, which they labeled A through E. Randall (1994) defined the Z Grid, which is an unstaggered grid for the vorticity, divergence, and mass. With a rectangular mesh, the Z Grid corresponds to the Arakawa–Lamb C Grid for the divergent component of the wind, and to the Arakawa–Lamb D Grid for

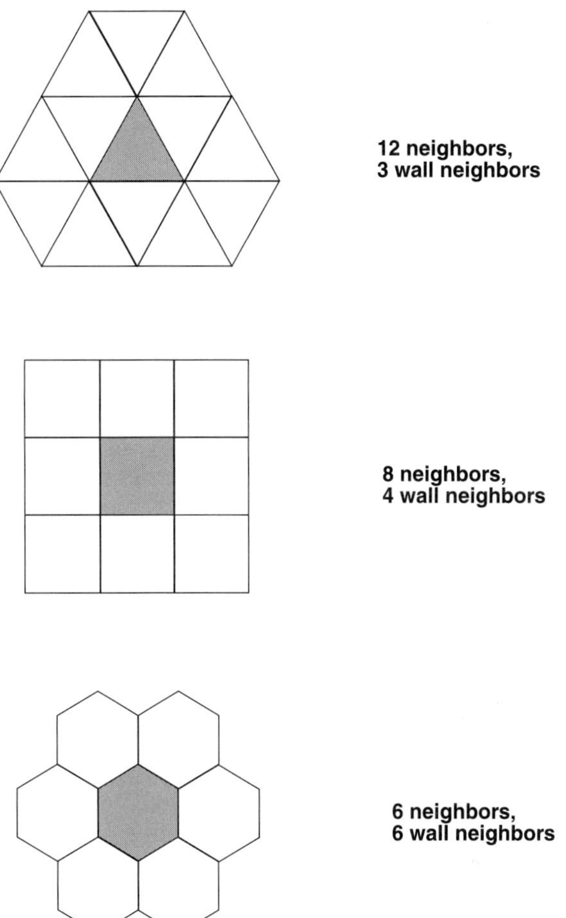

Figure 2 Cells neighboring a given cell (shaded) on triangular, square, and hexagonal grids. A "wall neighbor" is a neighbor that lies directly across a cell wall.

the rotational component. The spatial arrangements of the variables on the B, C, and Z Grids are shown in Fig. 3.

The analyses of Winninghoff (1968) and Arakawa and Lamb (1977) showed that geostrophic adjustment is best simulated on the C Grid, provided that the grid spacing is smaller than the radius of deformation. Although a modeler will try to choose a model's grid spacing to be smaller than the radius of deformation, this is not always possible; in atmospheric models, high internal modes can have small radii of deformation, and

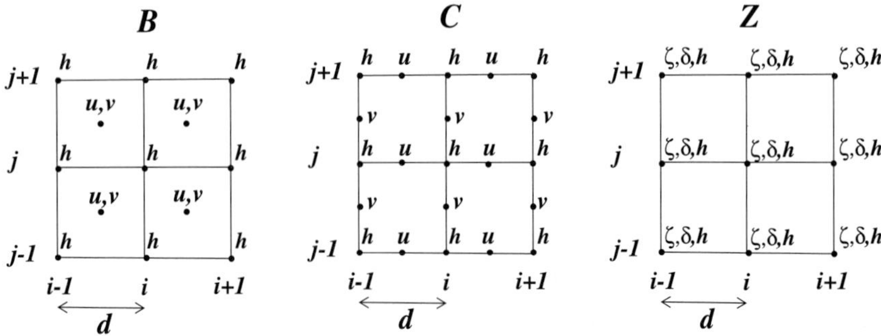

Figure 3 The spatial arrangements of the horizontal wind components (u and v) and the mass (h) on B Grid and C Grid. Also shown are the spatial arrangements of the vorticity, divergence, and mas on the Z Grid.

small radii of deformation are even more prevalent in ocean models because the ocean is weakly stratified. These problems are exacerbated near the poles, where the radius of deformation tends to be particularly small because the Coriolis parameter is large. When the grid spacing is larger than the radius of deformation, the C Grid gives poor results (Randall, 1994).

This is illustrated in Fig. 4, which shows the gravity-inertia wave frequency as a function of the horizontal wave numbers for the continuous shallow-water equations, and for the B, C, and Z Grids. The results are shown for a case in which the grid size is smaller than the radius of deformation, and for a case in which it is larger. The C Grid gives realistic results when the grid spacing is finer than the radius of deformation, but behaves very badly when the radius of deformation is smaller than the grid size. As shown by Randall (1994) and as illustrated in Fig. 4, the Z Grid behaves well regardless of the ratio of the grid size to the radius of deformation.

Figure 4 The nondimensional frequency (on the vertical axis), plotted as a function of the normalized horizontal wave numbers (the horizontal axes) for the continuous shallow-water equations, and for the B, C, and Z Grids, for the case in which the grid spacing is twice the radius of deformation (left column) and for the case in which the grid spacing is one-tenth of the radius of deformation (right column).

$\lambda/d=2$

$\lambda/d=0.1$

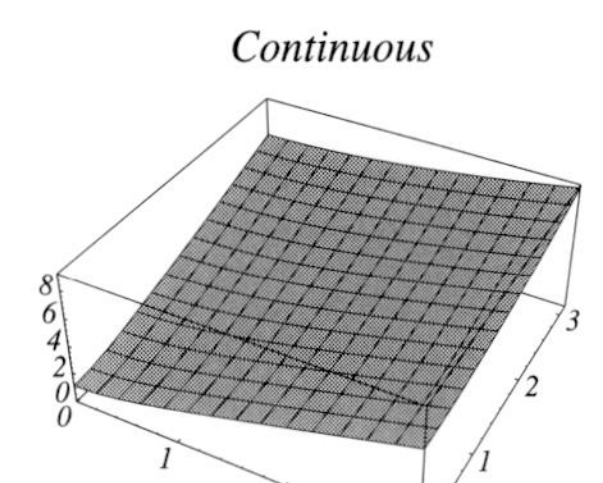

Continuous

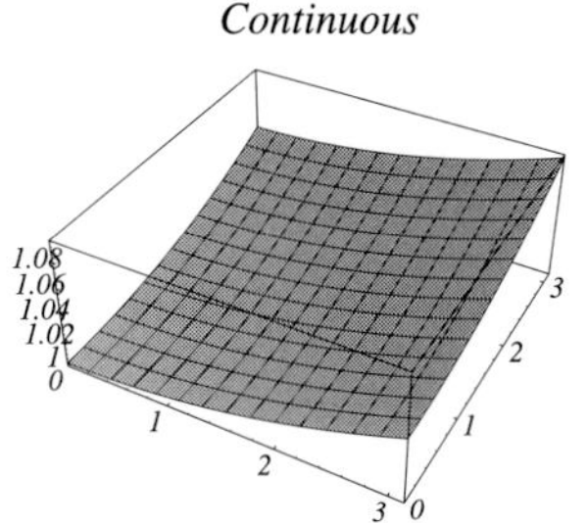

Continuous

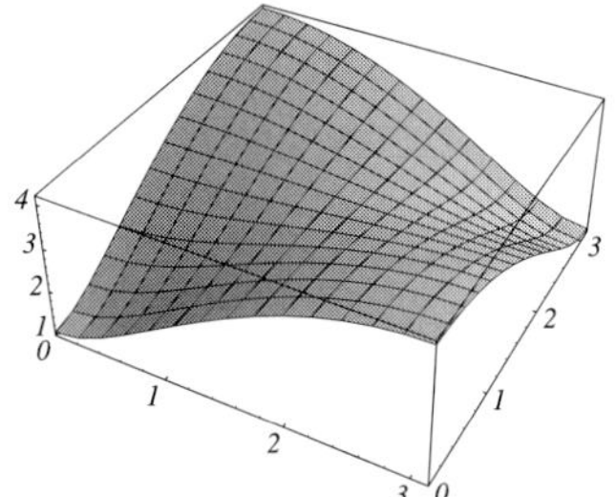

B Grid

B Grid

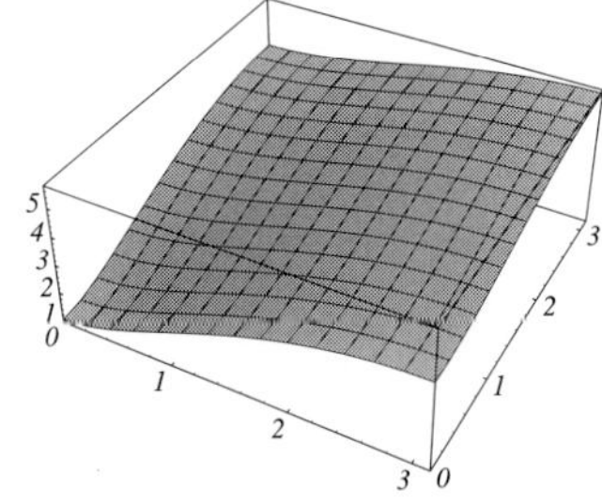

C Grid

C Grid

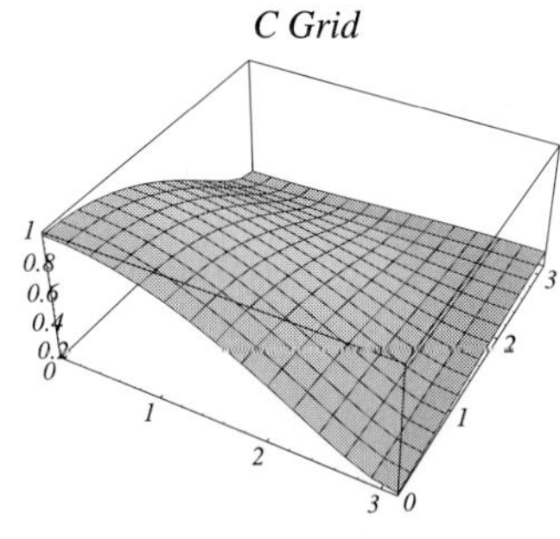

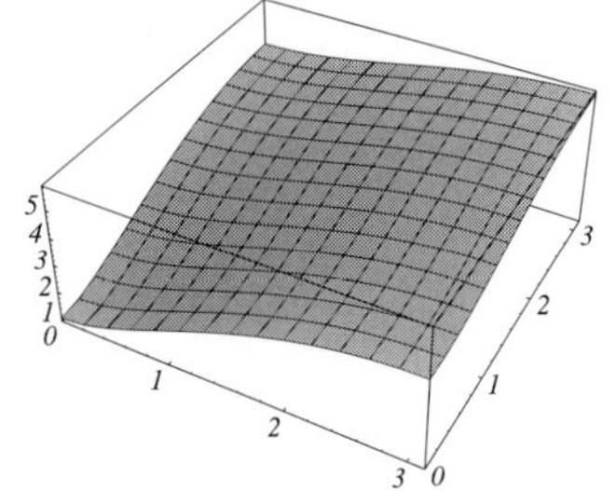

Z Grid

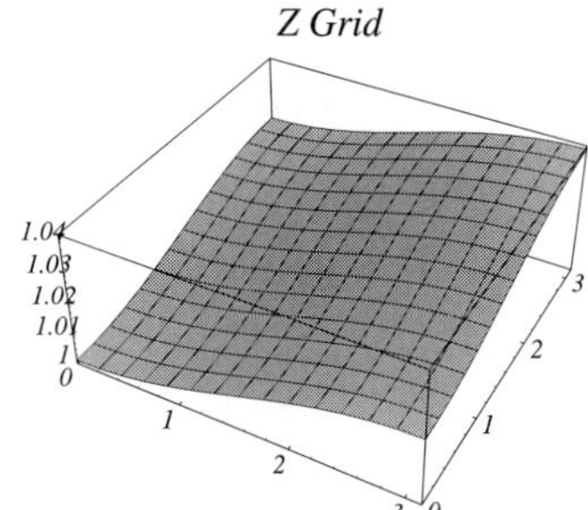

Z Grid

The reason for this is easy to see. The linearized shallow-water equations for divergence, vorticity, and mass are:

$$\frac{\partial \delta}{\partial t} - f\zeta + g\nabla^2 h = 0, \tag{1}$$

$$\frac{\partial \zeta}{\partial t} + f\delta = 0, \tag{2}$$

$$\frac{\partial h}{\partial t} + H\delta = 0. \tag{3}$$

The only differential operator appearing in these equations is the Laplacian. For reasons of symmetry, $\nabla^2 h$ is almost inevitably defined on the same points as h. Equations (1)–(3) therefore "want" an unstaggered grid, i.e., they want the Z Grid. Further discussion is given by Randall (1994).

The arguments given above carry over directly to geodesic grids, and so geostrophic adjustment is well simulated on a geodesic Z Grid for the same reasons that it is well simulated on a rectangular Z Grid.

If vorticity and divergence are used as the primary dynamical variables in a nonlinear model based on the Z Grid, then it is necessary to solve linear boundary-value problems for the stream function, ψ, and the velocity potential, χ, which are needed in the advection terms of the model:

$$\nabla^2 \psi = \zeta \quad \text{and} \quad \nabla^2 \chi = \delta. \tag{4}$$

Solutions to Eqs. (4) can be obtained by a variety of methods, but computational speed is an issue. We return to this point below.

III. A GEODESIC SHALLOW-WATER MODEL USING THE Z GRID

Masuda and Ohnishi (1986) integrated the shallow-water equations for vorticity, divergence, and mass on an unstaggered geodesic grid, derived from an icosahedron. Their model was thus based on a geodesic Z Grid, as discussed above. They found that their model gave good solutions to test problems, although they noted that for the cases they studied the solutions tended to reflect the geometry of the grid.

In the numerical results of Masuda and Ohnishi (1986), initial conditions which are symmetric across the equator slowly evolve to a state that is asymmetric across the equator, as a result of the flow interacting with the grid. This suggests that it is desirable to construct a spherical geodesic grid that is symmetric across the equator. Heikes and Randall (1995a) pointed out that this can be done simply by rotating all the faces in the Southern Hemisphere through $\pi/5$ radians; in this way, we obtain a symmetric polyhedron, which we call the *twisted icosahedron*. Grids based on this polyhedron can be called *twisted icosahedral grids*, or *twigs*. By recursively subdividing each triangular face and projecting the new vertices onto the unit sphere, we can generate polyhedra that progressively approximate a sphere.

Heikes and Randall (1995a,b) constructed a twig-based shallow-water model, using the vorticity, divergence, and mass equations without staggering. In test applications of finite-difference approximations to differential operators such as the Laplacian, Heikes and Randall (1995b) found that although the solutions converged uniformly (with second-order accuracy) in the mean-square sense used in the derivation of spectral transforms, local convergence failed at a few points on the sphere. An investigation revealed that, at these points, lines between cell centers do not intersect cell walls orthogonally, even in the limit of infinitely many points. As a result, operators defined in terms of outward-normal unit vectors on cell walls (e.g. the divergence, gradient, and Laplacian) do not converge to their differential counterparts in the limit of infinite resolution. This problem would not occur for the case of a hexagonal mesh on a plane; it arises from the projection of the mesh on the sphere. The problem was solved by slightly repositioning each of the cell centers, for each possible resolution, so as to minimize a global measure of the error; this "tweaking" of the grid is done once and for all, before the model is run, using a variational method. The changes are very small; to the eye, the tweaked grid is indistinguishable from the untweaked grid.

The shallow-water versions of the continuity, vorticity, and divergence equations can be discretized on the geodesic grid using straightforward "line-integral" methods similar to those discussed by Masuda and Ohnishi (1986). For reasons discussed by Sadourny *et al.* (1968), the horizontal discretization scheme conserves vorticity, kinetic energy, and enstrophy in the limit of two-dimensional nondivergent flow. Our free-surface shallow-water model conserves potential vorticity and potential enstrophy, as well as mass. The discretized pressure-gradient term of the free-surface shallow-water model does not give rise to any spurious sources or sinks of vorticity.

Heikes and Randall (1995a,b) took advantage of the nesting properties of successive geodesic grids to use multigrid methods (Fulton *et al.*, 1986) to solve Eqs. (4) for the streamfunction and velocity potential from the vorticity and divergence, respectively. This method proved to be fast enough so as to remove computational speed as an impediment to the use of the vorticity and divergence equations.

Heikes and Randall tested their model using the suite of seven test cases devised by Williamson *et al.* (1992). The results were compared with exact solutions where these are known, and also with the NCAR spectral shallow-water model, and with a C Grid shallow-water model based on the potential-enstrophy-conserving scheme of Arakawa and Lamb (1981). They obtained satisfactory results.

IV. SEMI-IMPLICIT TIME DIFFERENCING

Semi-implicit time-differencing can be implemented so that the gravity wave terms of a shallow-water model (or a three-dimensional model) permit a relatively long time step without computational instability. As already noted, semi-implicit time differencing is easily implemented in spectral models, and from the perspective of 2000 this appears to be one of the most important strengths of the spectral method when it is used with high resolution.

We have implemented semi-implicit time differencing in our geodesic-grid models. This requires solution of an elliptic system in which the unknowns are the "updated" or "end-of-time-step" values of the mass, temperature, and divergence fields. It turns out that the solution algorithm can be combined with that used to solve for the streamfunction and velocity potential; details are discussed by Ringler *et al.* (1998).

Whereas a semi-implicit time step requires about 20% more computation than a purely explicit time step, the semi-implicit scheme allows (at least) a fourfold increase in the length of the time step, so that a semi-implicit version of the model's dynamics is ultimately several times faster than an explicit version. A semi-implicit version of the model with approximately 200-km grid spacing runs contentedly with a time step of 20 min.

V. FLUX-CORRECTED TRANSPORT

Flux-corrected transport (FCT) schemes (Zalesak, 1979) are readily implemented on geodesic grids, in order to impose requirements such as sign preservation and/or monotonicity. We have implemented a multidimensional flux-corrected transport scheme, following the approach of Zalesak (1979), in order to maintain non-negative mass (necessary with the massless layer approach used in the isentropic coordinate model described below, and useful for moisture conservation in any model) as well as monotone advection for the PV and for passive tracers. The FCT schemes tend to dissipate second moments such as potential enstrophy, but they have no effects on the conservation of first moments such as potential vorticity. Details of the implementation will be reported elsewhere.

VI. A FULL-PHYSICS VERSION OF THE MODEL USING THE GENERALIZED SIGMA COORDINATE

To convert our geodesic shallow-water model into a full three-dimensional model suitable for the simulation of the atmospheric general circulation, we had to choose a vertical coordinate system and a vertical discretization scheme. We have also constructed and are currently testing versions of the model using the σ coordinate (Ringler *et al.*, 1998, 1999) with both Lorenz vertical staggering and Charney–Phillips vertical staggering (Moorthi and Arakawa, 1985; see also Arakawa's opening chapter, Chapter 1, in this volume). The σ-coordinate version with the Lorenz vertical staggering has recently been endowed with a full suite of physical parameterizations, so that it is now a full-fledged atmospheric general circulation model, although still an experimental one (Ringler *et al.*, 1999). The full-physics version actually makes use of a modified σ coordinate in which the PBL top is a coordinate surface (Suarez *et al.*, 1983). It is currently being tested, and a more complete discussion is given elsewhere (Ringler *et al.*, 2000). In Fig. 5 we show the simulated January-mean boundary-layer wind streamlines and precipitation field. These results were produced with a version of the model using 10,242 grid cells (roughly equivalent to 2° horizontal resolution) with 17 layers.

VII. A THREE-DIMENSIONAL VERSION OF THE MODEL WITH AN ISENTROPIC VERTICAL COORDINATE

Hsu and Arakawa (1990; hereafter HA) discussed the many advantages of the isentropic vertical coordinate for numerical modeling of the large-scale circulation of the atmosphere. These advantages include the following:

- Drastic reduction of the numerical difficulties associated with vertical advection

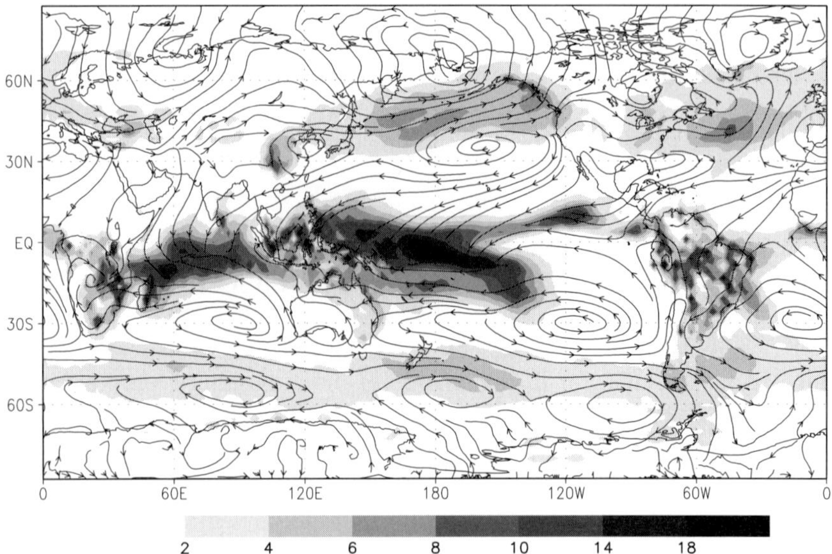

Figure 5 The simulated January-mean boundary-layer wind streamlines and precipitation field obtained in a simulation using a geodesic grid and the σ coordinate, with an embedded variable-depth boundary layer.

- Direct representation of the Ertel potential vorticity, and straightforward implementation of numerical schemes for its conservation
- Simple and natural definition of the available potential energy in a discrete numerical model

To deal with potential temperature surfaces that intersect the lower boundary, HA adopted the "massless layer" approach, originally suggested by Lorenz (1955), in which potential temperature surfaces that intersect the boundary are considered to be extended along the boundary with essentially zero mass between pairs of such surfaces. Figure 6 illustrates the massless layer approach. In the figure, two particular θ surfaces, denoted by θ_1 and θ_2, respectively, are shown intersecting the Earth's surface and then following along the Earth's surface toward the left.

We constructed a three-dimensional global atmospheric circulation model by combining the horizontal differencing scheme based on the geodesic grid, as discussed earlier, with a vertical discretization based on the isentropic vertical coordinate of HA, and including semi-implicit time differencing. The three-dimensional θ-coordinate version of the model conserves mass, the mass-weighted Ertel potential vorticity, potential temperature, and an arbitrary number of tracers. Because we followed the

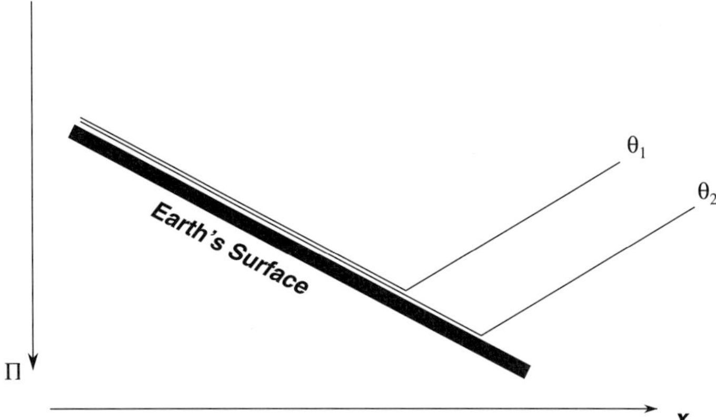

Figure 6 Sketch illustrating the intersections of two particular potential temperature surfaces with the Earth's surface. The horizontal axis represents a horizontal spatial coordinate, and the vertical axis is the Exner function, which increases downward. The Earth's surface is denoted by the thick line. The sketch shows a case in which the surface pressure varies in space.

methods of HA, the vertical differencing scheme of the model maintains total energy conservation under frictionless adiabatic flow.

Extensive tests of the isentropic-geodesic model were carried out using the Held–Suarez (1994) benchmark. The results demonstrated that the model is numerically stable and suitable for use in long simulations (years). Sample results are shown in Fig. 7.

VIII. FURTHER ANALYSIS OF THE ISENTROPIC COORDINATE

HA pointed out some difficulties with the θ coordinate, including these:

- Intersection of the potential temperature surfaces with the lower boundary
- Inability to represent dry statically unstable states of the atmosphere

The second of these is not important for large-scale modeling; generally speaking, any circulation that is influenced in an important way by large-scale dry static instability will not be compatible with the quasi-static approximation. The first difficulty is more serious, even though it is technical in nature. It gives rise to a number of problems, including one described by HA, which seems not to have been noticed earlier, and which

a

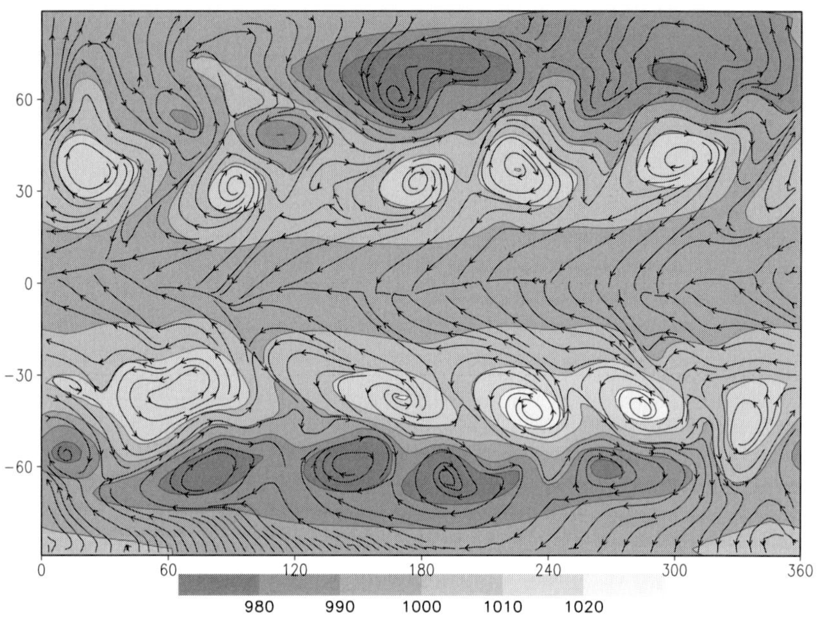

b

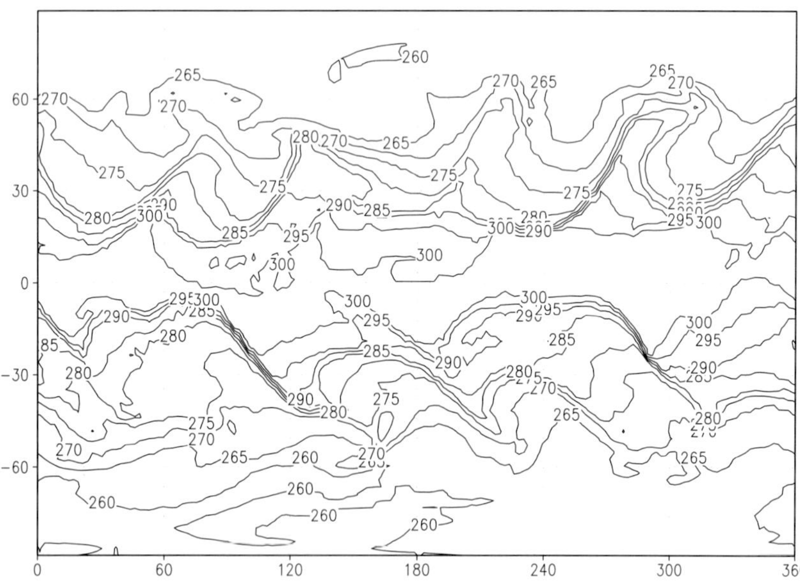

is sketched only briefly at the end of the main body of HA's paper: "In treating the intersections of the coordinate surfaces with the lower boundary, we have introduced massless layers along the lower boundary of the model.... This makes it difficult to define the surface air temperature.... Determining a sufficiently smooth distribution of the surface temperature is ... important in obtaining a sufficiently smooth distribution of the Montgomery potential, M_l. With the discrete hydrostatic equations ... almost discontinuous gradients of M_l (and therefore almost discontinuous geostrophic wind) appear for all l above the transitions between finite-mass and massless regions."

The nature of the problem described by HA can be understood as follows. The hydrostatic equation in isentropic coordinates is

$$\frac{\partial M}{\partial \theta} = \Pi. \tag{5}$$

Here $M(\theta)$ is the Montgomery potential, and Π is the Exner function, defined by

$$\Pi \equiv c_{\mathrm{p}} \left(\frac{p}{p_0} \right)^{\kappa}, \tag{6}$$

where c_{p} is the heat capacity of air at constant pressure, p_0 is a reference pressure (here taken to be 1000 mb), and κ is Poisson's constant. The lower boundary condition used with Eq. (5) is

$$M(0) = \phi_{\mathrm{S}}, \tag{7}$$

where ϕ is the geopotential and subscript S denotes the Earth's surface. For any isentropic surface that is locally in contact with the Earth's surface, $\Pi = \Pi_{\mathrm{S}}$, and so Eqs. (5) and (7) give

$$M(\theta) = \phi_{\mathrm{S}} + \theta \Pi_{\mathrm{S}}, \qquad \theta \le \theta_{\mathrm{S}}. \tag{8}$$

Here θ_{S} is the surface potential temperature; because we have massless layers, with many θ surfaces squeezed together along the lower boundary, the definition of θ_{S} may not be obvious. We define θ_{S} to be the warmest potential temperature for which $\Pi = \Pi_{\mathrm{S}}$.

Figure 7 Instantaneous fields obtained in Held–Suarez tests of the geodesic-isentropic model, with 10,242 cells and 48 levels. (a) The 950-mb wind streamlines and the surface pressure. (b) Surface temperature. Midlatitude baroclinic waves, including surface fronts, are clearly evident in the plots. Note the strong flow over the South Pole.

As a special case of Eq. (8), we can write the Montgomery potential at the Earth's surface as

$$M_S \equiv M(\theta_S) = \phi_S + \theta_S \Pi_S. \tag{9}$$

At higher levels,

$$M(\theta) = M(\theta_S) + \int_{\theta_S}^{\theta} \Pi(\theta')\, d\theta', \qquad \theta > \theta_S. \tag{10}$$

Here θ' is a dummy variable of integration. Equation (10) makes the simple point that the surface potential temperature influences the Montgomery potential at all levels. The horizontal pressure gradient force (hereafter HPGF) can be expressed in terms of the gradient of the Montgomery potential along θ surfaces:

$$\left(\frac{\partial \mathbf{V}}{\partial t} \right)_\theta \sim -\nabla_\theta M \equiv \mathrm{HPGF}. \tag{11}$$

Here $\mathbf{V}$ is the horizontal velocity vector and t is time.

We now introduce a discrete form of the hydrostatic equation, following HA. Consider a vertically discrete model using the θ coordinate with the massless layer approach, and with a total of N layers. The layers used are specified by choosing $N + 1$ constant layer-edge values of the potential temperature. The vertical index l, used to number the layers, is chosen to increase downwards, so that $l = N$ is the index of the lowest or "potentially coldest" layer. The "coldest" layer-edge potential temperature is chosen to be cold enough to ensure that layer N is massless over the entire globe. As we count backward to smaller values of l, we rise through a certain number of massless layers, until finally we encounter the lowest massy[1] layer in a given grid column; the index of this layer will be denoted by $l = L$. Note that L is expected to vary between grid columns and in time. An example is given below.

The vertically discrete form of the hydrostatic equation, as developed by HA, is

$$M_L = \phi_S + \Pi_S \theta_L \tag{12}$$

for the lowest massy layer, with index $l = L$. Equation (12) is equivalent to

$$\begin{aligned} M_L &= \phi_S + \theta_S \Pi_S + \Pi_S(\theta_L - \theta_S) \\ &= M_S + \Pi_S(\theta_L - \theta_S), \end{aligned} \tag{13}$$

---

[1] We were surprised to find that "massy" is a real word in the sense that it can be found in a standard dictionary.

which is a straightforward discrete approximation to Eq. (10). Note that Π_S is used here to approximate the mean value of Π over the interval $\theta_S \leq \theta \leq \theta_L$. This approximation has only first-order accuracy. As the vertical resolution of the model increases, $\theta_L \to \theta_S$, so that $M_L \to M_S$, $\Pi_L \to \Pi_S$, and Eq. (12) approaches (10). The fact that Eq. (13) involves θ_S appears to be a problem, because HA do not give a way to determine θ_S. The apparent problem is easily solved, however, because Eq. (12), which is algebraically equivalent to Eq. (13), does not involve θ_S.

HA determine the difference in Montgomery potential between a pair of layers using

$$M_l - M_{l+1} = \Pi_{l+1/2}(\theta_l - \theta_{l+1}) \qquad \text{for } l < L. \tag{14}$$

This corresponds to Eq. (5). By combining Eqs. (12) and (14), we find that

$$M_l = \phi_S + \theta_L \Pi_S + \sum_{i=l}^{L-1} \Pi_{i+1/2}(\theta_{i+1} - \theta_i) \qquad \text{for } l < L. \tag{15}$$

This corresponds to Eq. (10). Following HA, we use

$$\theta_l = \sqrt{\theta_{l+1/2}\,\theta_{l-1/2}} \qquad \text{for all } l. \tag{16}$$

To evaluate the errors arising from the vertically discrete hydrostatic equation, consider the following analytical distribution of pressure (Held and Suarez, 1994), p, with potential temperature and latitude, φ:

$$p(\theta, \varphi) = \begin{cases} p_0 & \text{for } \theta \leq \theta_S \\[2mm] p_0 \exp\left[\dfrac{\theta_S(\varphi) - \theta}{N(\varphi)}\right] & \text{for } \theta_S < \theta \end{cases}, \tag{17}$$

where

$$N(\varphi) = 24\cos^2\varphi + 34\sin^2\varphi \tag{18}$$

measures the rate with which pressure decreases with height as a function of latitude. The potential temperature measured along the Earth's surface is given by (Held and Suarez, 1994)

$$\theta_S(\varphi) = (1 - \sin^2\varphi)(302 - 45\sin^4\varphi)$$
$$+ \sin^2\varphi(257 + 45\sin^4\varphi). \tag{19}$$

Here the subscript S denotes the Earth's surface. Figure 8 shows various results obtained by using the analytically prescribed soundings given by

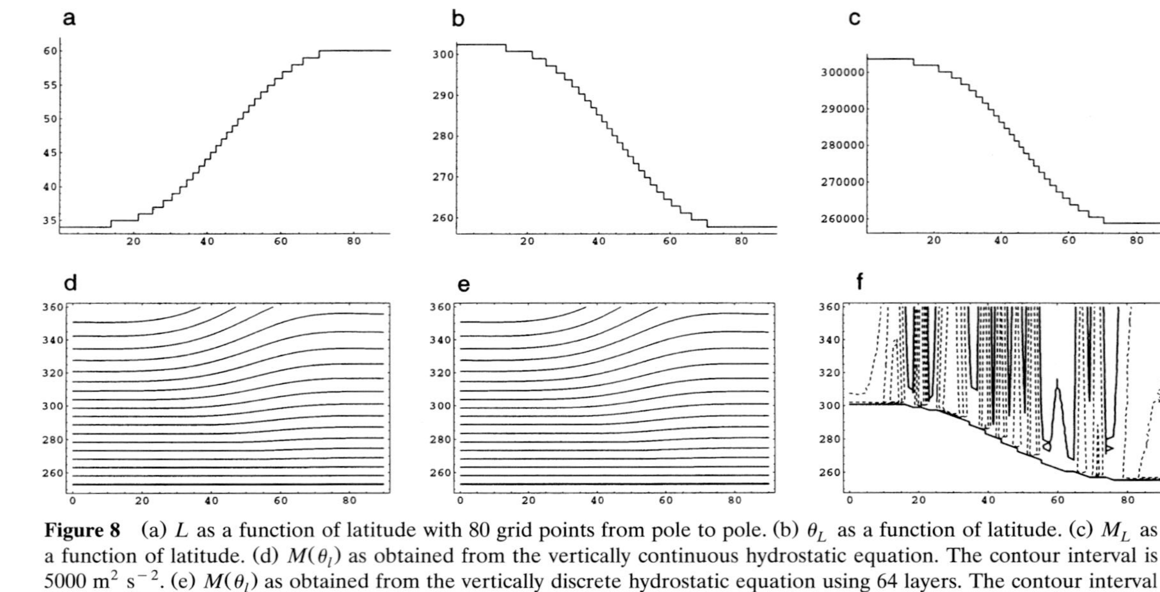

Figure 8 (a) L as a function of latitude with 80 grid points from pole to pole. (b) θ_L as a function of latitude. (c) M_L as a function of latitude. (d) $M(\theta_l)$ as obtained from the vertically continuous hydrostatic equation. The contour interval is 5000 m^2 s^{-2}. (e) $M(\theta_l)$ as obtained from the vertically discrete hydrostatic equation using 64 layers. The contour interval is 5000 m^2 s^{-2}. (f) The difference between the discrete and continuous solutions for $M(\theta_l)$. The contour interval is 1 m^2 s^{-2}. The thicker line is zero, and dashed lines are negative.

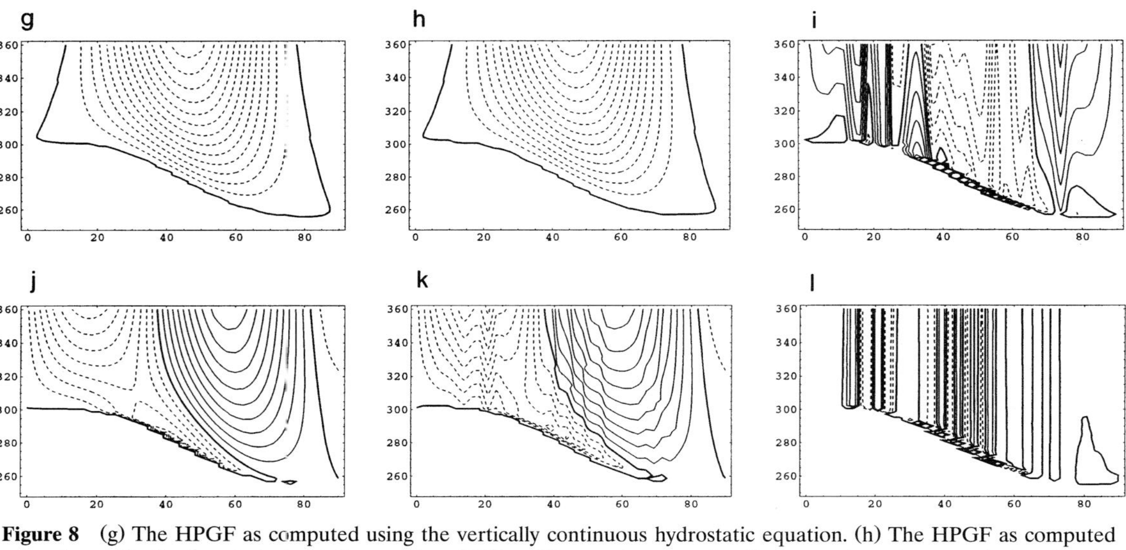

Figure 8 (g) The HPGF as computed using the vertically continuous hydrostatic equation. (h) The HPGF as computed using the vertically discrete hydrostatic equation. (i) The difference between the discrete and continuous solutions for the HPGF. (j) $\zeta_g(\theta_l)$ as computed using the vertically continuous hydrostatic equation. The contour interval is 2.5×10^{-6} s^{-1}. (k) $\zeta_g(\theta_l)$ as computed using the vertically discrete hydrostatic equation. (l) The difference between the discrete and continuous solutions for $\zeta_g(\theta_l)$. The contour interval is 5.0×10^{-7} s^{-1}.

Eqs. (17) and (19) to evaluate the pressure on various discrete levels and at various discrete latitudes. We combine these analytically prescribed pressures with the vertically discrete Eqs. (12) and (15). The results shown in Fig. 8 are computed with 40 points in latitude from equator to pole, and 64 layers for θ in the range 360–250 K. First, in Figs. 8a–c, we show the meridional distributions of L, θ_L, and M_L. These plots have a stairstep or sawtooth character. Next, in Fig. 8d, we show the latitude–height distribution of $M(\theta_l)$, computed using the analytic hydrostatic equation. Figure 8e shows the same field as computed using the vertically discrete equations, Eqs. (12) and (15). The results look smooth to the eye but the difference plot shown in Fig. 8f reveals the presence of computational noise. The noise originates in the stairstep structure of θ_L, which influences M_L through Eq. (12). The noisy structure of M_L contaminates the Montgomery potential throughout the atmospheric column, through integration of the hydrostatic equation, Eq. (14). Figures 8g–i show the corresponding results for the HPGF, and Figs. 8j–l show the corresponding results for the geostrophic vorticity, $\zeta_g(\theta_l)$, which is defined by

$$\zeta_g \equiv \nabla \cdot \left(\frac{1}{f} \nabla M \right). \tag{20}$$

The meridional differencing scheme used here to obtain the geostrophic vorticity from the Montgomery potential is

$$\zeta_g \cong \frac{1}{a^2 \cos \varphi_j} \frac{1}{\Delta \varphi} \left\{ \left[\frac{\cos \varphi_{j+1/2}}{f(\varphi_{j+1/2})} \left(\frac{M_{j+1} - M_j}{\Delta \varphi} \right) \right] \right.$$
$$\left. - \left[\frac{\cos \varphi_{j-1/2}}{f(\varphi_{j-1/2})} \left(\frac{M_j - M_{j-1}}{\Delta \varphi} \right) \right] \right\}. \tag{21}$$

The errors in the geostrophic vorticity due to vertical discretization are plainly evident in Fig. 8k.

Figure 9 shows the normalized error of the geostrophic vorticity, plotted as a function of the number of points in latitude. Results are plotted for 96, 64, and 32 layers, respectively. Here the normalized error of a variable x is defined by

$$\text{error} \equiv \frac{\sqrt{\sum_{j,k} \left(x_{j,k}^{\text{true}} - x_{j,k}^{\text{appx}} \right)^2}}{\sqrt{\sum_{j,k} \left(x_{j,k}^{\text{true}} \right)^2}}. \tag{22}$$

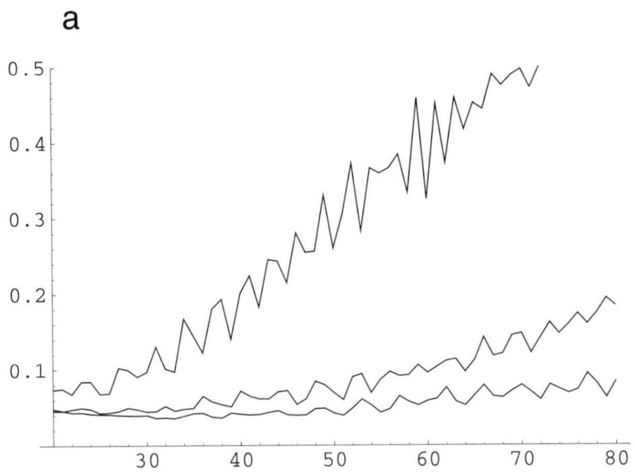

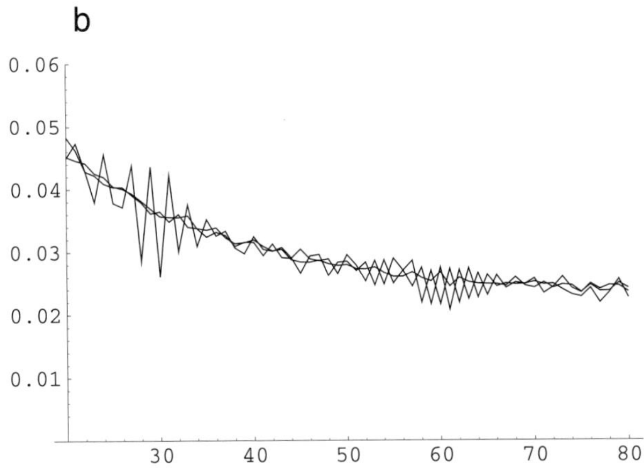

Figure 9 (a) The RMS errors of the geostrophic vorticity, plotted as a function of the number of points between the equator and pole, for three different vertical resolutions: 32, 64, and 96 layers. The RMS errors in $\zeta_g(\theta_l)$ are determined by comparing the results of numerical integration using Eq. (12) with analytic integration; these differences are then normalized with respect to the RMS of the analytically determined values of $\zeta_g(\theta_l)$, as shown in Eq. (22). As would be expected, the largest errors occur with 32 layers, and the smallest errors occur with 96 layers. Note, however, that the errors do not decrease monotonically as the meridional resolution is increased. (b) The corresponding results when $\zeta_g(\theta_l)$ is computed using the Montgomery potential as obtained by analytic integration of the hydrostatic equation. The errors in panel (b) come entirely from the horizontal discretization shown in Eq. (21).

Also shown in the figure are the errors when the Montgomery potential used in Eq. (21) is obtained through analytic integration of the hydrostatic equation, so that the errors arise entirely from the horizontal discretization given by Eq. (21). For the vertically discrete case (Fig. 9a), it appears that for a given vertical resolution, the normalized error is minimized for a particular horizontal resolution, and increases if the horizontal resolution is either greater or less than the optimal value. An explanation is given in the next subsection.

In summary, vertical discretization of the hydrostatic equation introduces numerical errors in the Montgomery potential that arise from the inability of the vertically discrete θ coordinate to accurately resolve the potential temperature of the air closest to the Earth's surface, i.e., θ_S. The errors can be reduced by increasing the vertical resolution for a given horizontal resolution, but even with very high vertical resolution (e.g., 96 layers) the errors are unacceptably large. For a given vertical resolution, the overall error can be reduced by increasing the horizontal resolution up to a point, but further increases in the horizontal resolution actually cause the error to increase substantially.

An interpretation of the source of the numerical errors is illustrated in Fig. 10. Here the vertical axis is the Exner function, and the horizontal axis is the potential temperature. For $\theta \leq \theta_S$, i.e., on the left side of the diagram, $\Pi = \Pi_S$. This is the realm of the massless layers. For $\theta > \theta_S$, we have $\Pi < \Pi_S$. Also indicated along the horizontal axis are $\theta_{L+1/2}$ and $\theta_{L-1/2}$, which are the layer-edge values of the potential temperature for layer L. By definition, layer L contains θ_S, the surface potential temperature. The figure shows that $\theta_{L+1/2} < \theta_S < \theta_{L-1/2}$; within the framework of the HA scheme, this is all that we know about θ_S, i.e., that it lies between certain bounds that are separated by a finite $(\delta\theta)_L \equiv \theta_{L+1/2} - \theta_{L-1/2}$. In a sense, the lowest massy layer is "partly filled," its edges are nominally defined by $\theta_{L+1/2}$ and $\theta_{L-1/2}$, but in fact no mass resides between $\theta_{L+1/2}$ and θ_S; the coldest values of θ are unoccupied.

First consider Fig. 10a, in which θ_S is cooler than θ_L. For this case, Eq. (12) can be made more accurate by replacing Π_S with a suitably chosen average value of the Exner function over the layer between the lower boundary and level L:

$$M_L = \phi_S + \theta_S \Pi_S + \frac{1}{2}(\Pi_S + \Pi_L)(\theta_L - \theta_S)$$

$$\text{for } \theta_S \leq \theta_L \quad \text{and} \quad \Pi_S \geq \Pi_L. \qquad (23)$$

Of course, we cannot use Eq. (23) to determine M_L unless θ_S and Π_L are known, and as already mentioned HA did not provide a way to determine

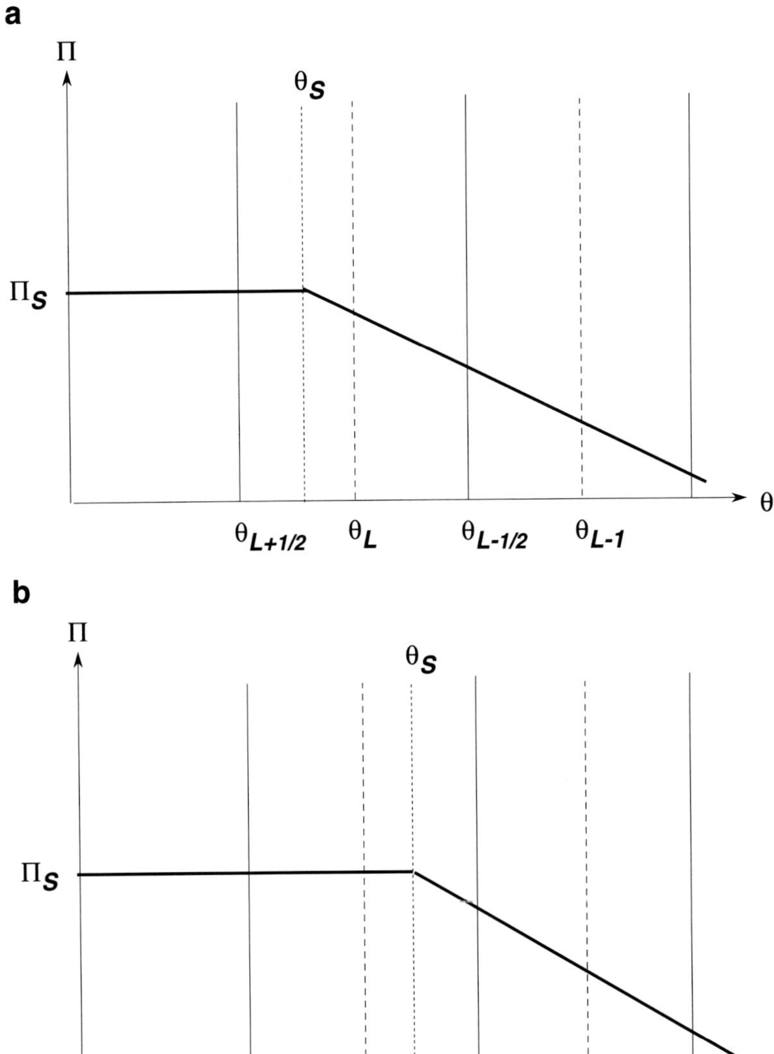

Figure 10 An interpretation of the cause of the numerical errors. In each panel, the vertical axis is Π, and the horizontal axis is θ. The two solid vertical lines in each panel denote layer edges, which are surfaces of constant θ, and the dashed line is a layer center, which is again a surface of constant θ, i.e., $\theta = \theta_L$. In each panel, the dotted lines denote $\theta = \theta_S$. In panel (a), θ_S is cooler than θ_L, while in panel (b) θ_S is warmer than θ_L.

θ_S. Therefore, another way to describe the problem is that θ_S is unknown, or rather is not known with sufficient accuracy since we do of course know that $\theta_{L+1/2} \leq \theta_S \leq \theta_{L-1/2}$. For this reason, we refer to the numerical errors illustrated in Figs. 8 and 9 as the "θ_S problem." For the situation illustrated in Fig. 10a, Eq. (14), when combined with Eq. (23), gives a perfectly satisfactory formula for M_{L-1}:

$$M_{L-1} = \phi_S + \theta_S \Pi_S + \frac{1}{2}(\Pi_S + \Pi_L)(\theta_L - \theta_S) + \Pi_{L-1/2}(\theta_L - \theta_{L-1})$$

$$\text{for } \theta_S \leq \theta_L \quad \text{and} \quad \Pi_S \geq \Pi_L. \qquad (24)$$

Now consider Fig. 10b, in which θ_S is warmer than θ_L. For this case, Eq. (12) gives the exact answer for M_L, which is actually on the Earth's surface so that $\Pi_S = \Pi_L$; we repeat it here for convenience:

$$M_L = \phi_S + \Pi_S \theta_L \quad \text{for } \theta_S \geq \theta_L \quad \text{and} \quad \Pi_S = \Pi_L. \qquad (25)$$

On the other hand, for the situation illustrated in Fig. 10b, Eq. (15) should be modified to give an accurate value of M_{L-1}:

$$M_{L-1} = \phi_S + \theta_S \Pi_S + \frac{1}{2}(\Pi_S + \Pi_{L-1})(\theta_{L-1} - \theta_S)$$

$$\text{for } \theta_S \geq \theta_L \quad \text{and} \quad \Pi_S = \Pi_L. \qquad (26)$$

Obviously, we need to know θ_S in order to use Eq. (26).

To confirm that the source of the numerical errors shown in Fig. 8 is in fact as explained above and as interpreted using Fig. 10, we repeat the computations used to generate Fig. 8, but this time we prescribe θ_S analytically using Eq. (19), and we use Eqs. (23)–(26) as described above. The results are shown in Fig. 11; the errors are greatly reduced, relative to those shown in Figs. 8 and 9. Note from Fig. 11d that with low vertical resolution, the errors in the geostrophic vorticity actually increase as the horizontal resolution is increased beyond a certain point. This shows that increasing the horizontal resolution with fixed vertical resolution can actually make the results worse. The vertical resolution essentially determines how many discrete θ surfaces can intersect the Earth's surface between the equator and pole. This number should be matched to the model's meridional resolution (A. Arakawa, personal communication, 1995). When the meridional resolution is fine, many θ surfaces must intersect the Earth's surface in order to define the meridionally resolved structure of the surface temperature. If not enough θ surfaces are provided, i.e., if the vertical resolution is inadequate, errors in the HPGF will

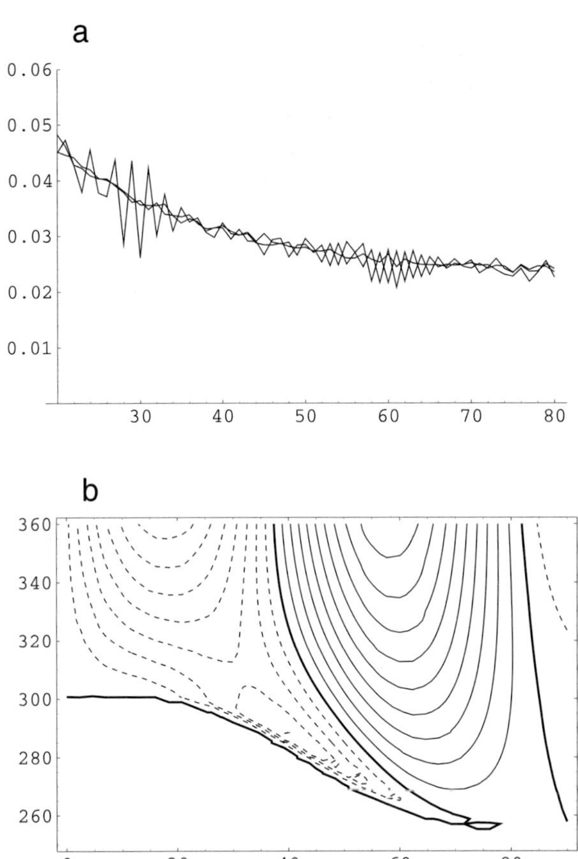

Figure 11 (a) Plot of the normalized error of the geostrophic vorticity [see Eqs. (22) and (20)], as obtained using the Montgomery potential derived from analytical integration of the hydrostatic equation, for three different vertical resolutions: 32, 64, and 96 layers. This panel is identical to Fig. 9b. (b) Geostrophic vorticity as obtained using the vertically discrete Eqs. (23)–(26) in place of Eqs. (12) and (15). These results can be compared with those shown in Figs. 8j and 8k.

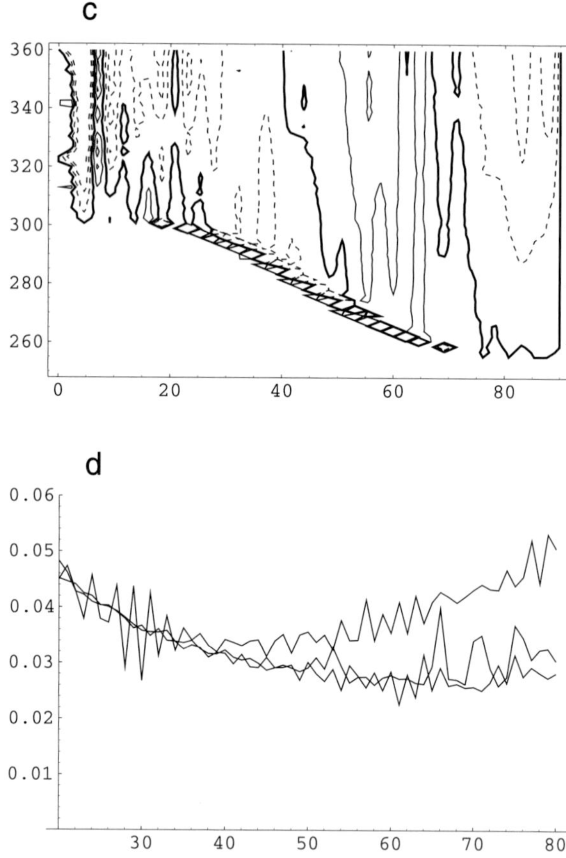

Figure 11 (c) Difference between the results plotted in panel (b) and those obtained with the analytically integrated hydrostatic equation. These results can be compared with those shown in Fig. 8l. (d) Normalized errors of the geostrophic vorticity when Eqs. (23)–(26) are used, for the same three vertical resolutions used in panel (a). These errors can be compared with those shown in panel a and also with those shown in Fig. 9a.

result and will become worse as the meridional resolution is further increased with fixed vertical resolution.

In summary, we have demonstrated that vertical discretization of the hydrostatic equation following the HA scheme introduces errors in the Montgomery potential, the HPGF, and the geostrophic vorticity, which decrease as the vertical resolution is increased, for a given horizontal resolution. The errors are minimized for a particular horizontal resolution, for a given vertical resolution.

Although we have shown that the θ_S problem can be mitigated by increasing the vertical resolution for a given horizontal resolution, the vertical resolution that would be required to reduce the problem to an acceptable level is excessive. A different approach is needed. We are currently exploring two such approaches, each of which involves a partial retreat from the pure isentropic coordinate.

The first possibility is to use a hybrid coordinate (HC), as proposed by various authors including Deaven (1976), Johnson and Uccellini (1983), Zapotocny *et al.* (1994), Zhu *et al.* (1992), and Konor and Arakawa (1997). Unlike schemes based on pure θ coordinates, HC schemes must include prognostic equations for the temperature, potential temperature, or a closely related state variable, in addition to prognostic equations for the vertical distribution of a density-like variable. As emphasized by Konor and Arakawa (1997), the design of HC schemes is strongly constrained by the requirement that the predicted distribution of mass in θ space be consistent with the predicted distribution of θ with pressure. Although the various HC schemes differ widely from each other, they all use something like a σ coordinate near the Earth's surface, which transitions either sharply or smoothly to a θ-coordinate aloft.

A second possibility, which we are currently investigating, is designed to permit use of the HA scheme with minimal modification. The approach is to maintain all layer edges (corresponding to the solid vertical lines in Fig. 10) and layer centers (corresponding to the dashed lines in Fig. 10) as surfaces of constant potential temperature, just as with a pure isentropic coordinates, but to predict $\bar{\theta}_l$, which is defined to be the *mass-weighted average* potential temperature for layer l, so that $\bar{\theta}_l$ can vary, just slightly, in time and space. Note the use of the overbar here to distinguish the predicted mass-weighted average potential temperature, $\bar{\theta}_l$, from the nominal "layer-center" value, θ_l (no overbar), which is independent of time and the horizontal coordinates. Further discussion will be given elsewhere.

IX. CONCLUSIONS

In this chapter, we have argued that geodesic grids, which were experimentally applied to the simulation of the global atmospheric circulation by Arakawa and colleagues in the 1960s, are now, 30 years later, emerging as the computational foundations of full-fledged atmospheric general circulation models. The vorticity and divergence equations are particularly well suited for use on geodesic grids. Use of the vorticity and divergence equations without horizontal staggering allows realistic simulation of geostrophic adjustment regardless of the ratio of the grid spacing to the radius of deformation. Multigrid solvers permit efficient inversion of the Laplacian as well as semi-implicit time differencing on geodesic grids. We believe that geodesic grids hold a great deal of promise for the future of atmospheric general circulation modeling.

We are now testing a full-physics geodesic general circulation model with the generalized σ coordinate. Preliminary results indicate that this approach can give realistic and computationally efficient simulations of the atmospheric general circulation.

For the future, the potential temperature coordinate of Hsu and Arakawa is very attractive, although the problems associated with the lower boundary condition still pose some challenges.

ACKNOWLEDGMENTS

The work reported here has been supported by the U.S. Department of Energy's CHAMMP and Climate Change Prediction Programs under grants DE-FG03-94ER61929 and DE-FG03-98ER62611, both to Colorado State University. Donald Dazlich assisted in the development of the full-physics geodesic model.

REFERENCES

Arakawa, A. (1966). Computational design for long-term numerical integration of equations of fluid motion: Two-dimensional incompressible flow. Part I. *J. Comput. Phys.* **1**, 119–143.

Arakawa, A., and V. R. Lamb (1981). A potential energy and enstrophy conserving scheme for the shallow water equations. *Mon. Wea. Rev.* **109**, 18–36.

Arakawa, A., and V. R. Lamb (1977). Computational design of the basic dynamical processes of the UCLA general circulation model. *In* "Methods in Computational Physics," Vol. 17, pp. 173–265. Academic Press, New York.

Augenbaum, J. M., and C. S. Peskin (1985). On the construction of the Voronoi mesh on a sphere. *J. Comput. Phys.* **14**, 177–192.

Baumgardner, J. R., and P. O. Frederickson (1985). Icosahedral discretization of the two-sphere. *SIAM Numer. Anal.* **22**, 1107–1115.

Browning, G. L., J. J. Hack, and P. N. Swarztrauber (1989). Comparison of three numerical methods for solving differential equations on the sphere. *Mon. Wea. Rev.* **117**, 1058–1075.

Deaven, D. G. (1976). A solution on boundary problems in isentropic coordinate models. *J. Atmos. Sci.* **33**, 1702–1713.

Dukowicz, J. K. (1984). Conservative rezoning (remapping) for general quadilateral meshes. *J. Comput. Phys.* **54**, 411–424.

Dukowicz, J. K., and J. W. Kodis (1987). Accurate conservative remapping (rezoning) for arbitrary Lagrangian–Eulerian computations. *SIAM J. Stat. Comp.* **3**, 305–321.

Fulton, S. R., P. E. Ciesielski, and W. H. Schubert (1986). Multigrid methods for elliptic problems: A review. *Mon. Wea. Rev.* **114**, 943–959.

Halem, M., and G. Russell (1973). A split-grid differencing scheme for the GISS model. *In* "Research Review, 1973, Part 2: Applications." Goddard Institute for Space Studies, NASA/GSFC.

Heikes, R. P., and D. A. Randall (1995a). Numerical integration of the shallow water equations on a twisted icosahedral grid. Part I: Basic design and results of tests. *Mon. Wea. Rev.* **123,** 1862–1880.

Heikes, R. P., and D. A. Randall (1995b). Numerical integration of the shallow water equations on a twisted icosahedral grid. Part II: Grid refinement, accuracy and computational performance. *Mon. Wea. Rev.* **123,** 1881–1887.

Held, I. M., and J. M. Suarez (1994). A proposal for the intercomparison of the dynamical cores of atmospheric general circulation models. *Bull. Am. Meteor. Soc.* **75,** 1825–1830.

Hsu, Y.-J. G., and A. Arakawa (1990). Numerical modeling of the atmosphere with an isentropic vertical coordinate. *Mon. Wea. Rev.* **118,** 1933–1959.

Jarraud, M., and A. J. Simmons (1983). The spectral technique. *In* "Seminar on Numerical Methods for Weather Prediction." pp. 1–59. European Centre for Medium Range Weather Prediction, Reading, UK.

Johnson, D. R., and L. W. Uccellini (1983). A comparison of methods for computing the sigma-coordinate pressure gradient force for flow over sloped terrain in a hybrid theta-sigma model. *Mon. Wea. Rev.* **111,** 870–886.

Konor, C. S., and A. Arakawa (1997). Design of an atmospheric model based on a generalized vertical coordinate. *Mon. Wea. Rev.* **125,** 1649–1673.

Lorenz, E. N. (1955). Available potential energy and the maintenance of the general circulation. *Tellus* **7,** 157–167.

Kurihara, Y. (1965). Numerical integration of the primitive equations on a spherical grid. *Mon. Wea. Rev.* **93,** 399–415.

Masuda, Y., and H. Ohnishi (1986). An integration scheme of the primitive equations model with an icosahedral-hexagonal grid system and its application to the shallow water equations. *In* "Short- and Medium-Range Numerical Weather Prediction" (T. Matsuno, ed.), pp. 317–326. Japan Meteorological Society, Tokyo.

McGregor, J. L. (1996). Semi-Lagrangian advection on conformal-cubic grids. *Mon. Wea. Rev.* **124,** 1311–1322.

Moorthi, S., and A. Arakawa (1985). Baroclinic instability with cumulus heating. *J. Atmos. Sci.* **42,** 2007–2031.

Nickovic, S. (1994). On the use of hexagonal grids for simulation of atmospheric processes. *Contrib. Atmos. Phys.* **67,** 103–107.

Phillips, N. A. (1957). A map projection system suitable for large-scale numerical weather prediction. *J. Meteor. Soc. Japan*, 75th Anniversary Volume, 262–267.

Popovic, J. M., S. Nickovic, and M. B. Gavrilov (1996). Frequency of quasi-geostrophic modes on hexagonal grids. *Meteor. Atmos. Phys.* **58,** 41–49, 1996.

Purser, R. J., and M. Rancic (1998). Smooth quasi-homogeneous gridding of the sphere. *Quart. J. Roy. Meteor. Soc.* **124,** 637–647.

Rancic, M., R. J. Purser, and F. Mesinger (1996). A global shallow water model using an expanded spherical cube: gnomonic versus conformal coordinates. *Quart. J. Roy. Meteor. Soc.* **122,** 959–981.

Randall, D. A. (1994). Geostrophic adjustment and the finite-difference shallow-water equations. *Mon. Wea. Rev.* **122,** 1371–1377.

Ringler, T. D., R. P. Heikes, and D. A. Randall (1998). Solving the primitive equations on a spherical geodesic grid: A technical report on a new class of dynamical cores, Atmospheric Science Paper No. 665. Colorado State University, Fort Collins.

Ringler, T. D., R. P. Heikes, and D. A. Randall (2000). Modeling the atmospheric general circulation using a spherical geodesic grid: A new class of dynamical cores. *Mon. Wea. Rev.* (in press).

Sadourny, R. (1972). Conservative finite-differencing approximations of the primitive equations on quasi-uniform spherical grids. *Mon. Wea. Rev.* **100,** 439–445.

Sadourny, R., and P. Morel (1969). A finite-difference approximation of the primitive equations for a hexagonal grid on a plane. *Mon. Wea. Rev.* **97,** 439–445.

Sadourny, R., A. Arakawa, and Y. Mintz (1968). Integration of the nondivergent barotropic equation with an icosahedral-hexagonal grid for the sphere. *Mon. Wea. Rev.* **96,** 351–356.

Suarez, M. J., A. Arakawa, and D. A. Randall (1983). Parameterization of the planetary boundary layer in the UCLA general circulation model: Formulation and results. *Mon. Wea. Rev.* **111,** 2224–2243.

Thacker, W. C. (1978). Comparison of finite-element and finite-difference schemes. Part II: Two-dimensional gravity-wave motion. *J. Phys. Oceanogr.* **8,** 680–689.

Thuburn, J. (1997). A PV-based shallow-water model on a hexagonal-icosahedral grid. *Mon. Wea. Rev.* **125,** 2328–2347.

Williamson, D. L. (1968). Integration of the barotropic vorticity equation on a spherical geodesic grid. *Tellus* **20,** 642–653.

Williamson, D. L. (1970). Integration of the primitive barotropic model over a spherical geodesic grid. *Tellus* **20,** 642–653.

Williamson, D. L., and P. J. Rasch (1994). Water vapor transport in the NCAR CCM2. *Tellus* **46A,** 34–51.

Williamson, D. L., J. B. Drake, J. J. Hack, R. Jakob, and P. N. Swarztrauber (1992). A standard test set for numerical approximations to the shallow-water equations in spherical geometry. *J. Comput. Phys.* **102,** 221–224.

Winninghoff, F. J. (1968). On the adjustment toward a geostrophic balance in a simple primitive equation model with application to the problems of initialization and objective analysis, Ph.D. Thesis. University of California, Los Angeles.

Zalesak, S. T. (1979). Fully multidimensional flux-corrected transport algorithms for fluids. *J. Comp. Phys.* **31,** 335–362.

Zapotocny, T. H., D. R. Johnson, and F. M. Reames (1994). Development and initial test of the University of Wisconsin global isentropic-sigma model. *Mon. Wea. Rev.* **122,** 2160–2178.

Zhu, Z., J. Thuburn, B. J. Hoskins, and P. H. Haynes (1992). A vertical finite-difference scheme based on a hybrid $\sigma\text{-}\theta\text{-}p$ coordinate. *Mon. Wea. Rev.* **120,** 851–862.

Chapter 18

A Coupled GCM Pilgrimage: From Climate Catastrophe to ENSO Simulations

Carlos R. Mechoso, Jin-Yi Yu, and Akio Arakawa
Department of Atmospheric Sciences
University of California, Los Angeles, California

I. INTRODUCTION

In his history of numerical modeling of the atmosphere, Akio Arakawa (see Chapter 1) recognizes a prelude and three phases. The most recent of these is the "Great Challenge" third phase, whose flag-bearer is the coupled atmosphere–ocean general circulation model (CGCM). Many of the difficulties found in modeling the atmosphere and ocean are further heightened in the coupled context, because interactions among processes in the components can lead to strong internal feedbacks. The current paper reviews our work in the early stages of the Great Challenge third phase, and presents the most important lessons learned along the road.

The first "C" in the acronym CGCM has both scientific and technical connotations. The former implies the intent to include the interactions

General Circulation Model Development

between processes represented in the atmospheric component (AGCM) and processes represented in the oceanic component (OGCM). The latter implies that the model codes have been traditionally built by linking independently developed AGCMs and OGCMs through an additional "coupling" component charged with ensuring that exchanges of information are made between the right locations at the right times. Here we are primarily concerned with the scientific connotations and only touch briefly on the technical ones.

This presentation is divided into seven sections. The first four after the introduction are organized as journeys in a "pilgrimage." Section II narrates the transition from the climate catastrophe obtained in our very first CGCM runs to a simulation that was realistic, but that suffered from a cold sea surface temperature (SST) bias and weak interannual variability at the equator. Section III presents our analyses of the latter results, our examination of the systematic errors of the CGCM, and the model revisions made for improvement. The influence of marine stratocumulus on the tropical climate is discussed in this section. Section IV analyzes a simulation with the revised model that succeeds in producing a realistic interannual variability at the equator. The analyses of results emphasize the structure of the simulated ENSO mode in several physical quantities such as SST, surface wind stress, and mean temperature in the upper ocean. Section V discusses the systematic errors that still linger in the CGCM, albeit the great improvements from previous versions, and presents our current strategy for research on assessing the importance of those errors and alleviating their impact. This section includes a short technical discussion on code optimization and parallelization. Section VII is an overview of lessons learned. We close in Section VIII with a discussion of future directions. Appendix A includes technical details on the data sets used for verification. Appendix B is a detour of our journey, in which a recent version of the CGCM was used to perform experimental predictions of the 1997–1998 ENSO event.

II. FIRST JOURNEY: FROM CATASTROPHE TO COLD BIAS AND WEAK INTERANNUAL VARIABILITY AT THE EQUATOR

A. MODEL DESCRIPTION

The UCLA atmospheric general circulation model is a state-of-the-art finite-difference model of the global atmosphere. The model predicts the horizontal wind, potential temperature, water vapor mixing ratio, planetary boundary layer (PBL) depth and surface pressure, and ground temperature and snow depth over land. The horizontal finite differencing of the primitive equations is based on the scheme designed by Takano and Wurtele (1982), which is the fourth-order version of the potential enstrophy and energy-conserving scheme for the shallow-water equations presented in Arakawa and Lamb (1981). The differencing of the thermodynamic energy and water vapor advection equations is also based on a fourth-order scheme.

The vertical coordinate used in the model is the modified σ coordinate of Suarez *et al.* (1983), in which the lowest model layer is the PBL. The vertical finite differencing is performed on a Lorenz-type grid following Arakawa and Lamb (1977) above 100 mb and Arakawa and Suarez (1983) below. This differencing is of second-order accuracy and designed to conserve the global mass integrals of potential temperature and total energy for adiabatic, frictionless flows. For the time integration of the momentum, thermodynamic energy, and water vapor advection equations, a leapfrog time-differencing scheme is used with a Matsuno step inserted periodically. To avoid the use of the extremely short time step required by the CFL condition near the poles, longitudinal averaging (which takes the form of a Fourier filter) is performed on selected terms in the prognostic equations to increase the effective longitudinal grid size. A more localized spatial filter is applied to the predicted PBL depths (Suarez *et al.*, 1983). In layers where an unstable stratification develops (potential temperature decreasing with height), a parameterization of sub-grid-scale dry convection results in a complete mixing of the horizontal momentum, potential temperature, and water vapor mixing ratio in the layers involved.

All work presented in this chapter was performed with the structure of the AGCM dynamics described in the preceding paragraphs of this subsection. This was not the case for the AGCM physics, which comprises the parameterization of physical processes. The following paragraphs describe this component as it was at the starting point of our journey.

Parameterization of PBL processes follows the mixed-layer approach of Suarez *et al.* (1983), which was pioneered by UCLA in the AGCM. In this parameterization, surface fluxes are calculated according to the bulk formula proposed by Deardorff (1972). Parameterization of cumulus convection, including its interaction with the PBL, follows Arakawa and Schubert (1974, hereafter A-S). The AGCM uses a finite time scale for adjustment of the cloud work function (Cheng and Arakawa, 1994). Compared to the case of instantaneous adjustment in the original A-S scheme, the cumulus parameterization produces approximately the same mass flux and precipitation, but a higher cloud incidence and larger time-averaged cumulus cloud amount. We have found that simulation of the outgoing long-wave radiation at the top of the atmosphere is significantly improved with the relaxed adjustment.

One could choose between two parameterizations of long-wave radiative transfer. One choice is the scheme developed by Katayama (1972) and modified by Schlesinger (1976). The other choice is the scheme developed by Harshvardhan *et al.* (1987, 1989). The two schemes consider the absorption by water vapor, carbon dioxide, and ozone. Harshvardhan's also includes the effects of the water vapor continuum, which is considered to play an important role in humid regions such as the tropics. The schemes also differ in the treatment of cloud radiative effects. The UCLA AGCM includes two types of clouds: (1) cumulus clouds associated with sub-grid-scale convection and (2) layer clouds associated with grid-scale supersaturation, which include stratocumulus associated with supersaturation in a sublayer within the PBL (Suarez *et al.*, 1983). Both schemes incorporate the effects of these two types of clouds in the radiation calculation except for cumulus clouds below 400 mb. In Katayama's scheme, clouds below 187.3 mb with temperatures warmer than 233 K are considered to be composed of water droplets and have an emissivity of 1. Clouds above 187.3 mb or colder than 233 K are considered to be composed of ice crystals and have an emissivity of 0.5. In Harshvardhan's scheme, cloud emissivities depend on pressure thickness and temperature in a functional form that has different coefficients for cumulus and layer clouds. In most cases, the emissivities of cumulus clouds are smaller with the former scheme than with the latter, while the emissivities of layer clouds are larger with the former scheme than with the latter, particularly in the upper troposphere. For the parameterization of short-wave radiative transfer, the AGCM included a scheme also developed by Katayama (1972).

The current oceanic component of the CGCM is the GFDL Modular Ocean Model (MOM) developed by Pacanowski, Dixon, and Rosati at GFDL. It is the successor to the code written by Cox (1984) based on Bryan (1969). MOM's code incorporates as options the parameterizations of vertical mixing by Pacanowski and Philander (1981) and by Mellor and Yamada (1982), as implemented by Rosati and Miyakoda (1988). In the former parameterization, the vertical diffusivity and viscosity are functions of the local Richardson number. In the upper 10 m of the model, the coefficient of vertical eddy viscosity has a minimum value to compensate for mixing by the high-frequency wind fluctuations that are absent from the monthly mean winds. The Mellor–Yamada (1982) level 2.5 closure scheme is used to model the turbulent transports of heat, salt, and momentum in terms of the larger scale fields. All results presented in this chapter were obtained with this option for turbulence mixing scheme in the OGCM.

The early versions of the CGCM included two additional features. First, the transfer coefficients for heat and moisture at the ocean surface were computed using a wind speed of 5 m s^{-1} whenever the PBL wind dropped below this level. This modification proved, however, to have a negligible impact on the simulated fields and was readily abandoned. The other feature, which is still part of the CGCM, is a parameterization of solar radiation penetration into the ocean according to the exponential formulation of Paulson and Simpson (1977).

Concerning the model resolutions, the earlier versions of the AGCM that were coupled to an OGCM had 9 layers in the vertical (with the top at 50 mb) and a horizontal resolution of 5° longitude by 4° latitude. The ocean model domain has been from 30°S to 50°N, and from 130°E to 70°W. There are 27 vertical layers, and the depth is constant at 4150 m. The longitudinal resolution is 1°; the latitudinal resolution varies gradually from 1/3° between 10°S and 10°N to almost 3° at 30°S and 50°N. In terms of the coupling, the surface wind stress and heat flux are calculated hourly by the AGCM, and their daily averages are passed to the OGCM. The SST is calculated hourly by the OGCM, and its value passed to the AGCM.

B. The Climate Catastrophe

Figure 1a presents the SST distribution in January from an experiment that started 1 year earlier with Katayama's long-wave radiation scheme (see Ma *et al.*, 1994, for details on model initialization). There are unrealistically warm SSTs over the entire tropics with a pool of water warmer than 30°C spreading over the central and eastern Pacific, and the equatorial cold tongue is conspicuously absent. The model seems to lock into this state, which is not a transient adjustment to coupling. The time evolution of SST at the equator in this experiment clearly suffers from severe climate drift or "climate catastrophe." Almost from the start, warm water surges eastward and eventually covers most of the equatorial Pacific Ocean. Consistently, regions of strong convective activity extend from the western to the central Pacific, where surface easterlies decrease to be gradually replaced by surface westerlies.

C. Overcoming the Catastrophe

The first modification made to the AGCM physics was the replacement of Katayama's long-wave radiation scheme with Harshvardhan's scheme. In contrast to Fig. 1a the results with this scheme in Fig. 1b are much more realistic, albeit systematically colder than in the observation. The tempera-

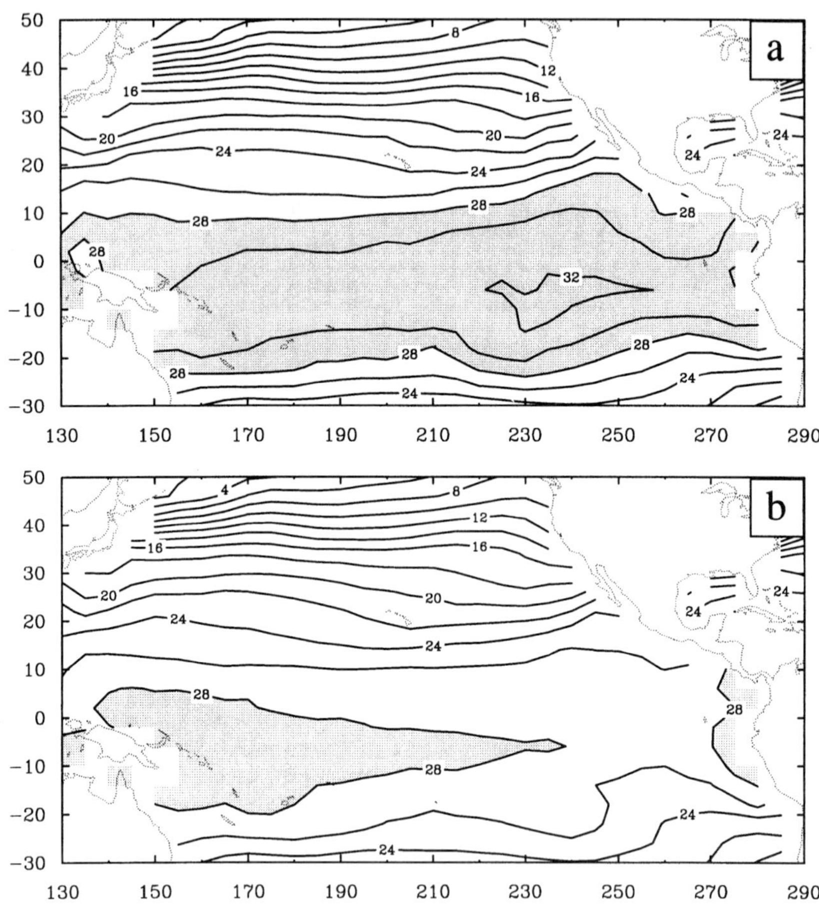

Figure 1   January-mean SST simulated using the (a) Katayama and (b) Harshvardhan long-wave radiation schemes. Values shown correspond to results 1 year after the integrations were started from identical initial conditions. Contour interval is 2°C. Values greater than 28°C are shaded.

ture gradient along the equator is maintained throughout the year and has realistic values. We will return to this point in Section III.A of this chapter. The experiment also simulates a realistic seasonal evolution of SST. The strengthening of the cold tongue starts in late July, which is about 2 months earlier than in the observed climatology. The wind stress at the ocean surface is reasonably well simulated.

Analyses based on the *uncoupled* AGCM with the same change of radiation schemes indicate that the mean SST change in the CGCM results from interactions between multiple processes in the atmosphere (Ma *et al.*, 1994). Since Harshvardhan's scheme assigns smaller emissivity to high clouds, the middle troposphere tends to become colder due to reduced greenhouse effect. This tends to destabilize the lower troposphere when SST is fixed, and enhance the overall cumulus activity, which in turn tends to dry the lower free atmosphere through cumulus-induced subsidence. It also tends to intensify the tropical direct circulation and, therefore, increase the surface wind speed in the tropics. The surface evaporation tends to increase and, when the AGCM is coupled with an OGCM, tropical SSTs become colder due to these two effects combined. The processes involved in this sensitivity are schematically shown by Arakawa in Chapter 23. On the other hand, the different parameterizations of vertical mixing in the OGCM did not help to overcome the climate catastrophe obtained when the Katayama scheme is used in the AGCM.

D. Interannual Variability

Robertson et al. (1995a,b) examined in detail the performance of the CGCM in multiyear simulations. The analyses included application of multichannel singular spectrum analysis (M-SSA) to time series of simulated fields at the equator to extract near periodicities and their associated spatiotemporal structures. (Also see Chapter 10, by Ghil and Robertson, which includes references on the M-SSA technique.) Robertson *et al.* (1995b) analyzed two simulations that differed in the inclusion or exclusion of the enhancements in transfer coefficient at low surface wind speeds and penetration of solar radiation into the upper ocean. They were able to isolate low-frequency modes with periods of either about 27 months (quasi-biennial or QB mode) or 48 months (quasi-quadrennial or QQ mode). The associated structures are characterized by predominantly standing oscillations in SST. These oscillations peak in the eastern Pacific, are accompanied by almost simultaneous zonal wind stress anomalies that peak to the west of the SST anomaly, and are preceded by thermocline anomalies that peak near the eastern edge of the basin.

The simulated modes of oscillation have three other important characteristics. First, the amplitude of the oscillations decays strongly during the first simulated decade to stabilize at a rather small value (0.25°C). Second, the ratio between the simulated magnitude of stress to SST anomalies reaches about 0.1 dyn cm^2 K^{-1}, which is about only half the magnitude in the observation. Third, the QB and QQ modes were not found to coexist in simulations with the same model version.

The characteristics of the QQ and QB mode pose many questions. One possible interpretation of the results is that largest amplitudes in the first years of simulation result from imbalances in initial conditions that are closer to the observation than to the model climatology. In this context, the weak amplitude of the modes in long simulations suggests that atmosphere–ocean interactions are also weak as expected in view of the cold SST bias and less favorable conditions for the development of feedbacks involving convection. The results are in agreement with the notion that the model climatology, which has substantial differences in the simulations, can have a large impact on the simulated interannual variability. We will come back to this point in Section IV.B of this chapter. The statement on the non-coexistence of the QQ and QB modes must be strongly qualified by the relatively short length of integrations.

III. SECOND JOURNEY: MODEL ANALYSES AND REVISIONS

A. Systematic Errors of CGCMS

There is no question that CGCMs face great challenges in attempting to model the tropical Pacific climate. The mean and annual variations of the climate in this part of the world are characterized by strong spatial and temporal asymmetries, which cannot be explained on the basis of variations in the configuration of the Earth–Sun system. The highest SSTs are not along the equator, but approximately at the locations of the Inter Tropical Convergence Zone (ITCZ) ($5°$–$10°$N). The oceanic seasonal march near the equator does not follow any simple model for the oceanic response to atmospheric forcing. For example, the maximum SST in March (to the west) *precedes* the onset of the period of eastward flow (to the east). In addition, north of the equatorial cold tongue the air accelerates as it moves over warmer waters and stratocumulus clouds form in the PBL.

The analysis of the seasonal cycle is a necessary first step in validating a CGCM, even from the point of view of interannual variability. Mechoso *et al.* (1995) examined the seasonal cycle over the tropical Pacific simulated by 11 coupled GCMs. Each of these models consists of a high-resolution ocean GCM of the tropical Pacific or global oceans coupled to a moderate or high-resolution atmospheric GCM without the use of flux correction. The results of this comparative study show that current state-of-the-art models share important successes and troublesome systematic errors in the simulation of the tropical Pacific climate:

1. Contemporary CGCMs succeed in simulating the zonal gradient in annual-mean SST at the equator over the central Pacific. The simulated equatorial cold tongue tends generally to be too strong, too narrow, and extend too far west. The mean SST is most poorly simulated in the coastal regions of the equatorial Pacific. East of 90°W simulated SST can be overestimated by up to 3°C. The zonal gradients of mean SST are much too large east of 110°W in most models, because the corresponding values go from being underestimated in the west to overestimated in the east. The mean SST along the equator is better simulated toward Indonesia than off the coast of Peru, but is generally warmer than that observed west of 150°W and has a zonal variation that can be somewhat erratic. The annual variation in zonal asymmetry of equatorial SST across the basin is believed to play a crucial role in determining the period and phase of ENSO (Meehl, 1990; Jin *et al.*, 1994).

2. No contemporary CGCM captures the relative amplitudes and phases of the two leading harmonics of the seasonal cycle in SST for the equatorial Pacific sufficiently well to reproduce the observed asymmetry in amplitude in the seasonal cycle that peaks in April. The timing of the warm phase of the cycle in April is generally more successfully captured than that of the cold phase, which arrives too early and decays too quickly. The transition to the cold phase is clearly much more rapid in the models in general, rather than in reality. The observed tendency for the seasonal cycle in SST to propagate westward is captured in many of the models, although—interestingly—there appears to be little correspondence between this success and a model's ability to simulate other aspects of the cold tongue.

3. Concerning the simulations of precipitation in the tropical Pacific, CGCMs fall roughly into two categories: (1) simulations that produce a rather persistent double ITCZ and (2) simulations in which the ITCZ shows a pronounced north–south seasonal migration. Neither of these behaviors is realistic, although the OLR data do show a limited meridional migration, and there is a hint of a double ITCZ in March–April (Waliser and Gauthier, 1993).

4. One of the most apparent systematic errors of contemporary CGCMs is that relatively warm water extends too far east near 10°S in every simulation. This feature, together with the tendency for the cold tongue to extend too far west along the equator in most models, makes simulated meridional gradients too large, while reducing zonal gradients. The overestimation of SSTs over the eastern Pacific south of the equator is generally associated with overestimated solar insolation.

5. Models that produce a realistic simulation of the seasonal cycle for the coupled atmosphere–ocean system seem to produce unrealistically

weak interannual variability (see Neelin *et al.*, 1992). The most recent results of CGCMs do not contradict the intriguing anticorrelation between the success in simulating the seasonal cycle and interannual variability of the coupled system. The extent to which such an anticorrelation has general validity, however, is unclear. For example, the existence of a cold tongue in the simulated SST field is rightly evaluated as a success of the simulation. Placing this success in its proper context is not easy since it is debatable whether the mechanisms responsible for the structure and evolution of the cold tongue are properly simulated.

To gain insight into the reasons for the model systematic errors, Yu and Mechoso (1999a) examine the errors in surface heat flux in the Tropical Pacific simulated by the AGCM component of the CGCM (Figs. 2a and 2b) and in SST simulated by the CGCM itself (Fig. 2c). Off the equator, the AGCM overestimates the heat flux out of the ocean in the northern and southern subtropics and along the coasts of Mexico and Central America. These errors are primarily due to excessive evaporation. The model also overestimates the heat flux into the ocean along the coasts of California and South America. These errors are primarily due to excessive downward short-wave radiation. The CGCM produces much smaller surface heat flux errors in those regions. However, the CGCM simulates cold and warm SST biases where the AGCM overestimates the heat fluxes from and into the ocean, respectively. The surface heat flux errors produced by AGCM deficiencies, therefore, tend to be compensated in the CGCM by SST errors. This compensation relationship suggests that interactions between surface heat flux and SST on monthly time scales off the equator are local.

Over the western equatorial Pacific, the AGCM produces realistic values of surface heat flux. The CGCM, on the other hand, produces a cold SST bias. This error is a manifestation of a spuriously elongated cold tongue. Since this phenomenon is primarily due to coupled atmosphere–ocean dynamical processes, there is not a straightforward connection between surface heat flux and SST errors. Over the eastern equatorial Pacific, ocean dynamics plays a key role in SST variations, and compensation between CGCM errors in SST and AGCM errors in surface heat flux is not found. Over the cold tongue, in particular, the deviations from the annual mean (hereafter "annual variations") of latent heat flux from the ocean surface are realistically simulated by the AGCM (not shown) but not by the CGCM (see Fig. 3b). Further analyses show that the annual variations in latent heat flux in both the observation and AGCM simulation are approximately in phase with those in surface or PBL wind speed. The variations in latent heat flux in the CGCM simulation, on the other

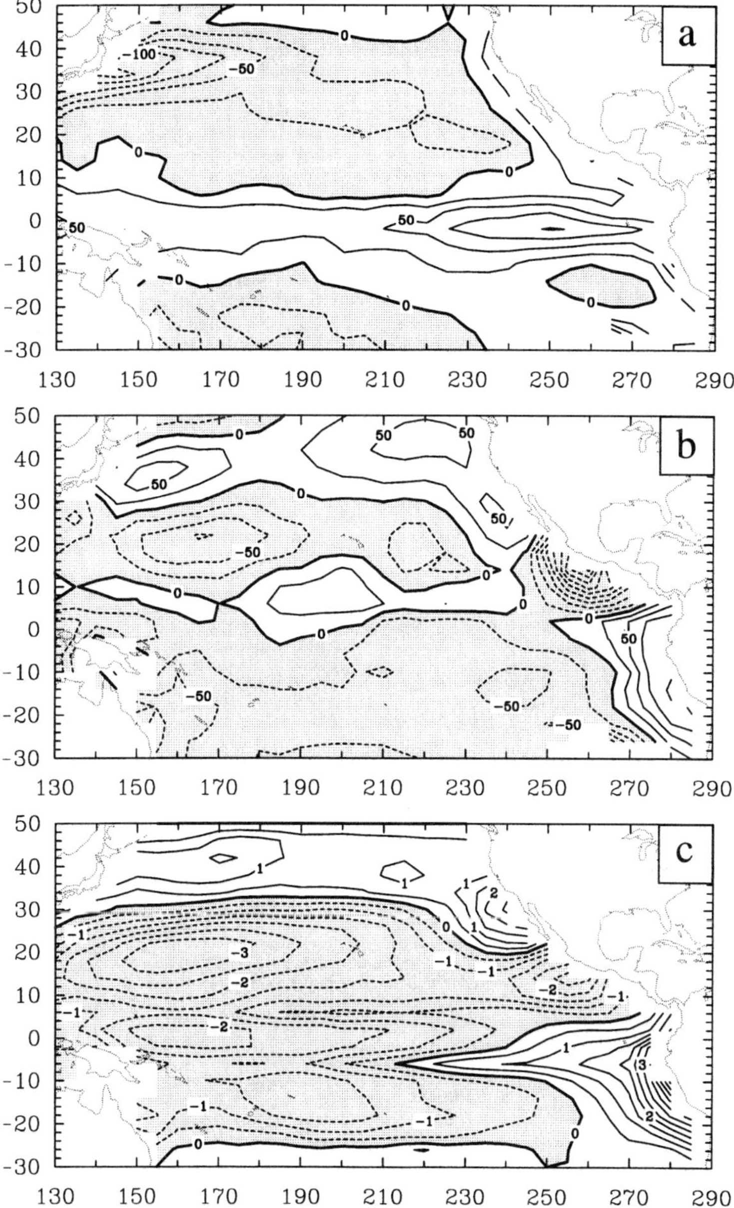

Figure 2 Annual-mean (a) surface heat flux from Oberhuber (1988), (b) errors in surface heat flux simulated by the AGCM, and (c) errors in SST simulated by the CGCM. Contour intervals are 25 W m^{-2} for (a) and (b), and 0.5°C for (c). Negative values are shaded. (Courtesy *Am. Meteor. Soc.*)

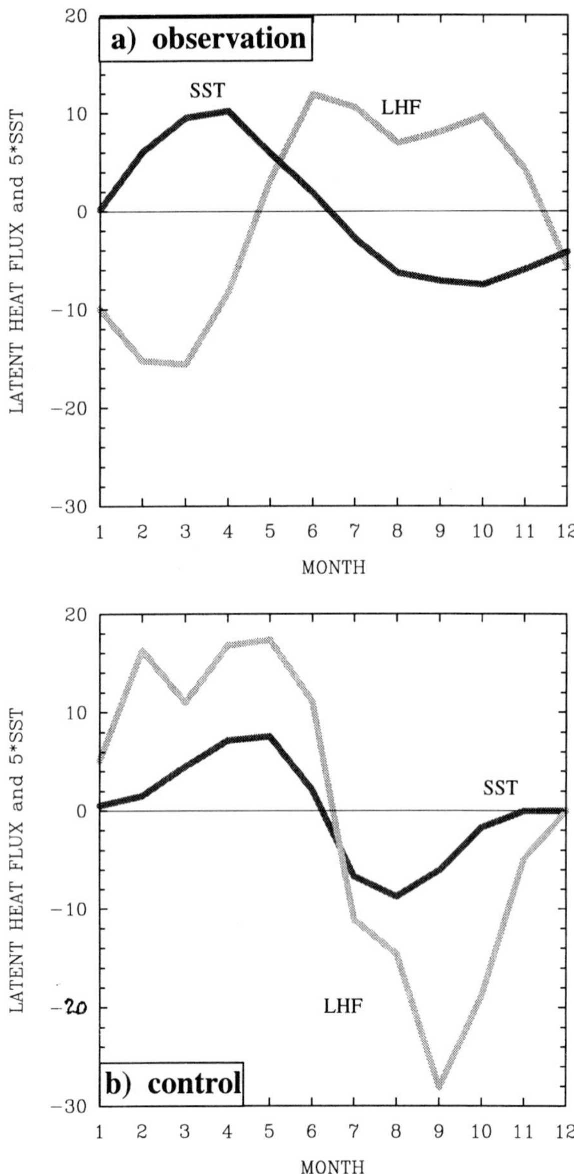

Figure 3 Annual variations of latent heat flux from the ocean surface (LHF) and SST averaged over the equatorial cold tongue (4°S–4°N; 120°–90°W) from (a) Oberhuber (1988) and (b) the CGCM. Units are W m^{-2} for HF and °C for SST, whose values are multiplied by 5. (Courtesy *Am. Meteor. Soc.*)

hand, are in phase with those in the humidity difference between ocean surface and overlying air. In addition, the annual variations in near-surface wind speed are too weak as part of the problem of unsuccessfully simulating a single ITCZ. This erroneous feature of the CGCM is associated with a spurious northerly wind simulated in the eastern equatorial Pacific during the first half of the calendar year when the zonal wind weakens.

The fairly realistic simulation of the annual variations in SST in the equatorial cold tongue by the CGCM shown in Fig. 3b agrees with the notion that annual variations of latent heat flux are not fundamental for the phenomenon in question. Mitchell and Wallace (1992) and Wang (1994) argue that annual variations in meridional wind stress at the equator play a key role in the annual variations of SST in the equatorial cold tongue, particularly in the initiation of its cold phase. A recent observational study (Nigam and Chao, 1996) supports this contention. The arguments of Mitchell and Wallace are based on the relationships between the strength of upwelling near the coast of equatorial South America and that of the meridional wind over the ocean. Over the equator in the eastern Pacific, meridional winds are always northward, but their magnitude is weaker than the annual mean in the earlier part of the year (January through May) with a minimum in March, and stronger in the latter part of the year (June through December) with a maximum in September. (The North American monsoon contributes to this evolution as surface convergence in the convective region over Central America is consistent with northerly winds over the eastern Pacific.) Accordingly, upwelling south of the equator is enhanced from September to next March and relaxed from March to September. In addition, the variations described are strongest along the northern fringe of the equatorial wave guide. This sets up a series of coupled-atmosphere ocean feedbacks that extends the cooling northward toward the equator.

The results presented in Yu and Mechoso (1999a) can also be interpreted as supportive of the links between North American monsoonal circulations and the initiation of the cold phase of the equatorial cold tongue. Even in CGCMs whose atmospheric components have a relatively low horizontal resolution, one would expect to resolve the major features in the seasonal evolution of tropical convection. In this context, the broad aspects of the annual variations of the equatorial cold tongue, particularly the existence and duration of its cold phase, may be relatively easy to simulate with a CGCM. In fact, 7 of the 11 models examined in Mechoso *et al.* (1995) show a realistic performance in this region. A comparable success for the warm phase is more elusive. Once again, these results are consistent with the notion that unrealistic features in the simulation of the eastern tropical Pacific climate by a CGCM may not be attributed to deficiencies in the representation of a single physical process, but are rather due to a combination of errors amplified by feedbacks of the coupled system.

B. Factors Contributing to Systematic Errors in the CGCM

Philander *et al.* (1996) point out that stratocumulus clouds along the coast of South America contribute significantly to the interhemispheric asymmetries of the climate in the eastern tropical Pacific, where they are a conspicuous feature. This is due to feedbacks, in which SST are cooled by the shielding effect of those clouds on solar radiation and the existence of those clouds is promoted by cooler SSTs. To investigate these ideas, Ma *et al.* (1996) performed a CGCM experiment in which the stratus cloud incidence was artificially enhanced to a constant 100% in a domain between extending from 10° to 30°S and from 90°W eastward to the Peruvian coast. Yu and Mechoso (1999b) extended the work of Ma *et al.* by considering the effect that annual variations in Peruvian stratocumulus can have on the climate of the eastern Pacific. Klein and Hartmann (1993) showed that these variations exist and are significant, with the area coverage of low-level clouds in that region varying from a minimum of about 40% in January–March to a maximum of about 70% in August–November. As an idealization of this behavior, Yu and Mechoso carried out CGCM experiments in which Peruvian stratocumulus clouds are artificially enhanced in either the first or the second half of the calendar year.

For the latter work, the AGCM was modified to have larger values of emissivity for layer clouds in the upper troposphere. In the uncoupled mode, this modification resulted in reduced surface evaporation and cumulus precipitation. The mechanism for this impact is opposite to that found when Katayama's long-wave radiation scheme was replaced by Harshvardhan's. When the AGCM is coupled to the OGCM, the corresponding SST field evidenced a smaller cold SST bias in the extratropics. In long runs the CGCM simulates a more realistic SST, and overall a more realistic evaporation.

The artificial increase in stratocumulus clouds along the Peruvian coast results in colder SSTs in that region due to the increased albedo. This cooling of SSTs extends northwestward toward the equator. All experiments with enhanced Peruvian stratocumulus produce strong asymmetry about the equator in association with a drastic reduction of SST in the Southern Hemisphere. They also produce stronger SST gradients along the equator in association with colder SSTs in the central to eastern sectors.

The intensity of cooling seems to depend more on the length than on the timing of the period in which stratocumulus are enhanced. An analysis of the results shows an intensification of the local Hadley circulation by the reduced absorption of insolation in the region of enhanced stratocumulus cloud incidence. The surface evaporation, which efficiently cools SSTs, is then enhanced at the northern boundary of the region due to the intensified advection of drier air by the lower branch of the Hadley circulation (Fig. 4). These experiments, therefore, demonstrate that Peruvian stratocumulus cloud decks are in fact crucial to the maintenance of the SST asymmetry about the equator in the eastern Pacific.

We next focus on the equatorial cold tongue. Figure 5 displays the annual variations of monthly-mean SSTs at the equator in the observation and enhanced stratocumulus experiments performed by Yu and Mechoso (1999b). (See Mechoso *et al.*, 1995, for the results obtained in the CGCM simulation.) In the observation (see Fig. 5a), there is a clear dominance of the semi-annual harmonic in the western sector and of the annual harmonic in the eastern sector. Here, values are positive from January through June and negative from July through December. These warm and cold phases of the cold tongue differ from each other in duration and speed of westward propagation. The cold phase lasts longer and propagates westward at a slower speed.

A comparison between the panels in Fig. 5 shows that the second-half-year experiment (see Fig. 5d) best captures the observed differences in duration and westward speed of propagation between the warm and cold phases of the cold tongue, including the onset, termination, and amplitude. The other experiments produce cold phases that are too short in duration or too weak in amplitude. The enhanced surface evaporation from May through December contributes to the longer duration of the cold phase of the cold tongue (see Fig. 6).

This success is primarily due to the more realistic portrayal of the meridional component of the PBL wind during the first and last months of the calendar year (not shown). In general, the second-half-year experiment produces stronger southerlies over the cold tongue during the second half of the year. In addition, the spurious northerlies that characterize the CGCM simulations during the first half of the year appear only in February and March with an amplitude of less than 1 m s^{-1}. The experiment also produces larger magnitudes of the zonal component of the wind from August through December than the CGCM simulation. The artificial enhancement of Peruvian stratocumulus in the second half of the year, therefore, contributes to produce more realistic PBL winds, particularly from the last few months of the year through the first couple of months of the following year. Consistently, the annual variations of surface

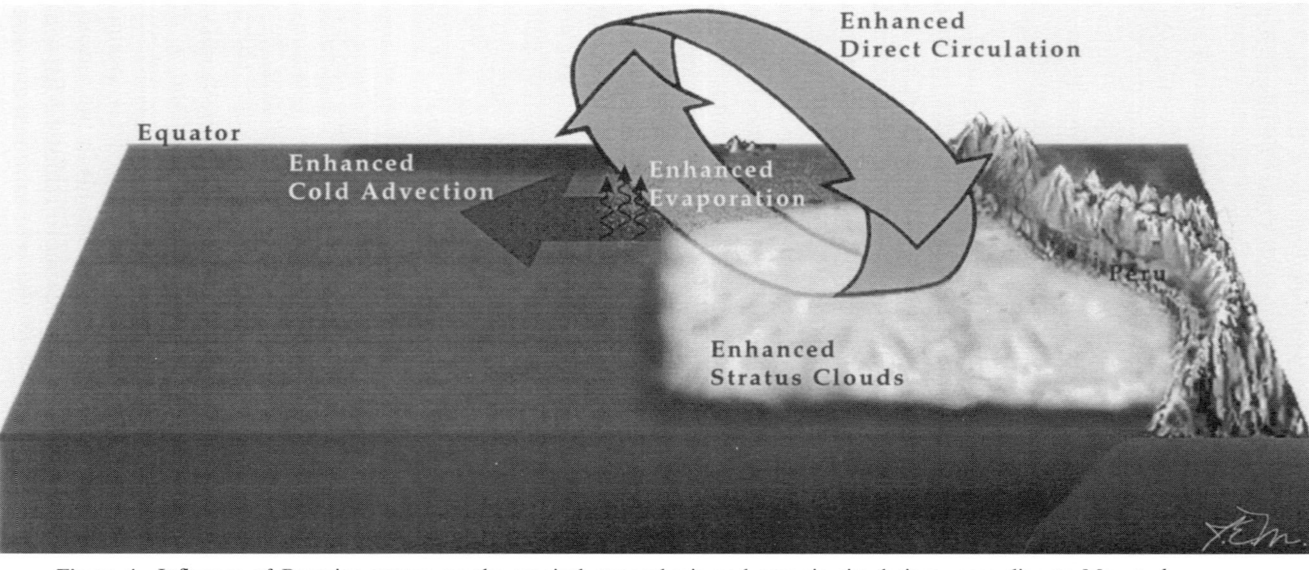

Figure 4 Influence of Peruvian stratus on the tropical atmospheric and oceanic circulations, according to Ma *et al.* (1996).

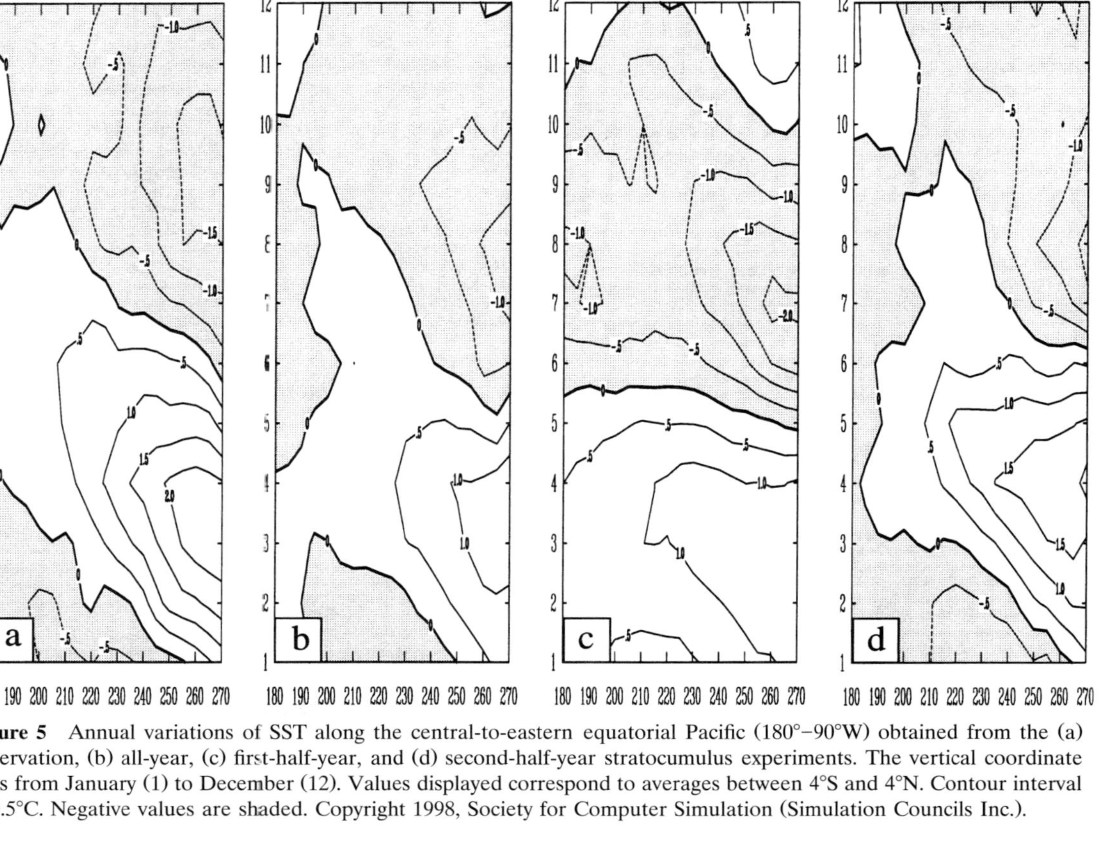

Figure 5 Annual variations of SST along the central-to-eastern equatorial Pacific (180°–90°W) obtained from the (a) observation, (b) all-year, (c) first-half-year, and (d) second-half-year stratocumulus experiments. The vertical coordinate runs from January (1) to December (12). Values displayed correspond to averages between 4°S and 4°N. Contour interval is 0.5°C. Negative values are shaded. Copyright 1998, Society for Computer Simulation (Simulation Councils Inc.).

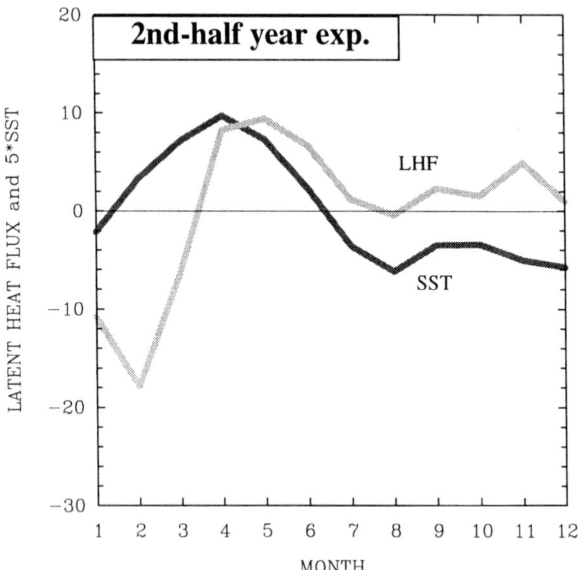

Figure 6 As in Fig. 3, except for the second-half-year stratocumlus experiment. (Courtesy *Am. Meteor. Soc.*)

evaporation over the cold tongue in the second-half-year experiment are very close to those in the observation, and much more realistic than in the CGCM simulation (contrast Figs. 3 and 4). A longer lasting cold phase of the cold tongue is thus obtained when Peruvian stratocumulus are enhanced in the second half of the year, and the temporal asymmetry in the annual variations of eastern equatorial Pacific SSTs is better captured by the model.

IV. THIRD JOURNEY: REALISTIC SIMULATION AT THE EQUATOR

A. MODEL IMPROVEMENTS

The most recent version of the AGCM includes the parameterization of gravity wave drag due to sub-grid-scale orography developed by Kim and Arakawa (1995). This scheme takes into account the drag enhancement due to the formation of nonlinear critical layers on the lee side of mountains. Based on statistical measures of the sub-grid-scale orography, given as inputs in addition to the mean and standard deviation of the surface height, the parameterization selectively enhances drag at the reference level and at levels immediately above when a nonlinear critical layer is expected to form. We have also extended the domain of the AGCM up to 1 mb with 15 layers in the vertical.

In view of the importance of the surface evaporation and stratocumulus cloud incidence in simulating SSTs, we dedicated an intense effort to revise the formulations of PBL moist processes, with an emphasis on the moisture exchange between the PBL and the lowest free-atmosphere layer. Whereas the PBL parameterization used in the AGCM explicitly predicts the height and depth of the stratocumulus sublayer in the PBL, the version of the AGCM used so far generally underpredicts the incidence of those clouds almost everywhere, particularly in the eastern subtropical Pacific of both hemispheres. This indicates that simulated PBL is generally too dry. Simulated surface evaporation, however, is not too small; in fact, it is too large compared to observation. These two concurrent deficiencies strongly suggest that the PBL in the AGCM is subject to too much drying effect, possibly due to the entrainment of drier air from above.

Even when the rate of *mass* entrainment to the PBL is known, however, computing the rate of *moisture* entrainment to the PBL is not straightforward due to vertical discretization. This is particularly true in models with relatively coarse vertical resolution since the rate of moisture entrainment to the PBL also depends on the specification of the water-vapor mixing ratio for the entraining air. In fact, Li and Arakawa (1997) find that the simulated stratocumulus cloud incidence is extremely sensitive to this specification. They demonstrate that the deficiencies of the AGCM mentioned above could be virtually eliminated by changing the specification to use the value extrapolated to the PBL top from above. In addition, enhancing the net entrainment of mass when the stratocumulus cloud layer is unstable helped to better simulate the geographical distribution of stratocumulus cloud incidence. Li and Arakawa (1997) show that this is greatly improved for a January simulation by the AGCM with these revisions incorporated.

B. Simulated Interannual Variability after Revisions

Figure 7 displays the annual-mean SST along the equator in the observation and the CGCM simulation before and after the PBL revisions. The zonal gradient in the observation is about 0.4°C per 10° longitude, and its sign alternates from positive in the west to negative in the central and again positive in the east. In the central region, the simulation before the revisions shows a realistic slope but a cold bias of about 2–3°C, which was the second strongest among the CGCMs analyzed by Mechoso *et al.* (1995). The success in capturing this slope is common to CGCMs, although

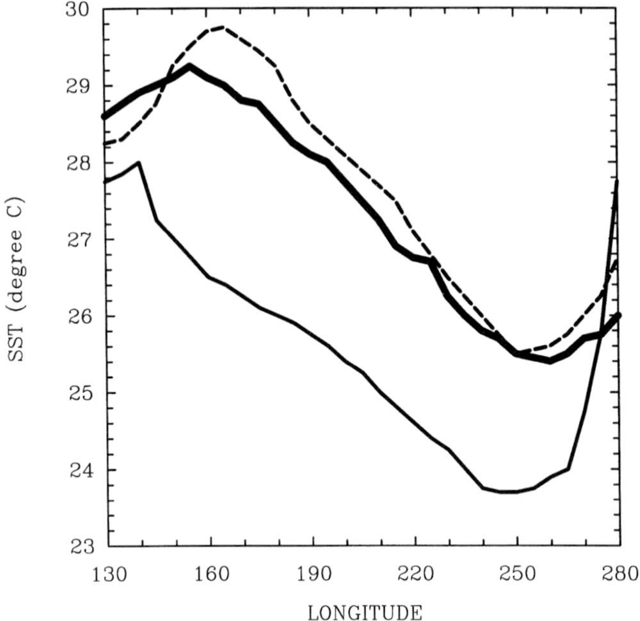

Figure 7   Annual-mean SST along the equatorial Pacific. The thick-solid line is the observed SST, thin-solid line is the SST produced by an earlier version of the UCLA CGCM, and thin-dashed line is the SST produced by the revised version of the CGCM. Values shown are averaged between 4°S and 4°N.

models generally underestimate the easterly surface wind stress at the equator. This may be another example of error compensation, whereby the effect on upwelling of a weaker wind is partially compensated by deficiencies in the mixing processes in the OGCM. The simulation after the revisions, on the other hand, is very realistic almost everywhere. One of the systematic errors of the CGCM has been alleviated, therefore, and the simulated cold tongue extends more realistically toward the South American coast. The annual variations of SST along the equator (not shown) also evidence improvements, particularly in the more realistic asymmetries between the warm and cold phases of the cold tongue.

The most spectacular improvement, however, is obtained in the simulated interannual variability at the equator. Figure 8 shows a longitude–time contour plot of anomalies in zonal wind stress, SST, and vertically averaged temperature in the upper 300 m of ocean (VAT) in the equatorial Pacific obtained in a multiyear integration with the revised version of the CGCM. The long-term monthly means (i.e., the annual cycles) are removed from all fields, and a low-pass filter is applied to remove the variability with time scales shorter than 1 year. The results in Fig. 8 correspond to the period from year 17 to year 38, which displays intense interannual variability. We can clearly see warm SST events approximately every 3–5 years (at years 18, 21, 25, 29, 33, 36). The largest SST anomalies are about 2°C. Figure 8 also shows large westerly wind stress anomalies to the west of the warm SST anomalies during the four major warming events. The magnitudes of the maximum wind stress anomalies are about 0.2 dyn cm^{-2}. VAT anomalies in the west lead those in the east, which peak near the coast ahead of SST anomalies. The frequency of simulated warm events and the relationships between simulated SST, wind stress, and VAT anomalies are very similar to those observed during El Niño events. The performance of the revised version of the CGCM is significantly improved in reference to the earlier version, in which interannual variability in equatorial SSTs has small magnitudes (~ 0.5°C).

The spatial structures and time evolution of SST anomalies for the warm event beginning in year 24 are displayed in Fig. 9. The warming of equatorial SST begins from the eastern sector in the earlier part of the year (Fig. 9a), and extends toward the central sector in the following months (Figs. 9b and 9c). A maximum in SST anomalies of about 2°C appears in October along the coast of South America. After this time, SST anomalies start to weaken in the eastern sector, but continue to develop in the central sector. Here, the warming continues until the following April and the event is over before the end of the second year. The spatial structure and time evolution of the simulated warm event are reasonably similar to those during an El Niño event, except that the model produces a weaker eastward propagation of the warm anomalies and tends to develop the strongest SST anomalies in the central equatorial Pacific.

To extract near periodicities in the SST averaged between 4°S and 5°N along the equatorial Pacific we apply the M-SSA method. The two leading modes produced by the method have principal components (PCs) and that are almost identical, except for a phase lag (not shown) and explain similar amounts of the SST variance (Fig. 10), which combined are more than 50% of the total. The similarity in PCs and explained variance between the two modes suggests that they represent an oscillatory phenomenon. The power spectra of the PCs peak sharply at about 48 months.

We next explore the possibility that the underlying dynamics of the simulated quasi-quadrennial oscillation involves ocean–atmosphere interactions. Our approach is based on applying the M-SSA method to combined variables in order to extract the multivariate structure associated with the oscillation. Specifically, we consider simultaneously the SST, zonal wind stress, and VAT fields between 20°S and 20°N. The data are preprocessed to amplify the signal-to-noise ratio. First, annual cycles are removed from the time series at each grid point. Second, the anomalies are low-pass

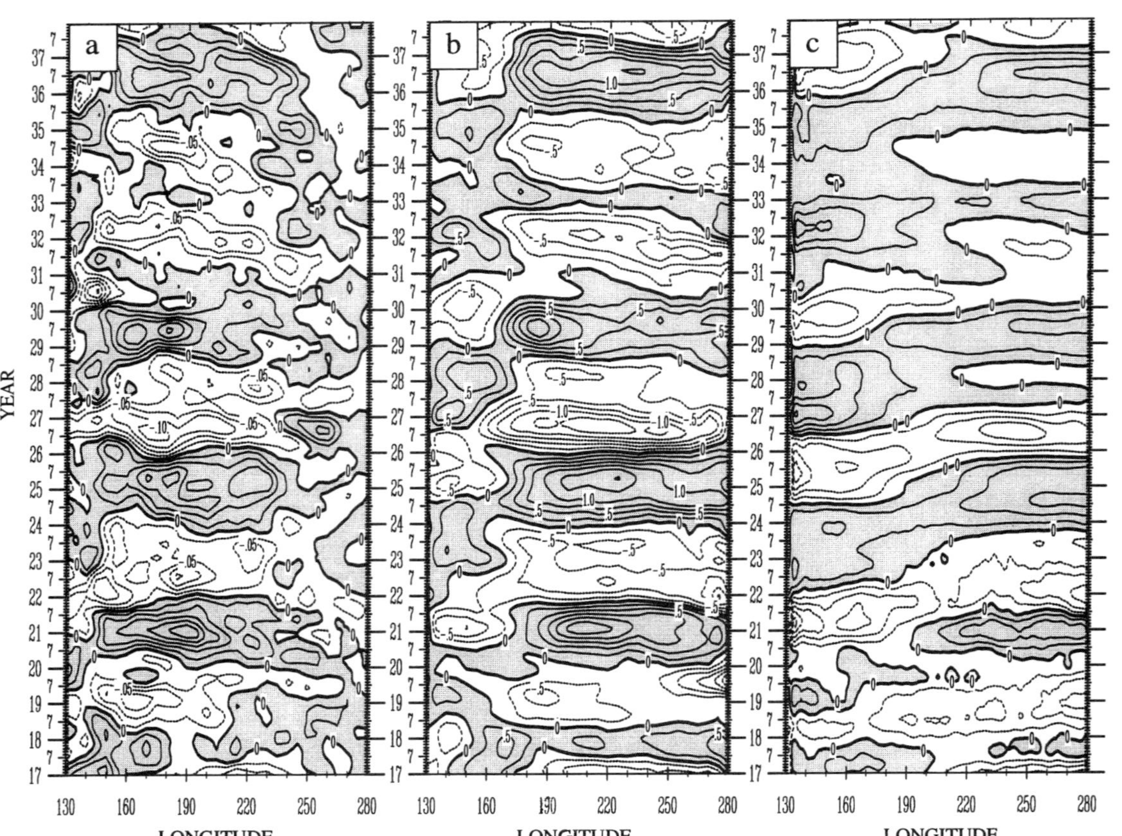

YEAR
a
b
c
LONGITUDE
LONGITUDE
LONGITUDE

filtered to remove variations with intraseasonal time scales. Finally, the resulting values for each field are scaled by appropriate constants to obtain nondimensional quantities with similar magnitudes.

The two leading modes of this combined M-SSA have similar eigenvalues (not shown), and together explain more than 40% of the variance in the combined variables. The PCs are also similar and their spectra show peaks at about 45 months. Figure 11 shows the eigenvectors of SST, zonal wind stress, and VAT obtained from the M-SSA. It is apparent that the oscillation of SST in the central-to-eastern equatorial Pacific is accompanied by large zonal wind stress variations to the west and VAT variations with a phase shift of about 90° across the basin. The pattern of these eigenvectors is similar to that of the corresponding fields shown in Fig. 8. Therefore, we identify the quasi-quadrennial oscillation with the simulated ENSO.

We also find that VAT variations at the equator are strongly anticorrelated with those at 10°N. Figure 12 displays the latitudinal variations of the VAT eigenvector associated with the leading oscillatory mode. This figure shows that negative (positive) VAT anomalies at the equator are accompanied by positive (negative) VAT anomalies at 10°N. This dipolar pattern of VAT is similar to that produced by a linear dynamical ocean model forced by wind stress obtained from observation (Zebiak, 1989). In that study, Zebiak pointed out that meridional transports of heat content develop during the ENSO cycle. Our results obtained with a full CGCM are supportive of the existence of such a connection. The largest differences between VAT at 10°N and at the equator occur approximately when SST anomalies change from positive to negative. This relationship suggests that the meridional displacement of VAT plays a role in the recurrence of ENSO events.

V. LESSONS LEARNED

Our research has identified several physical aspects crucial to the behavior of the coupled atmosphere–ocean system. In our first journey, we

---

Figure 8 Time–longitude plots along the equatorial Pacific of (a) zonal wind stress anomalies, (b) SST anomalies, and (c) vertically averaged temperature produced by the revised version of the CGCM. The anomalies are calculated by removing annual cycle from each quantity. A low-pass filtered was applied to remove variations with time scales shorter than 1 year. Positive values are shaded. Contour intervals are 0.025 dyn cm^{-2} for (a), and 0.25°C for (b) and (c).

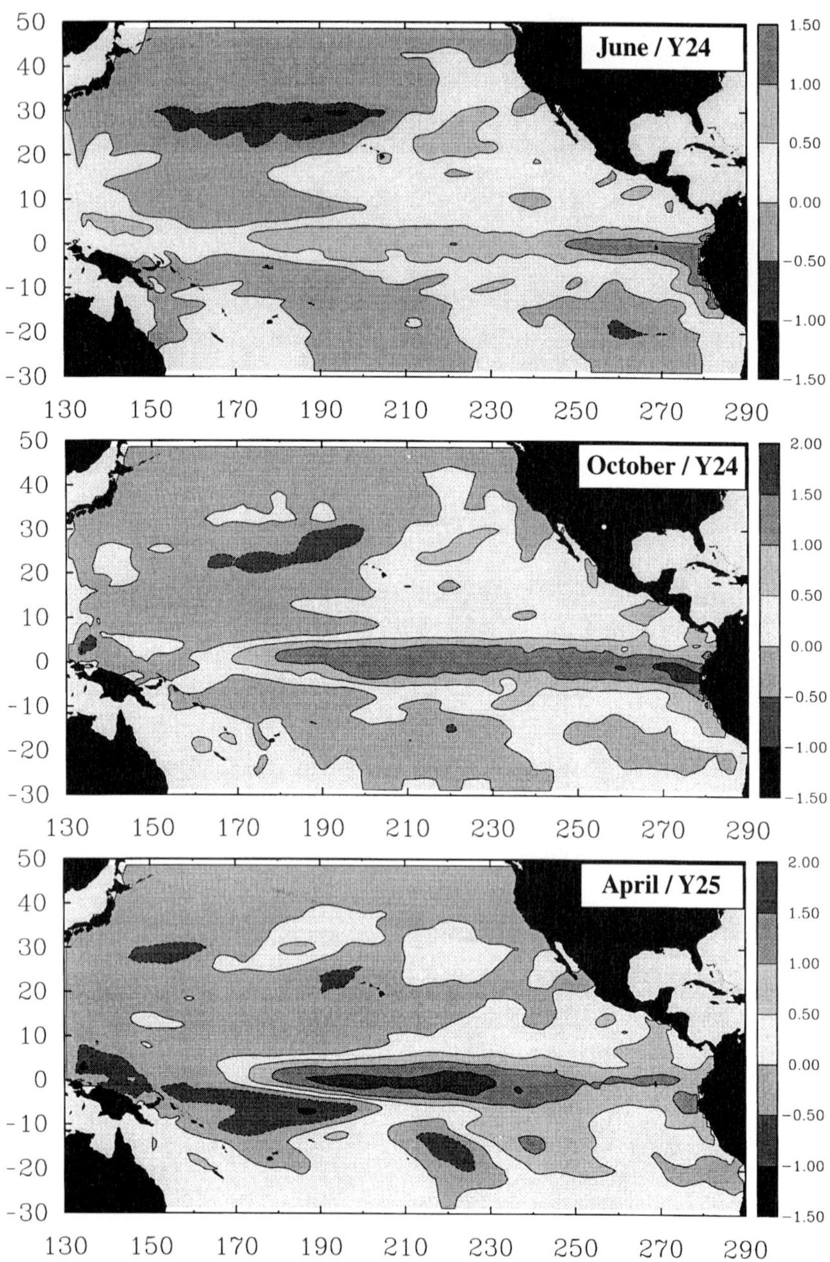

Figure 9   SST anomalies during the warm event produced by the revised version of the UCLA CGCM in year 24 of the simulation. The upper panel shows the anomalies in June of year 24, middle panel in October, and the lower panel in April of year 25. Contour interval is 0.5°C.

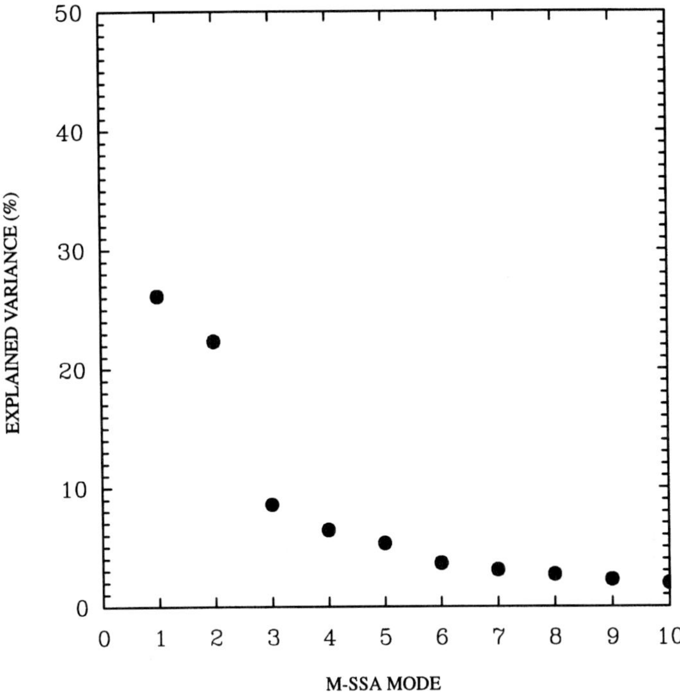

Figure 10 Percentage of variance explained by the leading 10 M-SSA modes of the simulated equatorial Pacific SST.

learned that SSTs simulated by the CGCM are extremely sensitive to the formulation of interacting physical processes, especially in the atmospheric component of the CGCM. For example, a relatively modest change in the long-wave radiation scheme can produce a drastically different mean SST. These sensitivities reflect the fact that the surface fluxes are consequence of complicated interactions between various processes in the atmosphere (and oceans). In the case of surface evaporation, the evaporated water vapor tends to increase PBL humidity, producing a strong negative feedback on the surface evaporation. Then, what really controls the amount of surface evaporation against the negative feedback is the amount of counteracting drying effects in the PBL, such as that due to the entrainment and/or horizontal advection of relatively dry air. In the case of Ma *et al.* (1994), the entrainment of free atmosphere air into the PBL is primarily responsible. In the case of Ma *et al.* (1996), on the other hand, horizontal advection of relatively dry air in the PBL is partially responsible for enhancing the surface evaporation. Similar situations can exist for the momentum and sensible heat fluxes.

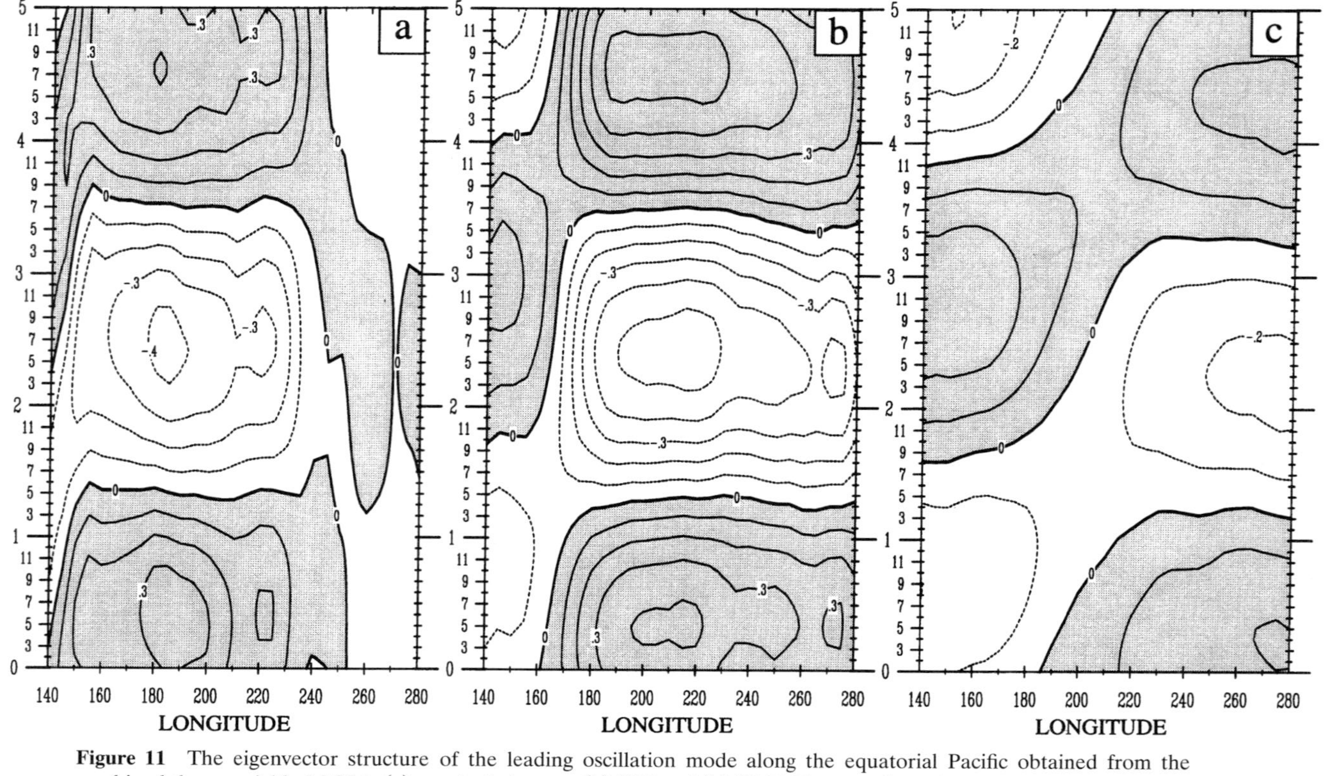

Figure 11 The eigenvector structure of the leading oscillation mode along the equatorial Pacific obtained from the combined three-variable M-SSA: (a) zonal wind stress, (b) SST, and (c) VAT. The coordinate is the 61-month lag used in M-SSA. Contour intervals are 0.1 dyn cm^{-2} for (a), 0.1°C for (b) and (c). Values shown in (a) are scaled by 10. Positive

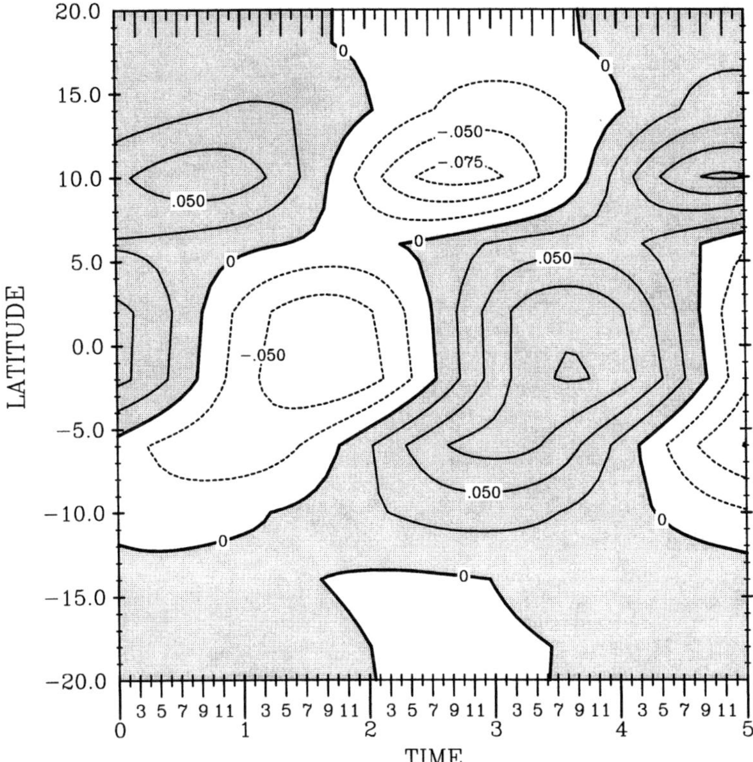

Figure 12 Time–latitude plot of VAT corresponding to the leading M-SSA mode. VATs are zonally averaged between 140°E and 80°W in the tropical Pacific. Contour interval is 0.25°C. Positive values are shaded. The time axis runs from January of year 0 to January of year 5.

In our second journey, we demonstrated that the climate in the eastern Pacific is highly sensitive to the characteristics of Peruvian stratocumulus. The annual variations of these clouds (1) affect the magnitude of the annual cycle over the equatorial cold tongue and (2) contribute to the temporally asymmetric features between the warm and cold phases of the cold tongue. The impact of Peruvian stratus clouds on equatorial SSTs is mainly through a direct circulation with sinking motion above the region of stratocumulus and rising motion north of the equator. The annual variations in cloud amount modulate the strength of this direct circulation and affect the intensity of surface winds over the equatorial cold tongue. The clouds' influence on the near-surface meridional wind affects the strength of the near-coastal upwelling, which is the major process determining the magnitude of the local annual cycle of SST. Peruvian stratocumulus also affect the annual variations of wind speed over the cold tongue, which in turn affect the local feedbacks between surface evaporation and SST. The temporal assymetry of the cold tongue can be captured by the model only when stratocumulus clouds off the coast of Peru are simulated at the appropriate times of the year.

The asymmetries between the warm and cold phases of the cold tongue, therefore, reflect the role of surface evaporation in determining the annual evolution of SST over the eastern equatorial Pacific. Results from this study suggest that Peruvian stratocumulus play an important role not only in the distribution but also in the annual variations of SSTs in the eastern equatorial Pacific. CGCMs have to properly simulate the relationship between the annual evolutions of SST and surface evaporation in order to capture the temporal asymmetry between the warm and cold phases of the cold tongue.

The results discussed in this paper are, of course, model dependent and may not necessarily occur in the same way in all models. They still imply that moist processes in the atmosphere can play central roles in the atmosphere–ocean system either directly or indirectly. Consequently, the quality of a CGCM crucially depends on how these processes are formulated. The same is true for an uncoupled AGCM although there the impact is less visible since the simulated results are strongly tied to the prescribed SST. In any event, we have learned that there are no quick fixes for the deficiencies of a CGCM. The development of better parameterizations and model simulations are possible only through improved understanding of model behavior.

In spite of the design improvements and simulation successes, model systematic errors have not been completely eradicated, particularly off the equator. Some of these errors are quite familiar. For example, SSTs are too warm south of the equator in the eastern Pacific, where there is still excessive short-wave heat flux at the ocean surface. In addition, the orientation of the South Pacific Convergence Zone (SPCZ) is too zonal. Nevertheless, there is a regional alleviation of the errors as their patterns do not extend so far into the eastern Pacific. The pertinacity of the systematic errors epitomizes the differences in modeling the atmosphere with prescribed SSTs and modeling the coupled atmosphere–ocean system in which SST is a prognostic variable. In principle, one cannot rule out that the ultimate cause for the poor CGCM performance with the Peruvian stratocumulus resides primarily in the AGCM. For example, the underestimation of subsidence along the coast of South America by compensation of rising motion associated with convection over the Andes or Amazonia might be sufficient to discourage a stratocumulus regime. In addition, the PBL winds along the coast of Peru in the AGCM tend to be directed inland in a configuration that is not suitable for coastal upwelling. On the other hand, one cannot rule out that the intensity of coastal upwelling simulated by the OGCM is too weak for consistency with a stratocumulus regime even with realistic wind forcing. In any event, whatever the source of errors, it is clear that their impact amplifies in the coupled system. Once those stratocumulus clouds disappear, SSTs warm up and the clouds do not return with the concurrent loss of interhemispheric SST asymmetries. The CGCM's ability to capture ENSO despite these difficulties is encouraging, but the implication of such a success ought to be evaluated in future studies.

VI. PRESENT AND FUTURE DIRECTIONS

A. THE PRESENT

The thrust of our current research is to improve the UCLA atmospheric general circulation model for better performance in simulations of the mean climate and its seasonal and interannual variations when the AGCM is used as a component of a coupled atmosphere–ocean general circulation model. In this regard, we are continuing our work on the development of improved representations of key physical processes that affect the mean climate and its variations, and on the identification of major sensitivities of the coupled atmosphere–ocean system. We chose to emphasize the better simulation of cloud and hydrological processes in view of the central role played by those processes in climate change through their interactions with dynamical, radiative, and PBL processes in the atmosphere, as well as with processes at and below the surface.

As a first step, we have introduced the prognostic closure of the A–S (1974) cumulus parameterization into the AGCM following the approach of Randall and Pan (1993). This closure relaxes the cloud-work-function (CWF) quasi-equilibrium by predicting the cloud-scale kinetic energy. One of the greatest advantages of this approach is that large-scale and cloud-scale processes do not have to be distinguished in calculating the time change of CWF. The relaxed adjustment tends to make the lifetime of simulated convective systems longer, which results in a higher cloud incidence and, therefore, a larger time-averaged cloud amount (Cheng and Arakawa 1994; Ma *et al.*, 1994).

We are also including vertical momentum and rainwater budgets and convective downdraft effects in the Arakawa–Schubert (1974) cumulus parameterization following Cheng and Arakawa (1994, 1997a,b). The outflow from downdrafts below cloud base significantly modifies the thermodynamic properties of the subcloud layer and, therefore, interactions between the atmosphere and oceans. Our next step will be to incorporate an explicit prediction of ice and liquid phases of water with detailed cloud microphysics interacting with radiation (Köhler *et al.*, 1997).

When this work is completed, the atmospheric component of the CGCM will have a version of the A-S cumulus parameterization that includes explicit budget calculations of vertical momentum and rainwater for each subensemble of updrafts and associated downdrafts. Liquid water and ice detrained from cumulus clouds within each model layer will be three-dimensionally advected, with detailed cloud microphysics interacting with radiation, to predict the spatial distribution of cloud mass. These model upgrades are expected to have important impacts on the simulations of cloud and humidity fields and, through their effects on radiation fields, on the performance of the CGCM.

We are also updating the OGCM component of the model, although at a considerably less intense pace than that of the AGCM. As a component of this effort, we have coupled the parallel version of the AGCM to the parallel version of the Bryan–Cox code developed at Los Alamos National Laboratory (LANL) and known as POP (Parallel Ocean Program). Further improvements on the original version of MOM made at LANL comprise the explicit prediction of surface pressure (instead of diagnosing a surface pressure balanced with the barotropic streamfunction), and a different data structure that is more suitable for a parallel computing environment.

B. Code Improvement

CGCMs face great technical challenges in view of their enormous demands on computer resources. The tendency toward higher resolutions and longer simulations requires highly optimized codes suitable for the most advanced supercomputers. We have dedicated a particularly intense effort to address these concerns. As a first step, the AGCM code was parallelized for massively parallel processors (Wehner *et al.*, 1995). To improve the efficiency of the code, we developed algorithms to load balance the different model components (Lou and Farrara, 1996). We also performed a number of single-node optimizations to increase the per-node performance on the CRAY T3E. As a result of these code upgrades, the AGCM now achieves an overall parallel efficiency of 50% on 256 nodes of a CRAY T3E-900, and an execution rate of about 40 Gflops on 512 nodes of that computer (Mechoso *et al.*, 1998). The AGCM running side by side with POP can currently achieve a performance of 40 Gflops when 324 nodes are assigned to the AGCM and 188 nodes to the OGCM (see Fig.

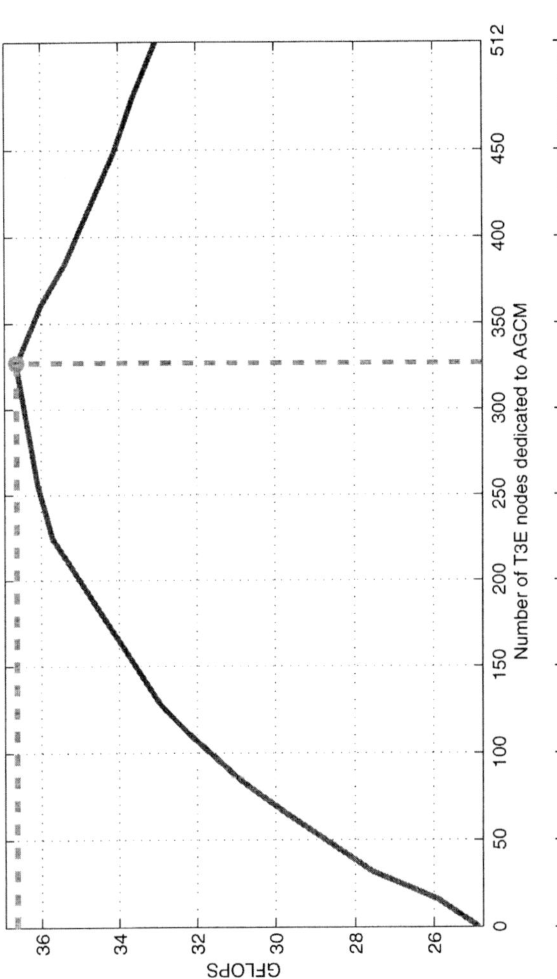

Figure 13 The expected Gflop rate after distributing 512 T3E nodes between the AGCM and OGCM codes. The upper x axis shows the number of T3E nodes dedicated to run the AGCM code and the lower x axis shows the number of T3E nodes dedicated to run the OGCM code. The expected maximum performance is 40.3 Gflops, dedicating 327 nodes to AGCM and 185 nodes to OGCM. Copyright 1999, Society for Computer Simulation (Simulation Councils, Inc.).

13). Our efforts with the AGCM code are aimed at the 100-Gflop level before the end of the millennium.

We have also developed an Earth System Information System (ESIS). ESIS follows a DBMS-centric approach, i.e., from the setup of a model run until the analysis and visualization of model output, all of the metadata of the model run is loaded into a database. The system is a web-based database application that has the following functions: (1) Sets up model runs, (2) loads model metadata and output into the database, (3) browses metadata information, (4) retrieves data sets, (5) analyzes and validates model output, and (6) visualizes data sets.

C. The Next-Generation UCLA AGCM

The advantage of an isentropic vertical (θ) coordinate in simulating potential vorticity advection is well recognized. Moreover, the simplest and most straightforward vertical grid for an isentropic coordinate is free from the problems the Lorenz grid has for a sigma-pressure (σ-p) type coordinate (e.g., Arakawa and Moorthi, 1988). The use of an isentropic vertical coordinate eliminates or reduces difficulties associated with the vertical advection of moisture because the coordinate acts as a quasi-Lagrangian coordinate before the onset of condensation. All of these features favor the use of an isentropic coordinate in a model that emphasizes realistic simulations of moisture and cloud fields, including thin layer clouds associated with fronts. Isentropic vertical coordinates, however, have both intrinsic and technical difficulties near the lower boundary. These difficulties are bypassed with a σ-type coordinate. The next version of the UCLA AGCM will have a new dynamical framework based on a generalized vertical coordinate (Arakawa *et al.*, 1994). This coordinate is essentially isentropic except near the lower boundary and in unstably stratified regions, where it switches automatically and smoothly to the σ-type coordinate (Konor and Arakawa, 1997).

We believe that the role of CGCMs as the fundamental tool for climate studies will increase even further as technological advances make them faster and friendlier to the users, and as societal demands for more accurate and detailed predictions increase even further. One approach that may be followed to address those demands will be to design multiscale, multigrid, integrated systems. Another approach may be based on simpler, physically based empirical parameterizations that will trade "complexity" for "high resolution of computation" in both space and time. Inevitably, the limits of resolution where the traditional CGCMs are applied will be reached, at least for individual components of the climate system. From now until then and beyond there will be more and exciting new phases in the numerical modeling of the atmosphere.

APPENDIX A

OBSERVATIONAL DATA

The SST climatology is taken from Alexander and Mobley (1976), which is used as the boundary condition for the AGCM outside the coupled domain. The surface heat flux, latent heat flux, and short-wave radiative flux are those compiled by Oberhuber (1988) from observational data mainly from the Comprehensive Ocean Atmosphere Datasets (COADS; Woodruff *et al.*, 1987) for the period from 1950 to 1979. The zonal and meridional components of the surface wind stress are those analyzed at Florida State University (FSU; Legler and O'Brien, 1984). All data are linearly interpolated to the grid of the AGCM to facilitate comparisons with simulated fields.

APPENDIX B

DETOUR: COUPLED GCM FORECASTS OF THE 1997–1998 EL NIÑO EVENT

The revised version of the UCLA CGCM is able to simulate El Niño-type variability with realistic amplitudes. We explored the potential of this CGCM in long-range prediction by focusing on the recent 1997–1998 El Niño event.

The initialization scheme used to generate the initial conditions for the ocean component of the CGCM is based on the model climatology from an extended run, and SST and wind stress anomalies from the CPC (Climate Prediction Center) assimilation data set and the FSU analyses, respectively. The initial conditions for the atmosphere are taken directly from the extended run of the CGCM. The ocean state at the end of the spin-up process is used as the initial condition for the ocean model. For the prediction of the 1997–1998 El Niño, the ocean model is integrated with observed wind stress and SST anomalies beginning in January 1994. We made several forecasts with different initialization times. Here we present the forecast in which the initial conditions for the OGCM correspond to June 15, 1997.

The predicted SST anomalies are shown in Fig. B.1. This figure shows a rapid intensification of the anomalies from the June initial condition to a

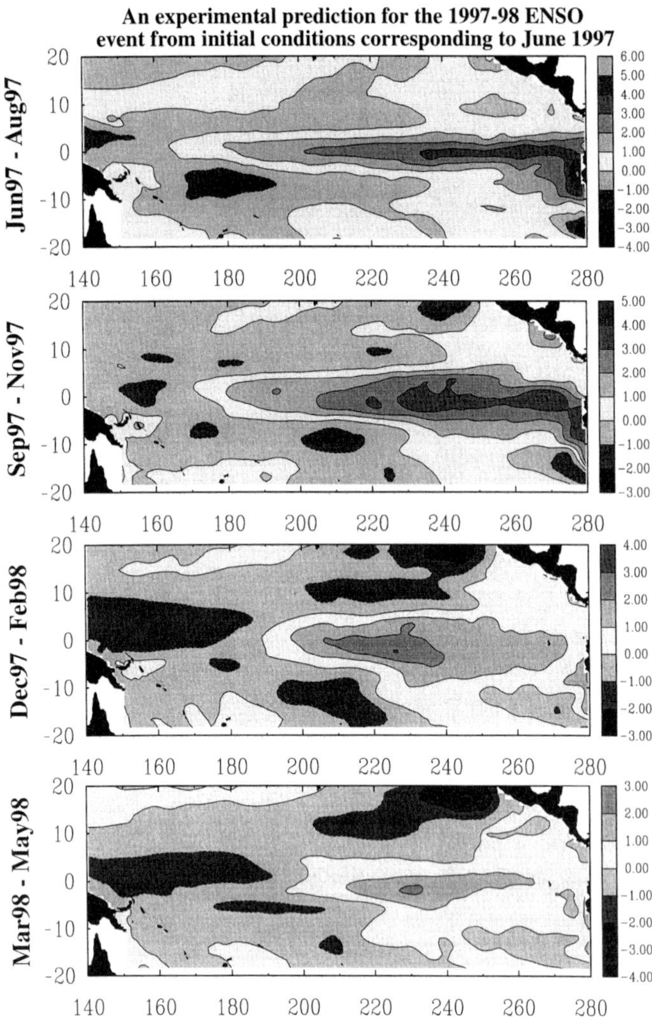

Figure B.1 An experimental prediction for the 1997–1998 ENSO event from initial conditions corresponding to June 1997.

September–October maximum. In the NIÑO3 region, SST anomalies increase by 1.5°C during this 3-month period. The largest SST anomalies were predicted to be more than 3°C in October 1997 over the eastern equatorial Pacific. The CGCM predicted that the 1997–1998 El Niño event would slowly decay after October 1997 and would terminate after the spring of 1998. A forecast initialized in July 1997 shows similar behavior.

The model is clearly able to capture the major features of the El Niño event targeted for prediction. This performance compares favorably with those of other CGCMs. The results of this experimental long-range prediction exercise, therefore, are very encouraging.

ACKNOWLEDGMENTS

It is a pleasure to recognize the contribution to this research made by our collaborators J. D. Farrara, C.-C. Ma, J. A. Spahr, and M. Fisher (now at the Hadley Center). This research was supported by NOAA GOALS under grant NA66GP0121, NSF under grant ATM96-30226, and DOE/CHAMMP under grant DE-FG03-91ER61214.

REFERENCES

Alexander, R. C., and R. L. Mobley (1976). Monthly average sea-surface temperatures and ice pack limits on a 1° global grid. *Mon. Wea. Rev.* **104,** 143–148.

Arakawa, A., and V. R. Lamb (1977). Computational design of the basic dynamical processes of the UCLA general circulation model. *In* "General Circulation Models of the Atmosphere" (Methods in Computational Physics), Vol. 17 (J. Chang, ed.), pp. 173–265. Academic Press, New York.

Arakawa, A., and V. R. Lamb (1981). A potential enstrophy and energy conserving scheme for the shallow-water equations. *Mon. Wea. Rev.* **109,** 18–36.

Arakawa, A., and S. Moorthi (1988). Baroclinic instability in vertically discrete systems. *J. Atmos. Sci.* **45,** 1688–1707.

Arakawa, A., and W. H. Schubert (1974). Interaction of a cumulus ensemble with the large-scale environment, Part I. *J. Atmos. Sci.* **31,** 674–701.

Arakawa, A., and M. J. Suarez (1983). Vertical differencing of the primitive equations in sigma coordinates. *Mon. Wea. Rev.* **111,** 34–35.

Arakawa, A., C. S. Konor, and C. R. Mechoso (1994). A generalized vertical coordinate and the choice of vertical grid for atmospheric models. *In* "Tenth Conference on Numerical Weather Prediction," Portland, OR, July 17–22, 1994, pp. 32–34.

Bryan, K. (1969). A numerical method for the study of the circulation of world oceans. *J. Comput. Phys.* **4,** 347–376.

Cheng, M.-D., and A. Arakawa (1994). Effects of including convective downdrafts and a finite cumulus adjustment time in a cumulus parameterization. *In* "Tenth Conference on Numerical Weather Prediction," Portland, OR, July 17–22, 1994, pp. 102–104. *Am. Meteor. Soc.*

Cheng, M.-D., and A. Arakawa (1997a). Inclusion of rainwater budget and convective downdrafts in the Arakawa–Schubert cumulus parameterization. *J. Atmos. Sci.* **54,** 1359–1378.

Cheng, M.-D., and A. Arakawa (1997b). Computational procedures for the Arakawa–Schubert cumulus parameterization, Tech. Rep. 101. General Circulation Modeling Group, Department of Atmospheric Sciences, UCLA.

Cox, M. D. (1984). A primitive equation three-dimensional model of the ocean, FDL Ocean Group Tech. Rep. 1. Princeton University.

Deardorff, J. W. (1972). Parameterization of the planetary boundary layer for use in general circulation models. *Mon. Wea. Rev.* **100,** 93–96.

Harshvardhan, R. Davies, D. A. Randall, and T. G. Corsetti (1987). A fast radiation parameterization for general circulation models. *J. Geophys. Res.* **92,** 1009–1016.

Harshvardhan, D. A. Randall, T. G. Corsetti, and D. A. Dazlich (1989). Earth radiation budget and cloudiness simulations with a general circulation model. *J. Atmos. Sci.* **46,** 1922–1942.

Jin, F.-F., J. D. Neelin, and M. Ghil (1994). El-Niño on the devil's staircase: Annual subharmonic steps to chaos. *Science* **264,** 70–72.

Katayama, A. (1972). A simplified scheme for computing radiative transfer in the troposphere. *In* "Numerical Simulation of Weather and Climate," Tech. Rep. 6. Dept. of Meteorology, UCLA.

Kim, Y.-J., and A. Arakawa (1995). Improvement of orographic gravity wave parameterization using a mesoscale gravity wave model. *J. Atmos. Sci.* **52,** 1875–1902.

Klein, S. A., and D. L. Hartmann (1993). The seasonal cycle of low stratiform clouds. *J. Climate* **6,** 11,587–11,606.

Köhler, M., C. R. Mechoso, and A. Arakawa (1997). Ice cloud formulation in climate modeling. *In* "Preprints, 7th Conference on Climate Variations," Long Beach, CA, Feb. 2–7, 1997, pp. 237–242. Amer. Meteor. Soc.

Konor, C. S., and A. Arakawa (1997). Design of an atmospheric model based on a generalized vertical coordinate. *Mon. Wea. Rev.* **125,** 1649–1673.

Legler, D. M., and J. J. O'Brien (1984). Atlas of tropical Pacific wind stress climatology 1971–1980. Department of Meteorology, Florida State University, Tallahassee.

Li, J.-L., and A. Arakawa (1997). Improved simulation of PBL moist processes with the UCLA GCM. *In* "7th Conference on Climate Variations," Feb. 2–7, 1997, Long Beach, CA pp. 35–40. Amer. Meteor. Soc.

Lou, J. Z., and J. D. Farrara (1996). Performance analysis and optimization on the UCLA parallel atmospheric general circulation model code. *In* "Proc. Supercomputing '96." Pittsburgh, PA, IEEE Computer Society.

Ma, C.-C., C. R. Mechoso, A. Arakawa, and J. D. Farrara (1994). Sensitivity of a coupled ocean–atmospheric model to physical parameterizations. *J. Climate* **7,** 1883–1896.

Ma, C.-C., C. R. Mechoso, A. W. Robertson, and A. Arakawa (1996). Peruvian stratus clouds and the tropical Pacific circulation: A coupled ocean–atmosphere GCM study. *J. Climate* **9,** 1635–1645.

Mechoso, C. R., L. A. Drummond, J. D. Farrara, and J. A. Spahr (1998). The UCLA AGCM in high performance computing environments. *In* "Proc. Supercomputing '98," Orlando, FL. IEEE Computer Society.

Mechoso, C. R., A. W. Robertson, N. Barth, M. K. Davey, P. Delecluse, P. R. Gent, S. Ineson, B. Kirtman, M. Latif, H. Le Treut, T. Nagai, J. D. Neelin, S. G. H. Philander, J. Polcher, P. S. Schopf, T. Stockdale, M. J. Suarez, L. Terry, O. Thual, and J. J. Tribbia (1995). The seasonal cycle over the tropical Pacific in general circulation models. *Mon. Wea. Rev.* **123,** 2825–2838.

Meehl, G. A. (1990). Seasonal cycle forcing of El Niño–southern oscillation in a global coupled ocean–atmosphere GCM. *J. Climate* **3,** 72–98.

Mellor, G. L., and T. Yamada (1982). Development of a turbulence closure model for geophysical fluid problems. *Rev. Geophys.* **20,** 851–875.

Mitchell, T. P., and J. M. Wallace (1992). On the annual cycle in equatorial convection and sea surface temperature. *J. Climate* **5,** 1140–1156.

Neelin, J. D., and co-authors. (1992). Tropical air-sea interaction in general circulation models. *Clim. Dyn.* **7,** 73–104.

Nigam, S., and Y. Chao (1996). Evolution dynamics of tropical ocean–atmosphere annual cycle variability. *J. Climate* **9,** 3187–3205.

Oberhuber, J. M. (1988). An atlas based on the COADS data set: The budgets of heat buoyancy and turbulent kinetic energy at the surface of the global ocean, Report 15. Max-Plank-Institut für Meteorologie, Germany.

Pacanowski, R., and S. G. H. Philander (1981). Parameterization of vertical mixing in numerical models of tropical oceans. *J. Phys. Oceanogr.* **11**, 1443–1451.

Paulson, E. A., and J. J. Simpson (1977). Irradiance measurements in the upper oceans. *J. Phys. Oceanogr.* **7**, 952–956.

Philander, S. G. H., D. Gu, D. Halpern, D. Lambert, N.-C. Lau, T. Li, and R. C. Pacanowski (1996). Why the ITCZ is mostly north of the equator. *J. Climate* **9**, 2958–2972.

Randall, D. A., and V. Pan (1993). Implementation of the Arakawa–Schubert cumulus parameterization with a prognostic cumulus kinetic energy. *In* "The Representation of Cumulus Convection in Numerical Models of the Atmosphere" (K. A. Emanuel and D. J. Raymond, eds.), pp. 137–144. Amer. Meteor. Soc.

Robertson, A. W., C.-C. Ma, C. R. Mechoso, and M. Ghil (1995a). Simulation of the tropical Pacific climate with a coupled ocean–atmosphere general circulation model. Part I: The seasonal cycle. *J. Climate* **8**, 1178–1198.

Robertson, A. W., C.-C. Ma, M. Ghil, and C. R. Mechoso (1995b). Simulation of the tropical Pacific climate with a coupled ocean–atmosphere general circulation model. Part II: The seasonal cycle. *J. Climate* **8**, 1199–1216.

Rosati, A., and K. Miyakoda (1988). A general circulation model for upper ocean simulation. *J. Phys. Oceanogr.* **18**, 1601–1626.

Schlesinger, M. E. (1976). A numerical simulation of the general circulation of atmospheric ozone, Ph.D. Thesis. Dept. of Atmos. Sci., UCLA.

Suarez, M. J., A. Arakawa, and D. A. Randall (1983). The parameterization of the planetary boundary layer in the UCLA general circulation model: Formulation and results. *Mon. Wea. Rev.* **111**, 2224–2243.

Takano, K., and M. G. Wurtele (1982). A fourth-order energy and potential enstrophy conserving difference scheme, *Report AFGL-TR-82-0205 (NTIS AD-A126626)*, Air Force Geophysics Laboratory, Hanscom AFB, MA.

Waliser, D. E., and C. Gauthier (1993). A satellite derived climatology of the ITCZ. *J. Climate* **6**, 2162–2174.

Wang, B. (1994). On the annual cycle in the tropical eastern central Pacific. *J. Climate* **7**, 1926–1942.

Wehner, M. F., A. A. Mirin, P. G. Eltgroth, W. P. Dannevik, C. R. Mechoso, J. D. Farrara, and J. A. Spahr (1995). Performance of a distributed memory, finite difference atmospheric general circulation model. *Parallel Comput.* **21**, 1655–1675.

Woodruff, S. D., R. J. Slutz, R. L. Jenne, and P. M. Stenrer (1987). A comprehensive ocean–atmosphere data set. *Bull. Am. Meteor. Soc.* **68**, 1239–1250.

Yu, J.-Y., and C. R. Mechoso (1999a). A discussion on the errors in the surface heat flux simulated by a coupled GCM. *J. Climate* **12**, 416–426.

Yu, J.-Y., and C. R. Mechoso (1999b). Links between annual variations of Peruvian stratocumulus clouds and of SST in the eastern equatorial. *Pacific. J. Climate* **12**, 3305–3318.

Zebiak, S. E. (1989). Oceanic heat content variability and El Niño cycles. *J. Phys. Oceanogr.* **19**, 475–486.

Chapter 19

Representing the Stratocumulus-Topped Boundary Layer in GCMs

Chin-Hoh Moeng and Bjorn Stevens
National Center for Atmospheric Research, Boulder, Colorado

I. INTRODUCTION

The most important role of the cloud-free PBL (planetary boundary layer) in large-scale circulations is to link the atmospheric motions to the Earth's surface through the turbulent fluxes of momentum, heat, and moisture. Most of the PBL parameterizations currently used in GCMs can predict these surface fluxes reasonably well, at least for simple PBL regimes that include no complications due to terrain, canopy, or precipitation.

The stratocumulus-topped boundary layer (STBL), however, influences large-scale circulations in more ways through its interaction with radiation. Because its fractional cloud cover and cloud albedo can significantly affect the amount of radiation reaching the Earth's surface, parameterizations for this PBL regime have to provide not only reasonable turbulent fluxes but also the cloud amount and cloud optical properties. Most of the existing parameterizations for STBL predict the turbulent fluxes and the cloud properties separately using different, uncoupled schemes. For example, the current version of NCAR Community Climate Model (CCM) predicts the PBL turbulent fluxes using a PBL turbulence model (Holtslag and Boville, 1993), which was developed specifically for cloud-free PBLs (Troen and Mahrt, 1986). The cloud cover is predicted using a slightly modified Slingo empirical cloud scheme (Slingo, 1987), which simply links the cloud amount to the inversion strength and heights, and humidity of the grid volume. The cloud cover is then used to calculate the radiation field. The parameterized turbulent transport, radiation, and cloud fields do not interact directly with each other as they do in nature.

In this chapter, we first review current understanding of the STBL, based mainly on previous observational and numerical studies of the subtropical marine STBL; survey the cloud and turbulence schemes currently used in GCMs; point out their problems; and show some possible simple improvements. Finally we present a current LES (large-eddy simulation) study that aims at further understanding of the STBL and developing a parameterization scheme that directly links the turbulent fluxes to the cloud-top radiative forcing. Here we treat the stratocumulus as solid clouds, namely, with a 100% cloud cover. Clouds with partial cloud cover in a GCM grid mesh are common and important phenomena to deal with (Randall, 1989), but are beyond the scope of this paper.

II. CURRENT UNDERSTANDING
OF THE STBL REGIME

Stratocumulus often occur over the subtropical oceans off the west coasts of the major continents throughout most of the northern summertime. Their favorite large-scale condition is strong subsiding motion over a cold ocean surface. This large-scale condition creates a very moist, shallow PBL. The lifting condensation level is below the top of the mixed layer so clouds that form in the upper part of the PBL are likely to spread out uniformly in the horizontal direction due to turbulent mixing. This cloud regime typically covers tens of thousands of square kilometers and, thus, can easily fill a few GCM grid meshes horizontally. Vertically, this cloud layer is often 100–500 m thick and thus can hardly be well resolved by a GCM vertical grid.

A. PHYSICAL PROCESSES

Turbulence in STBLs may be driven by many forcings, such as surface heating, cloud-top radiative and evaporative cooling, condensation warming, and wind shear, as sketched in Fig. 1. But the most typical and persistent stratocumulus-topped PBLs are believed to be driven mainly by cloud-top radiative cooling. Cloud drops emit long-wave radiation like a graybody. This results in a sharp increase in the long-wave radiation flux and, hence, a strong cooling at the cloud top. As the PBL air is cooled from the top, downdrafts then carry colder air into the PBL and hence are negatively buoyant. They can then drive turbulence just like thermal updrafts in the surface-heated PBL. But unlike the surface buoyancy flux in the surface-heated PBL, this cloud-top radiatively driven buoyancy forcing is difficult to measure or predict because it occurs in the thin, undulating interfacial layer where both entrainment and evaporation take place. All of these cloud-top physical processes (radiative cooling, entrainment warming, and evaporative cooling) together play an important role in determining the net amount of the cloud-top buoyancy forcing.

Entrainment brings in not only warm air but also dry air. Entrained warm air that mixes in with the radiatively cooled near-cloud-top air can make downdrafts warmer and hence decrease the cloud-top buoyancy. On the other hand, entrained dry air that mixes with cloudy air results in

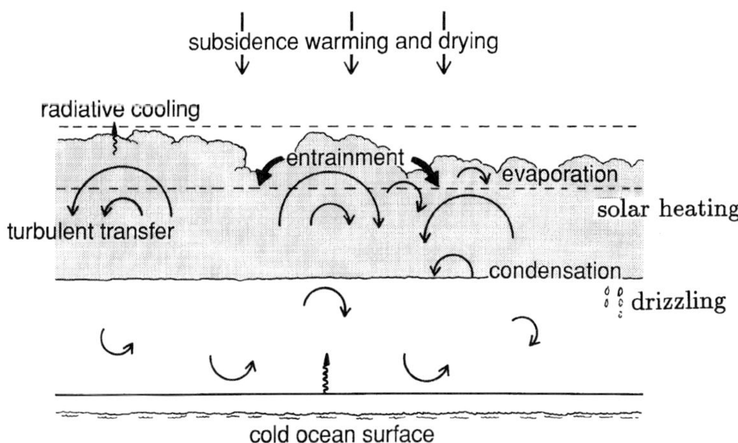

Figure 1 A sketch showing the various physical processes involved in the stratocumulus-topped PBL.

evaporation, which cools unsaturated downdrafts near the cloud top more rapidly and hence enhances the cloud-top buoyancy forcing.

Other physical processes, such as drizzle and solar heating, may also play important roles in regulating the turbulent fluxes and the cloud field. Some observational studies (e.g., Nicholls, 1984; Paluch and Lenschow, 1991) suggest that drizzle may cool the air just below the cloud layer, which can create a thin stably stratified layer near the cloud base and effectively cut off the moisture supply to the cloud layer from the ocean surface. Modeling studies (e.g., Wang and Wang, 1994; Stevens *et al.*, 1998) suggest that drizzle decreases the liquid water path, entrainment rate, buoyancy production, and turbulence intensity. Solar heating has been less often studied, but it is thought to lead to less turbulent kinetic energy production in the cloud layer and also raises the in-cloud saturation mixing ratio, leading to a thinner or disappearing of the cloud layer. Some numerical studies suggested that advecting the stratocumulus regime over to a warmer sea surface can change the dynamics of the STBL and result in the stratocumulus-to-cumulus transition in the subtropical marine boundary layer (Krueger *et al.*, 1995).

In summary, within the stratocumulus-topped PBL, turbulent transport, cloud microphysical processes, and radiation are strongly coupled. For instance, a larger turbulent moisture flux from the ocean surface supplies more moisture to the cloud layer, which modifies cloud microphysics; cloud microphysics are known to affect the radiation profile; and cloud-top radiation provides the main buoyancy forcing for turbulent fluxes. These interactions among turbulence, microphysics, and radiation occur on small time scales and spatial scales, and hence should be included in parameterization schemes. In other words, a reasonable parameterization scheme should consider all of these physical processes in a fully coupled mode. As Randall (1989) pointed out, the next generation of parameterizations requires solving for the cloud field, the radiation, and the turbulent fluxes simultaneously to ensure their full, direct coupling. As a first step toward such a parameterization, we need to know how to couple these processes. In Section IV we attempt to develop a simple parameterization that takes into account the fact that cloud-top radiative cooling drives turbulence within the stratocumulus-topped PBL.

B. Typical Profiles of the Thermodynamical Fields

A persistent STBL with no drizzle is usually well mixed; i.e., the horizontal means of the thermodynamic quantities that are moist-adiabatically conserved (such as the liquid water potential temperature, $\theta_1 \sim \theta -$

$(L/c_\mathrm{p})q_\mathrm{l}$, and the total water mixing ratio, $q_\mathrm{T} \equiv q_\mathrm{v} + q_\mathrm{l}$, where θ is the potential temperature, q_v vapor mixing ratio, and q_l liquid mixing ratio) do not vary much with height in the PBL. Within the solid cloud layer, potential temperature increases with height, and water vapor mixing ratio decreases with height, approximately following a moist adiabatic lapse rate. These mean thermodynamical profiles are sketched in Fig. 2. The well-mixed mean liquid water potential temperature implies that once the near cloud-top layer is cooled by cloud-top radiation or warmed by entrainment, turbulence rapidly mixes it; the whole process occurs in turbulence time scale, within minutes.

The total energy flux (e.g., the liquid water potential temperature flux $\rho_0 c_\mathrm{p}\overline{w\theta_\mathrm{l}}$ plus the radiation flux F_R) and the total moisture flux (the water vapor flux plus the liquid water flux) are often linear with height in the STBL. This linear feature is a consequence of statistical quasi-steadiness of the turbulent motion; in quasi-steady states the mean gradient of an adiabatically conserved quantity, such as the liquid water potential temperature, remains constant in time. Thus, from the mean conservation equation,

$$\frac{\partial}{\partial t}\frac{\partial \Theta_\mathrm{l}}{\partial z} = -\frac{\partial^2\left(\overline{w\theta_\mathrm{l}} + F_\mathrm{R}/\rho_0 c_\mathrm{p}\right)}{\partial z^2} = 0, \tag{1}$$

implying $\rho_0 c_\mathrm{p}\overline{w\theta_\mathrm{l}} + F_\mathrm{R}$ is linear with height for cloudy PBLs in quasi-steady states.

The buoyancy flux, which is a measure of the turbulence forcing (and hence turbulence intensity) in buoyancy-driven PBLs, is not linear with height, unlike that in the cloud-free convective PBL. This makes it difficult to measure the buoyancy flux profile in the field because measurements

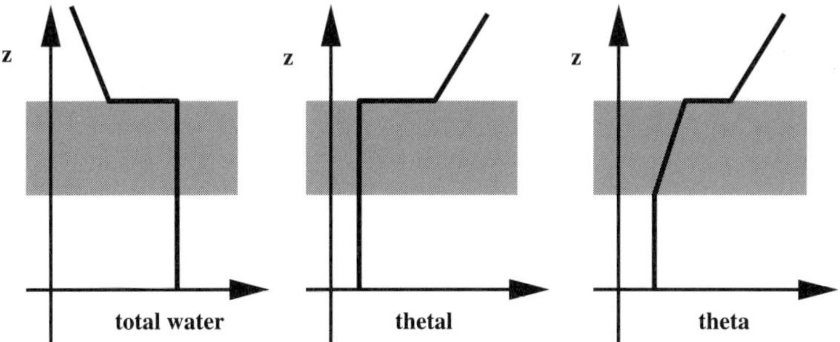

Figure 2 Typical profiles of the mean total mixing ratio, the mean liquid water potential temperature, and the mean potential temperature within a well-mixed STBL.

rely heavily on aircraft that commonly have only two to three flight legs within the STBL (in order to get enough samples in each leg). The nonlinear profile of the buoyancy flux makes it difficult to determine from a few levels of aircraft data in the vertical. For this reason, large-eddy simulations are often used to give detailed profiles of the buoyancy flux and other fluxes that are not linear in height.

Figure 3 shows the buoyancy flux, from 10 different LES codes, of an idealized STBL where cloud-top long-wave radiative cooling is the major driving force for the PBL turbulence. This study was carried out by the GCSS (GEWEX Cloud System Study) Boundary Layer Cloud Working Group and reported in Moeng *et al.* (1996). The 10 LES codes differ in their treatment of long-wave radiation, sub-grid-scale turbulence, sub-grid-scale condensation, and numerics. We note that although the absolute magnitudes of the flux differ among the LESs, the profile shapes are quite similar. In this case, the buoyancy flux is nearly zero at the surface (as expected because there is little surface heating in this case) and increases sharply across the cloud base due to condensation. The buoyancy flux remains large (and positive) within the cloud layer due to cloud-top cooling. The buoyancy flux drops sharply at the cloud top to a negative value, which is associated with entrainment. In this region, warm inversion

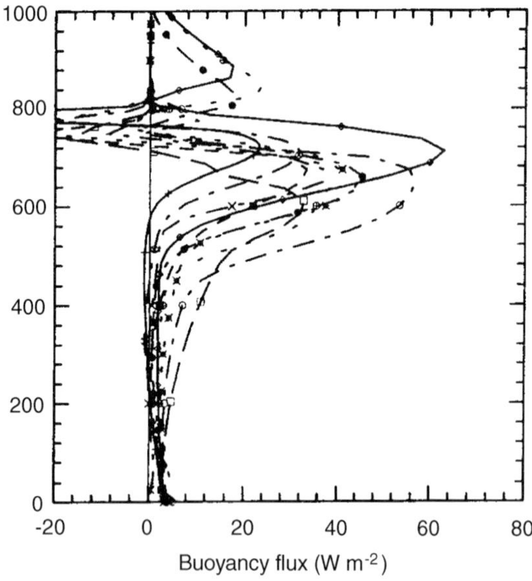

Figure 3   Buoyancy flux profile obtained from 10 LES codes. (From Moeng *et al.*, 1996.)

air ($\theta' > 0$) is carried down ($w' < 0$) into the cloud layer as a result of entrainment; this gives a negative buoyancy flux at the cloud top, which is often referred to as the entrainment buoyancy flux.

One important point to make here is that within the solid cloud layer the mean potential temperature Θ increases with height where its flux is positive. This indicates that the Θ field does not diffuse downgradiently. In other words, standard K-type turbulence models, which relate the fluxes to the mean gradients with eddy diffusivity, cannot apply to these nonconserved temperature variables in the solid cloud layer. However, as we will see in the next section, many GCMs use the above K-type turbulence model to solve for the prognostic variable Θ in both clear and cloudy layers. The same argument applies to the liquid water field where liquid water flux and the vertical gradient of the mean liquid water content are both positive in the solid cloud layer.

III. EXISTING STBL TURBULENCE AND CLOUD SCHEMES IN GCMs AND THEIR PROBLEMS

A. EXISTING MARINE STRATOCUMULUS PBL SCHEMES

Here we survey PBL and stratocumulus parameterizations used by 24 of the GCMs participating in the Atmospheric Model Intercomparison Project (AMIP). In all, more than 30 models participated in the project; we survey only those that are well documented and those that either have a PBL model or have 3 or more points below 800 mb. Because the physical packages within most GCMs are continually evolving (although not as rapidly as one might imagine), our survey represents a snapshot of the state of PBL and cloud parameterization circa 1995. The results of the survey are collected in Table I and described in more detail below. Further information about the AMIP project, and the models surveyed, can be found on the web at http://www-pcmdi.llnl.gov/pcmdi/modldoc/amip.

Approaches to PBL parameterization fall into two broad categories: bulk models and eddy diffusion models. Three GCMs, all stemming from the work of the UCLA group, parameterize the PBL using a mixed layer model, and allow the top of the PBL to be a coordinate surface of the GCM. The rest of the GCMs use K models, that is, an eddy diffusivity approach with fixed vertical levels. The eddy diffusivity approach states that the time rate of change of an arbitrary prognostic variable (such as Θ) can be represented by a diffusion-like process with a variable eddy diffu-

Table I
Summary of PBL Turbulence and Cloud Schemes in GCMs[a]

GCM	PBL scheme	Prognostic variables	Stratiform cloud	Levels below 800 hPa
BMRC	K(Ri)	No	$f(RH,\dots)$	5
CCC	K(Ri)	No	$f(RH)$	3
CCSR	K(Ri)	No	Statistical	5
CNRM	K(Ri)	No	$f(RH)$	4
COLA	K(Ri)	No	$f(RH,\dots)$	5
CSIRO	K(Ri)	No	$f(RH,\dots)$	3
CSU	MLM	N/A	N/A	N/A
DERF	K(Ri)	No	$f(RH,\dots)$	5
ECMWF	K(Ri)	No	$f(RH,\dots)$	5
GFDL	SMAG	No	$f(RH)$	4
GLA	$K - $ TKE	No	$f(RH)$	5
GSFC	$K - $ TKE	No	???	5
JMA	K(Ri)	No	$f(RH)$	6
LMD	?	No	Statistical	3
MGO	K(Ri)	No	$f(RH,\dots)$	5
MPI	$K - $ TKE $- PBL$	No	$f(RH,\dots)$	5
MRI	MLM	N/A	N/A	N/A
NCAR	$K - $ Prof	No	$f(RH,\dots)$	4
NMC	K(Ri)	No	$f(RH,\dots)$	5
NRL	K(Ri)	No	$f(RH,\dots)$	5
SUNY-Genesis	—	No	$f(RH,\dots)$	
UCLA	MLM	N/A	N/A	N/A
UGAMP	K(Ri)	No	$f(RH,\dots)$	5
UKMO	K(Ri) $-$ PBL	Yes	Statistical	4

[a] The different types of PBL models and cloud schemes are discussed in the text. Whether or not models use a prognostic thermodynamic variable that is conserved under moist adiabatic processes is indicated in the third column by "Yes" or "No."

sivity K dependent on the mean state of the flow, i.e.,

$$\frac{\partial \Theta}{\partial t} = \cdots + \frac{\partial}{\partial z}\left(K \frac{\partial \Theta}{\partial z}\right). \tag{2}$$

The K models differ in how they determine K. Fourteen of the GCMs surveyed let $K = f(\text{Ri})$, that is, they let K be a function of the gradient Richardson number Ri of the flow. Typically, this function is specified following Louis (1979) and Louis *et al.* (1981) or Mellor and Yamada level 2.0 (Mellor and Yamada, 1982). Three of the K models carry an additional equation for the turbulent kinetic energy (TKE), and then let $K = \lambda(\text{TKE})^{1/2}$ where λ is some length scale. One of the models, the NCAR model, uses the so-called K profile approach. This approach was developed by Troen and Mahrt (1986) and constrains the shape of the eddy diffusivity to a fixed profile within the PBL; the magnitude of the eddy diffusivity is related to the surface forcing of turbulence, i.e., surface buoyancy flux and surface stresses.

Only the K profile model and the mixed-layer models make explicit use of some measure of the PBL depth. Most of the K models make no explicit use of the PBL depth, although some prognostic TKE models use a turbulence length scale that depends on a diagnosed measure of PBL depth. Of the models surveyed, only one has more than 5 grid levels below 800 mb, and many have fewer than 5 grid levels below 800 mb. Five grid levels below 800 mb implies an average vertical grid spacing of 400 m, clearly too large to resolve processes that take place within a few hundred meters thick, such as cloud-top radiative cooling and entrainment. Even in simple, idealized, cloud-free PBLs (something which most of the K models were designed to represent), such coarse resolution has been shown to severely undermine the capacity of the model to adequately represent PBL processes (e.g., Ayotte *et al.*, 1996).

Apart from problems associated with poor vertical resolution, all but one of the K models surveyed suffer from an additional defect—one specific to cloud layers. Many GCMs use prognostic thermodynamic variables that are not conserved under moist adiabatic processes. They express the turbulent flux as $\overline{w\theta} = -K(\partial\Theta/\partial z)$ in the Θ equation. Such a turbulence model would mix Θ downgradiently so as to produce a well-mixed Θ profile in the whole turbulent layer. As discussed in Section II, a well-mixed Θ profile is an unrealistic feature in a solid cloud layer.

The methods used to diagnose stratiform cloudiness in the GCMs we surveyed were surprisingly uniform. Although a few of the models attempted to parameterize cloud fraction using statistical models, in conjunction with assumptions about the sub-grid distributions of temperature and moisture, most estimated the cloud fraction using simple functions of relative humidity, denoted as $f(RH)$ in Table I. We delineate between models that describe the cloud fraction solely in terms of relative humidity [$f(RH)$ models], and those that, following Slingo (1987), allow the cloud fraction to further depend on quantities like stability and height, [$f(RH,\ldots)$ models]. An example of this latter approach is the CCM3, in which cloud fraction c_1 (restricted to the level where inversion strength is strongest) is given by:

$$c_1 = \max[0., f(RH; a_1, b_1)f(\partial\Theta/\partial p; a_2, b_2)f(p; a_3, b_3)], \qquad (3)$$

where $f(x; a, b)$ is a linear function of some model field x with parameters a and b. The additional complexity of such models does not reflect a better understanding of the factors that control cloud fraction, but it does offer one the opportunity to tune the cloud schemes to produce observed low cloud climatologies irrespective of the peculiarities of an individual model climate. However, being flexible in tunning is not necessarily a desirable feature in a GCM.

The only GCMs that directly link the stratiform cloud and the PBL turbulence schemes are the ones that use mixed-layer models. It allows for, for example, cloud-top radiative cooling to drive the PBL turbulence directly and instantaneously. This approach, which appears well suited to well-mixed PBLs including typical STBLs, was developed by David Randall in the mid-1970s under the direction of A. Arakawa (Suarez *et al.*, 1983; Randall *et al.*, 1985). In the other models, radiation and PBL turbulence interact only indirectly, as the turbulence recognizes the existence of stratiform cloud layers only insofar as the cloud field leads to altered mean temperature profiles through radiation on the following GCM time step. The intimate coupling between radiation and turbulence production described in Section II is therefore ignored in the vast majority of GCMs used to study climate. It is our goal to develop a scheme that couples the radiation, turbulence, and cloud processes. We understand that most GCMs use the eddy diffusivity type of approach for turbulent transport; in other words, they use multiple grid levels to resolve the PBL structure. Converting this approach to the mixed-layer approach (which uses only one grid layer to represent the PBL) may require a major change to a GCM code. Thus, it would be most useful if such a new parameterization scheme were to be built on eddy diffusivity-type turbulence models, or some combination between the multiple grid level and mixed-layer model approaches.

B. Subtropical Stratocumulus in the CCM3

Despite intensive efforts devoted toward a better understanding of STBL, many (if not most) GCMs still poorly represent this PBL regime. And even in situations where the estimated cloud amount looks reasonable, agreement with observations is often spurious, as we shall see. Below we examine in more detail calculations from the NCAR CCM3 with AMIP boundary conditions. Our focus is on the Eastern Pacific stratocumulus regime during July over a 15-year integration. Our analysis of the CCM3 calculations illustrates a number of shortcomings, which may be evident in a broader class of models that use Slingo-like cloud schemes and eddy diffusivity PBL models. On the basis of our analysis, we are able to suggest some simple fixes that produce a more consistent simulation of the STBL, as well as outline some necessary steps toward a STBL parameterization that more realistically encodes our current understanding of the STBL.

In Fig. 4 we plot the 15-year averages of the low cloud amount, the PBL height, and the vertical velocity field near the surface. At first glance, the prediction of low-level cloudiness on the eastern edges of the subtropical oceans looks reasonable. Further inspection shows that while the spatial patterns of cloud fraction are in rough agreement with surface and satellite observations, there are some important and systematic differences. The regions of maximum cloud amount (which we focus on here) are underpredicted, and the model produces virtually no cloud just along the coasts; observations often suggest that large cloud fractions extend to the coast. Associated with this minimum in coastal low-level cloudiness are very large values of subsidence, which leads to a very shallow, almost-nonexistent PBL.

The oscillating pattern of low-level vertical velocity (Fig. 4c) is suggestive of the "ringing" associated with the spectral representation of the near-coast topography in the CCM3. This topographically induced modulation of the vertical velocity field appears to also modulate the patterns of the simulated low-level cloudiness. It may have amplified the strong subsidence along the coasts and led to the very low cloud amount in that region. This topographically induced vertical velocity field has also artificially enhanced the cloud amount shown in Fig. 4a. The effect of the vertical velocity field on the cloud field is illustrated in Fig. 5, which shows a vertical cross section of the time-averaged flow along a line from the Southern California coast to Hawaii. One of the maxima in stratocumulus amount is clearly associated with low-level rising motions due to "ringing." It is apparent that many of these low clouds are "frontal"-type clouds produced by the "frontal"-cloud parameterization, which depends in part on the strength of rising motions; it is not expected in this part of the globe. Modifying the CCM3 to prohibit "frontal" clouds from forming under strong inversions results in significantly less stratocumulus in the subtropical Eastern Pacific (cf. Figs. 6a and 6b), thereby illustrating that a significant component of the CCM3 subtropical stratocumulus is spurious.

Another undesirable feature in the representation of low clouds by the CCM3 is illustrated in Fig. 5b. This figure shows a cross-sectional view of eddy viscosity K and low-level cloud fraction. If one associates the depth of the PBL with the region of large values of K, and hence larger turbulent fluxes, then clearly much of the PBL cloud exists above—and disconnected from—the PBL. Because the PBL parameterization and the cloud parameterization are not coupled, cloud is allowed to form above the PBL. As a result, even the indirect coupling of the two schemes (through the mean state) is not properly represented.

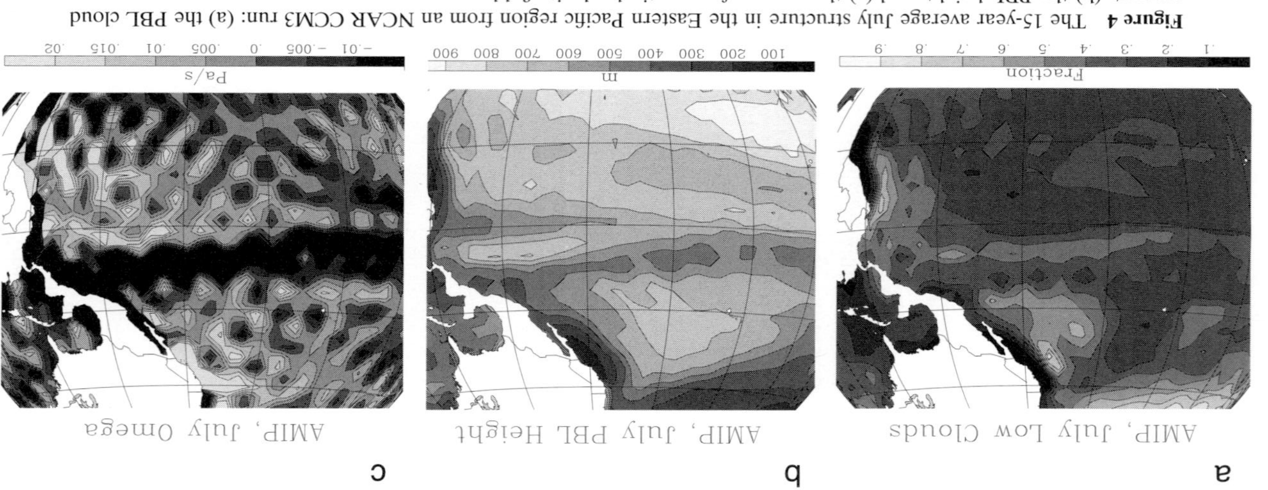

Figure 4 The 15-year average July structure in the Eastern Pacific region from an NCAR CCM3 run: (a) the PBL cloud amount, (b) the PBL height, and (c) the near-surface vertical velocity field.

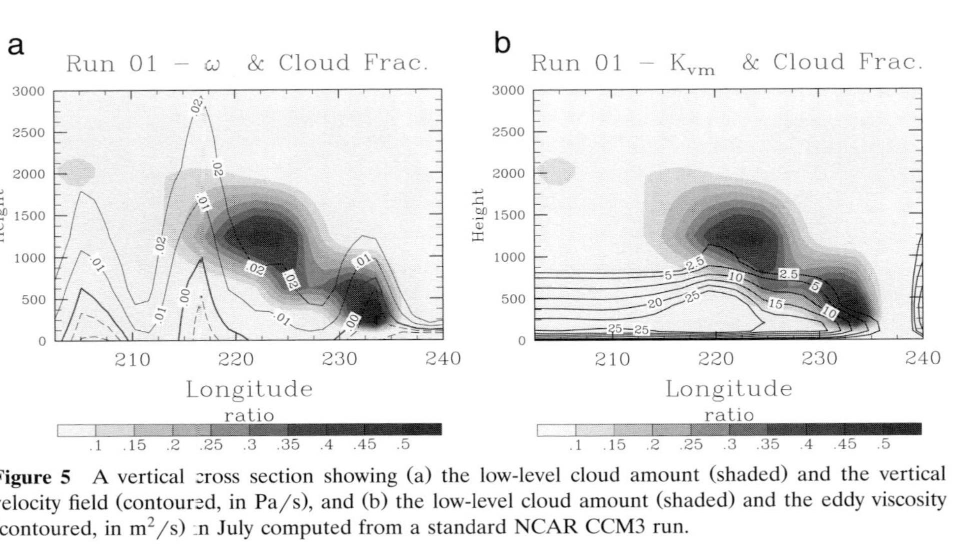

Figure 5 A vertical cross section showing (a) the low-level cloud amount (shaded) and the vertical velocity field (contoured, in Pa/s), and (b) the low-level cloud amount (shaded) and the eddy viscosity (contoured, in m^2/s) in July computed from a standard NCAR CCM3 run.

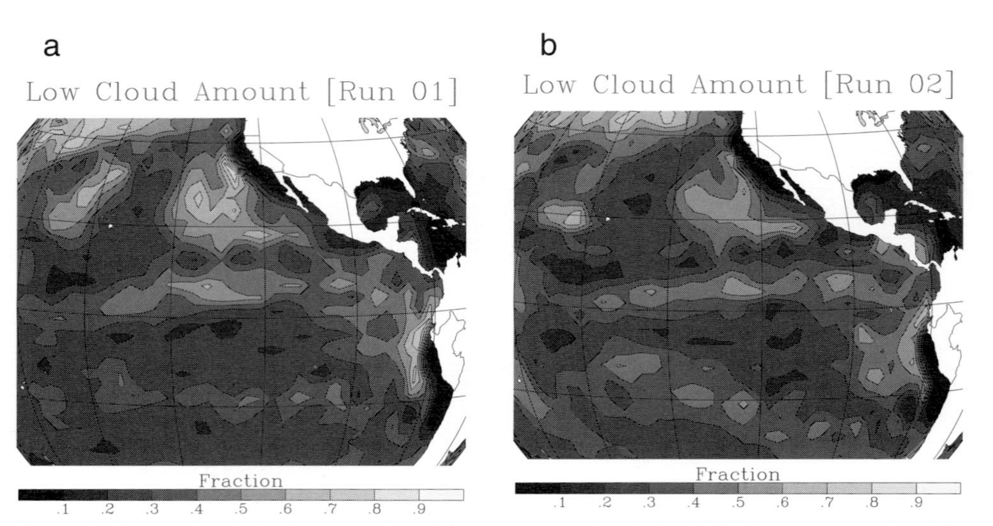

Figure 6 The low-level cloud amount from (a) a 1-year standard NCAR CCM run compared to (b) a run where no "frontal" clouds are allowed to form under an inversion.

To eliminate this unrealistic feature, we made the following simple improvement to the current cloud scheme. We allow for a relative-humidity-dependent fraction of cloud to form in any layer beneath the detected inversion if the relative humidity in the layer is larger than 80%. The resultant cloud parameterization uses three free parameters instead of the six used by the old scheme, and reflects our view that ad hoc models should attempt to minimize the number of unconstrained free parameters. Our changes lead to an increase in the cloud amount (cf. Figs. 6b and 7a), which is closer to observations. More importantly, the cloud layer is now forced to exist within the turbulent PBL as is evident in Fig. 7b. Still, the cloud amount close to the coasts is largely underestimated and also the parameterized eddy diffusivity is not directly linked to turbulence produced by cloud-top radiative cooling. But at least the indirect coupling between the turbulence and radiation is now properly represented.

In Fig. 8, we schematically illustrate how most GCMs treat the marine stratocumulus-topped PBL. In most GCMs the mean state (that is, mean temperature, moisture, and wind profiles) is used to predict the turbulent fluxes, which in turn modify the mean state at the next GCM time step. The mean state is also used to predict the cloud properties, such as the fractional cloud amount. The cloud properties, along with the mean state, are then used to calculate the radiation field, which again modifies the mean state. Therefore, the interactions incorporated in most GCMs (indicated by solid lines that connect the boxes in Fig. 8) among turbulence, radiation, and clouds are mediated by changes to the mean state, at consecutive GCM computational time steps, which are typically much longer than the turbulence time scale. In nature, all of the above physical processes are in a quasi-steady-state equilibrium with each other within a turbulence time scale.

Most GCMs do not consider many important interactions as shown by the dotted lines in Fig. 8. For example, cloud-top radiative cooling drives turbulent fluxes on the time scale of the turbulence, and these fluxes then help determine the cloud fraction and liquid water amount. As we discuss next, such interactions are beginning to be sufficiently well understood to allow for their incorporation into PBL-cloud parameterizations for use in GCMs.

IV. CURRENT EFFORT IN FURTHER UNDERSTANDING AND DEVELOPING PARAMETERIZATIONS OF THE STBL

Our understanding of the STBL described in Section II is mostly qualitative; to put our knowledge into parameterizations requires quantita-

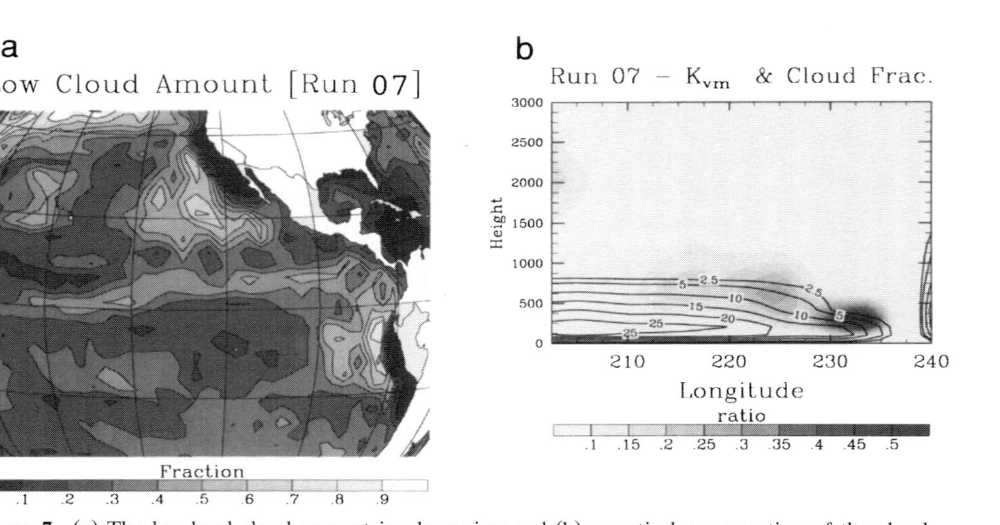

Figure 7 (a) The low-level cloud amount in plane view and (b) a vertical cross section of the cloud amount (shaded) and the eddy viscosity (contoured, in m^2/s) from an NCAR CCM run where some fractional amount of low-level clouds is allowed to form for all levels under the inversion if the relative humidity is larger than 80%.

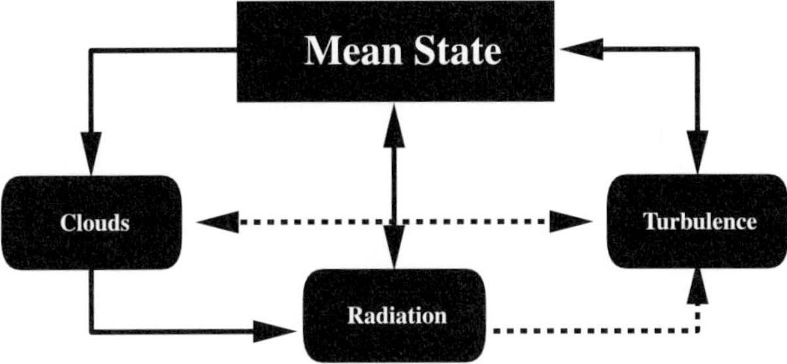

Figure 8 Interactions between the mean state, clouds, radiation, and turbulent fluxes within the STBL.

tive understanding about how all of these physical processes work together. One of our current efforts is to relate the turbulent fluxes to the cloud-top radiative forcing, which represents a major energy input that drives the PBL turbulence. The idea is similar to the treatment of the cloud-free convective PBL where most turbulence statistics can be reasonably described (or scaled) by its only energy source—the surface heat flux. Can turbulent fluxes within the STBL be directly related to the cloud-top radiative forcing?

To simplify the analysis, our first attempt is to use a prototype STBL where turbulence is driven only by cloud-top radiative cooling. This leads us to use the so-called smoke-cloud experiment, first suggested by Lilly (1968) and subsequently used by many investigators (e.g., Lilly and Schubert, 1980; Nieuwstadt and Businger, 1984; Moeng *et al.*, 1992; Bretherton *et al.*, 1999; Moeng *et al.*, 1999). Like stratocumulus, a smoke cloud also emits long-wave radiation like a graybody. This results in a sharp divergence of the long-wave radiation flux near the cloud top, leading to radiative cooling that can buoyantly drive turbulence as in the stratocumulus-topped PBL. Thus, a smoke-topped PBL shares the essential features of turbulence driven by radiative cooling, but without the additional complication of condensation/evaporation.

Our analysis is based on a smoke-cloud LES. The smoke-cloud case we considered here was designed for, and used in, the second model intercomparison study by the GCSS Boundary Layer Cloud Working Group (Bretherton *et al.*, 1999). This cloud case has no surface heating, no wind shear, and free-slip (i.e., no stress) bottom and top boundary conditions. In other words, long-wave radiative cooling at the cloud top is the only source of turbulence. A good understanding of this prototype smoke-filled PBL is a necessary first step toward better understanding of the more complicated stratocumulus-topped PBL.

Our sounding, large-scale conditions, and the treatment of long-wave radiation are all the same as those of the GCSS smoke-cloud case reported in Bretherton *et al.* (1999), except that we purposely choose a much weaker capping inversion to better resolve the cloud-top undulations. As shown later, our theory can only be examined through an LES that can resolve the cloud-top undulations. We also use a nested-grid LES code, developed by Sullivan *et al.* (1996), in which an outer (coarse) grid spans 3200 m in x and y and 1250 m in z, and a nested (finer) grid that is located in the entrainment zone and spans the same horizontal domain but only 250 m in z. The nested grid has $192 \times 192 \times 30$ grid points, so the grid size in the entrainment zone is about 16 m in the horizontal and about 8 m in the vertical. This, as shown from our flow visualization, seems sufficient to resolve the most important features in the entrainment zone.

Initially the smoke mixing ratio s is set to 1 below the inversion base and 0 above it. To mimic stratocumulus, which exhibits a sharp drop of liquid water field at the cloud top, we allow only smoke with concentration larger than 0.5 to be radiatively active. This is analogous to a real cloud case in which the water vapor field must exceed a certain threshold value before condensation occurs, so that smoke with concentration larger than 0.5 can be seen as the cloudy region and that less than 0.5 as the cloud-free region. The smoke field is then subject to cloud-top radiative cooling, which results in an increasingly unstable lapse rate near the cloud top during the initial spin-up of the simulated turbulence. When this thin layer becomes sufficiently unstable, convective overturning takes place and turbulence commences. After a spin-up period, which takes the first 30 min or so of the simulation time, vigorous turbulence, maintained by cloud-top radiative cooling, continuously mixes the liquid potential temperature field so that the thin unstable layer at the cloud top is no longer significant. We ran the LES for 150 simulation minutes, which is about 11 large-eddy-turnover times. The three-dimensional instantaneous flow fields were stored at 5-min intervals over the last 80 min of the simulation, so that 17 LES flow fields were available for analysis in this study.

A. LES Results

Figure 9 shows the temperature contours at the inversion and the flow velocity vectors from the LES. In the upper right corner, we also plot the grid mesh in the entrainment zone. Compared to the grid mesh, the

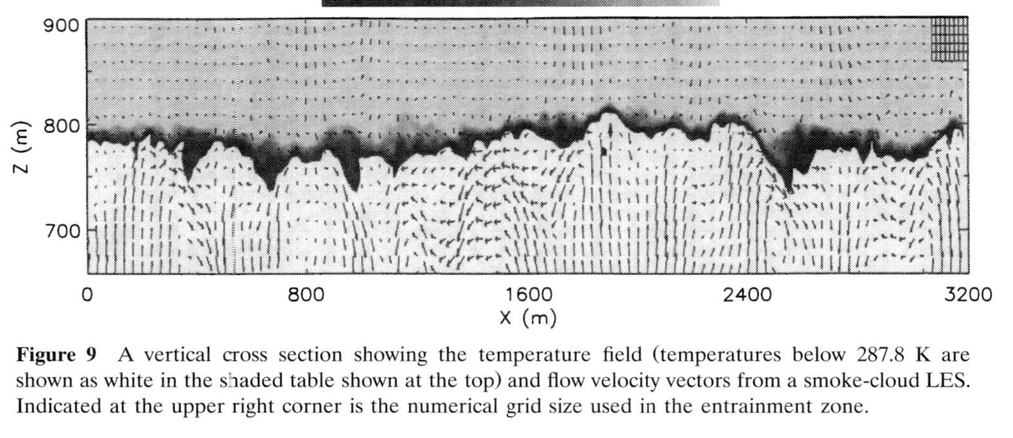

Figure 9 A vertical cross section showing the temperature field (temperatures below 287.8 K are shown as white in the shaded table shown at the top) and flow velocity vectors from a smoke-cloud LES. Indicated at the upper right corner is the numerical grid size used in the entrainment zone.

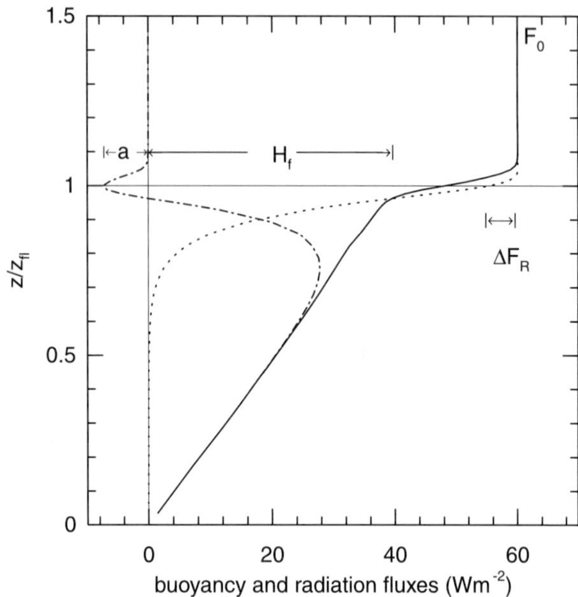

Figure 10   Vertical distributions of the buoyancy flux (dashed-dotted), radiation flux (dotted), and the total energy flux (solid) from the smoke-cloud LES. The symbol a indicates the amount of the entrainment buoyancy flux, H_f the extrapolated value of the linear H curve at the minimum buoyancy flux level, and ΔF_R is the amount of radiative flux jump above z_{fl}. Here $z_m \sim 0.95 z_{fl}$ and $z_{fl}^+ \sim 1.07 z_{fl}$.

cloud-top undulations appear to be well resolved as evidenced from the several entrainment events at $x \sim$ (400, 650, 1000, and 2600) m where wisps of warm air are being incorporated into the PBL.

For a smoke cloud, which has no latent heating, the buoyancy variable is adiabatically conserved; hence, the sum of the buoyancy flux and the radiative flux, i.e.,

$$H \equiv \rho_0 c_p \overline{w\theta_v} + F_R \equiv F_{sv} + F_R, \tag{4}$$

is linear with height in a statistically quasi-steady state. (For real clouds it is the sum of the liquid water potential temperature flux and the radiative flux that is linear with height as shown in Section II.) The buoyancy flux, radiative flux, and their sum H computed from the LES flow fields are shown in Fig. 10.

There are several points to note from Fig. 10: (1) The total energy flux H is linear with height in the quasi-steady state, but the buoyancy flux and radiative flux separately are not. (2) Some small amount of radiative flux divergence exists within the thin jump layer between the minimum buoyancy flux level, z_{fl}, and the top of the entrainment zone where turbulent fluxes vanish, z_{fl}^+. This radiation flux divergence must exist within the smoke-cloud region because no clear-air radiation is included in this LES. (3) If we define H_f as the value of H at z_{fl} by linearly extrapolating the mixed-layer H profile, as indicated in Fig. 10, we find that H_f is much less than F_0, which in this case is 60 W m^{-2}. There is clearly a large difference between H_f and F_0.

Determining the value of H_f is important because the cooling rate of the mean temperature field in the bulk of the PBL is $\sim \dfrac{1}{\rho_0 c_p} \dfrac{H_f}{z_i}$, not $\dfrac{1}{\rho_0 c_p} \dfrac{F_0}{z_i}$. So next we find out what determines H_f.

B. RELATING H_f TO RADIATION FLUX

As shown in Fig. 10, the total energy flux H in the smoke-cloud LES is linear with height only within the well-mixed layer, i.e., below z_m defined roughly at the level where the buoyancy flux crosses the zero line. But it can be approximated as

$$H \sim H_f \frac{z}{z_{fl}}, \tag{5}$$

because the layer between z_m and z_{fl} is very thin. Figure 10 shows that H_f is about 40 W m^{-2} and $F_{sv}(z_{fl})(\equiv \rho_0 c_p \overline{w\theta_v}$ at z_{fl}) about -8 W m^{-2}, so $H_f - F_{sv}(z_{fl}) < F_0$. What makes up the difference between $H_f - F_{sv}(z_{fl})$ and F_0? First, as mentioned before there is some radiative flux divergence above z_{fl}, denoted as ΔF_R, even though it is only a small portion of the total radiative flux jump across the cloud top. Our physical interpretation of this radiative flux divergence is motivated by Fig. 11, which plots the isosurface $s = 0.5$ of the simulated smoke field, thereby illustrating cloud-top undulations or hummocks. The z_{fl} level, indicated as a solid line in Fig. 11, is usually located near the middle of these cloud-top undulations. The contribution to ΔF_R comes from the local radiative flux divergence within each of the smoke-cloud hummocks that extend above z_{fl}.

It is also evident from Fig. 11 that some clear air exists below the z_{fl} level. As shown in Moeng *et al.* (1999), about half of the area at z_{fl} is occupied by clear air and the other half by smoke (or cloudy) air, on the average. For the following illustration, we assume that exactly half of the area is covered by clear air and the other half by smoke (or cloud) at

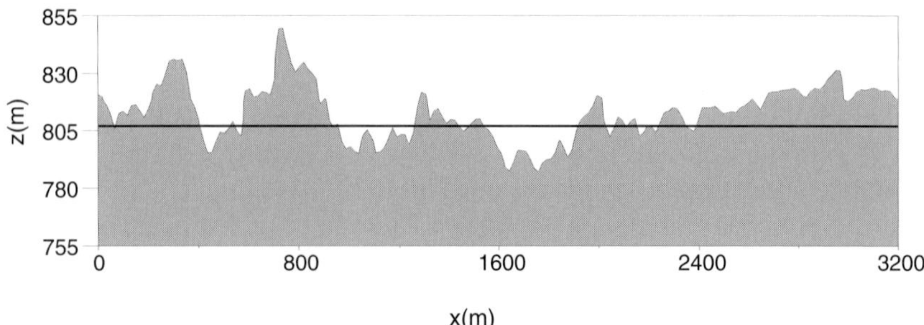

Figure 11 A vertical cross section showing the smoke-cloud-top undulations obtained from the LES. The shaded area has smoke concentrations larger than 0.5 and the solid line indicates the minimum buoyancy flux level, z_{fl}.

the z_{fl} level. Let the averaged value of the in-smoke-cloud radiative flux at z_{fl} be $\overline{F_c}$ while in the clear region the flux is F_0. Then the radiative flux at z_{fl} is

$$F_R(z_{fl}) = \frac{F_{R,\text{clear}}}{2} + \frac{F_{R,\text{cloud}}}{2} = \frac{F_0}{2} + \frac{\overline{F_c}}{2}, \tag{6}$$

hence

$$\Delta F_R \equiv F_0 - F_R(z_{fl}) = \frac{F_0 - \overline{F_c}}{2}. \tag{7}$$

If there was no clear air at the z_{fl} level, then the radiative flux there would be about $\overline{F_c}$. This means that the existence of clear air at z_{fl} makes the radiative flux there larger by $F_R(z_{fl}) - \overline{F_c} = (F_0 - \overline{F_c})/2$, which according to Eq. (7) is ΔF_R.

The sharp increase of the total energy flux H between z_m and z_{fl} shown in Fig. 10 is largely due to the existence of clear air in this layer. So, the difference between the actual H value at z_{fl} [i.e., $H(z_{fl})$] and the linearly extrapolated value H_f is $F_R(z_{fl}) - \overline{F_c}$, which is about ΔF_R. In other words,

$$H(z_{fl}) - H_f \sim \Delta F_R. \tag{8}$$

Because $F_0 \equiv F_R(z_{fl}) + \Delta F_R$ and $H(z_{fl}) \equiv F_{sv}(z_{fl}) + F_R(z_{fl})$ by definition, Eq. (8) gives

$$H_f \sim F_0 + F_{sv}(z_{fl}) - 2\Delta F_R. \tag{9}$$

The LES result in Fig. 10 shows $H_f \sim 40$ W m^{-2}, $F_{sv}(z_{fl}) \sim -8$ W m^{-2}, and $\Delta F_R \sim 5.5$ W m^{-2}. This approximately satisfies Eq. (9).

C. CLOSURE ASSUMPTIONS

To use Eq. (9), we need (1) a closure assumption for the entrainment flux $F_{sv}(z_{fl})$ and (2) a way to calculate ΔF_R. The latter issue was addressed in Moeng *et al.* (1999). They assumed a Gaussian distribution for the height of the undulations, which then allowed them to analytically derive a relationship between ΔF_R and the standard deviation σ_{z_i} of the cloud-top height fluctuations. They then empirically related σ_{z_i} to the interfacial Richardson number,

$$\mathrm{Ri} \equiv \frac{g}{T_0} z_i \Delta\Theta_v / w_*^2,$$

where $\Delta\Theta_v$ is the virtual potential temperature jump between z_{fl} and z_{fl}^+ and $w_* \equiv [2.5\int_0^{z_{fl}} F_{sv}\, dz]^{1/3}$. We do not repeat the derivation here.

One plausible closure assumption for the entrainment buoyancy flux (i.e., the flux at z_{fl}) is to assume that the entrainment flux is proportional to the averaged buoyancy flux over the whole PBL, as proposed by Deardorff (1976); i.e.,

$$F_{sv}(z_{fl}) = -A' \int_0^{z_{fl}} F_{sv}\, dz / z_{fl}. \tag{10}$$

This closure assumption is motivated by the fact that the entrainment buoyancy flux should depend on the turbulence intensity. For buoyancy-driven PBLs, turbulence intensity can be best measured by the layer-averaged buoyancy flux. The fidelity of the closure assumption of Eq. (10) was examined by Moeng *et al.* (1999). Their results, from several smoke-cloud LESs, are shown in Fig. 12; in all cases, except INV3-MFC, the proportionality constant A' is close to 0.5. The value $A' = 0.5$ is consistent with that proposed by Deardorff (1976), which gives the clear convective PBL limit where the entrainment buoyancy flux is about -0.2 of the surface buoyancy flux.

The run INV3-MFC has the strongest capping inversion (7 K jump in its initial temperature field) among all cases studied in Moeng *et al.* (1999) and it uses a monotone advection scheme. We have two reasons to doubt the accuracy of both INV3 results. First with such a strong capping inversion, the cloud-top undulations are not as well resolved as those of weaker capping inversion cases such as those shown in Fig. 9. Second, by

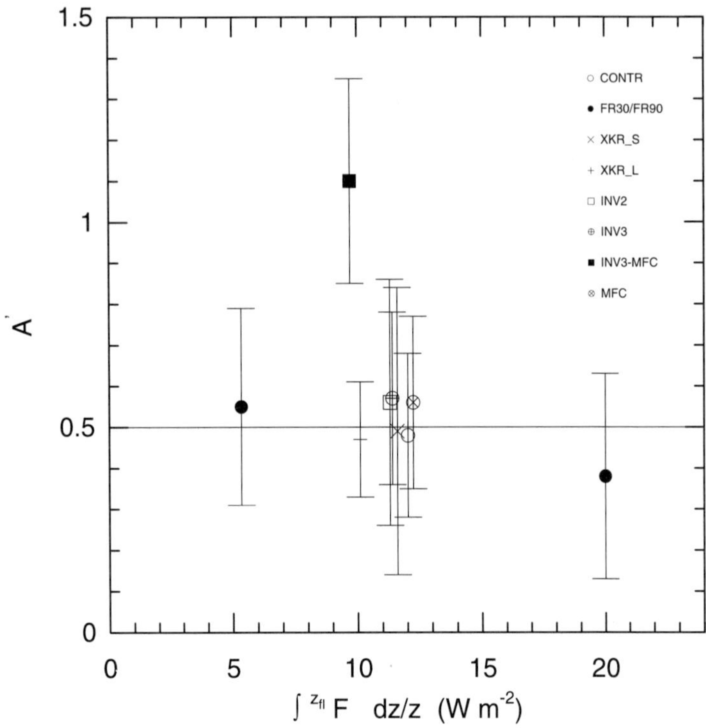

Figure 12   The ratio A' of the entrainment buoyancy flux to the layer-averaged buoyancy flux, as a function of the layer-averaged buoyancy flux, from several smoke-cloud LESs. (From Moeng *et al.*, 1999.)

comparing INV3-MFC and INV3 (which are the same except for the advection scheme used) in Fig. 12, we learn that the results of both strong capping inversion cases are strongly affected by numerics. These results suggest that the monotone advection scheme produces a larger numerical diffusion, hence more smoke in the entrainment zone, and hence a larger entrainment buoyancy flux. These effects are more pronounced with a stronger capping inversion. Finding out how numerics affects the entrainment buoyancy flux prediction obtained from LES is one of our current foci.

When the radiative cooling zone is thin enough, we may use a linear approximation to the buoyancy flux curve so that $F_{sv} \sim H$ in the mixed layer (i.e., approximating the dashed-dotted curve by the solid curve in

Fig. 10 within the mixed layer). Then, by geometry,

$$H_{\mathrm{f}} \sim 2 \int_0^{z_{\mathrm{fl}}} F_{\mathrm{sv}} \, dz / z_{\mathrm{fl}}. \tag{11}$$

This, along with the closure of Eq. (10), yields the following approximation:

$$F_{\mathrm{sv}}(z_{\mathrm{fl}}) \sim -(A'/2)H_{\mathrm{f}}. \tag{12}$$

We examined relationship (12) using the buoyancy flux profiles obtained from our LES and other smoke-cloud LESs reported in Lock and MacVean (1999), Lewellen and Lewellen (1998), and Stevens *et al.* (1999). We found that the proportionality constant $A'/2$ can be rather constant from the same LES code for different simulations, but varies among different LES codes. The LES results reported by the UKMO group (Lock and MacVean, 1999) gave about the same ratio as our LESs, i.e., $A'/2 \sim 0.25$; whereas the results from Lewellen and Lewellen (1998) and Stevens *et al.* (1999) showed a larger value $A'/2 \sim 0.5$. Stevens *et al.* (1999) also showed that this ratio can be sensitive to the sub-grid-scale Smagorinsky constant. We do not know at this moment whether the above variation of A' is physical or not because most of the above LESs dealt with strong capping inversion cases and might not resolve the cloud-top undulations well. This is an area that needs further investigation.

Nevertheless, for the purpose of illustration, if we use relationship (12) in (9), we obtain

$$H_{\mathrm{f}} \sim \frac{F_0 - 2\Delta F_{\mathrm{R}}}{1 + A'/2}. \tag{13}$$

Formula (13) links H_{f} to the net radiative forcing F_0. Then, from Eqs. (5), (10), and (12), we can calculate the total energy flux profile H, the layer-averaged buoyancy flux, and the entrainment buoyancy flux for a given radiative forcing F_0. This may also give the turbulence intensity and other turbulent statistics, such as eddy diffusivity.

After fully determining what processes control A', the next step is to extend this analysis to the real cloud case. This requires changing the dry adiabatically conserved thermodynamic variable to a moist adiabatically conserved one, e.g., from virtual potential temperature to liquid water potential temperature. Along with formulas that link the buoyancy flux to the fluxes of the moist adiabatically conserved variables (e.g., Randall, 1987), the above analysis can be easily applied to the real cloud case.

V. CONCLUSION

In this chapter, we first reviewed our current understandings of physical processes involved within the stratocumulus-topped PBL. Those understandings are mostly qualitative, and thus difficult to put into a parameterization format for GCM use.

We then described some of the existing parameterization schemes for the PBL turbulence and clouds. Most of these schemes do not treat radiation, turbulence, and clouds as a fully coupled system. Using a 15-year integration of the NCAR CCM3 as an example, we showed some unrealistic features resulting, in part, from the uncoupled treatments of radiation, turbulence, and clouds. We suggested some simple modifications to the NCAR CCM3 PBL schemes, which improved the model results somewhat, but these modifications do not incorporate the whole coupling. Our goal is to develop a scheme that fully couples the radiation, turbulence, and cloud processes in the STBL regime.

As a first step, we used large-eddy simulation of a prototype STBL (i.e., smoke-filled PBL) to show how turbulent fluxes can be directly linked to the cloud-top radiative forcing. We will apply this method to real cloud cases (clouds with condensation/evaporation) and more general large-scale conditions to obtain quantitative relationships among all the physical processes involved in the STBL.

ACKNOWLEDGMENTS

We thank Jim Hack, Jeff Kiehl, and Phil Rasch for their discussions on stratocumulus representations in NCAR CCM3, and Steve Krueger for his thorough review. NCAR is sponsored by the National Science Foundation. A part of this work is supported by DOE/ARM/PNL Project 97-104.

REFERENCES

Ayotte, K. W., P. P. Sullivan, A. Andren, S. C. Doney, A. A. M. Holtslag, W. G. Large, J. C. McWilliams, C.-H. Moeng, M. J. Otte, J. J. Tribbia and J. C. Wyngaard (1996). An evaluation of neutral and convective planetary boundary-layer parameterizations relative to large eddy simulations. *Bound.-Layer Meteor.* **79,** 131–175.

Bretherton, C. S., M. K. MacVean, P. Bechtold, A. Chlond, W. R. Cotton, J. Cuxart, H. Cuijpers, M. Khairoutdinov, B. Kosovic, D. Lewellen, C.-H. Moeng, P. Siebesma, B. Stevens, D. E. Stevens, I. Sykes, and M. C. Wyant (1999). An intercomparison of radiatively-driven entrainment and turbulence in a smoke cloud, as simulated by different numerical models. *Quart. J. Roy. Meteor. Soc.* **125,** 391–423.

Deardorff, J. W. (1976). On the entrainment rate of a stratocumulus-topped mixed layer. *Quart. J. Roy. Meteor. Soc.* **102,** 563–582.

Holtslag, A. A. M., and B. A. Boville (1993). Local versus nonlocal boundary-layer diffusion in a global climate model. *J. Climate* **6,** 1825–1842.

Krueger, S. K., G. T. McLean, and Q. Fu (1995). Numerical simulation of the stratus-to-cumulus transition in the subtropical marine boundary layer. Part I: Boundary-layer structure. *J. Atmos. Sci.* **52,** 2839–2850.

Lewellen, D. C., and W. S. Lewellen (1998). Large-eddy boundary layer entrainment. *J. Atmos. Sci.* **55,** 2645–2665.

Lilly, D. K. (1968). Models of cloud-topped mixed layers under a strong inversion. *Quart. J. Roy. Meteor. Soc.* **94,** 292–309.

Lilly, D. K., and W. H. Schubert (1980). The effects of radiative cooling in a cloud-topped mixed layer. *J. Atmos. Sci.* **37,** 482–487.

Lock, A. P., and M. K. MacVean (1999). The parameterization of entrainment driven by surface heating and cloud-top cooling. *Quart. J. Roy. Meteor. Soc.* **125,** 271–299.

Louis, J.-F. (1979). A parametric model of vertical eddy fluxes in the atmosphere. *Bound.-Layer Meteor.* **17,** 187–202.

Louis, J.-F., M. Tiedtke, and J.-F. Geleyn (1981). "A Short History of the PBL Parameterization at ECMWF, Workshop on Planetary Boundary Layer Parameterization, November 25–27, 1981. ECMWF.

Mellor, G. L., and T. Yamada (1982). Development of a turbulence closure model for geophysical fluid problems. *Rev. Geophys. Space Phys.* **20**(4), 851–875.

Moeng, C.-H., S. Shen, and D. A. Randall (1992). Physical processes within the nocturnal stratus-topped boundary layer. *J. Atmos. Sci.* **49,** 2384–2401.

Moeng, C.-H., W. R. Cotton, C. Bretherton, A. Chlond, M. Khairoutdinov, S. Krueger, W. S. Lewellen, M. K. MacVean, J. R. M. Pasquier, H. A. Rand, A. P. Siebesma, B. Stevens, and R. I. Sykes (1996). Simulation of a stratocumulus-topped planetary boundary layer: Intercomparison among different numerical codes. *Bull. Am. Meteor. Soc.* **77,** 261–278.

Moeng, C.-H., P. P. Sullivan, and B. Stevens (1999). Including radiative effects in an entrainment rate formula for buoyancy-driven PBLs. *J. Atmos. Sci.* **56,** 1031–1049.

Nicholls, S. (1984). The dynamics of stratocumulus: Aircraft observations and comparisons with a mixed layer model. *Quart. J. Roy. Meteor. Soc.* **110,** 783–820.

Nieuwstadt, F. T. M., and J. A. Businger (1984). Radiative cooling near the top of a cloudy mixed layer. *Quart. J. Roy. Meteor. Soc.* **110,** 1073–1078.

Paluch, I. R., and D. H. Lenschow (1991). Stratiform cloud formation in the marine boundary layer. *J. Atmos. Sci.* **48,** 2141–2157.

Randall, D. A. (1987). Turbulent fluxes of liquid water and buoyancy in partly cloudy layers. *J. Atmos. Sci.* **44,** 850–858.

Randall, D. A. (1989). Cloud parameterization for climate modeling: Status and prospects. *Atmos. Res.* **23,** 345–361.

Randall, D. A., J. A. Abeles, and T. G. Corsetti (1985). Seasonal simulations of the planetary boundary layer and boundary-layer stratocumulus clouds with a general circulation model. *J. Atmos. Sci.* **42,** 641–676.

Slingo, J. M. (1987). The development and verification of a cloud prediction scheme for the ECMWF model. *Quart. J. Roy. Meteor. Soc.* **113,** 899–927.

Stevens, B., W. R. Cotton, G. Feingold, and C.-H. Moeng (1998). Large-eddy simulations of strongly precipitating, shallow, stratocumulus-topped boundary layer. *J. Atmos. Sci.* **55,** 3616–3638.

Stevens, B., C.-H. Moeng, and P. P. Sullivan (1999). Large-eddy simulations of radiatively driven convection: Sensitivities to the representation of small scales. *J. Atmos. Sci.* **56,** 3963–3984.

Suarez, M. J., A. Arakawa, and D. A. Randall (1983). Parameterization of the planetary boundary layer in the UCLA general circulation model: Formulation and results. *Mon. Wea. Rev.* **111,** 2224–2243.

Sullivan, P. P., J. C. McWilliams, and C.-H. Moeng (1996). A grid nesting method for large-eddy simulation of planetary boundary-layer flows. *Bound.-Layer Meteor.* **80,** 167–202.

Troen, I., and L. Mahrt (1986). A simple model of the atmospheric boundary layer: Sensitivity to surface evaporation. *Bound.-Layer Meteor.* **37,** 129–148.

Wang, S., and Q. Wang (1994). Roles of drizzle in a one-dimensional third-order turbulence closure model of the nocturnal stratus-topped marine boundary layer. *J. Atmos. Sci.* **51,** 1559–1576.

Chapter 20

Cloud System Modeling

Steven K. Krueger

Department of Meteorology, University of Utah, Salt Lake City, Utah

I. INTRODUCTION

A. WHAT IS A CLOUD RESOLVING MODEL?

A cloud resolving model (CRM) is a 2-D or 3-D model that resolves cloud-scale motions while simulating a cloud system. The scales of motion that are explicitly represented thus depend on the cloud system of interest (Fig. 1). For example, to simulate a convective cloud system that contains both cumulus-scale and mesoscale circulations, a CRM would typically have a horizontal grid size of about 2 km, and a horizontal domain size of about 400 km.

It is interesting to compare the time scales and space scales of a CRM used to simulate a convective cloud system to those of global climate models (GCMs) (Table I). The scales are set by the characteristics of the dominant resolved eddies: cumulus clouds in CRMs, and baroclinic eddies in GCMs. The scales of cumulus clouds are about a hundredth of those of

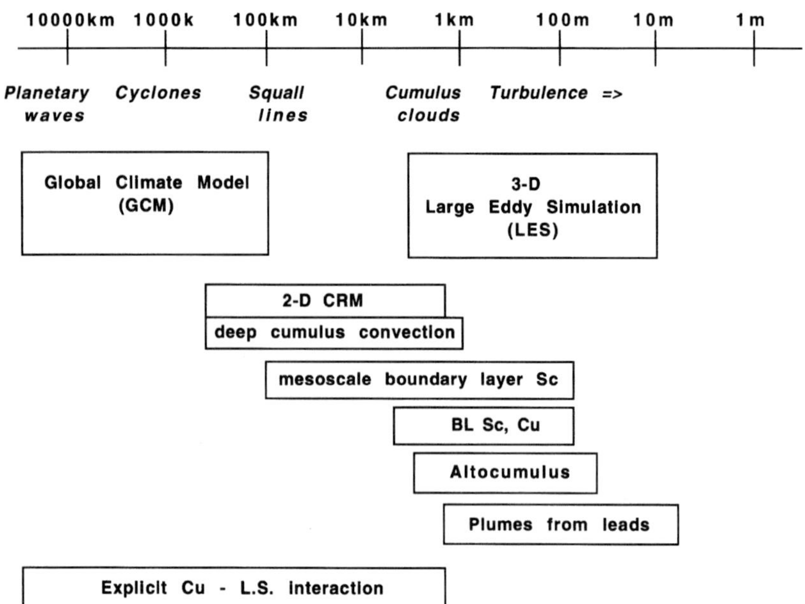

Figure 1 The scales of motion that are explicitly represented by various numerical models of the atmosphere.

Table I

CRMs and GCMs: A Scale Comparison

Aspect	CRM	GCM
Eddies	Cumulus clouds	Baroclinic eddies
Eddy time scale	3×10^3 sec	3×10^5 sec
Forcing time scale	3–4 days	365 days
Domain size	400 km	40,000 km
Horizontal grid size	2 km	200 km
Time step	10 sec	10^3 sec

baroclinic eddies. Thus, a 30-day CRM simulation is equivalent to a 10-year GCM simulation. CRMs have appropriately been called "cloud GCMs."

Both Fig. 1 and Table I make evident that cumulus clouds are not resolved by GCMs: they are sub-grid-scale. Therefore, their collective effects must be parameterized in a GCM. This is difficult. Because CRMs resolve cumulus clouds, a CRM can simulate an ensemble of cumulus clouds, and the results can be used to test or develop various aspects of cumulus parameterizations.

Although CRMs were originally developed for studying convective cloud systems (e.g., Soong and Ogura, 1980; Soong and Tao, 1980), 2-D versions have recently been applied to shallow convecting layers, such as the atmospheric boundary layer. In this case, the horizontal grid size is about 50 m, and the horizontal domain size about 5 km, and a more appropriate name is eddy-resolving model (ERM) because the large convective eddies are explicitly represented, even though the model is 2-D. Because ERMs are 2-D, they are intermediate in terms of physical realism (and computational cost) between the 1-D models used in GCMs to represent shallow convective layers and 3-D large-eddy simulation (LES) models.

In CRMs, the effects of scales of motion that are not resolved are parameterized using a turbulence closure, whereas the effects of scales larger than the domain size (the "large-scale forcing") must be specified. In addition, turbulent surface transfer, cloud microphysical, and radiative transfer processes are parameterized.

Given the large-scale forcing and the surface properties, a CRM is able to determine the evolution of a cloud system directly, to the extent that its representation of grid-scale dynamics and the parameterizations of its own sub-grid scale processes are accurate.

B. The University of Utah Cloud Resolving Model

Under the direction of Akio Arakawa, a 2-D (x-z) CRM was constructed at UCLA by Krueger (1985, 1988) to simulate cumulus ensembles. The model is based on the anelastic set of equations. It includes a third-moment ensemble-mean turbulence closure. Turbulent surface fluxes are diagnosed using flux-profile relationships based on Monin–Obukhov surface layer similarity theory (Businger *et al.*, 1971). Many improvements to the physics have been incorporated since 1985. The current University of Utah version includes the Coriolis force and the y component of the velocity and a diagnostic turbulent length scale (Xu and Krueger, 1991), a turbulent-scale condensation scheme (Chen 1991), a bulk ice-phase microphysics parameterization (Lord *et al.*, 1984; Krueger *et al.*, 1995a; Fu *et al.*, 1995), and an advanced solar and infrared radiation code (Fu, 1991; Fu *et al.*, 1995; Krueger *et al.*, 1995b).

The surface properties, including the surface temperature and ground wetness, can be specified as functions of time and horizontal position. The lateral boundary conditions are cyclic. The time-varying profiles of the large-scale horizontal advective tendencies of potential temperature and water vapor are prescribed. The time-varying profiles of the large-scale vertical advective tendencies of potential temperature and water vapor may also be prescribed, or they may be calculated using a specified large-scale vertical velocity profile and the predicted vertical gradients of potential temperature and water vapor. Large-scale forcing for the horizontal velocity may consist of specifying the time-varying profile of geostrophic velocity, or may be included as a nudging toward a specified time-varying profile.

C. What Is a CRM Good For?

CRMs have been extensively used for simulating convective cloud systems. How good are they for this purpose? Similar to GCMs, spatial resolution and sub-grid-scale physics are issues. Unlike GCMs, dimensionality and lateral boundary conditions (i.e., large-scale forcing) are additional issues. Recently, concerted efforts have been made to evaluate CRM simulations of convective cloud systems observed during GATE and TOGACOARE (e.g., Grabowski *et al.*, 1996; Xu and Randall, 1996; Wu *et al.*, 1998; Redelsperger *et al.*, 2000; Krueger *et al.*, 2000).

What is the appropriate horizontal grid size for simulating convective cloud systems? The philosophy is the same as for GCMs and LES models: Resolve the large eddies that do most of the transport. In deep convective cloud systems, these eddies are the tall convective towers (the "hot towers" of Riehl and Malkus, 1958). A horizontal grid size of 1–2 km is generally adequate to resolve these clouds.

Within the boundary layer (often called the subcloud layer), as well as within clouds, smaller turbulent eddies play a significant role. Turbulent eddies do most of the vertical transport in the boundary layer except in the vicinity of convective cloud systems, where cumulus-scale and mesoscale circulations are also important. Aircraft observations made during TOGA-COARE in the vicinity of active cumulus convection indicate that boundary layer motions at spatial scales of less than 2 km exhibit the characteristics of 3-D turbulence, whereas motions on larger scales have quite different characteristics (Williams *et al.*, 1996). This further justifies the use of a horizontal grid size of 1–2 km when using (2-D or 3-D) CRMs to simulate active cumulus convection. Thus, in a CRM, the turbulence closure parameterizes the vertical transport due to boundary layer turbulence as well as the horizontal transport into clouds (i.e., entrainment) due to in-cloud turbulence.

Is it appropriate to use 2-D CRMs to simulate convective cloud systems? The philosophy has been that 3-D CRMs need to be used if the 3-D structure of the cloud system is the object of the study (e.g., supercells). However, if the goal is to study the statistical (i.e., area-averaged) characteristics of a cloud system, especially in relation to the imposed large-scale forcing, then 2-D simulations appear to be adequate. They typically produce results that are statistically nearly identical to those from 3-D simulations except for larger temporal variability (e.g., Grabowski *et al.*, 1998).

Are the microphysical parameterizations now used in CRMs adequate to simulate convective cloud systems? McCumber *et al.* (1991) concluded that a three-ice-category (cloud ice, snow, graupel/aggregates) bulk parameterization allows a CRM to produce the characteristic radar reflectivity features of tropical squall lines. More recent studies (Redelsperger *et al.*, 2000) indicate that including the ice phase is needed to reproduce the observed dynamical structure of a tropical squall line.

Krueger *et al.* (1995a) found that the microphysics scheme used by Lord *et al.* (1984) underpredicted cloud ice amounts, and hence the horizontal extent of anvil clouds, in simulated tropical squall lines. These upper tropospheric stratiform clouds have a significant radiative impact on the global energy budget. Krueger *et al.* (1995a) modified the microphysics scheme so that it produces more cloud ice. Fu *et al.* (1995) compared the results of the modified scheme to the observations of cloud ice presented by Heymsfield and Donner (1990) and found good agreement.

In a recent CRM intercomparison of multiday simulations of convection during TOGACOARE, differences in microphysical and radiative transfer parameterizations produced significant differences in the cloud radiative forcings (Krueger *et al.*, 2000). Due to the uncertainties in many aspects of the microphysical parameterizations used by CRMs, it may be necessary in the short term to tune the parameterizations so that the top-of-atmosphere radiative fluxes predicted by the CRMs match those observed, similar to how GCMs have tuned their cloud schemes. Unfortunately, this tuning is not as simple or straightforward as it is for GCMs because CRM simulations of GATE and TOGACOARE cloud systems may be significantly affected by inaccuracies in the specification of the large-scale forcing, particularly by the lack of observations of large-scale horizontal advective tendencies of condensate.

As mentioned previously, CRMs have recently been used much like LES models to simulate shallow convective layers. Table II compares the features of 2-D CRMs that explicitly represent the large convective boundary layer eddies (called eddy-resolving models or ERMs) with those of 1-D turbulence closure models (TCMs) and 3-D LES models. Obviously, a 2-D model cannot explicitly represent the 3-D structure of the large turbulent eddies. However, intercomparisons of simulations of a stratocumulus-

Table II
Comparison of Convective Boundary Layer Models

	1-D Turbulence closure	2D CRM/ERM	3D LES
Explicitly represented circulations	None	Mesoscale and large convective eddies	Large turbulent eddies
Parameterized circulations	Clouds and turbulence	3-D turbulence	Small turbulent eddies
Cloud-regime-specific input required	Turbulent length scale Condensation scheme	None	None
Vertical domain	3 km	3 km	3 km
Horizontal domain	(none)	5 km	5×5 km
Grid size	50 m	50×50 m	$50 \times 50 \times 50$ m

topped convective boundary layer with four 2-D CRMs and eight 3-D LES codes by Moeng *et al.* (1996) indicate that the evolution, mean profiles, and scalar flux profiles produced by the two types of models are quite similar, although the turbulent kinetic energy, vertical velocity variance, horizontal velocity variance, and convective mass flux profiles are significantly different due to the differences in the resolved-eddy structure. In the 2-D CRMs, the resolved eddies are roll-like, whereas in the 3-D LES codes, they are plume-like. In addition, the x-component momentum flux profiles are much different due to the differences in large-eddy structure.

Two-dimensional CRMs used as ERMs are more general than 1-D TCMs for simulating shallow convective layers because the large eddies are explicitly represented. As a result, the ERMs do not require the cloud-regime-specific input (turbulent length scale and fractional cloudiness scheme) that the TCMs require (Krueger and Bergeron, 1994). This aspect of ERMs, combined with their computational economy compared to LES codes, makes ERMs an attractive intermediate level model. Their most notable use to date has been for simulating the Lagrangian evolution of the subtropical marine boundary layer (see Section IV).

D. Cloud Process Studies with the UCLA/UU CRM

Most of the studies done with the UCLA/UU (University of Utah) CRM have focused on deep convection over the Tropical oceans. Recently, several studies of shallow convective layers have also been performed. In this chapter, I describe several studies in which I have participated. These were selected to illustrate how a CRM can be used to better understand cloud processes. Two studies involve deep tropical convection. Section II discusses interactions between radiation and convection in tropical cloud clusters, and Section V deals with the enhancement of surface fluxes by tropical convection. The two intervening sections describe studies of shallow convective layers. Section III describes thin midlevel stratiform (altocumulus) clouds, and Section IV describes the stratocumulus-to-trade cumulus transition in the subtropical marine boundary layer. Section VI presents some results from a study of plumes generated by Arctic leads.

II. INTERACTIONS BETWEEN RADIATION AND CONVECTION IN TROPICAL CLOUD CLUSTERS

The results of this study were published in 1995 by Fu *et al.* (hereafter, FKL95). The goal was to better understand the interactions of infrared (IR) radiation and convection in tropical squall cloud clusters on the time scales and space scales of an individual cloud system (about 500 km and 12 hr). The life cycle of a tropical squall line was simulated over a 12-hr period using thermodynamic and kinematic initial conditions as well as large-scale advective forcing typical of a GATE Phase III squall cluster environment.

To study the impact of IR radiation, we performed three simulations: R1, with no radiative cooling; R2, with only clear-sky radiative cooling; and R3, with fully interactive radiative cooling. The differences between R2 and R1 thus represent the effects of *clear-sky IR radiative forcing*, whereas the differences between R3 and R2 reveal the effects of *IR cloud radiative forcing*. The radiative heating rate profiles for the three simulations are shown in Fig. 2.

The differences between the three simulations are primarily due to the differences in radiative forcing because of the highly organized nature of the simulated convective system. To ensure that such a system developed, the initial wind profile that we used was obtained from a multiday simulation described in Xu *et al.* (1992) during a period in which a long-lived squall line developed. In addition, we introduced a cold pool to initiate the squall line.

To illustrate the life cycle and structure of a typical simulated convective cloud system, Fig. 3 shows snapshots of the total hydrometeor mixing ratio, including cloud water, cloud ice, rain, snow, and graupel, based on the R3 simulation at three different times. In the incipient stage (Fig. 3a), the cloud system consists of isolated precipitating convective towers. In the

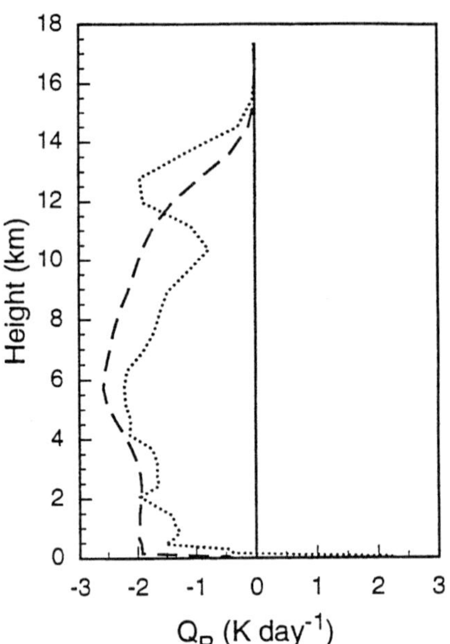

Figure 2 The time- and domain-averaged radiative heating rate profiles for the three simulations (R1, solid; R2, dashed; R3, dotted). [From Fu *et al.* (1995). Reprinted with permission from the American Meteorological Society.]

mature stage (Fig. 3b), new cumulonimbus cells grow at the leading edge of the mesoscale convective system, while older cells successively join the anvil cloud, which has a horizontal extent of about 180 km. At this time, significant precipitation covers a region about 140 km wide. In the dissipating stage (Fig. 3c), little precipitation remains, and the upper-tropospheric stratiform clouds are thinning.

Figure 4 illustrates the effects of clear-sky and cloud IR radiative forcing on the surface precipitation rate during simulations R1, R2, and R3. A comparison of R2 and R1 shows that the clear-sky IR radiative forcing increases the surface precipitation rate throughout the simulation. A comparison of R3 and R2 shows that the cloud IR radiative forcing has essentially no impact until the time of the peak surface precipitation rate (at 280 min), and acts to decrease the surface precipitation rate thereafter. Figure 5c in FKL95 (a time-height cross section of domain-averaged cloud radiative forcing) shows that the cloud radiative forcing does not become significant until about this time. Figure 4 in FKL95 (Hovmöller diagrams of cloud-top temperature for each simulation) shows that significant cloud radiative forcing does not occur until the anvil cloud becomes extensive.

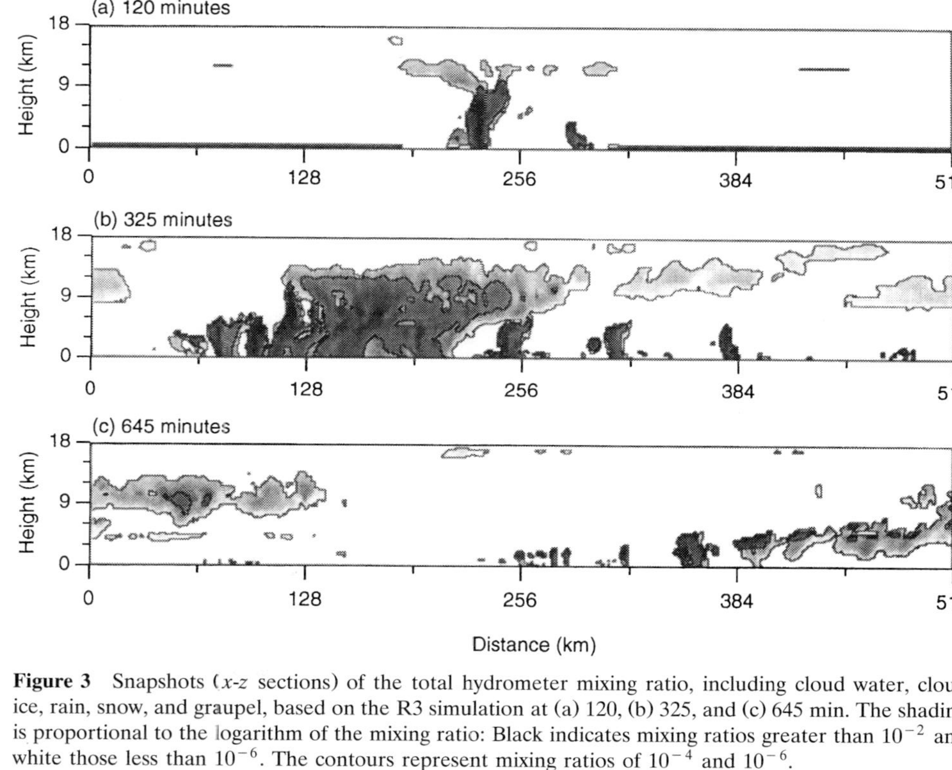

Figure 3 Snapshots (x-z sections) of the total hydrometer mixing ratio, including cloud water, cloud ice, rain, snow, and graupel, based on the R3 simulation at (a) 120, (b) 325, and (c) 645 min. The shading is proportional to the logarithm of the mixing ratio: Black indicates mixing ratios greater than 10^{-2} and white those less than 10^{-6}. The contours represent mixing ratios of 10^{-4} and 10^{-6}.

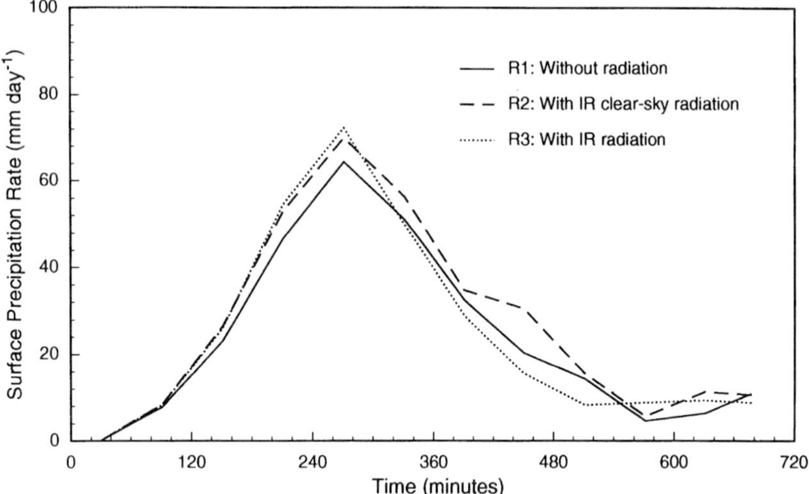

Figure 4 Time series of the surface precipitation rate for the three simulations.

The greater extent of the anvil clouds decreases the outgoing IR flux at the top of the atmosphere by as much as 20 W m^{-2}. This figure also shows that the anvil cloud becomes more extensive in R3 than in R2.

With fully interactive IR radiative heating, direct destabilization of the anvil clouds via IR cloud-top radiative cooling and cloud-base radiative warming (relative to the clear-sky IR cooling rate; see Fig. 2) generates more in-cloud turbulence and contributes to the longevity and extent of the anvil clouds. Figure 5 shows that the time- and domain-averaged cloud fraction at 12 km increases about 0.1 due to cloud radiative forcing, and that this is associated with a marked enhancement of the turbulent kinetic energy within the anvil cloud.

Figure 6 is a schematic illustration of the interactions of IR radiation and convection examined in FKL95's study. It shows that IR clear-sky radiation tends to destabilize the troposphere by cooling it, while IR cloud radiative forcing (due to the radiative effects of the anvil cloud) tends to stabilize the troposphere below 10 km by warming it (relative to IR clear-sky cooling). IR cloud radiative forcing also tends to destabilize the anvil cloud layer from 10 to 14 km, which thereby increases the extent and longevity of the anvil clouds, and thus feeds back to the IR cloud radiative forcing.

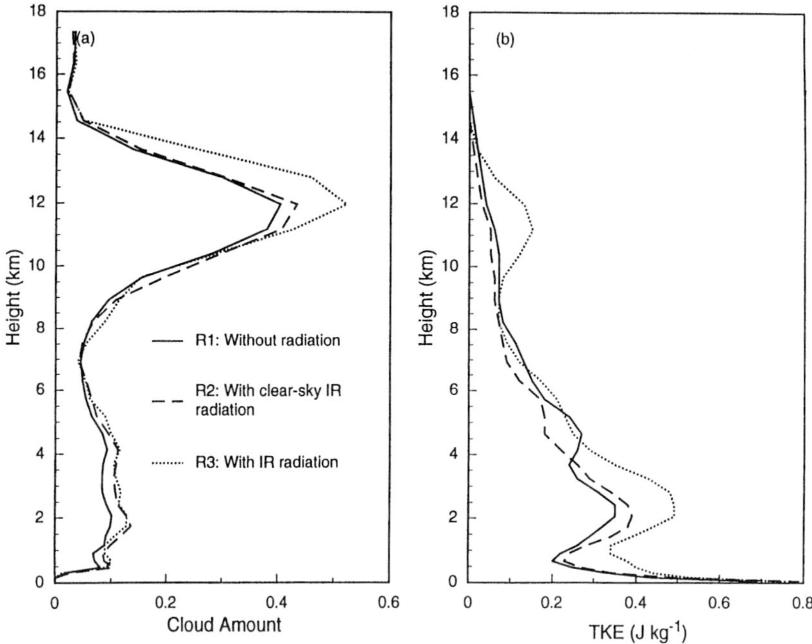

Figure 5 The time- and domain-averaged (a) cloud fraction and (b) turbulent kinetic energy (TKE) profiles for the three simulations. [From Fu *et al.* (1995). Reprinted with permission from the American Meteorological Society.]

III. THIN MIDLEVEL STRATIFORM (ALTOCUMULUS) CLOUDS

Altocumulus (Ac) and altostratus (As) clouds together cover approximately 22% of the Earth's surface (Warren *et al.*, 1986, 1988). Thus, they may play an important role in the Earth's energy budget through their effects on solar and infrared (IR) radiation. However, Ac clouds have been little investigated by either modelers or observational programs.

Heymsfield *et al.* (1991) examined two thin Ac clouds at about −30°C. They concluded that Ac clouds containing a convective structure are dynamically forced by radiative cooling effects. The radiative cooling causes sufficient negative buoyancy in cloud-top parcels to produce the observed downdraft velocities. They also pointed out that the absence of ice crystals implies a dearth of ice nuclei. Recently, Ryan (1996) noted that

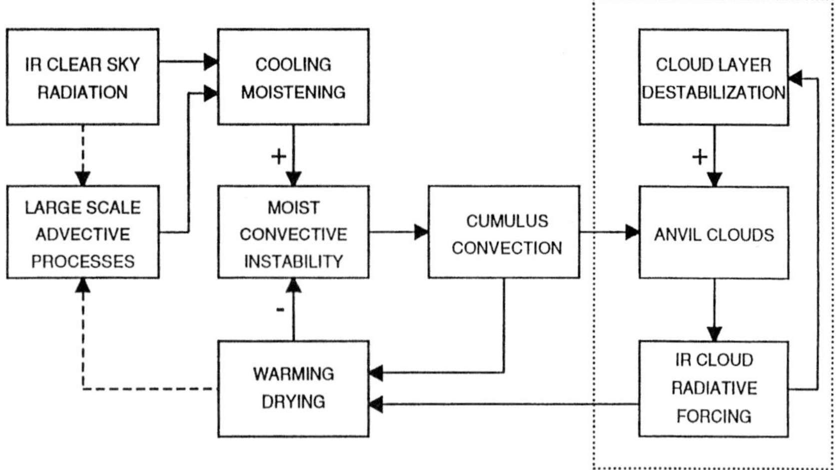

Figure 6 A schematic illustration of the mesoscale interactions of IR radiation and convection. The dashed lines indicate large-scale feedbacks that are not included in the CEM simulations. [From Fu *et al.* (1995). Reprinted with permission from the American Meteorological Society.]

observations indicate that middle-level stratiform clouds less than 1 km deep have either very few ice crystals or no ice crystals.

We studied the effects of radiation in simulated Ac cloud layers using the UCLA/UU CRM. We studied only liquid water clouds without precipitation because they represent most Ac cases. We also discussed the dynamics of the circulation occurring in the simulated Ac cloud layers. This study is described in more detail in Liu and Krueger (1997) and Liu (1998).

Starr and Cox (1985, hereafter SC85) simulated an Ac stratiformis cloud layer as a part of the same study in which they modeled a layer of cirrus. The profiles of temperature and moisture that we use are similar to those used by SC85 to simulate altocumulus. However, the initial supersaturation region thickness is reduced by one-half in our simulations. The model domain we used for our numerical simulations is 3.2 km long and 8.9 km high. The horizontal grid interval is 50 m. The vertical grid interval is 1 km from the surface to 5 km, 500 m from 5.0 to 5.5 km, and 25 m from 5.5 to 8.9 km.

Random perturbations are initially added to the potential temperature field in the supersaturation region to trigger the Ac circulations. The maximum magnitude of the perturbations is 0.1 K. The large-scale vertical velocity is zero.

We simulated a diurnal case which uses the radiative conditions at latitude 30°N on July 15. The starting time is midnight local time. The total simulation time is 36 hr. Solar radiation is significant between 6 and 18 hr, and after 29 hr.

In Fig. 7, the simulated field of liquid water mixing ratio (q_c) for the CRM simulation at 2 hr is displayed. This is a typical q_c snapshot for the simulation under nocturnal conditions. A cellular pattern is evident. The corresponding vertical velocity field is shown in Fig. 8. The updrafts are wider and weaker than the downdrafts. Comparison with Fig. 7 reveals

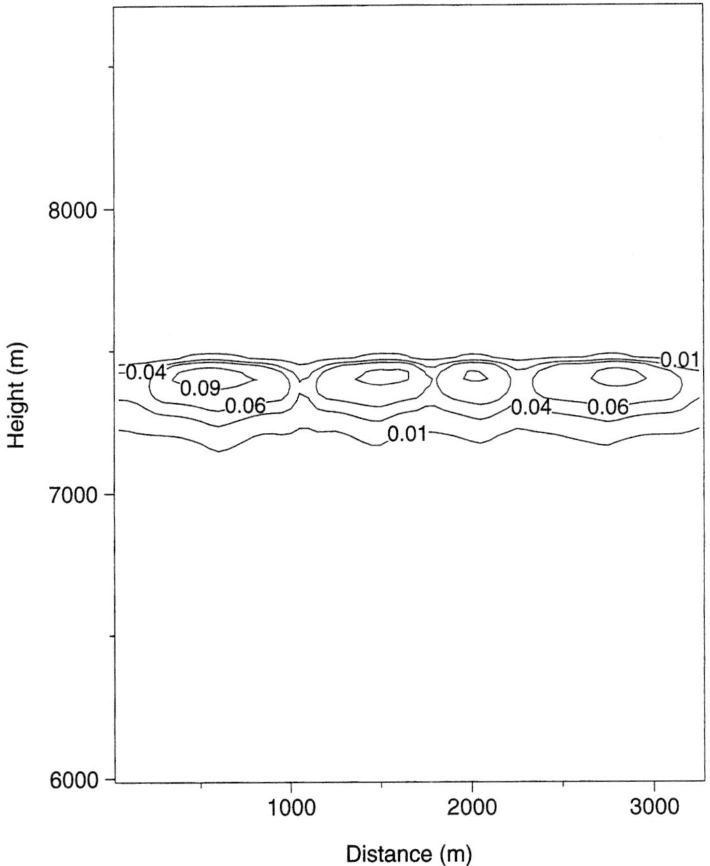

Figure 7 The liquid water mixing ratio field for the altocumulus cloud layer simulation at 2 hr.

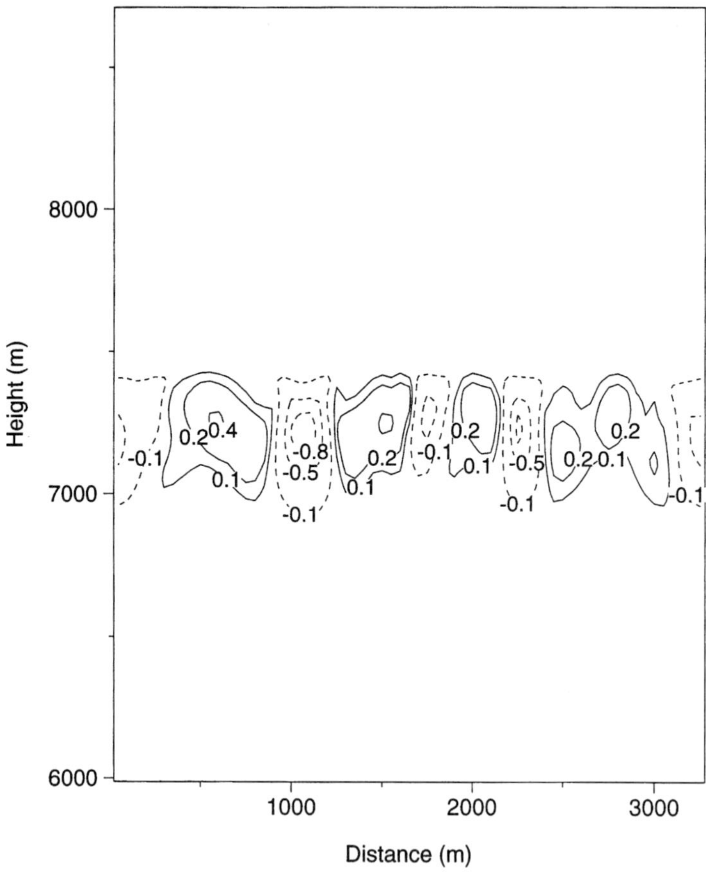

Figure 8   The vertical velocity field for the altocumulus cloud layer simulation at 2 hr.

that the updrafts are associated with regions of large q_c, and the downdrafts with regions of small q_c.

The evolution of the profiles of the horizontally averaged liquid water mixing ratio for the diurnal case is shown in Fig. 9. During the night, the cloud layer ascends and the cloud depth is almost constant. During the daytime, the cloud top height is nearly constant and the cloud depth decreases. After sunset, the cloud depth increases and the cloud layer again ascends. A decreased cloud depth during the daytime under diurnal conditions also occurred in the stratocumulus simulations of Bougeault (1985), Turton and Nicholls (1987), and Wyant *et al.* (1997), and in the marine stratocumulus observations of Hignett (1991).

The effects of radiation on the Ac layer depend on the radiative heating rate profile within the cloud layer. Figure 10 shows the evolution of the horizontally averaged total radiative heating rate for the diurnal case. During the night, there is strong cooling in the upper part of the cloud layer, and some heating in the lower part of the cloud layer. During the morning, we find that the maximum solar heating and IR cooling rates both occur near the cloud top, just like in a simulation of a stratocumulus-topped boundary layer under diurnal conditions (Krueger *et al.*, 1995b). In

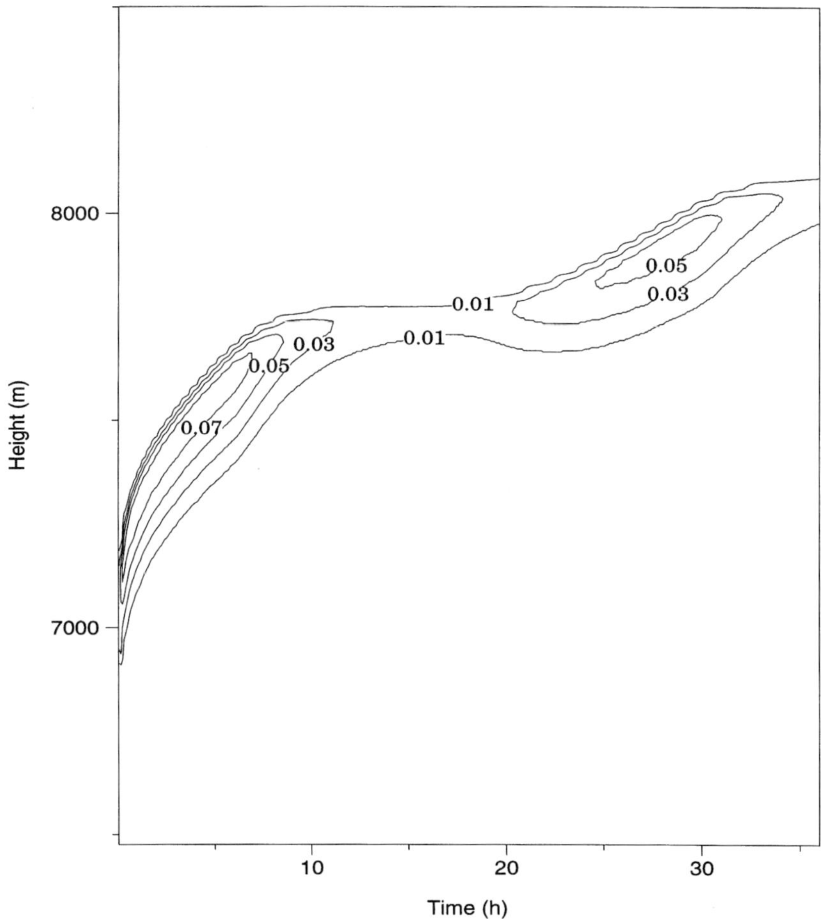

Figure 9 Time–height plot of the horizontally averaged liquid water mixing ratio for the altocumulus cloud layer simulation.

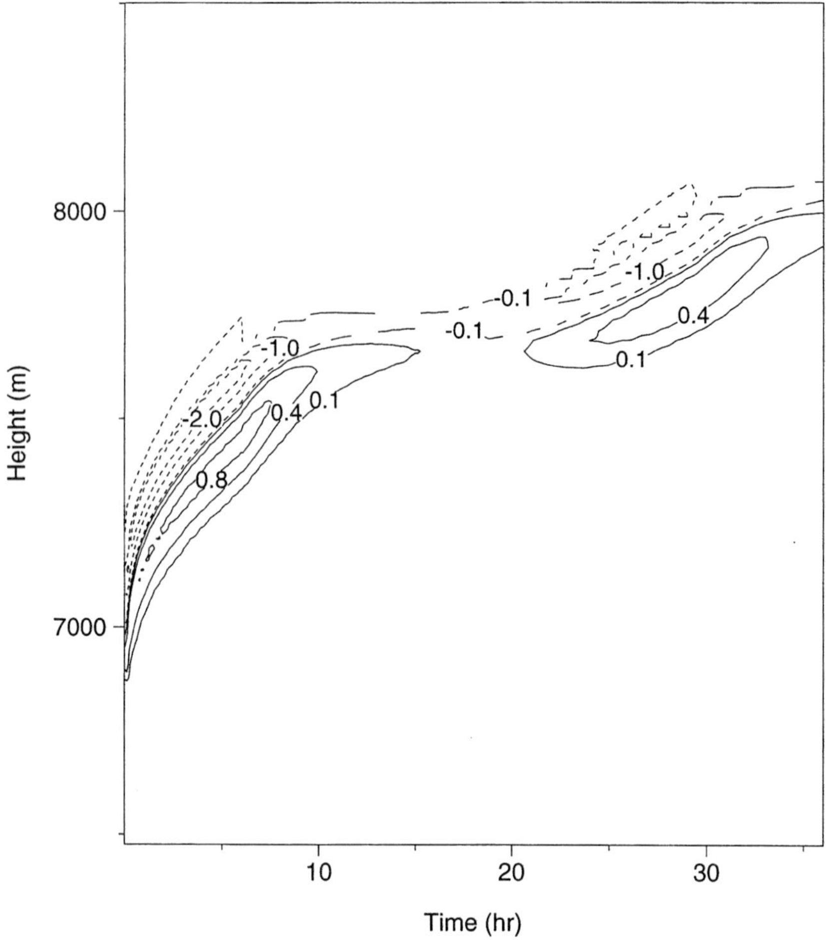

Figure 10　Time–height plot of the horizontally averaged total radiative heating rate for the altocumulus cloud layer simulation.

the Ac layer, this pattern produces net heating in most of the cloud layer during the morning, which decreases the cloud depth and cloud water mixing ratio. By the late afternoon, the cloud region is almost entirely radiatively cooling although the cooling rate is very small, which agrees with the thin afternoon Ac cloud examined by Heymsfield *et al.* (1991).

In Fig. 11 the profiles of the 6-hr averaged updraft and downdraft vertical velocities are shown for the diurnal case. The diurnal variation is obvious. The updraft and downdraft profile shapes are quite similar.

However, the downdraft peak values are larger. In the stratocumulus-topped boundary layer cases simulated by Krueger *et al.* (1995c) and Moeng *et al.* (1992), the updraft and downdraft profile shapes are also quite similar, but they have almost the same peak magnitudes.

The Ac cloud layer thickness and LWP (liquid water path) are controlled by the net cloud layer radiative heating and entrainment warming and drying. A positive feedback exists between LWP and IR radiative cooling, whereas a negative feedback exists between LWP and solar radiative heating. Typically, a negative feedback also exists between LWP and entrainment, but this depends on the relative humidity of the air above cloud top (Randall, 1984).

At the beginning of the diurnal simulation, the IR radiative cooling is large in the upper part of the Ac layer (Fig. 10). This quickly produces an active cloud layer circulation (Fig. 11) and a large entrainment rate (equal to the cloud top ascent rate in these simulations; see Fig. 9). But entrainment has a weak negative feedback on IR radiative cooling in this case. Therefore, only after several hours does the Ac structure and circulation become nearly steady.

After sunrise in the diurnal simulation, solar radiative heating cancels most of the IR radiative cooling in the upper part of the Ac layer, which means that the radiative destabilization is greatly reduced. The dramatic

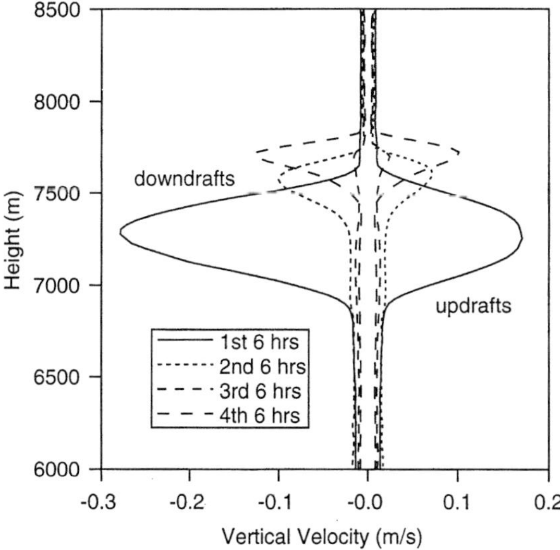

Figure 11   The profiles of the 6-hr-averaged updraft and downdraft vertical velocities for the altocumulus cloud layer simulation.

effect of this on the cloud layer circulation is shown in Fig. 11 and on the entrainment rate in Fig. 9. Solar radiation produces overall cloud layer heating during the morning (Fig. 10), which decreases the cloud depth and cloud water mixing ratio (Fig. 9). Both solar radiative heating and IR radiative cooling in the upper part of the Ac layer decrease quickly as the cloud thins. After most of the cloud liquid water has evaporated, IR cooling balances solar heating so that near-zero total radiative heating exists (Figs. 9 and 10). This allows the cloud to be maintained during the remainder of the afternoon. After sunset, the net cooling increases and the cloud water mixing ratio increases through the night (Fig. 9). At the same time, radiative destabilization increases and as a result so does the cloud circulation and the entrainment rate (Figs. 9 and 11).

These results suggest that actual Ac layers should display a diurnal variation in thickness. It is likely that some Ac layers will completely disappear during the day, and other Ac layers may simply become broken or scattered. Lazarus *et al.* (2000) examined the diurnal variation of Ac cloud amount over the southern Great Plains based on 10 years of edited synoptic cloud reports. They found a significant diurnal variation during the spring, summer, and fall, with maximum cloud amounts at 0600 local standard time and minimum amounts at 1800 local standard time. During winter there was no detectable diurnal variation. These observations are consistent with the CRM results.

IV. STRATOCUMULUS-TO-TRADE CUMULUS TRANSITION IN THE SUBTROPICAL MARINE BOUNDARY LAYER

In the eastern and equatorward quadrants of the subtropical high-pressure zones, the low level air flow (or trajectory) is generally equatorward and westward, across a progressively warmer sea surface and into regions of decreased subsidence. As air flows equatorward over the subtropical oceans, the initially stratocumulus-topped boundary layer (STBL) evolves into a trade cumulus boundary layer (TCBL). This stratocumulus-to-cumulus transition (SCT) involves both a radical decrease in cloud amount, from overcast stratocumulus to scattered cumulus, and a change in the boundary-layer structure and circulation from a well-mixed STBL to a two-layer TCBL with a well-mixed subcloud layer and a conditionally unstable cloud layer.

The physical processes responsible for the SCT have long been a puzzle. Randall (1980) and Deardorff (1980) suggested that cloud-top entrainment instability (CEI) is responsible for the SCT. They hypothesized that under certain conditions, entrainment leads to the generation of turbulence kinetic energy in a stratocumulus layer, which in turn leads to further entrainment and eventual breakup of the cloud layer. This process depends on rapid mixing and evaporative cooling of the entrained air.

However, subsequent observational studies found that STBLs occurred where the stratocumulus layers were predicted to break up due to CEI. More recent theoretical studies (e.g., Krueger, 1993) suggest that CEI does not occur as originally conceived because mixing of entrained air and cloudy air does not proceed rapidly enough.

The first attempts to model the SCT used versions of Lilly's (1968) cloud-topped, mixed-layer model. Schubert *et al.* (1979) presented solutions of a horizontally inhomogeneous version. In one of their experiments, boundary-layer air flows through regions of constant large-scale divergence but increasing SST. The parameters chosen were typical of the eastern North Pacific in July. In this experiment, the boundary layer warms, moistens, and deepens with time, and the turbulent fluxes increase. Wakefield and Schubert (1981) integrated the same model along climatological trajectories in the eastern North Pacific. This "Lagrangian" approach involves translating the model along the boundary-layer trajectory at a rate equal to the observed surface wind speed, while the lower and upper boundary conditions are continuously adjusted to their observed values along the trajectory. In neither of these studies was it possible to simulate the transition from a STBL to a TCBL because the boundary-layer model assumes that a mixed-layer structure always exists.

Moeng and Arakawa (1980) used 1-D second-moment turbulence closure, but in the framework of a 2-D $(y\text{-}z)$ mesoscale model with a horizontal domain of 1000 km and a horizontal grid interval of 10 km. The flow field within the model domain was that of the downward and equatorward branch of a Hadley cell. They found that the uniform stratus cloud layer was destroyed on the equatorward side of the domain, producing a mesoscale circulation of 30–50 km in size.

These studies of the SCT used 1-D approaches to represent the boundary-layer turbulence. The major drawback to using such approaches for simulating the SCT is their inability to predict the cloud fraction accurately. Most if not all 1-D models must be tuned for each type of boundary-layer cloudiness regime. In particular, boundary layers containing cumulus clouds are difficult to model using any 1-D turbulence closure, since the standard closure assumptions assume that the turbulence is nearly isotropic and nearly Gaussian. Cumulus convection has a large vertical velocity skewness and is thus highly non-Gaussian.

In 1-D bulk or turbulence closure models, the convective circulations and the small-scale turbulence are both parameterized. In contrast, a 3-D LES explicitly simulates the convective circulations. Only the small, sub-grid-scale eddies are parameterized. LES models have successfully simulated the STBL and the TCBL (e.g., Moeng, 1986; Sommeria, 1976; Siebesma and Cuijpers, 1995); however, LES models are too computationally expensive to use for the multiday simulations required to study the SCT.

To overcome the limitations of the 1-D and 3-D approaches, we used a 2-D $(x\text{-}z)$ numerical model, the UCLA/UU CRM, to simulate the SCT (Krueger *et al.*, 1995b). For simulating the SCT, a 2-D CRM offers the best combination of generality and economy in comparison to 1-D models and 3-D LES models. A 2-D CRM does not require the cloud-regime-specific input that 1-D models do, and is much less computationally expensive than LES models. Soong and Ogura (1980) were the first to use this type of model to simulate the TCBL.

We used the Lagrangian approach for our SCT simulations, which involves translating the domain along the boundary-layer trajectory at a rate equal to boundary-layer wind speed. This allows the CRM's domain to be small enough to allow high spatial resolution at a moderate computational expense. With this spatial resolution, the CRM can explicitly represent large convective boundary layer eddies. Turbulence other than the large eddies is parameterized using a third-moment turbulence closure. As in the Lagrangian studies described above, we used a July climatological boundary layer trajectory over the northeastern Pacific southwest of California.

We performed two simulations, SCT_1 and SCT_2. The boundary conditions and large-scale forcing are based on the climatological data presented by Betts *et al.* (1992). The surface wind speed was 7 m/s, the divergence was 3×10^{-6} s^{-1}, and the initial SST was 290.2 K. The SST increased 1.8 K/day along the trajectory.

These simulations did not include drizzle, the diurnal cycle, large-scale divergence changes, or mesoscale circulations. Each of these processes has been proposed as a necessary ingredient for a SCT. If a realistic SCT can occur in the simulations in the absence of these processes, it would suggest that these processes are not essential.

For these two simulations, the domain was 4.8 km wide and the horizontal grid size was 75 m. It is not possible to simulate mesoscale circulations with this horizontal domain size. In SCT_1, the domain was 2 km high and the vertical grid size was 50 m, whereas in SCT_2, the domain was 3 km high and the vertical grid size was 75 m.

In SCT_1, the boundary layer was initialized as a cloud-topped mixed layer in equilibrium with a SST of 290.2 K. This case was run for 120 hr. SCT_2 was run because the boundary layer's growth in SCT_1 was significantly affected by the top of the domain after about 84 hr. In SCT_2, the boundary layer was initialized with the SST and horizontally averaged profiles of potential temperature, water vapor mixing ratio, and liquid water mixing ratio of SCT_1 at 69 hr. This case was run for 75 hr (i.e., until hour 144 relative to the start of SCT_1).

The CRM included an advanced solar and infrared radiation code for these simulations (Krueger *et al.*, 1995b). The solar radiation was diurnally averaged so there was no diurnal cycle in the numerical simulation. The solar zenith angle (51.8°) was the mean *daytime* solar zenith angle for mid-July at 30°N.

The horizontally averaged cloud fraction versus height and time is shown in Fig. 12. The transition from overcast to broken and finally scattered clouds is evident. The cloud cover decreases from 100% at the start to about 20% at the end of 6 days. The results presented in Fig. 12 agree with the decrease in cloud cover observed along the climatological trajectory over increasing SST (Betts *et al.*, 1992). The simulated cloud top and base levels are also similar to those observed (as reported by Betts *et al.*, 1992). Riehl *et al.* (1951) presented the first set of soundings along the climatological trade wind trajectory northeast of Hawaii, which corre-

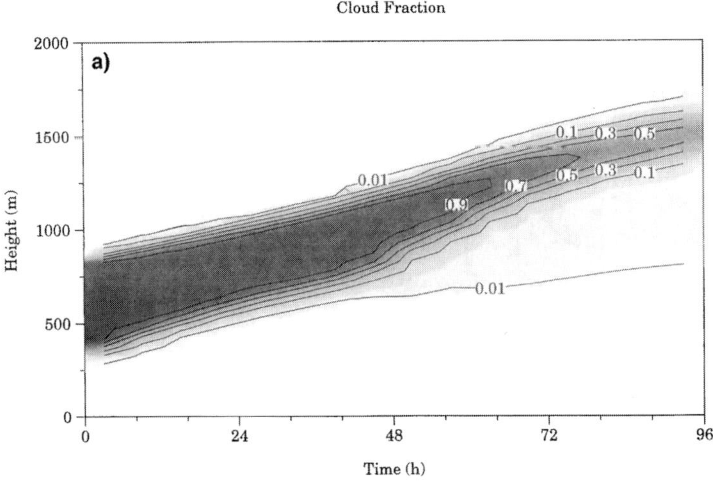

Figure 12 Horizontally averaged cloud fraction for (a) SCT_1 and (b) SCT_2. [From Krueger *et al.* (1995b). Reprinted with permission from the American Meteorological Society.]

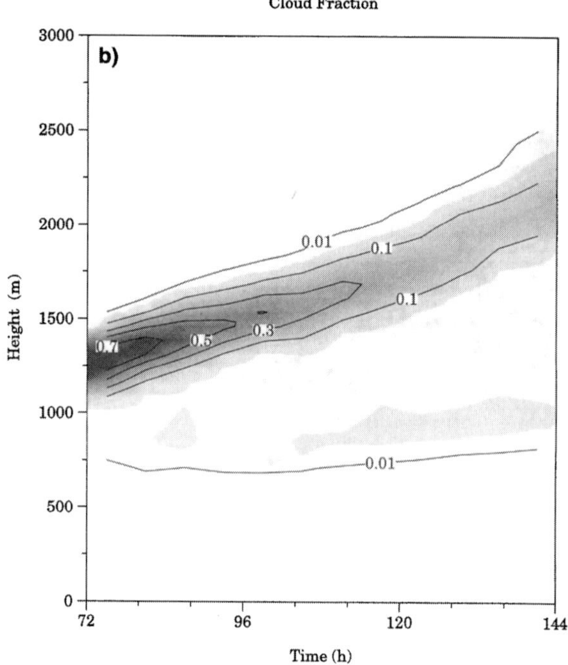

Figure 12   (*Continued*)

sponds to days 4–6 of SCT_2. Riehl *et al.* found that the cloud base slowly rises, while the inversion base rises from 900 to 800 mb, and inversion top from about 800 to 730 mb, in general agreement with Betts *et al.* (1992) and simulation SCT_2.

Examination of the 2-D cloud fields during SCT_1 (shown in Krueger *et al.*, 1995b) reveals that at 24 hr, the boundary layer is stratocumulus topped. By 48 hr, the stratocumulus-topped boundary layer has deepened significantly and some cumulus clouds are forming below it. At 72 hr, cumulus clouds are rising into a broken stratocumulus layer. This cumulus-under-stratocumulus cloud regime (WMO cloud type C_L8) is commonly observed at the location of Weather Ship N (30°N 140°W) during the summer (Klein *et al.*, 1994). The July average SST at Ship N is 295 K, which corresponds to hour 65 of SCT_1. After 72 hr in SCT_1 and SCT_2, the stratocumulus layer becomes less persistent, and by 120 hr, it occurs only in association with detraining cumulus clouds.

The 2-D velocity fields (shown in Krueger *et al.*, 1995b) portray the circulation changes during the SCT. At 24 hr, large eddies extend from the surface to cloud top. At 48 hr, some separation ("decoupling") between cloud and subcloud layer circulations is evident. At 72 hr, in the Cu-under-Sc regime, most of the eddies no longer extend through the entire boundary layer. This two-layer circulation pattern becomes progressively more evident as the stratiform cloud layer thins and breaks up, leaving only scattered cumulus clouds by 120 hr. The transition in the *circulation* pattern from a one-layer to a two-layer structure precedes the transition in the *cloud* pattern from overcast to scattered by 2–3 days. These changes in the boundary-layer circulation are associated with significant changes in the thermodynamic structure.

Figure 13 contains sequences of the horizontally averaged profiles of liquid water mixing ratio (q_c), total water mixing ratio (q_w), and liquid water static energy (s_l). Each sequence spans the entire 6 days of the two simulations. Days 0–3 are from SCT_1, and days 4–6 are from SCT_2. The simulated profiles of s_l and q_w, plus cloud base and cloud top height, are in generally good agreement with observations made during field studies of the trade cumulus boundary layer and the stratocumulus-topped boundary layer.

The q_c profiles (Fig. 13a) show that during the first 2 days, only an overcast stratocumulus cloud layer exists. After 2 days, a layer containing scattered cumulus clouds and very small q_c develops below the stratocumulus layer.

Figures 13b and 13c show the q_w and s_l profiles. Both q_w and s_l are conserved during moist adiabatic processes, and so their profiles are uniform with height in a mixed layer. During the first two days, both quantities exhibit mixed-layer profiles. Observations of the STBL indicate a mixed-layer structure capped by a strong temperature inversion and rapid decrease in water vapor mixing ratio in agreement with the simulation. Klein *et al.* (1994) present average July soundings at Weather Ship N where the Cu-under-Sc regime is commonly observed. These soundings do not exhibit a mixed-layer structure, in agreement with the corresponding simulation profiles at day 3.

By day 4, a two-layer structure resembling that of the TCBL has developed. Radiosonde soundings of the TCBL often exhibit a multilayer structure consisting of a mixed layer capped by an isothermal transition layer about 100 m thick in which the mixing ratio decreases. Above the transition layer, the conditionally unstable cloud layer extends to the base of the trade inversion. The mixing ratio decreases upward in the cloud layer. In the trade inversion, which caps the cloud layer, the temperature increases, and the mixing ratio decreases rapidly. The profiles for days 5 and 6 exhibit a mixed layer below the base of the cumulus clouds. Above a transition zone at about 750 m there is a stratified cloud layer. Due to averaging, the transition zone is not well defined. Across the transition

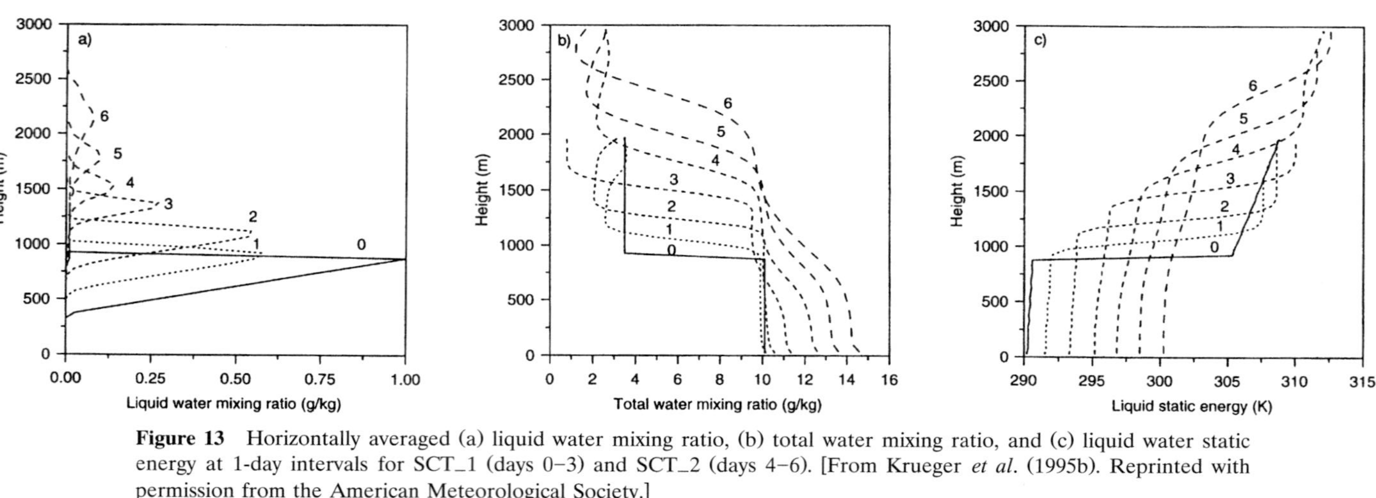

Figure 13 Horizontally averaged (a) liquid water mixing ratio, (b) total water mixing ratio, and (c) liquid water static energy at 1-day intervals for SCT_1 (days 0–3) and SCT_2 (days 4–6). [From Krueger *et al*. (1995b). Reprinted with permission from the American Meteorological Society.]

zone, q_w drops rapidly. Interestingly, the upper part of the cloud layer that contains the stratocumulus layer (see Fig. 13a) is relatively well-mixed in q_w. This feature may be the result of convective motions generated by cloud-top radiative and evaporative cooling that are restricted to the stratocumulus layer.

Figure 13c shows, as also found by Riehl *et al.* (1951), that the cloud layer stratification increases downstream in the trade wind region (days 4–6), while the inversion strength decreases.

Overall, the simulated SCT resembles observations. As noted previously, drizzle, the diurnal cycle, large-scale divergence changes, and mesoscale circulations were not included in the simulation, which therefore demonstrates that these processes are not essential for a SCT.

The results also suggest that there are three stages in the transition from the stratocumulus-topped boundary layer (STBL) to the trade cumulus boundary layer (TCBL). The simulated transition involves an intermediate stage, the "cumulus-under-stratocumulus" boundary layer (CUSBL). The CUSBL has a two-layer structure, like the TCBL, with a well-mixed subcloud layer and a stratified (partly mixed) cloud layer. The transition to a typical TCBL structure preceded the transition to a typical TCBL cloud fraction by about 2 days.

Krueger *et al.* (1995c) developed a convective updraft/downdraft partitioning scheme based on trajectory analysis and used it to analyze the boundary-layer circulation changes during the simulated SCT. They found that the "cumulus-under-stratocumulus" boundary layer, like the trade cumulus boundary layer, has an active subcloud layer circulation that is linked to the cloud layer by narrow cumulus updrafts. The circulation analysis also revealed that during the SCT, convective downdrafts originating near cloud top due to cloud-top cooling became less important, while convective (cumulus) updrafts became more important. This suggests that cloud-top entrainment instability does not play a significant role in the SCT.

A. DECOUPLING

The development of the CUSBL from the well-mixed STBL is due to the increasing frequency of "decoupling" of the subcloud and cloud layers. During decoupling, turbulent eddies are limited to the separate layers. Intermittent coupling of the two layers is achieved by cumulus convection. Decoupling and coupling are illustrated in Fig. 14, which shows a time series from SCT_1 of the horizontally averaged total water mixing ratio at two levels, one in the middle of the subcloud layer, and the other in the upper cloud layer. During decoupling, the subcloud layer is moistened by

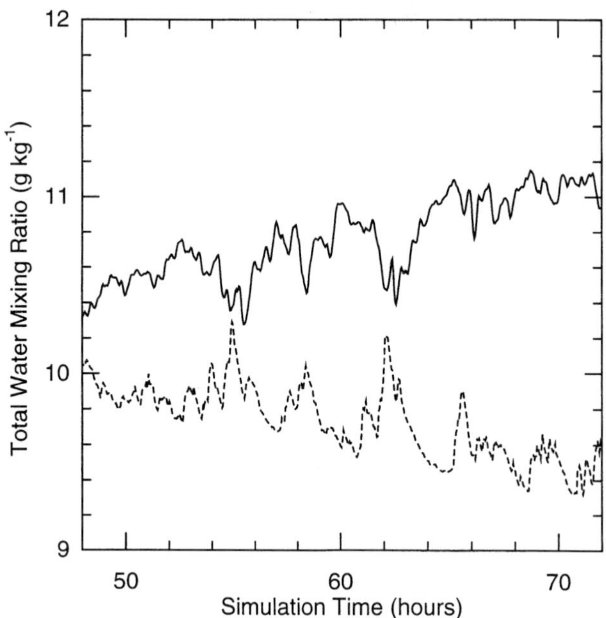

Figure 14   Time series from SCT_1 of the horizontally averaged total water mixing ratio in the middle of the subcloud layer (solid) and in the upper cloud layer (dashed).

surface fluxes while the upper cloud layer is dried by entrainment. During coupling, the opposite occurs, as cumulus updrafts carry moist subcloud-layer air into the cloud layer, while cumulus downdrafts and entrainment at the top of the subcloud layer transport dry air from the cloud layer into the subcloud layer.

Decoupling occurs when cloud-layer downdrafts are unable to penetrate to the surface due to deepening of the boundary layer. Recoupling occurs in the form of cumulus updrafts once the convective available potential energy of subcloud layer parcels increases sufficiently. This increase is due to surface fluxes that increase the moist static energy of the subcloud layer, and to radiative and evaporative cooling of the cloud layer.

Wyant *et al.* (1997) also studied the stratocumulus-to-trade cumulus transition using a similar numerical modeling approach, whereas Bretherton and Wyant (1997) examined the process of decoupling.

B. Summary

To simulate the stratocumulus-to-trade cumulus transition (SCT), a model must be able to represent a cumulus-under-stratocumulus cloud regime. A mixed-layer model can, at most, predict the onset of such a regime (by diagnosing decoupling). To be capable of simulating SCT, a 1-D turbulence closure model must include a realistic parameterization of cloud fraction and the associated buoyancy fluxes. Because a 2-D CRM explicitly represents the large convective eddies associated with cloud formation, it should not require the cloud-regime-specific input that 1-D models do. The results of 2-D CRM simulations of the SCT suggest that this is indeed the case.

V. ENHANCEMENT OF SURFACE FLUXES BY TROPICAL CONVECTION

Large-scale models typically diagnose the surface turbulent fluxes of sensible and latent heat over the ocean using the large-scale (i.e., area-averaged) near-surface temperature and water vapor mixing ratio and the speed of the large-scale wind vector. These fluxes may be called the "vector-mean" surface fluxes. Esbensen and McPhaden (1996) defined "mesoscale enhancement" as the difference between the vector-mean surface fluxes and the actual large-scale surface fluxes. In the absence of mesoscale circulations, there would be no mesoscale enhancement.

Based on buoy data, Esbensen and McPhaden (1996) found that mesoscale enhancement of evaporation can reach 30% of the total evaporation. They also showed that mesoscale enhancement is primarily due to mesoscale wind variability ("gustiness") and is associated with periods of significant precipitation.

Several multiday, large-domain simulations of tropical maritime convection were performed with the 2-D UCLA/UU CRM, as reported by Xu *et al.* (1992). Analyses of these simulations indicate that the characteristics of mesoscale enhancement as simulated by the model are quite similar to those observed (Zulauf and Krueger, 1997, 2000).

In the CRM, the bulk aerodynamic method of Deardorff (1972) is used to calculate the local (i.e., grid-point) latent and sensible heat fluxes at the surface, E and S. Using the notation of Esbensen and McPhaden (1996), the domain-averaged latent heat flux, $\langle E \rangle$, for example, is then

$$\langle E \rangle = \rho L_{\mathrm{v}} \langle C_{\mathrm{q}}(U, \cdot) U (q_{\mathrm{s}} - q) \rangle,$$

where ρ is the density, L_{v} is the latent heat of vaporization, C_{q} is the stability-dependent transfer coefficient, U is the wind speed, $\cdot$ represents additional dependences of C_{q}, q_{s} is the surface mixing ratio, q is the near-surface mixing ratio, and angle brackets indicate the domain average.

The corresponding vector-mean flux is

$$E_v = \rho L_v C_q(V, \langle \cdot \rangle) V (\langle q_{sfc} \rangle - \langle q \rangle),$$

where V is the magnitude of the average (or resultant) wind vector, which is predicted by large-scale models. By replacing V with the domain-averaged wind speed, $\langle U \rangle$, we obtain the scalar-mean flux,

$$E_s = \rho L_v C_q(\langle U \rangle, \langle \cdot \rangle) \langle U \rangle (\langle q_s \rangle - \langle q \rangle).$$

In the CRM simulations, the differences between V and $\langle U \rangle$ can be significant. We used V and $\langle U \rangle$ to calculate the vector-mean and scalar-mean latent and sensible heat fluxes, and compared them with the domain-averaged fluxes. We found that the vector-mean fluxes do a poor job of estimating the domain-averaged fluxes (i.e., that the mesoscale enhancement in the simulations is significant), but the scalar-mean fluxes track them extremely well.

For example, in one 11-day simulation (Q03), $\langle E \rangle = 88.2$ W m^{-2} and $E_v = 57.4$ W m^{-2}, whereas $E_s = 90.5$ W m^{-2}. The results are similar for the sensible heat fluxes: $\langle S \rangle = 17.5$ W m^{-2} and $S_v = 11.1$ W m^{-2}, whereas $S_s = 17.7$ W m^{-2}. The rms errors of the vector-mean and scalar-mean fluxes (relative to the domain-averaged fluxes) are 33.4 and 2.4 W m^{-2} for E_v and E_s, and 6.9 and 0.4 W m^{-2} for S_v and S_s, respectively. These are nearly the same as the mean errors.

We can approximate the scalar-mean fluxes, which are good estimates of the large-scale (i.e., domain-averaged) fluxes, if we can parameterize $\langle U \rangle$ in terms of V. Because V is always less than or equal to $\langle U \rangle$, we may introduce a gustiness speed U_g such that

$$\langle U \rangle^2 = V^2 + U_g^2.$$

The gustiness speed represents the contribution to the average wind speed from sub-grid-scale circulations (e.g., Mahrt and Sun, 1995). We will assume that U_g is primarily due to circulations driven by cumulus convection. However, boundary-layer convection also contributes to U_g, and formulations to parameterize this component of the gustiness speed have already been developed (e.g., Beljaars, 1995). (The CRM includes one.) Therefore, we separate U_g into components due to boundary layer convection, $U_{g,BL}$, and due to cumulus convection, $U_{g,CU}$, so that

$$\langle U \rangle^2 = V^2 + U_{g,BL}^2 + U_{g,CU}^2.$$

The variance of the cumulus-scale horizontal wind vector,

$$U_{\sigma,CU}^2 \equiv \langle u_{CU}'^2 \rangle + \langle v_{CU}'^2 \rangle,$$

where u'_{CU} and v'_{CU} are the cumulus-scale deviations from $\langle u \rangle$ and $\langle v \rangle$, the components of the average (or large-scale) wind vector, can be formally related to the gustiness speed due to cumulus convection:

$$U_{g,CU} = \alpha U_{\sigma,CU}.$$

Theoretical methods give a value of 0.8 for α (Jabouille *et al.*, 1996). In our simulations, α is usually between 0.75 and 0.85. These results suggest that a constant value of α is a good approximation. Then $U_{g,CU}$, and hence $\langle U \rangle$, can be obtained if $U_{\sigma,CU}$ can be parameterized.

The quantity $U^2_{\sigma,CU}$ should be related to a measure of the intensity of the cumulus convection, such as the updraft cloud mass flux at cloud-base level. The scatter plots in Fig. 15 show how $U^2_{\sigma,CU}$ is related to various measures of cumulus activity based on the results of several multiday CEM simulations. The quantities in the plots are 3-hr domain (512-km) averages. It is interesting that all six measures chosen are about equally correlated with $U^2_{\sigma,CU}$. In particular, the surface rainfall rate is as well correlated as any of the other more direct measures of cumulus activity. Jabouille *et al.* (1996) used observations to develop a similar parameterization based on the surface rainfall rate.

The linear fits displayed in Fig. 15 form the basis of a parameterization of $\langle U \rangle$ that augments V with a gustiness speed due to cumulus convection, which in turn can be linked via $U^2_{\sigma,CU}$ to one of several measures of cumulus activity. Many of these measures are available in large-scale models.

VI. PLUMES GENERATED BY ARCTIC LEADS

The interactions between sea ice, open ocean, atmospheric radiation, and clouds over the Arctic Ocean exert a strong influence on global climate. Uncertainties in the formulation of interactive air–sea–ice processes in global climate models (GCMs) result in large differences between the Arctic and global climates simulated by different models. In particular, the effects of leads on the atmosphere and the surface heat budget of the Arctic Ocean must be more accurately represented in climate models to allow possible feedbacks between leads and the sea ice thickness.

We are using the UCLA/UU Cloud Resolving Model to increase our understanding of (1) how atmospheric convective plumes emanating from leads affect the large-scale atmospheric budgets of sensible heat, water vapor, and condensate, and (2) how the contribution by such plumes to Arctic cloud cover, directly through the production of clouds and indirectly by increasing boundary-layer moisture, affects the surface heat budget of the Arctic Ocean.

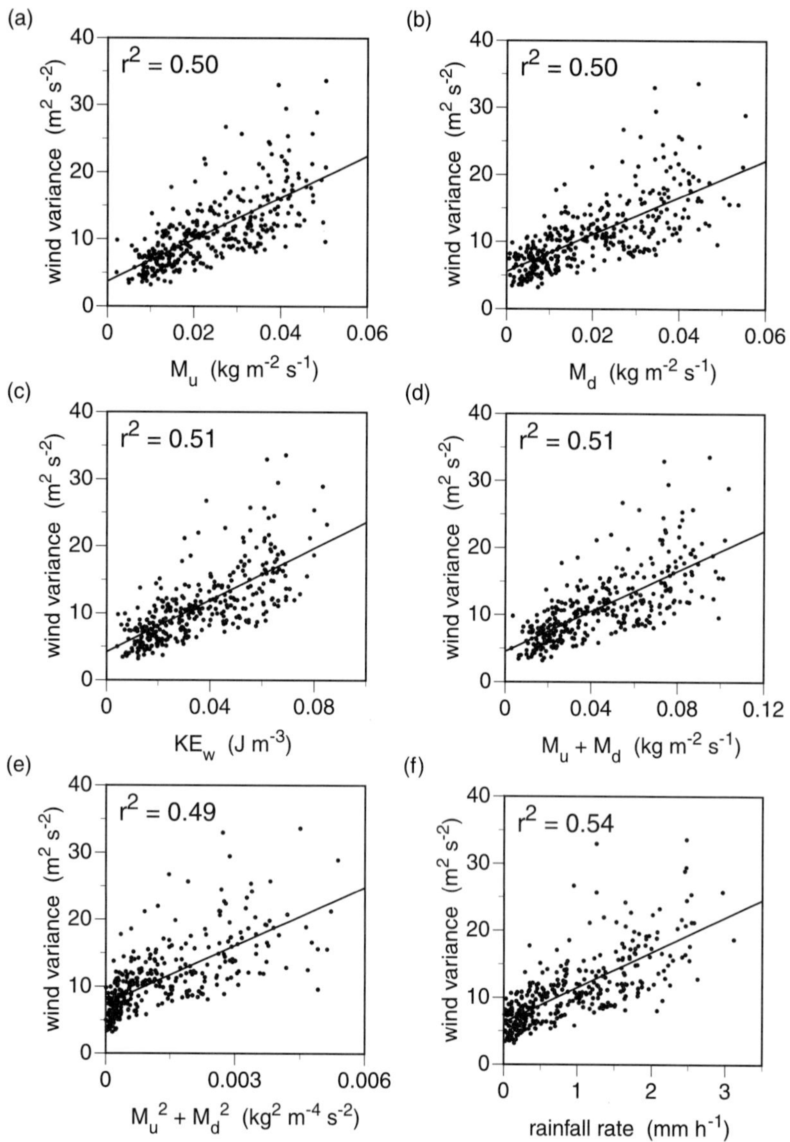

Figure 15 Scatterplots of large-scale cumulus activity versus $U^2_{\sigma,\mathrm{CU}}$ using consensus data for (a) updraft cloud mass flux $\langle M_u \rangle$, (b) downdraft cloud mass flux $\langle M_d \rangle$, (c) vertical component of kinetic energy $\langle KE_w \rangle$, (d) $\langle M_u \rangle + \langle M_d \rangle$, (e) $\langle M_u^2 \rangle + \langle M_d^2 \rangle$, and (f) surface precipitation rate $\langle P \rangle$. Also shown for each scatterplot is the least squares linear fit and r^2, the coefficient of determination.

Glendening and Burk (1992) and Glendening (1994) performed 3-D large-eddy simulations of the convective plumes generated by 200-m-wide leads with different geostrophic wind angles. We performed 2-D CRM simulations of the same cases to gauge the impacts of differences in model physics (Zulauf and Krueger, 1999).

The cases are characterized by specified ocean ($-2°C$) and ice ($-29°C$) temperatures, a stable, nearly isothermal, lower troposphere, a surface temperature of $-27°C$, and a geostrophic wind speed of 2.5 m s^{-1}. The CRM's vertical grid size was 4 m, and the horizontal grid size was 8 m. Each simulation was for 14 min.

Table III describes the setup of each of the four cases and summarizes the simulation results. Cases A, B, and C differ only in the geostrophic wind angle, and cases A and D differ only in the lead width.

Cases A, B, and C correspond to three of Glendening's cases. Figure 16 shows the time-averaged total turbulent kinetic energy for these cases. As found by Glendening, the plume character greatly depends on the geostrophic wind angle. For a geostrophic wind angle of 0 deg (case A), the geostrophic wind is parallel to the lead. For this case, the turbulent plume is symmetrical, erect, concentrated above the center of the lead, and reaches a height of 190 m. For a geostrophic wind angle of 15 deg (case B), the plume rises from the lee side of the lead. Further downstream it becomes elevated and reaches its maximum height of 115 m. For a geostrophic wind angle of 90 deg (case C), the geostrophic wind is perpendicular to the lead. For this case, the turbulent plume is similar to that for an angle of 15 deg, but the plume reaches only 65 m, and does not become completely detached from the surface.

The plume characteristics simulated by the 2-D CRM are qualitatively similar to those of Glendening's 3-D large-eddy simulations. The dependence of the plume height on geostrophic wind angle is in quantitative agreement with Glendening's results.

Due to the computational requirements of large-eddy simulations, Glendening was not able to study the dependence of plume characteristics

Table III

Simulations of Arctic Leads

Case	Domain size (m)	Lead width (m)	Geostrophic wind angle (deg)	$(F_s)_{lead}$ (W m^{-2})	$(LF_q)_{lead}$ (W m^{-2})	Plume height (m)
A	768	200	0	244	73	190
B	768	200	15	246	75	115
C	2304	200	90	243	72	65
D	768	400	0	264	80	270

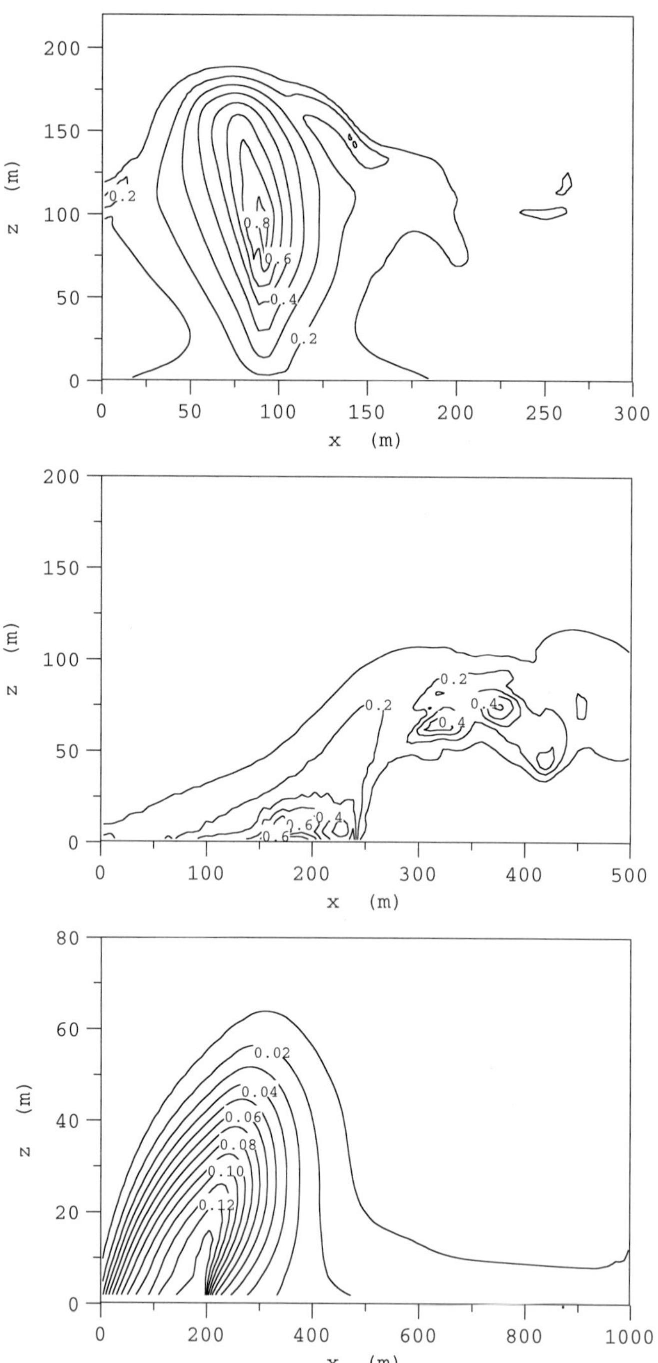

Figure 16   Mean total turbulent kinetic energy (m^2 s^{-2}) for cases A (top), B (middle), and C (bottom).

on lead width. However, it is possible to use a 2-D CRM to study this dependence in detail because such models are relatively economical. Table III shows that with a lead-parallel geostrophic wind, a plume generated by a 400-m-wide lead (case D) reaches 270 m, whereas one generated by a 200-m-wide lead (case A) reaches 190 m. The dependence on lead width is significant.

It would be interesting to use the 2-D CRM to study how cloud formation and interactive radiation affect the plume characteristics, and how cloudiness from a plume evolves and affects the radiative fluxes at the surface.

VII. CONCLUSIONS

Cloud-resolving models are numerical models that explicitly represent the cloud-scale circulations and their interactions with microphysical, radiative, and small-scale turbulent processes. Most CRMs are 2-D models, which allows them to simulate the mesoscale organization of cloud systems over multiday time periods. In contrast, 3-D large-eddy simulation models are limited to simulating large turbulent eddies over periods of several hours, and 1-D column models must parameterize the cloud-scale circulations, as well as the turbulence.

Two-dimensional CRMs have been used to study a variety of cloud types and systems, as this chapter illustrates. During the next decade, we can expect a number of improvements in CRMs. New observations of clouds will allow more thorough evaluations of the microphysical and radiative transfer parameterizations employed in CRMs. In addition, ever-increasing computational resources will allow 3-D CRMs and very-large-domain 2-D CRMs to become more commonly used. In the not-so-distant future, 3-D CRMs with domains covering an entire large-scale disturbance will become reality. This will allow cloud-resolving simulations of the interaction between cumulus ensembles and the large-scale circulation.

ACKNOWLEDGMENTS

This research was supported by AFOSR grant 91-0039, DOE Environmental Science Division grant DE-FG03-94ER61769, ONR grant N00014-91-J-1175, NASA grant NAG1-1718, and NSF grant OPP-9702583. Computing assistance was provided by NCAR SCD.

REFERENCES

Beljaars, A. C. M. (1995). The parameterization of surface fluxes in large scale models under free convection. *Quart. J. Roy. Meteor. Soc.* **121**, 255–270.

Betts, A. K., P. Minnis, W. Ridgway, and D. F. Young (1992). Integration of satellite and surface data using a radiative-convective oceanic boundary-layer model. *J. Appl. Meteor.* **31,** 340–350.

Bougeault, P. (1985). The diurnal cycle of the marine stratocumulus layer: A higher-order model study. *J. Atmos. Sci.* **42,** 2826–2843.

Bretherton, C. S., and M. C. Wyant (1997). Moisture transport, lower tropospheric stability and decoupling of cloud-topped boundary layers. *J. Atmos. Sci.* **54,** 148–167.

Businger, J. A., J. C. Wyngaard, Y. Izumi, and E. F. Bradley (1971). Flux-profile relationships in the atmospheric surface layer. *J. Atmos. Sci.* **28,** 181–189.

Chen, J.-M. (1991). Turbulence-scale condensation parameterization. *J. Atmos. Sci.* **48,** 1510–1512.

Deardorff, J. W. (1972). Parameterization of the planetary boundary layer for use in general circulation models. *Mon. Wea. Rev.* **100,** 93–106.

Deardorff, J. W. (1980). Cloud top entrainment instability. *J. Atmos. Sci.* **37,** 131–147.

Esbensen, S. K. and M. J. McPhaden (1996). Enhancement of tropical ocean evaporation and sensible heat flux by atmospheric mesoscale systems. *J. Climate* **9,** 2307–2325.

Fu, Q. (1991). Parameterization of radiative processes in vertically nonhomogeneous multiple scattering atmospheres. Ph.D. Dissertation, Dept. of Meteorology, University of Utah, Salt Lake City.

Fu, Q., S. K. Krueger, and K. N. Liou (1995). Interactions of radiation and convection in simulated tropical cloud clusters. *J. Atmos. Sci.* **52,** 1310–1328.

Glendening, J. W. (1994). Dependence of a plume heat budget upon lateral advection. *J. Atmos. Sci.* **51,** 3517–3530.

Glendening, J. W., and S. D. Burk (1992). Turbulent transport from an Arctic lead: A large-eddy simulation. *Bound.-Layer. Meteor.* **59,** 315–339.

Grabowski, W. W., X. Wu, and M. W. Moncrieff (1996). Cloud resolving modeling of tropical cloud systems during Phase III of GATE. Part I: Two-dimensional experiments. *J. Atmos. Sci.* **53,** 3684–3709.

Grabowski, W. W., X. Wu, M. W. Moncrieff, and W. D. Hall (1998). Cloud-resolving modeling of cloud systems during Phase III of GATE. Part II: Effects of resolution and the third spatial dimension. *J. Atmos. Sci.* **55,** 3264–3282.

Heymsfield, A. J., and L. J. Donner (1990). A scheme for parameterizing ice-cloud water content in general circulation models. *J. Atmos. Sci.* **47,** 1865–1877.

Heymsfield, A. J., L. M. Miloshevich, A. Slingo, K. Sassen, and D. O'C. Starr (1991). An observational and theoretical study of highly supercooled altocumulus. *J. Atmos. Sci.* **48,** 923–945.

Hignett, P. (1991). Observations of diurnal variation in a cloud-topped marine boundary layer. *J. Atmos. Sci.* **48,** 1474–1482.

Jabouille, P., J. L. Redelsperger, and J. P. Lafore (1996). Modification of surface fluxes by atmospheric convection in the TOGA-COARE region. *Mon. Wea. Rev.* **124,** 816–837.

Klein, S. A., D. L. Hartmann, and J. R. Norris (1994). On the relationships among low cloud structure, sea surface temperature, and atmospheric circulation in the summertime northeast Pacific. *J. Climate* **8,** 1140–1155.

Krueger, S. K. (1985). Numerical simulation of tropical cumulus clouds and their interaction with the subcloud layer. Ph.D. Dissertation, Dept. of Atmospheric Sciences, University of California, Los Angeles.

Krueger, S. K. (1988). Numerical simulation of tropical cumulus clouds and their interaction with the subcloud layer. *J. Atmos. Sci.* **45,** 2221–2250.

Krueger, S. K. (1993). Linear eddy modeling of entrainment and mixing in stratus clouds. *J. Atmos. Sci.* **50,** 3078–3090.

Krueger, S. K., and A. Bergeron (1994). Modeling the trade cumulus boundary layer. *Atmos. Res.* **33,** 169–192.

Krueger, S. K., Q. Fu, K. N. Liou, and H.-N. S. Chin (1995a). Improvements of an ice-phase microphysics parameterization for use in numerical simulations of tropical convection. *J. Appl. Meteor.* **34**, 281–287.

Krueger, S. K., G. T. McLean, and Q. Fu (1995b). Numerical simulation of the stratus-to-cumulus transition in the subtropical marine boundary layer. Part I: Boundary-layer structure. *J. Atmos. Sci.* **52**, 2839–2850.

Krueger, S. K., G. T. McLean, and Q. Fu (1995c). Numerical simulation of the stratus-to-cumulus transition in the subtropical marine boundary layer. Part II: Boundary-layer circulation. *J. Atmos. Sci.* **52**, 2851–2868.

Krueger, S. K., S. M. Lazarus, P. Bechtold, S. Chen, D. Cripe, L. Donner, W. Grabowski, M. Gray, D. Gregory, J. Gregory, F. Guichard, H. Jiang, D. Johnson, R. McAnelly, J. Petch, D. Randall, J.-L. Redelsperger, C. Seaman, H. Su, W.-K. Tao, X. Wu, and K.-M. Xu (2000). Intercomparison of multi-day simulations of convection during TOGA COARE using several cloud-resolving and single-column models. Submitted for publication.

Lazarus, S. M., S. K. Krueger, and G. G. Mace (2000). A cloud climatology of the Southern Great Plains ARM CART. Accepted for publication.

Lilly, D. K. (1968). Models of cloud-topped mixed layers under a strong inversion. *Quart. J. Roy. Meteor. Soc.* **94**, 292–309.

Liu, S. (1998). Numerical modeling of altocumulus cloud layers, Ph.D. Dissertation. Dept. of Meteorology, University of Utah, Salt Lake City.

Liu, S., and S. K. Krueger (1997). Effects of radiation in simulated altocumulus cloud layers. In "Preprints, Ninth Conference on Atmospheric Radiation," Long Beach, CA, pp. 330–334. Amer. Meteor. Soc.

Lord, S. J., H. E. Willoughby, and J. M. Piotrowicz (1984). Role of a parameterized ice-phase microphysics in an axisymmetric, nonhydrostatic tropical cyclone model. *J. Atmos. Sci.* **41**, 2836–2848.

Mahrt, L., and J. Sun (1995). The subgrid velocity scale in the bulk aerodynamic relationship for spatially averaged scalar fluxes. *Mon. Wea. Rev.* **123**, 3032–3041.

McCumber, M., W. K. Tao, J. Simpson, R. Penc, and S. T. Soong (1991). Comparison of ice-phase microphysical parameterization schemes using numerical simulations of convection. *J. Appl. Meteor.* **30**, 985–1004.

Moeng, C.-H. (1986). Large-eddy simulation of a stratus-topped boundary layer. Part I: Structure and budgets. *J. Atmos. Sci.* **43**, 2886–2900.

Moeng, C.-H., and A. Arakawa (1980). A numerical study of a marine subtropical stratus cloud layer and its stability. *J. Atmos. Sci.* **37**, 2661–2676.

Moeng, C.-H., S. Shen, and D. A. Randall (1992). Physical processes within the nocturnal stratus-topped boundary layer. *J. Atmos. Sci.* **49**, 2384–2401.

Moeng, C.-H., W. R. Cotton, C. Bretherton, A. Chlond, M. Khairoutdinov, S. K. Krueger, W. S. Lewellen, M. K. McVean, J. R. M. Pasquier, H. A. Rand, A. P. Siebesma, R. I. Sykes, B. Stevens (1996). Simulation of a stratocumulus-topped PBL: Intercomparison of different numerical codes. *Bull. Am. Meteor. Soc.* **77**, 216–278.

Randall, D. A. (1980). Conditional instability of the first kind upside-down. *J. Atmos. Sci.* **37**, 125–130.

Randall, D. A. (1984). Stratocumulus cloud-deepening through entrainment. *Tellus* **36A**, 446–457.

Redelsperger, J. L., P. R. A. Brown, F. Guichard, C. Hoff, M. Kawasima, S. Lang, T. Montmerle, K. Nakamura, K. Saito, C. Seman, W. K. Tao, and L. J. Donner (2000). A GCSS model intercomparison for a tropical squall line observed during TOGA-COARE. I: Cloud-resolving models. *Quart. J. Roy. Meteor. Soc.* **126**, 823–864.

Riehl, H., and J. S. Malkus (1958). On the heat balance of the equatorial trough zone. *Geophysica* **6**, 503–538.

Riehl, H., T. C. Yeh, J. S. Malkus, and N. E. La Seur (1951). The north-east trade of the Pacific Ocean. *Quart. J. Roy. Meteor. Soc.* **77**, 598–626.

Ryan, B. F. (1996). On the global variation of precipitating layer clouds. *Bull. Am. Meteor. Soc.*, **77**, 53–70.

Schubert, W. S., J. S. Wakefield, E. J. Steiner, and S. K. Cox (1979). Marine stratocumulus convection, Part II: Horizontally inhomogeneous solutions. *J. Atmos. Sci.* **36**, 1308–1324.

Siebesma, A. P., and J. W. M. Cuijpers (1995). Evaluation of parametric assumptions for shallow cumulus convection. *J. Atmos. Sci.* **52**, 650–666.

Sommeria, G. (1976). Three-dimensional simulation of turbulent processes in the undisturbed trade wind boundary layer. *J. Atmos. Sci.* **33**, 216–241.

Soong, S.-T., and W.-K. Tao (1980). Response of deep tropical cumulus clouds to mesoscale processes. *J. Atmos. Sci.* **37**, 2035–2050.

Soong, S.-T., and Y. Ogura (1980). Response of tradewind cumuli to large-scale processes. *J. Atmos. Sci.* **37**, 2016–2034.

Starr, D. O'C. and S. K. Cox (1985). Cirrus clouds. Part II: Numerical experiments on the formation and maintenance of cirrus. *J. Atmos. Sci.* **42**, 2682–2694.

Turton, J. D., and S. Nicholls (1987). A study of the diurnal variation of stratocumulus using a multiple mixed layer model. *Quart. J. R. Meteor. Soc.* **113**, 969–1009.

Wakefield, J. S., and W. S. Schubert (1981). Mixed-layer model simulation of Eastern North Pacific stratocumulus. *Mon. Wea. Rev.* **109**, 1952–1968.

Warren, S. G., C. J. Hahn, J. London, R. M. Chervin, and R. Jenne (1986). Global distribution of total cloud cover and cloud type amounts over land, NCAR Tech. Note TN-273 + STR.

Warren, S. G., C. J. Hahn, J. London, R. M. Chervin, and R. Jenne (1988). Global distribution of total cloud cover and cloud type amounts over ocean, NCAR Tech. Note TN-317 + STR.

Williams, A. G., H. Kraus, and J. M. Hacker (1996). Transport processes in the tropical warm pool boundary layer. Part I: Spectral composition of the fluxes. *J. Atmos. Sci.* **53**, 1187–1202.

Wu, X., W. W. Grabowski, and M. W. Moncrieff (1998). Long-term behavior of cloud systems in TOGA COARE and their interactions with radiative and surface processes. Part I: Two-dimensional modeling study. *J. Atmos. Sci.* **55**, 2693–2714.

Wyant, M. C., C. S. Bretherton, H. A. Rand, and D. E. Stevens (1997). Numerical simulations and a conceptual model of the stratocumulus to trade cumulus transition. *J. Atmos. Sci.* **54**, 168–192.

Xu, K.-M., and D. A. Randall (1996). Explicit simulation of cumulus ensembles with the GATE Phase III data: Comparison with observations. *J. Atmos. Sci.* **53**, 3710–3736.

Xu, K.-M., A. Arakawa, and S. K. Krueger (1992). The macroscopic behavior of cumulus ensembles simulated by a cumulus ensemble model. *J. Atmos. Sci.* **49**, 2402–2420.

Xu, K.-M., and S. K. Krueger (1991). Evaluation of cloudiness parameterizations using a cumulus ensemble model. *Mon. Wea. Rev.* **119**, 342–367.

Zulauf, M., and S. K. Krueger (1997). Parameterization of mesoscale enhancement of large-scale surface fluxes over tropical oceans. *In* "Preprints, 22nd Conference on Hurricanes and Tropical Meteorology," Fort Collins, CO, pp. 164–165. Amer. Meteor. Soc.

Zulauf, M. A., and S. K. Krueger (2000). Parameterization of mesoscale enhancement of large-scale surface fluxes over Tropical oceans. Submitted for publication.

Zulauf, M. A., and S. K. Krueger (1999). Two-dimensional numerical simulations of Arctic leads. *In* "Preprints, Fifth Conference on Polar Meteorology and Oceanography," Dallas, TX, pp. 404–408. Amer. Meteor. Soc.

Chapter 21

Using Single-Column Models to Improve Cloud-Radiation Parameterizations

Richard C. J. Somerville
Scripps Institution of Oceanography
University of California, San Diego
La Jolla, California

I. INTRODUCTION

For many years, the climate modeling community has recognized that the results of atmospheric or coupled general circulation models (GCMs) are especially sensitive to the treatment of certain sub-grid physical processes. In the early years of GCM development, modelers could justify highly idealized and even simplistic treatments of many poorly understood processes on practical grounds. After all, when the main GCM task was perceived as constructing a model to produce a mean climate bearing a qualitatively realistic resemblance to observations, then modelers could perhaps be forgiven for adopting cavalier attitudes toward some of the more thorny parameterization issues.

Vestiges of this era persist today. For example, parameterizing cloud amount as a simple function of grid-scale relative humidity, tuned to give a reasonable global-mean radiation budget at the top of the atmosphere, was the state of the art in GCMs until relatively recently. Nevertheless, numerous observational studies demonstrate that cloud amount is not a simple function of relative humidity. Climate modelers must come to terms with this reality. We are all entitled to believe in our own theories, but not in our own observational data.

Today, GCM parameterizations are held to a higher standard of verisimilitude. The climate modeling community today is part of a political debate, like it or not, because the prospect of anthropogenic climate change has become an element of the political calculus. It is this change in the relevance of climate modeling that has driven the research community toward empirical validation of parametric treatments of sub-grid processes. A prime example is the role of cloud feedbacks in determining the sensitivity of GCM climates to changes in the atmospheric concentration of carbon dioxide, other greenhouse gases, and to the direct and indirect effects of atmospheric aerosols.

Early GCMs had fixed clouds and hence a low sensitivity to prescribed carbon dioxide changes. When cloud amounts are parameterized on relative humidity, several positive feedbacks come into play. One is the reduction of cloud amount in a warmer climate, a typical GCM result. This change produces an augmented warming, because GCM clouds, like global average real clouds, have a larger contribution to the planetary albedo than to the planetary greenhouse effect. In more precise terms, their short-wave forcing dominates their long-wave forcing.

A separate class of positive feedbacks arises from an increase in average cloud altitude, which is also a characteristic GCM response to climatic warming. On average, higher clouds are less effective infrared emitters to space, because they are colder, and they are less effective reflectors of incoming solar photons, because they are optically thinner, than lower clouds. In combination with the reduction in cloud amount, the increase in cloud height has the effect of powerfully reinforcing the warming that gives rise to these changes in cloud amount. It is the GCMs which incorporate these effects most strongly, and which do not include compensating negative feedbacks, that produce the highest sensitivities to changes in greenhouse gas concentrations.

The GCMs with the lowest sensitivities, on the other hand, are those dominated by negative cloud feedbacks. One such feedback is due to hypothetical cloud microphysical processes. Chief among these is a hypothesized increase in cloud water content accompanying a warmer climate. If in fact a warmer climate is characterized by wetter clouds, then, other things being equal, the wetter clouds may produce a negative feedback by increasing the optical thickness and hence the albedo of lower clouds, or a positive feedback by increasing the infrared emissivity of optically thin higher clouds.

Which of these conflicting effects will dominate? Is it scientifically justifiable to focus on global average cloud feedbacks to the neglect of regional ones? How do these cloud feedbacks depend on season, or synoptic regime, or geographical locale? What are the dynamical and thermodynamic consequences of the diabatic processes that depend on cloud feedbacks? These questions form much of the research agenda of the climate community. Even if anthropogenic climate change were not such an important policy issue, cloud processes would play an important role in climate modeling, because they have such a pervasive influence on climate, as evidenced by recent GCM results (e.g., Lubin *et al.*, 1998).

II. SINGLE-COLUMN MODELING

Progress in attacking these questions depends on a multifaceted research strategy. It is now well recognized in the GCM and numerical weather prediction (NWP) communities that single-column models (SCMs) are tools that have a valuable role to play in testing and improving parameterizations by evaluating them empirically against field observations (e.g., see Randall *et al.*, 1996). This chapter explores the potential of SCMs in attacking the central questions of cloud-radiation feedbacks in climate.

The basic idea of the SCM is to force and constrain an isolated time-dependent atmospheric GCM column with estimates of observed advective flux convergences, then to compare the output with observations to judge the realism of the parameterizations. Because the SCM has one space dimension (vertical), it is very fast, and it is practical to explore large parameter spaces by making hundreds or even thousands of integrations, which is impossible with a full GCM. In ARM and TOGA-COARE, we have applied our SCM to investigate parameterizations of cloud-radiation processes.

Our SCM contains switch-selectable parameterizations based on current NWP and GCM practices. These include schemes from the NCAR CCM, UKMO, and ECHAM models, including several different versions of the Arakawa–Schubert cumulus convection routine and Sundqvist's computed cloud water budget. We apply this model to the ARM site in Oklahoma and to the TOGA-COARE site in the western tropical Pacific. The advantage of these sites is that we have extensive data in hand for both of them and can force the model with actual measurements. At other locations, we might have to force the model with NCEP or ECMWF analyses, which would require great care to be sure the NWP model used in the analysis does not contaminate the advective forcings too seriously. We have already extensively compared the cloud-radiation properties of these schemes against data. However, a great deal remains to be done with precipitation and other hydrological cycle components.

Our approach involves evaluating parameterizations directly against measurements from field programs, and using this validation to tune existing parameterizations and to guide the development of new ones. We use the single-column model (SCM) to make the link between observations and parameterizations. Surface and satellite measurements are both used to provide an initial evaluation of the performance of the different parameterizations. The results of this evaluation are then used to develop improved cloud-precipitation schemes, and finally these schemes are tested in GCM experiments (Lee *et al.*, 1997).

Our single-column model has evolved from the one described by Iacobellis and Somerville (1991a,b). The SCM now is a versatile and economical one-dimensional model, resembling a single column of a general circulation model (GCM). The SCM contains the full set of parameterizations of sub-grid physical processes that are normally found in a modern GCM. The SCM is applied at a specific site having a horizontal extent typical of a GCM grid cell. Since the model is one-dimensional, the advective terms in the budget equations are specified from observations.

The SCM is diagnostic rather than prognostic. Its input is an initial state, plus the time-dependent advection terms in the conservation equations, provided at all model layers. Its output is a complete heat and water budget for the study site, including temperature and moisture profiles, clouds and their radiative properties, diabatic heating terms, surface energy balance components, and hydrologic cycle elements, all specified as functions of time. For a more complete discussion of SCMs, see Randall *et al.* (1996) and Iacobellis and Somerville (1991a,b).

A typical configuration of our SCM has 16 vertical layers in the atmosphere and either a land-surface scheme with a surface energy balance calculation or an underlying ocean mixed-layer model. The standard version of the SCM uses a time step of 7.5 min. A diurnally varying solar signal dependent on the latitude and time of year is applied at the top of the model atmosphere. The standard SCM uses the long-wave radiation parameterization of Morcrette (1990) and the solar radiation parameterization of Fouquart and Bonnel (1980). The model is coded in a modular manner, however, and so alternative parameterizations can be substituted for the standard ones. When forced with observationally derived time-dependent horizontal advection terms, the model produces profiles of fluxes, diabatic heating rates, temperatures, etc., which are dynamically consistent with both the prescribed forcing and the model parameterizations.

The SCM is a convenient testbed for examining many aspects of the ways in which GCMs treat subgrid physical processes. For example, we have recently found strong sensitivity to vertical resolution in several test integrations in which we increased the number of layers substantially. Several possibilities are raised by this result. One is that the parameterizations are constructed around implicit assumptions as to how many layers are involved, so that they do not generalize to arbitrary vertical resolution and converge at sufficiently small vertical grid size. Another is that typical GCM and NWP vertical resolutions are simply inadequate for some aspects of parameterized sub-grid physics, although they may generally be satisfactory from the viewpoint of large-scale dynamics.

A variety of parameterizations representing cumulus convection, cloud prognostication, cloud-radiative properties, cloud water budgets, and precipitation can be tested in this manner and used for intercomparison. The cumulus convection parameterizations that we have used include those of Kuo (1974) (as modified by Anthes, 1977) and of Arakawa and Schubert (1974) (with downdrafts as specified by Kao and Ogura, 1987), including the relaxed version used in the GSFC GCM, and of Emanuel (1991). The cloud prediction schemes are those of Slingo (1987), Smith (1990), Sundqvist *et al.* (1989), and Tiedtke (1993). The cloud-radiative properties are specified from either McFarlane *et al.* (1992) or a combination of Stephens (1978) (water clouds) and Suzuki *et al.* (1993) (ice clouds). Cloud liquid water is a prognostic model variable when the Smith or Sundqvist or Tiedtke scheme is active, but not when the Slingo routine is being used. These parameterizations are switch-selectable in the present version of the model.

Because the domain of the model corresponds to a single column in a GCM or NWP model, and because the forcing is independent of processes occurring outside this domain, single-column models such as ours are sometimes referred to as *semi-prognostic*, although the terminology in this field is not yet standardized. Lacking a way to represent the feedback of processes inside the domain onto the large-scale fields that provide the forcing, these models are best applied in situations where the external forcing tends to change slowly relative to the time scale of the processes explicitly modeled. If their limitations are recognized clearly and their applications are chosen sensibly, however, single-column models can be powerful tools.

Semi-prognostic models may be looked on as a means of isolating the behavior of a model atmosphere over a single horizontal grid cell. Viewed in this way, they enable one to study the comparative merits and drawbacks of alternative parameterizations of physical processes, for example, or the sensitivity to errors in advective forcing. When detailed observational data are available, they provide a way to evaluate local model behavior comprehensively. In brief, they are a means of zooming in microscopically onto the model grid scale itself, to diagnose both the GCM physics and the actual atmosphere. By perturbing the advective forcing, for example, one can explore the dependence of the diabatic heating rates and other sub-grid parameterization outputs on the accuracy of the horizontal flux convergences. Similarly, for a given advective forcing, one can test how these model products depend on the choice of parameterizations. We have carried out many of these types of numerical experiments in a systematic manner, using observational data for verification. For a discussion of limitations of single-column modeling, including issues of sensitivity to errors in forcing data and consequences of constraining the SCM temperature and humidity profiles from departing too far from observations, see, e.g., Randall *et al.* (1996), Lee *et al.* (1997), Iacobellis and Somerville (1991a,b), and references therein.

III. PARAMETERIZATION VALIDATION AND SINGLE-COLUMN DIAGNOSTIC MODELS

A. DIAGNOSTIC MODELING

The single-column model is numerically integrated in time as an initial value problem that is forced and constrained by observational data. The input is an observed initial state, plus observationally derived estimates of the time-dependent advection terms in the conservation equations, provided at all model layers. Its output is a complete heat and water budget, including temperature and moisture profiles, clouds and their radiative properties, diabatic heating terms, surface energy balance components, and hydrologic cycle elements, all specified as functions of time. These model products can then be evaluated against observations. This validation provides a test of the realism of the model physical parameterizations and a means of evaluating proposed improvements. In addition, it allows an assessment of the sensitivity of the results to individual elements of the parameterizations.

Several years ago (Iacobellis and Somerville, 1991a,b), we described the theory and conceptual basis of single-column modeling. Virtually all the parameterizations in our model, however, have been replaced or substantially modified since these papers were written, as described below.

B. Model Structure

The atmospheric model is a single column divided into layers, and the fluxes of heat and moisture are determined for each of these layers. Like a GCM, the model includes parameterizations of solar and terrestrial radiation, shallow convection, deep cumulus convection, diffusion, distribution of surface fluxes, and cloud prediction. Unlike a GCM, the model is not global; instead, it is applied at a specific location.

The fundamental equations for the present model are the thermodynamic energy equation and the conservation equation for water vapor. These equations include terms representing the transport of heat and moisture due to horizontal and vertical advection. Resolution is variable and may be chosen to match available data and changed to test sensitivity to resolution. The model is applied to a grid area typically measuring 200 km on a side, and extending from the surface to the top of the atmosphere. Thus, the model volume closely resembles a single column in a GCM. Seasonal and diurnal variations are included in the model.

C. Solar Radiation

The solar radiation parameterization calculates the absorption of solar radiation within each of the model layers. The primary absorbers of solar radiation in the model atmosphere are water vapor and ozone. Clouds, as strong reflectors, play a major role in determining the fluxes of solar radiation. In this model both clouds and water vapor are interactive variables. The model uses a fixed profile of ozone concentration. At each time step, the parameterization uses the vertical distribution of water vapor and ozone concentrations, together with the cloud profile and the specified surface reflectivity, to compute the fluxes of solar radiation at each of the levels between the model layers. From these fluxes, the heating rates in each of the layers are calculated. As one example of using the single-column model to establish the relative merits of competing algorithms, we have recently compared the Fouquart and Bonnel (1980) solar radiation parameterization with that of Morcrette (1990).

D. Terrestrial Radiation

The model atmosphere is heated from below through the emission of infrared radiation from the surface. Within the atmosphere, infrared radiation is absorbed and radiated by water vapor, carbon dioxide, clouds, and ozone, in addition to other trace gases. The model parameterization uses the vertical profiles of these atmospheric constituents to calculate the net fluxes of terrestrial radiation at the levels between each of the layers. Our current terrestrial radiation parameterization is based on the one developed by Morcrette for the European Centre for Medium-Range Weather Forecasts (ECMWF) model. We are also incorporating the RRTM (Rapid Radiative Transfer Model), an improved long-wave scheme being developed by Anthony Clough and colleagues.

E. Horizontal Advection

The single-column model relies on analyzed observational data of the horizontal fields of temperature, humidity, and velocity. These are the source of the horizontal advection terms and are also used to infer the vertical velocities necessary for the computation of vertical advection and convection. Before the first such model was built and tested, a legitimate question was whether the observational data would be sufficient to guarantee the verisimilitude and realism of the model results. Our experience demonstrates that, except in data-sparse regions, current four-dimensional assimilation routines from operational numerical weather prediction can provide dynamically balanced fields, which, in conjunction with the parameterizations incorporated in the model, are accurate enough to yield realistic heat and moisture budgets. However, using an analysis based in part on an NWP model involves the risk that the advective forcing supplied to the SCM may be contaminated. Recently, we have shown that direct observational sounding data from the ARM and TOGA-COARE field experiments can be used to provide the needed horizontal flux convergence forcing terms, thus obviating the need to use model-dependent operational analyses. We have shown that this method provides accurate horizontal flux convergences and produces realistic budget closures in the model domain.

F. Convection

Recently, we have incorporated and extensively tested the cumulus convection schemes developed by Emanuel (1991) and Zhang and McFarlane (1995). The current model version now includes a choice of these modern and widely used cumulus convection schemes, as well as the "classical" schemes of Arakawa and Schubert (including the Kao–Ogura downdraft algorithm and the relaxed algorithm due to Max Suarez at GSFC), and Kuo (as modified by Anthes and Krishnamurti), in addition to simple algorithms such as moist convective adjustment. This type of software development is essential to our objective of being able to evaluate and intercompare competing parameterizations using observations.

G. Large-Scale Condensation

The model checks for large-scale condensation after all heat and moisture exchanges due to deep or shallow convection have been calculated. The model examines each layer in turn and computes the relative humidity. If the relative humidity is larger than a prescribed critical saturation relative humidity, water vapor is condensed out of the layer, and heat from the release of latent heat is added to the layer. All precipitation from large-scale condensation is ordinarily assumed to fall to the surface with no reevaporation in lower layers, although this simplification is not essential. The saturation relative humidity is set at less than unity (typically 0.95) to take into account possible sub-grid-scale saturation.

H. Cloud Prediction

The single-column diagnostic model, like many climate and forecasting models, is very sensitive to the specification of clouds. The standard version of the model employs a variant of a cloud parameterization scheme that empirically relates cloudiness to the large-scale model variables (Slingo, 1987). We have also incorporated a treatment of cloud optical properties adopted by the second-generation GCM of the Canadian Climate Centre, in which optical properties are based on cloud liquid water contents, which in turn are parameterized on temperature and pressure (McFarlane *et al.*, 1992).

Alternatively, cloud amount and optical properties in the model can be computed from cloud liquid water, which is carried in the model as a prognostic interactive variable, following the schemes of Smith (1990), Tiedtke (1993), or Sundqvist *et al.* (1989). In this case, we use algorithms such as those of Stephens (1978) or Suzuki *et al.* (1993) to compute cloud optical properties.

Recent observational work (e.g., Bower *et al.*, 1994) has provided promising suggestions for parameterizations that include the particle size distribution or effective radius of cloud liquid or ice particles. After the total water content, cloud particle size distribution is thought to be the single most important cloud microphysical parameter in determining cloud-radiative properties. It is straightforward to incorporate suggested functional forms, such as those of Bower *et al.*, into the single-column model algorithm, and to test the resulting radiative effects against measurements, and we are working along these lines now.

IV. MODEL EXPERIMENTS

A. Long-Term Experiments in the TOGA-COARE Region

As an example of the use of single-column models, we first present a series of seven "long-term" experiments from TOGA-COARE. The model configuration in each experiment differs only in the specification of the cumulus convection parameterization, the cloud prognostication scheme, and/or the parameterization of cloud-radiative properties. The parameterizations used in each experiment are shown in Table I.

In each experiment, the SCM was applied at nineteen $5° \times 5°$ sites within the bounded region of the western tropical Pacific extending from 137.5°E–172.5°E and 7.5°N–7.5°S. The size and location of the individual

Table I

Model Configurations

Exp. no.	Cumulus convection	Cloud scheme	Cloud-radiative properties
1	Kuo–Anthes	Slingo	McFarlane
2	Emanuel	Slingo	McFarlane
3	Arakawa–Schubert	Slingo	McFarlane
4	Kuo–Anthes	Smith	Stephens + Suzuki
5	Emanuel	Smith	Stephens + Suzuki
6	Arakawa–Schubert	Smith	Stephens + Suzuki
7	Kuo–Anthes	Sundqvist	Stephens + Suzuki

sites were selected to coincide with the ECMWF operational analysis data, which are used as model forcing in these experiments. Each experiment extends from 01 Nov 1992 to 28 Feb 1993 which corresponds to the IOP of TOGA-COARE. At each site the model was initialized at 01 Nov 1992 00LST and run for a period of 4 days. The model was then reinitialized and run again for the next 4 days. This process was repeated a total of 30 times with the last 4-day run extending from 25 Feb to 28 Feb 1993. These 30 runs were repeated at each of the nineteen $5° \times 5°$ sites. Long-term regional means during the TOGA-COARE IOP are formed by averaging results from the 570 (19 sites $\times$ 30 four-day periods) model runs for each configuration of model parameterizations.

The cloud-radiative forcing terms (Ramanathan *et al.* 1989; Senior and Mitchell, 1993) and other cloud-radiation variables produced by these SCM experiments can be examined to understand how the various parameterizations might affect the modeled climatological heat budget of the TOGA-COARE region. The cloud-radiative forcing represents the extra amount of radiation that is either emitted to space or absorbed by the earth–atmosphere system due to the presence of clouds. The total cloud forcing (CF) is composed of a short-wave component (CF_{SW}) and a long-wave component (CF_{LW}), or, in equation form:

$$CF = CF_{SW} + CF_{LW} = (Q - Q_c) - (F - F_c),$$

where Q is the net incoming solar radiation, F is the outgoing long-wave radiation, and the subscript c indicates clear-sky fluxes. If CF is positive, then the clouds have a warming effect on the earth–atmosphere system. These terms averaged over all 19 sites and over the 4-month time period for each of the seven SCM experiments are shown in Table II. The last row of Table II contains the observed values obtained from GMS-4 satellite measurements (Flament and Bernstein, 1993).

These results show that when the Slingo cloud scheme is used (experiments 1–3) model averages of CF_{SW}, CF_{LW}, OLR, albedo, and insolation are less sensitive to the choice of cumulus parameterization than when the Smith cloud scheme is employed (experiments 4–6). The effect of the treatment of cloud liquid water can be seen by comparing the results from experiments 1, 4, and 7. Here Kuo–Anthes convection was used in each case while the cloud prognostication scheme varied. The model results show large decreases in insolation and OLR and large increases (in magnitude) in albedo and both the short-wave and long-wave cloud forcing terms for experiments 4 and 7 compared to experiment 1.

The model results are much closer to the observed GMS-4 values when cloud liquid water is not an interactive variable (experiments 1–3). The

Table II
Cloud Forcing and Radiative Results

Exp. no.	CF_{SW} (W m^{-2})	CF_{LW} (W m^{-2})	CF (W m^{-2})	Albedo	OLR (W m^{-2})	Insolation (W m^{-2})
1	-100	56	-44	0.33	236	200
2	-101	57	-44	0.33	229	199
3	-108	65	-43	0.35	224	191
4	-121	66	-55	0.38	225	176
5	-137	76	-61	0.42	210	159
6	-144	82	-62	0.44	207	151
7	-138	71	-67	0.42	220	154
GMS	-108	62	-46	0.37	228	—

larger absolute values of the cloud forcing terms, the increase in albedo, and the decrease in OLR and insolation in experiments 4–7 all indicate that the model is either overestimating the cloud fraction and/or the cloud optical thickness when cloud liquid water is an interactive variable.

B. Short-Term Experiments in the IFA Region

The SCM is now applied over the site of the intensive flux array (IFA). In this series of experiments the horizontal advection terms necessary to force the SCM are derived from sounding measurements taken along the perimeter of the IFA. Thus, direct observations are used to force the model in these experiments as opposed to model assimilation products used in Section III.A. The SCM was run for several 5-day periods within the TOGA-COARE IOP of 01 Nov 92–28 Feb 93 using model configuration 7 (see Table I).

Figure 1 shows results that typify the SCM response during low precipitation conditions over the IFA. During these "dry" events, the SCM typically reproduced the observed temperature and humidity evolution and produced diagnostic radiative quantities that compared very well to both surface and satellite measurements. SCM results from a 5-day period that experienced significant precipitation are shown in Fig. 2. During these "wet" events, the SCM often underestimated the surface downwelling solar radiation and overestimated the planetary albedo while reproducing reasonably well the precipitation as measured by the satellite measurements of the Microwave Sounding Unit (MSU).

These model results from the IFA site are consistent with the long-term runs discussed earlier in that they both indicate that the model parameter-

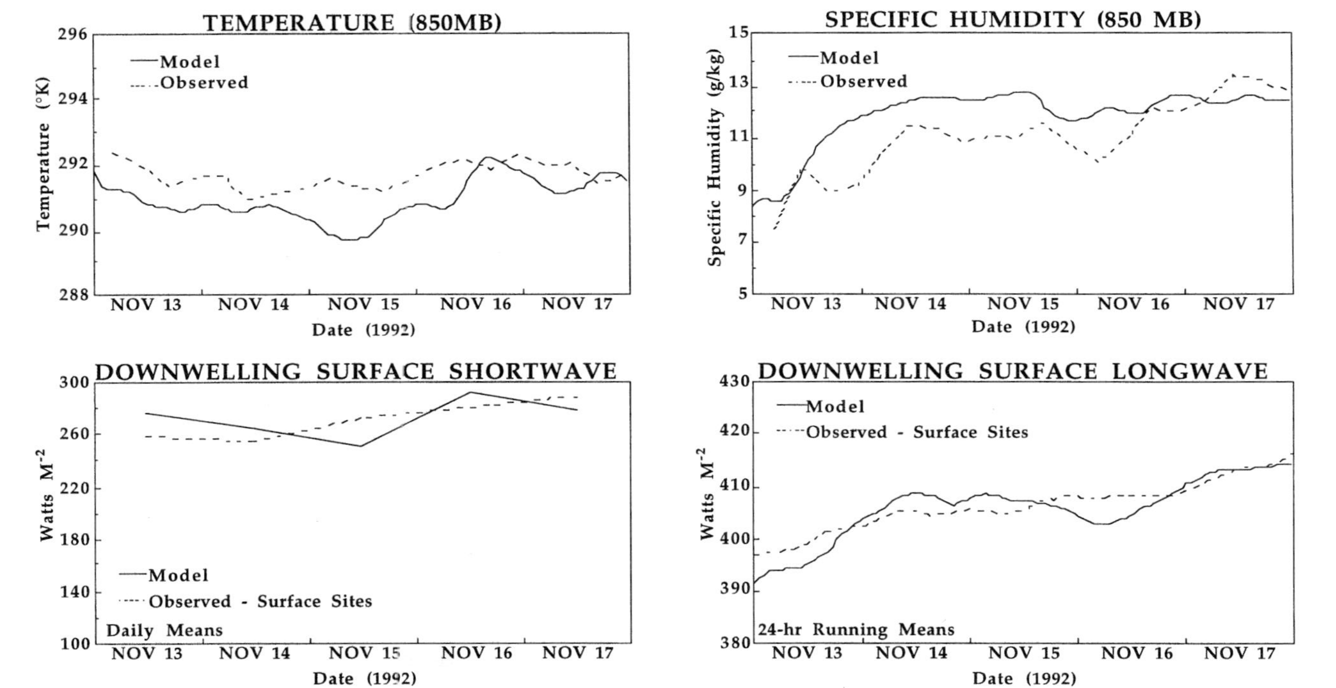

Figure 1 Model versus observation comparisons for low-precipitation conditions.

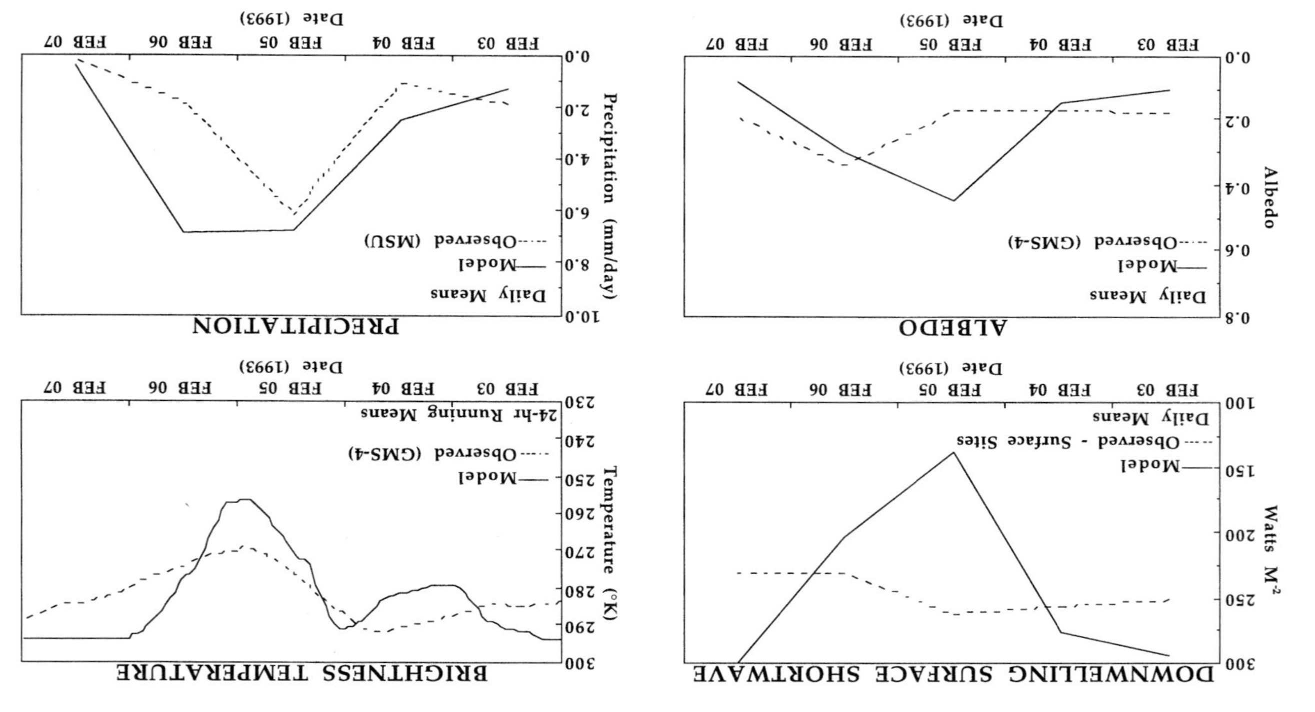

Figure 2 Model versus observation comparisons for significant precipitation conditions.

izations are overestimating the cloudiness and/or cloud optical thickness. Further analysis of the model results indicates that very large errors in cloud optical thickness would be necessary solely to explain the model discrepancies in albedo and surface short-wave. Thus, overestimation of the fractional cloud coverage is more likely to be the cause of the majority of the errors in albedo and surface short-wave. The fractional cloud cover for convective clouds in the Sundqvist cloud scheme is given by:

$$\text{CLD} = \xi\tau(1 + \text{RH})[1 + (\sigma_b - \sigma_t)/A]/(1 + B\xi\tau),$$

where CLD is fractional cloud amount, RH is relative humidity, σ_b and σ_t are the sigma level of cloud base and cloud top, respectively, ξ is a quantity proportional to the moisture convergence, τ is a convective time scale, and A and B are constants. In Sundqvist's scheme, the values of A and B are set to 0.3 and 2.5, respectively. We performed numerous SCM runs in the IFA region throughout the IOP period, each run with different values of A and B. A comparison of model albedo and surface short-wave to observations yielded "optimized" values of A and B of approximately 0.5 and 4.0.

The use of these optimized values of A and B improves the SCM estimates of albedo and downwelling short-wave radiation during precipitation events in the IFA region. A further and more substantial test of this optimization was performed by rerunning the long-term runs (SCM in configuration 7) with the optimized values of A and B. The model results from the "optimized" run and the original run are shown in Table III, along with measured values from the GMS-4 satellite. The results from the SCM run using the optimized values of A and B show a clear improvement over the original run when compared to GMS-4 measurements taken during the TOGA-COARE IOP.

Table III
Original versus Optimized

	CF_{SW} (W m^{-2})	CF_{LW} (W m^{-2})	CF (W m^{-2})	Albedo	OLR (W m^{-2})	Insolation (W m^{-2})
Original	−138	71	−67	0.42	220	154
Optimized	−112	59	−53	0.36	232	183
GMS	−108	62	−46	0.37	228	—

V. CONCLUSION

This illustrative study has shown that the SCM can be forced with direct observations of the horizontal advection terms, based on soundings on the COARE IFA boundary, rather than with estimates from inevitably model-dependent objective analyses, and can still produce realistic budgets, as confirmed by both surface and satellite measurements. The sensitivity test performed here with the "optimized" values of A and B is admittedly simple, and more rigorous analyses will be needed to yield a satisfactory and physically consistent cloud prognostication scheme for the western tropical Pacific. However, this illustrative study demonstrates the potential of the SCM as a useful and economical tool in parameterization development.

ACKNOWLEDGMENTS

The research reported in this paper has been sponsored in part by the U.S. Department of Energy under grant DE-FG03-90-ER61061, the National Science Foundation under grant ATM91-14109, the National Oceanic and Atmospheric Administration under grant NA36GP0372, and the National Aeronautics and Space Administration under grant NAG5-2238. I deeply appreciate the many essential contributions of my collaborators: Carolyn Baxter, Joannes Berque, Sam Iacobellis, Dana Lane, Wan-Ho Lee, and Boris Shkoller.

REFERENCES

Anthes, R. A. (1977). A cumulus parameterization scheme utilizing a one-dimensional cloud model. *Mon. Wea. Rev.* **105**, 270–286.

Arakawa, A., and W. H. Schubert (1974). Interaction of a cumulus cloud ensemble with the large-scale environment, Part I. *J. Atmos. Sci.* **31**, 674–701.

Bower, K. N., T. W. Choularton, J. Latham, J. Nelson, M. B. Baker, and J. Jenson (1994). A parameterization of warm clouds for use in atmospheric general circulation models. *J. Atmos. Sci.* **51**, 2722–2732.

Emanuel, K. A. (1991). A scheme for representing cumulus convection in large-scale models. *J. Atmos. Sci.* **48**, 2313–2335.

Flament, P., and R. Bernstein (1993). Images from the GMS-4 satellite during TOGA-COARE (November 1992 to February 1993), Technical Report 93-06. School of Ocean and Earth Science and Technology, University of Hawaii, Honolulu.

Fouquart, Y., and B. Bonnel (1980). Computation of solar heating of the Earth's atmosphere: A new parameterization. *Beitr. Phys. Atmos.* **53**, 35–62.

Iacobellis, S. F., and R. C. J. Somerville (1991a). Diagnostic modeling of the Indian monsoon onset, I: Model description and validation. *J. Atmos. Sci.* **48**, 1948–1959.

Iacobellis, S. F., and R. C. J. Somerville (1991b). Diagnostic modeling of the Indian monsoon onset, II: Budget and sensitivity studies. *J. Atmos. Sci.* **48**, 1960–1971.

Kao, J. C.-Y., and Y. Ogura (1987). Response of cumulus clouds to large-scale forcing using the Arakawa–Schubert cumulus parameterization. *J. Atmos. Sci.* **44,** 2437–2458.

Kuo, H. L. (1974). Further studies of the parameterization of the influence of cumulus convection on large-scale flow. *J. Atmos. Sci.* **31,** 1232–1240.

Lee, W.-H., S. F. Iacobellis, and R. C. J. Somerville (1997). Cloud radiation forcings and feedbacks: General circulation model tests and observational validation. *J. Climate* **10,** 2479–2496.

Lubin, D., B. Chen, D. H. Bromwich. R. C. J. Somerville, W.-H. Lee, and K. M. Hines (1998). The impact of cloud radiative properties on a GCM climate simulation. *J. Climate* **11,** 447–462.

McFarlane, N. A., G. J. Boer, J.-P. Blanchet, and M. Lazare (1992). The Canadian Climate Centre second-generation general circulation model and its equilibrium climate. *J. Climate* **5,** 1013–1044.

Morcrette, J.-J. (1990). Impact of changes to the radiation transfer parameterizations plus cloud optical properties in the ECMWF model. *Mon. Wea. Rev.* **118,** 847–873.

Price, J. F., R. A. Weller, and R. Pinkel (1986). Diurnal cycling: Observations and models of the upper ocean response to diurnal heating, cooling, and wind mixing. *J. Geophys. Res.* **91,** 8411–8427.

Ramanathan, V., B. R. Barkstrom, and E. F. Harrison (1989). Climate and the Earth's radiation budget. *Phys. Today* May, 22–32.

Randall, D. A., K.-M. Xu, R. C. J. Somerville, and S. Iacobellis (1996). Single-column models and cloud ensemble models as links between observations and climate models. *J. Climate* **9,** 1683–1697.

Senior, C. A., and J. F. B. Mitchell (1993). Carbon dioxide and climate: The impact of cloud parameterization. *J. of Climate* **6,** 393–418.

Slingo, J. M. (1987). The development and verification of a cloud prediction scheme for the ECMWF model. *Quart. J. Roy. Meteor. Soc.* **113,** 899–927.

Smith, R. N. B. (1990). A scheme for predicting layer clouds and their water content in a general circulation model. *Quart. J. Roy. Meteor. Soc.* **116,** 435–460.

Stephens, G. L. (1978). Radiation profiles in extended water clouds. II: Parameterization studies. *J. Atmos. Sci.* **35,** 2123–2132.

Sundqvist, H., E. Berge, and J. E. Kristjánsson (1989). Condensation and cloud parameterization studies with a mesoscale numerical weather prediction model. *Mon. Wea. Rev.* **117,** 1641–1657.

Suzuki, T., M. Tanaka, and T. Nakajima (1993). The microphysical feedback of cirrus cloud in climate change. *J. Meteor. Soc. Japan* **71,** 701–713.

Tiedtke, M. (1993). Representation of clouds in large-scale models. *Mon. Wea. Rev.* **121,** 3040–3061.

Zhang, G. J., and N. A. McFarlane (1995). Sensitivity of climate simulations to the parameterization of cumulus convection in the Canadian Climate Centre general circulation model. *Atmos.-Ocean* **33,** 407–446.

Chapter 22

Entropy, the Lorenz Energy Cycle, and Climate

Donald R. Johnson
Space Science and Engineering Center and
Department of Atmospheric and Oceanic Sciences
University of Wisconsin and
Division of Earth Sciences
Universities Space Research Association
Columbia, Maryland

I. INTRODUCTION

I appreciate the invitation to participate in this symposium in honor of Professor Arakawa. Of the stalwarts developing theory and models of the atmosphere and ocean, Professor Arakawa stands among the tallest. His contributions in combination with teaching in the classroom and mentoring of young scientists are singularly noteworthy and for which atmospheric scientists must be most grateful.

My personal interactions with Professor Arakawa have unfortunately been restricted to infrequent occasions, at least relative to his students and faculty colleagues gathered here. My interaction with Professor Arakawa through studying his scientific work has been more frequent (e.g., Arakawa and Lamb, 1977). Besides the delightful experience of summarizing Professor Yale Mintz's scientific contributions for the *Journal of Climate* (Johnson and Arakawa, 1996), we have developed a common interest in the study and modeling of atmospheric circulation using isentropic and/or generalized coordinates. Others (e.g., Bleck and Benjamin, 1993; Johnson *et al.*, 1993; Zapotocny *et al.*, 1994; Zhu, 1997; Zhu and Schneider, 1997) have also joined in the development and application of isentropic models for weather and climate prediction. The recent work of Arakawa and Hsu (1990), Hsu and Arakawa (1990), Arakawa *et al.* (1992, 1994), Arakawa and Konor (1994, 1996), Konor *et al.* (1994), and Konor and Arakawa (1997) in developing the numerics for isentropic and generalized coordinates provide a foundation that bodes well for future developments and application of weather and climate isentropic models. In recognition of Professor Arakawa's exceptional career, I am delighted to contribute this study in which classical concepts of entropy and the processes that change it in relation to the Lorenz energy cycle and climate are featured.

This study, which emphasizes the second law and entropy, expands the perspective needed to understand thermally forced circulation. Entropy concepts are fundamental to the study of large-scale exchange, monsoons, and the thermodynamic efficiency of differential heating in maintaining atmospheric circulation. Entropy concepts are essential for understanding reversible and irreversible thermodynamic processes. Entropy concepts are also important for understanding thermal equilibrium and disequilibrium, moist thermodynamics, phase changes, heat diffusion, and viscous dissipation. The study of entropy and its flux is also important for understanding radiative and chemical processes in relation to the earth system.

Within the climate state, a knowledge of the systematic entropy sources and sinks leads to direct insight into the scales and intensity of a mean mass circulation embedded within a stratified rotating atmosphere. Because entropy as an exact differential is uniquely related to atmospheric heating and cooling divided by temperature, the atmosphere must transport a net amount of entropy from heat sources to heat sinks. Within isentropic coordinates, the net entropy transport is realized through the condition that a branch of a mass circulation transports higher valued entropy from a source to the sink than is returned by a branch of a mass circulation transporting lower valued entropy from a sink to the source. This systematic exchange within the time-averaged mass circulation (Johnson *et al.*, 1985) and its relation to this net exchange of entropy and also dry static energy are described by Johnson (1989) in his study of "The Forcing and Maintenance of Global Monsoonal Circulations: An Isentropic Analysis."

These simple concepts, however, have been overshadowed by the emphasis on energy, the various forms of energy, and exchange processes in studies of the general circulation and climate. Entropy and energy are independent properties and, while complementary, they each, as this study emphasizes, provide unique insight into the maintenance of the general circulation and climate. The objective of this study is to link certain underlying concepts of the second law and entropy with the Lorenz energy cycle and the climate state. The understanding of the climate state as a thermally forced circulation is incomplete without including entropy and its exchange.

In bringing to the forefront the importance of entropy and its balance in modeling the climate state, certain key contributions, both past and present, are reviewed. Through this review, in-depth relations linking the classical Carnot cycle with the Lorenz energy cycle are set forth. From this background, the importance of the Lorenz energy cycle and the impact of spurious aphysical sources of entropy on the Lorenz energy cycle in simulating the climate state are developed. Finally, the relevance of mean entropy sources on the intensity of winter and summer time circulations is emphasized.

II. GLOBAL THERMODYNAMICS AND MONSOONAL CIRCULATIONS

In setting forth basic concepts concerning "the global thermodynamics of atmosphere motion," Dutton (1973) emphasizes the predictive aspect of entropy stemming from the second law. Subject to the constraint of isolation in the sense that the flux of energy through the bounding surfaces of the atmosphere vanishes, he equated the total energy of the actual atmosphere E to a corresponding total energy of a hydrostatic reference atmosphere E_0 with maximum entropy. He then noted that the reference state temperature T_0 was readily determined from the equality of the total energy E_0 with the volume integrated enthalpy $c_p T_0 M$ of the hydrostatic reference atmosphere. This was followed with the definition of entropic energy as the product of the reference state temperature T_0 and the difference of the entropy of the reference state S_0 minus that of actual state atmosphere S. He then established that the entropic energy $T_0(S_0 - S)$ was always positive from the condition that entropy of the reference state atmosphere always exceeded the entropy of the actual atmosphere. Dutton emphasized that entropic energy enjoyed a predictive component from the second law in that the ever-present natural processes of heat diffusion and kinetic energy dissipation perpetually force the atmospheric entropy S toward the state of equilibrium S_0 with maximum entropy. Although for different reasons, Dutton's concept of entropic energy is directly applicable to climate state modeling in that the boundary constraints of vanishing net energy flux for the "driftless climate state" and that of Dutton's reference atmosphere are equivalent (Johnson, 1989, 1997). As such, the reference state energy of a hydrostatic equilibrium maximum entropy atmosphere with uniform temperature everywhere is readily defined and invariant in time in a climate model without drift.

Later following Dutton's concepts, Johnson (1989) studied an unconstrained thermodynamic equilibrium state with T_0 uniform everywhere and a constrained equilibrium state with $T_\alpha(\theta, t)$ and $p_\alpha(\theta, t)$ uniform on isentropic surfaces. In his discussion of thermodynamic systems, Sommerfeld (1964) notes that a constrained equilibrium state can be defined when local differences of thermodynamic properties of a system are described by sufficiently small volume elements within which temperature and pressure are uniform, then the entropy of the constrained system may be taken to be the sum of the entropies of all volume elements. While the constrained equilibrium state has more entropy than the unconstrained state with maximum entropy, the state of uniform temperature and pressure $T_\alpha(\theta, t)$ and $p_\alpha(\theta, t)$ within the subdivided elements defined by the global extent of isentropic layers satisfies the condition for constrained thermodynamic equilibrium. Within the subdivided elements of equilibrium, internal energy assumes a minimum. For an equilibrium state in hydrostatic balance, the minimum in internal energy ensures a minimum in gravitational potential energy. Johnson proceded to show that the time rate of change of the constrained equilibrium state was determined by the isentropic areally integrated entropy source in conjunction with the mass continuity equation in which the equilibrium state pressure $p_\alpha(\theta, t)$ was equated to the isentropic areally averaged hydrostatic pressure $\bar{p}(\theta, t)$. In so doing, Johnson defined a reversible component of total energy created by the action of entropy sources in forcing the atmosphere from its equilibrium state, which is equivalent with the sum of Lorenz's (1955) available potential (A) and kinetic (K) energies. This reversible component, like Dutton's entropic energy, is contracted by the irreversible processes of heat diffusion and viscous dissipation. Increases of this reversible component of total energy occur through the covariance of entropy sources and the temperature departure $(T - T_\alpha)$. Consequently, the requirements for the driftless climate state with $\dot{E}$ and $\langle \dot{s} \rangle_\theta$ (the temporally areally averaged mass weighted averaged entropy source) equal to zero demand that Dutton's entropic energy and Johnson's (1989) reversible component of total energy for the climate state be invariant with time. In view of this requirement, Johnson (1997) concluded that a climate model will seek a state within which energy degraded by irreversible processes is balanced by differential heating that maintains the circulation against viscous dissipation, heat diffusion, and aphysical sources of entropy. From the existence of positive definite aphysical sources of entropy, Johnson concluded that in simulations of a driftless state, climate models will seek a state characterized by a general coldness relative to the true state.

In the maintenance of global monsoonal circulations and an energy phase space of reversible isentropic processes, the heating with respect to isentropic surfaces, just like for the generation of available potential energy, must occur at higher temperature (higher pressure) in tropical/subtropical latitudes and the cooling at lower temperature (lower pressure) in polar/extratropical latitudes (Johnson, 1989). The underlying premise for attributing the "coldness" in climate models that was developed earlier based on entropy balance (Johnson, 1997) will be developed utilizing the concepts of the reversible component of total energy. The result to be developed has its roots in recognizing the importance of the pattern of the efficiency factor distribution relative to the meridional variation of the isentropic layers set forth in Dutton and Johnson (1967). This pattern, which links isentropic layers of the cold upper polar troposphere with the warm lower tropical troposphere, provides a temperature structure that in combination with the systematic meridional variation of heating forces a mean isentropic Hadley circulation in each hemisphere. The Hadley circulations forced by the systematic variation of heating associated with the interception of beam solar radiation by the spherical earth and more uniform emission of terrestrial radiation to space involve cyclic processes in each hemisphere. There are also systematic monsoonal circulations forced by asymmetric differential heating associated with land–ocean contrasts. The evidence of this forcing is provided from the vertically averaged heating distributions filtered to emphasize wavelengths greater 10,000 km for the 4 months of January, April, July, and October 1979 presented in Fig. 1 (see color insert). Inspection of the heating distributions for January and July confirms that a distinct planetary scale of entropy sources and sinks related to the distribution of continents and oceans occurs in addition to the systematic meridional distribution.

A general characteristic of the patterns for the 4 months representing different seasons is the relatively uniform cooling in polar latitudes. The most prominent feature of the heating distributions for the 4 months is the major heat source over Asia and the Western Pacific in the boreal summer, which migrates to the Southern Hemisphere's tropical western Pacific in the austral summer. Note the reversal of the zonally varying heating distributions from winter to summer between the Northern Hemisphere continents and neighboring Pacific and Atlantic oceans of Asia and the Americas. In July, the heating over continents and cooling over oceans force the summer monsoons, whereas in January cooling over continents and heating over oceans force winter monsoons. A key feature of isentropic analysis is the correspondence of temporally, vertically integrated global transports of entropy and energy with the global scale of differential heating. Johnson (1989) discusses in more detail the seasonal variation of this planetary distribution of temporally, vertically integrated heat sources and sinks and global monsoonal circulations.

The point to be emphasized here is that for balance the systematic spatial distribution of heating requires a systematic transport of entropy from heat source to heat sink regions that maintains the atmosphere's circulation, and for these purposes a climate model's circulation. The trajectories from the entropy source to the entropy sink in a mean sense will be at higher potential temperature while the trajectories from the heat sink to the source will be at lower potential temperature. Within extratropical latitudes, the poleward trajectories embedded within the baroclinic waves will be displaced zonally relative to the equatorward trajectories. Within this structure the trajectories at higher potential temperature will transport more energy and entropy from the heat source to the heat sink than the trajectories at lower potential temperature return from the heat sink to the heat source (Otto-Bliesner and Johnson, 1982; Zillman and Johnson, 1985; Johnson *et al.*, 1985; and Johnson, 1989). Thus in maintaining balance, net energy and entropy are transported meridionally from lower to higher latitudes and also zonally from oceans to continents in winter with reversal of this latter transport in summer. This cyclic process viewed within the vertically integrated circulation requires correspondence of the scales of mean mass, energy, and entropy transport with the horizontal scales of differential heating (Wei *et al.*, 1983; Schaack *et al.*, 1990, 1991; Hoerling *et al.*, 1990; Zapotocny *et al.*, 1993; Schaack and Johnson, 1994). This cyclic process also involves work and generation of kinetic energy which balances dissipation (Johnson, 1989). This cyclic process in climate models which each mimics also incurs aphysical sources of entropy from the inadequacies of numerical modeling and parameterization.

The crucial thermodynamic issue here is to recognize that the mutual satisfaction of the energy and entropy constraints provides for the maintenance of the climate state circulation at a given level against irreversible dissipative processes. Furthermore, the temperature distribution of the climate model will adjust to a state within which the combination of temperature and differential entropy sources generates a sufficient amount of the reversible component of total energy by entropy decreasing processes to offset both positive definite physical and aphysical sources of entropy. With satisfaction of the boundary constraints of no net energy flux, the only means for the climate model to generate an additional amount of reversible component of total energy to offset positive definite aphysical sources over the multiannual time scale of the driftless climate state is for the equilibrium temperature of the model to be cold, thus increasing the efficiency of the differential heating (Johnson, 1997). The balance of the entropy of matter with no net gain within the atmosphere is achieved by a net removal of entropy by the greater outward flux of terrestrial radiative entropy flux than the incoming solar radiative flux.

While most key aspects of the reversible component of total energy are common with Lorenz's theory of available potential energy and maintenance of the atmospheric circulation against kinetic energy dissipation, there are some fundamental differences regarding entropy from Johnson's (1989) work that are not outwardly apparent from the governing equations for A plus K. Lorenz (1967) stated that the concept introduced in the theory of available potential and kinetic energy is a "separate concept from entropy since it involves the field of motion, while entropy depends only upon the thermodynamic state." From the combination of energy and entropy principles through the introduction of the constrained equilibrium state, Johnson (1989) established identical expressions for the maintenance of the atmospheric circulation as the ones derived by Lorenz. While there are other differences, this one difference where the emphasis is on the entropy provides insight into the thermodynamics of climate and the reasons for "coldness" in numerically simulated climate states.

The impact of aphysical sources of entropy and the entropy and energy constraints within a model simulated climate state are now extended to the reversible component of total energy and the Lorenz energy cycle. However, prior to this study which extends Johnson's (1997) analysis within the Lorenz energy cycle, some results are related to classical concepts that have their origins in the second law.

III. A HISTORICAL PERSPECTIVE CONCERNING ENTROPY AND CARATHÉODORY'S STATEMENT OF THE SECOND LAW

In his discussions of thermodynamic systems, Sommerfeld (1950, 1964, pp. 26–86) emphasizes that there are two parts to the second law. For the first part he states that "all thermodynamic systems possess a property which is called entropy. It is calculated by imagining that the state of the system is changed from an arbitrarily selected reference state to the actual state through a sequence of states of equilibrium and by summing up the quotients of the quantity of heat introduced at each step and the absolute temperature T." For the second part he emphasizes that "during real processes the entropy of an isolated system increases." He notes the equivalence of the Clausius form of the second law, which states that "heat cannot pass spontaneously from a lower to higher temperature level" and the Kelvin form that "it is impossible continuously to produce work by cooling only one body down to a temperature below the coldest part of its surroundings." He notes that this latter principle is normally quoted in the form expressed by Ostwald: "It is impossible to design a perpetual motion machine of the second kind."

At this point, note that none of the forms appears to be directly applicable to understanding the climate state and its maintenance against dissipative processes by differential heating. However, what appears less relevant became more relevant through Carathéodory's (1909) development.

In a series of lectures entitled "Natural Philosophy of Cause and Choice" delivered at Oxford, Born (1949) recalled earlier discussions with his "mathematical friend" Carathéodory concerning the basis of the second law of thermodynamics and concepts of idealized thermal machines that transform heat into work and vice versa (William Thompson, Lord Kelvin) or pump heat from one reservoir into another (Clausius). As a physicist, Born noted that these were new and strange concepts in that these idealizations lead to the underlying concepts for the second law. Yet while expressing "admiration for the men who invented these methods" he noted that they were obviously borrowed from engineering and thought they deviated too much from the ordinary methods of physics.

According to Born, as a result of his challenging discussions concerning the need for a more fundamental basis, Carathéodory produced a much more satisfactory statement of the second law from the mathematical consequence of the properties of the Pfaff equation known as the principle of Carathéodory (Margenau and Murphy, 1956). By the Pfaffian relation, entropy is defined to be a fundamental property of the atmosphere from transforming the heat added within the first law into a perfect differential by an integrating factor, namely, T^{-1}. Carathéodory's postulate is (Sommerfeld, 1964):

> In the neighborhood of every state which can be reached reversibly there exist states which cannot be reached along a reversible adiabatic path, or in other words, which can only be reached irreversibly or which cannot be reached at all.

In Born's own words, Carathéodory's postulate simply states "that there exist adiabatically inaccessible states in any vicinity of a given state." Born further comments (lecture delivered in 1948) that 40 years had passed (since 1909) and "still all textbooks reproduce the 'classical' method in lieu of Carathéordory's" and that "this state of affairs seems to me one of unhealthy conservatism."

Born concludes this particular lecture, entitled "Antecedence: Thermodynamics," by stating "why not apply the methods of Cauchy to thermal processes, by treating each volume element as a small thermodynamic system, and regarding not only strain, stress and energy, but also temperature and entropy as continuous functions in space," which he notes has of course been done but with limited success.

The perspective set forth in this study has roots in Sir Napier Shaw's (1930) view of the thermally forced circulation consisting of an overworld and an underworld in which higher and lower values of entropy are respectively transported from entropy sources to sinks in the overworld and from entropy sinks to sources in the underworld. Sir Napier Shaw (1923, 1930) does not indicate an awareness of Carathéodory's postulate in relation to entropy and the second law. From the nature of his discussions, Shaw (1923, 1930) was aware of the condition that entropy is a fundamental property with the term being introduced much earlier in 1865 by Clausius (Kutzbach, 1979). Presumably through Maxwell's influence and his development and use of the $t\phi$ (temperature–entropy) diagram for graphic representation of pressure and temperature obtained from observations by balloons, aeroplanes, or kites, Shaw (1923, 1925, 1930) developed arguments like Carathéodory's as evidenced from his deduction of cyclic atmospheric processes spanning each hemisphere. Shaw (1930) in effect emphasized that the mass and entropy exchange between the overworld and underworld regions of the troposphere was part of a cyclic process that could not be accounted for solely by reversible isentropic processes, that the entropy structure determined the natural level of an atmospheric parcel, and that there were inadmissible regions of the state of the system under reversible isentropic processes.

Sir Napier Shaw (1930) traces the life-history of an air parcel based on entropy and its changes due to heat added and removed as it ascends from the lower to upper equatorial troposphere through moist convection, travels poleward from equatorial to polar regions, descends to the lower troposphere, and then travels equatorward as a cold current, while obtaining warmth and moisture until it reaches its original equatorward position having acquired the temperature and humidity of its initial state. While others, most notably Hadley, had set forth the concept of a like circulation spanning each hemisphere, Ferrel and others (Lorenz, 1967) by this time had discounted the existence of simple direct thermodynamically driven mean hemispheric circulations based on the prevailing westerly winds in extratropical latitudes and the Coriolis force. Interestingly, Shaw must be recognized for reviving this perspective utilizing entropy concepts within which the time scales of a day for vertical ascent in tropical regions through moist convection and a fortnight for descent in polar regions through radiative cooling were identified. His conviction that such a circulation need exist was based on the knowledge that the entropy of air could only be changed by diabatic processes. Thus systematic tropical heating and polar cooling required poleward and equatorward branches of a mass circulation which transported higher values of entropy poleward and lower values of entropy equatorward. In considering the possibility of this circulation he relied on the principle that "the entropy of air tells us its proper place in the firmament... ." He was unaware of the basic hydrodynamic stability of the meridional circulations of a zonally symmetric state (Eliassen, 1951; Gallimore and Johnson, 1981). Still, by virtue of the exact differential nature of entropy sources leading to the definition of entropy, the requirement for transport of entropy envisaged by Sir Napier Shaw is substantiated unequivocally.

As a former student of Maxwell, the influence of Shaw's mentor concerning cyclic processes is evident in Figs. 80 and 81 of "The Air and Its Ways" (Shaw, 1923). Shaw acknowledges Maxwell's development of the "indicator diagram" in which area denotes work in the course of a cycle. It is from his analysis of cyclic processes involving entropy that Shaw (1923) concludes the atmosphere's efficiency as a heat engine is 25%. See his Fig. 81 (also Fig. 96 in Shaw, 1930) showing the cycle displayed in his temperature–entropy diagram with temperature as an abscissa and entropy as an ordinate. He attributes the basis for the temperature–entropy diagram to Sir Alfred Ewin and his work on the steam engine brake-horsepower efficiency. Shaw (1930) also acknowledges his mentor Maxwell for placing "scientific reason in a form which could be comprehended" and notes that for those "who wish to explore the real sources of the ideas of the thermodynamics of the atmosphere to keep by his side a copy of Maxwell's Theory of Heat."

At the same time as Shaw developed his views concerning cyclic processes, Sandström (1915, 1916) extended Bjerknes circulation theorem to include a thermodynamic cycle placing emphasis on heat supplied at low levels and extracted at high levels. Bjerknes (1916) also studied an idealized thermally forced circulation in a closed tube which in effect replicated the Carnot Cycle. Both Sandström and Bjerknes were criticized by Jeffreys (1925) concerning their statements regarding the relative height of heat source and sink. Jeffreys emphasized the importance of temperature gradients on level (geopotential) surfaces in creating motion. Later Godske (1936) and Jeffreys (1936) partially resolved differences. While all are partly right, no clear relation between differential heating and maintenance of atmospheric circulation emerged from this exchange (Johnson, 1989).

The development of isentropic analyses of atmospheric circulation emerged in the 1930s and 1940s, most notably through theory and analysis of potential vorticity (Rossby, 1940; Ertel, 1942; Kleinschmidt, 1955) and through synoptic studies of atmospheric circulation and air masses (Namais, 1940). Interestingly, Namais states that "the term entropy I should not attempt to define here, for it is a sort of mathematical adoption the synoptic meteorologist regards entropy as something proportional to potential temperature." Sir Napier Shaw (1930) had noted earlier that "meteorologists generally express aversion from any suggestion to introduce the idea of entropy into meteorological reasoning on the ground that it is an incomprehensible entity which suggests no physical reality and therefore confuses the argument." Quite possibly the failure of many meteorologists to embrace Shaw's entropy concepts was due to the view that the property was viewed as just another form of temperature conserved under vertical displacement.

To some extent, this situation still persists today, in view of the reluctance of atmospheric scientists to consider that entropy is the property most suited for studies determining the atmosphere's thermodynamic response to differential heating. Very simply, within the mean climate state there must be embedded circulations that systematically transport mass, entropy, and energy between regions of entropy sources and sinks in the form of global monsoonal circulations. Entropy and its conservation are also fundamental to dynamical theories employing potential vorticity.

Some appreciation of the application of entropy concepts to continuous media by physicists at the time of Born's lecture is provided by Tolman and Fine (1948). They state in an opening sentence of their paper that "In analyzing the thermodynamic behavior of physical chemical systems, it is essential to give due consideration to the production of entropy by irreversible processes that take place inside the system." They also state that these considerations are "very important" from both practical and theoretical points of views. From a practical point of view, one can evaluate the relative contribution of these irreversible processes to the inefficiency of a system. From a theoretical point of view, the proper inclusion of appropriate terms for the irreversible production of entropy makes it possible to utilize the second law in terms of an equation rather than as an inequality. Throughout their paper, they continually emphasize the importance of irreversible entropy production. The underlying basis for an entropy theorem as one of the fundamental equations of fluid dynamics stemming from Maxwell and Boltzmann's kinetic theory of gases is aptly presented by Sommerfeld (1964). Wulf and Davis (1952) utilized the second law and calculated the net rate of radiant heat absorbed divided by temperature in estimating the efficiency of the atmosphere. Lettau (1954) also used the second law but included all components of heat addition in estimating the entropy generated by irreversible processes. Spar (1949) contrasted the increase of kinetic energy with decrease of geopotential and internal energies from summer to winter. He noted that wintertime heat sources existed over the western Atlantic and western Pacific oceans, however, without gaining insight into the maintenance of atmosphere circulation. With the development of the theory of available potential energy by Lorenz (1955), nearly all diagnostic analysis turned to evaluation of the Lorenz energy cycle to estimate the intensity of the atmospheric circulation.

More recently, the importance of the second law has emerged with the treatment of the entropy of radiation and matter for moist atmospheres (e.g., Essex, 1987; Hauf and Höller, 1987; Callies and Herbert, 1988; Lesins, 1990; Ooyama, 1990; Peixoto *et al.*, 1991; Stephens and O'Brien, 1993, 1995; Li *et al.*, 1994; Li and Chýlek, 1994).

IV. THE CLASSICAL CONCEPT OF THE CARNOT CYCLE AND THE DRIFTLESS CLIMATE STATE

As a point of departure, the purpose in this section is to relate entropy, its balance, and energy dissipation within the driftless climate state to the classic Carnot cycle. Many texts discuss the Carnot cycle when introducing thermodynamics (e.g. Sears, 1953; Godske *et al.*, 1957; Carrington, 1994) in which heat is added and removed isothermally at two different temperatures with intermediate stages of adiabatic contraction and expansion.

Goody (1995) notes the need to understand the entropy balance and the introduction of Carnot efficiency in climate models. Even though outwardly the Carnot cycle only bears limited relevance to the atmosphere, there are certain basic similarities. Relative to isentropic surfaces, heat is added at higher temperature and removed at lower temperature in the generation of available potential energy. Thermodynamic work occurs through adiabatic expansion and contraction within the governing equations. Although this process transforms available potential to kinetic energy without a provision for "storage" of the reversible component of total energy (i.e., the sum of Lorenz's available potential and kinetic energies), the thermodynamic work within the driftless climate state equates with dissipation. The kinetic energy realized ultimately is removed by irreversible dissipative processes. The thermodynamic work also equates with the generation of available potential energy provided no irreversible diffusion of available potential energy occurs prior to its transformation to kinetic energy. Global integral relations will now be derived that relate to the classical Carnot cycle and its efficiency, after which a relation of entropy change to the Lorenz energy cycle will be established, keeping in mind that $D(E)$ in the following equation corresponds with work in the Carnot cycle.

The time rate of change of total energy as the sum of internal ($c_v T$), geopotential (ϕ), and kinetic (k) energies is expressed by

$$\dot{E} = G(E) - D(E), \tag{1}$$

where

$$G(E) = \int \rho J_\eta Q_m \, dV_\eta, \tag{2}$$

$$= \int \rho J_\eta \dot{s} T \, dV_\eta, \tag{3}$$

$$D(E) = -\int \rho J_\eta U \cdot F \, dV_\eta, \tag{4}$$

$$= \int J_\eta \varepsilon^2 \, dV_\eta. \tag{5}$$

For simplicity it is assumed that the work by viscous stresses is negligible. The relations between heat addition, its individual components, and entropy source are expressed by

$$Q_m = \left[\varepsilon^2 - \nabla \cdot \underset{\sim}{H}_r - \nabla \cdot \underset{\sim}{H}_s - \rho\frac{d}{dt}(Lq)\right]/\rho \tag{6}$$

$$= \dot{s}T, \tag{7}$$

where ε^2 is the positive dissipation function, $\underset{\sim}{H}_r$ is the radiative energy flux, $\underset{\sim}{H}_s$ is the sensible energy flux, L is the latent heat of energy, q is specific humidity, $\dot{s}$ is the entropy source, $\underset{\sim}{U}$ is the velocity, and $\underset{\sim}{F}$ is the frictional force. For generality the integral definitions are expressed in generalized coordinates where J_η is defined by $|\partial z/\partial \eta|$ and dV_η equals $dA_\eta d\eta$ (Johnson, 1980).

The time rate of change of entropy (S) is given by

$$\dot{S} = \int \rho J_\eta \dot{s}\, dV_\eta \tag{8}$$

where the time rate of change of specific entropy $\dot{s}$ is also expressed by (after Dutton, 1976)

$$\dot{s} = \left\{\left\{\varepsilon^2 - \left[\nabla \cdot \underset{\sim}{H}_r + \rho\frac{d}{dt}(Lq)\right]\right\}T^{-1}\right.$$
$$\left. + \nabla \cdot [k\nabla(\ln T)] + [k|\nabla \ln T|^2]\right\}/\rho. \tag{9}$$

Utilizing a factor of proportionality k, the internal energy flux $\underset{\sim}{H}_s$ by Fourier's hypothesis equals $(-k\nabla T)$. The purpose of expressing the entropy source in this form is to identify the positive definite sources of entropy from the dissipation function and heat conduction. More general forms of the entropy relations including expressions for irreversible sources of entropy for moist convective processes and radiative entropy are presented by Callies and Herbert (1988), Essex (1987), Hauf and Höller (1987), Stevens and O'Brien (1993, 1995), and others. While the details of these treatments are beyond the scope of this study, their results serve to point out that inaccuracies in the parameterization of these components of heat addition in climate models are aphysical sources of entropy and thus relevant to considerations of the "coldness of climate models" (Johnson, 1997).

Within the analysis carried out in this study, no provision will be made for the explicit representation of latent energy balance and its time rate of change from flux across the Earth's interface or condensation/evaporation within the atmosphere. See Johnson (1989) for governing equations where latent energy is included in the total energy balance with its integrated change globally determined by the net flux through the boundaries of the atmospheric system. By virtue of the driftless climate state, the implicit balance of water vapor is characterized by an equality of the net upward flux of water vapor across the Earth–atmosphere interface with the net condensation rate within the atmosphere. The effect of the latent energy within the atmosphere is realized as a source of dry entropy through condensation.

For clarity in relating the Carnot cycle with the Lorenz energy cycle, the only irreversible source of entropy considered in the following two sections is the positive dissipation of kinetic energy through viscous stresses. As such, within the driftless climate state the generation of available potential energy and its reversible transformation to kinetic energy by work are equal with kinetic energy dissipation.

Under the condition just noted, the combination of Eqs. (1) and (2) yields

$$\overline{D}(E) = \int \overline{\rho J_\eta Q_{\mathrm{m}}} \, dV_\eta, \tag{10}$$

where the time-area average of g and the temporally, areally mass weighted mean and deviation of f expressed in generalized coordinates (Dutton, 1976; Johnson, 1989) are given by

$$\bar{g} = \int_{t_1}^{t_2} \int_{A_\eta} g \, dA_\eta \, dt / A_\eta (t_2 - t_1),$$

$$\langle f \rangle_\eta = \overline{\rho J_\eta f} / \overline{\rho J_\eta}, \tag{11}$$

$$f^* = f - \langle f \rangle_\eta.$$

The time interval $(t_2 - t_1)$ is of sufficient length for a climate model to simulate a statistically stationary multiannual distribution of atmospheric properties subject to the constraints noted earlier that $\dot{E}$ and $\dot{S}$ vanish, i.e., the climate state without drift. Within this statistically stationary distribution, both the temporally, areally averaged boundary fluxes of energy at the top and the bottom of the atmosphere and the isentropic temporally, areally averaged entropy source ($\langle \dot{s} \rangle_\theta$) within the atmosphere must vanish (Johnson, 1997). Now Eq. (10) is expressed by

$$\overline{D}(E) = \int \overline{\rho J_\eta Q_{\mathrm{m}}^+} \, dV_\eta + \int \overline{\rho J_\eta Q_{\mathrm{m}}^-} \, dV_\eta, \tag{12}$$

$$= \int \overline{\rho J_\eta \dot{s}^+ T^+} \, dV_\eta + \int \overline{\rho J_\eta \dot{s}^- T^-} \, dV_\eta, \tag{13}$$

where for the categorization within the statistically stationary distribution

$$Q_{\mathrm{m}}^{+} = \beta Q_{\mathrm{m}},$$
$$Q_{\mathrm{m}}^{-} = (1 - \beta)Q_{\mathrm{m}},$$
$$\dot{s}^{+} = \beta \dot{s},$$
$$\dot{s}^{-} = (1 - \beta)\dot{s}, \tag{14}$$

given that

$$\beta = 1 \text{ for } \dot{s} > 0, \text{ also } Q_{\mathrm{m}} > 0,$$
$$\beta = 0 \text{ for } \dot{s} < 0, \text{ also } Q_{\mathrm{m}} < 0.$$

By this convention, the time space domain of entropy sources and entropy sinks are respectively and uniquely categorized into regions of heat addition and removal. Because energy dissipation is an irreversible source of entropy, the heat generated in Eq. (12) by this process is solely embedded in Q_{m}^{+}. As such, the difference in the magnitude of the integrals on the right is equal to $\overline{D}(E)$. From balance considerations, the energy source to maintain its supply against $\overline{D}(E)$ could simply be the heat added to the first integral in Eq. (12) through the positive dissipation ε^{2} in Eq. (6). This, however, is equivalent to the perpetual motion machine and thus precluded by the second part of the second law.

With the abbreviated notation that

$$\overline{Q^{+}} = \int \overline{\rho J_{\eta} \dot{s}^{+} T^{+}} \, dV_{\eta},$$
$$\overline{Q^{-}} = \int \overline{\rho J_{\eta} \dot{s}^{-} T^{-}} \, dV_{\eta}, \tag{15}$$

the energy dissipation expressed as a function of heat added is given by

$$\overline{D}(E) = \overline{Q}^{+} + \overline{Q}^{-} = \varepsilon^{+} \overline{Q}^{+}, \tag{16}$$

where the efficiency ε^{+} is expressed by

$$\varepsilon^{+} = \frac{\overline{Q}^{+} + \overline{Q}^{-}}{\overline{Q}^{+}}. \tag{17}$$

The energy dissipation expressed as a function of heat removed is given by

$$\overline{D}(E) = \varepsilon^{-} \overline{Q}^{-}, \tag{18}$$

where

$$\varepsilon^- = \frac{\overline{Q}^+ + \overline{Q}^-}{\overline{Q}^-} . \tag{19}$$

The first efficiency ε^+ just defined corresponds with the classical definition for a Carnot heat engine defined as the ratio for the work output $Q^+ + Q^-$ to the heat input Q^+. A negative value of the reciprocal ε^- corresponds with the coefficient of performance for a Carnot refrigerator defined as the negative ratio of Q^- to $Q^+ + Q^-$ (Sommerfeld, 1964).

From the mean value theorem applied to Eq. (15), the relation of the heat added and removed to entropy changes is now expressed by the following

$$\overline{Q}^+ = \tilde{T}^+ \int \overline{\rho J_\eta \dot{s}^+} \, dV_\eta = \tilde{T}^+ \overline{\tilde{S}}^+ , \tag{20}$$

$$\overline{Q}^- = \tilde{T}^- \int \overline{\rho J_\eta \dot{s}^-} \, dV_\eta = \tilde{T}^- \overline{\tilde{S}}^- , \tag{21}$$

where $\tilde{T}^+$ and $\tilde{T}^-$ in Eqs. (20) and (21) are unique mean values under the condition that $\rho J_\eta \dot{s}^+$ and $\rho J_\eta \dot{s}^-$ are everywhere positive and negative, respectively. The definitions of $\tilde{S}^+$ and $\tilde{S}^-$ follow from Eq. (8) as integrals of the entropy change in regions of heat added and removed, respectively. Here, it is important to recognize that the mean temperatures are unique and precise in the relation between the heat added and the entropy source.

For a climate state to be without drift, Johnson (1997) established that the isentropic temporally, areally averaged entropy source of matter must vanish, i.e., $\langle \dot{s} \rangle_\theta$ equals zero. Under this constraint,

$$\overline{\tilde{S}}^+ = -\overline{\tilde{S}}^- . \tag{22}$$

Note that $\langle \dot{s} \rangle_\theta$ equals zero requires $\overline{\tilde{S}}$ equal to zero; however, $\overline{\tilde{S}}$ equal to zero only requires the vertical integral of $\langle \dot{s} \rangle_\theta$ equal to zero. The substitution of Eqs. (20) and (21) into Eqs. (17) and (19) and use of Eq. (22) yields

$$\varepsilon^+ = \left[1 - \left(\tilde{T}^- / \tilde{T}^+ \right) \right] , \tag{23}$$

$$\varepsilon^- = \left[1 - \left(\tilde{T}^+ / \tilde{T}^- \right) \right] . \tag{24}$$

From the substitution of Eqs. (23) and (24) into Eqs. (16) and (18), the dissipation is expressed by

$$\overline{D}(E) = \left[1 - \left(\tilde{T}^- / \tilde{T}^+ \right) \right] \overline{Q}^+ \tag{25}$$

$$= \left[1 - \left(\tilde{T}^+ / \tilde{T}^- \right) \right] \overline{Q}^- . \tag{26}$$

Clearly the roles of both the first and second parts of the second law have been introduced through the entropy constraint in that the efficiency of the thermodynamic process in maintaining the atmospheric system against energy dissipation is readily defined. The crucial condition which in effect equates the maintenance of atmospheric circulation against dissipative processes by heat addition and removal is the condition that $\bar{\dot{S}}$ vanish with $\bar{\dot{S}}^+$ equal to $-\bar{\dot{S}}^-$. Since entropy is an exact differential and the constraint is uniquely determined from integrals of the Lagrangian source function in Eqs. (20) and (21), the equality of $\dot{S}^+$ and $\dot{S}^-$ requires a mass circulation that transports a precise amount of entropy from source to sink regions and a lesser amount from sink to source region.

Now the expressions for the energy dissipation, Eqs. (25) and (26), will be first subtracted from each other and then added, followed by substitution of Eqs. (20) and (21). The respective results from subtraction and then addition after some simplification are

$$0 = \left(\frac{\overline{Q}^+}{\overline{T}^-} + \frac{\overline{Q}^-}{\overline{T}^-} \right) = \bar{\dot{S}}^+ + \bar{\dot{S}}^- , \tag{27}$$

$$\overline{D}(E) = \frac{1}{2}(\tilde{T}^+ - \tilde{T}^-)\left(\frac{\overline{Q}^+}{\tilde{T}^+} - \frac{\overline{Q}^-}{\tilde{T}^-} \right)$$

$$= \frac{1}{2}(\tilde{T}^+ - \tilde{T}^-)\left(\bar{\dot{S}}^+ - \bar{\dot{S}}^- \right), \tag{28}$$

$$= (\tilde{T}^+ - \tilde{T}^-)\bar{\dot{S}}^+ = -(\tilde{T}^+ - \tilde{T}^-)\bar{\dot{S}}^- . \tag{29}$$

These results just derived for the driftless climate state with $\bar{\dot{S}} = 0$ correspond with the classical result for the Carnot cycle that the "quantities of heat absorbed and rejected are proportional to the corresponding temperature at which the two processes occur" (Sommerfeld, 1964; Sears, 1953). Furthermore, within the concept of the driftless climate state, the intensity of the energy dissipation is directly related to the difference of the mean value temperatures between the regions of entropy increase and decrease multiplied by the intensity of the entropy source (or sink).

An alternative relation in terms of the temperature deviations from a mean state with

$$\tilde{T}^+ = \bar{\tilde{T}} + \Delta\tilde{T}, \tag{30}$$

$$\tilde{T}^- = \bar{\tilde{T}} - \Delta\tilde{T}, \tag{31}$$

is given by

$$\overline{D}(E) = \Delta\tilde{T}\left(\overline{\tilde{S}}^+ - \overline{\tilde{S}}^-\right) \tag{32}$$

$$= 2\Delta\tilde{T}\overline{\tilde{S}}^+ = -2\Delta\tilde{T}\overline{\tilde{S}}^-, \tag{33}$$

where $\overline{\tilde{T}}$ is the mean temperature $(\tilde{T}^+ + \tilde{T}^-)/2$. Here, an increase in dissipation requires an increase in the difference of the mean value temperatures from their mean. Furthermore, since the balance of the entropy of matter within the driftless climate state occurs through the incoming entropy of solar radiation and exiting entropy of terrestrial radiation, in principle, the positive definite dissipation for the global state that can be supported in a state of balance is uniquely related to the magnitude of $\Delta\tilde{T}$ that develops. See Hauf and Höller (1987) for appropriate equations and Callies and Herbert (1988), Essex (1978), Lesins (1990), and Stephen and O'Brien (1993, 1995) for further discussions regarding the balance of entropy of matter in relation to radiative entropy exchange.

The results in Eqs. (28) and (29) correspond with the classical result of the Carnot cycle that "all reversible engines which exchange heat only at two temperatures T_1 and T_2 have equal efficiencies" (Sommerfeld, 1964, p. 29) and thus in this case are uniquely related to the intensity of the dissipation. Note within the defining equations for the atmosphere, the irreversible process of dissipation ε^2/ρ in Eq. (6) returns as heat added within the integral $\overline{Q}^+$. However, this does not preclude mathematical correspondence with the classical result of the Carnot cycle. With given heat addition the only degree of freedom to maintain balance with increased dissipation is for the difference of mean temperatures to increase. Since there is no accounting for the return of energy by irreversible processes by virtue of the categorization of the heat added $\overline{Q}^+$ and removed $\overline{Q}^-$ and since the conversion of available potential to kinetic energy is isentropic, in principle the cycle can be reversed such that the kinetic energy realized could be utilized to achieve a state with maximum total potential energy and no kinetic energy. Within the energy equations, an atmosphere with maximization of total potential energy by reversible isentropic processes from the reservoir of kinetic energy has the same probability as the maximization of kinetic energy from the reservoir of total potential energy (or available potential energy) by reversible processes. In the presence of differential heating by solar and terrestrial radiation in combination with earth rotation, and the balanced nature of atmospheric motion, a reservoir of both available potential and kinetic energies must exist; thus the probability of either extreme is zero.

Finally, it must be noted that the energy dissipation can be expressed by

$$D(\overline{E}) = (\Delta\tilde{T})^{+}\,\overline{\dot{S}}^{+} + (\Delta\tilde{T})^{-}\,\overline{\dot{S}}^{-}, \tag{34}$$

a covariant form of differential heating and temperature. The result in Eq. (34) for this simple system characterized by temperature departures is formally equivalent with the generation of the available potential energy (Lorenz, 1955), thus indicating that the entropy principle is an integral part of the Lorenz energy cycle. Here, ΔT may be viewed as a departure from the isobaric mean temperature multiplied by an assumed horizontally invariant static stability.

No attempt is made in this study to detail the energy transformations of the Lorenz energy cycle occurring within the climate state that have been studied extensively and summarized by Lorenz (1967) and others (e.g., Peixoto and Oort, 1992; Wiin-Nielsen and Chen, 1993). In this study, the reversible component of total energy is viewed as a property involved with the transport of entropy and dry static energy which is conserved during the reversible transformation of available potential to kinetic energy (Johnson, 1989). In the driftless climate state with $\langle \dot{s} \rangle_{\theta}$ zero, the reservoir of the reversible component of total energy is not increasing or decreasing, thus the rates of generation of A, the work involved with conversion of A to K and dissipation of K are equal. Thus, Eqs. (23) through (34) plus defining relations provide the simplest expressions relating the second law, Carnot efficiency, and the maintenance of the Lorenz energy cycle. Not only are the concepts related, but apart from the isobaric approximation set forth and utilized in diagnostic applications there is an exact physical and mathematical correspondence. The combination of mean value temperatures as derived with $\overline{\dot{S}}^{+}$ and $\overline{\dot{S}}^{-}$ being integrals of the Lagrangian source of entropy defined by an exact differential bears a unique and exact relation to the heat added $\overline{Q}^{+}$ and removed $\overline{Q}^{-}$.

Within the integral conditions just developed and the concept of the driftless climate state, the two parts of the second law are evident. The combination of entropy sources and sinks thermodynamically force the atmosphere from an equilibrium to an actual state, which in combination with the temperature distribution realized and energy transformations occurring within the reversible component of total energy maintain it against dissipative processes. In Eq. (34) the magnitudes of the temperature difference between the regions of heat addition and removal, the entropy source for matter, and the intensity of the energy dissipation place bounds on the intensity of atmosphere circulation. Within this context consider that the generation of the reversible component depends linearly on a temperature difference as defined by the Carnot cycle, while kinetic energy dissipation based on Lettau's (1959) geostrophic drag and surface roughness concepts for the planetary boundary layer depends on the cubic power of the surface geostrophic wind. With a cubic dependence of kinetic energy dissipation on the planetary boundary layer circulation, at some intensity level the magnitude of the temperature difference in combination with differential heat sources will be insufficient to maintain the energy dissipation. In constrast, if the intensity of the circulation and the dissipation is less than that required for balance, the temperature difference will increase with the result that the magnitude of the reversible component of total energy will also increase; thus ultimately the intensity of circulation and dissipation will increase until balance is attained. With recognition of irreversibility as the source of entropy, this basic result involving the Carnot cycle substantiates Johnson's (1997) conclusion that the only degree of freedom for a climate model to maintain a driftless state in the presence of positive definite aphysical sources of entropy would be for the model state to be biased cold.

V. THE CLIMATE STATE AND THE REVERSIBLE COMPONENT OF TOTAL ENERGY

In developing the concepts of global monsoonal circulations, Johnson (1989) derived the following expression for the reversible component of total energy:

$$\Delta E_\alpha = E - E_\alpha, \tag{35}$$

where the total energy (E), the total energy of a constrained equilibrium state (E_α) with equilibrium temperature $T_\alpha(\theta, t)$, the kinetic energy (K) of the atmosphere, and the available potential energy (A) of the atmosphere are respectively expressed as integrals in isentropic coordinates by

$$E = \int \rho J_\theta (c_v T + \phi + k)\, dV_\theta, \tag{36}$$

$$E_\alpha = \int \rho J_\theta (c_v T_\alpha + \phi_\alpha)\, dV_\theta, \tag{37}$$

$$K = \int \rho J_\theta k\, dV_\theta, \tag{38}$$

$$A = (E - E_\alpha) - K. \tag{39}$$

The definition of the total energy of the constrained equilibrium state E_α and its time rate of change formally corresponds with that of the reference state of available potential energy theory (Lorenz, 1955). However, a crucial point to recognize is that the derivation of the reversible component of total energy is based on an axiomatic concept of a constrained equilibrium state (Sommerfeld, 1964) with uniform temperature $T(\theta, t)$, in which by virtue of Poisson's equation, pressure and temperature are invariant within the global extent of incremental isentropic layers $d\theta$. There is no requirement for a virtual isentropic redistribution of mass to a horizontally invariant state in order to define Lorenz's reference state of minimum total potential energy. Rather the reversible component of total energy represents the energy realized from the thermodynamic forcing of the actual atmosphere from its equilibrium state by the systematic nature of entropy sources and sinks and the temperature differences that develop. Thus the issue of whether Lorenz's reference atmosphere was the proper one which the atmosphere attempted to reach relative to others involving momentum balance (Van Mieghem, 1956; Pfeffer *et al.*, 1965) and also the issue of appropriate modifications to the definition of a reference an atmosphere with orography are mute. As Sommerfeld (1964) notes, the entropy of a system can be taken to be the sum of the entropies of all volume elements in which equilibrium is constrained.

A subtle but important distinction between the entropy constraints $\bar{S}$ equal zero utilized in the prior development and the development with $\langle \dot{s} \rangle_\theta$ equal zero for the driftless climate state and reversible component of total energy stems from the differences in the implied temperature of equilibrium states. The implicit equilibrium state temperature for $\dot{S}$ equal zero is a uniform temperature $T_0(t_0)$ over the entire atmosphere, while the implicit equilibrium state for the reversible component of total energy with $\langle \dot{s} \rangle_\theta$ equal zero is $T_\alpha(\theta, t)$ (Johnson, 1989). The first corresponds with the temperature of Dutton's maximum entropy state, while the second as already noted corresponds with Lorenz's reference state. Both states of equilibrium require hydrostatic conditions and horizontal invariance along geopotential surfaces. Although not detailed in this study $\langle \dot{s} \rangle_\theta$ equal to zero mandates $\bar{S}$ equal 0; thus the flow of energy within the energy cycles of Dutton and Lorenz is identical from the condition that the time rate of change of reference state energies of both entropic and available potential energy vanish (Johnson, 1989).

The time rate of change of the reversible component of total energy is expressed by

$$\Delta \dot{E}_\alpha = \dot{E} - \dot{E}_\alpha = \dot{A} + \dot{K}, \tag{40}$$

$$= G(\Delta E_\alpha) - D(\Delta E_\alpha), \tag{41}$$

where the generation of ΔE_α is given by

$$G(\Delta E_\alpha) = \int \rho J_\theta \dot{s} T \varepsilon \, dV_\theta, \qquad (42)$$

and $D(\Delta E_\alpha)$ is simply equated to $D(E)$. The efficiency is defined by

$$\varepsilon = [1 - (T_\alpha/T)]. \qquad (43)$$

A substitution of Eq. (7) into Eq. (42) yields

$$G(\Delta E_\alpha) = \int \rho J_\theta Q_m \varepsilon \, dV_\theta, \qquad (44)$$

Under the condition for a climate state without drift with $\Delta \dot{E}_\alpha$ equal to zero, Johnson (1989, pp. 130–132) emphasized for a statistically stationary circulation that the dissipation by irreversible processes $D(\Delta E_\alpha)$ must be balanced by $G(\Delta E_\alpha)$ through the organization of entropy sources in relation to temperature and that the maintenance of the atmosphere's circulation against energy dissipation fundamentally rested on the entropy principle. To reinforce this result and provide more insight into the issues raised by Goody (1995) concerning the Carnot cycle and efficiency, consider the equality of the isentropic temporally, areally averaged integrands of Eqs. (2) and (3) now expressed by

$$\langle g(E) \rangle = \langle Q_m \rangle = \langle \dot{s} T \rangle, \qquad (45)$$

where for simplicity the subscript θ is suppressed for the average $\langle f \rangle_\theta$. A partitioning yields

$$\langle g(E) \rangle = \langle \dot{s} \rangle \langle T \rangle + \langle \dot{s}^* T^* \rangle. \qquad (46)$$

With $\langle \dot{s} \rangle$ equal to zero (Johnson 1997), $\langle g(E) \rangle$ is given by

$$\langle g(E) \rangle = \langle \dot{s}^* T^* \rangle. \qquad (47)$$

There is also a corresponding requirement from the condition $\langle \dot{s} \rangle$ equal zero that the isentropic temporally, areally averaged vertical diabatic mass flux vanish, i.e., the total upward balances the total downward mass transport within the statistically stationary state. This condition uniquely determines the irrotational component of a mass circulation that provides for net entropy transport from regions of entropy sources to entropy sinks (Johnson, 1989). Also note that within the covariance $\langle \dot{s}^* T^* \rangle$, for every Lagrangian incremental region with an entropy source $(\rho J_\eta \dot{s}^+) \, dA_\eta \, d\eta$, there is a corresponding Lagrangian entropy sink $(\rho J_\eta \dot{s}^-) \, dA_\eta \, d\eta$ imposed by the condition $\langle \dot{s} \rangle$ equal to zero. The net energy generated by the two regions in maintaining atmospheric circulation against dissipative processes is $(\rho J_\eta \dot{s}^+ T^+)\, dA_\eta\, d\eta$ plus $(\rho J_\eta \dot{s}^- T^-)\, dA\, d\eta$. Also since pressure raised to the R_d/c_p power is a unique function of temperature, the temperature difference of the two incremental Lagrangian regions also defines a unique pressure difference that in effect determines the kinetic energy produced by work involving isentropic expansion/contraction that is eventually dissipated by irreversibility processes. The net integral over all incremental volume elements equates with the total dissipation.

Now consider the corresponding generation of ΔE_α as given by the integrand of Eq. (44):

$$\langle g(\Delta E_\alpha)\rangle = \langle Q_m \varepsilon \rangle \tag{48}$$

$$= \langle \dot{s} T \varepsilon \rangle, \tag{49}$$

where ε is the efficiency defined in the Lorenz enegy cycle by Eq. (43). With the use of Eq. (43) and a partitioning, Eq. (49) becomes

$$\langle g(\Delta E_\alpha)\rangle = \langle \dot{s}\rangle\langle T\varepsilon\rangle + \langle \dot{s}^*(T - T_\alpha)^*\rangle. \tag{50}$$

Under the conditions that $\langle \dot{s}\rangle$ vanishes and $T_\alpha^*(\theta, t) \equiv 0$, $\langle g(\Delta E_\alpha)\rangle$ reduces to

$$\langle g(\Delta E_\alpha)\rangle = \langle \dot{s}^* T^*\rangle. \tag{51}$$

The exact equality of $\langle g(E)\rangle$ in Eq. (47) with $\langle g(\Delta E_\alpha)\rangle$ in Eq. (51) emphasizes the critical nature of introducing entropy through Eq. (7) and accurately simulating the joint distribution of the isentropic deviations of entropy sources and temperature from their mass weighted average in resolving the total energy balance of climate models. Now note, however, that $\langle g(\Delta E_\alpha)\rangle$ equal to $\langle g(E)\rangle$ need not be positive within all isentropic layers. For the atmospheric circulation to be maintained against dissipation, the integral with vertical mass weighting of $\langle \dot{s}^* T^*\rangle$ as expressed by Eq. (42) must be positive, i.e., the sum of the generation in layers with $\langle \dot{s}^* T^*\rangle$ positive overweighs the sum of the negative generation in all other layers, provided there are any layers of negative generation. Furthermore, under the condition for the driftless climate state with $\dot{E}$ zero, the integrals $G(E)$, $G(\Delta E_\alpha)$, $D(E)$, and $D(\Delta E_\alpha)$ representing the global balance of energy are equal. Thus if the isentropic deviations of temperature are identically zero everywhere as in the equilibrium state, the dissipation must vanish everywhere. The critical point to be noted here is that the conditions required for maintenance of the atmosphere's circulation against kinetic energy dissipation stem from the entropy principle whether viewed from the balance of total energy E or the reversible component of total energy ΔE_α.

VI. THE CLASSICAL CONCEPT OF EFFICIENCY IN RELATION TO $\langle g(E)\rangle$ AND $\langle g(\Delta E_\alpha)\rangle$

Within the set of equations utilized to illustrate the relation between the classical concept of the Carnot cycle and driftless climate state, the requirement for systematic differences in mean value temperatures in conjunction with the entropy sources led to the concept of a global efficiency as expressed by Eq. (23). Under the conditions just derived for $\langle g(E)\rangle$ and $\langle g(\Delta E)\rangle$, a definition of an efficiency corresponding with the Carnot cycle is now derived which, however, is a function of potential temperature. From the conditional relations in Eq. (14), $\langle \dot{s}\rangle$ and $\langle g(\Delta E_\alpha)\rangle$ (also $\langle g(E)\rangle$) are expressed by

$$\langle \dot{s}\rangle = \langle \dot{s}^+ + \dot{s}^-\rangle = \langle \dot{s}^+\rangle + \langle \dot{s}^-\rangle, \tag{52}$$

$$\langle g(\Delta E_\alpha)\rangle = \langle \dot{s}^+ T^{*+}\rangle + \langle \dot{s}^- T^{*-}\rangle, \tag{53}$$

where the plus and minus superscripts identify the time space domain in which entropy sources are everywhere either positive or negative. Now utilizing the condition that with $\langle \dot{s}^+\rangle$ equals minus $\langle \dot{s}^-\rangle$ with $\langle \dot{s}\rangle$ zero, Eq. (53) by the mean value theorem becomes

$$\langle g(\Delta E_\alpha)\rangle = \tilde{\varepsilon}^+ \tilde{T}^{*+}\langle \dot{s}^+\rangle \tag{54}$$

$$= \tilde{\varepsilon}^+ \langle \dot{s}^+ T^{*+}\rangle = \tilde{\varepsilon}^+ \langle Q_{\mathrm{m}}^+\rangle, \tag{55}$$

where the efficiency defined by the mean valued temperatures is defined by

$$\tilde{\varepsilon}^+ = \left[1 - \left(\tilde{T}^{*-}/\tilde{T}^{*+}\right)\right]. \tag{56}$$

A corresponding relation for the generation as a function of the entropy sink is expressed by

$$\langle g(\Delta E_\alpha)\rangle = \tilde{\varepsilon}^- \tilde{T}^{*-}\langle \dot{s}^-\rangle \tag{57}$$

$$= \tilde{\varepsilon}^- \langle \dot{s}^- T^{*-}\rangle = \tilde{\varepsilon}^- \langle Q_{\mathrm{m}}^-\rangle, \tag{58}$$

where

$$\tilde{\varepsilon}^- = \left[1 - \left(\tilde{T}^{*+}/\tilde{T}^{*-}\right)\right]. \tag{59}$$

As noted earlier, the negative reciprocal of $\tilde{\varepsilon}^-$ corresponds with coefficient of performance for a refrigerator.

Analogous with the earlier results, Eqs. (20) through (29), a subtraction followed by addition of Eqs. (55) and (58) with manipulation and substitution yields

$$0 = \frac{\langle Q_m^+ \rangle}{\tilde{T}^{*+}} + \frac{\langle Q_m^- \rangle}{\tilde{T}^{*+}}, \tag{60}$$

$$\langle g(\Delta E_\alpha) \rangle = (1/2)(\tilde{T}^{*+} - \tilde{T}^{*-})\left(\frac{\langle Q_m^+ \rangle}{\tilde{T}^{*+}} - \frac{\langle Q_m^- \rangle}{\tilde{T}^{*-}} \right) \tag{61}$$

$$= \frac{1}{2}(\tilde{T}^{*+} - \tilde{T}^{*-})(\langle \dot{s}^+ \rangle - \langle \dot{s}^- \rangle) \tag{62}$$

$$= (\tilde{T}^{*+} - \tilde{T}^{*-})(\langle \dot{s}^+ \rangle) = -(\tilde{T}^{*+} - \tilde{T}^{*-})\langle \dot{s}^- \rangle. \tag{63}$$

The equality of Eqs. (47), (51), and (61) through (63) reveals an equivalence between classical thermodynamics concepts and the Lorenz energy cycle in relation to the second law, entropy change, reversibility and irreversibility, and energy dissipation. This equivalence of $G(\Delta E_\alpha)$ with $D(\Delta E_\alpha)$ expressed in integral form utilizing Eq. (63) is

$$D(\Delta E_\alpha) = \int \overline{\rho J_\theta} \langle g(\Delta E_\alpha) \rangle \, dV_\theta \tag{64}$$

$$= \int \overline{\rho J_\theta} [\tilde{T}^{*+} - \tilde{T}^{*-}] \langle \dot{s}^+ \rangle \, dV_\theta \tag{65}$$

$$= -\int \overline{\rho J_\theta} [\tilde{T}^{*+} - \tilde{T}^{*-}] \langle \dot{s}^- \rangle \, dV_\theta, \tag{66}$$

The results expressed in Eqs. (65) and (66) extend to $G(E)$ in Eq. (2), and $D(E)$ in Eq. (5).

As noted earlier, the integrand of Eqs. (64) through (66) need not be positive within isentropic layers. However, the vertical integral of $\langle g(\Delta E_\alpha) \rangle$ must be positive. By virtue of the uniqueness of the mean valued temperatures $\tilde{T}^{*+}(\theta)$ and $\tilde{T}^{*-}(\theta)$, computations of the vertical distribution of the product of the temperature difference with $\langle \dot{s}^+ \rangle$ or $\langle \dot{s}^- \rangle$ utilizing Eqs. (65) and (66) would provide quantitative results regarding mean temperature differences and efficiencies as a function of potential temperature for estimating atmospheric dissipation, a quantity that cannot be evaluated directly. The equations just derived including the definition of mean valued temperatures $\tilde{T}^{*+}(\theta)$ and $\tilde{T}^{*-}(\theta)$ and efficiencies $\tilde{\varepsilon}^+(\theta)$ and $\tilde{\varepsilon}^-(\theta)$ also provide a unique set to relate the concept of Carnot efficiencies with the generation of available potential energy and the flow of energy through the Lorenz energy cycle. Although the form of efficiencies expressed by Eqs. (56) and (59) differ from Lorenz's isentropic efficiency, with the use of $T_\alpha(\theta, t)$, the Carnot efficiencies can be transformed into isentropic efficiencies (Johnson, 1989). In a discussion of thermodynamic efficiency, Lorenz (1967) noted that its determination was "the fundamental observational problem of atmospheric energetics." Clearly, $\langle g(E) \rangle$ and $\langle g(\Delta E_\alpha) \rangle$ as defined in Eqs. (47) and (51) yield identical results with $\langle g(\Delta E_\alpha) \rangle$ in Eqs. (55) and (58); thus a link of the efficiencies defined by the Lorenz energy cycle and the Carnot cycle has been established. The need expressed by Goody (1995) concerning the relation of Carnot efficiencies and the second law with available potential energy and the intensity of energy transformations has been satisfied. Reasons why this link has remained ill defined are several; the most probable being the very limited understanding of entropy, its relation to energy exchange, and failure to use isentropic coordinates.

VII. SOURCES OF ENTROPY IN THE MODELED CLIMATE STATE

The development heretofore has provided a perspective of the role of entropy and its sources in maintaining the atmosphere's circulation against dissipative processes, and related classical perspectives to the Lorenz energy cycle. In his previous study of the modeled climate state, Johnson (1997) determined from entropy balance that aphysical sources of entropy in the presence of the constrained equilibrium state with $\langle \dot{s} \rangle$ equal to zero requires the climate state to be cold. This requirement can only be established from an analysis of entropy balance and the condition that $\langle \dot{s} \rangle$ vanish. It cannot be established from the energy principle. This same constraint is now used within the concept of the reversible component of total energy to assess the increased dissipation by aphysical sources of entropy. This assessment will be made by comparing the magnitude of the differences of the generation of the reversible component of total energy between the model and the true climate state. Drawing on an earlier result that positive definite aphysical sources of entropy exist in climate models (Johnson, 1997, App. B), it will be established that an added component of generation is realized through the coldness of the model climate. This added component serves to balance the increase in dissipation stemming from positive definite aphysical sources of entropy and irreversibility.

The aim now is to utilize a functional representation of an entropy component $\Delta \dot{s}$ in terms of model departures of heat addition ΔQ and temperature ΔT from the true climate state and positive definite aphysical sources of entropy $\dot{s}_a$, and then determine a relationship between the two components. Because all climate models employ a common basic set of governing equations, all must obey the second law regardless of the form of the thermodynamic equation and coordinate system utilized. To establish an implicit representation of an isentropic transport equation and the isentropic temporally, areally averaged constraint for models in general, Johnson (1997) expressed the transport equation by

$$(\rho J_\eta)^{-1}\left[\frac{\partial}{\partial t_\eta}(\rho J_\eta f) + \nabla_\eta \cdot (\rho J_\eta \underset{\sim}{U} f) + \frac{\partial}{\partial \eta}(\rho J_\eta \dot{\eta} f)\right] = \dot{f}. \quad (67)$$

The generalized vertical coordinate η, which represents any of the convential meteorological coordinates (Johnson, 1980), must either be a monatomic increasing or decreasing function of height z. Here, J_η is $|\partial z/\partial \eta|$, and $\dot{f}$ is the model's Lagrangian source of f. The combination of terms on the left simply constitutes the Eulerian representation of the Lagrangian source within a model's coordinate system.

Equation (67) has been expressed in an invariant form in that the sum of the Eulerian components on the left constitutes the Lagrangian source of a property in all climate models regardless of coordinate system. Note that with substitution of Eq. (67) into Eq. (11),

$$\overline{\frac{\partial \eta}{\partial \theta}\left[\frac{\partial}{\partial t_\eta}(\rho J_\eta f) + \nabla_\eta \cdot (\rho J_\eta \underset{\sim}{U} f) + \frac{\partial}{\partial \eta}(\rho J_\eta \dot{\eta} f)\right]}/\overline{\rho J_\theta} = \langle \dot{f} \rangle, \quad (68)$$

the product of (ρJ_θ) and $(\rho J_\eta)^{-1}$ on the left, which weights the sum of the Eulerian components, simplifies to $(\partial \eta/\partial \theta)$; on the right $\langle \dot{f} \rangle$ remains the isentropic mass weighted Lagrangian source. Conceptually, the model's Eulerian transport calculations in η coordinates on the left multiplied by $(\partial \eta/\partial \theta)$ are being appropriately transformed as a sum, then averaged within the model's implicit isentropic structure, and then divided by $\overline{\rho J_\theta}$ to determine $\langle \dot{f} \rangle$. In the case of entropy, an aphysical source from spurious numerical dispersion/diffusion involves the "mixing" of energies from calculations of the Eulerian expansion on the left in relation to the calculation of the invariant Lagrangian entropy source on the right. The proof that numerical diffusion (Book *et al.*, 1981) is a positive definite source of entropy was developed for a simple box model in Appendix B (Johnson, 1997). Johnson concluded that in regions of strong temperature advection with a Courant number of 0.2 the corresponding anomalous heating rate would reach 3.16°C day^{-1}.

The entropy source within an atmospheric climate model is now defined by

$$\dot{s} = Q_{\mathrm{m}}/T + \dot{s}_{\mathrm{a}}, \tag{69}$$

where Q_{m}/T is the explicit entropy source determined from model computations and $\dot{s}_{\mathrm{a}}$ is the positive definite implicit aphysical source of entropy stemming from numerical diffusion/dispersion, Gibbs phenomena, inadequacies of parameterization, and other like processes that mix energy spuriously. With the use of a wavy underbar to express the true state for thermodynamic properties and processes, the following relation for the model's entropy source as a function of a true source, an explicit departure source $\Delta\dot{s}(\Delta Q_{\mathrm{m}}, \Delta T)$, and the implicit source $\dot{s}_{\mathrm{a}}$ is given by

$$\dot{s} = \underset{\sim}{\dot{s}} + \Delta\dot{s}(\Delta Q_{\mathrm{m}}, \Delta T) + \dot{s}_{\mathrm{a}}. \tag{70}$$

This relation in combination with Eq. (69) requires the model entropy source Q_{m}/T to be the sum of the true and systematic departure entropy sources given by

$$Q_{\mathrm{m}}/T = \underset{\sim}{\dot{s}} + \Delta\dot{s}(\Delta Q_{\mathrm{m}}, \Delta T), \tag{71}$$

where the true atmospheric entropy source corresponding with Eq. (7) is

$$\underset{\sim}{\dot{s}} = \underset{\sim}{Q}_{\mathrm{m}}/\underset{\sim}{T}. \tag{72}$$

The model errors in heat addition and temperature in relation to the actual and true states are, respectively,

$$Q_{\mathrm{m}} = \underset{\sim}{Q}_{\mathrm{m}} + \Delta Q_{\mathrm{m}}, \tag{73}$$

$$T = \underset{\sim}{T} + \Delta T, \tag{74}$$

where ΔQ_{m} and ΔT are departures. The combination of Eqs. (71) through (74) yields

$$Q_{\mathrm{m}}/T = \underset{\sim}{Q}_{\mathrm{m}}/\underset{\sim}{T} - \left(\underset{\sim}{Q}_{\mathrm{m}}/\underset{\sim}{T}\right)(\Delta T/T) + (\Delta Q_{\mathrm{m}}/T)$$

$$= \underset{\sim}{\dot{s}} - \underset{\sim}{\dot{s}}(\Delta T/T) + (\Delta Q_{\mathrm{m}}/T). \tag{75}$$

The explicit error source of entropy $\Delta\dot{s}$ following from Eqs. (71) and (75) is

$$\Delta\dot{s} = (\Delta Q/T) - \underset{\sim}{\dot{s}}(\Delta T/T). \tag{76}$$

Equations (69) through (76) considered together clarify the partitioning of entropy sources. Within the source $\Delta \dot{s}(\Delta Q_m, \Delta T)$, ΔQ_m represents explicit errors of simulated heat addition from erroneous parameterization of moist processes, radiation, sensible heating, and other processes, and ΔT represents explicit errors in simulated temperature. The aphysical source $\dot{s}_a$ is due to numerical dispersion/diffusion, Gibbs oscillations, and the inadequacies of parameterization, i.e., processes that serve to mix energy spuriously. However, the temperature errors that develop through integration from ΔQ_m and $\dot{s}_a$ interact from the nonlinearity of the relations. For example, a model's inability to simulate properly the temperature by virtue of spurious numerical dispersion/diffusion induces a departure entropy source $\Delta \dot{s}$ through division of the heat addition by an erroneous temperature. The temperature error also introduces errors in calculations of parameterized heat addition from the strong dependency of latent energy release and infrared radiation on temperature that lead to an erroneous specification of energy.

VIII. THE ENTROPY BALANCE

Now from the entropy balance for the driftless climate state that both the model and true state entropy source vanish, i.e., $\langle \dot{s} \rangle$ and $\langle \tilde{\dot{s}} \rangle$ equal zero, Johnson (1997) established from Eq. (70) that $\langle \Delta \dot{s} \rangle$ and $\langle \dot{s}_a \rangle$ are uniquely related by

$$\langle \dot{s}_a \rangle = -\langle \Delta \dot{s} \rangle \tag{77}$$

$$= \langle \tilde{\dot{s}}(\Delta T/T) \rangle - \langle (\Delta Q_m/T) \rangle. \tag{78}$$

The equality of $\langle \dot{s}_a \rangle$ and $-\langle \Delta \dot{s} \rangle$ has certain implications. Note that in Eq. (78) if a climate model simulates an unbiased distribution of heat addition and if aphysical sources of entropy $\dot{s}_a$ were to vanish, then the covariance of $\tilde{\dot{s}}$ and $(\Delta T/T)$ must vanish. By virtue of the systematic nature of $\tilde{\dot{s}}$ with polar cooling and tropical heating, the only means to ensure this condition would be for the temperature difference ΔT to vanish everywhere; i.e., the simulated temperature distribution would correspond with the true climate state. Alternatively, if ΔQ_m and ΔT were identically zero, both the systematic departure error and the positive definite source $\langle \dot{s}_a \rangle$ and $\langle \Delta \dot{s} \rangle$ must vanish with the result that the model and true climate state mean entropy sources become equal.

Now consider ΔT and ΔQ_m to be finite, however, under the condition that the aphysical sources of entropy vanish. Such a condition from Eqs.

(77) and (78) requires

$$0 = \langle (\dot{s}\Delta T - \Delta Q_{\mathrm{m}})/T \rangle \tag{79}$$

$$= \langle [\dot{s} - (\Delta Q_{\mathrm{m}}/\Delta T)](\Delta T/T) \rangle. \tag{80}$$

With ΔT finite, Eq. (80) requires the ratio of the error of heat addition ΔQ_{m} to the error in temperature ΔT to equal the true entropy source $\dot{s}$ everywhere. Since $\dot{s}$ is the true Lagrangian source, there is no means for ΔQ_{m} to induce an immediate response in temperature ΔT for a Lagrangian parcel. Furthermore, partitioning Eq. (80) into the products of mean and eddy components expressed by

$$\langle (\Delta Q_{\mathrm{m}}/\Delta T) \rangle \langle (\Delta T/T) \rangle = \langle [\dot{s} - (\Delta Q_{\mathrm{m}}/\Delta T)]^*(\Delta T/T)^* \rangle \tag{81}$$

reveals an inherent difficulty. Note for satisfaction of Eq. (81) that the mass weighted product on the left involving $\langle (\Delta Q_{\mathrm{m}}/\Delta T) \rangle$ and $\langle (\Delta T/T) \rangle$ would need to equal the mass weighted average of the covariant quantities on the right. Within these considerations, the possibility that ΔQ_{m} and ΔT are both finite while $\langle \dot{s}_{\mathrm{a}} \rangle$ vanishes is extremely unlikely. The condition that ΔT vanishes in Eq. (78) with $\langle \dot{s}_{\mathrm{a}} \rangle$ equal to the explicit error of heat addition $[-\langle (\Delta Q_{\mathrm{m}}/T) \rangle]$ is also not feasible from physical arguments based on the origins of the error. This leads to the end conclusion that if ΔQ_{m} is finite that both temperature errors and positive definite sources of entropy exist. The arguments just stated, however, do not preclude $\langle (\Delta Q_{\mathrm{m}}/T) \rangle$ equal to zero in Eq. (78) with $\langle \dot{s}_{\mathrm{a}} \rangle$ and $\langle \dot{s}(\Delta T/T) \rangle$ nonzero, although in actual fact this also is unlikely.

Since the functional dependence of $\Delta \dot{s}$ is nonlinear in Eq. (78) by virtue of the variation of T in the denominator of both terms on the right, with factoring and some manipulation, the reciprocal of temperature defined in Eq. (74) in relation to the true temperature and its error is expressed by

$$T^{-1} = \underset{\sim}{T}^{-1}[1 - (\Delta T/T)]. \tag{82}$$

The reciprocal of true temperature expressed as a mean and deviation by Eq. (11) is given by

$$\underset{\sim}{T}^{-1} = \langle \underset{\sim}{T} \rangle^{-1}[1 - (\underset{\sim}{T}^*/\underset{\sim}{T})]. \tag{83}$$

Their combination yields

$$T^{-1} = \langle \underset{\sim}{T} \rangle^{-1}[1 - (\underset{\sim}{T}^*/\underset{\sim}{T})][1 - (\Delta T/T)]. \tag{84}$$

Now substitution of Eq. (84) into Eq. (78) yields

$$\langle \dot{s}_{\mathrm{a}} \rangle = \langle \underset{\sim}{T} \rangle^{-1} \langle (\dot{s}\Delta T - \Delta Q_{\mathrm{m}})[1 - (\underset{\sim}{T}^*/\underset{\sim}{T})][1 - (\Delta T/T)] \rangle. \tag{85}$$

Within this relation consider that the temperature error ΔT is much less than the true departure temperature $\underset{\sim}{T}^*$. Consequently with $|\Delta T/\underset{\sim}{T}| \ll |\underset{\sim}{T}^*/\underset{\sim}{T}|$ and a partitioning, Eq. (85) is given by

$$\langle \dot{s}_a \rangle = \langle \underset{\sim}{T} \rangle^{-1} \left\{ \langle \dot{\underset{\sim}{s}}(\Delta T)^* \rangle - \langle \dot{\underset{\sim}{s}}\underset{\sim}{T}^* \rangle \langle (\Delta T/\underset{\sim}{T}) \rangle \right.$$
$$\left. - \langle (\dot{\underset{\sim}{s}}\underset{\sim}{T}^*)^*(\Delta T/\underset{\sim}{T})^* \rangle - \langle \Delta Q_m[1 - (\underset{\sim}{T}^*/\underset{\sim}{T})] \rangle \right\}. \qquad (86)$$

Under the simplification that $\langle \Delta Q_m[1 - (\underset{\sim}{T}^*/\underset{\sim}{T})] \rangle$ vanishes for purposes of assessing the impact of positive definite aphysical sources of entropy, the balance expressed by Eq. (86) reduces to

$$\langle \dot{s}_a \rangle = \langle \underset{\sim}{T} \rangle^{-1} \left\{ \langle \underset{\sim}{s}^*(\Delta T)^* \rangle - \langle \dot{\underset{\sim}{s}}^*\underset{\sim}{T}^* \rangle \langle \Delta T/\underset{\sim}{T} \rangle - \langle (\dot{\underset{\sim}{s}}^*\underset{\sim}{T}^*)^*(\Delta T/\underset{\sim}{T})^* \rangle \right\}.$$
$$(87)$$

Of the three terms, the last is higher in order than the first, in addition there is little expectation of a systematic covariance between $(\dot{\underset{\sim}{s}}^*\underset{\sim}{T}^*)^*$ and $(\Delta T/\underset{\sim}{T})^*$. As Johnson (1997) noted both of the first two terms were evident in Boer *et al.*'s (1991, 1992) results concerning the "coldness" of climate models. The isentropic layers that spanned the upper troposphere of polar latitudes and the lower troposphere of tropical latitudes were judged to be biased cold in the mean. However, the "coldness" in terms of temperature departure of the polar regions was greater than in equatorial latitudes. As such both terms are positive, the first from the covariance from $\dot{\underset{\sim}{s}}^*$ and $(\Delta T)^*$ and the second from the condition that $\langle \Delta T/\tilde{T} \rangle$ would be negative in combination with true generation of ΔE_α (also E) being positive as expressed by $\langle \dot{\underset{\sim}{s}}^*\underset{\sim}{T}^* \rangle$. These considerations combined with the result from 14 climate models in 35 different simulations that 104 out of 105 possible events as simulated in the upper troposphere of extratropical regions and the lower troposphere of tropical regions were judged to be cold (Boer *et al.*, 1991, 1992) and that the biased temperatures should be a maximum in the very layers that $\langle \dot{\underset{\sim}{s}}^*\underset{\sim}{T}^* \rangle$ is a maximum provide conclusive evidence for Johnson's conclusions. Positive definite aphysical sources of entropy are invariably present in any climate simulation, a condition that prevents $(\Delta T/T)$ from being zero. There is no direct means to remove $(\Delta T/T)$ since it is the temperature response of a cyclic process driven by differential aphysical entropy sources. Furthermore, with the satisfaction of boundary conditions for energy, any attempt to modify $(\Delta T/T)$ through adding ΔQ in one region must induce ΔQ of opposite sign in another region. Thus a corollary in accord with Carathéodory's postulate of the second law is:

> The presence of positive definite aphysical entropy sources $\dot{s}_a$ in the modeled climate state precludes simulating the true climate state and its structure.

IX. ENERGY BALANCE AND APHYSICAL SOURCES OF ENTROPY

Now the added component of model generation due to aphysical sources of entropy will be determined for both $\langle g(E) \rangle$ and $\langle g(\Delta E_\alpha) \rangle$ in order to contrast differences that emerge. The substitution of Eq. (69) into Eq. (45) yields

$$\langle g(E) \rangle = \langle Q_{\mathrm{m}} \rangle + \langle \dot{s}_{\mathrm{a}} T \rangle. \tag{88}$$

With use of Eq. (73), followed by expansion into products of means and deviations, Eq. (88) becomes

$$\langle g(E) \rangle = \langle \underset{\sim}{Q}_{\mathrm{m}} \rangle + \langle \Delta Q_{\mathrm{m}} \rangle + \langle \dot{s}_{\mathrm{a}} \rangle \langle T \rangle + \langle \dot{s}_{\mathrm{a}}^* T^* \rangle. \tag{89}$$

With the added component of the generation $\Delta g(E)$ as the difference between the model and true generation defined by

$$\langle \Delta g(E) \rangle = \langle g(E) \rangle - \langle g(\underset{\sim}{E}) \rangle, \tag{90}$$

where $\langle g(\underset{\sim}{E}) \rangle$ equals $\langle \underset{\sim}{Q}_{\mathrm{m}} \rangle$, the additional generation is given by

$$\langle \Delta g(E) \rangle = \langle \Delta Q_{\mathrm{m}} \rangle + \langle \dot{s}_{\mathrm{a}} \rangle \langle T \rangle + \langle \dot{s}_{\mathrm{a}}^* T^* \rangle. \tag{91}$$

If Eq. (91) were the only means to assess sources of error for $\langle g(E) \rangle$, the most likely choice would be to accept $\langle \Delta Q_{\mathrm{m}} \rangle$ as the primary source and discount $\langle \dot{s}_{\mathrm{a}} \rangle \langle T \rangle$ as a source. However, for consistency with the entropy analysis, $\langle \Delta Q_{\mathrm{m}} \rangle$ is equated to zero with the result that

$$\langle \Delta g(E) \rangle = \langle \dot{s}_{\mathrm{a}} \rangle \langle T \rangle, \tag{92}$$

where it is assumed that the covariance of $\dot{s}_{\mathrm{a}}^*$ and T^* vanishes, just as the covariance of kinetic energy dissipation and temperature is discounted in considering $g(\Delta E_\alpha)$. The result just derived corresponds with Eq. (39) of Johnson (1997) provided Eq. (92) is divided by $\langle g(E) \rangle$ as it is determined by $\langle \dot{s}^* T^* \rangle$.

Now substitute Eq. (69) in Eq. (48) and consider $\langle g(\Delta E_\alpha) \rangle$ expressed by

$$\langle g(\Delta E_\alpha) \rangle = \langle Q_{\mathrm{m}} \varepsilon \rangle + \langle \dot{s}_{\mathrm{a}} T \varepsilon \rangle. \tag{93}$$

An addition and subtraction of the efficiency of the true state yields

$$\langle g(\Delta E_\alpha) \rangle = \langle \underset{\sim}{Q}_{\mathrm{m}} \underset{\sim}{\varepsilon} \rangle + \langle \underset{\sim}{Q}_{\mathrm{m}} (\varepsilon - \underset{\sim}{\varepsilon}) \rangle + \langle \Delta Q_{\mathrm{m}} \varepsilon \rangle + \langle \dot{s}_{\mathrm{a}} T \varepsilon \rangle. \tag{94}$$

Now the difference of the efficiencies is expressed by

$$(\varepsilon - \underset{\sim}{\varepsilon}) = = [(\underset{\sim}{T}_\alpha/\underset{\sim}{T}) - (T_\alpha/T)]. \tag{95}$$

Now T_α defined as the sum of a true equilibrium temperature $\underset{\sim}{T}_\alpha$ and a departure ΔT_α is expressed by

$$T_\alpha = \underset{\sim}{T}_\alpha[1 + (\Delta T_\alpha/\underset{\sim}{T}_\alpha)]. \tag{96}$$

With use of Eqs. (82) and (96), the ratio T_α/T is expressed by

$$(T_\alpha/T) = (\underset{\sim}{T}_\alpha/\underset{\sim}{T})[1 + (\Delta T_\alpha/\underset{\sim}{T}_\alpha)][1 - (\Delta T/T)]. \tag{97}$$

A substitution of Eq. (97) into Eq. (95) with expansion and simplification yields

$$(\varepsilon - \underset{\sim}{\varepsilon}) = (\underset{\sim}{T}_\alpha/\underset{\sim}{T})\{(\Delta T/T)[1 + (\Delta T_\alpha/\underset{\sim}{T}_\alpha)] - (\Delta T_\alpha/\underset{\sim}{T})\}. \tag{98}$$

With the substitution of Eq. (98) into Eq. (94), $\langle g(\Delta E_\alpha)\rangle$ becomes

$$\begin{aligned}
\langle g(\Delta E_\alpha)\rangle = {} &\langle \underset{\sim}{Q}_m \underset{\sim}{\varepsilon}\rangle \\
&+ \langle \underset{\sim}{Q}_m(\underset{\sim}{T}_\alpha/\underset{\sim}{T})\{(\Delta T/T)[1 + (\Delta T_\alpha/\underset{\sim}{T}_\alpha)] - (\Delta T_\alpha/\underset{\sim}{T}_\alpha)\}\rangle \\
&+ \langle \Delta Q_m \varepsilon\rangle + \langle \dot{s}_a T\varepsilon\rangle.
\end{aligned} \tag{99}$$

by correspondence with Eq. (48), the generation of the mean reversible component of total energy for the true climate state is

$$\langle g(\Delta \underset{\sim}{E}_\alpha)\rangle = \langle \underset{\sim}{\dot{s}}\underset{\sim}{T}\underset{\sim}{\varepsilon}\rangle = \langle \underset{\sim}{Q}_m \underset{\sim}{\varepsilon}\rangle. \tag{100}$$

With the added component of the generation defined by

$$\langle \Delta g(\Delta E_\alpha)\rangle = \langle g(\Delta E_\alpha)\rangle - \langle g(\Delta \underset{\sim}{E}_\alpha)\rangle, \tag{101}$$

the combination of Eqs. (72) and (99) through (101) yields

$$\begin{aligned}
\langle \Delta g(\Delta E_\alpha)\rangle = {} &\langle \underset{\sim}{\dot{s}}\underset{\sim}{T}_\alpha\{(\Delta T/T)[1 + (\Delta T_\alpha/\underset{\sim}{T}_\alpha)] - (\Delta T_\alpha/\underset{\sim}{T}_\alpha)\}\rangle \\
&+ \langle \Delta Q_m \varepsilon\rangle + \langle \dot{s}_a T\varepsilon\rangle.
\end{aligned} \tag{102}$$

Noting that $T\varepsilon$ equals $(T - T_\alpha)$, expansion, averaging, and rearrangement yields

$$\begin{aligned}
\langle \Delta g(\Delta E_\alpha)\rangle = {} &\underset{\sim}{T}_\alpha\{[1 + (\Delta T_\alpha/\underset{\sim}{T}_\alpha)]\langle \dot{s}(\Delta T/T)\rangle \\
&- (\Delta T_\alpha/\underset{\sim}{T}_\alpha)\langle \dot{s}\rangle + \langle \Delta Q_m \varepsilon\rangle \\
&+ \langle \dot{s}_a\rangle(\langle T\rangle - T_\alpha) + \langle \dot{s}_a^* T^*\rangle\}.
\end{aligned} \tag{103}$$

Under the constraint for a driftless climate state with $\langle \dot{s} \rangle$ equal to zero, Eq. (103) reduces to

$$\langle \Delta g(\Delta E_\alpha) \rangle = \underline{T}_\alpha \{ [1 + (\Delta T_\alpha / \underline{T}_\alpha)] \langle \underline{\dot{s}}^*(\Delta T/T) \rangle \} + \langle \Delta Q_{\mathrm{m}} \varepsilon \rangle$$
$$+ \langle \dot{s}_{\mathrm{a}} \rangle (\langle T \rangle - T_\alpha) + \langle \dot{s}_{\mathrm{a}}^* T^* \rangle. \tag{104}$$

At this point, the comparison of $\langle \Delta g(E) \rangle$ and $\langle \Delta g(\Delta E_\alpha) \rangle$ from Eqs. (91) and (104) is interesting. Apart from the last term, the sources of error involve $\langle \Delta Q_{\mathrm{m}} \rangle$ and $\langle \Delta Q_{\mathrm{m}} \varepsilon \rangle$ and the seemingly unrelated terms $\langle \dot{s}_{\mathrm{a}} \rangle \langle T \rangle$ and $\langle \dot{s}^*(\Delta T/T) \rangle$. However, these terms are related through Eqs. (77) and (78). Note, with the substitution of Eq. (78) into Eq. (91) and some rearrangement along with the constraint of $\langle \dot{s} \rangle$ equal zero, the added component of generation of total energy $\langle \Delta g(E) \rangle$ is given by

$$\langle \Delta g(E) \rangle = \langle \Delta Q_{\mathrm{m}} \rangle + \langle \dot{s}^*(\Delta T/T) \rangle \langle T \rangle - \langle \Delta Q_{\mathrm{m}}/T \rangle \langle T \rangle \tag{105}$$

$$= \langle \underline{\dot{s}}^*(\Delta T/T) \rangle \langle T \rangle + \left\langle \Delta Q_{\mathrm{m}} \left(1 - \frac{\langle T \rangle}{T} \right) \right\rangle + \langle \dot{s}_{\mathrm{a}}^* T^* \rangle. \tag{106}$$

Under the simplification in Eq. (104) that $(\Delta T_\alpha / \underline{T}_\alpha) \ll 1$ and that the difference between $\langle T \rangle$ and T_α is negligible for these purposes,

$$\langle \Delta g(E) \rangle \cong \langle \Delta g(\Delta E_\alpha) \rangle. \tag{107}$$

The near equality of these results is revealing. Note that the induced error in $\langle g(E) \rangle$ from aphysical sources of entropy corresponding with the indirect error in $\langle \Delta g(\Delta E_\alpha) \rangle$ was determined without any recourse to introducing an equilibrium reference state through which either available potential energy or the reversible component of energy was defined. The result for $\langle \Delta g(E) \rangle$ was simply determined from the definition of the entropy source by Eq. (70).

Furthermore, since an equilibrium state involving T_α for total energy was never defined there was no provision for the departure error of the equilibrium state ΔT_α. The condition for a climate state without drift with $\langle \dot{s} \rangle$ equal to zero removes any degree of freedom for erroneous sources of entropy from the difference between T_α and $\langle T \rangle$. The equivalence for $\langle \Delta g(E) \rangle$ and $\langle \Delta g(\Delta E_\alpha) \rangle$ just derived removes any doubt that entropy and energy are fundamental independent considerations for understanding the maintenance of the atmospheric circulation. Tolman and Fine (1948) note that as long as entropy sources from irreversibility are included in the first law, the thermodynamic effects of the second law would be included in the governing equations. Still employing this strategy without an awareness of entropy and its exchange in the unraveling of causes to explain response, the effects of the second law and entropy balance remain unresolved. A classical example (Spar, 1949) is the inability to resolve the intensity and maintenance of atmospheric circulation from basic considerations of the balance of total energy apart from introducing the Lorenz energy cycle.

X. THE EXPECTED MAGNITUDES OF $\langle \Delta g(\Delta E_\alpha)\rangle$

There is every reason to expect local errors in heat addition $\Delta Q_{\rm m}$ both directly from the difficulties encountered in physical parameterization, and indirectly from numerical diffusion in simulating the long-range transport of energy and water substances. Indirect errors also stem from strong dependence of latent heat of phase changes and radiative flux on temperature and thus ΔT at the same time. However, Johnson (1997) noted that if the boundary conditions of energy flux are satisfied, then for every region with $\Delta Q_{\rm m}$ positive, there must be a corresponding region with $\Delta Q_{\rm m}$ negative provided the model equations satisfy energy conservation. Attempts to alleviate errors in $\Delta Q_{\rm m}$ are fraught with difficulties by virtue of global conservation under the condition that $\dot{E}$ vanish.

With the definition of efficiency from Eq. (43) and the constraint of $\langle \dot{s}\rangle$ equal to zero, Eq. (102) is expressed by

$$\langle \Delta g(\Delta E_\alpha)\rangle = \underset{\sim}{T}_\alpha[1 + (\Delta T_\alpha/T_\alpha)]\langle \underset{\sim}{\dot{s}}^*(\Delta T/T)\rangle + \langle \Delta Q_{\rm m}[1 - (T_\alpha/T)]\rangle$$

$$+ \langle \dot{s}_{\rm a}\rangle(\langle T\rangle - T_\alpha) + \langle s_{\rm a}^* T^*\rangle. \tag{108}$$

From the definition of temperature as the sum of a mass weighted average $\langle T\rangle$ and a deviation T^*, the reciprocal of temperature is expressed by

$$T^{-1} = \langle T\rangle^{-1}(1 - T^*T^{-1}). \tag{109}$$

Successive substitution of T^{-1} from Eq. (109) for T^{-1} on the right in Eq. (109) yields the second-order expression given by

$$T^{-1} = \langle T\rangle^{-1}[1 - (T^*/\langle T\rangle) + (T^{*2}/T\langle T\rangle)]. \tag{110}$$

Now the substitution for the reciprocal of temperature within Eq. (108) utilizing Eq. (110) yields

$$\langle \Delta g(\Delta E_\alpha)\rangle = \underset{\sim}{T}_\alpha\langle T\rangle^{-1}\langle\{[1 + (\Delta T_\alpha/\underset{\sim}{T}_\alpha)][\underset{\sim}{\dot{s}}^*(\Delta T)] - \Delta Q_{\rm m}\}$$

$$\times \{[1 - (T^*/\langle T\rangle) + (T^{*2}/T\langle T\rangle)]\}\rangle$$

$$+ \langle \Delta Q_{\rm m}\rangle + \langle \dot{s}_{\rm a}\rangle(\langle T\rangle - T_\alpha) + \langle s_{\rm a}^* T^*\rangle. \tag{111}$$

To isolate systematic and deviation components, let ΔT be the sum of the isentropic mass weighted average $\langle \Delta T \rangle$ and a deviation $(\Delta T)^*$ expressed by

$$\Delta T = \langle \Delta T \rangle + (\Delta T)^*. \tag{112}$$

The substitution of Eq. (112) into Eq. (111) and an expansion yields

$$\langle \Delta g(\Delta E_\alpha) \rangle = \underline{T}_\alpha \langle T \rangle^{-1} [1 + (\Delta \underline{T}_\alpha / \underline{T}_\alpha)] \Big\{ \langle \underline{\dot{s}}^* (\Delta T)^* \rangle$$

$$+ \langle T \rangle^{-1} \Big\{ \langle \Delta T \rangle \big[-\langle \underline{\dot{s}}^* T^* \rangle + \langle \underline{\dot{s}}^* (T^{*2}/T) \rangle \big]$$

$$- \langle s^* (\Delta T)^* T^* \rangle + \langle \underline{\dot{s}}^* \Delta T^* T^{*2}/T \rangle \Big\} \Big\}$$

$$+ \langle \Delta Q_{\mathrm{m}} \rangle [1 - (\underline{T}_\alpha / \langle T \rangle)]$$

$$+ \underline{T}_\alpha \langle T \rangle^{-2} \big[\langle (\Delta Q_{\mathrm{m}}) T^* \rangle - \langle \Delta Q_{\mathrm{m}} T^{*2}/T \rangle \big]$$

$$+ \langle \dot{s}_{\mathrm{a}} \rangle (\langle T \rangle - T_\alpha) + \langle s_\alpha^* T^* \rangle. \tag{113}$$

As noted previously there is every reason to expect ΔQ_{m} to be finite. However, the factor $[1 - (\underline{T}_\alpha / \langle T \rangle)]$ will be small with the near equality of $\underline{T}_\alpha$ and $\langle T \rangle$, thus the product of $\langle \Delta Q_{\mathrm{m}} \rangle$ and this factor will not dominate relative to other terms in the expression. In considering the order of the other terms involving ΔQ_{m}, one might suggest that the covariances of ΔQ_{m} with T^* and (T^{*2}/T) are negligible. This consideration may be questioned since the intensity of (ΔQ_{m}) due to erroneous parameterization of latent heat release during phase changes may be substantial. The extended result, which is simply presented in Eq. (113) deserves additional consideration.

Now for simplicity, ΔQ_{m} is assumed to vanish in order to assess the magnitude of error from positive definite aphysical sources. As discussed previously, there is an inconsistency in suggesting that ΔQ_{m} and ΔT are finite with $\langle \dot{s}_{\mathrm{a}} \rangle$ zero, but not with ΔQ_{m} equal to zero and ΔT and $\langle \dot{s}_{\mathrm{a}} \rangle$ finite.

Within the extended term of Eq. (113) involving $\langle \Delta T \rangle$, the first term $\langle \dot{s}^* T^* \rangle$ dominates the second term $\langle \dot{s}^* T^* (T^*/T) \rangle$ for two reasons: $T^*/\underline{T}$ is generally an order of magnitude less than unity and the function (T^{*2}/T) is positive everywhere while $\dot{s}^*$ is both positive and negative. Whatever the magnitude of the second and third terms, they are small relative to the systematic covariance of $\dot{s}^*$ and T^* from heating at high temperature in tropical/subtropical latitudes and cooling at low temperature in extratropical/polar latitudes. The fourth term shall be small relative to the first since $(\Delta T)^*$ is small relative to T^*. With $\langle \dot{s}^* T^* \rangle$

identically equal to $\langle g(\Delta E_\alpha)\rangle$ in Eq. (51), a division of Eq. (113) by this relation yields the ratio of the additional energy dissipation associated with aphysical processes to the generation needed to maintain a climate state. The result is

$$\langle \Delta g(\Delta E_\alpha)\rangle / \langle g(\Delta E_\alpha)\rangle = \underset{\sim}{T}_\alpha \langle T\rangle^{-1}[1 - (\Delta T_\alpha / T_\alpha)]$$

$$\times \left[\langle \underset{\sim}{\dot{s}}{}^*(\Delta T)^*\rangle\langle \dot{s}^* T^*\rangle^{-1} - \langle \Delta T\rangle\langle T\rangle^{-1}\right]$$

$$+ [\langle \dot{s}_a\rangle(\langle T\rangle - T_\alpha) + \langle \dot{s}_a^* T^*\rangle]\langle \dot{s}^* T^*\rangle^{-1},$$

$$(114)$$

where it has been assumed that the difference between $\langle \underset{\sim}{\dot{s}} T^*\rangle$ and $\langle \dot{s}^* T^*\rangle$ is negligible. Now with the additional assumptions that the covariance of the aphysical source of entropy and ε are negligible, $(\Delta T_\alpha / T_\alpha) \ll 1$ and the approximate equality of T_α and $\langle T\rangle$, the ratio of the additive component to the generation is given by

$$\langle \Delta g(\Delta E_\alpha)\rangle / \langle g(\Delta E_\alpha)\rangle = [\langle \dot{s}^*(\Delta T)^*\rangle\langle \dot{s}^* T^*\rangle^{-1} - \langle \Delta T\rangle\langle T\rangle^{-1}]. \quad (115)$$

As noted previously, the neglect of the two terms involving $\dot{s}_a$ is equivalent with the implicit neglect of the covariance of the kinetic energy dissipation and $T\varepsilon$.

In his assessment of magnitude of the temperature error expected from aphysical sources of entropy, Johnson (1997) simply evaluated the ratio of $\langle \dot{s}_a\rangle\langle \underset{\sim}{T}\rangle$ to $\langle \dot{s}^* T^*\rangle$. Note this formally corresponds with $\langle \Delta g(E)\rangle$ from Eq. (92) divided by $\langle g(E)\rangle$ from Eq. (47). Now as expressed by Eq. (115), the same ratio has been derived directly from the balances of both total energy and the reversible component of total energy through utilization of the entropy principle.

Utilizing Boer *et al.*'s (1991, 1992) results, Johnson (1997) noted that for a mean cold temperature bias of -10 K relative to a mean temperature of 250 K within the isentropic layers in which the differential heating is the greatest, the ratio of increased generation to the generation of the reversible component of total energy is ≈ 0.04. Clearly, if climate models have spurious sources of entropy such that their reversible dissipative processes of energy exceed that of the actual atmosphere by 4%, simulated cold biases of 10 K are expected and unavoidable provided that the simulated distribution of atmospheric heating is unbiased. Accurate modeling of the climate state poses extremely difficult challenges in view of the physical, dynamical, and numerical complexities. The result that an added generation is needed to balance the dissipative processes from positive definite aphysical entropy sources and that this added generation is realized through the coldness of the model climate state provides insight into the difficulties of replicating the temperature distribution of the climate state.

Johnson (1997) also discussed the issue of why the coldness simulated in all 14 climate models is almost universally preferred to occur in the upper troposphere of the extratropical/polar latitudes and the troposphere of tropical/subtropical latitudes. Dutton and Johnson (1967) and Johnson (1970, 1989, 1997) emphasized that the maintenance of the reversible component of total energy against irreversible dissipative processes requires that the heating occur at high temperature and cooling at low temperatures within isentropic layers. The solar and terrestrial distribution of radiation in combination with the Earth's spherical form demand that the net heating and relatively warm temperatures occur in the tropics and net cooling and cold temperatures occur in polar/extratropical latitudes. Insight into the structure of the efficiency distribution, temperature, and entropy sources relative to isentropic surfaces is extremely important for understanding generation and the maintenance of atmospheric circulation. The efficiency distribution in Fig. 2 (see color insert) portrays positive and negative efficiencies in low and high latitudes, respectively, which combined with systematic tropical heating and polar cooling determine the generation. Since the dominant process that maintains both the entropy balance and reversible component of total energy is proportional to the covariance of $\hat{s}^*$ and T^*, the biased temperature structure realized to offset aphysical entropy sources will be maximized on those isentropes where the covariance of $\hat{s}^*$ and T^* is a maximum. The largest magnitude of efficiencies is positioned within the red isentropic layers bounded by 300 and 325 K, which slope from the lower half of the tropical troposphere to the upper half of the extratropical/polar troposphere of each hemisphere. Furthermore, because the isentropic layers with greatest slope connect the tropical and polar regions of entropy sources and sinks, respectively, these are the primary layers that generate the reversible component of total energy regardless of the distribution of the irreversible entropy source.

Isentropic layers with the greatest slope in polar/extratropics latitudes are also the regions where spurious numerical diffusion in the exchange of entropy and dry static and latent energies is maximized in sigma models. Here relative to other regions of the atmosphere, vertical thermal wind shear within the baroclinic westerlies creates extreme vertical and horizontal gradients of potential temperature, which in combination with advection and vertical motion induces spurious numerical diffusion during meridional transport.

XI. THE MARCH OF THE SEASONS AND REVERSIBLE ISENTROPIC PROCESSES

In Johnson's (1989) concept of an energy phase space for reversible isentropic processes, the increase of the reversible component of total energy was expressed by

$$\Delta \dot{E}_\alpha = [G(E) - G(E_\alpha)] - D(\Delta E_\alpha), \qquad (116)$$

where the generation of the reversible component of total energy $G(\Delta E_\alpha)$ is the difference of the generation of the total energy in the actual atmosphere $G(E)$ minus the generation in the equilibrium atmosphere $G(E_\alpha)$. In this energy phase space, net heating $[G(E) > 0]$ expands the admissible region of isentropic thermodynamic trajectories while net cooling $[G(E) < 0]$ contracts the region. A decrease of the equilibrium state energy $[G(E_\alpha) < 0]$ though a sink of entropy expands the admissible region while an increase $[G(E_\alpha) > 0]$ through an entropy source contracts the region. Positive definite sources of entropy by viscous dissipation, heat diffusion, and all other irreversible sources of entropy $[D(\Delta E_\alpha) > 0]$ contract the region. With positive definite physical and aphysical sources of entropy which mix energy, the balance between the generation of A and dissipation within the Lorenz energy cycle of a climate model will not match that determined for the atmosphere. At first glance, this suggests that the efficiency of the climate model atmosphere must be less than the maximum efficiency defined by Carnot concepts (Godske *et al.*, 1957). In reality the opposite is the case. Since ideally the boundary flux of the energy of the climate model replicates that of the actual atmosphere, and since there is an additional positive definite aphysical source of entropy, thus additional dissipation; the climate model becomes more efficient by adjusting to a colder state in order for generation to balance increased dissipation (Johnson, 1997). With fixed heat addition (or cooling), the only degree of freedom to balance increased dissipation is for the difference of mean valued temperatures derived within the concept of the Carnot cycle to increase.

Now thermodynamic processes by which the efficiency of the actual atmosphere changes between summer and winter are examined. Consider the generation integral $G(\Delta E_\alpha)$ expressed by Eq. (42) as the sum of two components associated with mean and deviation entropy sources

$$G(\Delta E_\alpha) = G[\Delta E_\alpha(\langle \dot{s} \rangle)] + G[\Delta E_\alpha(\dot{s}^*)]. \qquad (117)$$

Since the aim in the following development is to relate the seasonal migration of the heat sources due to the meridional variation of incoming solar energy with changes in the reversible component of total energy, temporal averaging is eliminated and the mass weighted average in this section is simply defined to be an area mass weighted average. As such atmospheric integrals and averages vary continuously with respect to time. Utilizing Eqs. (42) and (43), and the Poisson relation, the components of generation associated with mean and deviation entropy sources in Eq. (117) are expressed by

$$G[\Delta E_\alpha(\langle \dot{s}\rangle)] = p_{00}^{-\kappa}\int \rho J_\theta \langle \dot{s}\rangle\langle \theta(p^\kappa - \bar{p}^\kappa)\rangle \, dV_\theta, \qquad (118)$$

$$G[\Delta E_\alpha(\dot{s}^*)] = p_{00}^{-\kappa}\int \rho J_\theta \langle \theta[\dot{s}^*(p^\kappa)^*]\rangle \, dV_\theta. \qquad (119)$$

In considering the relevance of the two integrals, clearly the positive generation by $\dot{s}^*$ and $(p^\kappa)^*$ expressed in Eq. (119) is primary in the maintenance of atmospheric circulation. In this form, the positive generation stems from heating at high pressure and cooling at low pressure with respect to isentropic surfaces (Dutton and Johnson, 1967). However, since in the mean with $\langle \dot{s}\rangle$ zero generation balances dissipation, the expansion/contraction of the reversible component of total energy requires that the mean entropy source $\langle \dot{s}\rangle$ in Eq. (118) be nonzero. With $\langle \dot{s}\rangle$ nonzero, the mean mass flux across an isentropic surface changes the mean pressure in the equilibrium state and thus the mean temperature $T_\alpha(\theta, t)$. As such, the span of the phase space of reversible isentropic processes is modified in accord with Carathéodory's postulate.

Johnson (1989, 1997) noted that throughout the annual cycle each hemisphere in the mean thermodynamically responds to differential heating independently. There is a dividing boundary between the two hemispheric regions that migrates with the sun and across which, in the zonally averaged sense, the net meridional transport of total energy and entropy vanish (Gallimore and Johnson, 1981; Townsend and Johnson, 1985; Johnson *et al.*, 1985; Johnson, 1989). Johnson (1989) then proceeded to discuss the differences in the intensity of isentropic Hadley mass circulations in each hemisphere for winter and summer seasons. He noted that the expansion of the phase space of reversible isentropic processes occurs during the transition fall season from the decrease of the constrained equilibrium energy $[G(E_\alpha) < 0]$ being greater than the decrease of the total energy $[G(E) < 0]$. As such $\Delta \dot{E}_\alpha$ is positive and the span of the phase space represented by ΔE_α increases. Within the context of Eq. (116), this expansion of the phase space of reversible isentropic processes leads to greater temperature contrasts and increased efficiency within the hemisphere, which in turn leads to greater $G(\Delta E_\alpha)$, and thus further increase in ΔE_α. The increase of the reversible component of total energy as the sum of A plus K provides for the increased vigor of the wintertime circulation and thus increased dissipation $D(E)$.

The phase space of reversible isentropic processes contracts during the spring season as the generation of the constrained equilibrium energy $[G(E_\alpha) > 0]$ exceeds the increase in generation of the total energy $[G(E) > 0]$, resulting in reduced ΔE_α and thus reduced A plus K. These considerations were developed earlier by Johnson (1989) from integral relations governing the equilibrium state in relation to systematic decreases of entropy during the fall season and increases during the spring season, plus empirical evidence. Within the theory of available potential energy as applied using the isobaric approximations, there is no provision for the systematic increase and decrease of $G(\Delta E_\alpha)$ from changes by $\langle \dot{s} \rangle$ nonzero. This degree of freedom is eliminated by the combination of an isobaric horizontally invariant static stability, a series expansion, and area averaging in expressing the generation as a covariance of $\langle Q^*T^* \rangle_p$. The winter increase and summer decrease in $G(A)$ are attributed to the increase and decrease in the meridional gradient of incoming solar energy associated with the onset of winter and summer, respectively, and the increased isobaric temperature gradient portrayed within the eddy covariance of temperature and heating deviations.

An alternative and more stringent relation for the role of systematic hemispheric entropy sources is now presented. With a partitioning into the product of vertically averaged mass weighted mean and deviation, Eq. (118) is expressed by

$$G[\Delta E_\alpha(\langle \dot{s} \rangle)] = p_{00}^{-\kappa} \int \rho J_\theta \{\langle \dot{s} \rangle\}_\theta \{\theta(p^\kappa - \bar{p}^\kappa)\}_\theta \, dV_\theta$$

$$+ p_{00}^{-\kappa} \int \rho J_\theta \{\langle \dot{s} \rangle^* \langle [\theta(p^\kappa - \bar{p}^\kappa)]\rangle^*\}_\theta \, dV_\theta. \quad (120)$$

where now the brackets with the subscript θ and the asterisks denote a vertical mass weighted average and its deviation. Since $\{\langle \dot{s} \rangle\}_\theta$ equals $\dot{S}/M$ and $\theta(p^\kappa - \bar{p}^\kappa)/p_{00}^\kappa$ equals $T\varepsilon$, Eq. (120) can be expressed by

$$G[\Delta E_\alpha(\langle \dot{s} \rangle)] = p_{00}^{-\kappa} \dot{S} \int \rho J_\theta \{\langle \theta(p^\kappa - \bar{p}^\kappa)\rangle\} \, dV_\theta/M$$

$$+ \int \rho J_\theta \{\langle \dot{s} \rangle^* \langle (T\varepsilon)\rangle^*\}_\theta \, dV_\theta. \quad (121)$$

where M is the mass of the atmosphere. Inspection of the first integral on the right in Eq. (121) reveals that both the vertical and horizontal averages of $\theta(p^{\kappa} - \bar{p}^{\kappa})$ can be removed without altering the value. With this step completed and use of a series expansion of $p^{\kappa} - \bar{p}^{\kappa}$ through the second-order

$$(p^{\kappa} - \bar{p}^{\kappa}) \cong \bar{p}^{\kappa}\left[(p'/\bar{p}) + \kappa(\kappa - 1)(p'/\bar{p})^{2}/2\right], \qquad (122)$$

along with the Poisson relation for the equilibrium atmosphere, Eq. (121) is expressed by

$$G\left[\dot{\Delta}E_{\alpha}(\langle \dot{s} \rangle)\right] \cong \kappa \dot{S} \int \rho J_{\theta} T_{\alpha}\left[(p'/\bar{p}) + (\kappa - 1)(p'/\bar{p})^{2}/2\right] dV_{\theta}/M$$

$$+ \int \rho J_{\theta}\{\langle \dot{s} \rangle^{*}\langle (T\varepsilon) \rangle^{*}\}_{\theta} \, dV_{\theta}. \qquad (123)$$

Within the first term of the series, the magnitude of $\kappa(p'/\bar{p})$ and sign in Eq. (122) are to a first-order approximation equivalent with the distribution of the efficiency. Inspection of the Northern Hemisphere distribution of the efficiency distribution in Fig. 2 reveals a dominance of magnitudes of negative efficiencies poleward of 30°N over the positive efficiencies in lower latitudes between the equator and 30°N. Thus, the integral of the first term of the series expansion must be negative. In the second term with $(\kappa - 1) < 0$ and the series being quadratic, this contribution is systematically negative throughout the hemisphere. Thus the contraction or expansion of the phase space of the reversible component of total energy corresponds inversely with the mean hemisphere source of entropy as expressed by the first integral in Eq. (123). Since the intensity of dissipation would lag the increase in the circulation, there is also an increase in the span of reversible processes. The key point here, however, is that without the action of a mean component, the eddy component would simply balance dissipation.

In the second integral of Eq. (123), $\langle \dot{s} \rangle^{*}$ as a function of θ is the vertical deviation of the horizontal mass weighted average. Preliminary calculations of this term indicate that its contribution to generation is positive from the condition that $\langle \dot{s} \rangle^{*}$ and $\langle T\varepsilon \rangle^{*}$ are positively correlated with the deviations of both being positive in lower isentropic layers and negative in upper isentropic layers, both globally and by hemisphere.

With the first integral as expressed by Eq. (121) being negative in the fall season with $\dot{S} < 0$, $G(\Delta E_{\alpha})$ is positive; in the spring season with $\dot{S} > 0$, $G(\Delta E_{\alpha})$ decreases. This result clearly isolates the unique dependence of seasonal variation of $G(\Delta E_{\alpha})$ and ΔE_{α} on systematic hemispheric increases and decreases of entropy. The result also represents a physical basis for increases/decreases of the reversible component of total energy and the span of reversible isentropic processes that captures a dependency on entropy change in accord with Carathéodory's postulate of the second law. A corollary following from Carathéodory's postulate would be:

> Considered within the joint global balance of energy and entropy, the extended phase space of the reversible component of total energy within a winter hemisphere can only be reached through a systematic decrease of entropy from summer to winter, and likewise the contracted phase space of the reversible component of total energy within a summer hemisphere can only be reached through a systematic increase of entropy from winter to summer.

The relevance of seasonal changes in relation to total energy, equilibrium state energy, and admissible regions of reversible isentropic processes is now summarized in Fig. 3 using results from Peixoto and Oort (1992). The results (Table I) from Oort and Peixoto (1983) present the mean atmospheric energy per unit surface area in units of 10^7 J m^{-2} for the summer and winter hemispheres. Although the available potential energy A has been evaluated using the isobaric quadratic approximation in lieu of an isentropic representation, for the purposes of illustrating seasonal values of the reversible component of total energy ΔE_α is assumed to be equal to the sum of A plus K in specifying hemispheric quantities.

Furthermore, to relate readily the increase of the winter (w) season values $(\Delta E_\alpha)_\mathrm{w}$ relative to the summer (s) season values $(\Delta E_\alpha)_\mathrm{s}$, all mean energy quantities are normalized relative to $(\Delta E_\alpha)_\mathrm{s}$ for ease of comparison. For example, in the winter, A plus K is 2.21 times larger than the corresponding summer value. See seasonal mean for each quantity in Table I as extracted from Oort and Peixoto (1983) and corresponding normalized values. As Lorenz (1967) emphasized, the seasonal changes and intensity of the atmosphere's circulation bear a direct relation with the intensity of available potential and kinetic energies being greater in winter than summer, while the relation with total potential energy $(I + \phi)$ is inverse.

Figure 3 (after Fig. 16 from Johnson, 1989) with kinetic energy as the abscissa and total potential energy as the ordinate presents an energy phase diagram illustrating the relations between total energy $(E = K + I + \phi)$, the reference state energy (E_α), the reversible component of total energy $(\Delta E_\alpha = E - E_\alpha)$, the total potential energy $(\Pi = I + \phi)$, and the kinetic energy (K) for summer and winter. Except for large differences between the summer and winter values of the total energy indicated by the break in the vertical axis, all values of energy quantities are scaled

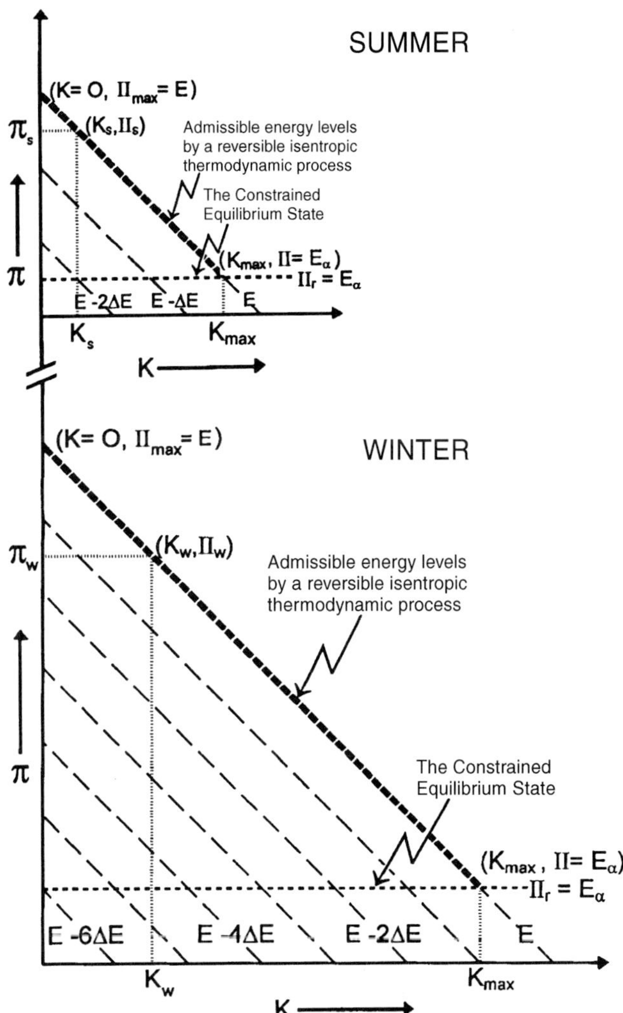

Figure 3 Energy phase-space diagram illustrating the relations between total energy (E), total potential energy (Π), kinetic energy (K), and the reversible component of total energy ($E - E_\alpha$) for summer and winter seasons. The abscissa is kinetic energy, the ordinate is total potential energy, and the family of sloping lines designates isopleths of total energy. The admissible region under reversible isentropic energy transformations between kinetic and available potential energies coincides with the bold dashed line, which is bounded by ($K_{max}, E = \Pi_r$) and ($K = 0, \Pi = E$). The length of the line l, which equals $\Delta E_\alpha / \sqrt{2}$, portrays the greater span of reversible isentropic processes in winter than in summer. The relative dimensions of the energy levels are scaled in accordance with the normalized values in Table I.

Table I
Estimates of Reversible Components[a]

	DJF		JJA	
	Amount	Ratio	Amount	Ratio
$A + K$	0.765	2.21	0.345	1.000
A	0.602	1.74	0.275	0.80
K	0.163	0.47	0.070	0.20
E	252.5	731.88	262.5	760.87
I	178.2	180.00	183.4	553.59
ϕ	69.0	92.20	71.0	205.80
LH	5.15	14.78	8.07	23.39
$K/(A + K)(\%)$	21.3		20.3	
$K/E(\%)$	0.065		0.027	

[a] Reversible components of total energy (A plus K), and available potential (A), kinetic (K), total (E), internal (I), gravitational potential (ϕ), and latent (LH) energies (10^7 J m^{-2}) for the winter (DJF) and summer seasons (JJA) as extracted from Tables 13.1 and Table 14.1 of Oort and Peixoto (1983). The ratios of kinetic (K) energy to the reversible component of total energy ($A + K$) and to total energy (E) are expressed in percent. The normalized ratio for all energies is with respect to the summer value of A plus K. The ratios of kinetic to total potential energy is expressed in percent.

appropriately in accord with the values in Table I. For example, the magnitude of the reversible component being proportional to the length of the heavy dashed line in winter is 2.2 times longer than length of the corresponding line in summer.

As Johnson (1989) discussed, at any given time the admissible level of reversible isentropic thermodynamic processes is restricted to occur along the heavy dashed line of the isopleth of total energy with bounds determined by the extreme values $[K_{\max} = (E - E_\alpha), (\Pi - E_\alpha) = 0]$ and $[K = 0, (\Pi - E_\alpha)_{\max} = (E - E_\alpha)]$. Although these extreme values are never realized, the contrast of the span of admissible regions between summer and winter is marked. Now, the processes that account for the geometric differences in the summer and winter diagrams utilizing the governing thermodynamic relations are readily determined.

By the Pythagorean theorem, the length of the isopleth of total energy E between the two maximum values is given by

$$l = \left[K_{\max}^2 + (\Pi - E_\alpha)_{\max}^2 \right]^{1/2}. \tag{124}$$

However, from the condition that

$$K_{max}^2 = (\Pi - E_\alpha)_{max}^2 \tag{125}$$

$$= (E - E_\alpha)^2, \tag{126}$$

the combination of Eqs. (35), (124), (125), and (126) yields the following relations:

$$\Delta E_\alpha = l/\sqrt{2}, \tag{127}$$

$$\Delta \dot{E}_\alpha = \left(\dot{E} - \dot{E}_\alpha\right) = \dot{l}/\sqrt{2}. \tag{128}$$

Thus the length l and its time rate of change $\dot{l}$ divided by $\sqrt{2}$ correspond with the magnitude of the reversible component of total energy ΔE_α and its time rate of change $\Delta \dot{E}_\alpha$. Note that changes in the reference state energy $\dot{E}_\alpha$ are in accord with $\dot{S}$. Inspection of Table I and Fig. 3 shows the greater length of l equal to $\sqrt{2}\,\Delta E_\alpha$ along E corresponds with the greater span in winter than in summer. In accord with Eq. (124), this greater length of l representing the greater span for reversible processes results from decreases of entropy ($\dot{S} < 0$) during the transition fall season. The restricted span for summer results from increases of entropy ($\dot{S} > 0$) during the transition spring season. Reversible transformations between A and K being isentropic restrict changes in A and K to a given isopleth of E, but bounded by the end points of l. Thus the heavy dashed isopleth determines permissible states of the system by isentropic processes. In accord with Carathéodory's postulate, any energy state not represented by the heavy dashed isopleth is not attainable unless diabatic processes occur. Here dissipation and heat diffusion as irreversible processes would not lead to changes in E, but would simply contract the phase space with $\dot{E}_\alpha > 0$.

Concerning Northern Hemisphere estimates of entropy, Dutton (1976) calculates 5.8×10^7 and 5.91×10^7 J/m^2 K for winter and summer, respectively. The winter reduction of a hemisphere's entropy relative to summer is evidenced by the formation of extensive polar air masses during the onset of winter, and an increase in static stability of the wintertime air masses over their summer counterparts (Johnson, 1989). The intensification of circulation in the wintertime is consistent with the increased efficiency of a thermodynamic system in producing ΔE_α by extracting entropy at a colder temperature and the greater temperature contrast in the presence of the greater meridional contrast of radiative entropy flux interacting with the state of the system. In contrast, the summer increase of entropy is manifested through the formation of warmer air masses, a decrease of static stability, lesser meridional contrast of radiative entropy flux, and reduced intensity of circulation. The decreased intensity of the circulation results from decreased efficiency, extracting heat at higher temperature and reduced meridional temperature contrast.

The perspective elucidated in this study is in accord with the classical concepts of the Carnot cycle, its efficiency, and Carathéodory's postulate. The perspective reconciles the role of entropy in the Lorenz energy cycle. The temperature distribution $T_\alpha(\theta, T)$ as a constrained equilibrium state and the global distribution of efficiency corresponding with the definition of the reversible component of total energy were developed axiomatically from thermodynamic concepts. As such, the underlying importance of the second law is recognized not only through entropy generation by irreversible processes, which like Dutton's contracts the phase space of the reversible component total energy, but it is also recognized through the contraction/expansion of this phase space by seasonally varying differential entropy sources and temperature. On this latter point, the phase space of the reversible component of total energy within the climate state exists as a thermally forced departure from an equilibrium state. Furthermore, theory based on a phase space of reversible isentropic processes both accommodates and requires the mutual existence of available potential and kinetic energy; both the temperature and pressure must vary on isentropic surfaces for entropy sources/sinks to expand the phase space of the reversible component of total energy through differential heating and thus balance dissipative processes. Their mutual existence is necessary to accommodate geostrophy and adjustment involving reversible isentropic processes including direct and indirect circulations. Kinetic energy is also essential for the maintenance of the long-range transport of energy and entropy that is required to link the far-field Lagrangian entropy sources and sinks. There is, however, a fundamental difference between the balances of entropy and energy within the driftless climate state. There is no net source of energy within the driftless climate state since the net boundary flux of energy vanishes. In contrast, the atmosphere as a natural system generates entropy. This entropy generated, however, is removed through the condition that the long-wave entropy flux to space exceeds the short-wave entropy flux received by the atmosphere.

Future studies should mutually consider both the radiative entropy and entropy of matter in conjunction with the radiative energy and energy of matter. The radiative flux of entropy is extremely relevant to the entropy of matter throughout the atmosphere due to its importance in vertical exchange of both energy and entropy of matter by virtue of selective absorption and emission of H_2O, CO_2, and other trace gases and also in removing the entropy created by irreversible processes within the atmosphere. The more general relations expressed in briefest detail here combined with the recent works detailing radiative entropy and its exchange (eg. Essex, 1987; Hauf and Höller, 1987; Callies and Herbert, 1988; Lesins, 1990; Stephens and O'Brien, 1993; Li *et al.*, 1994; Li and Chýlek, 1994) offer a means to address the balance of entropy and energy of the modeled climate state.

XII. CONCLUSIONS AND ADDITIONAL CONSIDERATIONS

Among the scientists engaged in modeling of atmospheric circulation, Professor Arakawa is foremost in emphasizing the appropriate conservation of atmospheric properties. He is also foremost in ensuring that the numerics employed satisfy appropriate constraints concerning mass, momentum, energy, entropy, enstrophy, and circulation. He has also emphasized comparisons of the dynamics of discrete systems with that of the continuous system and the need to maintain integral constraints of physical importance. His emphasis on these matters in modeling atmospheric circulation accounts in part for his exceptional success.

In selecting a topic for this symposium, my familiarity with Professor Arakawa's exemplary efforts to advance modeling of the global atmospheric circulation and the climate state and his new directions in developing a generalized but largely isentropic coordinate model led me to select thermodynamics with a focus on entropy. By virtue of my own interests, the study largely employs analysis in isentropic coordinates. Very simply, in studies of the thermodynamics of the atmosphere's circulation that involve mass, energy, and entropy, the use of isentropic coordinates for analysis provides a perspective that is simple and straightforward in isolating the atmosphere's thermodynamic response to differential heating.

The theoretical analysis set forth in this study is an extension of earlier results setting forth the role of entropy and its change with respect to the reversible component of total energy within global monsoonal circulations (Johnson, 1989). The concept of global monsoonal circulations emerged from a long-standing interest in the Lorenz energy cycle, entropy, and the role of differential heating in maintaining atmospheric circulation. Global monsoonal circulations (Johnson, 1989) must exist within the time-averaged climatic response to differential heating in which the scales of mass, energy, and entropy transport correspond with the scales of the vertically integrated source/sink distribution of entropy. The mean transport is unique in the sense that trajectories collectively through a systematic component must transport mass and entropy from heat source to heat sink in higher valued isentropic layers and from heat sink to heat source in lower valued isentropic layers. Furthermore, since the quasi-horizontal transport occurs through the mean mass transport, the total transport of entropy is entirely free of eddy modes, whereas the energy transport is essentially free of eddy modes (Johnson, 1989). In studying thermodynamic processes, Carrington (1994) comments "how the use of the most suitable thermodynamic coordinates can make the analysis of a process quite simple. Ill-chosen thermodynamic coordinates, conversely, will create untold complications."

Lorenz (1967) argued that the theory of available potential energy embraced concepts that differed from entropy in that the motion field was involved, presumably through the isentropic redistribution of the mass of the atmosphere to a reference state and the generation of kinetic energy. While it is true that the transformation between A and K involves the motion field, the combination of A plus K as the reversible component of total energy removes these considerations. Furthermore, with the combination of kinetic and internal energy as the sum of mean and random molecular motion along with gravitational potential energy to define total energy (Johnson and Downey, 1982), the thermodynamics of mass, total energy, and entropy are then the only relevant properties.

Johnson (1989) also established that, within the time-averaged state of global monsoonal circulations, generation of kinetic energy occurred primarily through the thermally forced irrotational component of the time-averaged isentropic mass transport, as it is determined by the product of the Montgomery streamfunction and the Laplacian of a potential function $\psi_m \nabla_\theta^2 \overline{X}$, where the Laplacian of $\overline{X}$ equates with $\nabla_\theta \cdot \overline{(\rho J_\theta \underset{\sim}{U})}$. Within the driftless climate state, the divergence of the isentropic mass transport is directly related to the vertical derivative of the isentropic mass transport $\partial(\overline{\rho J_\theta \dot\theta})/\partial\theta$ [also $\partial(\overline{\rho J_s \dot s})/\partial s$]. Although the generation of kinetic energy by this process within the atmosphere does not equate directly with pressure work of expansion $\langle p\nabla \cdot \underset{\sim}{U} \rangle$, their global integrals are equal (Johnson and Downey, 1982). As such, within the time-averaged state, the energy transformation between A and K globally becomes directly linked with the atmospheric distribution of differential heating and entropy sources and sinks.

In his textbook "Principles of Atmosphere Physics and Chemistry," Goody (1995) comments that general circulation models should obey entropy balance in which the global entropy decrease by differential heating balances the entropy increase by irreversible processes, "but it is not clear that most do so." He also states that the kinetic energy and total potential energy provide no a priori reason "why all of the available potential energy cannot end up as kinetic energy." Goody's comment here points to a typical misconception of many—that available potential energy, once generated, is the "spring which drives atmospheric circulations." Goody also states that "the equations should contain a Carnot efficiency if they are to satisfy the second law of thermodynamics" and notes that a limitation of available potential energy is "that it does not bring in the thermodynamic efficiency of the system, and refers to the maximum energy available, constrained by geometrics and energetical considerations."

In his classic work setting forth the theory of available potential energy, Lorenz (1955) references Margules (1903) who discussed an initial atmosphere not in equilibrium but which "tends to attain a condition of stable equilibrium" with entropy uniform horizontally and increasing vertically. Lorenz develops his theory by extending Margules's concepts of the available kinetic energy for storms to the atmosphere globally. In developing the relations of differential heating and efficiency in the maintenance of the general circulation, Lorenz (1967) also discusses the concept of an equilibrium state in considerable detail including classical views of thermodynamic efficiency. His approximate expression for available potential energy as the variance of temperature with respect to isobaric surfaces is in accord with Jeffreys's (1925) emphasis on the importance of temperature differences on level surfaces as the thermodynamic energy source for kinetic energy. Tolman and Fine (1948), Wulf and Davis (1952), and others utilized entropy in their studies of efficiency.

Attention must be given to the exceedingly important role that diagnostic studies of the transformation of available potential energy into kinetic energy have played in understanding the maintenance of the atmosphere's circulation. Available potential energy theory, which emerged just after baroclinic instability theory (Charney, 1947; Eady, 1949) and was developed in the late 1940s, emphasized the importance of amplifying baroclinic waves as the mechanism for poleward eddy transport of angular momentum and energy in the satisfaction of balance requirements. Emphasis was placed on the generation of zonal available potential energy by the systematic meridional variation of differential heating followed by the transformation of the zonal available potential energy into eddy available potential energy through poleward eddy energy transport with subsequent, if not, coincident transformation to eddy kinetic energy. The zonal kinetic energy was maintained against dissipation by the inertial transformation of eddy into zonal kinetic energy. Within this perspective, the link of the generation of available potential energy with the irreversible process of dissipation and entropy increase was circuitous. The central focus being addressed at that time was to understand why there was no apparent direct response in the form of kinetic energy generation from differential heating within systems dominated by geostrophy. At the same time, however, Riehl and Fultz (1957, 1958) did establish that relative to the wind maximum in a dishpan, there was a direct response. Also see Palmén and Newton (1969).

The developments set forth here in this analysis relating entropy and climate do not negate the utility of the Lorenz energy cycle. On the contrary, the developments in this study place new importance on the use of the Lorenz energy cycle for modeling and diagnosis of the climate state. The developments also place importance on understanding the role of entropy and its source/sink distribution in maintaining the atmosphere's circulation against irreversible processes. Apart from the actual atmospheric state, the impact of aphysical sources of entropy in model simulations has been ascertained and relations with the classical Carnot cycle have been established. Within the perspective of the reversible component energy and its equivalence with the Carnot cycle, the emphasis is on the thermodynamics of an actual atmospheric state being forced from an equilibrium state by differential entropy sources. The result is generation of a reversible component of total energy, where in the case of $\langle \dot{s} \rangle$ zero, the entropy gained through heat added equals the entropy lost by the heat removed.

In contrasting efforts by Lettau (1954), Dutton (1973), and Pearce (1978), Johnson noted that estimates using Lorenz's definition of the reference state available potential energy ensured that the reversible component is isolated and accurately bounded when, through entropy sources and sinks, the total energies of the actual and equilibrium states increase or decrease. Johnson also noted of all the candidates for equilibrium temperature distribution that $T_\alpha(\theta, t)$ is the one that determines the sharpest delineation of reversible thermodynamic trajectories in the sense that with expansion or contraction of the envelope, no component of total energy is included that is unable to engage in reversible exchange between kinetic and total potential energies. Still, the failure to relate available potential energy to classical concepts of the second law and entropy constitutes a stumbling block to the theoretical understanding of climate and also to the realization of the full potential of applications of the Lorenz energy cycle for modeling the thermodynamic forcing of climate. For the most part, climate modelers have focused on the balance of total potential energy with little attention given to the generation of the available potential energy and the Lorenz energy cycle, apparently without realizing that under the condition of the driftless climate state, the critical issue of thermodynamic forcing with respect to total potential and available potential energy are one and the same, i.e., with $\langle \dot{s} \rangle$ equal zero the covariance of isentropic deviations of entropy sources and temperature directly determines the thermodynamic work that maintains the climate circulation. That these are fundamental requirements from the combination of the first and second laws in relation to the Carnot cycle has been unequivocally documented in this analysis.

The entropy constraints of $\dot{S}$ and $\langle \dot{s} \rangle$ equal to zero are at the heart of this analysis, in which the generation of available potential energy is related to the Carnot cycle. Not only are these necessary conditions for simulation of a driftless climate state, but the satisfaction of these constraints in conjunction with the categorization of the time space domain into distinct regions of heat added and heat removed leads to unique and exact definitions of mean valued temperatures corresponding with heat addition and heat removal. Within this context the balance equations derived emulate the Carnot cycle in which the product of heat added and the thermodynamic efficiency prescribed by mean valued temperatures determine the thermodynamic work. The underlying reason in a mean sense that the generation of available potential energy is equal to work rests on definition of entropy with $\dot{s}$ equal to Q/T. As developed earlier, with $\langle \dot{s} \rangle$ equal zero, and with incremental heat addition $\rho J_\theta Q^+ \, dV_\theta$ specified by $\rho J_\theta \dot{s}^+ T^+ \, dV_\theta$ at some location and a corresponding amount of heat removed $\rho J_\theta Q^- \, dV_\theta$ by $\rho J_\theta \dot{s}^- T^- \, dV_\theta$ at another location, a unique amount of work is specified by virtue of the exact differential defining entropy as a physical property, and the condition that the temperature difference $T^+ - T^{-1}$ by Poisson's relation defines a unique pressure difference. The simplicity of the Carnot cycle provides fundamental insight into why temperature deviations and heat added and removed relative to isentropic surfaces maintain the atmosphere circulation against dissipation. The integral relations with $\dot{S}$ equal zero for the driftless climate state correspond with the balance requirements for Dutton's entropic energy where in this case the generation and idealized work would balance the energy dissipation by all irreversible sources of entropy. The integral relations with $\langle \dot{s} \rangle$ equal zero for the driftless climate state correspond with the Lorenz energy cycle in which the generation and work balances kinetic energy dissipation.

The complexity of ensuring $\langle \dot{s} \rangle$ equal to zero in modeling the driftless climate state or its proper value during the march of the seasons is exceedingly complicated and deserves special attention. This requirement places an internal constraint on the sum of mean component entropy sources/sinks within each isentropic layer. Here relative to the stratification of the troposphere, stratosphere, mesosphere, and thermosphere, the mean entropy increase from latent heat release and convergence of sensible heat flux balances the net decrease of entropy by the net divergence of energy associated with spectral absorption/emission of radiatively active gases. Ensuring the proper balance of mean entropy sources relative to seasonal variations of the thermal structure throughout the troposphere, stratosphere, mesosphere, and thermosphere poses challenges for the future in the modeling of the climate state and advancing understanding of the global nature of atmosphere circulations including solar atmosphere interactions.

Related to these considerations in the modeling of climate and entropy exchange are the complications introduced by positive definite aphysical sources of entropy involving the mixing of energy from numerics, Gibbs phenomena, and inadequacies of parameterization identified previously with respect to the reversible component of total energy.

In the remaining comments, attention is focused on certain inconsistencies that stem from using the isentropic versus isobaric approximation to ascertain the relevance of component diabatic processes and their maintenance of atmospheric circulation. The approximations are also misleading conceptually since dependence on the addition of heat with respect to pressure as embodied in the classical Carnot cycle has been eliminated. Two examples bear evidence of these considerations.

Within the theory of available potential energy as applied using the isobaric approximations, there is no provision for the systematic increase and decrease of $G(\Delta E_\alpha)$ from changes with $\dot{S}$ and $\langle \dot{s} \rangle$ nonzero [see Eq. (123)]. The winter increase and summer decrease in $G(A)$ are merely attributed to the increase and decrease in the meridional gradient of incoming solar energy associated with the onset of winter and summer, respectively, without any appreciation of the thermodynamic implications of the isentropic areally averaged mean entropy source $\langle \dot{s} \rangle$.

The second example involves a serious misconception concerning the role of sensible heat addition at the Earth's interface in the generation of available potential energy. Palmén and Newton (1969), among others have noted that when "cold air overruns the warm surface waters east of the Gulf Stream" (also the Kuroshio Current) that the heat added leads to "a decrease of available potential energy." This results from the condition that the isobaric temperature deviation of the cold air relative to the mean isobaric temperature is negative, thus $\langle Q_{\mathrm{m}}^* T^* \rangle_p$ in these regions is negative. This interpretation violates the second law, and the thermodynamics of the Carnot cycle in that heat added to high pressure relative to its extraction by radiation at lower pressure generates a reversible component of total energy which ultimately produces work within the cyclic nature of atmospheric circulation. Concerning sensible heat addition, the Carnot efficiency and the efficiency defined isentropically at the Earth's surface are positive; therefore, the generation of available potential energy regardless of magnitude must be positive.

In studies of the generation of available potential energy for storms over a relatively small area of 5.4×10^{13} m^2 (Gall and Johnson, 1971; Johnson, 1970), the erroneous nature of the isobaric calculation estimates a negative generation of -7.6 W m^{-2} whereas the isentropic estimates a positive generation of 6.1 W m^{-2} within a cold air outbreak over the gulf stream. Since the wintertime net entropy sources over the Pacific and Atlantic oceans are rather extensive as polar air masses migrate equatorward and are modified into subtropical air masses, sensible heat flux from the ocean to the atmosphere in extratropical and even subtropical latitudes represents a substantial fraction of the heat addition to the atmosphere, all at relatively high pressure. In the Southern Hemisphere, isentropic estimates of the generation of available potential energy are also positive in regions where Antarctic air masses flowing equatorward over the Southern Ocean are modified by sensible heat flux from the ocean (Bullock and Johnson, 1972). Boer (1989) calls attention to the nontrivial difference between exact and approximate versions of the energy budget equations. Siegmund (1994) finds from global assimilated data that the largest differences between the isobaric approximate and isentropic calculations of the generation of available potential energy occurs in the lower tropopause of the extratropics with the estimates being opposite in sign and the magnitude of the negative isobaric estimate being 10 times the positive isentropic estimate of generation. These differences are due in large part to the reasons noted by Gall and Johnson (1971). Here the relevance of thermodynamic efficiency, the Carnot cycle and the impact of sensible heat addition at high pressure needs to be recognized in climate model development.

There is also the related matter that the upward flux of sensible heat from the ocean to the atmosphere in northern latitudes of the Atlantic Ocean is responsible of the formation of cold water and forcing of the downward vertical branch of the global conveyor belt within the ocean (Broeker, 1991). This action through a cyclic process is linked with the transports of warm waters by the gulf stream from southern to northern latitudes within the upper ocean and reverse transport of cold waters in the interior of the Atlantic. The net effect of this action is to decrease the entropy of the ocean and increase the entropy of the atmosphere, both of which play a critical role in maintaining the conveyer belt within the ocean. Similar exchanges ring the Antarctic continent and the surrounding sea ice, which affect the circumpolar circulation of the Southern Ocean, the global conveyor belt, and the bottom waters of the global ocean. Here as well as in the North Atlantic there are complications due to the increase in density from the salt released during the formation of sea ice. The coupling of the energy and entropy exchange between the atmosphere and ocean in these regions in determining the climate state is exceedingly important.

A comparison of available potential energy estimates from isobaric and isentropic formulas reveals differences ranging from 40 to 100%, (Dutton and Johnson, 1967). Although these computations were based on cross-sectional data along 75°W for 1958, they provide insight into the systematic dependence of available potential energy on the determination of mean lapse rate structure used for the isobaric approximation.

Accurate modeling of the climate state is an extremely difficult challenge now and in the future. There must be a recognition that accuracies of model simulated climate states are limited and that this limit is a function of numerics and inadequacies of parameterization that induce the mixing of energy and thus irreversibility. The relevance of thermodynamic efficiency and the Carnot cycle needs to be recognized on these matters, particularly within the context of consistencies in the parameterization of component diabatic processes and conservation in the development of climate models of dry and/or moist entropy when approrpriate. In terms of thermodynamics, the necessity of parameterizations in climate models for nearly all of the processes of heat addition introduces additional degrees of freedom in terms of tunable parameters which frequently vary in time and space. Johnson (1997) discusses the numerous additional degrees of freedom introduced by "tuning" of the parameterized components of heat addition and the risks of masking underlying deficiencies without gaining fundamental insight (Oreskes *et al.*, 1994).

Consider that calculations of the generation of the available potential energy have generally been limited to a global statistic involving the integral of isobaric approximations of $\langle Q_{\mathrm{m}} \varepsilon \rangle$ with little direct insight into the role of component diabatic processes and the associated global response. The relations developed herein concerning the Carnot cycle and the second law provide direct insight into the relation between Lagrangian sources/sinks of entropy, temperature, and work. For example, straightforward calculations of $\langle \dot{s}^{+} \rangle$, $\langle \dot{s}^{-} \rangle$, $\langle Q^{+} \rangle$, and $\langle Q^{-} \rangle$ as a function of potential temperature enable $\tilde{T}^{+}(\theta, t)$ and $\tilde{T}^{-}(\theta, t)$ to be determined, from which dissipation may be estimated. The mean value temperatures so defined in conjunction with $\langle \dot{s}^{*} T^{*} \rangle$ provide information on whether the generation as a function of potential temperature is associated with large differences in temperature with relatively weak entropy sources and sinks or whether the generation is from a small difference in temperature coupled with intense entropy sources and sinks. As different parameterizations of heating are examined in climate models, the analysis just described will provide information on how the temperature as a response couples with the heating distribution to maintain atmospheric circulation. There must also be recognition that the component entropy sources from latent and sensible heating are intense locally and small scale relative to the global extent of the removal of entropy by infrared emission to space. With regard to the coupled nature of entropy and entropy transport with heat sources and sinks, scaling arguments fail in attempting to isolate scales of differential heating corresponding with certain scales of atmospheric circulation.

Current challenges involve improving the accuracies of all aspects of the hydrological cycle, cloud radiation interaction, and the fluxes of radiation and sensible heat. Future challenges involve understanding and improving the accuracy of chemical exchanges and processes, biosphere interaction, and increased spectral resolution of the radiative flux of energy including the spectral absorptivity and emissivity of water substances, chemical constituents, and aerosols. Here moist thermodynamic processes involving cloud generation, condensation and evaporation, phase changes, the indirect role of cloudiness in photodisassociation, photosynthesis, and other like processes in which the presence or absence of sunlight acts as a switch are particularly important. All of these processes involve reversibility and irreversible sources of entropy from physics and numerics in modeling the climate state. In efforts to understand the impact of these processes, attention must be focused on differential heating and the related atmospheric transport of mass, entropy, and energy. Very simply, the governing equations are nonlinear, the actual forcing of change and response may be separated by thousands of miles, and, more importantly, the response in terms of the state of the system may stem from a single process or the interaction of several processes.

The theory developed herein within the concept of global monsoonal circulations and the Lorenz energy cycle provides a basis for examining mass, entropy, and energy transport in relation to entropy sources and sinks by component processes. The study of the climate state is incomplete without considering the Lorenz energy cycle in relation to entropy sources and sinks and second law. Eventually when such analyses are coupled with studies of the mass, entropy, and energy transport and carried out for actual and modeled climate states, new insight will be gained in understanding reversible and irreversible processes as they impact atmospheric circulation and the climate state.

ACKNOWLEDGMENTS

I express appreciation to Dr. Ronald Taylor who alerted me to Born's (1949) lectures at Oxford concerning Carathéodory. I also gratefully acknowledge the assistance of Mr. Todd

Schaack, Drs. Allen Lenzen and Tom Zapotocny in the preparation of this study, Mrs. Judy Mohr in preparation of this manuscript, and Krista Ommodt in the preparation of figures. Appreciation is also extended to the UCLA Department of Atmospheric Sciences and the organizing committee for honoring Professor Arakawa and the invitation received. Support for this study was provided by DOE grant DE-FG02-92ER61439 and NASA grants NAG5-1330 and NAG5-4398.

REFERENCES

Arakawa, A., and Hsu, Y.-J. G. (1990). Energy conserving and potential-entrophy dissipating schemes for the shallow water equations. *Mon. Wea. Rev.* **118,** 1960–1969.

Arakawa, A., and Lamb, V. (1977). Computational design of the basic dynamical processes of the UCLA general circulation model. *In* "Methods in Computational Physics" (J. Chang, ed.), Vol. 17, pp. 173–265. Academic Press, San Diego.

Arakawa, A., and Konor, C. S. (1994). A generalized vertical coordinate and the choice of vertical grid for atmospheric models. *In* "Life Cycles of Extratropical Cyclones," Vol. III, Proceedings of an International Symposium, Bergen, Norway, pp. 259–264. Geophysical Institute, University of Bergen.

Arakawa, A., and Konor, C. S. (1996). Vertical differencing of the primitive equations based on the Charney–Phillips grid in hybrid σ-θ vertical coordinates. *Mon. Wea. Rev.* **124,** 511–528.

Arakawa, A., Mechoso, C. R., and Konor, C. S. (1992). An isentropic vertical coordinate model: Design and application to atmospheric frontogenesis studies. *Meteor. Atmos. Phys.* **50,** 31–45.

Arakawa, A., Konor, C. S., and Mechoso, C. R. (1994). A generalized vertical coordinate and the choice of vertical grid for atmospheric models. *In* "Preprints, 10th Conf. on Numerical Weather Prediction," Portland, OR, pp. 32–34. Amer. Meteor. Soc.

Bjerknes, V. (1916). Über thermodynamische Maschinen die unter Mitwirkung der Schwerkraft arbeiten. *Abh. K. Sächs. Gesellsch. Wissensch., Leipzig, Math.-Phys. Kl.* **30,** 37–65.

Bleck, R., and Benjamin, S. (1993). Regional weather prediction with a model combining terrain-following and isentropic coordinates. Part I. *Mon. Wea. Rev.* **121,** 1770–1785.

Boer, G. J. (1989). An exact and approximate energy equations in pressure coordinates. *Tellus,* Series A, **41,** 97–108.

Boer, G. J., Arpe, K., Blackburn, M., Déqué, M., Gates, W. L., Hart, T. L., Le Treut, H., Roeckner, E., Sheinin, D. A., Simmonds, I., Smith, R. N. B., Tokioka, T., Wetherald, R. T., and Williamson, D. (1991). An intercomparison of the climates simulated by 14 atmospheric general circulation models, WCRP Rep. 58, WMO/TD-425, *World Meteorol. Organ.,* Geneva.

Boer, G. J., Arpe, K., Blackburn, M., Déqué, M., Gates, W. L., Hart, T. L., Le Treut, H., Roeckner, E., Sheinin, D. A., Simmonds, I., Smith, R. N. B., Tokioka, T., Wetherald, R. T., and Williamson, D. (1992). Some results from an intercomparison of the climates simulated by 14 atmospheric general circulation models. *J. Geophys. Res.* **97,** 12771–12786.

Book, D. L., Boris, J. P., and Zalesak, S. T. (1981). Flux-corrected transport. *In* "Finite-Difference Techniques for Fluid Dynamics Calculations" (D. L. Book, ed.), pp. 29–55. Springer-Verlag, Berlin.

Born, M. (1949). "Natural Philosophy of Cause and Change," Clarendon Press, Oxford.

Broecker, W. (1991). The great ocean conveyor. *Oceanography* **4,** 79–89.

Bullock, B. R., and Johnson, D. R. (1972). The generation of available potential energy by sensible heating in Southern Ocean cyclones. *Quart. J. Roy. Meteor. Soc.* **98,** 459–518.

Callies, U., and Herbert, F. (1988). Radiative processes and non-equilibrium thermodynamics. *J. Appl. Math. Phys.* **39,** 242–266.

Carathéodory, C. (1909). Untersuchungen über die Grundlagen der Thermodynamik. *Mathematische Annalen,* **67,** 355–386.

Carrington, G. (1994). "Basic Thermodynamics." Oxford University Press, UK.

Charney, J. G. (1947). The dynamics of long waves in a baroclinic westerly current. *J. Meteorol.* **4,** 135–162.

Dutton, J. A. (1973). The global thermodynamics of atmospheric motion. *Tellus* **25,** 89–110.

Dutton, J. A. (1976). "The Ceaseless Wind." McGraw-Hill, New York.

Dutton, J. A., and Johnson, D. R. (1967). The theory of available potential energy and a variational approach to atmospheric energetics. *In* "Advances in Geophys" (H. E. Landberg and J. Van Mieghem, eds.), Vol. 12, pp. 333–436. Academic Press, New York.

Eady, E. T. (1949). Long waves and cyclone waves. *Tellus* **1,** 35–42.

Eliassen, A. (1951). Slow thermally or frictionally controlled meridional circulation in a circular vortex. *Astrophys. Norv.* **5,** 19–60.

Ertel, H. (1942). Ein neuer hydrodynamicher Wirbelsatz. *J. Meteorol.* **Z59,** 277–281.

Essex, C. (1987). Global thermodynamics, the Clausius inequality and entropy radiation. *Geophys. Astrophys. Fluid Dyn.* **38,** 1–13.

Gall, R. L., and Johnson, D. R. (1971). The generation of available potential energy by sensible heating. A case study. *Tellus* **23,** 465–482.

Gallimore, R. G., and Johnson, D. R. (1981). The forcing of the meridional circulation of the isentropic zonally averaged circumpolar vortex. *J. Atmos. Sci.* **38,** 583–599.

Godske, C. L. (1936). Note on an apparent discrepancy between Bjerknes–Sandström and H. Jeffreys. *Quart. J. Roy. Meteor. Soc.* **62,** 446–449.

Godske, C. L., Bergeron, J., Bjerknes, J., and Bundgaard, R. C. (1957). "Dynamic Meteorology and Weather Forecasting." Am. Meteorol. Soc., Boston, Massachusetts and Carnegie Institution of Washington, Washington, DC.

Goody, R. (1995). "Principles of Atmospheric Physics and Chemistry." Oxford University Press, UK.

Hauf, T., and Höller, H. (1987). Entropy and potential temperature. *J. Atmos. Sci.* **44,** 2887–2901.

Hoerling, M. P., Schaack, T. K., and Johnson, D. R. (1990). Heating distributions for January and July simulations of the NCAR community climate model. *J. Climate* **3,** 417–434.

Hsu, Y.-J. G., and Arakawa, A. (1990). Numerical modeling of the atmosphere with an isentropic vertical coordinate. *Mon. Wea. Rev.* **118,** 1933–1959.

Jeffreys, H. (1925). On fluid motions produced by differences of temperature and humidity. *Quart. J. Roy. Meteor. Soc.* **51,** 347–356.

Jeffreys, H. (1936). Note on Dr. Godske's paper. *Quart. J. Roy. Meteor. Soc.* **62,** 449–450.

Johnson, D. R. (1970). The available potential energy of storms. *J. Atmos. Sci.* **27,** 727–741.

Johnson, D. R. (1980). A generalized transport equation for use with meteorological coordinate systems. *Mon. Wea. Rev.* **108,** 733–745.

Johnson, D. R. (1989). The forcing and maintenance of global monsoonal circulations: An isentropic analysis. *In* "Advances in Geophys" (B. Saltzman, ed.) Vol. 31, pp. 43–316. Academic Press, San Diego.

Johnson, D. R. (1997). 'General coldness of climate models' and the second law: Implications for modeling the earth system. *J. Climate* **10,** 2826–2846.

Johnson, D. R., and Arakawa, A. (1996). On the scientific contributions and insight of Professor Yale Mintz. *J. Climate* **9,** 3211–3224.

Johnson, D. R., and Downey, W. K. (1982). On the energetics of open systems. *Tellus* **34,** 458–470.

Johnson, D. R., Townsend, R. D., and Wei, M.-Y. (1985). The thermally coupled response of the planetary scale circulation to the global distribution of heat sources and sinks. *Tellus* **37A**, 106–125.

Johnson, D. R., Zapotocny, T. H., Reames, F. M., Wolf, B. J., and Pierce, R. B. (1993). A comparison of simulated precipitation by hybrid isentropic-sigma and sigma models. *Mon. Wea. Rev.* **121**, 2088–2114.

Kleinschmidt, E. (1955). Die Entstehung einer Höhenzyklone über Nordamerika. *Tellus* **7**, 96–110.

Konor, C. S., and Arakawa, A. (1997). Design of an atmospheric model based on a generalized vertical coordinate. *Mon. Wea. Rev.* **125**, 1649–1673.

Konor, C. S., Mechoso, C. R., and Arakawa, A. (1994). Comparison of frontogenesis simulations with θ and σ vertical coordinates and design of a general vertical coordinate model. *In* "Life Cycles of Extratropical Cyclones, Vol. II, Proceedings of an International Symposium, Bergen, Norway, pp. 464–468. Geophysical Institute, University of Bergen.

Kutzbach, G. (1979). "The Thermal Theory of Cyclones." Am. Meteor. Soc.

Lesins, G. B. (1990). On the relationship between radiative entropy and temperature distributions. *J. Atmos. Sci.* **47**, 795–803.

Lettau, H. (1954). A study of the mass, momentum, and energy budget of the atmosphere. *Arch. Meteorol. Geophys. Bioklimatol. Ser. A* **7**, 133–157.

Lettau. H. (1959). Wind profile, surface stress and geostrophic drag coefficients in the atmospheric surface layer. *In* "Advances in Geophysics" Vol. 6, pp. 241–257. Academic Press, New York.

Li, J. and Chýlek, P. (1994). Entropy in climate models. Part II: Horizontal structure of atmospheric entropy production. *J. Atmos. Sci.* **51**, 1702–1708.

Li, J., Chýlek, P., and Lesins, G. B. (1994). Entropy in climate models. Part I: Vertical structure of atmospheric entropy production. *J. Atmos. Sci.* **51**, 1691–1701.

Lorenz, E. N. (1955). Generation of available potential energy and the intensity of the general circulation. *Tellus* **7**, 157–167.

Lorenz, E. N. (1967). The nature and theory of the general circulation of the atmosphere, No. 218, TP115. World Meteorol. Organ., Geneva.

Margenau, H., and Murphy, G. M. (1956). "The Mathematics of Physics and Chemistry," 2nd ed. D. Van Nostrand Company, New York.

Margules, M. (1903). Über die Energie der Stürme. *Jahrb. Zentralanst. Meteorol.* **40**. *In* "The Mechanics of the Earth's Atmosphere," 3rd Coll. (Trans. by Abbe, C. (1910), pp. 533–595. Smithsonian Inst., Washington, DC.

Namais, J. (1940). "An Introduction to the Study of Air Mass and Isentropic Analysis." Am. Meteorol. Soc.

Oort, A. H., and Peixoto, J. P. (1983). Global angular momentum and energy balance requirements from observations. *Adv. Geophys.* **25**, 355–490.

Ooyama, K. (1990). A thermodynamic foundation for modeling the moist atmosphere. *J. Atmos. Sci.* **47**, 2580–2593.

Oreskes, N., Shrader-Frechette, K., and Belitz, K. (1994). Verification, validation, and confirmation of numerical models in the earth sciences. *Science* **263**, 641–646.

Otto-Bliesner, B. L., and Johnson, D. R. (1982). Thermally forced mean mass circulation in the Northern Hemisphere. *Mon. Wea. Rev.* **110**, 916–932.

Palmén, E., and Newton, C. W. (1969). "Atmospheric Circulation Systems." Academic Press, New York.

Pearce, R. P. (1978). On the concept of available potential energy. *Quart. J. Roy. Meteor. Soc.* **104**, 737–755.

Peixoto, J. P., and Oort, A. H. (1992). "Physics of Climate." American Institute of Physics, New York.

Peixoto, J. P., Oort, A. H., Almeida, M. D., and Tome, A. (1991). Entropy budget of the atmosphere. *J. Geophys. Res.* **96** (D6), 10981–10988.

Pfeffer, R. L., Mardon, D., Sterbenz, P., and Fowlis, W. (1965). A new concept of available potential energy. In "Dynamics of Large-Scale Atmospheric Processes," Proc. Int. Symp. Dyn. Large-Scale Processes Atmos., Moscow, pp. 179–189. World Meteorol. Organ., Geneva.

Riehl, H., and Fultz, D. (1957). Jet stream and long waves in a steady rotating dishpan experiment. *Quart. J. Roy. Meteor. Soc.* **83,** 215–231.

Riehl, H., and Fultz, D. (1958). The general circulation in a steady rotating dishpan experiment. *Quart. J. Roy. Meteor. Soc.* **84,** 389–417.

Rossby, C.-G. (1940). Planetary flow patterns in the atmosphere. *Quart. J. Roy. Meteor. Soc.* **66 (Suppl.),** 68–87.

Sandström, J. W. (1915). The origin of the wind. *Mon. Wea. Rev.* **43,** 161–163.

Sandström, J. W. (1916). Meteorologische Studien in schwedischen Hochgebirge. *Goeteborgs K. Vetensk. Vitterhets-Samh. Handl., Sect. 4* **22,** No. 2.

Schaack, T. K., and Johnson, D. R. (1994). January and July global distributions of atmospheric heating for 1986, 1987, and 1988. *J. Climate* **7,** 1270–1285.

Schaack, T. K., Johnson, D. R., and Wei, M.-Y. (1990). The three-dimensional distribution of atmospheric heating during the GWE. *Tellus* **42A,** 305–327.

Schaack, T. K., Lenzen, A. J., and Johnson, D. R. (1991). Global heating distributions for January 1979 calculated from GLA assimilated and simulated model-based data sets. *J. Climate* **4,** 395–406.

Sears, F. W. (1953). "An Introduction to Thermodynamics, the Kinetic Theory of Gases, and Statistical Mechanics." Addison-Wesley, Reading, MA.

Shaw, Sir Napier (1923). "The Air and Its Ways." Cambridge Univ. Press, London.

Shaw, Sir Napier (1925). Note in the graphics representation of pressure and temperature obtained from observations by balloons, aeroplanes and kites. *Reprinted in* "Selected Meteorological Papers of Sir Napier Shaw." F. R. S. Macdonalds & Co. Ltd.

Shaw, Sir Napier (1930). "Manual of Meteorology, The Physical Processes of Weather," Vol. III. Cambridge Univ. Press, London.

Siegmund, P. (1994). The generation of available potential energy, according to Lorenz' exact and approximate equations. *Tellus* **46A,** 566–582.

Sommerfeld, A. (1950). "Mechanics of Deformable Bodies," Vol. 2 (Translation German to English by G. Kuerti.) Academic Press, New York.

Sommerfeld, A. (1964). "Thermodynamics and Statistical Mechanics: Lectures on Theoretical Physics," Vol. 5 (Translation German to English by J. Kestin; F. Bopp and J. Mixner, eds.). Academic Press, New York.

Spar, J. (1949). Energy changes in the mean atmosphere. *J. Meteorol.* **6,** 411–415.

Stephens, G. L., and O'Brien, D. M. (1993). Entropy and climate. I: ERBE observations of the entropy production of the earth. *Quart. J. Roy. Meteor. Soc.* **119,** 121–152.

Stephens, G. L., and O'Brien, D. M. (1995). Entropy and climate. II. Simple models. *Quart. J. Roy. Meteor. Soc.* **121,** 1773–1796.

Tolman, R. C., and Fine, P. C. (1948). On the irreversible production of entropy. *Rev. Mod. Phys.* **20,** 51–77.

Townsend, R. C., and Johnson, D. R. (1985). A diagnostic study of the isentropic zonally averaged mass circulation during the First GARP Global Experiment. *J. Atmos. Sci.* **42,** 1565–1579.

Van Mieghem, J. (1956). The energy available in the atmosphere for conversion into kinetic energy. *Beitr. Phys. Atmos.* **29,** 4–17.

Wei, M.-Y., Johnson, D. R., and Townsend, R. D. (1983). Seasonal distributions of diabatic heating during the First GARP Global Experiment. *Tellus* **35A,** 241–255.

Wiin-Nielsen, A. and Chen, T-C. (1993). "Fundamentals of Atmospheric Energetics." Oxford University Press, New York.

Wulf, O. R., and Davis, L. Jr. (1952). On the efficiency of the engine driving the atmosphere circulation. *J. Meteorol.* **9,** 79–82.

Zapotocny, T. H., Johnson, D. R., and Reames, F. M. (1993). A comparison of regional isentropic-sigma and sigma model simulations of the January 1979 Chicago blizzard. *Mon. Wea. Rev.* **121,** 2115–2135.

Zapotocny, T. H., Johnson, D. R., and Reames, F. M. (1994). Development and initial test of the University of Wisconsin global isentropic-sigma model. *Mon. Wea. Rev.* **122,** 2160–2178.

Zillman, J. W., and Johnson, R. D. (1985). Thermally forced mean circulations in Southern Hemisphere. *Tellus* **37A,** 56–76.

Zhu, Z. (1997). Precipitation and water vapor transport simulated by a hybrid-coordinate GCM. *J. Climate* **10,** 988–1003.

Zhu, Z., and Schneider, E. K. (1997). Improvement in stratosphere simulation with a hybrid σ-θ coordinate GCM. *Quart. J. Roy. Meteor. Soc.* **123,** 2095–2114.

Chapter 23

Future Development of General Circulation Models

Akio Arakawa

Department of Atmospheric Sciences
University of California, Los Angeles, California

I. INTRODUCTION: THE BEGINNING OF THE "GREAT CHALLENGE" THIRD PHASE

This chapter, which is a continuation of the first chapter of this book, is based on the lecture with the same title I presented at the end of the AA Fest: Symposium on General Circulation Model Development: Past, Present, and Future, held at UCLA, January 20–22, 1998. As in the first chapter, I present a personal perspective, this time on the future development of general circulation models (GCMs), and point out some modeling issues that should be considered more seriously than they have been in the past.

Numerical modeling of the atmosphere has gone through the second phase of its history (see Fig. 1 of Chapter 1), during which its scope was

General Circulation Model Development

"magnificently" expanded. Numerical models including GCMs are now indispensable tools for studying and predicting a variety of atmospheric phenomena.

In spite of the expansion, however, the geographical distribution of sea surface temperature (SST) was prescribed as an external condition in most GCM applications during the second phase. From the point of view of climate simulations, this assumes the most important part of the answer and, therefore, it might have hidden crucial deficiencies that may exist in the model. For example, calculation of the surface heat flux does not have to be very accurate when SST is prescribed, as long as the heat flux adjusts the low-level air temperature toward the prescribed SST with a relatively short time scale. Calculation of the vertical distribution of radiative heating/cooling also does not have to be very accurate, at least for the tropical troposphere, as long as the cumulus parameterization adjusts the temperature lapse rate toward a realistic value. Thus, parameterizations of planetary boundary layer (PBL) and radiation/cloud processes did not appear to be very demanding problems in the second phase.

Even a successful simulation of tropical precipitation does not necessarily verify the cumulus parameterization used in the model. In the subtropics, the radiative cooling approximately determines the amount of the adiabatic warming in the descending branch of the Hadley circulation (see Fig. 1). Then the amount of adiabatic cooling and, therefore, the amount of condensation heating in the ascending branch are more or less determined, while its position is strongly controlled by the prescribed SST distribution. Indeed, two GCMs with cumulus parameterizations constructed using completely different reasoning can produce equally realistic gross features of tropical precipitation.

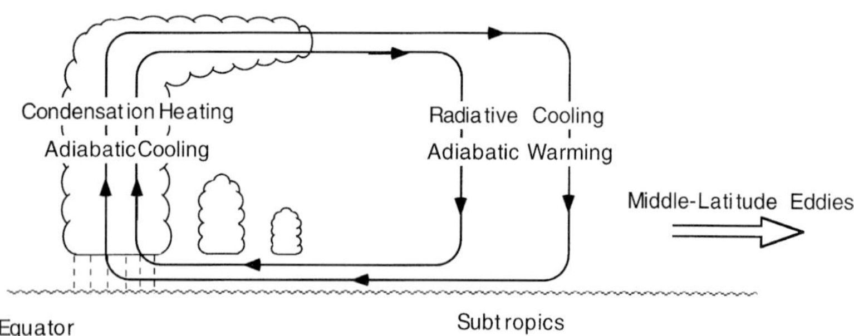

Figure 1 A schematic figure showing the balance of warming/cooling in the Hadley circulation.

When I was working with earlier generations of the UCLA GCM, I was amazed that the total precipitation in the equatorial region always came up within a reasonable range almost regardless of the way in which moist convection was treated. This was even true for the "precipitation" estimated from the convective heating simulated by the Generation I UCLA GCM (Mintz–Arakawa "dry" model), in which moisture was not predicted. (In the model, the radiative cooling in the subtropics was guaranteed to be realistic because it was based on an empirical formulation. See Section V of Chapter 1 in this volume.) A good simulation of the gross features of tropical precipitation, therefore, can tell us relatively little about the validity of the cumulus parameterization used for the simulation. (Simulation of the partition of precipitation between the continents and oceans is a different story.)

When the atmospheric GCM is coupled with an oceanic GCM, however, the situation can be entirely different. Any deficiencies in the model may now appear as errors in the simulated SST and can cause a serious climate drift of the coupled system.

In my opinion, the "great challenge" third phase of numerical modeling of the atmosphere has already begun (see Fig. 1 of Chapter 1). The opening of this phase is partly stimulated by the recent development of coupled atmosphere–ocean GCMs, as well as a clearer identification of important issues in climate change, such as the global warming and ENSO prediction problems. The opening is also characterized by the increased use of large-eddy simulation models (LESs) and cloud-resolving models (CRMs) to evaluate parameterizations in larger scale models.

In the past several years, improvements in the performance of coupled atmosphere–ocean GCMs have been remarkable. As of 1995, most coupled GCMs could successfully simulate the zonal gradient of the equatorial SST over the central Pacific (see Mechoso *et al.*, 1995). Many models even simulate the annual cycle of the equatorial SST realistically. This is in a sharp contrast to the situation only 3 years earlier (see Neelin *et al.*, 1992). We may then wonder if something revolutionary in our modeling capability might have happened during those 3 years simultaneously at different institutions. Of course, this is not likely to be the case. More likely, the improvement was a consequence of relatively minor changes, or "tuning," of the models, which might have been done even on different aspects of the models at different institutions.

Indeed, the mean SSTs simulated by coupled atmosphere–ocean models have been found to be extremely sensitive to those aspects that are relatively poorly formulated in the models, such as

- The vertical redistribution of moisture from the planetary boundary layer (PBL) to the free atmosphere due to entrainment/diffusion processes

- The amount of each cloud type (e.g., high, low, deep, shallow, liquid-water, ice, stratiform, cumuliform)
- The emissivity, reflectivity, and absorptivity of each cloud type

These sensitivities arise because surface fluxes are not determined by the processes near the surface alone. Instead, they are the consequences of complicated interactions between various processes in the atmosphere (and oceans). Let us consider surface evaporation as an example. To calculate surface evaporation $E \equiv (F_q)_S$ over a wet surface, it is conventional to use a formula such as

$$(F_q)_S = \rho C_E |\mathbf{v}_B|(q^*_S - q_B), \tag{1}$$

where F denotes the turbulent flux, q is the specific humidify, the subscript S denotes the surface, ρ is the density, C_E is the surface transfer coefficient for water vapor, $\mathbf{v}$ is the horizontal velocity, the subscript B denotes a representative value near the surface or for the entire PBL, and q^*_S is the saturation value of q at the surface. [In Eq. (20) of Chapter 1, subscript g is used instead of S.] If the evaporated water vapor is locally stored within the PBL, q_B increases through evaporation. Equation (1) then indicates that the surface evaporation is subject to a negative feedback, or an adjustment toward zero, due to the change of q_B. We may then ask an important question: In nature (and in GCMs), what controls the amount of E against this eminent adjustment?

To answer this question, we first note that the value of q_B is almost in a quasi-equilibrium beyond the time scale of adjustment, which should be on the order of a day. Then, for a longer time scale, the amount of E is approximately determined by the counteracting drying effects in the PBL, such as the dilution of moisture due to entrainment/diffusion processes through the PBL top and the horizontal advection of drier air within the PBL. Figure 2 illustrates this situation. These drying effects then "force"

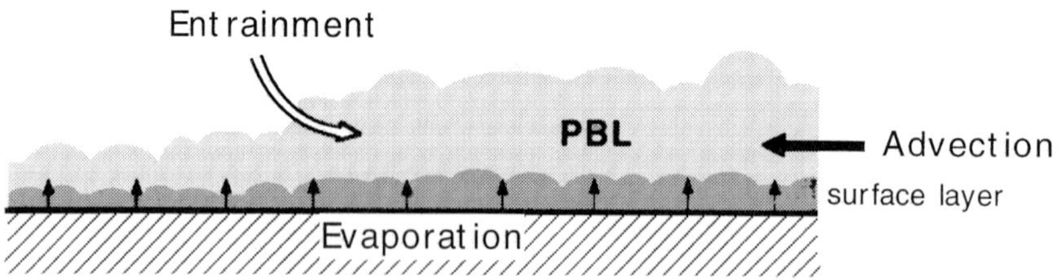

Figure 2   A schematic figure showing the balance of moistening and drying effects on PBL humidity.

the surface evaporation against the adjustment. From this point of view, $q_S^* - q_B$ in Eq. (1) is a consequence of E, rather than its cause.

Note that the quasi-equilibrium referred to here is for the PBL specific humidity, q_B, which is an intensive quantity, not for the total amount of moisture in the PBL, which is an extensive quantity. This difference is important because, for example, the entrainment of mass adds moisture to the PBL (unless the entraining air is completely dry), increasing the total amount of moisture in the PBL, while the PBL specific humidity is decreased as the drier entrained air is mixed with the PBL air.

The argument given above for the surface evaporation is analogous to the quasi-equilibrium argument of Arakawa and Schubert (1974) for cumulus parameterization, in which a measure of conditional instability is assumed to be in a quasi-equilibrium beyond the moist-convective adjustment time scale. Then the intensity of cumulus activity is approximately determined by the destabilizing effects due to large-scale processes, which "force" the cumulus activity against the adjustment, not by the degree of existing instability. See Fig. 15 of Chapter 1 for an illustration of this situation.

Ma *et al.*'s (1994) experiment with a coupled atmosphere–ocean GCM is a good demonstration of the sensitivity of the surface evaporation to various atmospheric processes. In this experiment, the emissivity of high-level clouds was reduced by changing the radiation scheme. Then the mean SST simulated by the coupled GCM became colder. This may seem obvious since the reduced emissivity should give less greenhouse effect below. We have found, however, that the decrease in SSTs is not directly through a change in the radiation flux at the surface. Instead, as shown in Fig. 3, cumulus convection and surface evaporation play central roles in

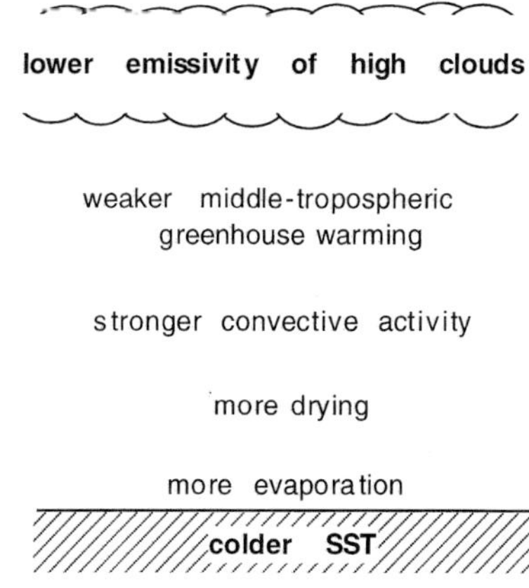

Figure 3　Processes involved in the influence of a lower emissivity of high clouds on SST.

this decrease. See Chapter 18 in this book for more details of this experiment.

These results are of course model dependent. For example, if the cumulus parameterization used in the GCM does not directly recognize the upper level radiative cooling, or if it does not produce the low-level drying, the result can be totally different. Resolving these model dependencies is exactly among the objectives of the "great challenge" third phase.

Although simpler models will continue to be useful for many purposes, I anticipate two general trends in the development of advanced GCMs during the third phase: the expansion of internal processes and the increase of resolution. During the expansion of internal processes, more and more processes and their interactions are included in the model. The typical example is the inclusion of oceanic processes through coupling the atmospheric GCM with an oceanic GCM. Other examples are the inclusions of more detailed ground, cryosphere, biosphere, microphysical, and chemical processes. The expected benefit of this trend is, of course, a broader applicability of the GCMs. However, besides the obvious practical drawbacks from the increased complexity, this trend may tend to degrade the quality of simulated climatology as the newly added processes eliminate negative feedbacks in current models with prescribed external conditions. This has usually been the case when the atmospheric GCM is coupled with an oceanic GCM.

Due to the use of higher resolution, which is the other anticipated trend in the third phase, mesoscale processes, or even individual deep clouds, will be at least partly resolved by the model. Besides a decrease of truncation errors, the expected benefit of this trend is a reduced need for "sub-grid-scale" parameterizations. Again there can be the following concerns with this trend: Not all computational errors decrease as the resolution increases, and the nature of "sub-grid-scale" processes changes as the resolution changes while they still need to be parameterized. Correspondingly, we may have to "retune" or even reformulate the physics package as the resolution changes substantially.

In the following sections, I discuss selected aspects of future general circulation modeling with these general trends in mind. Section II discusses the problem of choosing dynamics equations, including the possibility of abandoning the primitive equations. Sections III and IV discuss discretization problems, with an emphasis on choosing the vertical grid, vertical coordinate, and horizontal grid in Section III, and on advection schemes in Section IV. Section V discusses the formulations of PBL and stratiform cloud processes and representation of the effects of surface irregularity. Section VI discusses future problems in formulating moist convection and cloud processes. Finally, Section VII gives closing remarks.

II. CHOICE OF DYNAMICS EQUATIONS

Practically all of the GCMs developed during the "magnificent" second phase use the primitive equations. With the spherical and pressure coordinates, these equations may be written as

$$\frac{Du}{Dt} - \left(2\Omega + \frac{u}{a\cos\varphi}\right)v\sin\varphi + \frac{1}{a\cos\varphi}\frac{\partial\phi}{\partial\lambda} = F_\lambda, \tag{2}$$

$$\frac{Dv}{Dt} + \left(2\Omega + \frac{u}{a\cos\varphi}\right)u\sin\varphi + \frac{1}{a}\frac{\partial\phi}{\partial\varphi} = F_\varphi, \tag{3}$$

$$\frac{\partial\phi}{\partial p} = -\alpha, \tag{4}$$

$$\frac{1}{a\cos\varphi}\left[\frac{\partial u}{\partial\lambda} + \frac{\partial(v\cos\varphi)}{\partial\varphi}\right] + \frac{\partial\omega}{\partial p} = 0, \tag{5}$$

$$c_p\frac{D\theta}{Dt} = \left(\frac{p_0}{p}\right)^{R/c_p}Q, \tag{6}$$

with

$$\frac{D}{Dt} \equiv \frac{\partial}{\partial t} + \frac{u}{a\cos\varphi}\frac{\partial}{\partial\lambda} + \frac{v}{a}\frac{\partial}{\partial\varphi} + \omega\frac{\partial}{\partial p}, \tag{7}$$

$$\theta \equiv \alpha\frac{p}{R}\left(\frac{p_0}{p}\right)^{R/c_p}. \tag{8}$$

Here all symbols are standard. Approximations used in these equations are as follows:

1. Neglecting the acceleration and friction terms in the vertical component of the momentum equation—the "quasi-hydrostatic approximation"
2. Neglecting the terms involving $2\Omega\cos\varphi$—the traditional approximation
3. Approximating the radial distance from the Earth's center by the mean radius of the Earth, a—the shallow-atmosphere approximation

4. Neglecting the uw/r and vw/r terms in the horizontal momentum equation and the u^2/r and v^2/r terms in the vertical momentum equation—the "small-curvature approximation"

The advantage of using all four of these approximations as a package is the maintenance of the potential vorticity and absolute angular momentum conservation laws in the following approximate forms:

$$\frac{D}{Dt}\frac{(\mathbf{k}2\Omega\sin\varphi + \nabla\times\mathbf{v})\cdot\nabla\theta}{\rho} = 0 \text{ for adiabatic frictionless flow,} \qquad (9)$$

where $\mathbf{v}$ is the horizontal velocity and ∇ is the three-dimensional gradient operator, and

$$\frac{D}{Dt}[(\Omega a\cos\varphi + u)a\cos\varphi] = 0 \text{ for frictionless flow when } \partial p/\partial\lambda = 0.$$
$$(10)$$

Among the four approximations listed above, justification for the traditional approximation is perhaps the weakest, and we may ask ourselves if this is about the time to consider abandoning that approximation. We can argue that the approximation is valid when

$$(2\Omega\cos\varphi)^2 \ll N^2, \qquad (11)$$

where $2\Omega\cos\varphi$ is the frequency of the inertial oscillation in a vertical west–east plane and N is the Brunt–Väisälä frequency. This condition is normally well satisfied except when the heat of condensation significantly reduces the effective Brunt–Väisälä frequency. Using the p (or σ) coordinate, White and Bromley (1995) derived a generalized quasi-hydrostatic system of equations, with a complete representation of the Coriolis force, that is "dynamically consistent" in the sense that the system maintains the conservation laws, Eqs. (9) and (10), in more general forms.

An alternative approach for including the complete Coriolis force is the use of the original nonhydrostatic system of equations even for global models, as in the new U.K. Meteorological Office unified model (Cullen *et al.*, 1994). This approach is analogous to the transition from the quasi-geostrophic models to the primitive equation models, which occurred near the beginning of the "magnificent" second phase. The reason for this transition, instead of a transition to the intermediate models (McWilliams and Gent, 1980), such as the balanced model, was computational convenience rather than a higher accuracy of the solutions with the primitive equations. An analogous situation may exist in the transition to nonhydrostatic models. It introduces new problems, however, i.e., initialization of vertical velocity and simulation of hydrostatic adjustment (see Section III.B), as the transition to the primitive equation models introduced the problems of initializing horizontal velocity and simulation of geostrophic adjustment.

Perhaps the greatest advantage we might gain through the use of a nonhydrostatic GCM is the possibility of unifying models for different scales by developing unified models (see Fig. 1 of Chapter 1 in this book). For this purpose, however, models' physics packages also need to be unified. We should try to do this anyway, as I will discuss later, although it is a much more difficult task than changing the dynamics equation.

Even when we continue to use the package of approximations commonly used for the primitive equations, we have two options: the use of the momentum equations or the use of vorticity/divergence equations. A choice between these two systems of equations can be based on computational convenience, as in Ritchie *et al.* (1995) with the use of a semi-Lagrangian scheme and in Heikes and Randall (1995) with the use of an icosahedral global grid; or it can be based on the more substantial reason of improving the discrete dynamics, as in Randall (1994) (see Section III of this chapter and Randall's chapter in this book, Chapter 17).

III. DISCRETIZATION PROBLEMS: CHOICE OF VERTICAL GRID, VERTICAL COORDINATE, AND HORIZONTAL GRID

A. INTRODUCTION

In numerical modeling of the atmosphere, either a finite-difference method or a spectral method is commonly used for horizontal discretization, while finite-difference methods are used almost exclusively for vertical and time discretization, although finite-element methods have sometimes been used. See Mesinger (1997) for reviews of these subjects.

From my point of view, the following computational problems are among the issues that need to be considered during the "great challenge" third phase.

1. *Computational aspects of the dynamical system that cannot be automatically improved by increasing the resolutions.* Control is needed of "computational noise" of any kind, which may influence larger scale processes thermodynamically, as well as dynamically, through the collective effects of associated phase changes of water and cloud effects on radiation.

2. *Advection.* A greater emphasis on the Lagrangian accuracy is needed, while not sacrificing the Eulerian conservation, due to the increased importance of simulating the distributions of moisture and other atmospheric constituents. The Lagrangian accuracy is especially important for the vertical advection of the water vapor mixing ratio, because its magnitude relative to the saturation value is crucial in determining the existence of clouds.

I discuss selected problems relevant to item 1 in the rest of Section III and the problem of item 2 in Section IV.

B. Choice of Vertical Grid in the σ Coordinate

Most existing GCMs, with a few exceptions including models originally developed by Krishnamurti (1969) and Robert *et al.* (1972), use the Lorenz grid (Lorenz, 1960; L Grid), rather than the Charney–Phillips grid (Charney and Phillips, 1953, CP Grid), with a pressure-based vertical coordinate such as the σ coordinate (Phillips, 1957) or its variation. As indicated in Section VII of Chapter 1, this does not seem to be the best choice for the reasons I discuss in the rest of this subsection. Figure 4 shows these grids with the σ coordinate.

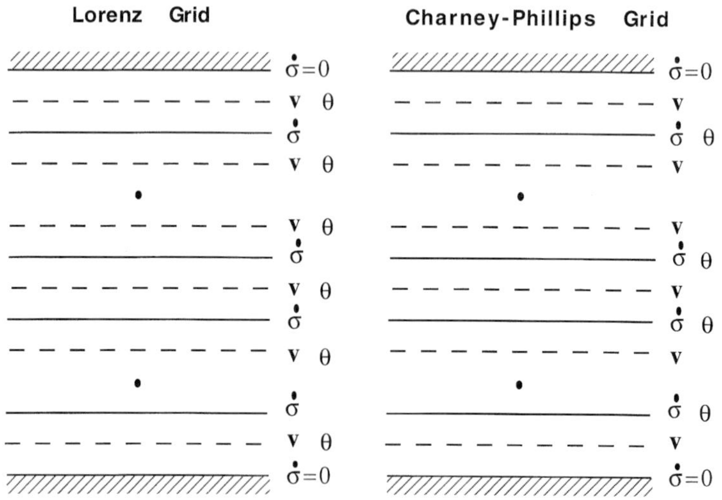

Figure 4   The Lorenz and Charney–Phillips grids with the σ coordinate.

The advantage of the L Grid for nongeostrophic models such as the primitive equation models has been well recognized. With this grid, all three-dimensional prognostic variables are carried for each model layer to which the continuity equation representing the mass budget is applied. With the L Grid, therefore, it is easier to keep track of the budget of mass-weighted quantities, such as the total energy and the total potential enthalpy.

The problem with the L Grid is due to the existence of an extra degree of freedom in the vertical distribution of θ that cannot be controlled by any baroclinic (or internal) modes. By comparing the number of degrees of freedom in θ and that in $\dot{\sigma}$ (or ω in the p coordinate) shown in Fig. 4, it is clear that the L Grid allows the existence of such an extra degree of freedom in θ while the CP Grid does not. This extra degree of freedom introduces a vertical "computational mode" in the solutions, characterized by a zigzag vertical structure of θ, which has no counterpart in the solutions of the continuous equation (Tokioka, 1978; Arakawa, 1983; Schneider, 1987; Arakawa and Moorthi, 1988; Arakawa, 1988; Leslie and Purser, 1992; Clark and Hanes, 1994; Fox-Rabinovitz, 1994; Cullen and James, 1994; Hollingsworth, 1995; Arakawa and Konor, 1996). It is important to note that this problem with the L Grid is a property of the grid rather than of a particular scheme. An increase in order of accuracy does not reduce the error due to this mode because a computational mode cannot become "accurate" since there is no corresponding true solution with which to compare it.

Figure 5, taken from Arakawa and Konor (1996), shows the time evolutions of potential temperature perturbations associated with standing inertia-gravity waves generated by an unbalanced initial condition, simulated with the L Grid (upper panel) and with the CP Grid (lower panel). The result with the L Grid clearly demonstrates the existence of the computational mode, which is not vertically propagating.

The extra degree of freedom in θ with the L Grid also gives an extra degree of freedom in the potential vorticity, which can cause a spurious realization of baroclinic instability for short waves (Arakawa and Moorthi, 1988). Figures 6a and 6b, also taken from Arakawa and Konor (1996), show the geopotential height and potential temperature fields interpolated to the 900-mb level for disturbances that developed from initially random disturbances in a baroclinically unstable basic state, simulated using an L Grid model (Fig. 6a) with the σ coordinate and a CP Grid model (Fig. 6b) with the σ coordinate. The spurious growth of small-scale perturbations with the L Grid is apparent in Fig. 6a. These perturbations are dynamically active since their growth is due to spurious energy conversion, rather than spurious energy generation as in the usual computational instabilities.

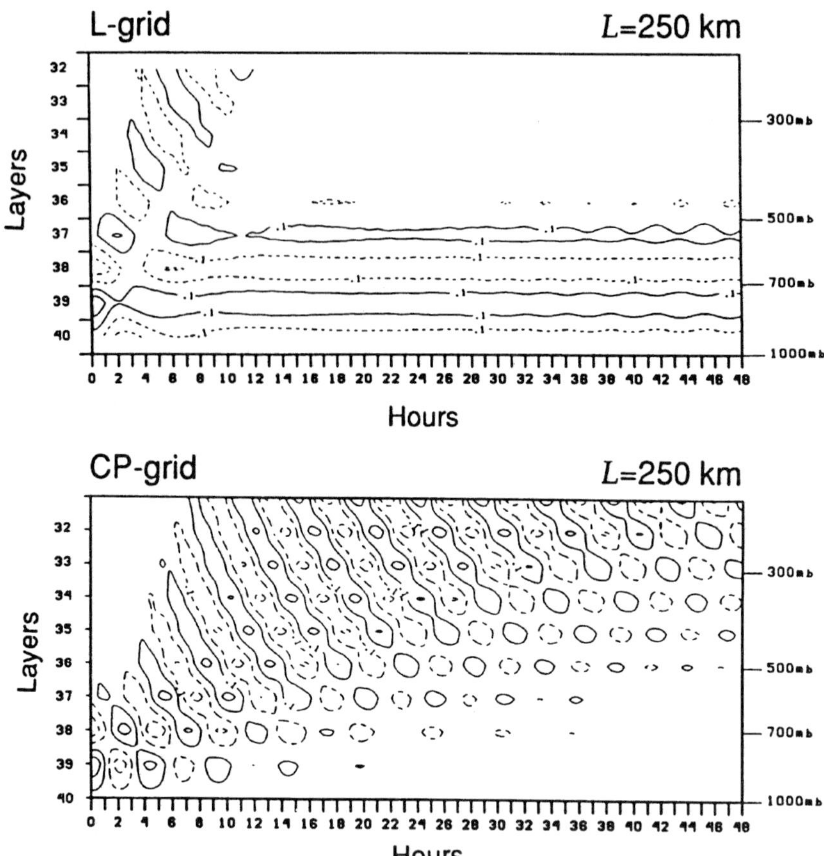

Figure 5   Time evolution of potential temperature perturbations associated with standing inertia-gravity waves generates by an unbalanced initial condition, simulated with the L Grid (upper panel) and with the CP Grid (lower panel). (From Arakawa and Konor, 1996. Courtesy of the American Meterological Society.)

Arakawa and Konor (1996) further showed that even large-scale fields simulated with the L Grid are very sensitive to the form of horizontal diffusion and the magnitude of its coefficient, while those with the CP Grid are not.

The use of an appropriate vertical grid is also crucial in nonhydrostatic models (e.g., Cullen and James, 1994). The spurious growth of small-scale perturbations with the L Grid is expected to occur also in a nonhydrostatic large-scale model. Moreover, just as the way in which geostrophic adjust-

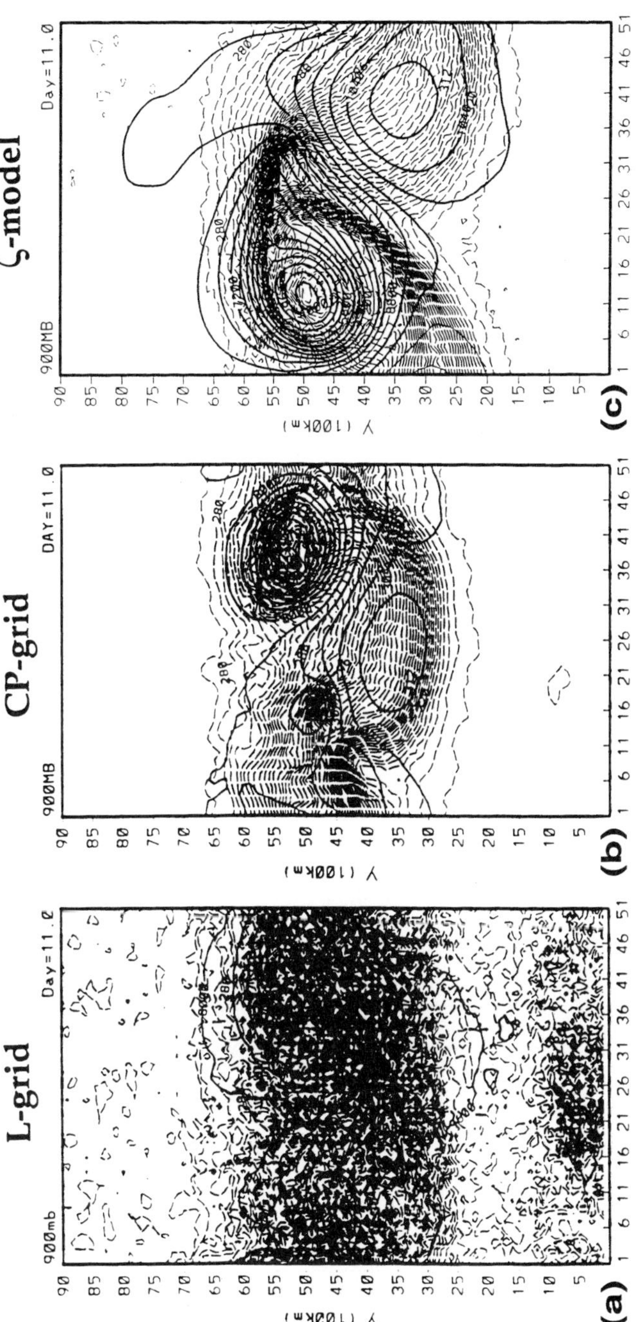

Figure 6 The geopotential height (solid lines) and potential temperature (dashed lines) fields interpolated to the 900-mb level for disturbances developed from initially random disturbances in a baroclinically unstable basic state, simulated by (a) an L Grid model with the σ coordinate, (b) a CP Grid model with the σ coordinate, and (c) the hybrid θ-σ model described in Section III.D. (From Arakawa and Konor, 1996, and Konor and Arakawa, 1997. Courtesy of the American Meteorological Society.)

ment takes place depends on the horizontal grid, the way in which hydrostatic adjustment takes place depends on the vertical grid. Hydrostatic adjustment occurs through the dispersion of vertically propagating sound waves due to the buoyancy force (Lamb, 1932, Section 309; Bannon, 1995). Since the effective buoyancy is almost zero in the L Grid for marginally resolvable vertical scales due to the inevitable vertical averaging, hydrostatic adjustment does not properly operate for such scales with the L Grid.

C. Isentropic Vertical Coordinates

An alternative to the CP Grid with a pressure-based coordinate is the use of an isentropic vertical coordinate such as the θ coordinate. In the p or σ coordinate, what determines the thermal structure of the model atmosphere is the space distribution of θ, whereas in the θ coordinate it is the space distribution of p; and it is most natural to define p in a θ-coordinate model at the interfaces of model layers (see Fig. 7) as the mass continuity equation is applied to each model layer. Thus, p in the θ coordinate is analogous to θ in the p or σ coordinate. The grid shown in Fig. 7 is then analogous to the CP Grid rather than the L Grid shown in Fig. 4. Therefore, the problems with the L Grid mentioned in Section III.B do not exist with the most naturally selected grid in the θ coordinate.

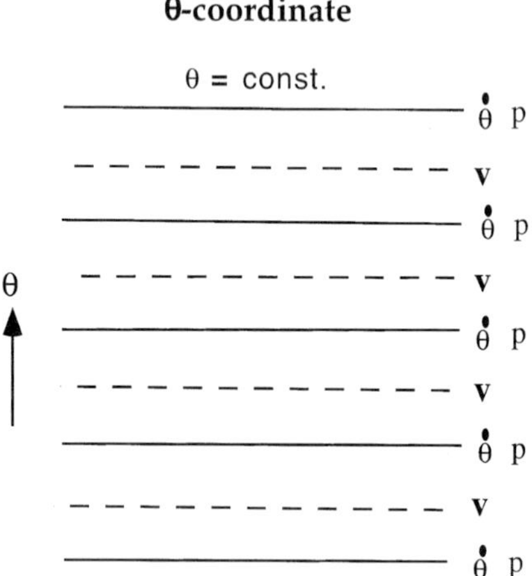

Figure 7   A grid for a θ-coordinate model analogous to the CP Grid with the σ coordinate.

There are also well-recognized advantages in the use of isentropic vertical coordinates. Those advantages are consequences of the following two properties of isentropic surfaces:

1. The horizontal pressure gradient force is irrotational when the curl is taken along an isentropic surface so that it generates no isentropic vorticity $\mathbf{k} \cdot \nabla_\theta \times \mathbf{v}$ as long as the surface does not intersect the Earth's surface.
2. The mass flux across an isentropic surface depends only on diabatic processes so that isentropic surfaces are material surfaces under adiabatic processes.

One of the important consequences of these two properties of isentropic surfaces is that the (quasi-static version of) Ertel's potential vorticity can be expressed as $(2\Omega \sin\varphi + \mathbf{k} \cdot \nabla_\theta \times \mathbf{v})\Delta\theta/\rho\Delta z$, where Δz is the thickness of the layer bounded by two isentropic surfaces, $\theta = \theta_0$ and $\theta = \theta_0 + \Delta\theta$, where θ_0 is a constant. Thus, with an isentropic vertical coordinate, the potential vorticity of each model layer is analogous to the shallow-water potential vorticity, $(2\Omega \sin\varphi + \mathbf{k} \cdot \nabla \times \mathbf{v})/h$, where h is the fluid depth. This is an advantage in developing a discrete model with the θ coordinate since the development can be more closely guided by schemes tested for the shallow-water equations. For example, SICK (symmetric instability of computational kind, see Chapter 1) may appear when a scheme for the vector-invariant form of the shallow-water equations is applied to each layer of a three-dimensional model with the σ coordinate, while it does not seem to appear when the same is done with the θ coordinate.

Property 2 listed earlier gives an additional important advantage to the θ coordinate; i.e., the three dimensionality matters only through the pressure gradient force when adiabatic, leading to simpler expressions for angular momentum and energy budgets and for wave-mean flow interactions. Furthermore, with the θ coordinate, the thermodynamic equation is used to simply equate the "vertical velocity" $\dot\theta$ to Q/Π, where Q is heating and Π is the Exner function, $c_\mathrm{p}(p/p_0)^{R/c_\mathrm{p}}$. It thus eliminates the space-discretization problem for the thermodynamic equation.

In the context of this article, the virtual elimination of the discretization problem for the vertical advection terms is perhaps the most important advantage of the θ coordinate because we still have difficulties with advection schemes. These difficulties, as we discuss in the next section, are especially serious for the vertical advection of moisture, which strongly influences clouds and associated radiation processes in the model. The use of the θ coordinate is advantageous even for simulating condensation processes, which are not adiabatic, because moisture is transported along an isentropic surface in surrounding unsaturated regions or before the onset of condensation (see Johnson *et al.*, 1993).

Another advantage of the θ coordinate is that coordinate surfaces normally do not directly cross frontal zones because they tend to form along isentropic surfaces. Furthermore, unlike the case of the height and pressure-based coordinates, the mass of a layer between two isentropic surfaces can become arbitrarily small. Thus, a θ coordinate model can resolve fronts, automatically providing higher resolution where it is needed. This should be advantageous in simulating frontal cloud systems.

There are, of course, disadvantages with the θ coordinate. Obviously it cannot be used if θ is not monotonic in height. Moreover, allowing an arbitrarily small mass for model layers in a θ coordinate creates computational difficulties. Obviously the scheme for the continuity equation for predicting the mass of such layers must be positive definite. Even then, values of the prognostic variables can easily be unrealistic for layers with small mass. In addition, a high vertical resolution concentrated in a frontal zone allows very shallow inertia-gravity waves to be trapped in the zone, which are essentially local inertial oscillations. Because these waves satisfy the dispersion relation at least approximately, they are not computational modes although their generation can be computational. In the model developed by Hsu and Arakawa (1990) and Konor and Arakawa (1997), these difficulties are overcome by introducing a diffusion term designed to be effective only where the vertical resolution is very high.

Another problem with the θ coordinate arises when implementing diabatic effects. With this coordinate, the vertical mass flux crossing a coordinate surface is determined by the thermodynamic equation, $\dot{\theta} = Q/\Pi$. (This by itself is an advantage rather than a disadvantage because the mass flux has a clear physical meaning.) It is then natural that Q is computed at the interfaces of the model layers, where $\dot{\theta}$ is defined. On the other hand, the mass budget of the model is expressed by the continuity equation applied to each model layer. This situation makes formulation of the heat budget of the model, including the latent heat budget, somewhat awkward.

Still another problem with the θ coordinate appears near the lower boundary. In addition to the technical difficulties associated with coordinate surfaces intersecting the lower boundary, there is a problem in determining the values of θ at the lower boundary. In a θ-coordinate model, the values of θ at the lower boundary can be determined by an extrapolation from above and, therefore, their distribution along the boundary depends strongly on the vertical resolution. Although higher vertical resolution is needed as the horizontal resolution increases in any model (Lindzen and Fox-Rabinovitz, 1989), it is especially important to keep this in mind in designing a θ-coordinate model since potential temperature advection along the lower boundary is crucially important for quasi-geostrophic dynamics (e.g., Held *et al.*, 1995) and nongeostrophic surface frontogenesis.

D. HYBRID θ-σ COORDINATES

An approach to allow a sufficiently high resolution for the potential temperature distribution along the lower boundary without a very high vertical resolution is the use of a hybrid θ-σ coordinate, which introduces a σ-coordinate domain near the lower boundary and couples it with a θ-coordinate domain above. Historically, the objective of using this kind of coordinate was to eliminate computational difficulties associated with the intersections of isentropic surfaces with the lower boundary. It seems that we now have a stronger reason not to extend the θ-coordinate domain all the way down to the boundary, as discussed in the last subsection. Recent examples of such models include those described by Zapotocny *et al.* (1994) based on Johnson and Uccellini (1983), Bleck and Benjamin (1993) based on Bleck (1978, 1979), Zhu *et al.* (1992), and Konor and Arakawa (1997).

Suppose that a hybrid vertical coordinate is defined by $\zeta = F(\theta, \sigma)$, where the function $F(\theta, \sigma)$ is chosen to be always monotonic in height. Recall that θ is predicted in a σ-coordinate model to determine the thermal structure, whereas p is predicted in a θ-coordinate model for that purpose. Then, in a hybrid-coordinate model that has a smooth transition between the σ domain and the θ domain, both p and θ must be predicted. If these predictions are done independently, $[\partial F(\theta, \sigma)/\partial t]_\zeta = 0$ generally does not hold and, therefore, predicted surfaces of constant $F(\theta,\sigma)$ may not coincide with surfaces of constant ζ. This indicates that, even though $F(\theta, \sigma)$ is monotonic in height, ζ does not necessarily remain monotonic and crossing of coordinate surfaces may occur in the prediction.

In the model constructed by Konor and Arakawa (1997), $[\partial F(\theta, \sigma)/\partial t]_\zeta = 0$ is guaranteed by the vertical mass flux equation, which reduces to $\dot{\zeta} = Q/\Pi$ when $\zeta = \theta$ and $\dot{\zeta} = \omega$ when $\zeta = p$. They chose an expression for $F(\theta, \sigma)$ that automatically becomes essentially a function of σ near the lower boundary and wherever static stability is approximately zero or negative, and quickly but smoothly becomes essentially θ away from those domains. Figure 8 schematically shows a typical structure of the hybrid coordinate they used. Figure 6c shows the geopotential height and potential temperature fields interpolated to 900 mb for disturbances that

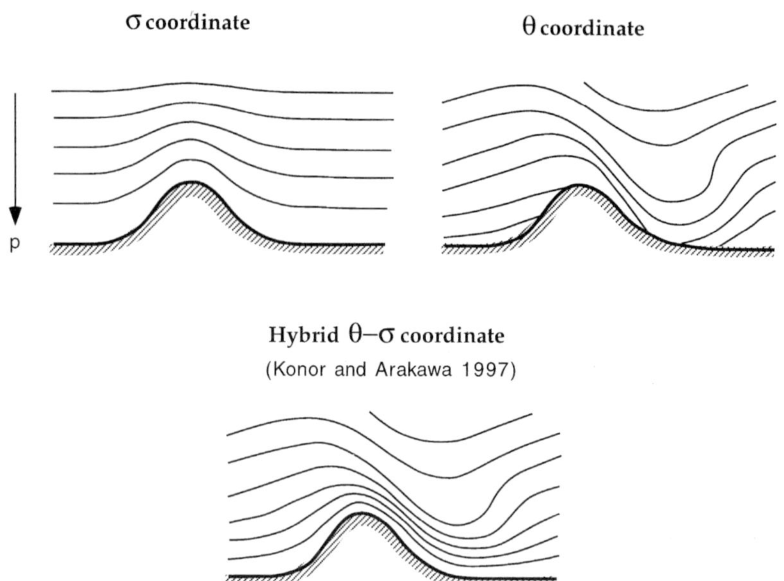

Figure 8 A schematic showing coordinate surfaces in the hybrid θ-σ coordinate used by Konor and Arakawa (1997) and its comparisons with those in the σ and θ coordinate.

developed from initial conditions similar to those used for Figs. 6a and 6b. In this figure, we see no sign of the spurious growth of short waves.

Constant σ surfaces are approximately material surfaces near the lower boundary, even more so than isentropic surfaces are under the existence of surface heating. Thus the inclusion of a σ-coordinate domain near the surface even strengthens advantage 2 mentioned in Section III.C, if the σ domain is shallow enough so that the advantage of the θ coordinate is maintained for the majority of the model domain.

It is not my intention to say that all future GCMs should use a hybrid θ-σ coordinate. I believe, however, that the possibility of developing advanced climate models based on this kind of coordinate should be further explored. I also note the merit of the η coordinate based on step mountains (Mesinger, 1984; Mesinger *et al.*, 1988; Janjic, 1990, 1994), which has been successfully used in regional NWP models. For more details of this subject, see Chapter 13 by Mesinger.

E. UPPER AND LOWER BOUNDARY CONDITIONS

GCMs and NWP models almost always assume that there is no vertical mass flux across the upper boundary. This rigid condition is often used even when the upper boundary is not at the top of the atmosphere. With this condition, however, spurious reflection of upward-propagating waves takes place at the boundary. The same problem exists, as Lindzen *et al.* (1968) pointed out, even with the upper boundary formally placed at the top of the atmosphere because in practice discrete models cannot have enough vertical resolution all the way to infinity. At present, no simple but fully justifiable way of handling the upper boundary is available for general use in numerical models.

The lower boundary condition is a statement that the Earth's surface is a material surface. While there is no doubt about the validity of this condition expressed as $w_s = \mathbf{v}_s \cdot \nabla z_s$, where subscript s denotes the Earth's surface, we generally have $\overline{w}_s \neq \overline{\mathbf{v}}_s \cdot \nabla \overline{z}_s$, where the overbar denotes the area average over the horizontal grid interval. Then the effective height of the surface to be used in the lower boundary condition is not $\overline{z}_s$ but $\tilde{z}_s$ satisfying $\overline{w}_s = \overline{\mathbf{v}}_s \cdot \nabla \tilde{z}_s$. This kind of consideration led Wallace *et al.* (1983) to use an "envelope orography," which is higher than $\overline{z}_s$ depending on the sub-grid-scale variance of z_s. A more general treatment of the kinematical lower boundary condition is needed, together with physical problems associated with irregular terrain (see Section V.D).

F. CHOICE OF HORIZONTAL GRID

Throughout the entire history of the UCLA GCM (with the exception of Generation II), it has been required that the discrete momentum equations be equivalent to the discrete vorticity equation that conserves enstrophy and energy if the flow is frictionless, two dimensional, and nondivergent (see Chapter 1). From the additional point of view of geostrophic adjustment, the C Grid has been chosen for recent generations of the UCLA GCM. The problem of choosing a horizontal grid, however, is not a settled problem since the C Grid also has disadvantages (see Chapter 1).

Figure 9 illustrates an alternate possibility. If we are concerned with only nondivergent motions, the D Grid shown in Fig. 9a, which was used in the Generation I UCLA GCM, is the best. If we are concerned with only pure (i.e., with no Coriolis force effect) gravity waves, the C Grid shown in

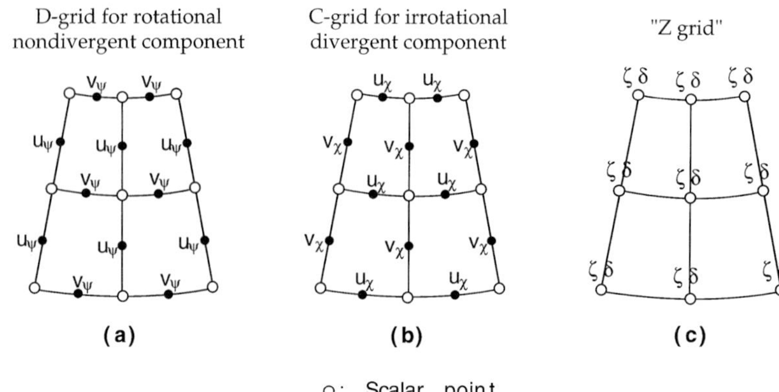

Figure 9   Illustration of the Z Grid by Randall (1995).

Fig. 9b is the best. Here ψ and χ are the streamfunction and velocity potential, respectively, defined at the open-circle points in the figure. Randall (1994) pointed out that if the momentum equations are abandoned in favor of the vorticity/divergence equations, the simple unstaggered grid shown in Fig. 9c, which he called the Z Grid, gives very satisfactory dispersion properties and, therefore, is expected to simulate geostrophic adjustment very well.

Covering the entire globe by a quasi-uniform resolution is also an unsettled problem. Constructing a global model using an icosahedral global grid, originally accomplished by Sadourny *et al.* (1968) and Williamson (1968) and later by Masuda and Ohnishi (1987), has been revived by Heikes and Randall (1995) (see Randall's chapter in this book, Chapter 17).

IV. DISCRETIZATION PROBLEMS: ADVECTION SCHEMES

A. INTRODUCTION

The problem of discretizing the advection equation, or the advection terms in other prognostic equations, is still one of the unsettled problems in numerical modeling of the atmosphere. A number of approaches, methods and techniques have been proposed and used for this problem, with terminology such as "Eulerian," "Lagrangian," "semi-Lagrangian," "advective form," "flux (divergence) form," "flux correction," "constancy," "conservation," "dispersion," "dissipation," "overshooting/ under-shooting," "positive-definite," "monotonic," and "shape-preserving." It is beyond the scope of this chapter to review these subjects even briefly. Instead, toward the end of the section, I present the major concern I have with this problem.

In any case, discretization of the advection equation is based on one of the following three forms:

1. Eulerian advective form:

$$\frac{\partial q}{\partial t} + \mathbf{V} \cdot \nabla q = 0, \tag{12}$$

   for which "constancy" is automatic
2. Eulerian flux (divergence) form:

$$\frac{\partial(mq)}{\partial t} + \nabla \cdot (m\mathbf{V}q) = 0, \tag{13}$$

   for which "conservation" is automatic
3. Lagrangian form:

$$\frac{Dq}{Dt} = 0, \tag{14}$$

   for which "stability" is (almost) automatic.

Here q is a quantity per unit mass to be advected, $\mathbf{V}$ is the advecting velocity, m is the pseudo-density (i.e., the mass per unit horizontal area per unit increment of the vertical coordinate) predicted by the continuity equation, and D/Dt is the material time derivative. "Constancy" means that if initially $q = $ constant, it remains so in time. "Conservation" means that $\overline{mq}$ does not change in time, where the overbar denotes the area-average over a closed domain. [Note that "conservation" here is that of the first moment, $\overline{mq}$, not that of the second moment $\overline{mq^2}$ as in (potential) enstrophy conservation or energy conservation (e.g., Arakawa and Lamb, 1981).] Finally, "stability" here means the boundedness of predicted q.

"Constancy" is perhaps one of the minimum requirements for any advection scheme. It is not automatically satisfied, however, in a scheme based on the flux (divergence) form, Eq. (13), which is a combination of the advection equation, Eq. (12), and the continuity equation, unless the scheme becomes equivalent to the discrete continuity equation used in the model when q is identically 1. Any reasonable Eulerian scheme should be able to be rewritten from the advective form to the flux (divergence) form, or vice versa, although actual computations are done using one of the two. Lagrangian (or semi-Lagrangian) schemes satisfy "constancy" but usually not "conservation." The time step of any explicit Eulerian schemes are restricted by the Courant-Friedrich-Levy (CFL) stability condition while such a restriction does not exist in Lagrangian (or semi-Lagrangian) schemes.

B. Computational Mode in Discrete Advection Equations

Most of the difficulties in discretizing the advection equation are associated with multidimensionality, nonuniformity (of the current), and nonlinearity. Problems, however, can exist even without these features. The existence of a computational mode is an example. As in Section III.B, a "computational mode" refers to a mode in the solutions of a finite-difference equation that has no counterpart in the solutions of the original differential equation. Because there is no corresponding true solution to compare with, a computational mode cannot be made more "accurate" by increasing the resolution or using a higher order scheme.

The existence of a computational mode in time with the leapfrog time differencing is well known. When the frequency is given, however, the relevant computational mode is in space, which commonly exists in most finite-difference schemes for the advection equation. The mode is especially visible in solutions with space-centered nondissipative schemes. To see the existence of a computational mode following Matsuno (1966), let us consider Eq. (12) in its simplest case of one-dimensional advection equation with a constant current U given by

$$\frac{\partial q}{\partial t} + U\frac{\partial q}{\partial x} = 0. \tag{15}$$

When the space derivative is replaced by the usual second-order centered finite difference, the relation between ν and $k\Delta x$ becomes as shown by the heavy half-sine curve in Fig. 10. Here ν is the frequency, k is the wave number, and Δx is the grid size. Unlike the continuous case, there are two wave numbers for a given frequency. As the grid size approaches zero for a given frequency, only the smaller wave number approaches the true wave number. The other wave number then represents a spurious mode, the computational mode in space. The group velocity associated with this mode is negative when $U > 0$, i.e., against the current. When the order of

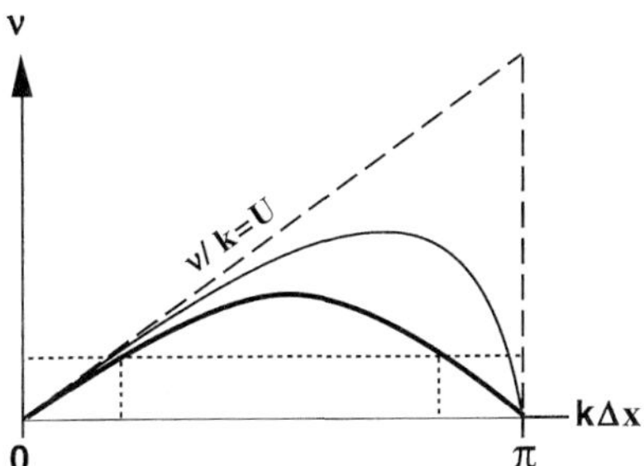

Figure 10 Dispersion relations for solutions of the advection equation, Eq. (15), with centered space finite differencing. See text for further explanation.

accuracy is raised to 4, for example, the relation becomes as shown by the thin solid line in Fig. 10, with a faster group velocity associated with the computational mode. When the order of accuracy is further increased to infinity, corresponding to the use of a spectral model, the relation becomes as shown by the dashed lines in Fig. 10. The computational mode still exists, now with the wave number equal to $\pi/\Delta x$ (i.e., wavelength equal to $2\Delta x$), with an infinite group velocity. This is a well-known problem with a spectral method applied to the Eulerian advection equation.

Note that, in the simple cases presented above with a uniform current and centered space finite differences, conservation of q^2 is automatic when time is continuous. Thus controlling a computational mode is a separate problem from the problem of conserving the second moment, such as the problem of enstrophy conservation in the nondivergent vorticity equation. In practice, the computational mode can only be handled by sacrificing the exact conservation of the second moment (while there is no justification for not conserving the first moment.) In the case of the nondivergent vorticity equation, however, enstrophy conservation helps the situation by preventing (or reducing) the spurious energy cascade to small scales, which may generate the computational mode. Also, the method of conserving the second moment for a nonlinear system can be applied in a modified way to guarantee that the deviation from conservation is dissipation, rather than generation, of the second moment (Takacs, 1985; Arakawa and Hsu, 1990; Hsu and Arakawa, 1990).

C. SEMI-LAGRANGIAN SCHEMES

Following the pioneering work of Robert (e.g., Robert, 1981, 1982), semi-Lagrangian schemes are becoming increasingly popular in numerical modeling of the atmosphere. When the semi-implicit method is used for the terms responsible for the existence of fast-moving waves in a Eulerian scheme, the advective current restricts the time interval satisfying the CFL condition. Removing this restriction is the major motivation for the Lagrangian approach. Semi-Lagrangian schemes consider trajectories whose arrival points coincide with grid points, which are fixed in space, while values at the departure points are determined by interpolations from the grid points. Staniforth and Côté (1991) presented an excellent review of semi-Lagrangian schemes for atmospheric models. These schemes combined with the semi-implicit method are used in an increasing number of NWP models and GCMs (e.g., Tanguay *et al.*, 1989; Bates *et al.*, 1993; Williamson and Olson 1994, 1998a,b; Ritchie *et al.*, 1995; Moorthi *et al.*, 1995; Chen *et al.*, 1997).

There seems to be, however, some confusion about the merit of semi-Lagrangian schemes in the literature. Leslie and Dietachmayer (1997) pointed out that:

> Over the years there have been a number of studies comparing the relative merits of semi-Lagrangian and Eulerian schemes. These studies, which continue to appear in the literature up to the present, almost invariably conclude that semi-Lagrangian schemes are superior in accuracy, and produce less noise, than Eulerian schemes.... Such conclusions are not justified because they have compared semi-Lagrangian and Eulerian schemes of different orders of accuracy. Typically, the semi-Lagrangian schemes tested have employed cubic spatial interpolation and therefore are third order in space, whereas the Eulerian schemes have usually been second order (and sometimes fourth order) in space.

I might add that a number of Eulerian schemes share the same order of accuracy but perform quite differently, especially in multi-dimensional nonlinear problems. Thus deriving a general conclusion on the difference through a casual comparison of solutions with a particular semi-Lagrangian scheme and those with a particular Eulerian scheme cannot be justified.

Indeed, for the one-dimensional advection with a uniform current given by Eq. (15), a family of semi-Lagrangian schemes is identical to a family of Eulerian schemes when the Courant number is smaller than one (e.g., Staniforth and Côté, 1991). Figure 11 shows the solutions of Eq. (15) with a family of semi-Lagrangian schemes and a family of Lax–Wendroff type Eulerian schemes (see Takacs, 1985) for $\mu = 0.5$ using an initial condition in which q is nonzero only at a single grid point. Here μ is the Courant number $U\Delta t/\Delta x$. [Recall that the solution of Eq. (15) for an arbitrary initial condition is a superposition of the solutions for initial conditions concentrated at single grid points such as those considered here.] Solutions with each pair of semi-Lagrangian and Eulerian schemes are identical.

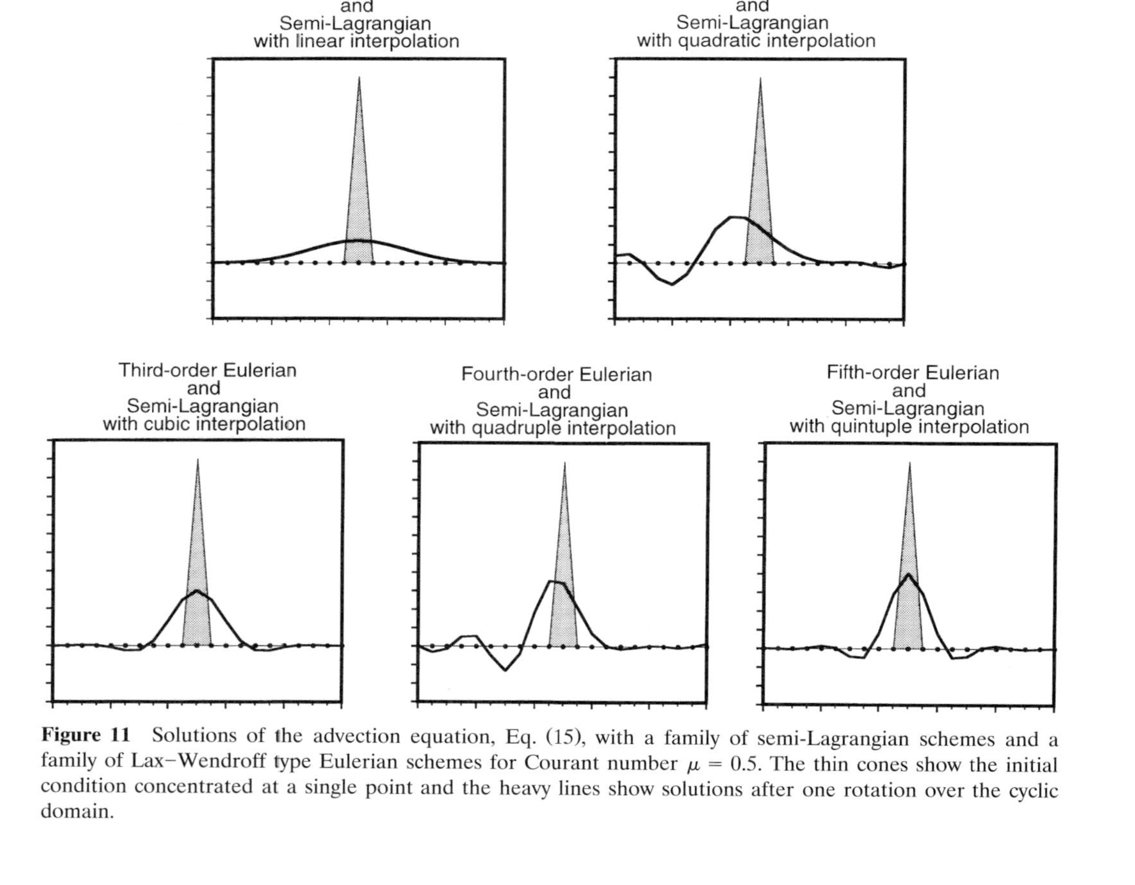

Figure 11   Solutions of the advection equation, Eq. (15), with a family of semi-Lagrangian schemes and a family of Lax−Wendroff type Eulerian schemes for Courant number $\mu = 0.5$. The thin cones show the initial condition concentrated at a single point and the heavy lines show solutions after one rotation over the cyclic domain.

From this figure, we can clearly see the point raised by Leslie and Dietachmayer (1997): Comparing a semi-Lagrangian scheme with an Eulerian scheme of different order of accuracy is almost meaningless.

When the Courant number is not small, the time interval must be reduced for Eulerian schemes to satisfy the CFL condition; this is not necessary for semi-Lagrangian schemes because they automatically shift the grid points used at the departure level further upstream as the Courant number increases. This is the only true difference between semi-Lagrangian schemes and Eulerian schemes. This does not mean, however, the solution produced using a semi-Lagrangian scheme with a large Courant number is accurate. Figure 12 shows solutions of Eq. (15) on a stretched grid with a semi-Lagrangian scheme using a cubic-spline interpolation, again for an initial condition concentrated at a single point. The motivation for considering a stretched grid is to mimic the situation in which the mass of model layers varies in height, as is the case in almost all multilevel atmospheric models. As shown in Figs. 12a and 12c, the single time-step solutions for $t = T$ are very sensitive to the gross Courant number UT when it is larger than one and they are very different from the multiple time-step solutions shown in Figs. 12b and 12d for the same T but with a smaller time interval, $\Delta t = T/10$. We also see that conservation (of the first moment) is severely violated in Figs. 12a and 12c, while it is not so in 12b and 12d.

This kind of problem in classical semi-Lagrangian schemes can be alleviated in a noninterpolating semi-Lagrangian scheme (Ritchie, 1986; see also Smolarkiewicz and Rasch, 1991; Smolarkiewicz and Pudykiewicz, 1992) as well as in the "flux form semi-Lagrangian scheme" (Lin and Rood, 1996), all of which can be interpreted as hybrid semi-Lagrangian and Eulerian schemes. For further discussions on semi-Lagrangian schemes, see Mesinger (1997).

D. An Inherent Difficulty in Discretizing the Advection Equation

In my mind, an inherent difficulty in a discrete advection equation is in defining what we want in the solutions. To illustrate this point, let us consider solutions similar to those in Fig. 12. Figure 13 shows three hypothetical solutions: a solution with a perfect Lagrangian accuracy both in magnitude and phase (Fig. 13a), a solution satisfying conservation with

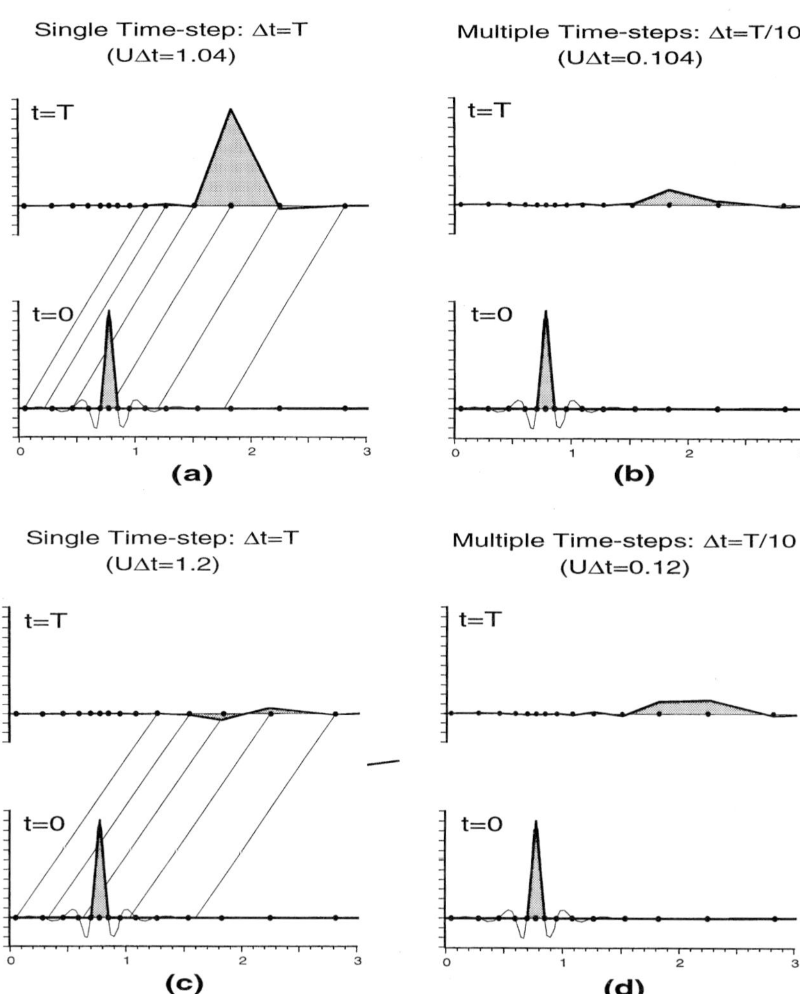

Figure 12 Solutions of the advection equation, Eq. (15), on a stretched grid with a semi-Lagrangian scheme using a cubic-spline interpolation for an initial condition concentrated at a single point. Upper and lower panels are for slightly different gross Courant numbers based on the unit space increment. Left and right panels show single and multiple time-step solutions for $t = T$, respectively.

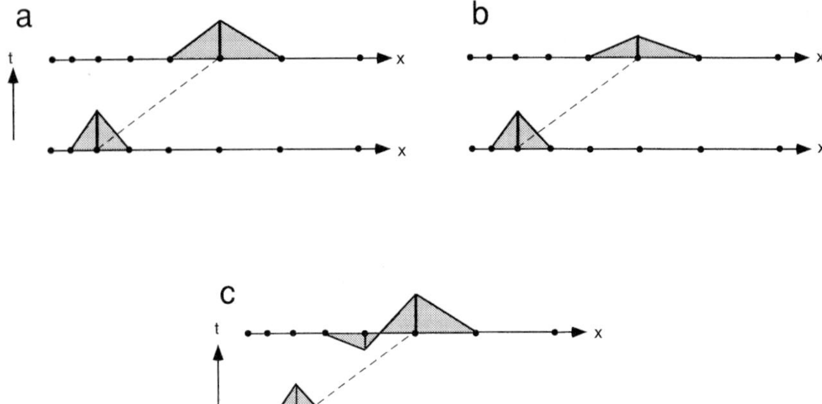

Figure 13 Three hypothetical solutions of the advection equation, Eq. (15), on a stretched grid: (a) a solution with a perfect Lagrangian accuracy both in magnitude and phase, (b) a solution satisfying conservation with no dispersion error, and (c) a solution satisfying conservation with a perfect Lagrangian accuracy for the major peak but with dispersion error.

no dispersion error (Fig. 13b), and a solution satisfying conservation with a perfect Lagrangian accuracy for the major peak but with dispersion error (Fig. 13c). All of these solutions are hypothetical and cannot be obtained in practice. Still they illustrate the problem of defining what we ultimately want in the solutions in view of their impact on the performance of the entire model.

If the quantity advected is the specific humidity, the solution of Fig. 13a does not conserve the total water content, while the solution of Fig. 13b gives excessive drying and, therefore, less cloudiness. Both of these cases are simple translations and, therefore, they are "shape preserving." The solution of Fig. 13c seems to be the optimum, but there is a question of whether we can tolerate such a large distortion of the shape. This kind of consideration makes me feel that the use of a grid fixed in space has inherent difficulties for both Eulerian and semi-Lagrangian schemes. This is one of the major reasons why I am in favor of the quasi-Lagrangian vertical coordinate, such as an isentropic coordinate, at least as one of the promising options.

V. PARAMETERIZATIONS OF PBL AND STRATIFORM CLOUD PROCESSES AND REPRESENTATION OF THE EFFECTS OF SURFACE IRREGULARITY

A. Various Approaches in PBL Parameterization

Around 1970, I had the feeling that parameterization of PBL processes was one of the weakest aspects of the GCMs that existed at that time. When I shared that feeling with Jim Deardorff, he looked a little surprised and said, "Then you should speak up!" He of course knew the status of the PBL parameterizations in those GCMs. At that time, however, general circulation modeling was considered the job of large-scale dynamicists, whose interest in the PBL usually did not go much beyond the classical Ekman layer with a constant viscosity coefficient. In that sense, large-scale dynamicists and boundary-layer meteorologists lived in different worlds.

Mintz shared the same feeling with me and invited Deardorff to UCLA for a 6-month period to think about what a GCM can do with the PBL. An outcome of this visit was the paper by Deardorff (1972), which had a strong impact on later development of the UCLA GCM (see Section IX.C of Chapter 1). Although I didn't necessarily speak up and I didn't work on PBL problems myself, I encouraged some of my students to perform PBL-related research, including David Randall (1976), Chin-Hoh Moeng (1979), and Steven Krueger (1985).

Things have changed significantly since 1970, both in general circulation modeling and boundary-layer meteorology. Most importantly, general circulation modelers and boundary-layer meteorologists now live in the same world. I still feel that, in view of their crucial roles in the climate system, the formulation of PBL processes remains among the weakest aspects of the existing GCMs. This is especially true for the stratocumulus-topped PBL.

It is not my intention here to discuss PBL modeling in any depth or breadth. For such discussions, see chapters by Chin-Hoh Moeng (Chapter 19) and Steve Krueger (Chapter 20) in this book. I would like to emphasize some points, however, with which I am especially concerned.

Browning (1974) wrote to the directors of centers concerned with global modeling to ascertain what they regard as the priority issues on which to concentrate when parameterizing the large-scale effects of clouds and cloud-related processes. He reports, "Many respondents referred to inadequacies in the present schemes for representing cloud-cover, mainly, but not entirely, in the context of boundary-layer clouds." In my point of view, the problem with boundary-layer clouds is more than technical. It is in the basic approach traditionally followed in most GCMs, in which the formation of "boundary-layer" clouds is treated as a process passive to *dry* turbulent processes without considering cloud effects on turbulence.

One of the main purposes of a PBL parameterization is to calculate turbulent fluxes of momentum, heat, and water vapor at the surface. Unless we are working with a single-layer model of the atmosphere, however, the prognostic equations of GCMs need the divergence of those fluxes. Thus, a PBL parameterization must provide vertical profiles of turbulent fluxes, including their vertical extent: the PBL depth.

Moreover, surface fluxes themselves cannot be determined without considering the processes throughout the entire PBL. Calculation of surface fluxes is usually done with bulk aerodynamical formulas such as Eq. (1). For the potential temperature, θ, the formula may be written as

$$(F_\theta)_S = \rho C_H |\mathbf{v}_B| (\theta_S - \theta_B). \tag{16}$$

There are a number of ambiguities, however, in defining $|\mathbf{v}_B|$, θ_S, and θ_B. Thus Eq. (16) "can be applied in a model only after several somewhat arbitrary choices have been made" and the formula is "really nothing more than the definition of C_H" (Zhang *et al.*, 1996). Once C_H is defined by Eq. (16), we can proceed to empirically determine its magnitude with the aid of similarity theory (e.g., Businger *et al.*, 1971). Equation (16) can then be used in prognostic models to obtain the flux $(F_\theta)_S$.

This last step, however, is only computational and it does not necessarily reflect the cause-and-effect relationship. If the height for θ_B is chosen within the surface layer, for which the storage of θ is negligible for time scales longer than the order of, say, 10 min, an argument similar to that presented in Section I can be applied to this situation; i.e., the surface flux of θ is approximately determined by counteracting effects on θ_B. The counteracting effect in this case is the flux of θ through the top of the surface layer, $(F_\theta)_{S+}$. Thus, for those time scales, it is more appropriate to view Eq. (16) as an equation that diagnoses θ_B from $(F_\theta)_S$ determined by $(F_\theta)_{S+}$.

The common way of formulating $(F_\theta)_{S+}$ and the profile of F_θ above the surface layer is through the K closure. For F_θ, for example, the closure assumes

$$F_\theta = -\rho K \frac{\partial \theta}{\partial z}, \tag{17}$$

with a specified diffusion coefficient, K, typically as a function of the local Richardson number (e.g., Louis, 1979). As in the case of Eq. (16), however, Eq. (17) is really nothing more than the definition of K; but the situation is much worse here since F_θ in the convective PBL does not depend on the local gradient of θ except near the surface.

Higher order single-point closures, which generalize the K closure, are attractive in principle and can be very useful in basic research outside the GCMs (see Chapter 20 in this book). In large-scale models, the level 2.5 version of the Mellor–Yamada model (Mellor and Yamada, 1974, 1982) is particularly popular (e.g., Miyakoda and Sitrus, 1977, 1990; Sitrus and Miyakoda, 1990). Without a very high vertical resolution, however, it is difficult to justify the use of a single-point closure especially when the PBL has a stratocumulus sublayer.

Another approach in generalizing the K closure is the "nonlocal" K schemes or the K-profile parameterization (Troen and Mahrt, 1986; Holtslag and Boville, 1993; Vogelezang and Holtslag, 1996; Large *et al.*, 1994). This is a bulk approach that includes a formulation of the nonlocal effect of the surface heat flux on the transport using a diagnosed height of the PBL top. It remains to be seen, however, how this approach can be generalized to a stratocumulus-topped PBL.

Finally, there are bulk models based on a well-mixed single-layer PBL. This approach was proposed by Deardorff (1972), in which surface fluxes are determined by equations formally similar to the bulk aerodynamical formulas but with the mean values over the entire PBL replacing the surface-layer values. Crucial in this formulation is the PBL depth, which is predicted from the mass budget equation for the PBL with parameterizations of the mass source due to entrainment and the mass sink due to cumulus mass flux (see Section IX.C of Chapter 1).

Although Deardorff (1972) did not consider cloud effects on turbulence, the model's framework allows inclusion of a Lilly (1968)-type mixed-layer model for stratocumulus clouds when the predicted PBL top is higher than the condensation level. When such a stratocumulus sublayer exists, the radiative cooling concentrated near the cloud top drives in-cloud moist turbulence, making the character of the entire PBL dramatically different from that of a cloud-free PBL. The profiles of F_θ (or F_S, where s is the dry static energy) and F_q are particularly different with and without clouds, while the moist static energy and the mixing ratio of total water can be even more well mixed with clouds. Thus, the bulk models allow us to treat cloud-free and cloud-topped PBLs within the same framework. It is also possible to incorporate the effect of cloud top entrainment instability (CTEI; Lilly, 1968; Randall, 1980; Deardorff, 1980; see also Moeng *et al.*, 1995) on stratocumulus cloud layers. Randall (1976) and Suarez *et al.*

(1983) implemented such a model in the UCLA GCM (see Section IX.D of Chapter 1). Recent revisions of the model have produced extremely encouraging simulations of the global distributions of stratocumulus cloud incidence for all seasons (Li *et al.*, 1999). Their results indicate that the CTEI does tend to decrease the cloud incidence unless counteracted by other effects.

Although the assumption of a well-mixed PBL might be considered an oversimplification, I consider it a useful idealization at this stage, allowing us to concentrate on the processes crucially important for a stratocumulus-topped PBL. What we need in the future seems to be a generalization of the assumption, rather than abandoning it. The paper by Randall *et al.* (1992) suggests a way to relax this idealization by considering three internally consistent layers: a ventilation layer near the surface, an entrainment layer near the top, and a convective layer in between. A simple second-order closure, including prediction of the bulk turbulence kinetic energy (TKE), is used in their model.

Prediction of the bulk TKE makes the formulation of surface fluxes more reasonable than the use of the traditional bulk aerodynamic formula. Stull (1994) suggested replacing the wind speed in such a formula by a velocity scale that is a measure of the intensity of the turbulence. Zhang *et al.* (1996) used the square root of the vertically averaged kinetic energy over the PBL depth for this purpose and replaced θ_B in Eq. (1) by the vertical mean of θ over the entire PBL depth. They diagnostically tested the new formulation against observations with very encouraging results.

B. IMPLEMENTATION OF PBL PROCESSES IN A VERTICALLY DISCRETE MODEL

We now face the problem of implementing a variable-depth PBL model in a vertically discrete GCM. Figures 14a and 14c correspond to the structures in Generation III and Generations IV and V of the UCLA GCM, respectively (see Section IX.D and Fig. 12 of Chapter 1). Figure 14c illustrates the modified σ-coordinate used by Suarez *et al.* (1983), in which the PBL top is a coordinate surface. As Arakawa pointed out in Chapter 1, the major advantage of using such a coordinate is that the PBL properties are expected to be "similar" along a coordinate surface (see the solid and dashed curves in Figs. 14c and 14d. Thus this coordinate uses vertical self-similarity to increase the homogeneity along a coordinate surface. It especially makes the formulation of processes concentrated near the PBL

Standard σ-coordinate

(a) Coarse Resolution

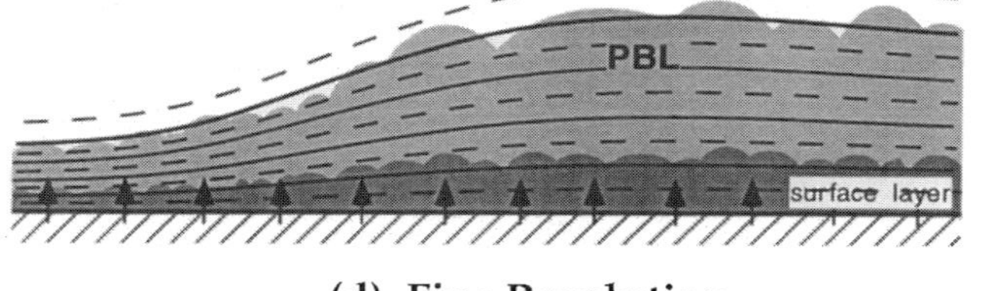

(b) Fine Resolution

Modified σ-coordinate

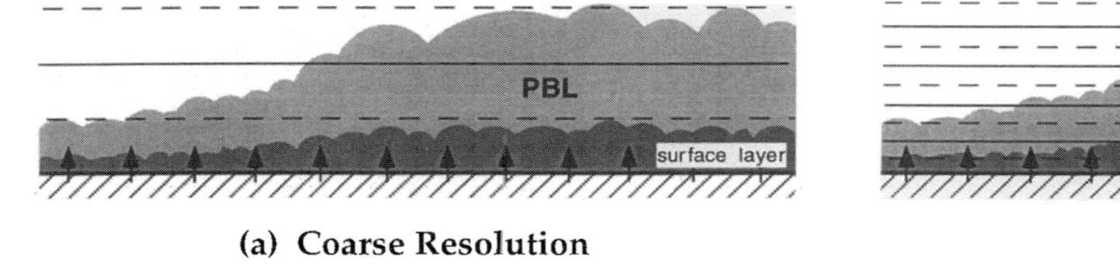

(c) Coarse Resolution

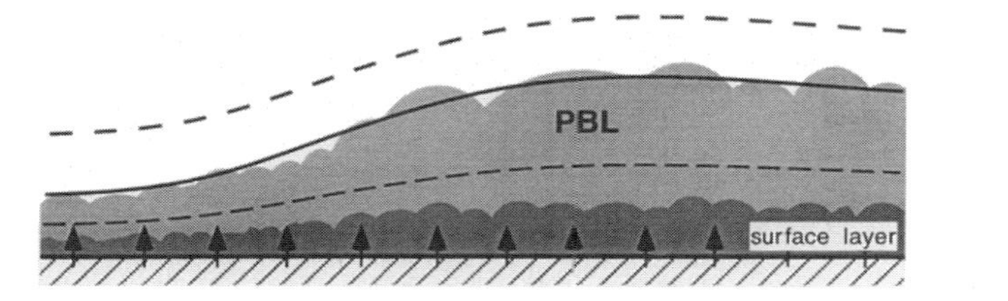

(d) Fine Resolution

Figure 14 Structures of low (left panels) and high (right panels) vertical resolution models with the standard σ coordinate (upper panels) and the modified σ coordinate used by Suarez *et al.* (1983), in which the PBL top is a coordinate surface (lower panels).

top much more tractable. Moreover, the PBL top becomes a material surface if the entrainment rate is zero. All of these, especially the last, are formally analogous to the θ coordinate, in which coordinate surfaces are material surfaces if heating is zero. Even if heating is nonzero, the mass flux across a θ-coordinate surface has a clear physical meaning, i.e., heating. In the modified σ coordinate, the mass flux through the coordinate surface representing the PBL top also has a clear physical meaning, i.e., entrainment minus cumulus mass flux. Figures 14b and 14d illustrate the structures of high-resolution models with the standard σ coordinate and the modified σ coordinate.

When the PBL does not have a well-defined top, such as the "daytime" PBL in evening, the definitions of the modified σ coordinate become ambiguous. In such a situation, the coordinate can be viewed as an arbitrarily chosen coordinate.

C. Unsolved Problems in Modeling Stratiform Clouds

Besides what I have already mentioned regarding the PBL stratocumulus clouds, a number of problems still exist when modeling stratiform clouds. It is not my intention here to review such problems, but rather to point out two problems that are crucially important for physically based prediction of cloud amount: the transition of PBL-associated clouds from stratiform to cumuliform and the decay and breakup processes of upper level stratiform clouds.

The cloud regime transition of PBL-associated clouds was one of my old concerns (Arakawa, 1975). Recently, how this transition takes place has become clearer through observations (e.g., Bretherton, 1992) and numerical simulation studies (e.g, Krueger *et al.*, 1995a,b), and the classical picture shown in Fig. 13 of Chapter 1 in this book needs a modification. The picture should include a transition zone that has cumuliform clouds within the PBL topped by stratiform clouds near the PBL top. Formulation of the transition zone with this multiple-layer vertical structure in a GCM will be quite a challenge.

The first step for GCMs in predicting cloudiness is to explicitly predict the three-dimensional distributions of the liquid and ice phases of water, which may be generated by large-scale ascents or detrainment from cumulus clouds. Formulating the decay processes of those clouds is then crucial for physically based prediction of cloudiness.

What matters in the decay process of these clouds, however, is more than cloud microphysics. Due to the existence of radiative cooling near cloud top and radiative warming near cloud base, we expect generation of in-cloud convection, which tends to make the cloud layer well mixed, as is found by Köhler *et al.* (1997) for ice clouds using a cloud-resolving model. In their results, the convection transports cloud ice upward and thus clouds tend to expand upward. The convection also enhances snowfall, which tends to make the cloud base higher. Sublimation from snowfall may form new clouds below, which tend to rise and merge the mother clouds above. In this way, ice clouds can be more persistent than expected without the radiation-driven in-cloud convection.

Formulating processes responsible for the breakup of a cloud layer is also important because the effect of a given mass of cloud water on radiation depends on the shape of clouds: primarily horizontal area for cloudiness and primarily vertical extent for optical depth. Eventually, as in the case of PBL stratiform clouds, implementing all of these processes in a vertically discrete model will require special considerations.

D. PROCESSES ASSOCIATED WITH IRREGULAR SURFACE

Here I discuss only orographic irregularity, although in practice irregularities in vegetation and soil properties should also be considered. Corresponding to the turbulence scale, convection scale, and mesoscale of atmospheric motions (see Fig. 2 of Chapter 1), I separate the irregularity into the same three scales, which can influence the atmosphere in the following ways:

1. *Turbulence scale.* This scale of the irregularity is considered in most GCMs through the dependency of the bulk aerodynamical transfer coefficient on the roughness length of the surface. Ambiguities exist, however, as to what the length really means especially for sensible heat transfer and evaporation.

2. *Convection scale.* To my knowledge, the effect of orographic irregularity in this spectral range is completely ignored in all GCMs. The irregularity generates horizontal inhomogeneity in the PBL, which may force or trigger either dry or moist convection. The current parameterization schemes for convection, however, are not constructed to recognize such inhomogeneity in the PBL (see Section VI.B, objective 5).

3. *Mesoscale.* Most large-scale models now include a parameterization of stationary internal gravity waves generated by sub-grid-scale orography in this spectrum range following Palmer *et al.* (1986), McFarlane (1987), and Lindzen (1988). For a review of the existing schemes and development of an improved scheme, see Kim and Arakawa (1995). These formulations, however, ignore any effects of condensation processes on those waves.

Some models include an "envelope orography" formulation (see Section III.E), which represents the blocking effect on flow near the surface by irregular orography. Presumably this effect can coexist with the gravity wave effect, because waves involved in the blocking effect must be highly trapped near the surface, indicating that they represent a different range of the wave spectrum.

Irregular orography has the obvious geometrical effect on the fractional cloudiness of stratocumulus clouds. In addition, although completely ignored in the existing GCMs, irregular orography can have an important thermal effect due to inhomogeneous short-wave absorption during the daytime and inhomogeneous long-wave cooling during the nighttime, with or without stratiform clouds. There is an indication that this effect is crucial for simulating the cloud regimes over irregular mountains in winter (Li *et al.*, 1999).

VI. CUMULUS PARAMETERIZATION

A. INTRODUCTION

Cumulus parameterization is the problem of formulating the collective effects of cumulus convection in terms of large-scale variables. In Fig. 15, the upper half of the loop represents the effects of large-scale processes on cumulus-convective processes, while the lower half represents those of cumulus-convective processes on large-scale processes. For the loop to be closed in a model that does not resolve cumulus-convective processes, it

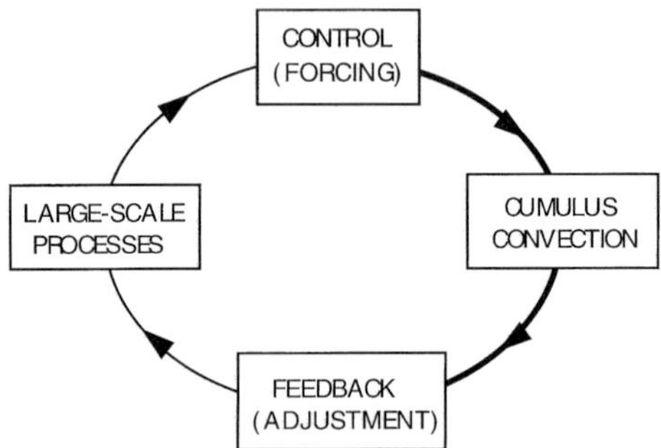

Figure 15   A loop representing the effects of large-scale processes on cumulus-convective processes (upper half) and those of cumulus-convective processes on large-scale processes (the lower half). The right half shown by the heavy segments represents cumulus parameterization.

must include the right half shown by the heavy segments, a formulation of which is precisely the problem of cumulus parameterization. We refer to the upper half of the loop as "control" or "large-scale forcing" and the lower half as "feedback" or "adjustment," although there is no need to identify which is the first and which is the second. These terminologies are convenient only when we are concentrating on the parameterization problem represented by the right half of the loop.

Because cumulus parameterization is an attempt to formulate the collective effect of cumulus clouds without predicting individual clouds, it is a closure problem, in which we seek a limited number of equations that govern the statistics of a system with huge dimensions. The core of the parameterization problem is, therefore, in the choice of appropriate closure assumptions.

A question then arises: To what extent can the collective effects of cumulus convection be formulated in terms of large-scale variables? This is the question of the "parameterizability" of cumulus convection. It is the most fundamental question in the cumulus parameterization problem because the logical structure of a cumulus parameterization should be constructed based on one's understanding of cumulus parameterizability.

Recently, the concept of quasi-equilibrium seems to be more widely accepted as the principal closure than before (e.g., Emanuel *et al.*, 1994; Randall *et al.*, 1997; Neelin, 1997; see also Emanuel's chapter in this book, Chapter 8). At the dawn of the new phase in the history of numerical modeling of the atmosphere, we should ask ourselves the parameterizability question more seriously than we have in the past to plan for the future directions of our approach to this problem. Parameterizability, however, should depend on the objective of the parameterization, which I discuss in the next subsection.

B. The Objectives of Cumulus Parameterization

The objectives of cumulus parameterization keep expanding as the scope of GCMs does. I classify the objectives into two categories: classical objectives and nonclassical or future objectives. The quantities to be determined by cumulus parameterization are listed, roughly in order of difficulty in my judgment.

Objective 1: Vertically integrated cumulus heating and drying (which are approximately equal to surface precipitation)

These quantities measure the overall intensity of cumulus activity. The question of to what extent the overall intensity can be determined from large-scale conditions still represents the core of the parameterizability problem.

> *Objective 2:* Vertical distributions of cumulus heating (and cooling) and drying (and moistening)

These are the quantities GCMs demand a cumulus parameterization to provide at each time step. This objective, which includes objective 1, can be considered as the classical objective of cumulus parameterization. With the pressure coordinate, the large-scale potential temperature and water vapor budget equations can be written as

$$c_p\left[\frac{\partial \overline{T}}{\partial t} + \left(\frac{p}{p_0}\right)^{R/c_p}\left(\overline{\mathbf{v}}\cdot\nabla + \overline{\omega}\frac{\partial}{\partial p}\right)\overline{\theta}\right] = L\underline{C_1} + c_p B_1 + Q_R(\equiv Q_1), \quad (18)$$

$$L\left[\frac{\partial \overline{q}}{\partial t} + \left(\overline{\mathbf{v}}\cdot\nabla + \overline{\omega}\frac{\partial}{\partial p}\right)\overline{q}\right] = -L\underline{C_2} + LB_2(\equiv -Q_2), \quad (19)$$

where an overbar denotes the average over a large-scale horizontal area, Q_1 and Q_2 are the *apparent heat source* and *apparent moisture sink* (Yanai *et al.*, 1973), respectively, C denotes the effects of (net) condensation and associated transports, B denotes the effects of boundary-layer turbulent flux, subscripts 1 and 2 denote the effects on the time changes of temperature and water vapor mixing ratio, respectively, and Q_R is the radiation heating. All other symbols are standard. Let us temporarily assume that B_1, B_2, and Q_R have already been determined (see objectives 5 and 6 below). Then, for a given large-scale state, the four underlined terms in Eqs. (18) and (19), i.e., C_1, C_2, and the two time derivative terms, are unknowns.

> *Objective 3:* Deviation from the quasi-equilibrium due to relaxed adjustment

Many cumulus parameterization schemes, including that of Arakawa and Schubert (1974, AS hereafter), are implemented as adjustment schemes. Suppose that quasi-equilibrium states, to which adjustment is made, are explicitly defined in a scheme. The scheme still has to decide whether the adjustment will be made fully at each time step, or only partially so that the adjustment is relaxed as in the Betts–Miller scheme (Betts, 1986, 1996; Betts and Miller, 1986, 1993).

It has been pointed out that the use of a finite adjustment time scale has a dynamical impact (Emanuel, 1993a; Neelin and Yu, 1994; Yu and Neelin, 1994; Emanuel *et al.*, 1994). Here I show some simple evidence that the choice of adjustment time scale matters in other ways also. In Fig. 16, the results of partial adjustments (right figures) are compared with those of full adjustments (left figures), for a steady large-scale forcing starting from $t = 0$ (upper figures) and an impulse-type large-scale forcing concentrated at the time step immediately after $t = 0$ (lower figures). The ordinate is the quantity to be adjusted toward zero, such as the excess cloud work function over its critical value, ΔA. The tick marks on the abscissa show the time levels at which the adjustment is performed. The heavy lines show the change of ΔA due to the large-scale forcing and the dotted lines show the adjustment, which is assumed to be $-\Delta A$ in the full adjustment and $-\Delta A/2$ in the partial adjustment. With steady large-scale forcing, there is practically no difference in the actual amount of adjustment between the full and partial adjustments after an initial transition period. This indicates that the choice of adjustment time scale is not important if the adjustment time scale is sufficiently smaller than the time scale of the large-scale forcing (Arakawa, 1969; see Eq. (25) of Chapter 1). With the impulse-like large-scale forcing, there is also no difference in the *total* amount of the adjustment (the sum of the lengths of the dotted lines). This again indicates that the choice of adjustment time scale is not important in determining the total effect of cumulus activity per episode of large-scale forcing. These conclusions are valid for the classical objectives,

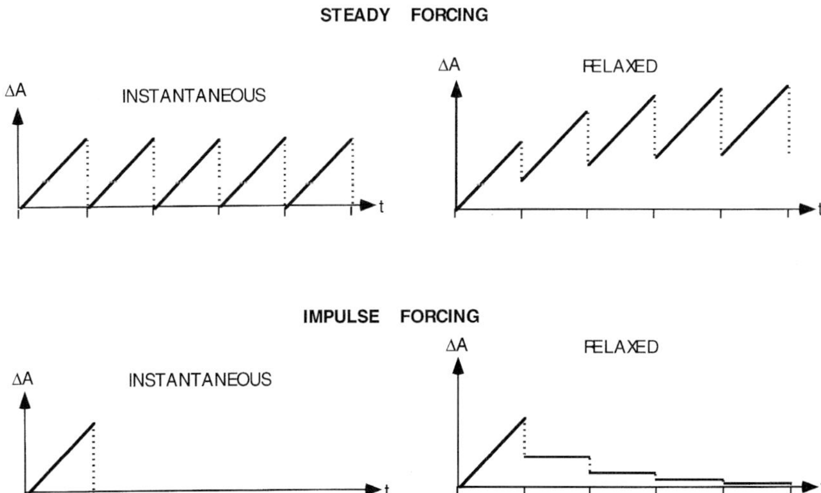

Figure 16 Illustration of an impact of the relaxed adjustment, in which only one-half of the required amount of adjustment is performed at each time. See text for further explanation.

which are determination of the quantities that depend only on the (total) amount of adjustment, such as surface precipitation and cumulus heating and drying averaged over a few hours or longer.

The sensitivity to the adjustment time scale can be quite different for nonclassical objectives. In the case of steady large-scale forcing, for example, the mean value of ΔA is larger with partial adjustment than with full adjustment. When A is the cloud work function of Arakawa and Schubert (1974) or the convective available potential energy (CAPE), a larger value of ΔA means a steeper temperature lapse rate and/or a higher humidity.

In the case of impulse-like large-scale forcing, on the other hand, the main difference is in the duration of the adjustment, or the duration of cumulus activity, which is longer with the relaxed adjustment. We then expect in GCM simulations that time-averaged cloudiness can be significantly increased with relaxed adjustment. This has been confirmed through actual GCM simulations (Cheng and Arakawa, 1994; Ma *et al.*, 1996).

The use of finite adjustment time scales, or soft adjustment (e.g., Emanuel 1991, 1993b), has recently become common. For the AS scheme, for example, the relaxed AS scheme (Moorthi and Suarez, 1992) and the prognostic AS scheme (Randall and Pan, 1993; Pan and Randall, 1998) have been developed, besides the use of a simple partial adjustment described above.

Objective 4: Generation, transport, and mixing of liquid and ice phases of water

In Eq. (19), only the vapor phase of water is considered. More generally, Eqs. (18) and (19) should be replaced by

$$c_p \left[\frac{\partial \overline{T}}{\partial t} + \left(\frac{p}{p_0} \right)^{R/c_p} \left(\overline{\mathbf{v}} \cdot \nabla + \overline{\omega} \frac{\partial}{\partial p} \right) \overline{\theta} \right]$$

$$= L_c \underline{C_1} + L_f \underline{F_1} - L_s \underline{S_1} + c_p B_1 + Q_R, \tag{20}$$

$$\left[\frac{\partial \overline{q_v}}{\partial t} + \left(\overline{\mathbf{v}} \cdot \nabla + \overline{\omega} \frac{\partial}{\partial p} \right) \overline{q_v} \right] = -\underline{C_2} + \underline{S_2} + B_2, \tag{21}$$

$$\left[\frac{\partial \overline{q_l}}{\partial t} + \left(\overline{\mathbf{v}} \cdot \nabla + \overline{\omega} \frac{\partial}{\partial p} \right) \overline{q_l} \right] = \underline{C_3} - \underline{F_3} + B_3, \tag{22}$$

$$\left[\frac{\partial \overline{q_i}}{\partial t} + \left(\overline{\mathbf{v}} \cdot \nabla + \overline{\omega} \frac{\partial}{\partial p} \right) \overline{q_i} \right] = \underline{F_4} - \underline{S_4} + B_4 \tag{23}$$

Here subscripts 2, 3, and 4 correspond to subscripts v, l, and i, which denote the vapor, liquid, and ice phases, respectively; C, F, and S denote condensation, freezing, and sublimation, respectively, including the effects of convective-scale transports and associated mixing, respectively; L_c, L_f, and L_s denote the latent heat for condensation, freezing, and sublimation, respectively; and, as in Eqs. (18) and (19), B and Q_R denote the effect of turbulent transport and the radiation heating, respectively, both of which are temporarily assumed to be already determined (again see objectives 5 and 6 below). At this point all of the underlined terms are unknowns.

In Eqs. (20) through (23), the separations of liquid water into cloud water and rain and of ice into cloud ice and snow are avoided for simplicity. Even so, these equations are complicated. Unfortunately, the past level of guidance via diagnostic analysis of observed budgets cannot be expected for this system, at least in the near future.

Obviously, the inclusion of realistic cloud microphysics is crucial for this objective. There is an important prerequisite, however: The parameterization must be able to realistically determine the mass circulation associated with cumulus convection, including the vertical distributions of detrainment from updrafts and downdrafts.

To meet the classical objectives, there is no need to separately determine σ and w in the equation $m/\rho = \sigma w$, where m is the mass flux, ρ is the density, σ is the fractional area covered by the updraft or downdraft, and w is the vertical velocity. To include the budget of in-cloud rainwater, for example, we must determine w through the vertical component of the momentum equation applied to convective-scale motions (Cheng and Arakawa, 1997). This determination becomes even more important for objective 6 given below.

Objective 5: Interactions with the subcloud layer

For this objective, the mass exchange between cloud and subcloud layers must be determined. In addition, formulations of the effects of vertical and horizontal inhomogeneities in the subcloud layer on updrafts and the role of downdrafts in producing such inhomogeneities in the subcloud layer are included in this objective. In introducing such effects, it is probably important to simultaneously include the effects of the large rate of entrainment near cloud base found by Lin and Arakawa (1997a,b).

For objectives 1 through 4, the B terms in Eqs. (18) and (19) or those in Eqs. (20) through (23) are assumed to be determined separately from the C terms (or the C, F, and S terms). To accomplish objective 5, we must combine the B terms with the C terms (or the C, F, and S terms) so that the distinction between cumulus-convective and PBL processes becomes ambiguous. This is true especially for relatively shallow cumulus convection.

Objective 6: Interactions with radiation

So far, we have assumed that the Q_R terms in Eqs. (18) and (19) are determined separately from the other terms. To accomplish objective 6, however, we must combine the Q_R term with the other terms so that the distinction between radiative processes and cumulus-convective and PBL processes becomes almost meaningless, especially for the cumulonimbus-anvil and stratocumulus-topped PBL problems.

Also, the separate determination of σ and w in $m/\rho = \sigma w$ becomes crucial for determining the cloud cover of the convective clouds. Generally, the response of m to large-scale processes is primarily through changes in σ rather than changes in w (see Robe and Emanuel, 1996; Emanuel and Bister, 1996; and Emanuel's chapter in this book, Chapter 8).

Needless to say, achieving objective 4 is an important prerequisite for objective 6. Knowing the mass of cloud water, however, is not sufficient to determine its effect on radiation, as pointed out in Section V.C.

Objective 7: Momentum transports

In spite of the past efforts on this problem (e.g., Moncrieff, 1981, 1992; Wu and Yanai, 1994), I list it here because there are still a number of difficulties involved with parameterizing the effects of momentum transports in GCMs. The difficulties arise mainly because momentum is not a conservative property and its transport can be either downgradient or upgradient depending on how cumulus convection is organized. For this objective, therefore, the parameterization must determine how cumulus convection is organized into a mesoscale system, particularly whether the organization is quasi-two-dimensional squall-line type or not and, if yes, the orientation of organization as well.

Objective 8: Horizontal transports

This is a problem so far completely ignored in the cumulus parameterization problem but it will have to be considered as resolution increases. See Section VI.D for a related discussion.

1. Parameterizability for the Classical Objectives under an Idealized Situation

Consider a horizontal area—large enough to contain an ensemble of cumulus clouds but small enough to cover only a fraction of a large-scale

> disturbance. The existence of such an area is the basic assumption of this paper. (Arakawa and Schubert, 1974)

This is the assumption introduced at the beginning of AS to treat a cumulus ensemble as a locally homogeneous statistical entity of implicit clouds. This idealization was intentionally introduced to discuss how a parameterization could be constructed under the simplest condition we can imagine. Unless parameterizability of cumulus convection is clear under this condition, it will never be clear for conditions that are more realistic.

For the classical objectives of cumulus parameterization, the "feedback" shown in Fig. 15 is represented by C_1 and C_2 in Eqs. (18) and (19). If we interpret observed and subsequently spatially smoothed values of $\mathbf{v}$, θ, and q as $\bar{\mathbf{v}}$, $\bar{\theta}$, and $\bar{q}$, and if we have a time sequence of observed $\bar{\theta}$ and $\bar{q}$, then all terms on the left sides of Eqs. (18) and (19), including the time derivatives, are known. Then we can calculate Q_1 and Q_2 as residuals, from which C_1 and C_2 can be inferred. This procedure, which is standard in diagnostic studies of cumulus ensembles, follows the thin lower-left segment of the loop in Fig. 15 backwards. Because the path does not include the heavy segments of the loop, the values of C_1 and C_2 obtained in this way have nothing to do with closures necessary for parameterization. In spite of this, or perhaps because of this, such diagnostic studies provide useful information on cumulus ensembles in nature.

In the cumulus parameterization problem, as pointed out in Section VI.B, all of the four underlined terms in Eqs. (18) and (19) are unknowns. Thus, we need *at least* two types of assumptions to close the problem. Arakawa and Chen (1987; see also Arakawa, 1993, and Section X.E in Chapter 1) called an assumption that constrains the C_1 and C_2 terms a Type II closure and an assumption that constrains the two tendency terms a Type I closure. A Type II closure is a consequence of "static control" on the intensive properties of clouds, such as cloud temperature, cloud water mixing ratio, and cloud vertical velocity, through the choice of a cloud model. This control effectively determines cloud heating (C_1) and cloud drying (C_2) *per unit mass flux at cloud base* from given vertical profiles of large-scale temperature and humidity. A Type I closure, on the other hand, is a consequence of "dynamic control" on the statistical behavior of a cumulus ensemble responding to large-scale processes, either directly through a quasi-equilibrium assumption or indirectly through a formulation of the cumulus adjustment process. Together with the Type II closure, Type I closure effectively determines the extensive measures of cumulus activity, such as cloud mass flux and cloud-covered area, from given large-scale dynamical, parameterized radiation and parameterized PBL processes.

For objective 1, i.e., when the vertical distributions of C_1 and C_2 do not have to be determined, the simple requirement of (moist static) energy conservation given by

$$\int_0^{p_s}(C_1 - C_2)\, dp = 0 \tag{24}$$

is the only Type II closure needed. Thus, no cloud model is necessary for this objective.

To meet objective 2, AS considers a spectral cloud ensemble model and expresses the two unknown one-dimensional functions of height, C_1 and C_2, in terms of a single unknown one-dimensional variable, $m_B(\lambda)$, where m_B is the cloud-base mass flux and λ is a parameter identifying cloud types. This decrease in the degrees of freedom through the spectral cloud ensemble model represents the Type II closure of AS. The Type I closure of AS, on the other hand, is through the cloud work function quasi-equilibrium assumption applied to each cloud type, which constrains the coupling of the time changes of the temperature and humidity profiles. In a model that has N levels above cloud base, there are $4N$ unknowns (the two tendency terms, C_1 and C_2 at N levels) in $2N$ equations [Eqs. (18) and (19) applied to N levels], leaving $2N$ unknowns to be determined by closure assumptions. AS then consider N cloud types and introduce N Type II closures and N Type I closures. (Here, for the sake of argument, the subcloud layer values are fixed.) Thus, AS distributed the necessary closures evenly into the two types.

In the original version of AS, downdraft effects are neglected. To include these effects, we must consider both the updraft and downdraft mass fluxes for each cloud type; but we may avoid the increase in the degrees of freedom by assuming that downdrafts are slaved to updrafts for each cloud type so that both mass fluxes are still determined by $m_B(\lambda)$ of updrafts. This is what is done in most existing formulations of downdraft effects including Cheng and Arakawa (1997).

The partition of closure assumptions into these two types can be used to characterize the logical structure of a cumulus parameterization. Suppose that quasi-equilibrium assumptions are introduced for temperature and humidity profiles separately, as in the moist-convective adjustment scheme of Manabe *et al.* (1965) and the Betts–Miller scheme (Betts, 1986, 1996; Betts and Miller, 1986, 1993), rather than for the coupling of temperature and humidity profiles as in AS. Then, there will be $2N$ Type I closures, leaving no room for Type II closures through a cloud model. In this case, clouds are entirely passive to large-scale processes without their own physics except for satisfying Eq. (24). If we consider a generalized spectral cloud ensemble model, on the other hand, as in Ding and Randall (1998), in which clouds can start not only from the PBL but from any levels, there will be $N(N + 1)/2$ cloud types to be considered. If each of these cloud types is dynamically controlled by the large-scale processes independently, we would need $N(N + 1)/2$ Type I closures, which is larger than $2N$ when $N > 3$. It is conceivable, however, that coupling between different cloud types provides a partial closure so that the dynamical control by the large-scale processes is only on some bulk properties of the cumulus ensemble. The question here is whether the concept of quasi-equilibrium should be expanded from cloud–environment interactions to cloud–cloud interactions.

As we have seen, there are still ambiguities in dividing the closure assumptions into different types even for the idealized conditions. Eliminating these ambiguities in the logical structure is perhaps one of the very basic goals when we attempt to develop a future cumulus parameterization scheme.

2. Parameterizability with a Coarse Horizontal Resolution

We now begin to consider parameterizability for more realistic (and, therefore, inevitably more complicated) situations. In a model with a coarse horizontal resolution, the complication is mainly due to the existence of mesoscale organization of cumulus convection. The results of cloud-resolving model simulations by Xu *et al.* (1992) do show fluctuations in cumulus activity not fully modulated by large-scale forcing, especially when the low-level vertical shear is strong. These fluctuations obviously cannot be parameterized using a quasi-equilibrium assumption. This situation represents the limit of "diagnostic parameterizability." The only way to overcome this limit seems to be by introducing more prognostic equations for cumulus activity using a higher order closure.

Randall and Pan (Randall and Pan, 1993; Pan and Randall, 1998) developed a prognostic closure for the AS parameterization, in which convective-scale bulk kinetic energy K is predicted for each cloud type using

$$\frac{dK}{dt} = m_{\mathrm{B}} A - \frac{K}{\tau_{\mathrm{D}}}, \tag{25}$$

where m_{B} is the cloud-base mass flux as defined earlier, A is the cloud work function, which can be determined from the vertical profiles of temperature and humidity, and τ_{D} is the dissipation time scale for K, all of which are defined for each cloud type. Following Arakawa and Xu

(1990) and Xu (1991), the prognostic closure uses a parameter α defined by

$$K = \alpha m_\mathrm{B}^2. \tag{26}$$

Because K includes convective-scale horizontal kinetic energy, which tends to be large when the convection is organized on the mesoscale (see Xu *et al.*, 1992), we can crudely interpret properly scaled α as a measure of the mesoscale organization of convection. If α is known, m_B can be diagnosed from K using Eq. (26). Based on m_B thus determined for all cloud types, the AS spectral cumulus ensemble model can determine the vertical distributions of C_1 and C_2. The vertical profiles of temperature and humidity can then be predicted using Eqs. (18) and (19), and thus the cloud work function A can be updated. Using this A in Eq. (25), K can be predicted. This is the procedure of the prognostic closure of AS proposed by Randall and Pan (1993) and Pan and Randall (1998).

Although it is not needed in actual predictions, we may interpret the above procedure using the cloud work function tendency equation,

$$\frac{dA}{dt} = Jm_\mathrm{B} + F, \tag{27}$$

where Jm_B and F represent the cumulus and large-scale effects on dA/dt, representing cumulus adjustment and large-scale forcing, respectively. The factor J, which symbolically represents the kernel of the integral equation in AS, is assumed to be negative (i.e., the cumulus effect is to decrease A) so that we may write $J = -|J|$. Assuming α and J are constants and eliminating A and K between Eqs. (25), (26), and (27), we obtain

$$2\alpha\frac{d^2 m_\mathrm{B}}{dt^2} + \frac{\alpha}{\tau_\mathrm{D}}\frac{dm_\mathrm{B}}{dt} + |J|m_\mathrm{B} = F. \tag{28}$$

This is a forced damping oscillation equation for m_B. In the limit as $\alpha \to 0$, Eq. (28) gives $m_\mathrm{B} = F/|J|$. This is the equilibrium solution of Eq. (27). Thus the procedure described above can be considered as a generalization of the Type I closure used in AS, although it is not easy to tell how realistic the generalization is due to the uncertainties in the magnitudes of α and τ_D. This is why Pan and Randall (1998) claim that they can predict m_B but not necessarily K or A. The prognostic closure, however, does have an advantage: Eq. (27) is not used in predictions so that large-scale and cumulus-scale processes do not have to be separated in that equation. Although this is to some extent true in any relaxed adjustment or soft quasi-equilibrium formulation, the particular formulation described above is useful in discussing a possible future direction of cumulus parameterization (see Section VI.D).

3. Parameterizability with a Fine Resolution

This is the case when the grid scale is not sufficiently larger than the convective scale. This is typical of the situations in mesoscale models and, therefore, problems in mesoscale modeling will eventually become problems in GCMs. Molinari and Dudek (1992) pointed out that "In mesoscale models, explicit clouds may form at a grid point while essentially similar clouds are simultaneously being parameterized." Consequently, "double counting" by parameterized and unparameterized forms and "scale misrepresentation" due to spurious competition between parameterized and unparameterized forms may occur in those models. Molinari and Dudek (1992; see also Molinari, 1993, and Frank, 1993) proposed a "hybrid approach," in which a fraction of convective sources of condensate determined by a parameterization is added to the grid-scale cloud-water and rainwater equations. From my point of view, this should be done anyway even in GCMs to meet objective 4 given in the last subsection.

With a high resolution (or to a lesser extent with almost any resolution), we have a more fundamental problem because "grid-scale" and "sub-grid-scale" are not well separable; thus there are ambiguities even in interpreting what grid point values represent. There are at least three ways of defining those values:

1. *As the space means over the sub-grid area representing the environment of convective clouds.* In this case, the processes to be parameterized are the effects of clouds on their environment, such as the cumulus-induced subsidence and cumulus detrainment effects.
2. *As the space means over the entire gridbox, including clouds and their environment.* In this case, processes to be parameterized are the sub-grid eddy effects due to the existence of clouds.
3. *As the ensemble means over infinite realizations at that grid point.* In this case, processes to be parameterized are ensemble mean effects.

These three definitions are equivalent to each other under the idealized conditions considered earlier. Although these conditions are assumed in most existing schemes at least implicitly, they cannot be easily justified for realistic situations. Then, which of the three definitions given above should be chosen?

Definition 1 is perhaps the most common among the existing schemes, including AS, although the separation of a sub-grid area into the area covered by convective clouds and that of the environment is ambiguous mainly due to the existence of associated stratiform clouds.

Definition 2 is also not free of this kind of ambiguity because the separation of a sub-grid area into cloud-covered and environmental areas also has to be made in formulating the eddy effects. In any event, separating a variable into the mean over a grid size and the deviation from it does not make much sense when the mean itself lacks significance.

Definition 3 is conceptually superior to the others since no distinction is made between grid-scale and sub-grid-scale processes, and thus all physical processes can be treated in a single package. A question remains: What conditions should be fixed over the "infinite realizations"? Some way of separating nonparameterized and parameterized processes is still needed though it does not have to depend on the grid size.

C. Future Directions

Discussion in the last subsection suggests possible future directions of cumulus parameterization. First, we should try to abandon the concept of "sub-grid-scale" parameterization because the separation between "grid-scale" and "sub-grid-scale" is artificial. Having two formulations of condensation processes based on this artificial separation, one in the cumuliform clouds as "sub-grid-scale" condensation, which can only be in a statistical equilibrium in a model, and the other in stratiform clouds as "grid-scale" condensation, which can be highly transient, makes the situation worse than ambiguous. We can think of two extreme approaches to eliminate this situation.

1. *Resolve everything*! In this approach, individual clouds are treated explicitly through an application of a cloud-resolving model to the entire globe. No cumulus parameterization is needed except for the purpose of interpreting the results. This approach may be possible for a limited purpose some time in the future; but even then, it is not likely to be feasible for all GCM applications.

2. *Parameterize everything*! Due to differential radiative heating/cooling between cloud base and cloud top, and possibly due to shear effects, even "stratiform clouds" have internal convection and turbulence, which influence the thickness, area coverage, precipitation, and lifetime of clouds, as discussed in Section V.C. In this approach, "stratiform clouds" are considered as a horizontally extended manifestation of convective processes.

In my point of view, these two approaches are complementary. The following is a summary of my perspective on the future goals of cumulus parameterization:

1. *From "diagnostic parameterization" to "prognostic parameterization."* More prognostic equations governing the higher moments and life cycles of convective processes are used. The extreme case in this direction is the "resolve everything" approach.

2. *From "cumulus parameterization" to "unified cloud parameterization."* A single physics package is used to treat all types of clouds, without artificially separating them into "grid-scale" and "sub-grid-scale," using a unified cloud model and finding an appropriate statistical closure valid for all cloud types. The extreme case in this direction is the "parameterize everything" approach.

3. *From "single-column parameterization" to "multiple-column parameterization."* Neglecting the convergence of horizontal transports cannot be justified if the horizontal distance for averaging is not sufficiently larger than the horizontal scale of convection. This is especially important when mesoscale processes are resolved but not fully resolved.

4. *From "deterministic parameterization" to "nondeterministic parameterization."* Nondeterministic components due to the existence of nonparameterizable parts of small-scale processes are included in the parameterized results.

All of these are long-term goals and, at this stage, they only suggest the direction of future development. In the following, I outline one of the short-term scenarios I have in mind as an example, which is admittedly immature and requires clarification of many details.

A good starting candidate for goal 1 is a model similar to that used in the prognostic AS discussed in Section VI.C. The model should have additional prognostic equations, however, at least one for an explicit prediction of cloud-base mass flux, $m_{\rm B} \equiv (\rho \sigma w)_{\rm B}$, replacing the diagnostic determination using Eq. (26). In this way, the parameter α, which is a measure of the mesoscale organization of convection, is effectively predicted. The prediction of $m_{\rm B}$ can be done, after properly determining the magnitude of w, by predicting σ as an adjustment to a quasi-equilibrium as in Emanuel (1991). The predictions of K, which includes the horizontal kinetic energy, and the predictions of $m_{\rm B}$, which is vertical mass flux, must be done consistently, including the physics that determines the degree of mesoscale organization of convection such as shear effects. All of these should be done for each properly defined cloud type with interactions between them.

Developing such a model is a challenge and its success will depend heavily on the guidance provided by the extensive use of cloud-resolving models. An example of methods to identify cloud types in CRM simulations can be found in Lin and Arakawa (1997b), in which cloud parcels are classified in terms of their eventual heights. Another possibility for classifying cloud types (or cloud regimes) is to follow an empirical statistical approach, such as the use of canonical correlation or rotated EOF analysis of the vertical profiles of simulated C_1 and C_2, as Chen (1989; see also Arakawa and Chen, 1987; Betts, 1996), Liu (1995; see also Arakawa, 1993) and Lin and Arakawa (2000) did for observed Q_1 and Q_2. In view of goal 2, the analysis should cover a variety of situations, including the cases in which clouds are dominantly stratiform. Development of a low-order mesoscale model based on an EOF analysis of CRM simulated data has already been proposed by Palmer (1996), although his approach is purely empirical, whereas what I have in mind is semi-empirical, maintaining the logical structure of the AS (or the prognostic AS) parameterization.

I realize that future parameterizations for use in a high-resolution model, in which mesoscale processes are partially resolved, should almost inevitably use a multiple-column (i.e., horizontally nonlocal) parameterization in view of goal 3. Even before the problem of parameterizing horizontal transports is solved, however, we should try to depart from single-column parameterization. A simple way to do this is through using horizontal (weighted) averages of grid point values as inputs to the (prognostic) parameterization scheme when the grid size is smaller than a properly chosen length, say, 500 km. It then introduces a region in the spectral domain in which predictions by the GCM (or NWP model) and predictions by the prognostic parameterization are both performed. Naturally, these two types of predictions should be done in an interactive way. This strategy is again very similar to what Palmer (1996) proposed with the EOF mesoscale model, saying, "Large-scale (i.e., large mesoscale) EOFs should be partially resolved by NWP model—so coupling not confined only to smallest resolvable scale of NWP models." Figure 17 is essentially the same as one of his figures, but modified for the present purpose.

Finally, in reference to goal 4, I simply quote from Lorenz (1969):

> In a typical model atmosphere, it is assumed that only the statistical properties of the smaller-scale motions influence the larger scales, and that at any instant these statistical properties are determined by the larger-scale motions....Effectively a system consisting of only the larger scales is assumed to be deterministic....Errors in the small-scale statistics ought to appear in more realistic models.

VII. CONCLUSIONS

In numerical modeling of the atmosphere, we must formulate complicated atmospheric processes with a limited number of degrees of freedom.

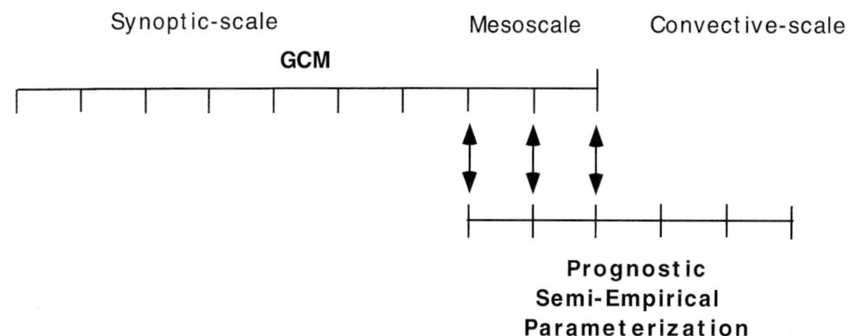

Figure 17 Coupling of a GCM with a prognostic parameterization of smaller scale processes. (After Palmer, 1996, with modification)

This is an enormous task. Although the last two phases in the history have been epoch making and magnificent (see Chapter 1 in this book), the new "great challenge" third phase will require more than simple extrapolations of what we have achieved in the past.

The emphasis of our efforts in the third phase will be on formulations of the interactions of various processes shown in Fig. 18, by the dashed arrows for the second phase and the solid arrows for the third phase, indicating that our understanding of these interactions has not yet become "solid." In the past, much modeling effort has been spent on synthesizing what we knew about individual boxes in the left panel of Fig. 18. Modeling efforts during the third phase will pose new problems for basic research on the interactions between those boxes. After the successful completion of the third phase, the atmospheric sciences will become a truly unified science, represented by a single box for the atmosphere in the right panel of Fig. 18.

One of the trends in the new phase is toward unified models (see Fig. 1 of Chapter 1 in this book), or more precisely speaking, toward *unification of modeling efforts,* because many of the modeling difficulties we have now are common to mesoscale, regional NWP, global NWP, and global climate models. Extensive use of large-eddy simulation and cloud-resolving models (LESs and CRMs) will become more indispensable for the future progress of our modeling capability.

In conclusion, I quote Bjerknes (1904) as I did at the beginning of my first chapter:

> The problem is of huge dimensions ... I am convinced that it is not too soon to consider this problem as the objectives of our researches.

I will not be able to see the completion of the "great challenge" third phase, but I am happy to see at least its beginning.

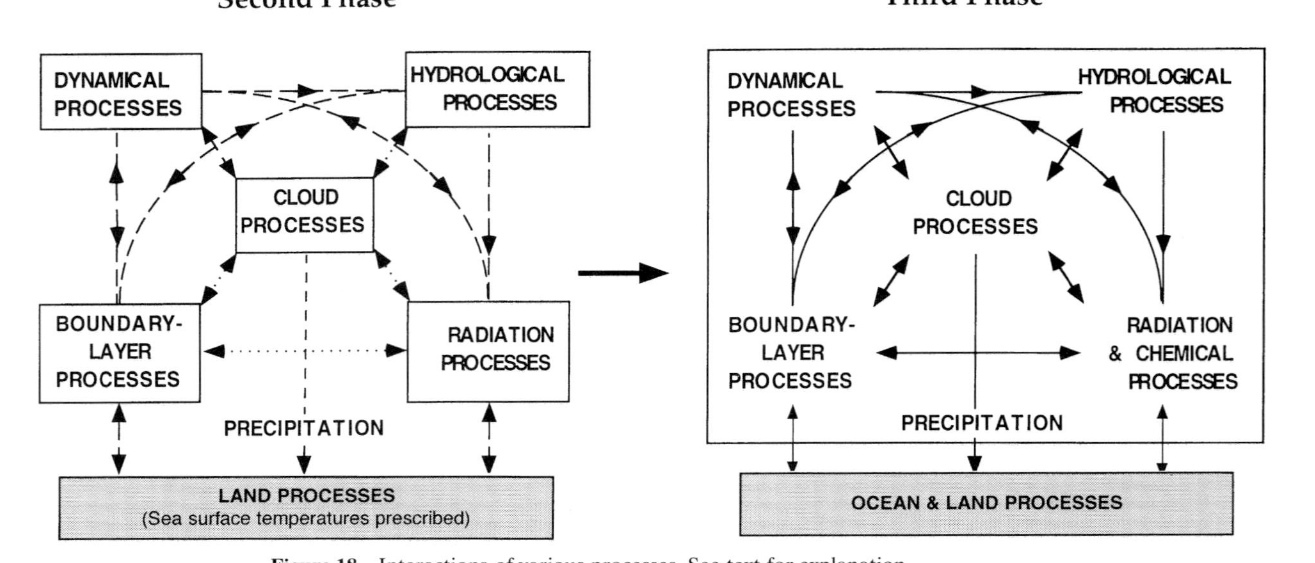

Figure 18 Interactions of various processes. See text for explanation.

ACKNOWLEDGMENTS

I would like to thank again all participants and sponsors of the AA Fest, particularly Kayo Ide, David Randall, and Roger Wakimoto for their wonderful organization of the symposium.

Preparation of this article was supported by the NSF grant ATM-96139, NASA grant NAG 5-4420, and DOE grant DE-FG03-91ER61214. I greatly appreciate the help provided by Prof. David Randall and Drs. John Farrara and Celal Konor in revising the original manuscript.

REFERENCES

Arakawa, A. (1969). Parameterization of cumulus clouds. *In* "Proceedings of the WMO/IUGG Symposium on Numerical Weather Prediction," Tokyo, 1968, pp. IV-8-1–IV-8-6. Japan Meteorological Agency.

Arakawa, A. (1975). Modelling clouds and cloud processes for use in climate model. The Physical Basis of Climate and Climate Modelling, GARP Publication Series No. 16, pp. 183–197.

Arakawa, A. (1983). Vertical differencing of filtered models. *In* "Numerical Methods for Weather Prediction," pp. 183–206. European Centre for Medium Range Weather Forecasts, Reading, UK.

Arakawa, A. (1988). Finite-difference methods in climate modeling. *In* "Physically-Based Modeling and Simulation of Climate and Climate Change" (M. Schlesinger, ed.), Part I, pp. 79–168. Kluwer Academic Publishers, The Netherlands.

Arakawa, A. (1993). Closure assumptions in the cumulus parameterization problem. *In* "The Representation of Cumulus Convection in Numerical Models of the Atmosphere" (K. A. Emanuel and D. J. Raymond, eds.), pp. 1–16. American Meteorological Society.

Arakawa, A., and J.-M. Chen (1987). Closure assumption in the cumulus parameterization problem. *In* "Short- and Medium-Range Numerical Weather Prediction" (T. Matsuno, ed.), Special Volume of *J. Meteor. Soc.* Japan, pp. 107–131.

Arakawa, A. and Y.-J. G. Hsu (1990). Energy conserving and potential-enstrophy dissipating schemes for the shallow water equations. *Mon. Wea. Rev.* **118,** 1960–1969.

Arakawa, A., and C. S. Konor (1996). Vertical differencing of the primitive equations based on the Charney-Phillips grid in hybrid σ-p vertical coordinates. *Mon. Wea. Rev.* **124,** 511–528.

Arakawa, A., and V. R. Lamb (1981). A potential enstrophy and energy conserving scheme for the shallow water equations. *Mon. Wea. Rev.* **109,** 18–36.

Arakawa, A., and S. Moorthi (1988). Baroclinic instability in vertically discrete systems. *J. Atmos. Sci.,* **45,** 1688–1707.

Arakawa, A., and W. H. Schubert (1974). Interaction of a cumulus cloud ensemble with the large-scale environment. Part I. *J. Atmos. Sci.* **31,** 674–701.

Arakawa, A., and K.-M. Xu (1990). The macroscopic behavior of simulated cumulus convection and semi-prognostic tests of the Arakawa–Schubert cumulus parameterization. *In* "Proceedings of the Indo-US Seminar on Parameterization of Sub-Grid Scale Processes in Dynamical Models of Medium Range Prediction and Global Climate," Pune, India. IITM.

Bannon, P. R. (1995). Hydrostatic adjustment: Lamb's problem. *J. Atmos. Sci.* **52,** 1743–1752.

Bates, J. R., S. Moorthi, and R. W. Higgins (1993). A global multi-level atmospheric model using a vector semi-Lagrangian finite-difference scheme. Part I: Adiabatic formulation. *Mon. Wea. Rev.* **121,** 244–263.

Betts, A. K. (1986). A new convective adjustment scheme. Part I: Observational and theoretical basis. *Quart. J. Roy. Meteor. Soc.* **112,** 677–691.

Betts, A. K. (1996). The parameterization of deep convection. *In* "The Physics and Parameterization of Moist Convection" (R. T. Smith, ed.), pp. 255–279. Kluwer Academic Publishers, The Netherlands.

Betts, A. K., and M. J. Miller (1986). A new convective adjustment scheme. Part II: Single column tests using GATE wave. BOMEX, ATEX and arctic airmass data sets. *Quart. J. Roy. Meteor. Soc.* **112,** 693–709.

Betts, A. K., and M. J. Miller (1993). The Betts–Miller scheme. *In* "The Representation of Cumulus Convection in Numerical Models of the Atmosphere" (K. A. Emanuel and D. J. Raymond, eds.), pp. 107–121. American Meteorological Society.

Bjerknes, V., 1904: Das Problem der Wettervorsage, betrachtet vom Standpunkte der Mechanik und der Physik. *Meteor. Z.,* **21,** 1–7. (Translation by Yale Mintz)

Bleck, R. (1978). Finite-difference equations in generalized vertical coordinates. Part I: Total energy conservation. *Beitr. Phys. Atmos.* **51,** 360–372.

Bleck, R. (1979). Finite-difference equations in generalized vertical coordinates. Part II: Potential vorticity conservation. *Beitr. Phys. Atmos.* **52,** 95–105.

Bleck, R., and S. Benjamin (1993). Regional weather prediction with a model combining terrain-following and isentropic coordinates. Part I. *Mon. Wea. Rev.* **121,** 1770–1785.

Bretherton, C. S. (1992). A conceptual model of the stratocumulus-trade-cumulus transition in the subtropical oceans. *In* "Proc. 11th Int. Conf. on Clouds and Precipitation," Vol. I, Montreal, Quebec, Canada, pp. 374–377. International Commission on Clouds and Precipitation and International Association of Meteorology and Atmospheric Physics.

Browning, K. A. (1974). Survey of perceived priority issues in the parameterizations of cloud-related processes. *Quart. J. Roy. Meteor. Soc.* **120,** 483–487.

Businger, J. A., J. C. Wyngaard, Y. Izumi, and E. F. Bradley (1971). Flux-profile relationships in the atmospheric surface layer. *J. Atmos. Sci.* **28,** 181–189.

Charney, J. G., and N. A. Phillips (1953). Numerical integration of the quasi-geostrophic equations for barotropic and simple baroclinic flows. *J. Meteor.* **10,** 71–99.

Chen, J.-M. (1989). Observational study of the macroscopic behavior of moist-convective processes, Ph.D. Thesis. Department of Atmospheric Sciences, UCLA.

Chen, M., R. B. Rood, and L. L. Takacs (1997). Impact of a semi-Lagrangian and an Eulerian dynamical core on climate simulations. *J. Climate* **10,** 2374–2389.

Cheng, M.-D., and A. Arakawa (1994). Effects of including convective downdrafts and a finite cumulus adjustment time in a cumulus parameterization. *In* "Tenth Conference on Numerical Weather Prediction," Portland, OR, July 17–22, 1994, pp. 102–104.

Cheng, M.-D., and A. Arakawa (1997). Inclusion of rainwater budget and convective downdrafts in the Arakawa–Schubert cumulus parameterization. *J. Atmos. Sci.* **54,** 1359–1378.

Clark, P. D., and P. Haynes (1994). Equatorial inertial instability: Effects of vertical finite differencing and radiative transfer. *J. Atmos. Sci.* **51,** 2101–2109.

Cullen, M. J. P., and J. James (1994). A comparison of two different vertical grid staggerings. *In* "Preprints, 10th Conference on Numerical Weather Prediction," Portland, OR, July 18–22, 1994, pp. 38–40. Amer. Meteor. Soc., Boston, MA.

Cullen, M. J. P., T. Davies, M. H. Mawson, J. A. James, and S. C. Coulter (1994). Some issues in numerical methods for the next generation of NWP and climate models. *Forecasting Research*, Sci. Paper No. 26. U.K. Meteor. Office, 35 pp.

Deardorff, J. W. (1972). Parameterization of the planetary boundary layer for use in general circulation models. *Mon. Wea. Rev.* **100,** 93–106.

Deardorff, J. W. (1980). Cloud top entrainment instability. *J. Atmos. Sci.* **37,** 329–350.

Ding, P., and D. A. Randall (1998). A cumulus parameterization with multiple cloud bases. *J. Geophys. Res.* 103, No. D10, 11,341–11,352.

Emanuel, K. A. (1991). A scheme representing cumulus convection in large-scale models. *J. Atmos. Sci.* **48**, 2313–2335.

Emanuel, K. A. (1993a). A cumulus representation based on the episodic mixing model: The importance of mixing and microphysics in predicting humidity. *In* "The Representation of Cumulus Convection in Numerical Models of the Atmosphere" (K. A. Emanuel and D. J. Raymond, eds.), pp. 185–192. American Meteorological Society.

Emanuel, K. A. (1993b). The effect of convective response to WISHE modes. *J. Atmos. Sci.* **50**, 1763–1775.

Emanuel, K. A., and M. Bister (1996). Moist convective velocity and buoyancy scales. *J. Atmos. Sci.* **53**, 3276–3285.

Emanuel, K. A., J. D. Neelin, and C. S. Brethren (1994). On large-scale circulations in convecting atmosphere. *Quart. J. Roy. Meteor. Soc.* **120**, 1111–1143.

Fox-Rabinovitz, M. S. (1994). Computational dispersion properties of vertically staggered grids for atmospheric models. *Mon. Wea. Rev.* **122**, 377–392.

Frank, W. M. (1993). A hybrid parameterization with multiple closures. *In* "The Representation of Cumulus Convection in Numerical Models of the Atmosphere" (K. A. Emanuel and D. J. Raymond, eds.), pp. 151–154. American Meteorological Society.

Heikes, R., and D. A. Randall (1995). Numerical integration of the shallow-water equations on a twisted icosahedral grid. Part I: Basic design and results of tests. *Mon. Wea. Rev.* **123**, 1862–1880.

Held, I. M., R. T. Pierrehumbert, S. T. Garner, and K. L. Swanson (1995). Surface quasi-geostrophic dynamics. *J. Fluid. Mech.* **282**, 1–20.

Hollingsworth, A. (1995). A spurious mode in the "Lorenz" arrangement of F and T which does not exist in the "Charney–Phillips" arrangement. ECMWF Technical Memorandum No. 211.

Holtslag, A. A. M., and B. A. Boville (1993). Local versus nonlocal boundary-layer diffusion in a global climate model. *J. Climate* **6**, 1825–1842.

Hsu, Y-J. G., and A. Arakawa (1990). Numerical modeling of the atmosphere with an isentropic vertical coordinate. *Mon. Wea. Rev.* **118**, 1933–1959.

Janjic, Z. I. (1990). The step mountain coordinate: Physical package. *Mon. Wea. Rev.* **118**, 1429–1443.

Janjic, Z. I. (1994). The step mountain Eta coordinate model: Further development of the convection, viscous sublayer, and turbulence closure schemes. *Mon. Wea. Rev.* **122**, 927–945.

Johnson, R. D., and L. W. Uccellini (1983). A comparison of methods for computing the sigma-coordinate pressure gradient force for flow over sloped terrain in a hybrid theta-sigma coordinate model. *Mon. Wea. Rev.* **111**, 870–886.

Johnson, R. D., T. H. Zapotocny, F. M. Reames, B. J. Wolf, and R. B. Pierce (1993). A comparison of simulated precipitation by hybrid isentropic-sigma and sigma models. *Mon. Wea. Rev.* **121**, 2088–2114.

Kim, Y. J., and A. Arakawa (1995). Improvement of orographic gravity wave parameterization using a mesoscale gravity wave model. *J. Atmos. Sci.* **52**, 1875–1902.

Köhler, M., C. R. Mechoso, and A. Arakawa (1997). Ice cloud formulation in climate modeling. *In* "7th Conference on Climate Variations," Long Beach, CA, February 2–7, 1997, pp. 237–242. American Meteorological Society.

Konor, C. S., and A. Arakawa (1997). Design of an atmospheric model based on a generalized vertical coordinate. *Mon. Wea. Rev.* **125**, 1649–1673.

Krishnamurti, T. N. (1969). An experiment in numerical prediction in equatorial latitudes. *Quart. J. Roy. Meteor. Soc.* **95**, 594–620.

Krueger, S. K. (1985). Numerical simulation of tropical cumulus clouds and their interaction with the subcloud layer, Ph.D. Thesis. Department of Atmospheric Sciences, UCLA.

Krueger, S. K., G. T. McLean, and Q. Fu (1995a). Numerical simulation of the stratus-to-cumulus transition in the subtropical marine boundary layer. Part I: Boundary-layer structure. *J. Atmos. Sci.* **52**, 2839–2850.

Krueger, S. K., G. T. McLean, and Q. Fu (1995b). Numerical simulation of the stratus-to-cumulus transition in the subtropical marine boundary layer. Part II: Boundary-layer circulation. *J. Atmos. Sci.* **52**, 2851–2868.

Lamb, H. (1932). "Hydrodynamics," 6th ed. Dover Publications, New York.

Large, W. G., J. C. McWilliams, and S. C. Doney (1994). Oceanic vertical mixing: A review and a model with a nonlocal boundary layer parameterization. *Rev. Geophys.* **32**, 363–403.

Leslie, L. M., and G. S. Dietachmayer (1997). Comparing schemes for integrating Euler equations. *Mon. Wea. Rev.* **125**, 1687–1691.

Leslie, L. M., and R. J. Purser (1992). A comparative study of the performance of various vertical discretization schemes. *Meteor. Atmos. Phys.* **50**, 61–73.

Li, J.-L. F., C. R. Mechoso, and A. Arakawa (1999). Improved moist processes with the UCLA GCM. *In* "10th Symposium on Global Change Studies," Dallas, Texas, January 10–15, 1999, pp. 423–426, American Meteorological Society.

Lilly, D. K. (1968). Models of cloud-topped mixed layers under a strong inversion. *Quart. J. Roy. Meteor. Soc.* **94**, 292–309.

Lin, C.-C., and A. Arakawa (2000). Empirical determination of convective cloud regimes. Accepted by *J. Atmos. Sci.*

Lin, C.-C., and A. Arakawa (1997a). The macroscopic entrainment processes of simulated cumulus ensemble. Part I. Effective sources of entrainment. *J. Atmos. Sci.* **54**, 1027–1043.

Lin, C.-C., and A. Arakawa (1997b). The macroscopic entrainment processes of simulated cumulus ensemble. Part II. Testing the entraining-plume model. *J. Atmos. Sci.* **54**, 1044–1053.

Lin, S.-J., and R. B. Rood (1996). Multidimensional flux-form semi-Lagrangian transport schemes. *Mon. Wea. Rev.* **124**, 2046–2070.

Lindzen, R. S. (1988). Supersaturation of vertically propagating internal gravity waves. *J. Atmos. Sci.* **45**, 705–711.

Lindzen, R. S., and M. Fox-Rabinovitz (1989). Consistent vertical and horizontal resolution. *Mon. Wea. Rev.* **117**, 2575–2583.

Lindzen, R. S., E. S. Batten, and J. W. Kim (1968). Oscillations in atmospheres with tops. *Mon. Wea. Rev.* **96**, 133–140.

Liu, Y.-Z. (1995). The representation of the macroscopic behavior of observed moist-convective processes, Ph.D. Thesis. Department of Atmospheric Sciences, UCLA.

Louis, J.-F. (1979). A parametric model of vertical eddy fluxes in the atmosphere. *Bound.-Layer Meteor.* **17**, 187–202.

Lorenz, E. N. (1960). Energy and numerical weather prediction. *Tellus* **12**, 364–373.

Lorenz, E. N. (1969). The predictability of a flow which possesses many scales of motion. *Tellus* **21**, 289–307.

Ma, C.-C, C. R. Mechoso, A. Arakawa, and J. Farrara (1994). Sensitivity of a coupled ocean–atmosphere model to physical parameterizations. *J. Climate* **7**, 1883–1896.

Ma, C.-C., C. R. Mechoso, A. W. Robertson, and A. Arakawa (1996). Peruvian stratus clouds and the tropical Pacific circulation: A coupled ocean–atmosphere GCM study. *J. Climate* **9**, 1635–1645.

Manabe, S., J. Smagorinsky, and R. F. Strickler (1965). Simulated climatology of a general circulation model with a hydrological cycle. *Mon. Wea. Rev.* **93**, 769–798.

Masuda, Y., and H. Ohnishi (1987). An integration scheme of the primitive equation model with an icosahedral-hexagonal grid system and its application to the shallow water equations. *In* "Short- and Medium-Range Numerical Weather Prediction" (T. Matsuno, ed.), pp. 317–326. Special Volume, *J. Meteor. Soc. Japan.*

Matsuno, T. (1966). False reflection of waves at the boundary due to use of finite differences. *J. Meteor. Soc. Japan* **44**, 145–157.

McFarlane, N. A. (1987). The effect of orographically excited gravity-wave drag on the general circulation of the lower stratosphere and troposphere. *J. Atmos. Sci.* **44**, 1775–1800.

McWilliams, J. C., and P. R. Gent (1980). Intermediate models of planetary circulations in the atmosphere and ocean. *J. Atmos. Sci.* **37**, 1657–1678.

Mechoso, C. R., A. W. Robertson, and coauthors (1995). The seasonal cycle over the tropical Pacific in coupled ocean–atmosphere general circulation models. *Mon. Wea. Rev.* **123**, 2825–2838.

Mellor, G. L., and T. Yamada (1974). A hierarchy of turbulence closure models for planetary boundary layer. *J. Atmos. Sci.* **31**, 1791–1806.

Mellor, G. L., and T. Yamada (1982). Development of a turbulence closure model for geophysical fluid problems. *Rev. Geophys. Space Phys.* **20**, 851–875.

Mesinger, F. (1984). A blocking technique for representation of mountains in atmospheric models. *Riv. Meteor. Aeronautica* **44**, 195–202.

Mesinger, F. (1997). Dynamics of limited-area models: Formulation and numerical methods. *Meteor. Atmos. Phys.* **63**, 3–14.

Mesinger, F., Z. I. Janjic, S. Ničković, D. Gavrilov, and D. G. Deaven (1988). The step-mountain coordinate: Model description and performance for cases of Alpine lee cyclogenesis and for a case of an Appalachian redevelopment. *Mon. Wea. Rev.* **116**, 1493–1518.

Miyakoda, K., and J. Sitrus (1977). Comparative integrations of global models with various parameterized processes of sub-grid scale vertical transports: Description of parameterizations. *Beit. Phys. Atmos.* **50**, 445–487.

Miyakoda, K., and J. Sitrus (1990). Subgridscale physics in 1-month forecasts. Part II: Systematic error and blocking forecasts. *Mon. Wea. Rev.* **118**, 1065–1081.

Moeng, C.-H. (1979). Stability of a turbulent layer cloud within the planetary boundary layer, Ph.D. Thesis. Department of Atmospheric Sciences, UCLA.

Moeng, C.-H., D. H. Lenschow, and D. A. Randall (1995). Numerical investigations of the roles of radiative and evaporative feedbacks in stratocumulus entrainment and breakup. *J. Atmos. Sci.* **52**, 2869–2883.

Molinari, J. (1993). An overview of cumulus parameterization in mesoscale models. *In* "The Representation of Cumulus Convection in Numerical Models of the Atmosphere" (K. A. Emanuel and D. J. Raymond, eds.), pp. 155–158. American Meteorological Society.

Molinari, J., and M. Dudek (1992). Parameterization of convective precipitation in mesoscale numerical models: a critical review. *Mon. Wea. Rev.* **120**, 326–344.

Moncrieff, M. W. (1981). A theory of organized steady convection and its transport properties. *Quart. J. Roy. Meteor. Soc.* **107**, 29–50.

Moncrieff, M. W. (1992). Organized convective systems. Archetypal dynamical models, mass and momentum flux theory, and parameterization. *Quart. J. Roy. Meteor. Soc.* **118**, 819–850.

Moorthi, S., and M. Suarez (1992). Relaxed Arakawa-Schubert: A parameterization of moist convection for general circulation models. *Mon, Wea. Rev.* **120**, 978–1002.

Moorthi, S., R. W. Higgins, and J. R. Bates (1995). A global multilevel atmospheric model using a vector semi-Lagrangian finite-difference scheme. Part II: Version with physics. *Mon. Wea. Rev.* **123**, 1523–1541.

Neelin, J. D. (1997). Implications of convective quasi-equilibrium for the large-scale flow. *In* "The Physics and Parameterization of Moist Convection" (R. T. Smith, ed.), pp. 413–446. Kluwer Academic Publishers, The Netherlands.

Neelin, J. D., and J. Yu (1994). Models of tropical variability under convective adjustment and the Madden–Julian oscillation. Part I: Analytical results. *J. Atmos. Sci.* **51**, 1876–1894.

Neelin, J. D., M. Latif, and coauthors (1992). Tropical air–sea interaction in general circulation models. *Climate Dyn.* **7**, 73–104.

Palmer, T. N. (1996). On parameterizing scales that are only somewhat smaller than the smallest resolved scales, with application to convection and orography. *In* "Workshop on New Insights and Approaches to Convective Parameterization," November 4–7, 1996, pp. 328–337. ECMWF, Reading, UK.

Palmer, T. N., G. J. Shutts, and R. Swinbank (1986). Alleviation of a systematic westerly bias in circulation and numerical weather prediction models through an orographic gravity-wave drag parameterization. *Quart. J. Roy. Meteor. Soc.* **112**, 1001–1039.

Pan, D.-M., and D. A. Randall (1998). A cumulus parameterization with a prognostic closure. *Quart. J. Roy. Meteor. Soc.* **124**, 949–981.

Phillips, N. A. (1957). A coordinate system having some special advantages for numerical forecasting. *J. Meteor.* **14**, 184–185.

Randall, D. A. (1976). The interaction of the planetary boundary layer with large-scale circulations, Ph.D. Thesis. Department of Atmospheric Sciences, UCLA.

Randall, D. A. (1980). Conditional instability of the first kind upside-down. *J. Atmos. Sci.* **37**, 125–130.

Randall, D. A. (1994). Geostrophic adjustment and the finite-difference shallow-water equations. *Mon. Wea. Rev.* **122**, 1371–1377.

Randall, D. A., and D. M. Pan (1993). Implementation of the Arakawa–Schubert cumulus parameterization with a prognostic closure. *In* "The Representation of Cumulus Convection in Numerical Models of the Atmosphere" (K. A. Emanuel and D. J. Raymond, eds.), pp. 137–144. American Meteorological Society.

Randall, D. A., Q. Shao, and C.-H. Moeng (1992). A second-order bulk boundary-layer model. *J. Atmos. Sci.* **49**, 1903–1923.

Randall, D. A., D. M. Pan, P. Ding, and D. G. Cripe (1997). Quasi-equilibrium. *In* "The Physics and Parameterization of Moist Convection" (R. T. Smith, ed.), pp. 359–385. Kluwer Academic Publishers, The Netherlands.

Ritchie, H. (1986). Eliminating the interpolation associated with the semi-Lagrangian scheme. *Mon. Wea. Rev.* **114**, 135–146.

Ritchie, H., C. Temperton, A. Simmons, M. Hortal, T. Davies, D. Dent, and M. Hamrud (1995). Implementation of the semi-Lagrangian method in a high-resolution version of the ECMWF forcast model. *Mon. Wea. Rev.* **123**, 489–514.

Robe, F. R., and K. A. Emanuel (1996). Moist convective scaling: some inferences from three-dimensional cloud ensemble simulations. *J. Atmos. Sci.* **53**, 3265–3275.

Robert, A. (1981). A stable numerical integration scheme for the primitive meteorological equations. *Atmos.-Ocean* **19**, 35–46,

Robert, A. (1982). A semi-Lagrangian and semi-implicit numerical integration scheme for the primitive meteorological equations. *J. Met. Soc. Japan* **60**, 319–325.

Robert, A. J., J. Henderson, and C. Turnbull (1972). An implicit time integration scheme for baroclinic models of the atmosphere. *Mon. Wea. Rev.* **100**, 329–335.

Sadourny, R., A. Arakawa, and Y. Mintz (1968). Integration of the nondivergent barotropic vorticity equation with an icosahedral-hexagonal grid for the sphere. *Mon. Wea. Rev.* **96**, 351–356.

Schneider, E. K. (1987). An inconsistency in vertical discretization in some atmospheric models. *Mon. Wea. Rev.* **115**, 2166–2169.

Sitrus, J., and K. Miyakoda (1990). Subgridscale physics in 1-month forecasts. Part I: Experiment with four parameterization packages. *Mon. Wea. Rev.* **118**, 1043–1064.

Smolarkiewicz, P. K., and J. A. Pudykiewicz (1992). A class of semi-Lagrangian approximations for fluids. *J. Atmos. Sci.* **49**, 2082–2096.

Smolarkiewicz, P. K., and P. Rasch (1991). Monotone advection on the sphere: An Eulerian versus a semi-Lagrangian approach. *J. Atmos. Sci.* **48**, 793–810.

Staniforth, A., and J. Côté (1991). Semi-Lagrangian integration schemes for atmospheric models—a review. *Mon. Wea. Rev.* **119**, 2206–2223.

Stull, R. B. (1994). A convective transport theory for surface fluxes. *J. Atmos. Sci.* **51**, 3–22.

Suarez, M. J., A. Arakawa, and D. A. Randall (1983). The parameterization of the planetary boundary layer in the UCLA general circulation model: formulation and results. *Mon. Wea. Rev.* **111**, 2224–2243.

Takacs, L. L. (1985). A two-step scheme for the advection equation with minimized dissipation and dispersion errors. *Mon. Wea. Rev.* **113**, 1050–1065.

Tanguay, M., A Simard, and A. Staniforth (1989). A three-dimensional semi-Lagrangian scheme for the Canadian regional finite element forecast model. *Mon. Wea. Rev.* **117**, 1861–1871.

Tokioka, T. (1978). Some consideration on vertical differencing. *Meteor. Soc. Japan* **56**, 89–111.

Troen, I., and L. Mahrt (1986). A simple model for the atmospheric boundary layer: Sensitivity to surface evaporation. *Bound.-Layer Meteor.* **37**, 129–148.

Vogelezang, D. H. P., and A. A. M. Holtslag (1996). Evaluation and model impacts of alternative boundary-layer height formulations. *Bound.-Layer Meteor.* **81**, 245–269.

Wallace, J. M., S. Tibaldi, and A. J. Simmons (1983). Reduction of systematic forecast errors in the ECMWF model through the introduction of an envelope orography. *Quart. J. Roy. Meteor. Soc.* **109**, 683–717.

White, A. A., and R. A. Bromley (1995). Dynamically consistent, quasi-hydrostatic equations for global models with a complete representation of the Coriolis force. *Quart. J. Roy. Meteor. Soc.* **121**, 399–418.

Williamson, D. L. (1968). Integration of the barotropic vorticity equation on a spherical geodesic grid. *Tellus* **20**, 642–653.

Williamson, D. L., and J. G. Olson (1994). Climate simulations with a semi-Lagrangian version of the NCAR Community Climate model. *Mon. Wea. Rev.* **122**, 1594–1610.

Williamson, D. L., and J. G. Olson (1998a). A comparison of semi-Lagrangian and Eulerian polar climate simulations. *Mon. Wea. Rev.* **126**, 991–1000.

Williamson, D. L., and J. G. Olson (1998b). A comparison of semi-Lagrangian and Eulerian tropical climate simulations. *Mon. Wea. Rev.* **126**, 1001–1012.

Wu, X., and M. Yanai (1994). Effects of vertical wind shear on the cumulus transport of momentum: Observations and parameterization. *J. Atmos. Sci.* **51**, 1640–1660.

Xu, K.-M. (1991). The coupling of cumulus convection with large-scale processes, Ph.D. Thesis. Department of Atmospheric Sciences, UCLA.

Xu, K.-M., A. Arakawa, and S. K. Krueger (1992). The macroscopic behavior of cumulus ensembles simulated by a cumulus ensemble model. *J. Atmos. Sci.* **49**, 2402–2420.

Yanai, M., S. Esbensen, and J. Chu (1973). Determination of bulk properties of tropical cloud clusters from large-scale heat and moisture budgets. *J. Atmos. Sci.* **30**, 611–627.

Yu, J., and J. D. Neelin (1994). Models of tropical variability under convective adjustment and the Madden–Julian oscillation. Part II: Numerical results. *J. Atmos. Sci.* **51,** 1895–1914.

Zapotocny, T. H., D. R. Johnson, and F. M. Reames (1994). Development and initial test of the University of Wisconsin global isentropic-sigma model. *Mon. Wea. Rev.* **122,** 2160–2178.

Zhang, C., D. A. Randall, C.-H. Moeng, M. Branson, K. A. Moyer, and Q. Wang (1996). A surface flux parameterization based on the vertically averaged kinetic energy. *Mon. Wea. Rev.* **124,** 2521–2536.

Zhu, Z., J. Thuburn, B. J. Hoskins, and P. H. Haynes (1992). A vertical finite-difference scheme on a hybrid σ-θ-p coordinate model. *Mon. Wea. Rev.* **120,** 851–862.

Index

INTERNATIONAL GEOPHYSICS SERIES

EDITED BY

RENATA DMOWSKA
Division of Applied Sciences
Harvard University
Cambridge, Massachusetts

JAMES R. HOLTON
Department of Atmospheric Sciences
University of Washington
Seattle, Washington

H. THOMAS ROSSBY
Graduate School of Oceanography
University of Rhode Island
Narragansett, Rhode Island

Volume 1   BENO GUTENBERG. Physics of the Earth's Interior. 1959*
Volume 2   JOSEPH W. CHAMBERLAIN. Physics of the Aurora and Airglow. 1961*
Volume 3   S. K. RUNCORN (ed.). Continental Drift. 1962*
Volume 4   C. E. JUNGE. Air Chemistry and Radioactivity. 1963*
Volume 5   ROBERT G. FLEAGLE AND JOOST A. BUSINGER. An Introduction to Atmospheric Physics. 1963*
Volume 6   L. DEFOUR AND R. DEFAY. Thermodynamics of Clouds. 1963*
Volume 7   H. U. ROLL. Physics of the Marine Atmosphere. 1965*
Volume 8   RICHARD A. CRAIG. The Upper Atmosphere: Meteorology and Physics. 1965*
Volume 9   WILIS L. WEBB. Structure of the Stratosphere and Mesosphere. 1966*
Volume 10   MICHELE CAPUTO. The Gravity Field of the Earth from Classical and Modern Methods. 1967*
Volume 11   S. MATSUSHITA AND WALLACE H. CAMPBELL (eds.). Physics of Geomagnetic Phenomena (In two volumes). 1967*
Volume 12   K. YA KONDRATYEV. Radiation in the Atmosphere. 1969*
Volume 13   E. PAL MEN AND C. W. NEWTON. Atmosphere Circulation Systems: Their Structure and Physical Interpretation. 1969*
Volume 14   HENRY RISHBETH AND OWEN K. GARRIOTT. Introduction to Ionospheric Physics. 1969*

* Out of print.

Volume 15 C. S. RAMAGE. Monsoon Meteorology. 1971*

Volume 16 JAMES R. HOLTON. An Introduction to Dynamic Meteorology. 1972*

Volume 17 K. C. YEH AND C. H. LIU. Theory of Ionospheric Waves. 1972*

Volume 18 M. I. BUDYKO. Climate and Life. 1974*

Volume 19 MELVIN E. STERN. Ocean Circulation Physics. 1975

Volume 20 J. A. JACOBS. The Earth's Core. 1975*

Volume 21 DAVID H. MILLER. Water at the Surface of the Earth: An Introduction to Ecosystem Hydrodynamics. 1977

Volume 22 JOSEPH W. CHAMBERLAIN. Theory of Planetary Atmospheres: An Introduction to Their Physics and Chemistry. 1978*

Volume 23 JAMES R. HOLTON. An Introduction to Dynamic Meteorology, Second Edition. 1979*

Volume 24 ARNETT S. DENNIS. Weather Modification by Cloud Seeding. 1980

Volume 25 ROBERT G. FLEAGLE AND JOOST A. BUSINGER. An Introduction to Atmospheric Physics, Second Edition. 1980

Volume 26 KUG-NAN LIOU. An Introduction to Atmospheric Radiation. 1980

Volume 27 DAVID H. MILLER. Energy at the Surface of the Earth: An Introduction to the Energetics of Ecosystems. 1981

Volume 28 HELMUT G. LANDSBERG. The Urban Climate. 1991

Volume 29 M. I. BUDKYO. The Earth's Climate: Past and Future. 1982*

Volume 30 ADRIAN E. GILL. Atmosphere-Ocean Dynamics. 1982

Volume 31 PAOLO LANZANO. Deformations of an Elastic Earth. 1982*

Volume 32 RONALD T. MERRILL AND MICHAEL W. MCELHINNY. The Earth's Magnetic Field. Its History, Origin, and Planetary Perspective. 1983

Volume 33 JOHN S. LEWIS AND RONALD G. PRINN. Planets and Their Atmospheres: Origin and Evolution. 1983

Volume 34 ROLF MEISSNER. The Continental Crust: A Geophysical Approach. 1986

Volume 35 M. U. SAGITOV, B. BODKI, V. S. NAZARENKO, AND K. G. TADZHIDINOV. Lunar Gravimetry. 1986

Volume 36 JOSEPH W. CHAMBERLAIN AND DONALD M. HUNTEN. Theory of Planetary Atmospheres, 2nd Edition. 1987

Volume 37 J. A. JACOBS. The Earth's Core, 2nd Edition. 1987

Volume 38 JOHN R. APEL. Principles of Ocean Physics. 1987

Volume 39 MARTIN A. UMAN. The Lightning Discharge. 1987

Volume 40 DAVID G. ANDREWS, JAMES R. HOLTON, AND CONWAY B. LEOVY. Middle Atmosphere Dynamics. 1987

Volume 41 PETER WARNECK. Chemistry of the Natural Atmosphere. 1988

Volume 42 S. PAL ARYA. Introduction to Micrometeorology. 1988

Volume 43 MICHAEL C. KELLEY. The Earth's Ionosphere. 1989

Volume 44 WILLIAM R. COTTON AND RICHARD A. ANTHES. Storm and Cloud Dynamics. 1989

Volume 45 WILLIAM MENKE. Geophysical Data Analysis: Discrete Inverse Theory, Revised Edition. 1989

Volume 46 S. GEORGE PHILANDER. El Niño, La Niña and the Southern Oscillation. 1990

Volume 47 ROBERT A. BROWN. Fluid Mechanics of the Atmosphere. 1991

Volume 48 JAMES R. HOLTON. An Introduction to Dynamic Meteorology, Third Edition. 1992

* Out of print.